LIBRARY/NEW ENGLAND INST. OF TECHNOLOGY
3 0147 1003 8161 8

D1609187

TA 345.5 .A98 G73 2011

Graham, Rick, 1958-

Mastering AutoCAD Civil 3D 2012

DATE DUE

NOV 0 8 2017			

Demco, Inc. 38-293

Masterir
AutoCAD® Civi

Autodesk
Official Training Guide

NEW ENGLAND INSTITUTE
OF TECHNOLOGY
LIBRARY

Mastering
AutoCAD® Civil 3D® 2012

Autodesk
Official Training Guide

Richard Graham

Louisa Holland

Wiley Publishing, Inc.

NEW ENGLAND INSTITUTE
OF TECHNOLOGY
LIBRARY

Senior Acquisitions Editor: Willem Knibbe
Development Editor: Lisa Bishop
Technical Editor: Lisa Pohlmeyer
Production Editor: Elizabeth Campbell
Copy Editor: Liz Welch
Editorial Manager: Pete Gaughan
Production Manager: Tim Tate
Vice President and Executive Group Publisher: Richard Swadley
Vice President and Publisher: Neil Edde
Book Designer: Maureen Forys, Happenstance Type-O-Rama
Compositor: Jeff Lytle, Happenstance Type-O-Rama
Proofreader: Jen Larsen
Indexer: Nancy Guenther
Project Coordinator, Cover: Katherine Crocker
Cover Designer: Ryan Sneed

Copyright © 2011 by Wiley Publishing, Inc., Indianapolis, Indiana
Published simultaneously in Canada

ISBN: 978-1-118-01681-7
ISBN: 978-1-118-12448-2 (ebk.)
ISBN: 978-1-118-12447-5 (ebk.)
ISBN: 978-1-118-12449-9 (ebk.)

No part of this publication may be reproduced, stored in a retrieval system or transmitted in any form or by any means, electronic, mechanical, photocopying, recording, scanning or otherwise, except as permitted under Sections 107 or 108 of the 1976 United States Copyright Act, without either the prior written permission of the Publisher, or authorization through payment of the appropriate per-copy fee to the Copyright Clearance Center, 222 Rosewood Drive, Danvers, MA 01923, (978) 750-8400, fax (978) 646-8600. Requests to the Publisher for permission should be addressed to the Legal Department, Wiley Publishing, Inc., 10475 Crosspoint Blvd., Indianapolis, IN 46256, (317) 572-3447, fax (317) 572-4355, or online at http://www.wiley.com/go/permissions.

Limit of Liability/Disclaimer of Warranty: The publisher and the author make no representations or warranties with respect to the accuracy or completeness of the contents of this work and specifically disclaim all warranties, including without limitation warranties of fitness for a particular purpose. No warranty may be created or extended by sales or promotional materials. The advice and strategies contained herein may not be suitable for every situation. This work is sold with the understanding that the publisher is not engaged in rendering legal, accounting, or other professional services. If professional assistance is required, the services of a competent professional person should be sought. Neither the publisher nor the author shall be liable for damages arising herefrom. The fact that an organization or Web site is referred to in this work as a citation and/or a potential source of further information does not mean that the author or the publisher endorses the information the organization or Web site may provide or recommendations it may make. Further, readers should be aware that Internet Web sites listed in this work may have changed or disappeared between when this work was written and when it is read.

For general information on our other products and services or to obtain technical support, please contact our Customer Care Department within the U.S. at (800) 762-2974, outside the U.S. at (317) 572-3993 or fax (317) 572-4002.

Wiley also publishes its books in a variety of electronic formats and by print-on-demand. Not all content that is available in standard print versions of this book may appear or be packaged in all book formats. If you have purchased a version of this book that did not include media that is referenced by or accompanies a standard print version, you may request this media by visiting http://booksupport.wiley.com. For more information about Wiley products, visit us at www.wiley.com.

Library of Congress Cataloging-in-Publication Data is available from the publisher.

TRADEMARKS: Wiley, the Wiley logo, and the Sybex logo are trademarks or registered trademarks of John Wiley & Sons, Inc. and/or its affiliates, in the United States and other countries, and may not be used without written permission. AutoCAD and AutoCAD Civil 3D are registered trademarks of Autodesk, Inc. All other trademarks are the property of their respective owners. Wiley Publishing, Inc. is not associated with any product or vendor mentioned in this book.

10 9 8 7 6 5 4 3 2 1

Dear Reader,

Thank you for choosing *Mastering AutoCAD® Civil 3D® 2012*. This book is part of a family of premium-quality Sybex books, all of which are written by outstanding authors who combine practical experience with a gift for teaching.

Sybex was founded in 1976. More than 30 years later, we're still committed to producing consistently exceptional books. With each of our titles, we're working hard to set a new standard for the industry. From the paper we print on, to the authors we work with, our goal is to bring you the best books available.

I hope you see all that reflected in these pages. I'd be very interested to hear your comments and get your feedback on how we're doing. Feel free to let me know what you think about this or any other Sybex book by sending me an email at `nedde@wiley.com`. If you think you've found a technical error in this book, please visit `http://sybex.custhelp.com`. Customer feedback is critical to our efforts at Sybex.

    Best regards,

    Neil Edde
    Vice President and Publisher
    Sybex, an Imprint of Wiley

Acknowledgments

First and foremost, thanks to James Wedding, who is the pioneer of this book. Along with his team through the years, it has come a long way. We have big shoes to fill with this project and we are humbled to be a part of it now.

Thanks to the whole team at Wiley: Willem Knibbe for listening to our pleas and giving us encouragement all the way; Pete Gaughan, who promptly answered our technical questions; the editors, Lisa Bishop, Elizabeth Campbell, and Liz Welch, for making sure that what we wrote made grammatical sense.

Thanks to Lisa Pohlmeyer, who meticulously combed through our datasets making sure they worked and offered great suggestions to make the book even better.

Thanks to Matt Anderson of Autodesk for his help and encouragement in the face of our incessant questions. Matt, you are the best!

—*Rick Graham & Louisa Holland*

Thanks to my coauthor Louisa Holland for her expertise, enthusiasm, and attitude toward this project. I learned a lot from you and hope that this new partnership will carry forward to volumes in the future.

Thanks to my wife Melony for her love and encouragement and for letting me stay up all hours of the night so I could meet my deadlines. I owe you a cruise.

Thanks to John Huenke of Cornerstone Development and James R. Holley & Associates, Inc. for the use of the subdivision dataset; to Kevin Clark for the GIS dataset; to Matt Anderson for the MicroStation dataset; and to the general twittersphere population that follows me and has answered many questions.

And finally, thank you for purchasing this book.

—*Rick Graham*

There are so many people without whom this book would not be possible. Thanks to Rick Graham for being a supportive and knowledgeable collaborator. Thanks to all my clients, past and present, for teaching me more about engineering than I ever could learn at university. I'd like to thank all my Tweeps for answering questions and listening to me kvetch. I'd like to thank my parents for not letting me become an architect.

For their technical and dataset contributions I want to thank Rick Larson, Brad Hollister, and Eric Arneson of WisDOT, Ladd Nelson of Carlson Software, and Steve Biver of Eagle Point's left earlobe.

I'd like to thank my coworkers at MasterGraphics, especially Russ Nicloy and Ron Butzen, for being super cool about letting me divide my focus from my day job.

Most of all I'd like to thank my husband Mark, who kept me sane and caffeinated during the writing process.

—*Louisa Holland*

About the Authors

Richard "Rick" Graham has been a part of numerous Civil firms over the years. He has worked closely with the AutoCAD Civil 3D development team to help shape future versions of the product, is president of the AutoCAD Civil 3D User Group, speaks at Autodesk University, and is a respected moderator on Autodesk's Civil 3D discussion boards as well as blogger and social network-ite.

Louisa "Lou" Holland is a LEED-accredited Civil Engineer from Milwaukee, WI. She has trained users on Eagle Point Software and AutoCAD since 2001, and on AutoCAD Civil 3D since 2006. She has worked extensively with the Wisconsin Department of Transportation and various consultants on AutoCAD Civil 3D implementations. Louisa is an Autodesk Approved Instructor (AAI), an AutoCAD Civil 3D Certified Professional, and a regular speaker at Autodesk University, Autodesk User Group International, and other industry events.

Contents at a Glance

Introduction . . . . *xxiii*

- **Chapter 1** • The Basics of AutoCAD Civil 3D 1
- **Chapter 2** • Survey 41
- **Chapter 3** • Points 71
- **Chapter 4** • Surfaces 103
- **Chapter 5** • Parcels 161
- **Chapter 6** • Alignments 205
- **Chapter 7** • Profiles and Profile Views 245
- **Chapter 8** • Assemblies and Subassemblies 309
- **Chapter 9** • Basic Corridors 347
- **Chapter 10** • Advanced Corridors, Intersections, and Roundabouts 383
- **Chapter 11** • Superelevation 459
- **Chapter 12** • Cross Sections and Mass Haul 477
- **Chapter 13** • Pipe Networks and Part Builder 505
- **Chapter 14** • Storm and Sanitary Analysis 573
- **Chapter 15** • Grading 611
- **Chapter 16** • Plan Production 647
- **Chapter 17** • Interoperability 675
- **Chapter 18** • Quantity Takeoff 711
- **Chapter 19** • Styles 731
- **Appendix A** • The Bottom Line 835
- **Appendix B** • AutoCAD Civil 3D Certification 867

Index *871*

Contents

Introduction . . . . *xxiii*

Chapter 1 • The Basics of AutoCAD Civil 3D 1
The Interface 1
   Toolspace 1
   Panorama 17
   Ribbon 17
Labeling Lines and Curves 18
   Coordinate Line Commands 18
   Direction-Based Line Commands 21
Creating Curves 26
   Standard Curves 26
   Re-creating a Deed Using Line and Curve Tools 30
   Best Fit Entities 31
   Attach Multiple Entities 34
   The Curve Calculator 34
   Adding Line and Curve Labels 35
Using Transparent Commands 36
   Standard Transparent Commands 37
   Matching Transparent Commands 38
The Underlying Engine 38
   Managing Civil 3D Information 39
The Bottom Line 39

Chapter 2 • Survey 41
Setting Up the Databases 41
   Survey Database Defaults 41
   The Equipment Database 43
   The Figure Prefix Database 44
   The Linework Code Set Database 46
   The Main Event: Your Project's Survey Database 47
   "Mommy, Where Does Survey Data Come From?" 51
   Under the Hood in Your Survey Database 52
   Other Survey Features 60
   The Coordinate Geometry Editor 63
Using Inquiry Commands 66
The Bottom Line 68

Chapter 3 • Points ... 71
Anatomy of a Point ... 71
   A Quick Word on Styles ... 71
   COGO Points vs. Survey Points ... 72
Creating Basic Points ... 72
   Point Settings ... 72
   Importing Points from a Text File ... 74
   Converting Points from Land Desktop, SoftDesk, and Other Sources ... 77
   Getting to Know the Create Points Dialog ... 81
Basic Point Editing ... 85
   Physical Point Edits ... 86
   Panorama and Prospector Point Edits ... 86
   Point Groups: Don't Skip This Section! ... 87
   Changing Point Elevations ... 90
Description Keys: Field to Civil 3D ... 91
   Creating a Description Key Set ... 93
Point Tables ... 97
User-Defined Properties ... 98
The Bottom Line ... 100

Chapter 4 • Surfaces ... 103
Understanding Surface Basics ... 103
Creating Surfaces ... 104
   Free Surface Information ... 105
   Surface Approximations ... 111
   Surface from GIS Data ... 114
Refining and Editing Surfaces ... 117
   Surface Properties ... 118
   Surface Additions ... 121
Surface Styling and Analysis ... 136
   Contouring Basics ... 137
   Slopes and Slope Arrows ... 141
   Visibility Checker ... 143
Comparing Surfaces ... 144
   Simple Volumes ... 145
   Volume Surfaces ... 146
Labeling the Surface ... 149
   Contour Labeling ... 150
   Surface Point Labels ... 151
Point Cloud Surfaces ... 154
   Let Point Clouds Reign! ... 154
The Bottom Line ... 159

Chapter 5 • Parcels 161

Creating and Managing Sites 161
    Best Practices for Site Topology Interaction 161
    Creating a New Site 165
Creating a Boundary Parcel 167
Creating a Wetlands Parcel 168
Creating a Right-of-Way Parcel 170
Creating a Cul-de-sac 172
Creating Subdivision Lot Parcels Using Precise Sizing Tools 174
    Attached Parcel Segments 174
    Precise Sizing Settings 175
    Slide Line – Create Tool 177
    Swing Line – Create Tool 179
Using the Free Form Create Tool 180
Editing Parcels by Deleting Parcel Segments 182
Best Practices for Parcel Creation 185
    Forming Parcels from Segments 185
    Parcels Reacting to Site Objects 186
    Constructing Parcel Segments with the Appropriate Vertices 192
Labeling Parcel Areas 193
Labeling Parcel Segments 196
    Labeling Multiple Parcel Segments 197
    Labeling Spanning Segments 199
    Adding Curve Tags to Prepare for Table Creation 200
    Creating a Table for Parcel Segments 202
The Bottom Line 203

Chapter 6 • Alignments 205

Alignment Concepts 205
    Alignments and Sites 205
    Alignment Entities 206
Creating an Alignment 207
    Creating from a Line, Arc, or Polyline 208
    Creating by Layout 211
    Best Fit Alignments from Lines and Curves 215
    Reverse Curve Creation 217
    Creating with Design Constraints and Check Sets 219
Editing Alignment Geometry 222
    Grip-Editing 223
    Tabular Design 224
    Component-Level Editing 225
    Understanding Alignment Constraints 226
    Changing Alignment Components 229

Alignments as Objects.. 230
   Renaming Objects... 230
   The Right Station.. 233
   Assigning Design Speeds 235
   Labeling Alignments 236
   Alignment Tables 240
The Bottom Line... 243

Chapter 7 • Profiles and Profile Views 245
The Elevation Element 245
   Surface Sampling 246
   Layout Profiles.. 251
   Editing Profiles 261
   Matching Profile Elevations 267
   Profile Display.. 270
   Profile Labels... 270
Profile Views... 281
   Creating Profile Views During Sampling.............................. 281
   Creating Profile Views Manually 283
   Splitting Views 283
Profile Utilities... 290
   Superimposing Profiles 290
   Object Projection... 291
Editing Profile Views... 294
   Profile View Properties.. 294
   Labeling Styles 304
The Bottom Line... 307

Chapter 8 • Assemblies and Subassemblies 309
Subassemblies... 309
   The Corridor Modeling Catalog...................................... 310
Building Assemblies... 312
   The Pre-Cooked Assemblies.. 312
   Creating a Typical Road Assembly 313
   Getting the Most from Subassembly Help 319
   Jumping into Help 320
   Commonly Used Subassemblies 323
   Editing an Assembly 325
   Creating Assemblies for Nonroad Uses 327
Specialized Subassemblies... 331
   Using Generic Links.. 331
   Daylighting with Generic Links..................................... 334
   Working with Daylight Subassemblies 335
Advanced Assemblies... 340
   Offset Assemblies.. 340
   Marked Points and Friends.. 340

| Assembling Your Assemblies. 342
    Storing a Customized Subassembly on a Tool Palette 342
    Storing a Completed Assembly on a Tool Palette 343
The Bottom Line. 344

Chapter 9 • Basic Corridors 347

Understanding Corridors 347
Creating a Simple Corridor. 348
    Baseline. 349
    Regions. 349
    Frequency. 350
    Rebuilding Your Corridor 354
    Troubleshooting Corridor Problems 355
Corridor Feature Lines 358
Understanding Targets 361
    Using Target Alignments and Profiles 362
    Editing Sections 367
Creating a Corridor Surface 369
    The Corridor Surface 369
    Creation Fundamentals 370
    Adding a Surface Boundary 372
Performing a Volume Calculation 377
Non-Road Corridors 378
The Bottom Line. 381

Chapter 10 • Advanced Corridors, Intersections, and Roundabouts. 383

Getting Creative with Corridor Models 383
Using Alignment and Profile Targets to Model a Roadside Swale 384
    Corridor Utilities. 384
Multiregion Baselines 390
Modeling a Cul-de-sac 392
    Using Multiple Baselines. 392
    Establishing EOP Design Profiles 393
    Putting the Pieces Together. 394
    Troubleshooting Your Cul-de-Sac 398
Intersections: The Next Step Up 401
    Using the Intersection Wizard 402
    Manually Modeling an Intersection 412
    Creating an Assembly for the Intersection. 414
    Adding Baselines, Regions, and Targets for the Intersections 415
    Troubleshooting Your Intersection 421
    Checking and Fine-Tuning the Corridor Model 423
    Refining a Corridor Surface 429
Using an Assembly Offset 433
Using a Feature Line as a Width and Elevation Target 440

Roundabouts: The Mount Everest of Corridors 443
    Drainage First 444
    Roundabout Alignments 445
    Center Design 450
    Profiles for All 451
    Tie It All Together 453
    Finishing Touches 454
The Bottom Line 457

Chapter 11 • Superelevation 459

Getting Ready for Super 459
    Design Criteria Files 461
    Ready Your Alignment 464
    Super Assemblies 464
Applying Superelevation to the Design 469
    Start with the Alignment 469
    Transition Station Overlap 471
Superelevation Views 473
The Bottom Line 475

Chapter 12 • Cross Sections and Mass Haul 477

The Corridor 477
Lining Up for Samples 479
    Creating Sample Lines along a Corridor 481
    Editing the Swath Width of a Sample Line Group 483
Creating the Views 483
Creating a Single-Section View 484
It's a Material World 488
    Creating a Materials List 489
    Creating a Volume Table in the Drawing 490
    Adding Soil Factors to a Materials List 491
    Generating a Volume Report 493
A Little More Sampling 493
Annotating the Sections 495
Mass Haul 496
    Taking a Closer Look at the Mass Haul Diagram 497
    Create a Mass Haul Diagram 498
    The Create Mass Haul Diagram Dialog Explained 500
The Bottom Line 503

Chapter 13 • Pipe Networks and Part Builder 505

Planning a Typical Pipe Network 505
The Part Catalog 507
    The Structures Domain 508
Part Builder 512

    Part Builder Orientation 513
        Understanding the Organization of Part Builder 514
        Adding a Part Size Using Part Builder 516
        Sharing a Custom Part 518
        Adding an Arch Pipe to Your Part Catalog 518
    Part Rules. 519
        Structure Rules 519
        Pipe Rules. 521
        Creating Structure and Pipe Rule Sets 524
    Parts List 526
    Exploring Pipe Networks 526
    Pipe Network Object Types 527
    Creating a Sanitary Sewer Network 527
        Creating a Pipe Network with Layout Tools 528
        Establishing Pipe Network Parameters. 528
        Using the Network Layout Creation Tools 529
        Creating a Storm Drainage Pipe Network from a Feature Line 536
    Changing Flow Direction 539
    Editing a Pipe Network. 539
        Editing Your Network in Plan View 540
        Making Tabular Edits to Your Pipe Network. 544
        Context Menu Edits 545
        Editing with the Network Layout Tools Toolbar 546
    Creating an Alignment from Network Parts. 548
    Drawing Parts in Profile View 550
        Vertical Movement Edits Using Grips in Profile 552
        Removing a Part from Profile View 554
        Showing Pipes That Cross the Profile View 555
        Exploring the Tools on the Pipe Network Tab 558
    Adding Pipe Network Labels. 560
        Creating a Labeled Pipe Network Profile Including Crossings 561
        Pipe Labels. 562
        Structure Labels 563
        Special Profile Attachment Points for Structure Labels 563
    Creating an Interference Check between a Storm and Sanitary Pipe Network. 564
    Creating Pipe Tables 567
        Exploring the Table Creation Dialog 570
        Changes to a Pipe Network and Pipe Tables 570
        The Table Panel Tools 570
    The Bottom Line. 571

Chapter 14 • Storm and Sanitary Analysis **573**
    Getting Started on the CAD Side. 573
        Water Drop. 573
        Catchments 575
        Exporting Pipes to SSA 584

Storm and Sanitary Analysis 586
    Guided Tour of SSA 586
    Hydrology Methods 588
    From Civil 3D, with Love 599
    Make It Rain 604
    Running Reports from SSA 606
The Bottom Line 610

Chapter 15 • Grading 611

Working with Grading Feature Lines 611
    Accessing Grading Feature Line Tools 611
    Creating Grading Feature Lines 612
    Editing Feature Line Horizontal Information 620
    Editing Feature Line Elevation Information 625
    More Feature Line Editing Tools 628
    Labeling Feature Lines 636
Grading Objects 637
    Creating Gradings 637
    Editing Gradings 640
    Creating Surfaces from Grading Groups 641
The Bottom Line 645

Chapter 16 • Plan Production 647

Preparing for Plan Sets 647
Prerequisite Components 647
Using View Frames and Match Lines 648
    The Create View Frames Wizard 648
    Creating View Frames 655
    Editing View Frames and Match Lines 657
Creating Plan and Profile Sheets 659
    The Create Sheets Wizard 659
    Managing Sheets 665
Creating Section Sheets 668
    Creating Section View Groups 668
    Creating Section Sheets 670
Supporting Components 671
The Bottom Line 674

Chapter 17 • Interoperability 675

Data Shortcuts 675
    Publishing Data Shortcut Files 676
    Using Data Shortcuts 681
Playing Nicely in the Sandbox 690
    Earlier Versions of Civil 3D or Land Desktop 691
    Playing With Other Formats 697

An Introduction to Map 3D 703
  Where Is It? 704
  Setup 704
  Queries 706
The Bottom Line 709

Chapter 18 • Quantity Takeoff 711

Pay Item Files 711
  Pay Item Favorites 712
  Searching for Pay Items 714
Keeping Tabs on the Model 715
  AutoCAD Objects as Pay Items 715
  Pricing Your Corridor 717
  Pipes and Structures as Pay Items 720
  Highlighting Pay Items 725
Inventorying Your Pay Items 726
The Bottom Line 730

Chapter 19 • Styles 731

Civil 3D Templates 731
  Importing Styles 732
  Drawing Settings 734
  Object Settings 743
Get Started with Object Styles 745
  Frequently Seen Tabs 746
  Basic Object Styles 747
  Linear Object Styles 752
  Surface Styles 755
  Pipe and Structure Styles 765
Label Styles 773
  General Note Labels 785
  Point Label Styles 787
  Line and Curve Labels 790
  Pipe and Structure Labels 794
Profile and Alignment Labels 798
  Label Sets 798
  Alignment Labels 799
Advanced Style Types 815
  Table Styles 815
  Profile View Styles 817
  Section View Styles 827
  Code Set Styles 831
The Bottom Line 834

Appendix A • The Bottom Line 835
Chapter 1: The Basics of AutoCAD Civil 3D 835
Chapter 2: Survey. 837
Chapter 3: Points 839
Chapter 4: Surfaces 841
Chapter 5: Parcels. 844
Chapter 6: Alignments 846
Chapter 7: Profiles and Profile Views 848
Chapter 8: Assemblies and Subassemblies 850
Chapter 9: Basic Corridors 851
Chapter 10: Advanced Corridors, Intersections, and Roundabouts 852
Chapter 11: Superelevation. 854
Chapter 12: Cross Sections and Mass Haul 855
Chapter 13: Pipe Networks and Part Builder. 855
Chapter 14: Storm and Sanitary Analysis 857
Chapter 15: Grading. 858
Chapter 16: Plan Production. 861
Chapter 17: Interoperability 862
Chapter 18: Quantity Takeoff. 864
Chapter 19: Styles. 865

Appendix B • AutoCAD Civil 3D Certification 867

Index 871

Introduction

AutoCAD Civil 3D was introduced in 2004 as a trial product. Designed to give the then–Land Development desktop user a glimpse of the civil engineering software future, it was a sea change for AutoCAD-based design packages. Although there was need for a dynamic design package, many seasoned Land desktop users wondered how they'd ever make the transition.

Over the past few years, AutoCAD Civil 3D series have evolved from the wobbly baby introduced on those first trial discs to a mature platform used worldwide to handle the most complex engineering designs. With this change, many engineers still struggle with how to make the transition. The civil engineering industry as a whole is an old dog learning new tricks.

We hope this book will help you in this journey. As the user base grows and users get beyond the absolute basics, more materials are needed, offering a multitude of learning opportunities. While this book is starting to move away from the basics and truly become a *Mastering* book, we hope that we are headed in that direction with the general readership. We know we cannot please everyone, but we do listen to your comments—all toward the betterment of this book.

Designed to help you get past the steepest part of the learning curve and teach you some guru-level tricks along the way, *Mastering AutoCAD Civil 3D 2012* is the ideal addition to any AutoCAD Civil 3D user's bookshelf.

Who Should Read This Book

The *Mastering* book series is designed with specific users in mind. In the case of *Mastering AutoCAD Civil 3D 2012*, we expect you'll have some knowledge of AutoCAD in general and some basic engineering knowledge as well. A basic understanding of AutoCAD Civil 3D will be helpful, although there are explanations and examples to please everyone. We expect this book should appeal to a large number of AutoCAD Civil 3D users, but we envision a few primary users:

Beginning Users Looking to Make the Move to Using AutoCAD Civil 3D These people understand AutoCAD and some basics of engineering, but they are looking to learn AutoCAD Civil 3D on their own, broadening their skill set to make themselves more valuable in their firms and in the market.

AutoCAD Civil 3D Users Looking for a Desktop Reference With the digitization of the official help files, many users still long for a book they can flip open and keep beside them as they work. These people should be able to jump to the information they need for the task at hand, such as further information about a confusing dialog or troublesome design issue.

Users Looking to Prepare for the Autodesk Certification Exams   This book focuses on the elements you need to pass the Associate and Professional exams with flying colors, and includes margin icons to note topics of interest. Just look for the icon.

Classroom Instructors Looking for Better Materials   This book was written with real data from real design firms. We've worked hard to make many of the examples match the real-world problems we have run into as engineers. This book also goes into greater depth than many basic texts, allowing short classes to review the basics and leave the in-depth material for self-discovery, while longer classes can cover the full material presented.

This book can be used front-to-back as a self-teaching or instructor-based instruction manual. Each chapter has a number of exercises and most (but not all) build on the previous exercise. You can also skip to almost any exercise in any chapter and jump right in. We've created a large number of drawing files that you can download from www.sybex.com/go/masteringcivil3d2012 to make choosing your exercises a simple task.

What You Will Learn

This book isn't a replacement for training. There are too many design options and parameters to make any book a good replacement for training from a professional. This book teaches you to use the tools available, explores a large number of the options available, and leaves you with an idea of how to use each tool. At the end of the book, you should be able to look at any design task you run across, consider a number of ways to approach it, and have some idea of how to accomplish the task. To use one of our common analogies, reading this book is like walking around your local home-improvement warehouse. You see a lot of tools and use some of them, but that doesn't mean you're ready to build a house.

What You Need

Before you begin learning AutoCAD Civil 3D, you should make sure your hardware is up to snuff. Visit the Autodesk website and review graphic requirements, memory requirements, and so on. One of the most frustrating things that can happen is to be ready to learn, only to be stymied by hardware-related crashes. AutoCAD Civil 3D is a hardware-intensive program, testing the limits of every computer on which it runs.

We also strongly recommend using either a wide format or dual-monitor setup. The number of dialogs, palettes, and so on make AutoCAD Civil 3D a real estate hog. By having the extra space to spread out, you'll be able to see more of your design along with the feedback provided by the program itself.

You need to visit www.sybex.com/go/masteringcivil3d2012 to download all of the data and sample files. Finally, please be sure to visit the Autodesk website at www.autodesk.com to download any service packs that might be available.

The Mastering Series

The *Mastering* series from Sybex provides outstanding instruction for readers with intermediate and advanced skills, in the form of top-notch training and development for those already working in their field and clear, serious education for those aspiring to become pros. Every *Mastering* book includes:

- Real-world scenarios ranging from case studies to interviews that show how the tool, technique, or knowledge presented is applied in actual practice
- Skill-based instruction, with chapters organized around real tasks rather than abstract concepts or subjects
- A self-review section called The Bottom Line, so you can be certain you're equipped to do the job right

What Is Covered in This Book

This book contains 19 chapters and two appendices:

Chapter 1, "The Basics of AutoCAD Civil 3D," introduces you to the interface and many of the common dialogs in AutoCAD Civil 3D. This chapter looks at the Toolbox and some underused Inquiry tools as well. We also explore various tools for creating linework.

Chapter 2, "Survey," looks at the Survey Toolspace and the unique toolset it contains for handling field surveying and fieldbook data handling. We also look at various surface and surveying relationships.

Chapter 3, "Points," introduces AutoCAD Civil 3D points and the various methods of creating them. We also spend some time discussing the control of AutoCAD Civil 3D points with description keys and groups.

Chapter 4, "Surfaces," introduces the various methods of creating surfaces, using free and low-cost data to perform preliminary surface creation. Then we investigate the various surface edits and analysis methods. We wrap up the chapter with a look at point clouds and their use.

Chapter 5, "Parcels," describes the best practices for keeping your parcel topology tight and your labeling neat. It examines the various editing methods for achieving the desired results for the most complicated plats.

Chapter 6, "Alignments," introduces the basic AutoCAD Civil 3D horizontal control element. This chapter also examines using layout tools that maintain the relationships between the tangents, curves, and spiral elements that create alignments.

Chapter 7, "Profiles and Profile Views," looks at the sampling and creation methods for the vertical control element. We also examine the editing and element level control. In addition, we explore how profile views reflect the required format for your design and plans.

Chapter 8, "Assemblies and Subassemblies," looks at the building blocks of AutoCAD Civil 3D cross-sectional design. We discuss the available tool catalogs and show you how to build full design sections for use in any design environment.

Chapter 9, "Basic Corridors," introduces the basics of corridors—building full designs from horizontal, vertical, and cross-sectional design elements. We look at the various components to understand them better before moving to a more complex design set.

Chapter 10, "Advanced Corridors, Intersections, and Roundabouts," looks at using corridors in more complex situations. We discuss building surfaces, intersections, and other areas of corridors that make them powerful in any design situation.

Chapter 11, "Superelevation," takes a close look at the tools used to add superelevation to roadways. This functionality has changed greatly in the last few years, and you will have a chance to use the new Axis of Rotation subassemblies that can pivot from several design points.

Chapter 12, "Cross Sections and Mass Haul," looks at slicing sections from surfaces, corridors, and pipe networks using alignments and the mysterious sample-line group. Working with the wizards and tools, we show you how to make your sections to order. We explore Mass Haul to demonstrate the power of AutoCAD Civil 3D for creation of the Mass Haul diagrams.

Chapter 13, "Pipe Networks and Part Builder," gets into the building blocks of the pipe network tools. We look at modifying an existing part to add new sizes and then building parts lists for various design situations. We then work with the creation tools for creating pipe networks, and plan and profile views to get your plans looking like they should.

Chapter 14, "Storm and Sanitary Analysis," is a first look at the hydrology and hydraulic design tools included with AutoCAD Civil 3D 2012. We introduce the new catchment objects in AutoCAD Civil 3D and the best workflow to export data to this analysis tool.

Chapter 15, "Grading," examines both feature lines and grading objects. We look at creating feature lines to describe critical areas and then using grading objects to describe mass grading. We also explore using the basic tools to calculate some simple volumes.

Chapter 16, "Plan Production," walks you through the basics of creating view frame groups, sheets, and templates used to automate the drawing sheet process.

Chapter 17, "Interoperability," looks at the various ways of sharing and receiving data. We describe the data-shortcut mechanism for sharing data between AutoCAD Civil 3D users. We also consider other methods of importing and exporting, such as XML and DGN.

Chapter 18, "Quantity Takeoff," shows you the ins and outs of assigning pay items to pipes, corridor codes, blocks, and areas. You learn how to set up new pay items and generate quantity takeoff reports.

Chapter 19, "Styles," is devoted to object and label styles. We start by examining what makes a good AutoCAD Civil 3D template. You learn to navigate the Text Component Editor and how to master style conundrums you may come across.

Appendix A, "The Bottom Line," gathers together all the Master It problems from the chapters and provides a solution for each.

Appendix B, "AutoCAD Civil 3D Certification," points you to the chapters in this book that will help you master the objectives for each exam.

How to Contact the Authors

We welcome feedback from you about this book and/or about books you'd like to see from us in the future. You can reach us by emailing c3d.rickgraham@yahoo.com. For more information about our work, please visit our respective websites/blogs SimplyCivil3D.wordpress.com (Rick) and www.engineeringbird.com (Louisa).

Sybex strives to keep you supplied with the latest tools and information you need for your work. Please check their website at www.sybex.com, where we'll post additional content and updates that supplement this book if the need arises. Enter **Civil 3D** in the Search box (or type the book's ISBN—**9781118016817**) and click Go to get to the book's update page.

Thanks for purchasing *Mastering AutoCAD Civil 3D 2012*. We appreciate it, and look forward to exploring AutoCAD Civil 3D with you!

Chapter 1

The Basics of AutoCAD Civil 3D

Before we get into the "mastering" of AutoCAD Civil 3D, it is important to understand the basics. There are numerous dialogs, ribbons, menus, and icons to pore over. They might seem daunting at first glance, but as you use them, you will gain familiarity with their location and use. In this chapter, you will explore the interface and learn terminology that will be used throughout this book.

In addition, we will introduce the Lines and Curves commands, which offer a plethora of options for drawing lines and curves accurately.

In this chapter, you will learn to:

- Find any Civil 3D object with just a few clicks
- Modify the drawing scale and default object layers
- Modify the display of Civil 3D tooltips
- Navigate the Ribbon's contextual tabs
- Create a curve tangent to the end of a line
- Label lines and curves

The Interface

If you have used Civil 3D 2010 or 2011, the interface for Civil 3D 2012 is basically the same. If you are coming into Civil 3D 2012 from an earlier release, then this part of the chapter is for you. The context-sensitive Ribbon is one of the biggest differences you will encounter. The tools within Civil 3D can now be accessed via the Ribbon. Toolspace and the general look and feel of the Civil 3D interface make this release easy to use. Figure 1.1 shows the Civil 3D palette sets along with the AutoCAD tool palettes and context-sensitive Ribbon displayed in a typical environment.

Toolspace

Toolspace is one of the unique Civil 3D palette sets. Toolspace can have as many as four tabs to manage user data. These tabs are as follows:

- Prospector
- Settings
- Survey
- Toolbox

Using a Microsoft Windows Explorer–like interface within each, these tabs drive a large portion of the user control and data management of Civil 3D.

FIGURE 1.1
Civil 3D in a typical environment. Toolspace is docked on the left, and tool palettes float over the drawing window. The Ribbon is at the top of the workspace.

Prospector

Prospector is the main window into the Civil 3D object model. This palette, or tab, is where you go seeking data; it also shows points, alignments, parcels, corridors, and other objects as one concise, expandable list. In addition, in a project environment this window is where you control access to your project data, create references to shared project data, and observe the check-in and check-out status of a drawing. Finally, you can also use Prospector to create a new drawing from the templates defined in the Drawing Template File Location branch in your AutoCAD Options dialog. Prospector has the following branches:

- Open Drawings
- Projects (only if the Vault client is installed)
- Data Shortcuts
- Drawing Templates

> **MASTER AND ACTIVE DRAWING VIEWS**
>
> If you can't see the Projects or Drawing Templates branch in Figure 1.1, look at the top of the Prospector pane. There is a drop-down menu for operating in Active Drawing View or Master View mode. Selecting Active Drawing View displays only the active drawing and data shortcuts. Master View mode, however, displays the Projects, the Drawing Templates, and the Data Shortcuts branches, as well as the branches of all drawings that are currently open.

In addition to the branches, Prospector has a series of icons across the top that toggle various settings on and off. Some of the Civil 3D icons from previous versions have been removed, and their functionality has been universally enabled for Civil 3D 2012. Let's take a closer look at those icons:

Item Preview Toggle   Turns on and off the display of the Toolspace item preview within Prospector. These previews can be helpful when you're navigating drawings in projects (you can select one to check out) or when you're attempting to locate a parcel on the basis of its visual shape. In general, however, you can turn off this toggle—it's purely a user preference.

Preview Area Display Toggle   When Toolspace is undocked, this button moves the preview area from the right of the tree view to beneath the tree view area.

Panorama Display Toggle   Turns on and off the display of the Panorama window (which we'll discuss in a bit). To be honest, there doesn't seem to be a point to this button, but it's here nonetheless.

Help   This should be obvious, but it's amazing how many people overlook this icon.

> **HAVE YOU LOOKED IN THE HELP FILE LATELY?**
>
> The AutoCAD Civil 3D development team in Manchester, New Hampshire, has worked hard to make the Help files in Civil 3D top-notch and user friendly. The help files should be your first line of support!

Open Drawings

This branch of Prospector contains the drawings currently open in Civil 3D. Each drawing is subdivided into groups by major object type, such as points, point groups, surfaces, and so forth. These object groups then allow you to view all the objects in the collection. Some of these groups are empty until objects are created. You can learn details about an individual object by expanding the tree and selecting an object.

Within each drawing, the breakdown is similar. If a collection isn't empty, a plus sign appears next to it, as in a typical Windows Explorer interface. Selecting any of these top-level collection names displays a list of members in the preview area. Right-clicking the collection name allows you to select various commands that apply to all the members of that collection. For example, right-clicking the Point Groups collection brings up the menu shown in Figure 1.2.

FIGURE 1.2
Context-sensitive menus in Prospector

In addition, right-clicking the individual object in the list view offers many commands unique to Civil 3D: Zoom To Object and Pan To Object are typically included. By using these commands, you can find any parcel, point, cross section, or other Civil 3D object in your drawing almost instantly.

Many longtime users of AutoCAD have resisted right-clicking menus for their daily tasks since AutoCAD 14. In other AutoCAD products this may be possible, but in Civil 3D you'll miss half the commands! This book focuses on the specific options and commands for each object type during discussions of the particular objects.

Projects

The Projects branch of Prospector will only be visible if you are using Vault. This branch allows you to sign in and out of Vault, review what projects are available, manage the projects you sort through for information, check out drawings for editing, and review the status of drawings as well as that of individual project–based objects.

Data Shortcuts

A data shortcut identifies the path to a specific object, in a specific drawing. Many users have found data shortcuts to be ideal in terms of project collaboration for two reasons: flexibility and simplicity.

Drawing Templates

The Drawing Templates branch is added more as a convenience than anything else. You can still create new drawings via the standard File ➤ New option, but by using the Drawing Templates branch, you can do the same thing without leaving Prospector. The Drawing Templates branch searches the file path specified in your AutoCAD Options dialog and displays a list of all the DWT files it finds. You can customize this path to point to a server or other folder, but by default it's a local user-settings path. Right-clicking the name of a template presents you with the options shown in Figure 1.3.

FIGURE 1.3
Creating a new drawing from within the Drawing Templates branch of Prospector. The templates shown here are located in the folder set in your AutoCAD Options dialog.

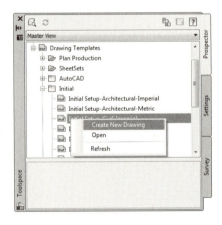

Civil 3D is built on both AutoCAD and AutoCAD Map, so Civil 3D 2012 comes with a variety of templates. However, most users will want to select one of the top few templates, which start with _Autodesk Civil 3D and then have some descriptive text. These templates have been built on the basis of customer feedback to provide Civil 3D with a varying collection of object styles. These templates give you a good starting point for creating a template that meets your needs or the needs of your firm.

Settings

The Settings tab of Toolspace is where you can adjust how Civil 3D objects look and how the Civil 3D commands work. You use this tab to control styles, labels, and command settings for each component of Civil 3D. This book starts by looking at the top level of drawing settings and a few command settings to get you familiar, and then covers the specifics for each object's styles and settings in their respective chapters.

Drawing Settings

Starting at the drawing level, Civil 3D has a number of settings that you must understand before you can use the program efficiently. Civil 3D understands that the end goal of most users is to prepare construction documents on paper. To that end, most labeling and display settings are displayed in inches for imperial users and millimeters for metric users instead of nominal units like many other AutoCAD objects. Because much of this is based on an assumed working scale, let's look at how to change that setting, along with some other drawing options:

1. Open the file `Basic Site.dwg` from this book's companion web page, www.sybex.com/go/masteringcivil3d2012.

2. Switch to the Settings tab.

3. Right-click the filename, and select Edit Drawing Settings to display the dialog shown in Figure 1.4.

Figure 1.4
The Drawing Settings dialog

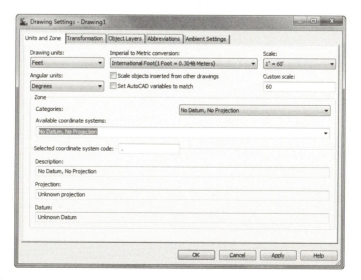

Each tab in this dialog controls a different aspect of the drawing. Most of the time, you'll pick up the object layers, abbreviations, and ambient settings from a companywide template. However, the drawing scale and coordinate information change for every job, so you'll visit the Units And Zone and the Transformation tabs frequently.

Units And Zone Tab

The Units And Zone tab lets you specify metric or imperial units for your drawing. You can also specify the conversion factor between systems. In addition, you can control the assumed plotting scale of the drawing. The drawing units typically come from a template, but the options for scaling blocks and setting AutoCAD variables depend on your working environment. Many engineers continue to work in an arbitrary coordinate system using the settings as shown earlier, but using a real coordinate system is easy! For example, setting up a drawing for the Harrisburg, Pennsylvania area, you'd follow this procedure:

1. Select USA, Pennsylvania from the Categories drop-down menu on the Units And Zone tab.
2. Select NAD83 Pennsylvania State Planes, South Zone, US Foot from the Available Coordinate Systems drop-down menu. You could have also typed **PA83-SF** in the Coordinate System Code box.

There are literally hundreds, if not thousands, of available coordinate systems. These are established by international agreement; because Civil 3D is a worldwide product, almost any recognized surveying coordinate system can be found in the options. Once your coordinate system has been established, you can change it on the Transformation tab if desired.

This tab also includes the options Scale Objects Inserted From Other Drawings and Set AutoCAD Variables To Match. In Figure 1.4, both are unchecked to move forward.

The Scaling option has been problematic in the past because many firms work with drawings that have no units assigned and therefore scale incorrectly. But you can experiment with this setting as you'd like. The Set AutoCAD Variables To Match option attempts to set the AutoCAD variables AUNITS, DIMUNITS, INSUNITS, and MEASUREMENT to the values placed in this dialog. You can learn about the nature of these variables via the help files. Because of some inconsistencies between coordinate-based systems and the AutoCAD engine, sometimes these variables must be approximated. Again, you won't typically set this flag to True; you should experiment in your own office to see if it can help you.

Transformation Tab

With a base coordinate system selected, you can now do any further refinement you'd like using the Transformation tab (Figure 1.5). The coordinate systems on the Units And Zone tab can be refined to meet local ordinances, tie in with historical data, complete a grid to ground transformation, or account for minor changes in coordinate system methodology. These changes can include the following:

Apply Sea Level Scale Factor   Takes into account the mean elevation of the site and the spheroid radius that is currently being applied as a function of the selected zone ellipsoid.

Grid Scale Factor   Based on a 1:1 value, a user-defined uniform scale factor, a reference point scaling, or a prismoidal transformation in which every point in the grid is adjusted by a unique amount.

Reference Point Can be used to set a singular point in the drawing field via pick or via point number, local northing and easting, or grid northing and easting values.

Rotation Point Can be used to set the reference point for rotation via the same methods as the reference point.

Specify Grid Rotation Angle Enter an amount or set a line to North by picking an angle or deflection in the drawing. You can use this same method to set the azimuth if desired.

FIGURE 1.5
The Transformation tab

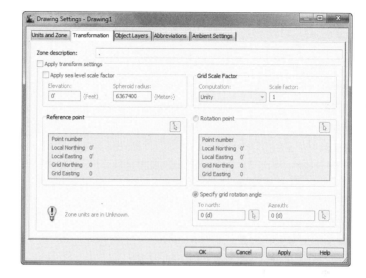

Most engineering firms work on either a defined coordinate system or an arbitrary system, so none of these changes are necessary. Given that, this tab will be your only method of achieving the necessary transformation for certain surveying and geographic information system (GIS)–based and land surveying–based tasks.

Object Layers Tab

Setting object layers to your company standard is a major part of creating the feel you're after when using Civil 3D in your office. The nearly 50 objects described here make up the entirety of the Civil 3D modeling components and the objects you and other users will deal with daily.

Let's see how to change a parameter in the Object Layers tab. First, click the Layer column in the Catchment row, as shown in Figure 1.6. Then in the Layer Selection dialog, select _CATCHMENT and click OK.

> **ONE OBJECT AT A TIME**
>
> Note that this procedure only changes the Catchment object. If you want to change the standard of all the objects, you need to adjust the Catchment Labeling, Catchment Table, Profile, Profile View, Profile View Labeling, and so on. To do this, it's a good idea to right-click in the grid view and select Copy All. You can then paste the contents of this matrix into Microsoft Excel for easy formatting and reviewing.

FIGURE 1.6
Changing the Layer setting for the Catchment object

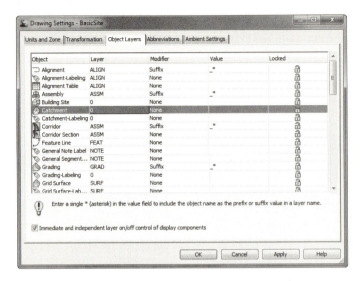

One common question that surrounds the Object Layers tab is the check box at the lower left: Immediate And Independent Layer On/Off Control Of Display Components. What the heck does that mean? Relax—it's not as complicated as it sounds.

Many objects in Civil 3D are built from underlying components. Take an alignment, for example. It's built from tangents, curves, spirals, extension lines, and so on. Each of these components can be assigned its own layer—in other words, the lines could be assigned to the LINES layer, curves to the CURVES layer, and so on. When this check box is selected, the *component's* layer exerts some control. In the example given, if the alignment is assigned to the ALIGN layer and the box is selected, turning off (not freezing) the LINES layer will make the line components of that alignment disappear. Deselect this control, and the LINES layer's status won't have any effect on the visibility of the alignment line components.

Finally, it's important to note that this layer control determines the object's parent layer *at creation*. Civil 3D objects can be moved to other layers at any time. Changing this setting doesn't change any objects already in place in the drawing.

Abbreviations Tab

You could work for years without noticing the Abbreviations tab. The options on this tab allow you to set the abbreviations Civil 3D uses when labeling items as part of its automated routines. The prebuilt settings are based on user feedback, and many of them are the same as the settings from Land Desktop, the last-generation civil engineering product from Autodesk.

Changing an abbreviation is as simple as clicking in the Value field and typing a new one. Notice that the Alignment Geometry Point Entity Data section has a larger set of values and some formulas attached. They are more representative of other label styles, and we'll visit the label editor in Chapter 19, Styles.

Ambient Settings Tab

The Ambient Settings tab can be daunting at first. The term *ambient* means "surround" or "surrounding," and these settings control many of the math, labeling, and display features, as well as the user interaction surrounding the use of Civil 3D. Being familiar with the way this tab works will help you further down the line, because almost every other setting dialog in the program works like the one shown in Figure 1.7.

You can approach this tab in the following ways:

Top to Bottom Expand one branch, handle the settings in that branch, and then close it and move to the next.

Print and Conquer Expand all the branches using the Expand All Categories button found at the lower right.

FIGURE 1.7
The Ambient Settings tab with the General branch expanded

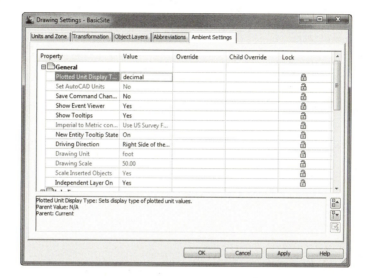

> **DRAWING PRECISION VS. LABEL PRECISION**
>
> You can create label styles (discussed in Chapter 19, "Styles") to annotate objects using precision, units, or specifications other than those set in the Ambient or Command Settings dialog. Establish settings to reflect how you'd like to input and track your data, not necessarily how you'd like to label your data.

The Ambient Settings for Direction offer the following choices:

- Unit: Degree, Radian, and Grad
- Precision: 0 through 8 decimal places
- Rounding: Round Normal, Round Up, and Truncate

- ◆ Format: Decimal, two types of DDMMSS, and Decimal DMS
- ◆ Direction: Short Name (spaced or unspaced) and Long Name (spaced or unspaced)
- ◆ Capitalization
- ◆ Sign
- ◆ Measurement Type: Bearings, North Azimuth, and South Azimuth
- ◆ Bearing Quadrant

From this list, it becomes clear where these settings apply to the tools discussed in this chapter. When you're using the Bearing Distance transparent command, for example, these settings control how you input your quadrant, your bearing, and the number of decimal places in your distance.

Explore the other categories, such as Angle, Lat Long, and Coordinate, and customize the settings to fit how you work.

At the bottom of the Ambient Settings tab is a Transparent Commands category. These settings control how (or if) you're prompted for the following information:

Prompt For 3D Points Controls whether you're asked to provide a z elevation after x and y have been located.

Prompt For Y Before X For transparent commands that require x and y values, this setting controls whether you're prompted for the y-coordinate before the x-coordinate. Most users prefer this value set to False so they're prompted for an x-coordinate and then a y-coordinate.

Prompt For Easting Then Northing For transparent commands that require Northing and Easting values, this setting controls whether you're prompted for the Easting first and the Northing second. Most users prefer this value set to False, so they're prompted for Northing first and then Easting.

Prompt For Longitude Then Latitude For transparent commands that require longitude and latitude values, this setting controls whether you're prompted for Longitude first and Latitude second. Most users prefer this set to False, so they're prompted for Latitude and then Longitude.

After you have expanded the branches, right-click in the middle of the displayed options and select Copy To Clipboard. Then paste the settings to Excel for review, as you did with the Object Layers tab.

> ### Sharing the Workload
>
> The print and conquer approach makes it easy to distribute multiple copies to surveyors, land planners, engineers, and so on and let them fill in the changes. Then, creating a template for each group is a matter of making their changes. If you're asking end users who aren't familiar with the product to make these changes, it's easy to miss one. Working line by line is fairly foolproof.

After you decide how to approach these settings, get to work. The settings are either drop-down menus or text boxes (in the case of numeric entries). Many of them are self-explanatory and common to land-development design. Let's look at these settings in more detail (see Figure 1.7).

Plotted Unit Display Type   Remember, Civil 3D knows you want to plot at the end of the day. In this case, it's asking you how you would like your plotted units measured. For example, would you like that bit of text to be 0.25" tall or ¼" high? Most engineers are comfortable with the Leroy method of text heights (L80, L100, L140, and so on), so the decimal option is the default.

Set AutoCAD Units   This displays whether or not Civil 3D should attempt to match AutoCAD drawing units, as specified on the Units And Zone tab.

Save Command Changes To Settings   This setting is incredibly powerful but a secret to almost everyone. By setting it to Yes, your changes to commands will be remembered from use to use. This means if you make changes to a command during use, the next time you call that Civil 3D command, you won't have to make the same changes. It's frustrating to do work over because you forgot to change one out of the five things that needed changing, so this setting is invaluable.

Show Event Viewer   Event Viewer is Civil 3D's main feedback mechanism, especially when things go wrong. It can get annoying, however, and it takes up valuable screen real estate (especially if you're stuck with one monitor!), so many people turn Show Event Viewer off. We recommend leaving it on and pushing it to the side if needed.

Show Tooltips   One of the cool features that people remark on when they first use Civil 3D is the small pop-up that displays relevant design information when the cursor is paused on the screen. This includes things such as Station-Offset information, Surface Elevation, Section information, and so on. Once a drawing contains numerous bits of information, this display can be overwhelming; therefore, Civil 3D offers the option to turn off these tooltips universally with this setting. A better approach is to control the tooltips at the object type by editing the individual feature settings. You can also control the tooltips by pulling up the properties for any individual object and looking at the Information tab.

Imperial To Metric Conversion   This displays the conversion method specified on the Units And Zone tab. The two options currently available are US Survey Foot and International Foot.

New Entity Tooltip State   You can also control tooltips on an individual object level. For instance, you might want tooltip feedback on your proposed surface but not on the existing surface. This setting controls whether the tooltip is turned on at the object level for new Civil 3D objects.

Driving Direction   This specifies the side of the road that forward-moving vehicles use for travel. This setting is important in terms of curb returns and intersection design.

Drawing Unit, Drawing Scale, and Scale Inserted Objects   These settings were specified on the Units And Zone tab but are displayed here for reference and so that you can lock them if desired.

Independent Layer On   This is the same control that was set on the Object Layers tab.

The settings that are applied here can also be applied at the object levels. For example, you may typically want elevation to be shown to two decimal places, but when looking at surface elevations, you might want just one. The Override and Child Override columns give you feedback about these types of changes. See Figure 1.8.

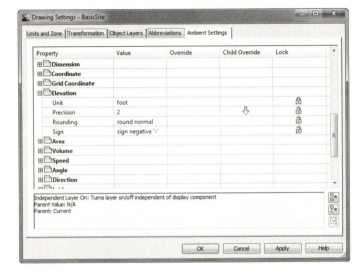

FIGURE 1.8
The Child Override indicator in the Elevation values

The Override column shows whether the current setting is overriding something higher up. Because you're at the Drawing Settings level, these are clear. However, the Child Override column displays a down arrow, indicating that one of the objects in the drawing has overridden this setting. After a little investigation through the objects, you'll find the override is in the Edit Feature Settings of the Profile view, as shown in Figure 1.9.

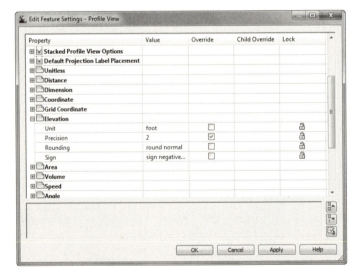

FIGURE 1.9
The Profile Elevation Settings and the Override indicator

Notice that in this dialog, the box is checked in the Override column. This indicates that you're overriding the settings mentioned earlier, and it's a good alert that things have changed from the general Drawing Settings to this Object Level setting.

But what if you don't want to allow those changes? Each Settings dialog includes one more column: Lock. At any level, you can lock a setting, graying it out for lower levels. This can be handy for keeping users from changing settings at the lower level that perhaps should be changed at a drawing level, such as sign or rounding methods.

Object Settings

If you click the Expand button next to the drawing name, you see the full array of objects that Civil 3D uses to build its design model. Each of these has special features unique to the object being described, but there are some common features as well. Additionally, the General collection contains settings and styles that are applied to various objects across the entire product.

The General collection serves as the catchall for styles that apply to multiple objects and for settings that apply to *no* objects. For instance, the Civil 3D General Note object doesn't really belong with the Surface or Pipe collection. It can be used to relate information about those objects, but because it can also relate to something like "Don't Dig Here!" it falls into the General category. The General collection has three components (or branches):

Multipurpose Styles   These styles are used in many objects to control the display of component objects. The Marker Styles and Link Styles collections are typically used in cross-sectional views, whereas the Feature Line Styles collection is used in grading and other commands. Figure 1.10 shows the collection of multipurpose styles and some of the marker styles that ship with the product.

FIGURE 1.10
General multipurpose styles and some marker styles

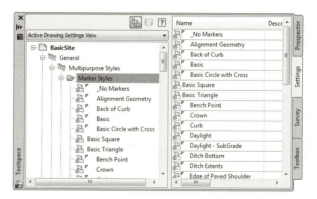

Label Styles   The Label Styles collection allows Civil 3D users to place general text notes or label single entities outside the parcel network while still taking advantage of Civil 3D's flexibility and scaling properties. With the various label styles shown in Figure 1.11, you can get some idea of their usage.

Because building label styles is a critical part of producing plans with Civil 3D, Chapter 19, "Styles," looks at how to build a new basic label and some of the common components that appear in every label style throughout the product.

Commands Almost every branch in the Settings tree contains a Commands folder. Expanding this folder, as shown in Figure 1.12, shows you the typical long, unspaced command names that refer to the parent object.

FIGURE 1.11
Line label styles

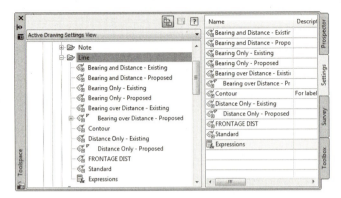

FIGURE 1.12
Surface command settings in Toolspace

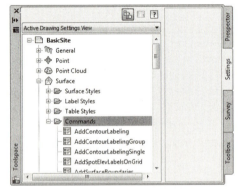

SURVEY

The Survey palette is displayed optionally and controls the use of the survey, equipment, and figure prefix databases. Survey is an essential part of land-development projects. Because of the complex nature of this tab, all of Chapter 2, "Survey," is devoted to it.

TOOLBOX

The Toolbox is a launching point for add-ons and reporting functions. To access the Toolbox, from the Home tab in the Ribbon, select Toolspace ➤ Palettes ➤ Toolbox. Out of the box, the Toolbox contains reports created by Autodesk, but you can expand its functionality to include your own macros or reports. The buttons on the top of the Toolbox, shown in Figure 1.13, allow you to customize the report settings and add new content.

THE INTERFACE | 15

Figure 1.13
The Toolbox palette with the Edit Toolbox Content button

Real World Scenario

A Toolbox Built Just for You

You can edit the Toolbox content and the Report Settings by selecting the desired tool, right-clicking, and then executing. Don't limit yourself to the default reports that ship in the Toolbox, though. Many firms find that adding in-house customizations to the Toolbox gives them better results and is more easily managed at a central level than by customizing via the AutoCAD custom user interface (CUI) and workspace functionality.

Let's add one of the sample Civil 3D Visual Basic Application (VBA) macros to a new Toolbox:

1. Click the Edit Toolbox Content button (shown in Figure 1.13) to open the Toolbox Editor in Panorama.

2. Click the button shown here to add a new root category.

3. Click the Root Category1 toolbox that appears. The name will appear in the preview area, where you can edit it. Change the name to **Sample Files**, and press ↵.

4. Right-click the Sample Files toolbox, and select New Category as shown here.

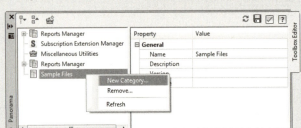

5. Expand the Sample Files toolbox to view the new category, and then click the name to edit it in the preview area. Change the name to **VBA**, and press ↵.
6. Right-click the VBA category, and select New Tool.
7. Expand the VBA category to view the new tool, and then click the name to edit it in the preview area. Change its name to **Pipe Sample**.
8. Change the description to **Sample VBA**.
9. Working down through the properties in the preview area, select VBA in the drop-down menu in the Execute Type field.
10. Click in the Execute File field, and then click the More button.
11. Browse to `C:\Program Files\Autocad Civil 3D 2012\Sample\Civil 3D API\COM\Vba\Pipe\`, and select the file `PipeSample.dvb`.
12. Click Open.
13. Click in the Macro Name text field, and type **PipeSample** as shown here.

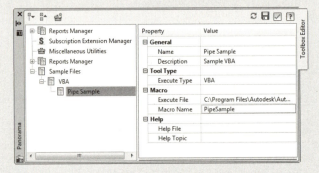

14. Click the green check box at the upper right to dismiss the editor.
15. You will be asked, "Would you like to apply those changes now?" Select Yes.

You've now added that sample VBA macro to your Toolbox. By adding commonly used macros and custom reports to your Toolbox, you can keep them handy without modifying the rest of your Civil 3D interface or programming buttons. It's just one more way to create an interface and toolset for the way you work.

Panorama

The Panorama window is Civil 3D's feedback and tabular editing mechanism. It's designed to be a common interface for a number of different Civil 3D–related tasks, and you can use it to provide information about the creation of profile views, to edit pipe or structure information, or to run basic volume analysis between two surfaces. For an example of Panorama in action, change to the View tab, and then select Palettes ➢ Event Viewer. You'll explore and use Panorama more during this book's discussion of specific objects and tasks.

> **RUNNING OUT OF SCREEN REAL ESTATE?**
>
> It's a good idea to turn on Panorama using this technique and then drag it to the side so you always see any new information. Although it's possible to turn it off, doing so isn't recommended—you won't know when Civil 3D is trying to tell you something! Place Panorama on your second monitor (now you see why you need to have a second monitor, don't you?), and you'll always be up-to-date with your Civil 3D model.
>
> And in case you missed it, you were using Panorama when you added the sample VBA macro in the previous exercise.

Ribbon

As with AutoCAD, the Ribbon is the primary interface for accessing Civil 3D commands and features. When you select an AutoCAD Civil 3D object, the Ribbon displays commands and features related to that object. If several object types are selected, the Multiple contextual tab is displayed. Use the following procedure to familiarize yourself with the Ribbon:

1. Open the `BasicSite.dwg`, which you will find at www.sybex.com/go/masteringcivil3d2012.
2. Select one of the parcel labels (the labels in the middle of the lot areas).
3. Notice that the Labels & Tables, General Tools, Modify, and Launch Pad tabs are displayed, as shown in Figure 1.14.
4. Select a parcel line and notice the display of the Multiple contextual tab.
5. Use the Esc key to cancel all selections.
6. Reselect a parcel line. Select the down arrow next to the Modify panel. Using the pin at the bottom-left corner of the panel, pin the panel open.
7. Select the Properties command in the General Tools panel to open the AutoCAD Properties palette. Notice that the Modify panel remains opened and pinned.

> **STYLES AND MORE STYLES**
>
> Civil 3D uses styles to change the look and feel of objects and labels. Throughout this book, you will see many styles. For a better look at styles, refer to Chapter 19.

FIGURE 1.14
The context-sensitive Ribbon

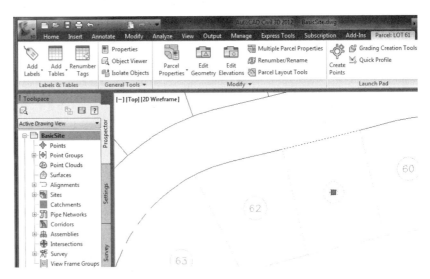

Labeling Lines and Curves

You can draw lines many ways in an AutoCAD-based environment. The tools found on the Draw panel of the Home tab create lines that are no more intelligent than those created by the standard AutoCAD Line command. How the Civil 3D lines differ from those created by the regular Line command isn't in the resulting entity, but in the process of creating them. Figure 1.15 shows the available line commands.

Note that you can switch between any of the Line commands without exiting the command. For example, if your first location is a point object, use Line By Point Object; then, without leaving the command, go back to the Lines/Curves menu and choose any Line or Curve command to continue creating your linework. You can also press the Esc key once, while in a Lines/Curves menu command, to resume the regular Line command.

Coordinate Line Commands

The next few commands discussed in this section help you create a line using Civil 3D points and/or coordinate inputs. Each command requires you to specify a Civil 3D point, a location in space, or a typed coordinate input. These Line tools are useful when your drawing includes Civil 3D points that will serve as a foundation for linework, such as the edge of pavement shots, wetlands lines, or any other points you'd like to connect with a line.

Line Command

The Create Line command on the Draw panel of the Home tab issues the standard AutoCAD Line command. It's equivalent to typing **line** on the command line or clicking the Line tool on the Draw toolbar.

Create Line By Point # Range Command

The Create Line By Point # Range command prompts you for a point number. You can type in an individual point number, press ↵, and then type in another point number. A line is drawn connecting those two points. You can also type in a range of points, such as **640-644**. Civil 3D draws a line that connects those lines in numerical order—from 640 to 641, and so on (see Figure 1.16). This order won't give you the desired linework for edge of asphalt, for example.

Figure 1.15
Line creation tools

Figure 1.16
A line created using 640-644 as input

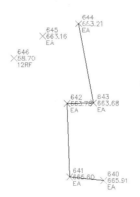

Alternatively, you can enter a list of points such as **640, 643, 644** (Figure 1.17). Civil 3D draws a line that connects the point numbers in the order of input. This approach is useful when your points were taken in a zigzag pattern (as is commonly the case when cross-sectioning pavement), or when your points appear so far apart in the AutoCAD display that they can't be readily identified.

FIGURE 1.17
A line created using 640, 643, 644 as input

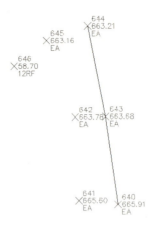

CREATE LINE BY POINT OBJECT COMMAND

The Create Line By Point Object command prompts you to select a point object. To select a point object, locate the desired start point and click any part of the point. This tool is similar to using the regular Line command and a Node object snap (also known as osnap).

CREATE LINE BY POINT NAME COMMAND

The Create Line By Point Name command prompts you for a point name. A *point name* is a field in Point properties, not unlike the point number or description. The difference between a point name and a *point description* is that a point name must be unique. It is important to note that some survey instruments name points rather than number points as is the norm.

To use this command, enter the names of the points you want to connect with linework.

CREATE LINE BY NORTHING/EASTING AND CREATE LINE BY GRID NORTHING/EASTING COMMANDS

The Create Line by Northing/Easting and Create Line By Grid Northing/Easting commands let you input northing (y) and easting (x) coordinates as endpoints for your linework. The Create Line By Grid Northing/Easting command requires that the drawing have an assigned coordinate system. This command can be useful when working with known surveyed points or features in a state plane coordinate system (SPCS).

CREATE LINE BY LATITUDE/LONGITUDE COMMAND

The Create Line By Latitude/Longitude command prompts you for geographic coordinates to use as endpoints for your linework. This command also requires that the drawing have an assigned coordinate system. For example, if your drawing has been assigned Delaware State Plane NAD83 US Feet and you execute this command, your Latitude/Longitude inputs are translated into the appropriate location in your state plane drawing. This command can be useful when you are drawing lines between waypoints collected with a standard handheld GPS unit.

Direction-Based Line Commands

The next few commands help you specify the direction of a line. Each of these commands requires you to choose a start point for your line before you can specify the line direction. You can specify your start point by physically choosing a location, using an osnap, or using one of the point-related line commands discussed earlier.

CREATE LINE BY BEARING COMMAND

The Create Line By Bearing command will likely be one of your most frequently used line commands.

This command prompts you for a start point, followed by prompts to input the Quadrant, Bearing, and Distance values. You can enter values on the command line for each input, or you can graphically choose inputs by picking them on screen. The glyphs at each stage of input guide you in any graphical selections. After creating one line, you can continue drawing lines by bearing, or you can switch to any other method by clicking one of the other Line By commands on the Draw panel (see Figure 1.18).

FIGURE 1.18
The tooltips for a quadrant (top), a bearing (middle), and a distance (bottom)

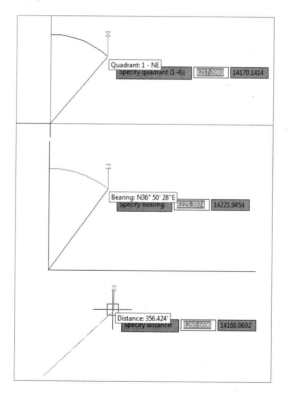

CREATE LINE BY AZIMUTH COMMAND

The Create Line By Azimuth command prompts you for a start point, followed by a north azimuth, and then a distance (Figure 1.19).

FIGURE 1.19
The tooltip for the Create Line by Azimuth command

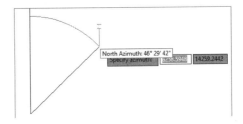

CREATE LINE BY ANGLE COMMAND

The Create Line By Angle command prompts you for a turned angle and then a distance (Figure 1.20). This command is useful when you're creating linework from angles right (in lieu of angles left) and distances recorded in a traditional handwritten field book (required by law in many states).

FIGURE 1.20
The tooltip for the Create Line By Angle command

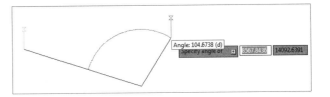

CREATE LINE BY DEFLECTION COMMAND

By definition, a *deflection angle* is the angle turned from the extension of a line from the backsight extending through an instrument. Although this isn't the most frequently used surveying tool in this day of data collectors and GPSs, on some occasions you may need to create this type of line. When you use the Create Line By Deflection command, the command line and tooltips prompt you for a deflection angle followed by a distance (Figure 1.21). In some cases, deflection angles are recorded in the field in lieu of right angles.

FIGURE 1.21
The tooltips for the Create Line By Deflection command

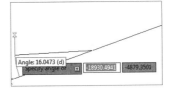

CREATE LINE BY STATION/OFFSET COMMAND

To use the Create Line By Station/Offset command, you must have a Civil 3D Alignment object in your drawing. The line created from this command allows you to start and/or end a line on the basis of a station and offset from an alignment.

You're prompted to choose the alignment and then input a station and offset value. The line *begins* at the station and offset value. On the basis of the tooltips, you might expect the line to be drawn from the alignment station at offset zero and out to the alignment station at the input offset. This isn't the case.

When prompted for the station, you're given a tooltip that tracks your position along the alignment, as shown in Figure 1.22. You can graphically choose a station location by picking in the drawing (including using your osnaps to assist you in locking down the station of a specific feature). Alternatively, you can enter a station value on the command line.

FIGURE 1.22
The Create Line By Station/Offset command provides a tooltip for you to track stationing along the alignment.

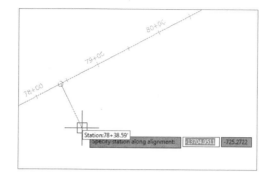

Once you've selected the station, you're given a tooltip that is locked on that particular station and tracks your offset from the alignment (see Figure 1.23). You can graphically choose an offset by picking in the drawing, or you can type an offset value on the command line.

FIGURE 1.23
The Create Line By Station/Offset command provides a tooltip that helps you track the offset from the alignment.

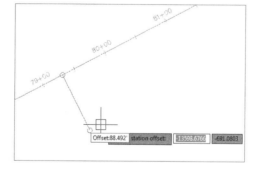

CREATE LINE BY SIDE SHOT COMMAND

The Create Line By Side Shot command lets you occupy one point, designate a backsight, and draw a line that has endpoints relative to that point. The occupied point represents the setup of your surveying station, whereas the second point represents your surveying backsight. This tool may be most useful when you're creating stakeout information or re-creating data from field notes. If you know where your crew set up and you have their side-shot angle measurements but you don't have electronic information to download, this tool can help. To specify locations relative to your occupied point, you can specify the angle, bearing, deflection, or azimuth on the command line or pick locations in your drawing. In some cases, it is more appropriate to supply a survey crew with handwritten notes regarding backsights, foresights, angles right, and distances rather than upload the same information to a data collector.

While the command is active, you can toggle between angle, bearing, deflection, and azimuth by following the command-line prompts.

When you're using the Create Line By Side Shot command, you're given a setup glyph at your occupied point, a backsight glyph, and a tooltip to track the angle, bearing, deflection, or azimuth of the side shot (see Figure 1.24). You can toggle between these options by following the command-line prompts.

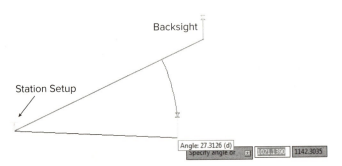

FIGURE 1.24
The tooltip for the Create Line By Side Shot command tracks the angle, bearing, deflection, or azimuth of the side shot.

CREATE LINE EXTENSION COMMAND

The Create Line Extension command is similar to the AutoCAD Lengthen command. This command allows you to add length to a line or specify a desired total length of the line.

You are first prompted to choose a line. The command line then displays the prompt `Specify distance to change, or [Total]`. The distance you specify is added to the existing length of the line. The command draws the line appropriately and provides a short summary report on that line. The summary report in Figure 1.25 indicates that the beginning line length was 100" and that an additional distance of 50" was specified with the Line Extension command. It is important to note that in some cases, it may be more desirable to create a line by a turned angle or deflection of 180 degrees so as not to disturb linework originally created from existing legally recorded documents.

FIGURE 1.25
The Create Line Extension command provides a summary of the changes to the line.

```
Select line object:
Specify distance to change, or [Total]: 50
-------------------------------------------------------------
                            LINE DATA
-------------------------------------------------------------
Begin . . . . . North: 1216.9486'        East: -617.9729'
End . . . . . . North: 1279.6199'        East: -481.6927'
              Distance: 150.000'       Course: N65° 18' 13"E
```

If instead you specify a total distance on the command line, the length of the line is changed to the distance you specify. The summary report shown in Figure 1.26 indicates that the beginning of the line was the same as in Figure 1.26 but with a total length of only 100".

FIGURE 1.26
The summary report on a line where the command specified a total distance

```
Command: _AeccLineExtension
Select line object:
Specify distance to change, or [Total]: t
Specify total distance, or [Change]: 100
-------------------------------------------------------------
                            LINE DATA
-------------------------------------------------------------
Begin . . . . . North: 1216.9486'        East: -617.9729'
End . . . . . . North: 1258.7295'        East: -527.1194'
              Distance: 100.000'       Course: N65° 18' 13"E
```

Create Line From End Of Object Command

The Create Line From End Of Object command lets you draw a line tangent to the end of a line or arc of your choosing. Most commonly, you'll use this tool when re-creating deeds or other survey work where you have to specify a line that continues a tangent from an arc (see Figure 1.27).

FIGURE 1.27
The Create Line From End Of Object command lets you add a tangent line to the end of an arc.

Create Line Tangent From Point Command

The Create Line Tangent from Point command is similar to the Create Line From End Of Object command, but Create Line Tangent From Point allows you to choose a point of tangency that isn't the endpoint of the line or arc (see Figure 1.28).

FIGURE 1.28
The Create Line Tangent From Point command can place a line tangent at the midpoint of an arc (or line).

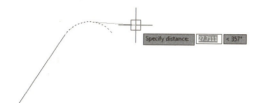

Create Line Perpendicular From Point Command

Using the Create Line Perpendicular From Point command, you can specify that you'd like a line drawn perpendicular to any point of your choosing. In the example shown in Figure 1.29, a line is drawn perpendicular to the endpoint of the arc. This command can be useful when the distance from a known monument perpendicular to a legally platted line must be labeled in a drawing.

FIGURE 1.29
A perpendicular line is drawn from the endpoint of an arc, using the Create Line Perpendicular From Point command.

Creating Curves

Curves are an important part of surveying and engineering geometry. In truth, curves are no different from AutoCAD arcs. What makes the curve commands different than the basic AutoCAD commands isn't the resulting arc entity but the inputs used to draw the arc. Civil 3D wants you to provide directions to the arc commands using land surveying terminology rather than with generic Cartesian parameters.

Figure 1.30 shows the Create Curves menu options.

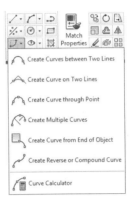

FIGURE 1.30
Create Curves commands

Standard Curves

When re-creating legal descriptions for roads, easements, and properties, engineers, surveyors, and mappers often encounter a variety of curves. Although standard AutoCAD arc commands could draw these arcs, the AutoCAD arc inputs are designed to be generic to all industries. The following curve commands have been designed to provide an interface that more closely matches land surveying, mapping, and engineering language.

CREATE CURVE BETWEEN TWO LINES COMMAND

The Create Curve Between Two Lines command is much like the standard AutoCAD Fillet command, except that you aren't limited to a radius parameter. The command draws a curve that is tangent to two lines of your choosing. This command also trims or extends the original tangents so their endpoints coincide with the curve endpoints. The lines are trimmed or extended to the resulting PC (point of curve, which is the beginning of a curve) and PT (point of tangency, or the end of a curve). You may find this command most useful when you're creating foundation geometry for road alignments, parcel boundary curves, and similar situations.

The command prompts you to choose the first tangent and then the second tangent. The command line gives the following prompt:

```
Select entry [Tangent/External/Degree/Chord/Length/Mid-Ordinate/
miN-dist/Radius]<Radius>:
```

Pressing ↵ at this prompt lets you input your desired radius. As with standard AutoCAD commands, pressing T changes the input parameter to Tangent, pressing C changes the input parameter to Chord, and so on.

As with the Fillet command, your inputs must be geometrically possible. For example, your two lines must allow for a curve of your specifications to be drawn while remaining tangent to both. Figure 1.31 shows two lines with a 25″ radius curve drawn between them. Note that the tangents have been trimmed so their endpoints coincide with the endpoints of the curve. If either line had been too short to meet the endpoint of the curve, then that line would have been extended.

FIGURE 1.31
Two lines using the Create Curve Between Two Lines command

CREATE CURVE ON TWO LINES COMMAND

The Create Curve On Two Lines command is identical to the Create Curve Between Two Lines command, except that the Create Curve On Two Lines command leaves the chosen tangents intact. The lines aren't trimmed or extended to the resulting PC and PT of the curve.

Figure 1.32, for example, shows two lines with a 25″ radius curve drawn on them. The tangents haven't been trimmed and instead remain exactly as they were drawn before the Create Curve On Two Lines command was executed.

FIGURE 1.32
The original lines stay the same after you execute the Create Curve On Two Lines command.

CREATE CURVE THROUGH POINT COMMAND

The Create Curve Through Point command lets you choose two tangents for your curve followed by a pass-through point. This tool is most useful when you don't know the radius, length, or other curve parameters but you have two tangents and a target location. It isn't necessary that the pass-through location be a true point object; it can be any location of your choosing.

This command also trims or extends the original tangents so their endpoints coincide with the curve endpoints. The lines are trimmed or extended to the resulting PC and PT of the curve.

Figure 1.33, for example, shows two lines and a desired pass-through point. Using the Create Curve Through Point command allows you to draw a curve that is tangent to both lines and that passes through the desired point. In this case, the tangents have been trimmed to the PC and PT of the curve.

FIGURE 1.33
The first image shows two lines with a desired pass-through point. In the second image, the Create Curve Through Point command draws a curve that is tangent to both lines and passes through the chosen point.

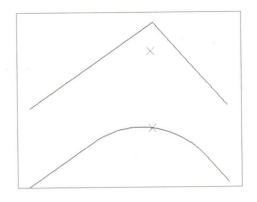

CREATE MULTIPLE CURVES COMMAND

The Create Multiple Curves command lets you create several curves that are tangentially connected. The resulting curves have an effect similar to an alignment spiral section. This command can be useful when you are re-creating railway track geometry based on field survey data.

The command prompts you for the two tangents. Then, the command line prompts you as follows:

    Enter Number of Curves:

The command allows for up to 10 curves between tangents.

One of your curves must have a flexible length that's determined on the basis of the lengths, radii, and geometric constraints of the other curves. Curves are counted clockwise, so enter the number of your flexible curve:

    Enter Floating Curve #:
    Enter the length and radii for all your curves:
    Enter curve 1 Radius:
    Enter curve 1 Length:

The floating curve number will prompt you for a radius but not a length.

As with all other curve commands, the specified geometry must be possible. If the command can't find a solution on the basis of your length and radius inputs, it returns no solution (see Figure 1.34).

FIGURE 1.34
Two curves were specified with the #2 curve designated as the floating curve.

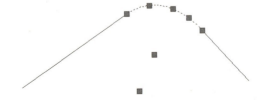

Create Curve From End Of Object Command

The Create Curve From End Of Object command enables you to draw a curve tangent to the end of your chosen line or arc.

The command prompts you to choose an object to serve as the beginning of your curve. You can then specify a radius and an additional parameter (such as Delta or Length) for the curve or the endpoint of the resulting curve chord (see Figure 1.35).

FIGURE 1.35
A curve, with a 25″ radius and a 30″ length, drawn from the end of a line

Create Reverse Or Compound Curves Command

The Create Reverse Or Compound Curves command allows you to add additional curves to the end of an existing curve. Reverse curves are drawn in the opposite direction (i.e., a curve to the right tangent to a curve to the left) from the original curve to form an S shape. In contrast, compound curves are drawn in the same direction as the original curve (see Figure 1.36). This tool can be useful when you are re-creating a legal description of a road alignment that contains reverse and/or compound curves.

FIGURE 1.36
A tangent and curve before adding a reverse or compound curve (left); a compound curve drawn from the end of the original curve (right); and a reverse curve drawn from the end of the original curve (bottom)

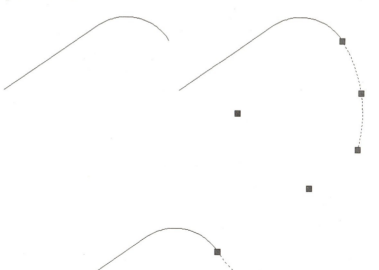

Re-creating a Deed Using Line and Curve Tools

This exercise will help you apply some of the tools you've learned so far to reconstruct the overall parcel that will be used as the sample exercises for the majority of the book.

```
From Point of Beginning
South 44 degrees 54 minutes 15 seconds West 68.64 feet to a point
North 07 degrees 05 minutes 24 seconds East 217.80 feet to a point
North 72 degrees 12 minutes 10 seconds East 4.23 feet to a point
North 05 degrees 53 minutes 27 seconds East 201.09 feet to a point
South 86 degrees 32 minutes 10 seconds East 121.22 feet to a point
North 03 degrees 25 minutes 51 seconds West 168.78 feet to a point
North 14 degrees 38 minutes 58 seconds East 283.16 feet to a point
North 07 degrees 19 minutes 22 seconds West 79.64 feet to a point
North 07 degrees 04 minutes 00 seconds West 205.45 feet to a point
South 46 degrees 24 minutes 36 seconds West 121.05 feet to a point
South 48 degrees 31 minutes 20 seconds West 414.66 feet to a point
North 49 degrees 29 minutes 56 seconds West 50.80 feet to a point
North 48 degrees 37 minutes 57 seconds East 150.29 feet to a point
North 05 degrees 39 minutes 50 seconds East 497.28 feet to a point
North 84 degrees 20 minutes 01 seconds East 290.33 feet to a point
North 05 degrees 20 minutes 48 seconds West 195.08 feet to a point
North 76 degrees 46 minutes 10 seconds East 701.96 feet to a point
South 23 degrees 42 minutes 48 seconds East 130.68 feet to a point
South 20 degrees 13 minutes 35 seconds East 526.50 feet to a point
South 76 degrees 04 minutes 14 seconds West 379.96 feet to a point
South 13 degrees 22 minutes 41 seconds East 320.08 feet to a point
South 12 degrees 36 minutes 45 seconds East 159.86 feet to a point
South 12 degrees 21 minutes 15 seconds East 274.32 feet to a point
South 61 degrees 15 minutes 09 seconds West 272.81 feet to a point
North 06 degrees 15 minutes 30 seconds West 131.45 feet to a point
South 72 degrees 12 minutes 22 seconds West 301.60 feet to a point
South 06 degrees 58 minutes 04 seconds East 206.04 feet to a point
Returning to Point of Beginning
The resulting enclosure should be: 26.25 acres (more or less)
```

Follow these steps:

1. Open the `Deed Create Start.dwg` file, which you can download from this book's web page at www.sybex.com/masteringcivil3d2012.

2. Turn off Dynamic Input by pressing F12, or by toggling the icon off at the status bar.

3. From the Draw panel on the Home tab, select the Line drop-down and choose the Create Line By Bearing command.

4. At the `Select first point:` prompt, select any location in the drawing to begin the first line.

5. At the `>>Specify quadrant (1-4):` prompt, enter **3** to specify the SW quadrant, and then press ↵.

6. At the >>Specify bearing: prompt, enter **44.5415**, and press ↵.

7. At the >>Specify distance: prompt, enter **68.64**, and press ↵.

8. Repeat steps 4 through 6 for the rest of the courses.

9. Press Esc to exit the Create Line By Bearing command.

10. The finished linework should look like Figure 1.37. There will be an error of closure of 10.0016″. Typically, rounding errors can cause an error in closure. Perhaps reworking the deed holding a different rounding value would improve your results. Consult your office survey expert about how this would be handled in house, and refer to Chapter 2 for more information about traverse adjustment and similar tools.

11. Save your drawing. You'll need it for the next exercise.

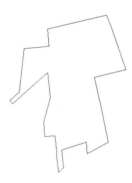

Figure 1.37
The finished linework

Best Fit Entities

Although engineers and surveyors do their best to make their work an exact science, sometimes tools like the Best Fit Entities are required.

Roads in many parts of the world have no defined alignment. They may have been old carriage roads or cart paths from hundreds of years ago that evolved into automobile roads. Surveyors and engineers are often called to help establish official alignments, vertical alignments, and right-of-way lines for such roads on the basis of a best fit of surveyed centerline data.

Other examples for using Best Fit Entities include property lines of agreement, road rehabilitation projects, and other cases where existing survey information must be approximated into "real" engineering geometry (see Figure 1.38).

Figure 1.38
The Create Best Fit Entities menu options

CREATE BEST FIT LINE COMMAND

The Create Best Fit Line command under the Best Fit drop-down on the Draw panel takes a series of Civil 3D points, AutoCAD points, entities, or drawing locations and draws a single best-fit line segment from this information. In Figure 1.39, for example, the Create Best Fit Line command draws a best-fit line through a series of points that aren't quite collinear. Note that the best-fit line will change as more points are picked.

FIGURE 1.39
A preview line drawn through points that aren't quite collinear

Once you've selected your points, a Panorama window appears with a regression data chart showing information about each point you chose, as shown in Figure 1.40.

FIGURE 1.40
The Panorama window lets you optimize your best fit.

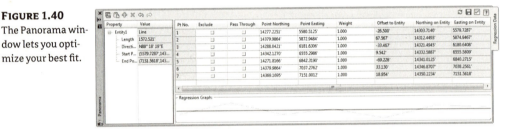

This interface allows you to optimize your best fit by adding more points, selecting the check box in the Pass Through column to force one of your points on the line, or adjusting the value under the Weight column.

CREATE BEST FIT ARC COMMAND

The Create Best Fit Arc command under the Best Fit drop-down works identically to the Create Best Fit Line command, except that the resulting entity is a single arc segment as opposed to a single line segment (see Figure 1.41).

FIGURE 1.41
A curve created by best fit

Create Best Fit Parabola Command

The Create Parabola command under the Create Best Fit Entities option works in a similar way to the line and arc commands just described. This command is most useful when you have a Triangulated Irregular Network (also known as TIN) sampled or surveyed road information and you'd like to replicate true vertical curves for your design information.

After you select this command, the Parabola By Best Fit dialog appears (see Figure 1.42).

FIGURE 1.42
The Parabola By Best Fit dialog

You can select inputs from entities (such as lines, arcs, polylines, or profile objects) or by picking on screen. The command then draws a best-fit parabola on the basis of this information. In Figure 1.43, the shots were represented by AutoCAD points; more points were added by selecting the By Clicking On The Screen option and using the Node osnap to pick each point.

FIGURE 1.43
The best-fit preview line changes as more points are picked.

Once you've selected your points, a Panorama window appears, showing information about each point you chose. Also note the information in the right pane regarding K-value, curve length, grades, and so forth.

In this interface (shown in Figure 1.44), you can optimize your K-value, length, and other values by adding more points, selecting the check box in the Pass Through column to force one of your points on the line, or adjusting the value under the Weight column.

FIGURE 1.44
The Panorama window lets you make adjustments to your best-fit parabola.

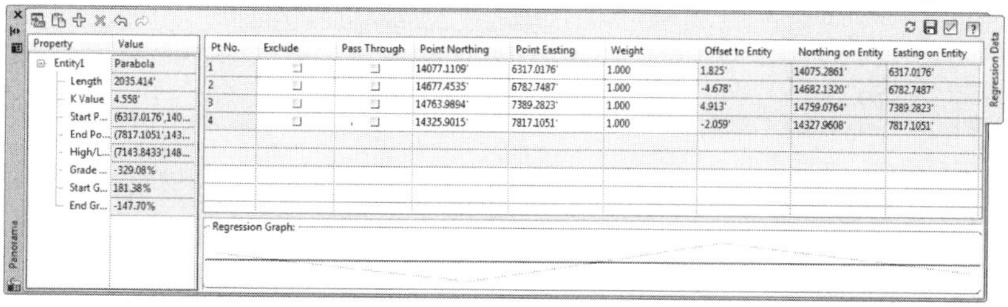

34 | **CHAPTER 1** THE BASICS OF AUTOCAD CIVIL 3D

Attach Multiple Entities

The Attach Multiple Entities command (found on the Home tab and extended Draw panel pull-down) is a combination of the Line From End Of Object command and the Curve From End Of Object command. This command is most useful for reconstructing deeds or road alignments from legal descriptions when each entity is tangent to the previous entity. Using this command saves you time because you don't have to constantly switch between the Line From End Of Object command and the Curve From End Of Object command (see Figure 1.45).

Figure 1.45
The Attach Multiple Entities command draws a series of lines and arcs so that each segment is tangent to the previous one.

The Curve Calculator

Sometimes you may not have enough information to draw a curve properly. Although many of the curve-creation tools assist you in calculating the curve parameters, you may find an occasion where the deed you're working with is incomplete.

The Curve Calculator found in the Curves drop-down on the Draw panel helps you calculate a full collection of curve parameters on the basis of your known values and constraints. The units used in the Curve Calculator match the units assigned in your Drawing Settings.

The Curve Calculator can remain open on your screen while you're working through commands. You can send any value in the Calculator to the command line by clicking the button next to that value (see Figure 1.46).

The button at the upper left of the Curve Calculator inherits the arc properties from an existing arc in the drawing, and the drop-down menu in the Degree Of Curve Definition selection field allows you to choose whether to calculate parameters for an arc or a chord definition.

The drop-down menu in the Fixed Property selection field also gives you the choice of fixing your radius or delta value when calculating the values for an arc or a chord, respectively (see Figure 1.47). The parameter chosen as the fixed value is held constant as additional parameters are calculated.

Figure 1.46
The Curve Calculator

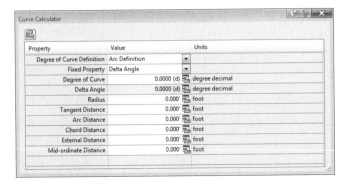

CREATING CURVES | 35

FIGURE 1.47
The Fixed Property drop-down menu gives you the choice of fixing your radius or delta value.

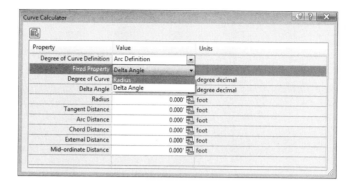

As explained previously, you can send any value in the Curve Calculator to the command line using the button next to that value. This ability is most useful while you're active in a curve command and would like to use a certain parameter value to complete the command.

Adding Line and Curve Labels

CERT OBJECTIVE

Although most robust labeling of site geometry is handled using Parcel or Alignment labels, limited line- and curve-annotation tools are available in Civil 3D. The line and curve labels are composed much the same way as other Civil 3D labels, with marked similarities to Parcel and Alignment Segment labels.

Our next exercise leads you through labeling the deed you re-created earlier in this chapter:

1. Continue working in the `Deed Create Start.dwg` file.

2. Click the Labels button in the Labels & Tables panel on the Annotate tab. The Add Labels dialog appears, as shown in Figure 1.48.

3. Choose Line And Curve from the Feature drop-down menu.

4. Choose Multiple Segment from the Label Type drop-down menu. The Multiple Segment option places the label at the midpoint of each selected line or arc.

5. Confirm that Line Label Style is set to Bearing Over Distance and that Curve Label Style is set to Distance-Radius And Delta.

6. Click the Add button.

FIGURE 1.48
The Add Labels dialog, set to Multiple Segment Labels

> **Where Is Delta?**
>
> In the Text Component Editor for a curve label, the value that most people would refer to as a delta angle is called the General Segment Total Angle. To insert the Delta symbol in a label, type \U+0394 in the Text Editor window on the right side of the Text Component Editor dialog.

7. At the Select Entity: prompt, select each line and arc that you drew in the previous exercise. A label appears on each entity at its midpoint, as shown in Figure 1.49.

Figure 1.49
The labeled linework

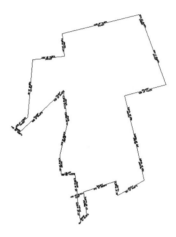

8. Save the drawing—you'll need it for the next exercise.

Using Transparent Commands

In many cases, the "Create Line By…" commands in the Draw panel are the standard AutoCAD Line commands combined with the appropriate transparent commands.

A transparent command behaves somewhat similarly to an osnap command. You can't click the Endpoint button and expect anything to happen—you must be active inside another command, such as a line, an arc, or a circle command.

The same principle works for transparent commands. Once you're active in the Line command (or any AutoCAD or Civil 3D drawing command), you can choose the Bearing Distance transparent command and complete your drawing task using a bearing and distance.

As stated earlier, the transparent commands can be used in any AutoCAD or Civil 3D drawing command, much like an osnap. For example, you can be actively drawing an alignment and use the Northing/Easting transparent command to snap to a particular coordinate, and then press Esc once and continue drawing your alignment as usual.

While a transparent command is active, you can press Esc once to leave the transparent mode but stay active in your current command. You can then choose another transparent command if you'd like. For example, you can start a line using the Endpoint osnap, activate the Angle

Distance transparent command, draw a line-by-angle distance, and then press Esc, which takes you out of angle-distance mode but keeps you in the Line command. You can then draw a few more segments using the Point Object transparent command, press Esc, and finish your line with a Perpendicular osnap.

You can activate the transparent commands using keyboard shortcuts or using the Transparent Commands toolbar. Be sure you include the Transparent Commands toolbar (shown in Figure 1.50) in all your Civil 3D and survey-oriented workspaces.

Figure 1.50
The Transparent Commands toolbar

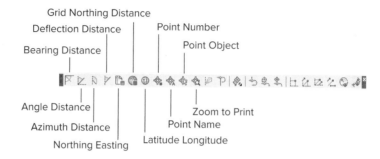

The six profile-related transparent commands will be covered in Chapter 7, "Profiles and Profile Views."

Standard Transparent Commands

The transparent commands shown in Table 1.1 behave identically to their like-named counterparts from the Draw panel (discussed earlier in this chapter). The difference is that you can call up these transparent commands in any appropriate AutoCAD or Civil 3D draw command, such as a line, polyline, alignment, parcel segment, feature line, or pipe-creation command.

Table 1.1: The transparent commands

Tool Icon	Menu Command
⊠	Angle Distance
⊠	Bearing Distance
⟋	Azimuth Distance
⊳	Deflection Distance
▥	Northing Easting
◉	Grid Northing Easting

TABLE 1.1: The transparent commands *(CONTINUED)*

TOOL ICON	MENU COMMAND
	Latitude Longitude
	Point Number
	Point Name
	Point Object
	Zoom To Point
	Side Shot
	Station Offset

Matching Transparent Commands

You may have construction or other geometry in your drawing that you'd like to match with new lines, arcs, circles, alignments, parcel segments, or other entities.

While actively drawing an object that has a radius parameter, such as a circle, an arc, an alignment curve, or a similar object, you can choose the Match Radius transparent command and then select an object in your drawing that has your desired radius. Civil 3D draws the resulting entity with a radius identical to that of the object you chose during the command. You'll save time using this tool because you don't have to first list the radius of the original object and then manually type in that radius when prompted by your circle, arc, or alignment tool.

The Match Length transparent command works identically to the Match Radius transparent command except that it matches the length parameter of your chosen object.

The Underlying Engine

Civil 3D is part of a larger product family from Autodesk. During its earliest creation, various features and functions from other products were recognized as important to the civil engineering community. These included the obvious things such as the entire suite of AutoCAD drafting, design, modeling, and rendering tools as well as more esoteric options such as Map's GIS capabilities. An early decision was made to build Civil 3D on top of the AutoCAD Map product, which in turn is built on top of AutoCAD.

This underlying engine provides a host of options and powerful tools for the Civil 3D user. AutoCAD and Map add features with every release that change the fundamental makeup of how Civil 3D works. With the introduction of workspaces in 2006, users can now set up Civil 3D to display various tools and palettes depending on the task at hand. Creating a workspace is like having a quick-fix bag of tools ready: preliminary design calls for one set of tools, and final plan production calls for another.

Workspaces are part of a larger feature set called the *custom user interface* (referred to as CUI in the help documentation and online). As you grow familiar with Civil 3D and the various tool palettes, menus, and toolbars, be sure to explore the CUI options that are available from the Workspace toolbar.

You may have noticed that when you start typing in Civil 3D, it shows a list of the commands and set variables that begin with that letter, and as you type further, it refines that list. This is AutoCAD's new autocomplete feature. If you do not want to use this feature, you can type **AUTOCOMPLETE** and set the command to **OFF**.

Managing Civil 3D Information

The Manage tab contains many of the management tools available in Civil 3D. Many of these tools are continuations of the basic AutoCAD tools, but they are worth some discussion.

- The Data Shortcuts panel contains all the tools related to using data shortcuts. You will learn more about data shortcuts in chapter 17. "Interoperability".

- The Customization panel has the tools for manipulating the user interface and tool palettes via the Customize dialog box. You can also import and export your customized user interface (CUI). And for the real hackers, you can set aliases; for example, if you wanted to change C (which is the default keyboard shortcut for circle) to COPY, you can take care of that here.

- The Applications panel allows you to run specialized third-party applications, as well as manage Lisp files. And you thought that Lisp was dead!

- The CAD Standards panel allows someone such as a CAD Manager to set compliances for layering, allow importing of third-party drawing file layers, and, via a macro, change their layers to your company's standards automatically.

- The Action Recorder panel contains all the tools for recording and playback of keystrokes. It has been around for some time but is probably not used much in the Civil 3D world. You can record keystrokes and play them back to remedy the repetitive keystrokes one might use over and over.

- The Styles panel is new to Civil 3D 2012 and is a welcome addition. You can now import styles via a dialog box instead of the old ways of accomplishing this. And the Purge tool will look at all the styles in your drawing and allow you to remove ones that are not in use. You will learn more about these tools in Chapter 19, "Styles."

The Bottom Line

Find any Civil 3D object with just a few clicks. By using Prospector to view object data collections, you can minimize the panning and zooming that are part of working in a CAD program. When common subdivisions can have hundreds of parcels or a complex corridor can have dozens of alignments, jumping to the desired one nearly instantly shaves time off everyday tasks.

Master It Open `BasicSite.dwg` from `www.sybex.com/go/masteringcivil3d2012`, and find parcel number 18 without using any AutoCAD commands or scrolling around on the drawing screen.

Modify the drawing scale and default object layers. Civil 3D understands that the end goal of most drawings is to create hard-copy construction documents. By setting a drawing scale and then setting many sizes in terms of plotted inches or millimeters, Civil 3D removes much of the mental gymnastics that other programs require when you're sizing text and symbols. By setting object layers at a drawing scale, Civil 3D makes uniformity of drawing files easier than ever to accomplish.

Master It Change `BasicSite.dwg` from the 100-scale drawing to a 40-scale drawing.

Modify the display of Civil 3D tooltips. The interactive display of object tooltips makes it easy to keep your focus on the drawing instead of an inquiry or report tools. When too many objects fill up a drawing, it can be information overload, so Civil 3D gives you granular control over the heads-up display tooltips.

Master It Within the same BasicSite drawing, turn off the tooltips for the Road A alignment.

Navigate the Ribbon's contextual tabs. As with AutoCAD, the Ribbon is the primary interface for accessing Civil 3D commands and features. When you select an AutoCAD Civil 3D object, the Ribbon displays commands and features related to that object. If several object types are selected, the Multiple contextual tab is displayed.

Master It Using the Ribbon interface, access the Alignment Style Editor for the Proposed Alignment style. (Hint: it's used by the Road A alignment.)

Create a curve tangent to the end of a line. It's rare that a property stands alone. Often, you must create adjacent properties, easements, or alignments from their legal descriptions.

Master It Create a curve tangent to the end of the first line drawn in the first exercise that meets the following specifications:

Radius: 200.00"

Arc Length: 66.580"

Label lines and curves. Although converting linework to parcels or alignments offers you the most robust labeling and analysis options, basic line- and curve-labeling tools are available when conversion isn't appropriate.

Master It Add line and curve labels to each entity created in the exercises. Choose a label that specifies the bearing and distance for your lines and length, radius, and delta of your curve.

Chapter 2

Survey

Civil 3D supports a collaborative workflow in many aspects of the design process, but especially in the survey realm. Accurate data starts outdoors. A survey that has been consistently and correctly coded in the field can save hours of drafting time. Nobody knows the site better than the folks who were out there freezing in their boots, sloshing through the project location. Surveyors can collect line information such as swales, curbs, or even pavement markings and communicate this digitally to the data collectors.

Civil 3D can often eliminate the need for third-party survey software. Civil 3D can download and process survey data directly from a data collector. To enter data in a manner that is easily digested by Civil 3D, your survey process can incorporate the information from this chapter.

In this chapter, you will learn to:

- Properly collect field data and import it into Civil 3D
- Set up description key and figure databases
- Create and edit field book files
- Manipulate your survey data

Setting Up the Databases

Before any project-specific data is imported, there is a bit of onetime setup that will improve the translation between the field and the office.

Your survey database defaults, equipment database, linework code set, and the figure prefix database should be in place before you import your first survey. You can find the location of these files by going to the Survey tab in Toolspace and clicking the Survey User Settings button in the upper-left corner. The dialog shown in Figure 2.1 opens.

It is common practice to place these files on a network server so your organization can share them. Change the paths in this dialog and they will "stick" regardless of which drawing you have open.

Survey Database Defaults

When you first set up your survey settings, it is a good idea to create a test database for setting the survey database defaults. To create the test database, right-click Survey Databases on the Survey tab and select New Local Survey Database. Name the new database **Test**, and click OK to continue. Right-click the new Test database and select Edit Survey Database Settings.

Once you set your desired defaults for units, precision, and other options, click the Export Settings To A File button. Then save the settings to the folder specified in the Survey User Settings dialog (see Figure 2.2).

FIGURE 2.1
Survey User Settings dialog

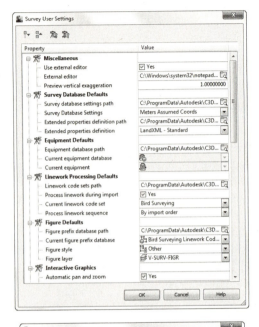

FIGURE 2.2
Survey Database settings

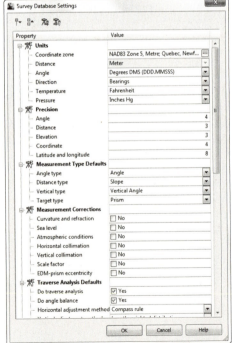

The Survey User Settings dialog contains these settings:

Units This section is where you set your master coordinate zone for the database. If you insert any information in the database into a drawing with a different coordinate zone set, the program will automatically translate that data to the drawing coordinate zone. Your coordinate zone units will lock the distance units in the Units section. You can also set the angle, direction, temperature, and pressure here.

Precision This section is where you define and store the precision information of angles, distance, elevation, coordinates, and latitude and longitude.

Measurement Type Defaults This section lets you define the defaults for measurement types, such as angle type, distance type, vertical type, and target type.

Measurement Corrections This section is used to define the methods (if any) for correcting measurements. Some data collectors allow you to make measurement corrections as you collect the data, so that needs to be verified, because double correction applications could lead to incorrect data.

Traverse Analysis Defaults This section is where you choose how you perform traverse analyses and define the required precision and tolerances for each. There are four types of 2D traverse analyses: Compass Rule, Transit Rule, Crandall Rule, and Least Squares Analysis.

There are also two types of 3D traverse analyses: Length Weighted Distribution and Equal Distribution.

Least Squares Analysis Defaults This section is where you set the defaults for a least squares analysis. You only need to change settings here if least squares analysis is the method you will use for your horizontal and/or vertical adjustments.

Survey Command Window The Survey Command window is the interface for manual survey tasks and for running survey batch files. This section lets you define the default settings for this window.

Error Tolerance Set tolerances for the survey database in this section. If you perform an observation more than one time and the tolerances set within are not met, an error will appear in the Survey Command window and ask you what action you want to take.

Extended Properties This section defines settings for adding extended properties to a survey LandXML file. Extended properties are useful for certain types of surveys, including Federal Aviation Administration (FAA)–certified surveys.

To delete your test database, you will need to close the database from the Survey tab. To do so, locate the database in the Survey tab. Right-click on it and select Close Survey Database. Using Windows Explorer, you can then browse to the working folder containing the database and delete it. There is no way to delete a survey database from within the software.

The Equipment Database

The equipment database is where you set up the various types of survey equipment that you are using in the field. Doing so allows you to apply the proper correction factors to your traverse analyses when it is time to balance your traverse. Civil 3D comes with a sample piece of equipment for you to inspect to see what information you will need when it comes time to create your equipment. The Equipment Properties dialog provides all the default settings for the sample equipment in the equipment database. Expand the Equipment Databases ➢ Sample branches, right-click Sample, and select Manage Equipment Database to access this dialog in Toolspace, as shown in Figure 2.3.

FIGURE 2.3
Equipment Database Manager

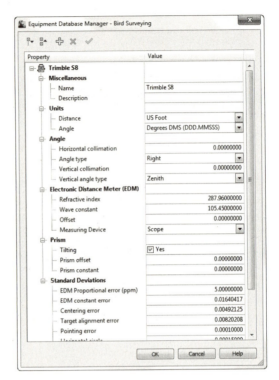

You will want to create your own equipment entries and enter the specifications for your particular total station. Add a new piece of equipment to the database by clicking the plus sign at the top of the Equipment Database Manager window. If you are unsure of the settings to enter, refer to the user documentation that you received when you purchased your total station.

The Figure Prefix Database

The figure prefix database is used to translate descriptions in the field to lines in CAD. These survey-generated lines are called *figures*. If a description matches a listing in the figure prefix database, the figure is assigned the properties and style dictated therein (see Figure 2.4).

The Figure Prefix Database Manager contains these columns:

Breakline   This column specifies whether the figure should be used as a breakline during surface creation. To dispel common misconceptions about this option, you should know that a figure flagged as a breakline still needs to be added to the surface definition by the user (i.e., no automatic breaklines appear in a surface), and a figure not flagged as a breakline can still be added to the surface. This option sets the Breakline Option to Yes automatically when you right-click the Figures listing and select Create Breaklines.

Figure 2.4
Figure Prefix Database Manager

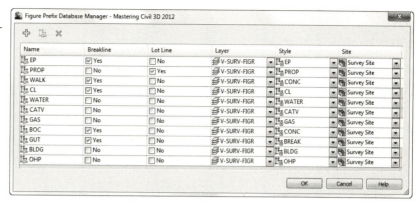

Lot Line   This column specifies whether the figure should behave as a parcel segment. If a closed area is formed with figures of this kind, a parcel is also formed.

Layer   This column specifies the layer that the figure will reside on when inserted into the drawing. If the layer already exists in the drawing, the figure will be placed on that layer. If the layer does not exist in the drawing, the layer will be created and the figure placed on the newly created layer.

Style   This column specifies the style to be used for each figure. The style contains the linetype and color of the figure. If the style exists in the current drawing, the figure will be inserted using that style. If the style does not exist in the drawing, the style will be created with the standard settings and the figure will be placed in the drawing according to the settings specified with the newly created style.

Site   This column specifies which site the figures should reside on when inserted into the drawing. As with previous settings, if the site exists in the drawing, the figure will be inserted into that site. If the site does not exist in the drawing, a site will be created with that name and the figure will be inserted into the newly created site.

Next you'll explore these settings in a practical exercise. You'll use the styles created in the previous exercise:

1. Open the drawing `figure prefix library.dwg`, which you can download from this book's webpage, www.sybex.com/go/masteringcivil3d2012.

2. In the Survey tab of Toolspace, right-click Figure Prefix Databases and select New. The New Figure Prefix Database dialog opens.

3. Enter **Mastering Civil 3D** in the Name text box, and click OK to dismiss the dialog. Mastering Civil 3D will now be listed under the Figure Prefix Databases section in the Survey tab on Toolspace.

4. Right-click the newly created Mastering Civil 3D figure prefix database and select Manage Figure Prefix Database. The Figure Prefix Database Manager will appear.

5. Select the white + symbol in the upper-left corner of the Figure Prefix Database Manager to create a new figure prefix.

6. Click the SAMPLE name and rename the figure prefix to **EP** (for Edge of Pavement).
7. Check the No box in the Breakline column to change it to Yes so the figure will be treated as a breakline. Leave the box in the Lot Line column unchecked, so the figure will not be treated as a parcel segment.
8. Under the Layer column, select V-SURV-FIGR.
9. Under the Style column, select EP.
10. Under the Site column, leave the name of the site set to Survey Site.
11. Complete the Figure Prefixes table with the values shown in Table 2.1.
12. Click OK to dismiss the Figure Prefix Database Manager.

TABLE 2.1: Figure settings

NAME	BREAKLINE	LOT LINE	LAYER	STYLE	SITE
PROP	No	Yes	V-FIGURE	PROP	PROPERTY
WALK	Yes	No	V-FIGURE	CONC	SURVEY SITE
CL	Yes	No	V-FIGURE	CL	SURVEY SITE
WATER	No	No	V-FIGURE	WATER	SURVEY SITE
CATV	No	No	V-FIGURE	CATV	SURVEY SITE
GAS	No	No	V-FIGURE	GAS	SURVEY SITE
BOC	Yes	No	V-FIGURE	CONC	SURVEY SITE
GUT	Yes	No	V-FIGURE	BREAK	SURVEY SITE
BLDG	No	No	V-FIGURE	BLDG	SURVEY SITE
OHP	No	No	V-FIGURE	OHP	SURVEY SITE

The Linework Code Set Database

The linework code set (Figure 2.5) lists what designators are used to start, stop, continue, or add additional geometry to lines. For example, the B code that is typically used to begin a line can be replaced by a code of your choosing, a decimal (.) can be used for a right-turn value, and a minus sign (–) can be used for a left-turn value. Linework code sets allow a survey crew to customize their data collection techniques based on methods used by various types of software not related to Civil 3D.

FIGURE 2.5
The Edit Linework Code Set dialog

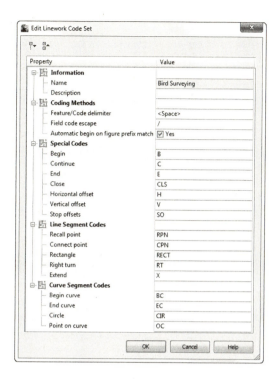

> **YOU DO NOT NEED LINEWORK CODES IN YOUR SURVEY TO GENERATE LINES!**
>
> There is a setting in the linework code set called Automatic Begin On Figure Prefix Match (as you can see in Figure 2.5). If you select this option, lines will start when a shot description matches one of your figure prefixes (such as EP or CL in Figure 2.4), with no additional coding needed.
>
> Depending on your survey department's method for picking up linework, Automatic Begin On Figure Prefix Match can be a good thing or a bad thing. If the survey crew has fastidiously collected each line with its own name, such as CL1, CL2, EP1, EP2, and so on with no repeats, it can work well. If they use start and stop codes but reuse line names throughout the survey, it can produce figures that connect unintentionally.

The Main Event: Your Project's Survey Database

Now that you know how to get everything set up, you are probably eager to get some real, live data into your drawing.

First, set your survey working folder to your desired survey storage location. The Civil 3D survey database is a set of external Microsoft SQL Server Compact database files that reside in your survey working folder. By default, this folder is located in `C:\Civil 3d Projects\`.

To set the survey working folder, right-click on Survey Databases and select Set Working Folder, as shown in Figure 2.6.

FIGURE 2.6
Set the survey working folder.

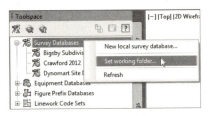

In the current version of Civil 3D, this is a different, independent setting from the working folder for data shortcuts. Most folks want this on a network location, tucked neatly into the survey folder for the project.

Keep in mind that the survey database is version specific. You can open a 2011 database in 2012, but once it has been converted, there is no going back.

To create a survey database, you can either right-click and select New Local Survey Database as mentioned earlier, or you can select Import Survey Data from the Home tab's Create Ground Data panel.

The contents of a survey database are organized into the following categories:

Import Events Import events provide a framework for viewing and editing specific survey data, and they are created each time you import data into a survey database. The default name for the import event is the same as the imported filename. The Import Event collection contains the networks, figures, and survey points that are referenced from a specific import command, and provides an easy way to remove, re-import, and reprocess survey data in the current drawing.

Networks A survey network is a collection of related data that is collected in the field. The network consists of setups, control points, noncontrol points, known directions, observations, setups, and traverses. A network must be created in a survey database before any data can be imported. A survey database can have multiple networks. For example, you can use different networks for different phases of a project. If working within the same coordinate zone, some users have used one survey database with many networks for every survey that they perform. This is possible because individual networks can be inserted into the drawing simply by dragging and dropping the network from the Survey tab in Toolspace into the drawing. You can hover your cursor over any Survey Network component in the drawing to see information about that component and the survey network. You can also right-click any component of the network and browse to the observation entry for that component.

Network Groups Network groups are collections of various survey networks within a survey database. These groups can be created to facilitate inserting multiple networks into a drawing at once simply by dragging and dropping.

Figures Figures are the linework created by codes and commands entered into the raw data file during data collection. The figure names typically come from the descriptor or description of a point.

Figure Groups Similar to network groups, figure groups are collections of individual figures. These groups can be created to facilitate quick insertion of multiple figures into a drawing.

Survey Points One of the most basic components of a survey database, points form the basis for each and every survey. Survey points look just like regular Civil 3D point objects, and their visibility can be controlled just as easily. However, one major difference is that a survey point cannot be edited within a drawing. Survey points are locked by the survey database, and the only way of editing is to edit the observation that collected the data for the point. This provides the surveyor with the confidence that points will not be accidentally erased or edited. Like figures, survey points can be inserted into a drawing by either dragging and dropping from the Survey tab of Toolspace or by right-clicking Surveying Points and selecting the Points ➢ Insert Into Drawing option.

Survey Point Groups   Just like network groups and figure groups, survey point groups are collections of points that can be easily inserted into a drawing. When these survey point groups are inserted into the drawing, a Civil 3D point group is created with the same name as the survey point group. This point group can be used to control the visibility or display properties of each point in the group.

In the following exercise, you'll create an import event and import an ASCII file with survey data. The survey data includes linework.

1. Create a new drawing from the _AutoCAD Civil 3D (Imperial) NCS.dwt template file.

2. From the Home tab, in the Create Ground Data panel of the Ribbon, click Import Survey Data to open the Import Survey Data wizard.

3. Click Create New Survey Database.

4. Enter **Roadway** as the name of the folder in which your new database will be stored. Click OK. Roadway is now added to the list of survey databases.

5. Click Edit Survey Database Settings and verify the units are set to US Foot. Click OK to dismiss the dialog.

6. Click Next.

7. Set the Data Source Type drop-down to Point File.

8. Click the Add Files button on the right side of the Selected File text box, and browse to the `import points.txt` file, which you can download from this book's webpage, www.sybex.com/go/masteringcivil3d2012.

9. Click Open and the name will be listed in the dialog as shown in Figure 2.7.

FIGURE 2.7
Select the correct file type, file, and format in the Import Survey Data wizard.

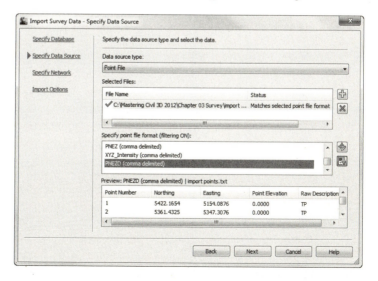

10. Under Specify Point File Format, scroll through the list until you find PNEZD (Comma Delimited). Highlight the format. The preview will show that the correct data type is selected. Click Next to continue.

11. Click Create New Network. Enter the name **Roadway** as the name of the network. Click OK. Click Next to continue.

12. Verify that your import options match what is shown in Figure 2.8. Click Finish.

FIGURE 2.8
The Import Options page in the Import Survey Data wizard

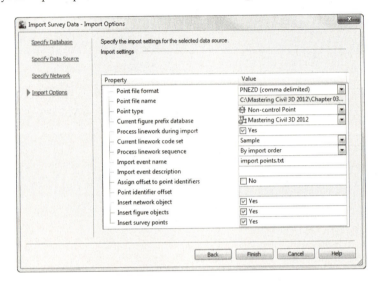

The data is imported and the linework is drawn; however, the building is missing the left side. The following steps will resolve this issue:

1. Continue working in the drawing from the previous exercise. In the Survey tab, select Survey Databases, and then select Roadway, Networks, Roadway, and Non-Control Points.

2. Right-click and select Edit to bring up the Non-Control Points Editor palette in Panorama.

3. Scroll to the bottom of the point list and notice the last line in the file describing point number 34.

4. Move your cursor to the right of the description and type **CLS**, as shown in Figure 2.9. This is the default Close Figure command.

5. Click the check box in the upper right of the palette to apply changes and save your edits. A warning dialog will appear. Click Yes to apply your changes.

6. Select Survey Databases, choose Roadway, click Import Events, and select import points.txt. Right-click import points.txt, and select Process Linework to bring up the Process Linework dialog.

7. Click OK to reprocess the linework with your updated point description. The building figure line and your drawing should look something like Figure 2.10.

FIGURE 2.9
Editing the import event to add the CLS command and close the building geometry

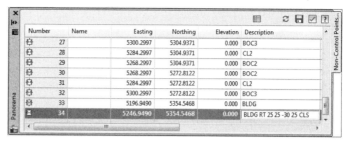

FIGURE 2.10
After editing and reprocessing the linework

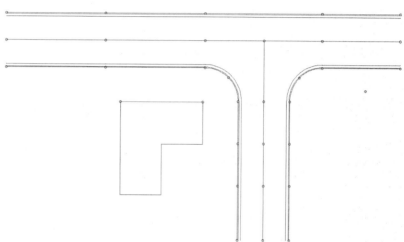

EDITING THE SURVEYED POINT?

Many surveyors will cringe at this new ability to easily modify the drawing data without updating the source files from the field survey. We think this is an improvement in that you can create drawings that accurately reflect personal field observations without modifying the legal record that is the original survey data.

"Mommy, Where Does Survey Data Come From?"

The type of file you will end up bringing into Civil 3D largely depends on your data collector. Civil 3D can import a number of different file formats, but file types vary in what information they can place in the survey database:

ASCII Text File   Most commonly, this is a PNEZD file format (Point Number, Northing, Easting, Elevation, Description). This type of file can produce linework, but every shot is listed as a noncontrol point. The file extension on this type of file may be shown as .txt, .csv, or something else.

LandXML   Contains raw data, points, linework, and even point groups and figure groups. LandXML is the most efficient method for transferring the entirety of a survey database if needed by an outside organization.

FBK (Field Book Format) Can contain raw data, linework, and setup information. FBK is a specially formatted text-based file that was the only format option for creating linework in legacy Autodesk products. In the current release, FBK is no longer king, but it is still a widely used format in many data collectors.

Let's look at some of the manufacturers in the data collector market:

Trimble Trimble has two major offerings: the Trimble TCS Survey Controller and the Trimble SCS900 Site Controller. Both of these data collectors interact directly with Civil 3D via a freely available download called Trimble Link from the Trimble website. This download will add a Trimble Ribbon tab to Civil 3D and allow uploading and downloading of data directly using both data collectors as well as Trimble Site Vision software. Trimble Link will take a JOB file (the default file format for Trimble), convert it to an FBK transparently, create a new survey network in a new or existing database, and import the FBK into that newly created network, all in one step. You can download the software from www.trimble.com/link_ts.asp?Nav=Collection-63438.

TDS Now a division of Trimble, Survey Pro is one of the most popular data collection software packages on the market today. The TDS RAW or RW5 file format is the format on which other data collector manufacturers base their RAW data files. To convert from a TDS RAW data file to an FBK, users must either purchase TDS ForeSight DXM or use TDS Survey Link, which is included in your Civil 3D install. You can initiate Survey Link by typing **STARTSURVEYLINK** on the command line, or by selecting Survey Data Collection Data Link from the Create Ground Data panel on the Home tab.

Leica The Leica System 1200 data collectors can be integrated with Civil 3D via the use of Leica X-Change. Leica X-Change allows for the import of Leica data and conversion to the FBK format, as well as export options for Leica System 1200 collectors. You can download the software from www.leica-geosystems.com/corporate/en/downloads/lgs_page_catalog.htm?cid=239.

MicroSurvey MicroSurvey has an advantage not shared by any other data collector in this list: It has the ability to directly export an FBK from the data collector. However, if users demand a conversion process, the MicroSurvey RAW data file is based on TDS RAW data, and it can be converted using the included TDS Survey Link.

Carlson Like Trimble, Carlson offers a freely downloadable plug-in for Civil 3D. Carlson Connect allows for direct import and export to Carlson SurvCE data collectors as well as a conversion option for drawings containing Carlson point blocks. You can download the software by going to http://update.carlsonsw.com/updates.php and selecting Carlson Connect as the product you are seeking.

Under the Hood in Your Survey Database

Once you import survey data, Civil 3D will help you make sense of it by organizing the data into groups, as shown in Figure 2.11:

Control Points Control points are typically points in your data that have a high confidence factor. These are usually benchmarks or other known points used to lock the survey into its geographical location.

Non-Control Points Noncontrol points are also stored in the survey database. Keyed-in points; GPS-collected points, such as those located by real-time kinematic (RTK) GPS; and any point brought in through an ASCII file will appear as noncontrol points.

Directions The direction from one point to another must be manually entered into the data collector for the direction to show up later in the survey network for editing. The direction can be as simple as a compass shot between two initial traverse points that serves as a rough basis of bearings for a survey job.

Setups The setup is typically where the meat of the data is found, especially when working with conventional survey equipment. Every setup, as well as the points (side shots) located from that setup, can be found. Setups will contain two components: the station (or occupied point) and the backsight. Setups, as well as the observations located from the setup, can be edited. The interface for editing setups is shown in Figure 2.12. Angles and instrument heights can also be changed in this dialog.

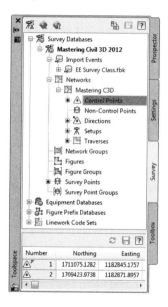

FIGURE 2.11
A typical survey database network with data

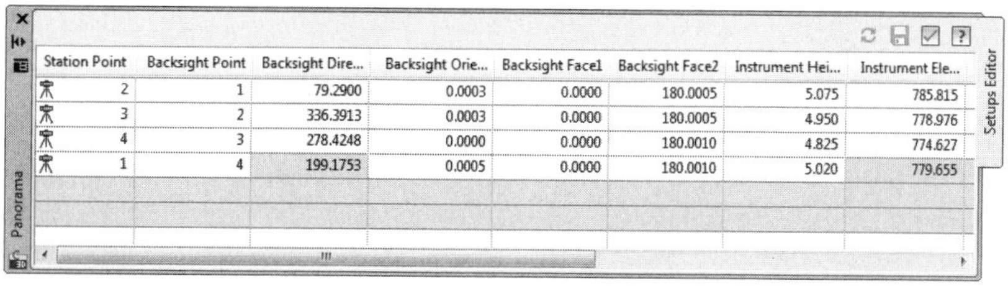

FIGURE 2.12
Setups and observations can be changed in the Setups Editor.

Station Point	Backsight Point	Backsight Dire...	Backsight Orie...	Backsight Face1	Backsight Face2	Instrument Hei...	Instrument Ele...
2	1	79.2900	0.0003	0.0000	180.0005	5.075	785.815
3	2	336.3913	0.0003	0.0000	180.0005	4.950	778.976
4	3	278.4248	0.0000	0.0000	180.0010	4.825	774.627
1	4	199.1753	0.0005	0.0000	180.0010	5.020	779.655

Traverses

The Traverses section is where new traverses are created or existing ones are edited. These traverses can come from your data collector, or they can be manually entered from field notes via the Traverse Editor, as shown in Figure 2.13. You can view or edit each setup in the Traverse Editor, as well as the traverse stations located from that setup.

FIGURE 2.13
The Traverse Editor

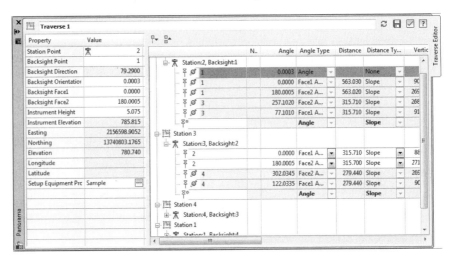

Once you have defined a traverse, you can adjust it by right-clicking its name and selecting Traverse Analysis. You can adjust the traverse either horizontally or vertically, using a variety of methods. The traverse analysis can be written to text files to be stored, and the entire network can be adjusted on the basis of the new values of the traverse, as you'll do in the following exercise:

1. Create a new drawing from the _AutoCAD Civil 3D (Imperial) NCS.dwt template file.
2. Navigate to the Survey tab of Toolspace.
3. Right-click Survey Databases and select New Local Survey Database. The New Local Survey Database dialog opens.
4. Enter **Traverse** as the name of the folder in which your new database will be stored.
5. Click OK to dismiss the dialog. The Traverse survey database is created as a branch under the Survey Databases branch.
6. Expand the Traverse branch, right-click Networks, and select New. The New Network dialog opens.
7. Expand the Network branch in the dialog if needed. Name your new network **Traverse Practice**.
8. Click OK. The Traverse Practice network is now listed as a branch under the Networks branch of the Traverse survey database in Prospector.
9. Right-click the Traverse Practice network and select Import ➤ Import Field Book.

10. Select the file `traverse.fbk`, which you can download from this book's webpage (www.sybex.com/go/masteringcivil3d2012) and click Open. The Import Field Book dialog opens.

11. Make sure you have checked the boxes shown in Figure 2.14. You will be inserting the points into the drawing this time.

FIGURE 2.14
The Import Field Book dialog

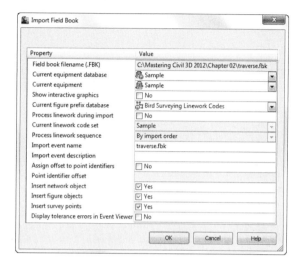

12. Click OK. Save the drawing for the next exercise.

Inspect the data contained within the network. You have one control point—point 2—that was manually entered into the data collector. There is one direction, and there are four setups. Each setup combines to form a closed polygonal shape that defines the traverse. Notice that there is no traverse definition. In the following exercise, you'll create that traverse definition for analysis:

1. Continue working in the drawing from the previous exercise. Right-click Traverses under the Traverse Practice network and select New to open the New Traverse dialog.

2. Name the new traverse **Traverse1**.

3. Enter **2** as the value for the Initial Station and **1** (if necessary) for the Initial Backsight.

   The traverse will now pick up the rest of the stations in the traverse and enter them in the next box.

4. Enter **2** as the value for the Final Foresight if necessary (the closing point for the traverse). Your New Traverse dialog box will look like Figure 2.15. Click OK.

5. Right-click Traverse1 in the bottom portion of Toolspace. Select Traverse Analysis.

6. In the Traverse Analysis dialog, ensure that Yes is selected for Do Traverse Analysis and Do Angle Balance.

7. Select Least Squares for both the horizontal and vertical adjustment methods.

8. Select 30,000 for both the horizontal and vertical closure limit 1:X.
9. Leave Angle Error Per Set at the default.
10. Make sure the option Update Survey Database is set to Yes.
11. The Traverse Analysis dialog will look like Figure 2.16. Click OK.

FIGURE 2.15
Defining a new traverse

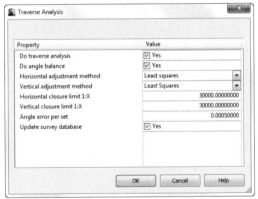

FIGURE 2.16
Specify the adjustment method and closure limits in the Traverse Analysis dialog.

The analysis is performed, and four text files are displayed that show the results of the adjustment. Note that if you look back at your survey network, all points are now control points, because the analysis has upgraded all the points to control point status.

Figure 2.17 shows the Traverse1 Raw Closure.trv and Traverse 1 Vertical Adjustment.trv files that are generated from the analysis. The raw closure file shows that your new precision is well within the tolerances set in step 8. The vertical adjustment file describes how the elevations have been affected by the procedure.

The third text file is shown in three separate portions. The first portion is shown in Figure 2.18. This portion of the file displays the various observations along with their initial measurements, standard deviations, adjusted values, and residuals. You can view other statistical data at the beginning of the file.

FIGURE 2.17
Traverse analysis results

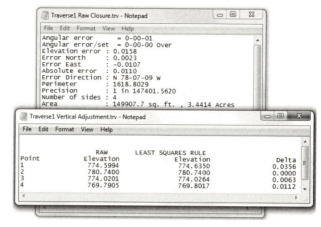

FIGURE 2.18
Statistical and observation data

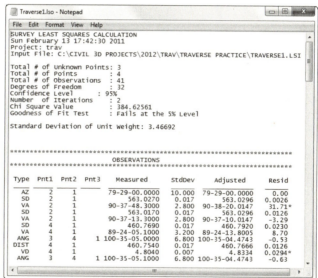

Figure 2.19 shows the second portion of this text file and displays the adjusted coordinates, the standard deviation of the adjusted coordinates, and information related to error ellipses displayed in the drawing. If the deviations are too high for your acceptable tolerances, you will need to redo the work or edit the field book.

Figure 2.20 displays the final portion of this text file—Blunder Detection/Analysis. Civil 3D will look for and analyze data in the network that is obviously wrong and choose to keep it or throw it out of the analysis if it doesn't meet your criteria. If a blunder (or bad shot) is detected, the program will not fix it. You will have to edit the data manually, whether by going out in the field and collecting the correct data or by editing the FBK file.

FIGURE 2.19
Adjusted coordinate information

ADJUSTED COORDINATES

Point	Northing	Easting	Elevation
1	13740905.9355	2157152.4444	774.6350
3	13740513.3864	2156723.9875	774.0264
4	13740471.0583	2157000.1693	769.8017

Standard Deviations - Adjusted Coordinates

	STANDARD DEVIATIONS		
Point	North	East	Elevation
1	0.026865	0.008341	0.003265
3	0.008517	0.015068	0.003629
4	0.020368	0.017793	0.003828

Least Squares Error Ellipses
at 95% Confidence Level

Point	Semi-Major Axis	Semi-Minor Axis	NE-Axis Azimuth
1	0.2438418757	0.0607871778	169-29-00
3	0.1432969460	0.0580918243	67-58-16
4	0.2347534324	0.0571615924	40-38-54

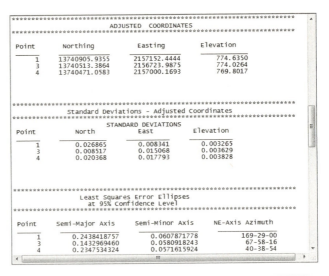

FIGURE 2.20
Blunder analysis

Blunder Detection/Analysis

Type	Pnt1	Pnt2	Pnt3	Adjusted	Resid	Redun	Reliability Estimate	Tests Marg	Ext
AZ	2	1		79-29-00.0000	0.000	0.000	<None>	F	F
SD	2	1		563.030	0.003	0.840	-0.003	P	P
VA	2	1		90-38-20.0147	31.715	0.818	-38.789	F	F
SD	2	1		563.030	0.013	0.840	-0.015	P	P
VA	2	1		90-37-10.0147	-3.285	0.818	4.018	P	P
SD	4	1		460.792	0.023	0.844	-0.027	P	P
VA	4	1		89-24-13.8005	8.701	0.805	-10.802	P	F
ANG	3	4	1	100-35-04.4743	-0.526	0.703	0.747	P	P
DIST	4	1		460.767	0.013	0.844	-0.015	P	P
VD	4	1		4.833	0.029	0.797	-0.037	F	F
ANG	3	4	1	100-35-04.4743	-0.626	0.703	0.890	P	P
SD	3	4		279.439	0.002	0.830	-0.002	P	P
VA	3	4		90-51-35.8216	-11.378	0.804	14.144	P	P
ANG	2	3	4	122-03-35.8498	-4.150	0.754	5.502	P	P
SD	3	4		279.439	0.002	0.830	-0.002	P	P
VA	3	4		90-52-35.8216	18.622	0.804	-23.148	P	F
ANG	2	3	4	122-03-35.8498	0.850	0.754	-1.127	P	P

OTHER METHODS OF MANIPULATING SURVEY DATA

Often, it is necessary to edit the entire survey network at one time. For example, rotating a network to a known bearing or azimuth from an assumed one happens quite frequently. To find this hidden gem of functionality, change to the Modify tab and choose Survey from the Ground Data panel. Select the Survey tab, and choose Translate Database from the Modify panel (shown in Figure 2.21) to manipulate the location of a network.

FIGURE 2.21
The elusive yet indispensible Translate Database command

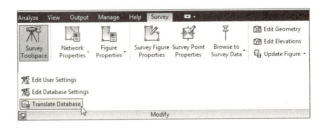

Real World Scenario

MANIPULATING THE NETWORK

Translating a survey network can move a network from an assumed coordinate system to a known coordinate system, it can rotate a network, and it can adjust a network from assumed elevations to a known datum.

1. Create a new drawing from the NCS Extended Imperial template file.
2. In the Survey tab of Toolspace, right-click and select New Local Survey Database. The New Local Survey Database dialog opens. Enter **Translate** in the text box. This is the name of the folder for the new database.
3. Click OK, and the Translate database will now be listed under the Survey Databases branch on the Survey tab.
4. Select Networks under the new Translate branch. Right-click and select New to open the New Network dialog. Enter **Translate** as the name of this new network. Click OK to dismiss the dialog.
5. Right-click the Translate network, and select Import ➢ Import Field Book.
6. Navigate to the file traverse.fbk which you can download from this book's webpage, www.sybex.com/go/masteringcivil3d2012, and click Open. The Import Field Book dialog opens. Click OK to accept the default options.
7. Update the Point Groups collection in Prospector so that the points show up in the drawing if needed.
8. Draw a vertical polyline directly from point 3 (the point to the southwest), to the north of point 3 (toward the middle of the loop). It can be any length, but be sure to use the object snap node at point 3.
9. Change to the Modify tab and choose Survey from the Ground Data panel to open the Survey tab.
10. Choose Translate Database from the Modify panel drop-down menu.
11. For the purposes of this exercise, you will leave the points on their same coordinate system, but change the bearing of the line defined from point 3-point 2 to due north. Elevations will remain unchanged.
12. In the first window, type **3** as the Number value. This is the Base Point number (the number that you will be rotating the points around). Click Next.
13. In the next window, click the Pick In Drawing button in the lower-left corner to specify the new angle.
14. Using osnaps, pick point 3 and then point 2 (to the northwest) for your reference angle.

15. When you see the prompt Specify New Direction, pick point 3; somewhere along the orthogonal polyline you'll be prompted to specify the second point to define the new angle, and pick point 3 again. Click Next.
16. In the next window, click the Pick In Drawing button on the lower left to pick point 3 as the destination point. This will essentially negate any translation features and provide you with only a rotation.
17. Leave the Elevation Change box empty. If you were raising or lowering the elevations of the network, this box is where you would enter the change value.
18. Click Next to review your results, and click Finish to complete the translation.
19. Go back to the drawing and inspect your points. Point 2 should now be due north of point 3.
20. Close the drawing without saving. Even though you did not save the drawing, the changes were made directly to the database and the network can be imported into any drawing.

Once figures are part of your drawing, you may have some cleanup to do. To edit a figure, select any one and use the commands from the context Ribbon that appears.

Be aware that reimporting your linework will result in your edits getting blown away. Reprocessing your linework after a modification is made can sometimes result in duplicate figures.

Using a Dirty Word, "Explode"

Most of the time in Civil 3D, exploding objects is a huge no-no. Exploding an object removes the underlying Civil 3D intelligence and reduces objects to base AutoCAD components. In this author's opinion, survey figures are the exception to the no-exploding rule. An exploded survey figure will revert to a 3D polyline and will become divorced from the survey database that created it. Creating an independent polyline can be useful if your survey figures are going to be the basis for grading feature lines at a later time.

Other Survey Features

Other components of the survey functionality included with Civil 3D 2012 are the Astronomic Direction Calculator, the Geodetic Calculator, Mapcheck reports, and Coordinate Geometry Editor. All of these features are accessed from the Survey flyout on the Ground Data panel of the Analyze tab.

The Astronomic Direction Calculator, shown in Figure 2.22, is used to calculate sun shots or star shots. For the obscure art of these type of observations, Civil 3D has all the ephemeris data built in, making this a very convenient tool.

The Geodetic Calculator is used to calculate and display the latitude and longitude of a selected point, as well as its local and grid coordinates. It can also be used to calculate unknown points. If you know the grid coordinates, the local coordinates, or the latitude and longitude of a point, you can enter it in the Geodetic Calculator and create a point at that location. Note that the Geodetic Calculator only works if a coordinate system is assigned to the drawing in the Drawing Settings dialog. In addition, any transformation settings specified in this dialog will be reflected in the Geodetic Calculator, shown in Figure 2.23.

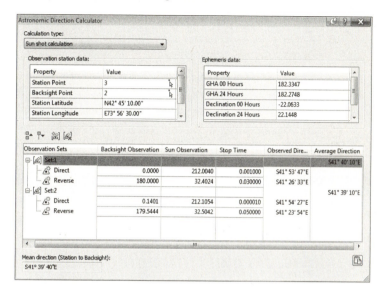

FIGURE 2.22
The Astronomic Direction Calculator

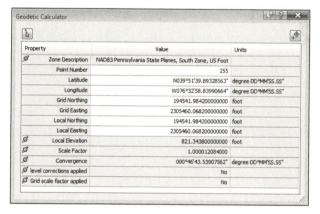

FIGURE 2.23
The Geodetic Calculator

Remember those labels you added to your lines in Chapter 1? Put them to work in a Mapcheck Report. The Mapcheck Report computes closure based on line, curve, or parcel segment labels.

1. Open the `Mapcheck-start.dwg` file, which you can download from this book's webpage, www.sybex.com/go/masteringcivil3d2012.

2. Change to the Analyze tab, and select Survey ➢ Mapcheck from the Ground Data panel to display the Mapcheck Analysis palette.

3. Click the New Mapcheck button at the top of the menu bar.

4. At the Enter name of mapcheck: prompt, type **Record Deed**.

5. At the Specify point of beginning (POB) prompt, choose the north endpoint of the line representing the east line of the parcel (the longest line in the file).

6. Working clockwise, select each parcel label one at a time. Be sure not to skip the small segment in the southwest portion of the site.

7. In the northwest portion of the site, you will encounter a label whose bearing is flipped. You will use the Mapcheck Reverse command to rectify this in your mapcheck.

8. At the Select a label or [Clear/Flip/New/Reverse] prompt, type **R** and then press ↵ to reverse direction. The mapcheck glyph will now appear in the correct location along the segment, as shown in Figure 2.24.

FIGURE 2.24
The Mapcheck glyph verifies your input is correct.

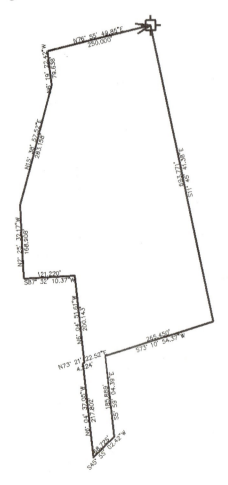

9. Select the last line label along the north of the parcel.
10. Press ↵ to complete the mapcheck entry. The completed parcel should have 12 sides.
11. Select the output view as shown in Figure 2.25 to verify closure.

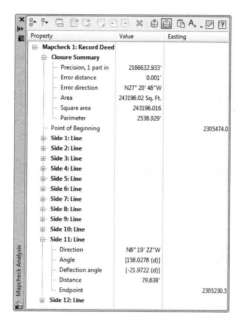

FIGURE 2.25
The completed deed in the Mapcheck Analysis palette

The Coordinate Geometry Editor

The new Coordinate Geometry Editor (Figure 2.26) is a powerhouse tool that makes creating and evaluating 2D boundaries easier than before. The functionality introduced with this feature supplants entering parcel data one segment at a time using the Line By Bearing And Distance command. Traverse analysis can be performed on manually entered segments, polylines, or COGO point objects without needing to define them in a survey database.

Boundary data can be entered into the Coordinate Geometry Editor using a mix of methods. As shown in the first line of the traverse seen in Figure 2.26, you can use formulas to enter data. In the example shown, multiple segments with the same bearing have been consolidated into a single entry.

If a value is unknown, such as the distance in line 6 of Figure 2.26, you can enter a **U**. Civil 3D will calculate the unknown value when you generate the traverse report.

To enter data using points, use the Pick COGO Points In Drawing button to select the points in the direction of the traverse side. Note that the direction and distance are entered independently of each other, so you will need to repeat the selection for each column of the table. You may also copy and paste between columns.

If a line has been entered in error, the Coordinate Geometry Editor offers a variety of tools for fixing problems. To remove a line of the table, highlight the row, right-click, and select Delete Row, as shown in Figure 2.27.

FIGURE 2.26
Your new best friend, the Coordinate Geometry Editor

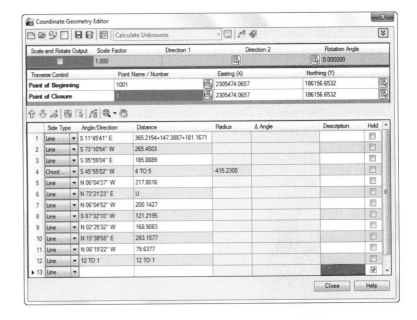

FIGURE 2.27
Removing unwanted traverse data

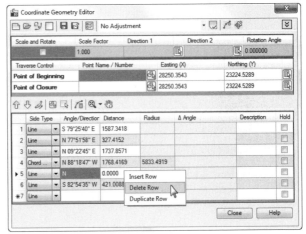

Similar to the glyphs you saw in the Mapcheck command, the Coordinate Geometry will appear in the graphic showing the side directions and point of closure, as seen in Figure 2.28.

When you want to run a traverse report, set the report type you wish to run from the top of the Coordinate Geometry Editor. If you have unknowns in your traverse, your only option will be to calculate the unknown values. Click the Display Report button to view the results of your entries. Depending on the type of adjustment you chose, your results should resemble Figure 2.29.

FIGURE 2.28
Temporary graphics or "glyphs" to help you identify your boundary

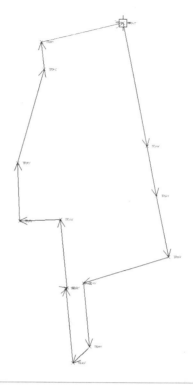

FIGURE 2.29
Traverse report created by the Coordinate Geometry Editor

Traverse Report

Closure

Total Traverse Length	2538.928
Error in Closure	0.001
Closure is one part in	1866445.9895
Error in North(Y)	0.0013
Error in East(X)	0.0004
Direction of Error	N 19°00'01" E

Traverse Control

	Point Name	Northing	Easting
Point of Beginning	1001	186156.6532	2305474.0657
Point of Closure	1	186156.6532	2305474.0657

Input Data

Side	Angle/Direction	Distance	Radius	#Delta Angle	Description
1	S 11°45'41" E	365.2154			
2	13 TO 14	13 TO 14			
3	S 11°45'41" E	181.1671			
4	S 73°10'54" W	265.4503			
5	S 05°59'04" E	185.8889			
6	S 45°55'02" W	68.7261	-393.6557		
7	N 06°04'37" W	217.8016			

Using Inquiry Commands

A large part of a surveyor's work involves querying lines and curves for their length, direction, and other parameters.

The Inquiry commands panel (Figure 2.30) is on the Analyze tab, and it makes a valuable addition to your Civil 3D and survey-related workspaces. Remember, panels can be dragged away from the Ribbon and set in the graphics environment much like a toolbar.

FIGURE 2.30
The Inquiry commands panel

The Inquiry tool (shown in Figure 2.31) provides a diverse collection of commands that assist you in studying Civil 3D objects. You can access the Inquiry tool by clicking the Inquiry Tool button on the Inquiry panel.

FIGURE 2.31
Choosing an Inquiry type from the Inquiry tool palette

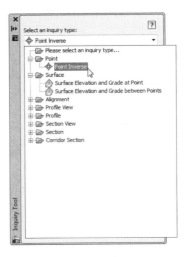

To use the Point Inverse option in the Inquiry commands, select the Point Inverse option as shown in Figure 2.32.

You can key in the point number or use the Pick In CAD icon to select the points you wish to examine.

The other Inquiry commands that are specific to Civil 3D are also handy to the survey process.

The List Slope tool provides a short command-line report that lists the elevations and slope of an entity (or two points) that you choose, such as a line or feature line.

The Line And Arc Information tool provides a short report about the line or arc of your choosing (see Figure 2.33). This tool also works on parcel segments and alignment segments. Alternatively, you can type **P** for points at the command line to get information about the apparent line that would connect two points on screen.

FIGURE 2.32
Point Inverse results

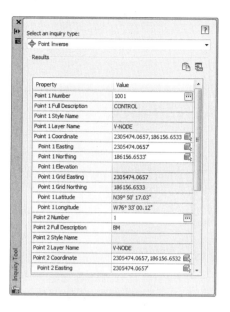

FIGURE 2.33
Command-line results of a line inquiry and arc inquiry

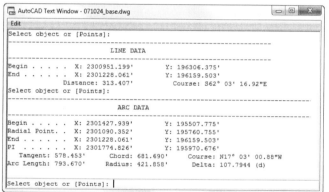

> **DON'T GET BURNED BY AUTOCAD ANGLES**
>
> Do not set base AutoCAD angular units to Surveyors Units. This setting may seem perfectly logical if you have been using base AutoCAD prior to using Civil 3D. However, this angular setting will end up confusing you more than helping you.
>
> When looking for information about a line, use the Line and Arc tool mentioned in this section rather than the base AutoCAD **LIST** command. The Civil 3D Line and Arc tool (**CGLIST**) works on more object types and is not affected by rotated coordinate systems.

When entering angles in base AutoCAD, the full **N50d10'10"E** is needed to denote a bearing. In Civil 3D commands, **N50.1010E** will denote the same thing and is much faster to type. However, as every surveyor knows, there is a huge difference between 50.1010° and 50°10'10".

A setting you will want to change in your Civil 3D template is the angular entry method for general angles, as shown here. This setting mainly affects the Angle-Distance Transparent command. You will learn more about Civil 3D templates in Chapter 19, "Styles."

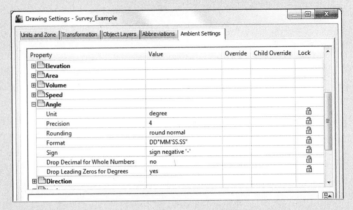

Keep your base AutoCAD angular units set to Decimal Degrees to help you differentiate when you are in base AutoCAD angular entry and Civil 3D's more surveyor-friendly angular entry.

The Angle Information tool lets you pick two lines (or a series of points on the screen). It provides information about the acute and obtuse angles between those two lines. Again, this also works for alignment segments and parcel segments.

The Continuous Distance tool provides a sum of distances between several points on your screen, or one base point and several points.

The Add Distances tool is similar to the Continuous Distance command, except the points on your screen do not have to be continuous.

The Bottom Line

Properly collect field data and import it into Civil 3D.   You learned best practices for collecting data, how the data is translated into a usable format for the survey database, and how to import that data into a survey database. You learned what commands draw linework in a raw data file, and how to include those commands into your data collection techniques so that the linework is created correctly when the field book is imported into the program.

Master It   In this exercise, you'll create a new drawing and a new survey database and import the MASTER_IT_C3.txt file into the drawing. The format of this file is PNEZD (comma delimited).

Set up description key and figure databases. Proper setup is key to working successfully with the Civil 3D survey functionality.

Master It In this exercise, you'll create a figure prefix database and a description key set.

Create a new description key set and the following description keys using the default styles. Make sure all description keys are going to V-Node:

- CL*
- EOP*
- TREE*
- BM*

Create a figure prefix database called MasterIt containing the following codes:

- CL
- EOP

Create and edit field book files. You learned how to create field book files using various data collection techniques and how to import the data into a survey database.

Master It In this exercise, you'll create a new drawing and survey database. Translate the database based on:

- Base Point **1**
- Rotation Angle of **10.3053°**

Manipulate your survey data. You learned how to use the traverse analysis and adjustments to create data with a higher precision.

Master It In this exercise, you'll use the survey database and network from the previous exercises in this chapter. You'll analyze and adjust the traverse using the following criteria:

- Use the Compass Rule for Horizontal Adjustment.
- Use the Length Weighted Distribution Method for Vertical Adjustment.
- Use a Horizontal Closure Limit of 1:25,000.
- Use a Vertical Closure Limit of 1:25,000.

Chapter 3

Points

The foundation of any civil engineering project is the simple point, frequently referred to as shots. Most commonly, points are used to identify the location of existing features, such as trees and property corners; topography, such as ground shots; or stakeout information, such as road geometry points. However, points can be used for much more. This chapter will both focus on traditional point uses and introduce ideas to apply the dynamic power of point editing, labeling, and grouping to other applications.

In this chapter, you will learn to:

- ◆ Import points from a text file using description key matching
- ◆ Create a point group
- ◆ Export points to LandXML and ASCII format
- ◆ Create a point table

Anatomy of a Point

Civil 3D *points* (see Figure 3.1) are intelligent objects that represent x, y, and z locations in space. Each point has a unique number and, optionally, a unique name that can be used for additional identification and labeling.

FIGURE 3.1
A typical point object showing a marker, a point number, an elevation, and a description

A Quick Word on Styles

Separating the point functionality discussed in this chapter from the styles that make them look the way they do is difficult. Chapter 19, "Styles," will go into the nitty-gritty of creating and manipulating point styles and label styles. In this chapter, we will work with styles that are already part of a drawing. This is true for points, labels, and tables.

COGO Points vs. Survey Points

In Chapter 2, "Survey," we imported survey data that contained points. Points brought in through the methods described in Chapter 2 are referred to as *survey points*. In this chapter, we will import points from a delimited text file and place them in CAD using the point creation tools. Points created in this manner are referred to as *COGO points*. Figure 3.2 shows the subtle differences between points brought in as COGO points (top) and points brought in through a database (bottom).

FIGURE 3.2
The context-sensitive Ribbon reflects similarities and differences between COGO points (top) and survey points (bottom).

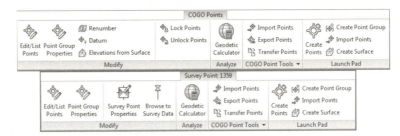

The differences between COGO points and survey points are subtle but important to note. A COGO point is unlocked by default—meaning it can readily be edited. A survey point, on the other hand, must be unlocked if a user wishes to edit the point. A survey point stays tied to the database that it came from, whereas a COGO point maintains no tie to the originating text file or object it was created against. Other than their origin, both COGO points and survey points obey the principles outlined in this chapter.

Creating Basic Points

You can create points many ways, using the Points menu in the Create Ground Data panel on the Home tab. Points can also be imported from text files or external databases or converted from AutoCAD, Land Desktop, or SoftDesk point objects.

Point Settings

Point settings are our first glimpse at what is known as *command settings*. All Civil 3D objects have command settings tucked away in the Settings tab of Toolspace. In the case of points, it is handy to have these settings readily available for on-the-fly modifications. Whether or not the changes you make on-the-fly are remembered the next time you create points depends on your template settings (see Chapter 19 for more information on feature settings in the template).

Access the Create Points toolbar by going to the Home tab, and in the Create Ground Data panel, select Points ➢ Point Creation Tools. Expand the toolbar by clicking the chevron button on the far-right side.

DEFAULT LAYER

For most Civil 3D objects, the object layer is established in the drawing settings. In the case of points, the default object layer is set in the command settings for point creation and can be changed in the Create Points dialog (see Figure 3.3).

FIGURE 3.3
Verify the point object layer before creating points.

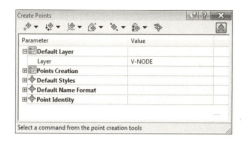

Prompt for Elevations, Names, and Descriptions

When creating points in your drawing, you have the option of being prompted for elevations, names, and descriptions (see Figure 3.4). In many cases, you'll want to leave these options set to Manual. The command line will ask you to assign an elevation, name, and description for every point you create.

FIGURE 3.4
You can change the elevation, point name, and description settings from Manual to Automatic.

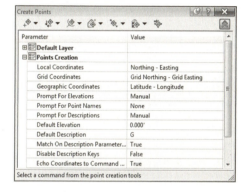

If you're creating a batch of points that have the same description or elevation, you can change the Prompt toggle from Manual to Automatic and then provide the description and elevation in the default cells. For example, if you're setting a series of trees at an elevation of 10′, you can establish settings as shown in Figure 3.5.

FIGURE 3.5
Default settings for placing tree points at an elevation of 10′

Note that these settings only apply to points created from this toolbar. The settings do not affect the elevation or description of points imported from a file.

What's in a Name?

Users often confuse a point's *name* with *description*. The name is a unique, alphanumeric sequence that can be used in lieu of a point number. Description refers to the all-important code given to a point out in the field. Most survey software today does not use alphanumeric point identification, so this option is usually set to None in Civil 3D. If you do have a point file that uses a name instead of point number, you will need to create a custom point format, as described later in this chapter.

Importing Points from a Text File

One of the most common means of creating points in your drawing is to import an external text file (see Figure 3.6).

FIGURE 3.6
The Import Points and the Point File Formats – Create Group dialogs

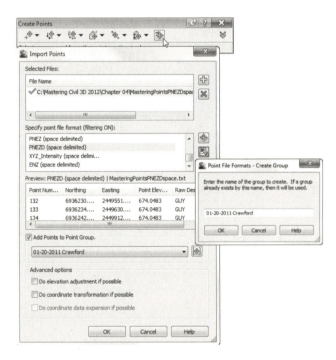

To add a file to your Import Points dialog, click the plus sign to browse. You can add multiple files at once if they are in the same point format (i.e., PNEZD (comma delimited)). The import process supports most text formats as well as Microsoft Access Database (MDB) files.

When your file is listed in the top of the dialog box, a green check mark will indicate that Civil 3D can parse the information. Be careful, though, as Civil 3D does not know the difference between a Northing and an Easting or a point number and an elevation. You still need to select the correct file format.

The file format filter is there to help you. Civil 3D recognizes how the file is delimited (i.e., tab, comma, space) and only shows you the formats that apply. If you don't want the help, you can turn the filtering off by clicking the Filter icon.

If the file format you need is not available, or you wish to use the adjustment and transformation capabilities, you can do so by clicking the Manage Formats button.

You can make an elevation adjustment if the point file contains additional columns for thickness, Z+, or Z–. You can add these columns as part of a custom format. See the *Civil 3D 2012 Users Guide* section "Using Point File Format Properties to Perform Calculations" for more details.

You can perform a coordinate system transformation if a coordinate system has been assigned both to your drawing (under the drawing settings) and as part of a custom point format. In this case, the program can also do a coordinate data expansion, which calculates the latitude and longitude for each point.

A common use of the formats is to create a point file format for importing a name-based point file. In the following example, you will create a new file format to accommodate names (instead of point numbers):

1. Open the Import Points dialog box and click the Manage Points icon. (You can also access this functionality by going to the Settings tab and choosing Point and then Point File Formats.)

2. Click New, choose the User Point File option, and click OK.

3. Name the format **Name-NEZD**. Toggle on the Delimited By option and place a comma in the field.

4. Click the first <unused> column heading. Select Name from the Column Name drop-down menu and click OK.

5. Click the next <unused> column and select Northing from the Column Name drop-down. Leave Invalid Indicator and Precision fields as default and click OK.

6. Repeat the process for Easting, Point Elevation, and Raw Description.

7. To test the format, click Load. Select the file `Test Format.txt` and click Open.

8. Click Parse. If the format has been created successfully, you will see the file preview as shown in Figure 3.7.

9. Click OK to complete the format and close the Point File Format dialog. Dismiss the Import Points dialog. The new file format will remain in case you need it.

FIGURE 3.7
A completed and tested new point file format

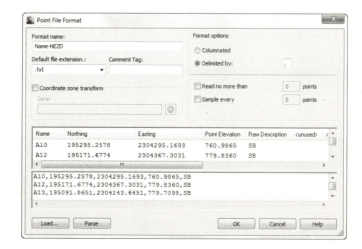

Importing a Text File of Points

In this exercise, you'll learn how to import a TXT file of points into Civil 3D:

1. Open the Mastering Points.dwg file, which you can download from this book's web page at www.sybex.com/go/masteringcivil3d2012.

2. Open the Import Points dialog by selecting Points From File in the Import panel of the Insert tab.

3. Change the Format field to PNEZD (Comma Delimited).

4. Click the file folder's plus (+) button to the right of the Source File field, and navigate out to locate the Mastering_C3D_Points.txt file.

5. Click the Create Point group icon. Name the point group **Survey 05-05-2011** and click OK.

6. Leave all the other boxes unchecked.

7. Click OK. You may have to use Zoom Extents to see the imported points. (Hint: Double-click your middle mouse wheel for zooming extents.)

LandXML and Points: A Match Made in Heaven

LandXML is a file format specifically made to share the type of data Civil 3D and its counterparts create. For points, it works particularly well, since it will also carry point group information. Import LandXML files from the Import panel on the Insert tab. You'll learn much more about LandXML in Chapter 17, "Interoperability."

Thinking Ahead: Assigning Point Numbers

Point numbers are assigned using the Point Identity settings in the Create Points dialog or the point file from which they originated. To list available point numbers, enter **ListAvailablePointNumbers** on the command line, or select any point to open the context-sensitive Ribbon and choose COGO Point Tools ➢ List Available Point Numbers.

Converting Points from Land Desktop, SoftDesk, and Other Sources

Civil 3D contains several tools for migrating legacy point objects to the current version. The best results are often obtained from an external point list, such as a text file, LandXML, or an external database. However, if you come across a drawing that contains the original Land Desktop, SoftDesk, AutoCAD, or other types of point objects, tools and techniques are available to convert those objects into Civil 3D points.

A Land Desktop point database (the Points.mdb file found in the COGO folder in a Land Desktop project) can be directly imported into Civil 3D in the same interface in which you'd import a text file.

Land Desktop point objects, which appear as AECC_POINTs in the AutoCAD Properties palette, can also be converted to Civil 3D points (see Figure 3.8). Upon conversion, this tool gives you the opportunity to assign styles, create a point group, and more.

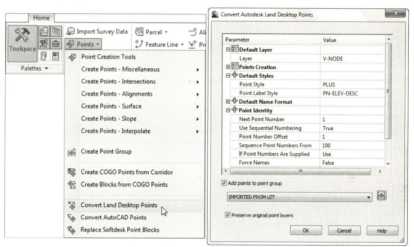

FIGURE 3.8
The Convert Land Desktop Points option (left) opens the Convert Autodesk Land Desktop Points dialog (right).

Occasionally, you'll receive AutoCAD point objects drawn at elevation from aerial topography information or other sources. It's also not uncommon to receive SoftDesk point blocks from outside surveyors. Both of these can be converted to or replaced by Civil 3D points by using the Points drop-down on the Create Ground Data panel of the Home tab (see Figure 3.8).

USING AUTOCAD ATTRIBUTE EXTRACTION TO CONVERT OUTSIDE PROGRAM POINT BLOCKS

Occasionally, you may receive a drawing that contains point blocks from a third-party program such as Eagle Point or Carlson. These point blocks may look similar to SoftDesk point blocks, but the block attributes may have been rearranged and you can't convert them directly to Civil 3D points using Civil 3D tools.

Whenever possible, your best course of action would be to request the source survey in text format or LandXML. If that is not possible, the exercise that follows will pull the data you need without losing the information from the attributes.

1. Open the file Mystery Plat.dwg. First, examine the blocks you are given to determine the names of the blocks with which you are working. A glance at AutoCAD properties will give you an idea of your block name and what attributes you are extracting. In this example, we are seeing Carlson survey blocks, which use the SRVPNO1 as an anchor for its attributes.

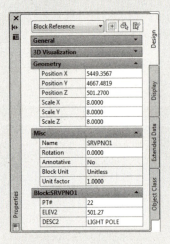

2. Use the Data Extraction tool by typing **EATTEXT** in the command line to launch the Data Extraction Wizard.

3. Select the radio button to create a new data extraction and click Next. The Save Extraction As page appears, prompting you to name and save this extraction. Give the extraction a meaningful name, and save it in the appropriate folder. Click Next.

4. Confirm that the drawings to be scanned for attributed blocks are on the list:

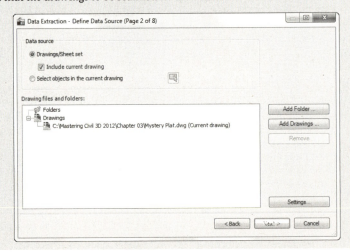

CREATING BASIC POINTS | **79**

5. In the Select Objects screen of the Data Extraction Wizard, uncheck Display All Object Types. Check Display Blocks With Attributes Only and Display Objects Currently In-Use Only; doing so will filter out most unneeded blocks. Eliminate the other types of attributed blocks by deselecting their boxes. Click Next.

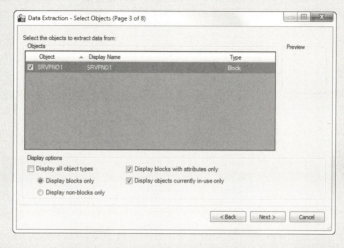

6. On the next screen, select these properties: the point number (PT#), elevation (ELEV2), and description (DESC2), as well as Position X and Position Y. In this case, we are relying on the attribute ELEV2 instead of the Position Z; you do not need both. Click Next.

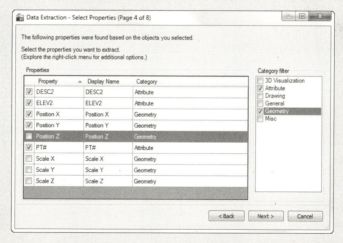

7. On the Refine Data screen, rearrange the columns into a PNEZD format by clicking and dragging the column headers into place. Uncheck Show Count Column and Show Name Column. Click Next.

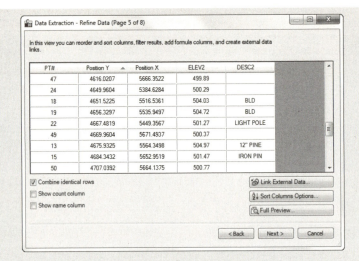

8. On the Choose Output screen, set Output Data to External File, and save your extraction as an CSV file in a logical place.

9. Locate the file in Windows. Right-click on it and select Open With ➤ Notepad. Remove the first line of text (the header information). Save and close the file.

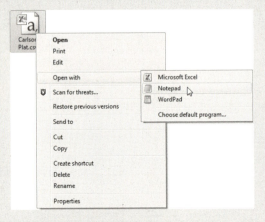

10. In a drawing that contains your Civil 3D styles, use the Import Points tool in the Create Points dialog to import the CSV file.

CREATING BASIC POINTS | 81

CONVERTING POINTS

In this exercise, you'll convert Land Desktop point objects and AutoCAD point entities into Civil 3D points:

1. Open the `Convert LDT Points.dwg` file, which you can download from this book's webpage.

2. Use the List command or the AutoCAD Properties palette to confirm that most of the objects in this drawing are `AECC_POINTs`, which are points from Land Desktop. Also, note a cluster of cyan-colored AutoCAD point objects in the western portion of the site.

3. Run the Land Desktop Point Conversion tool by switching to the Home tab and choosing Points ➤ Convert Land Desktop Points. Note that the Convert Autodesk Land Desktop Points dialog allows you to choose a default layer, point creation settings, and styles.

4. Place a check mark next to Add Points To Point Group. Click the Create A New Point Group button. Name the group **Converted from LDT** and click OK.

5. Click OK to complete the conversion process. Civil 3D scans the drawing looking for Land Desktop point objects.

6. Once Civil 3D has finished the conversion, zoom in on any of the former Land Desktop points. The points should now be `AECC_COGO_POINTs` in both the List command and in the AutoCAD Properties palette, confirming that the conversion has taken place. The Land Desktop points have been replaced with Civil 3D points, and the original Land Desktop points are no longer in the drawing.

7. In Prospector, expand the Point Groups category. Notice there is a symbol indicating that the point group needs to be updated. Right-click on Point Groups and select Update.

8. Zoom in on the cyan AutoCAD point objects.

9. Run the AutoCAD Point Conversion tool by choosing Points ➤ Convert AutoCAD Points from the Home tab.

10. The command line reads `Select AutoCAD Points`. Use a crossing window to select all the cyan-colored AutoCAD points.

11. At the command-line prompt, enter a description of **GS** (Ground Shot) for each point.

12. Zoom in on one of the converted points, and confirm that it has been converted to a Civil 3D point. Also, note that the original AutoCAD points have been erased from the drawing.

Getting to Know the Create Points Dialog

CERT
OBJECTIVE

In Civil 3D 2012, you can find point-creation tools directly under the Points drop-down on the Create Ground Data panel of the Home tab as well as in the Create Points toolbar. The toolbar is *modeless,* which means it stays on your screen even when you switch between tasks. Figure 3.9 shows the toolbar with the point creation methods labeled.

FIGURE 3.9
The Create Points toolbar

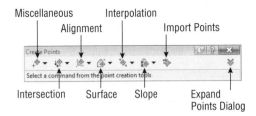

As you place points using these tools, a few general rules apply to all of them. If you place a point on an object with elevation, the point will automatically inherit the elevation of the object. If you use the surface options, the point will automatically inherit the elevation of the surface you choose.

Miscellaneous Point-Creation Options The options in the Miscellaneous category are based on manually selecting a location or on an AutoCAD entity, such as a line, pline, and so on. Some common examples include placing points at intervals along a line or polyline, as well as converting SoftDesk points or AutoCAD entities (see Figure 3.10).

FIGURE 3.10
Miscellaneous point-creation options

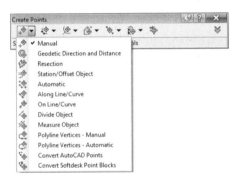

Intersection Point-Creation Options The options in the Intersection category allow you to place points at a certain location without having to draw construction linework. For example, if you needed a point at the intersection of two bearings, you could draw two construction lines using the Bearing Distance transparent command, manually place a point where they intersect, and then erase the construction lines. Alternatively, you could use the Direction/Direction tool in the Intersection category (see Figure 3.11).

FIGURE 3.11
Intersection point-creation options

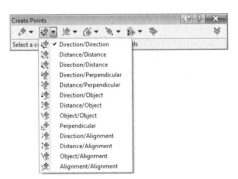

Alignment Point-Creation Options The options in the Alignment category are designed for creating stakeout points based on a road centerline or other alignments. You can also set Profile Geometry points along the alignment using a tool from this menu. See Figure 3.12.

FIGURE 3.12
Alignment point-creation options

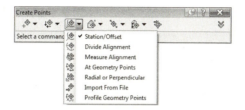

> **AUTOMATIC – OBJECT: UNMASKING THE MYSTERY**
>
> The description option in the point settings, Automatic – Object, can only be used when placing points along an alignment. For example, when placing points using the At Geometry Points option, the point will inherit the alignment's name, the station value of the point, and the type of geometry as its description.
>
> For all other point placement options, Automatic – Object will behave exactly the same as Automatic.

Surface Point-Creation Options The options in the Surface category let you set points that harvest their elevation data from a surface. Note that these are points, not labels, and therefore aren't dynamic to the surface. You can set points manually, along a contour or a polyline, or in a grid. See Figure 3.13.

FIGURE 3.13
Surface point-creation options

Interpolation Point-Creation Options The Interpolation category lets you fill in missing information from survey data or establish intermediate points for your design tasks. For example, suppose your survey crew picked up centerline road shots every 100', and you'd like to interpolate intermediate points every 25'. Instead of doing a manual slope calculation, you could use the Incremental Distance tool to create additional points (see Figure 3.14).

FIGURE 3.14
The Interpolation point-creation options

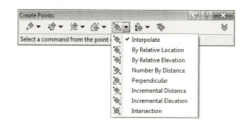

Another use would be to set intermediate points along a pipe stakeout. You could set a point for the starting and ending invert, and then set intermediate points along the pipe to assist the field crew.

Slope Point-Creation Options   The Slope category allows you to set points between two known elevations by setting a slope or grade. Similar to the options in the Interpolation and Intersection categories, these tools save you time by eliminating construction geometry and hand calculations (see Figure 3.15).

FIGURE 3.15
Slope point-creation options

CREATING POINTS

In this exercise, you'll learn how to create points along a parcel segment and along a surface contour:

1. Open `Mastering Point Creation.dwg`, which you can download from this book's webpage. Note the drawing includes an alignment, a series of parcels, and an existing ground surface.

2. Open the Create Points dialog by selecting Points ➢ Point Creation Tools in the Create Ground Data panel on the Home tab.

3. Click the chevron icon on the right to expand the dialog.

4. Expand the Points Creation option. Change the Prompt For Elevations value to None and the Prompt For Descriptions value to Automatic by clicking in the respective cell, clicking the down arrow, and selecting the appropriate option. Enter **LOT** for Default Description (see Figure 3.16). This will save you from having to enter a description and elevation each time. Because you're setting stakeout points for rear lot corners, you will disregard elevation for now.

FIGURE 3.16
Point-creation settings in the Create Points dialog

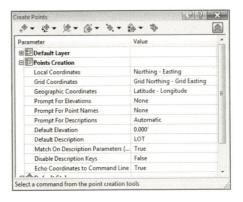

5. Select the Automatic tool under the Miscellaneous category (the first button on the top left of the Create Points dialog).

6. Select all five of the blue property lines in the drawing. Press ↵.

7. Press Esc to exit the command. A point is placed at each property corner and at the endpoints of each curve.

8. Select the Measure Object tool under the Miscellaneous category to set points along the boundary for Property 10. After selecting the object, this tool prompts you for starting and ending stations (press ↵ twice to accept the measurements), offset (press ↵ to accept 0), and an interval (enter **25**). Press Esc to exit the command. A point is placed at 25′ intervals along the property boundary.

9. Next you'll experiment with the Direction-Direction option from the Intersection Point flyout. Click the Direction-Direction icon and click the southeast endpoint of the "floating" parcel line. Click the opposite endpoint to establish the direction of the line. Press ↵ to specify a zero offset.

10. Starting from the southeastern corner of Property 6, establish the second direction by clicking each endpoint of the segment. Press ↵ to specify a zero offset. A point is generated where the two lines would intersect if they were to be extended. Press Esc to exit the command.

11. In the point settings, change Prompt For Descriptions to Automatic – Object.

12. From the Alignments flyout, choose At Geometry Points. Select the green centerline alignment. You are now prompted for a profile. Select Layout (1) from the drop-down and click OK.

13. Press ↵ twice to confirm the station values along which you will place points. Press Esc to exit the command. You should now see points with the name of the alignment and station values.

14. Return to the Point Settings options and change Prompt For Elevations to Manual and Default Description to EG. The next round of points you'll set will be based on the existing ground elevation.

15. Select the Along Polyline/Contour tool in the Surface category to create points every 25′ along the driveway near HOUSE2.

16. Experiment with the plethora of point placement tools available to you!

Basic Point Editing

Despite your best efforts, points will often be placed in the wrong location or need additional editing after their initial creation. It's common for property-corner points to be rotated to match a different assumed benchmark or for points used in a grading design to need their elevations adjusted.

Physical Point Edits

Points can be moved, copied, rotated, deleted, and more using standard AutoCAD commands and grip edits. Rotate a point with the special Point-Rotation grip, as shown in Figure 3.17.

FIGURE 3.17
The new top grip allows label modifications (left); the center grip allows marker modifications (right).

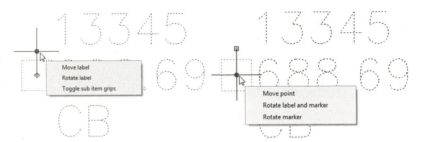

Panorama and Prospector Point Edits

Edit/List Points

You can access many point properties through the Point Editor in Panorama. Choose a point (or points), and then select Edit/List Points from the contextual Ribbon. Panorama brings up information for the selected point(s) (see Figure 3.18).

FIGURE 3.18
Edit points in Panorama

Point Nu...	Northing	Easting	Point Elev...	Name	Raw Description	Full Description	Description For
4	196006.3211'	2305168.6365'	821.711'		TP4	TP4	
11	196006.4760'	2305168.5780'	821.615'		OPUS 3109-11	OPUS 3109-11	
708	196040.0894'	2305170.0278'	821.314'		CL DYL	CL DYL	$*
709	196009.9786'	2305179.6236'	821.945'		CL DYL	CL DYL	$*
710	195992.5522'	2305186.0707'	822.370'		CL DYL	CL DYL	$*
711	195964.3460'	2305197.2031'	822.867'		CL	CL	$*
742	195979.0835'	2305179.1747'	822.310'		EM	EM	
743	195999.1276'	2305171.5996'	821.936'		EM	EM	
744	196012.7283'	2305166.3933'	821.588'		EM DW	EM DW	
745	196015.4935'	2305161.0280'	821.718'		EM DW	EM DW	
746	196036.3229'	2305158.2661'	821.153'		EM DW	EM DW	
747	196036.1508'	2305157.8396'	821.407'		EM DW	EM DW	
748	196036.8479'	2305174.6820'	821.425'		EM	EM	
749	196021.1647'	2305188.1071'	821.291'		EM	EM	
750	196020.2548'	2305188.0515'	821.290'		EM	EM	
751	196010.7712'	2305195.5571'	821.282'		EM	EM	
752	196007.1122'	2305202.3786'	821.337'		FC GUTTER	FC GUTTER	

You can access a similar interface in the Prospector tab of Toolspace by highlighting the Points collection (see Figure 3.19).

> **ZOOM TO POINT**
>
> Right-click on a point number as shown in Figure 3.19 and notice the Zoom To option. This is especially handy for quickly finding important points as denoted by field crews.

FIGURE 3.19
Prospector lets you view your entire Points collection at once.

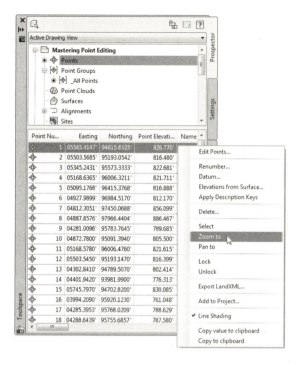

Point Groups: Don't Skip This Section!

CERT OBJECTIVE

Working with point groups is one of the most powerful techniques you will learn from this chapter. Want to turn all your points off without touching layers? Make a point group! Want to move last week's survey up by the blown instrument height difference? Make a point group! Want to show all your Topo shots as a dot rather than an X? Want to prevent invert shots from throwing off your surface model? Point group! Point group!

A point group is a collection of points that has been filtered for a certain criterion. You can use any property of the points, such as description, elevation, and point number, or you can select points in the drawing.

Civil 3D creates the _All Points group for you, which contains every point in the drawing. It cannot be renamed, deleted, or have its properties modified to exclude any points. Create point groups for a collection of points you might wish to separate from others, as shown in Figure 3.20.

FIGURE 3.20
An example of useful point groups in Prospector

Point groups can (and should!) be created upon import of a text file, as shown back in Figure 3.6. That way, if a problem comes to light about that group of points (such as incorrect instrument height) they can be isolated and dealt with apart from other points. Create a new point group by right-clicking on the main point group category and pick New.

In this exercise, you'll learn how to use point groups to separate points into usable categories:

1. Open the drawing `Mastering Point Groups.dwg`, which you can download from this book's webpage.

2. In Prospector, right-click Point Groups and select New.

3. On the Information tab, name the point group **Vegetation**.

4. Set Point Style to Tree.

5. Set Point Label Style to Description Only.

6. Switch to the Include tab and place a check mark next to With Raw Descriptions Matching.

7. In the Raw Descriptions Matching field, type **TREE*, SHRUB*, TL***. The asterisk acts as a wildcard to include points that may have additional information after the description. We are adding multiple descriptions by separating them with a comma, as shown in Figure 3.21.

FIGURE 3.21
The Include tab of the Vegetation point group

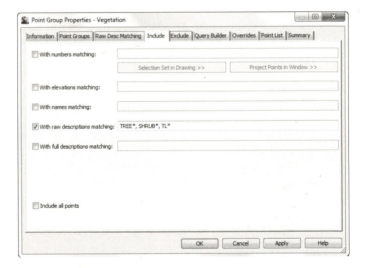

8. Switch to the Overrides tab. Place a check mark next to Style and Point Label Style, as shown in Figure 3.22. Doing so ensures that the point group will control the styles instead of the description keys.

9. Switch to the Point List tab and examine the points that have been picked up by the group. Only points beginning with SHRUB, TREE, and TL should appear in the list. Click OK.

10. Again, right-click on Point Groups and select New.

11. On the Information tab, name the group **NO DISPLAY**, as shown in Figure 3.23. Set both Point Style and Point Label Style to <none>.

FIGURE 3.22
Overrides forces the styles to conform by point group rather than description key.

FIGURE 3.23
Most drawings should contain a NO DISPLAY point group with styles set to <none>.

12. Switch to the Include tab. At the bottom of the dialog box, put a check mark next to Include All Points.
13. Switch to the Overrides tab and place a check mark next to Style and Point Label Style.
14. Click OK. All the points are hidden from view as a result of the point group.
15. Create another point group called **Topo**. Set Point Style to Basic X (BLACK) and Point Label Style to Elevation And Description.
16. Switch to the Exclude tab and place a check mark next to With Elevations Matching. Type **<1** in the accompanying field. Place a check mark next to With Raw Descriptions Matching. Type **INV*, HYD*** in the accompanying field, as shown in Figure 3.24. Click OK.

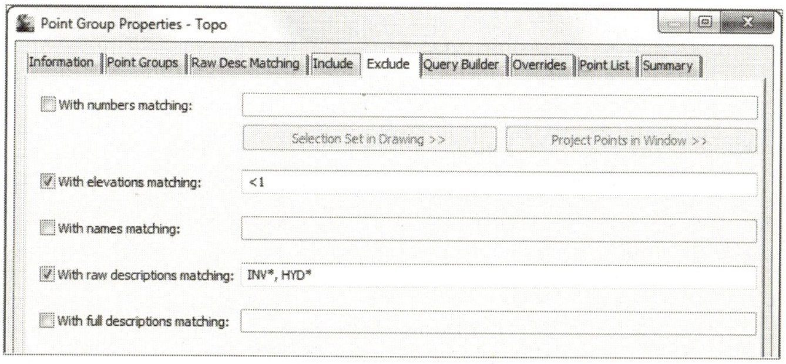

FIGURE 3.24
Use Exclude to create a Topo point group.

Best Practice: Control Point Display Using Point Groups Rather than Layers

Civil 3D drawings will have many layers in them. It is much easier to switch the display of the point groups rather than create layer states for each point visibility scenario.

A point can belong to more than one group at once. For instance, a water valve cover with elevation may be in a Topo group, a Utilities group, and the _All Points group. In these cases, the order in which the point group is displayed in Prospector determines which point group a point is "listening to" for its properties.

In this exercise, we will walk you through an example of how point group display order works:

1. Open the drawing `Mastering Point Display.dwg`, which you can download from this book's webpage.

2. In Prospector, right-click Point Groups and select Properties (see Figure 3.25).

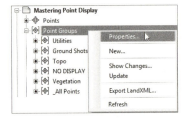

Figure 3.25
Select Point Groups ➢ Properties to change point group display precedence.

3. Using the arrows on the far right of the Point Groups listing, move Vegetation to the top of the list. Move NO DISPLAY so that it is listed directly below Vegetation, as shown in Figure 3.26.

4. Click OK. Notice that only the Vegetation group is visible.

5. Experiment with changing the order of the point groups using the properties.

Figure 3.26
The order in which the point groups appear in this list controls precedence.

Changing Point Elevations

Points placed on or along an object that has elevation will automatically inherit the object's elevation. Points placed with tools in the Surface flyout will automatically inherit the elevation of a surface model. If you have chosen to place points with manual elevation entry and press ↵ when prompted to specify an elevation, the elevation will be null (no elevation).

You are never stuck with a COGO point's elevation. They can be changed individually or as a group using the Panorama window. Additional tools are available for manipulating points (see Figure 3.27), in the contextual panel that opens when you select a point object.

Elevations From Surface is an extremely handy tool for forcing points to a surface elevation (see Figure 3.28).

FIGURE 3.27
Point-editing commands in the Ribbon

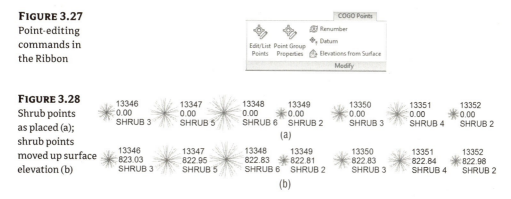

FIGURE 3.28
Shrub points as placed (a); shrub points moved up surface elevation (b)

When you change the datum, you are most likely going to move a group of points' elevations. Right-click on the name of the point group in Prospector and select Edit Points. Panorama will appear for your point-editing delight.

Use Windows keyboard tricks to control which points are selected for modification. Ctrl+A will select all points in the Panorama listing, as shown in Figure 3.29. When you are done selecting points, right-click to choose Datum. The command line will prompt you to specify the change in elevation you require.

FIGURE 3.29
Right-click to access point modification tools from Panorama.

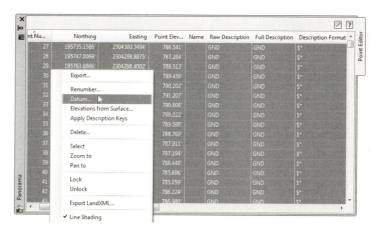

Description Keys: Field to Civil 3D

Description keys bridge the gap between the field and the office. *The Description Key Set* is a listing of field descriptions and how they should look and behave once they are imported into Civil 3D.

For example, a surveyor may code in **BM** to indicate a benchmark. When the file is imported into Civil 3D either through the Survey database (as we examined in Chapter 2) or by import from a text file (as you did earlier in this chapter), it will be checked against the description key set. If BM exists in the list as it is in Figure 3.30, then styles, format, layer, and several other parameters are set.

FIGURE 3.30
Description key set

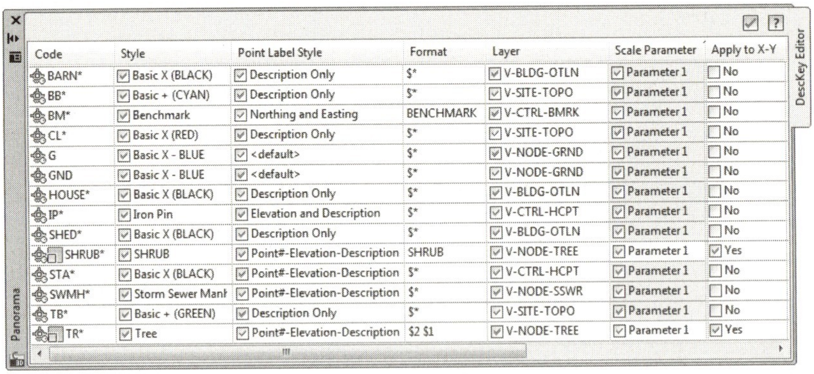

Code	Style	Point Label Style	Format	Layer	Scale Parameter	Apply to X-Y
BARN*	☑ Basic X (BLACK)	☑ Description Only	$*	☑ V-BLDG-OTLN	☑ Parameter 1	☐ No
BB*	☑ Basic + (CYAN)	☑ Description Only	$*	☑ V-SITE-TOPO	☑ Parameter 1	☐ No
BM*	☑ Benchmark	☑ Northing and Easting	BENCHMARK	☑ V-CTRL-BMRK	☑ Parameter 1	☐ No
CL*	☑ Basic X (RED)	☑ Description Only	$*	☑ V-SITE-TOPO	☑ Parameter 1	☐ No
G	☑ Basic X - BLUE	☑ <default>	$*	☑ V-NODE-GRND	☑ Parameter 1	☐ No
GND	☑ Basic X - BLUE	☑ <default>	$*	☑ V-NODE-GRND	☑ Parameter 1	☐ No
HOUSE*	☑ Basic X (BLACK)	☑ Description Only	$*	☑ V-BLDG-OTLN	☑ Parameter 1	☐ No
IP*	☑ Iron Pin	☑ Elevation and Description	$*	☑ V-CTRL-HCPT	☑ Parameter 1	☐ No
SHED*	☑ Basic X (BLACK)	☑ Description Only	$*	☑ V-BLDG-OTLN	☑ Parameter 1	☐ No
SHRUB*	☑ SHRUB	☑ Point#-Elevation-Description	SHRUB	☑ V-NODE-TREE	☑ Parameter 1	☑ Yes
STA*	☑ Basic X (BLACK)	☑ Point#-Elevation-Description	$*	☑ V-CTRL-HCPT	☑ Parameter 1	☐ No
SWMH*	☑ Storm Sewer Manh	☑ Point#-Elevation-Description	$*	☑ V-NODE-SSWR	☑ Parameter 1	☐ No
TB*	☑ Basic + (GREEN)	☑ Description Only	$*	☑ V-SITE-TOPO	☑ Parameter 1	☐ No
TR*	☑ Tree	☑ Point#-Elevation-Description	$2 $1	☑ V-NODE-TREE	☑ Parameter 1	☑ Yes

An interesting fact to note about description keys is that they take over styles and layers set elsewhere. For example, if your point placement options have a layer set but you place a point that matches a description key with a layer set to something different, the description key set "wins." In the point group creation examples, we set the Overrides tab to have Style and Point Label Style selected (Figure 3.22). Those settings wrestle control of the styles away from the description key and into the hands of the point group.

Code The raw description or field code entered by the person collecting or creating the points. The code works as an identifier for matching the point with the correct description key. Click inside this field to activate it, and then type your desired code. Wildcards are useful when more information is added to the shot in addition to the field code.

Point Style The point style that will be applied to points that meet the code criteria. Check the box, and then click inside the field to activate a style-selection dialog. By default, styles set here will take precedence over styles set elsewhere. For more information on creating or modifying point styles, see Chapter 19.

Point Label Style The point label style that will be applied to points that meet the code criteria. Check the box, and then click inside the field to activate a style-selection dialog. By default, styles set here will take precedence over styles set elsewhere. For more information on creating or modifying point styles, see Chapter 19.

Format The Format column can convert a surveyor's shorthand into something that is more drafter-friendly. In Civil 3D terms, the Format column converts the raw description to the full description. The default of $* means the raw description and full description will have the same value. In Figure 3.30, Format will convert all codes starting with BM to a full description of BENCHMARK.

If a survey crew is consistent in coding, even fancier formats can be used. The code should always come first, but the crew can use a space to indicate a parameter.

Consider the example raw description: TREE 30 PINE. TREE is the code, or $0. Parameter 1 is 30, or $1 in the Format field. PINE is the second bit of information after the code referred to as parameter 2, or $2. Based on the example description key set in Figure 3.30, this would translate to a full description of PINE 30. You can have up to nine parameters after the code if your survey crew is feeling verbose.

Layer   The layer that will be applied to points that meet the code criteria. Click inside this field to activate a layer-selection dialog. The layer set here will take precedence over layer defaults set in the point command settings or the point creation tools.

Scale Parameter   The Scale parameter is used to tell Civil 3D which bit of information after the code will be used to scale the symbol. By default it is checked on, but it won't do anything unless Apply To X-Y is also selected. Once you enable Apply To X-Y, you can change which parameter contains scale information.

In our example, TREE 30 PINE, 30 is the Scale parameter.

Fixed Scale Factor   Fixed Scale Factor is an additional scale multiplier that can be applied to the symbol size. The most frequent use of Fixed Scale Factor is to convert a field measurement of inches to feet. If the 30 in our example represents a canopy measurement and is meant to be feet, no fixed scale factor is needed. However, if the 30 represents inches (i.e., a trunk diameter), we would need to turn on Fixed Scale Factor and set the value to 0.0833.

Use Drawing Scale   In most cases, you will leave this option unchecked. By default, marker styles dictate that they will grow or shrink based on the annotative scale of the drawing. Generally, this setting is not needed unless you want to scale your point symbol based on a parameter in addition to the scale factor.

Apply To X-Y   If you wish to scale symbols based on information in the field code, you need to turn this option on by placing a check mark in the box. This option works with the marker style and the Scale parameter to increase the size of an item to a scale indicated by the surveyor in the raw description.

Apply To Z   In most cases, you will leave this option unchecked. Most marker symbols are 2D blocks, so checking this on will have no effect on the point. If your marker symbol consists of a 3D block, it will be stretched by the parameter value, which is rarely needed.

Rotation Parameter, Fixed Rotation, and Rotation Direction   These options work similarly to the scale factor parameter except they dictate the rotation of a symbol. They are not widely used, however, since it is often more time-effective to have the drafter rotate the points in CAD than to have the surveyor key in a rotation.

Creating a Description Key Set

Description key sets appear on the Settings tab of Toolspace under the Point branch. You can create a new description key set by right-clicking the Description Key Sets collection and choosing New, as shown in Figure 3.31.

Figure 3.31
Creating a new description key set on the Settings tab of Toolspace

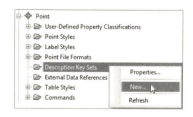

In the resulting Description Key Set dialog, give your description key set a meaningful name, and click OK. You'll create the actual description keys in another dialog.

Creating Description Keys

To enter the individual description key codes and parameters, right-click your description key set, as illustrated in Figure 3.32, and select Edit Keys. The DescKey Editor in Panorama appears.

To enter new codes, right-click a row with an existing key in the DescKey Editor, and choose New or Copy from the context menu, as shown in Figure 3.33.

> **Using Wildcards**
>
> The asterisk (*) acts as a wildcard in many places in Civil 3D. Two of the most common places to use a wildcard are the DescKey Editor and the Point Group Properties dialog. Whereas a DescKey code of TREE flags any points created with that raw description, a DescKey code of TREE* also picks up raw descriptions of TREE1, TREE2, TREEMAPLE, TREEOAK, and so on. You can use the wildcard the same way in the Point Group Properties dialog when specifying items to include or exclude.

> **Point Groups or Description Keys?**
>
> After reading the last two sections, you're probably wondering which method is better for controlling the look of your points. This question has no absolute answer, but there are some things to take into consideration when making your decision.
>
> Point groups are useful for both visibility control and sorting. They're dynamic and can be used to control the visibility of points that already exist in your drawing.
>
> Both can be standardized and stored in your Civil 3D template.
>
> Your best bet is probably a combination of the two methods. For large batches of imported points or points that require advanced rotation and scaling parameters, description keys are the better tool. For preparing points for surface building, exporting, and changing the visibility of points already in your drawing, point groups will prove most useful.

Figure 3.32
Editing a description key set

FIGURE 3.33
Creating or copying a description key

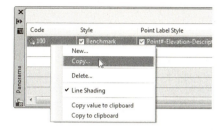

Activating a Description Key Set

Once you've created a description key set, you should verify the settings for your commands so that Civil 3D knows to match your newly created points with the appropriate key.

The Commands ➢ CreatePoints branches are stored on the Settings tab of Toolspace under the Point branch. Edit these command settings by right-clicking, as shown in Figure 3.34.

In the Edit Command Settings dialog, ensure that Match On Description Parameters is set to True and that Disable Description Keys is set to False, as shown in Figure 3.35.

FIGURE 3.34
Right-click Create-Points and choose Edit Command Settings.

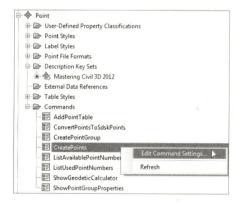

FIGURE 3.35
Do these settings look familiar? Verify that Disable Description Keys is set to False.

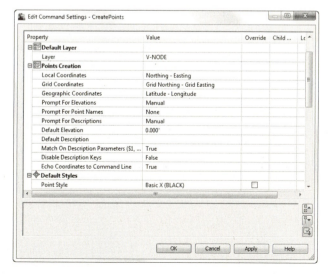

It is not uncommon to have multiple description key sets in your template for multiple clients or external survey firms that you work with. If you have multiple description key sets, they are all active, but if a set has a duplicate key, the first one Civil 3D runs across will take precedence. For example, if one set uses FL for flowline but a second set uses FL for fence line, the second occurrence of the FL key gets ignored.

You can control the search order from the Settings tab of Toolspace by right-clicking on Description Keys Sets and selecting Properties. Figure 3.36 shows the Description Key Sets Search Order Listing dialog box. Use the arrows on the right side of the dialog box to set the order. The set listed first takes first priority, then the second, and so on. Note that the listing in the Settings tab may not reflect the true listing in the properties.

FIGURE 3.36
The Description Key Sets Search Order dialog

Real World Scenario

WORKING WITH LAYERS AND DESCRIPTION KEYS TOGETHER

It's common for surveyors to import points, apply description keys, and use the LAYISO command to isolate a group of points and create two-dimensional linework or breaklines. The following exercise walks you through the steps to apply this concept effectively:

1. Open the Description Keys and Layers.dwg file, which you can download from this book's webpage.

2. Choose Points From File from the Import panel of the Insert tab. The Import Points dialog appears.

3. Be sure the Format is PNEZD (space-delimited), and then click the white plus sign and select Survey.txt (which you can download from this book's webpage).

4. Select Open and then click OK to exit the dialogs and review the results.

5. In Prospector, the groups will need to be updated. Right-click on Point Groups Category to update all of them at once. Click the Top Of Bank group to highlight it.

6. At the bottom of Toolspace you will see the listing of points. Locate point 3002 in the listing (it should be the first one).

7. Right-click on 3002 and select Zoom To. Point3002 is a Top Of Bank shot, which we will use with the Layer Isolate tool.
8. From the Home tab, on the Layers panel click Layer Isolate. Select one of the points labeled Top Of Bank and press ↵. Notice that the layer is isolated. (Note: Use the REGEN command if you are still seeing other points not on the Top Of Bank layer.)
9. Open the Description Key Sets branch of the Points category on the Settings tab of Toolspace.
10. Right-click on Mastering Civil 3D 2012 and select Edit Keys. The DescKey Editor will open in Panorama.
11. Review the Layer settings for both the BOTB* and TOPB* codes as selected in the DescKey Editor.

Point Tables

You've seen some of the power of dynamic point editing; now let's look at how those dynamic edits can be used to your advantage in point tables.

Most commonly, you may need to create a point table for survey or stakeout data; it could be as simple as a list of point numbers, northing, easting, and elevation. These types of tables are easy to create using the standard point-table styles and the tools located in the Points menu under the Add Tables option.

1. Open the `Point Table.dwg` file, which you can download from this book's webpage. This file will appear empty, but it isn't. First, we will review reordering point group properties to change the display of points.
2. In Prospector, expand Point Groups. You will see that NO DISPLAY is listed on the top, which means the styles set in its properties are taking over the other point groups. Right-click on Point Groups and select Properties.
3. In the listing, move the group Trees To Be Removed to the top by using the arrows on the right. Click OK.
4. Use Zoom Extents to see the cluster of trees you are working with.
5. On the Ribbon, switch to the Annotate tab. Click Add Tables ➢ Add Point Table.
6. Verify that Table Style is set to Tree Removal. Click the Point Group icon. Select Trees To Be Removed and deselect Split Table. When all your settings match those in Figure 3.37, click OK.
7. Click anywhere in the graphic to place the table.

Figure 3.37
Point Table Creation options

User-Defined Properties

Standard point properties include items such as number, easting, northing, elevation, name, description, and the other entries you see when examining points in Prospector or Panorama. But what if you'd like a point to know more about itself?

It's common to receive points from a soil scientist that list additional information such as groundwater elevation or infiltration rate. Surveyed manhole points often include invert elevations or flow data. Tree points may also contain information about species or caliber measurements. All this additional information can be added as user-defined properties to your point objects. You can then use user-defined properties in point labeling, analysis, point tables, and more.

 **Real World Scenario**

How Can Civil 3D Work with Soil Boring Data?

In the following example you will add user-defined properties to some soil boring points and leverage point groups to work with the data. The skills you learn in this example can be applied to multiple soil boring values for the purposes of creating subsurface data.

1. Open the file `Soil Borings.dwg`. This file contains an existing ground surface along with several point groups, including one containing soil boring points. Take a moment to examine the soil boring points and the current elevation listing.

2. From the Settings tab, in the Point collection, right-click on User-Defined Property Classifications and select New. Name the new classification **Soil Borings** and click OK.

3. Expand the User-Defined Property Classifications area (if it is not already) and right-click on Soil Borings. Select New.

4. Name the new property **Watertable Elevation**. Set Property Field Type to Elevation. Deselect Default Value and click OK.

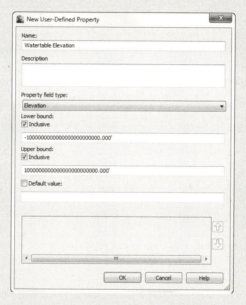

5. Jump back to the Prospector tab. Highlight the main Point Groups listing. At the very bottom of Toolspace you will see a listing of all the point groups, as seen here. Set the classification as shown.

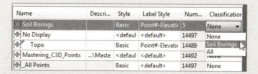

6. Right-click the Soil Borings point group and select Edit Points. Scroll over in Panorama until you locate the new classification column. This is the information you added in the previous step.

7. Add the Watertable Elevation entries in Panorama, as shown here. Dismiss Panorama when you're done.

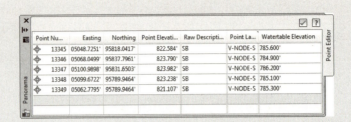

8. Right-click on the Soil Borings group and select Properties. Switch to the Overrides tab.

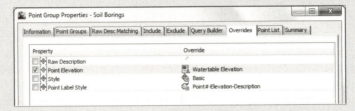

9. Place a check mark next to Point Elevation on the Overrides tab. Click the tiny pencil icon twice, or until it turns into the user-defined property icon.

10. Click the field next to the icon to set the value to Watertable Elevation.

11. Click OK to dismiss the Point Group Properties dialog. Notice that the elevation labels for the five points are listed as the watertable elevations.

The Bottom Line

Import points from a text file using description key matching. Most engineering offices receive text files containing point data at some point during a project. Description keys provide a way to automatically assign the appropriate styles, layers, and labels to newly imported points.

Master It Create a new drawing from _AutoCAD Civil 3D (Imperial) NCS.dwt. Revise the Civil 3D description key set to use the parameters listed here:

Code	Point style	Point label style	Format	Layer
GS	Basic	Elevation Only	Ground Shot	V-NODE
GUY	Guy Pole	Elevation and Description	Guy Pole	V-NODE
HYD	Hydrant (existing)	Elevation and Description	Existing Hydrant	V-NODE-WATR
TOP	Basic	Point#-Elevation-Description	Top of Curb	V-NODE
TREE	Tree	Elevation and Description	Existing Tree	V-NODE-TREE

Import the `Concord_PNEZD_SpaceDelim.txt` file from the data location, and confirm that the description keys made the appropriate matches by looking at a handful of points of each type. Do the trees look like trees? Do the hydrants look like hydrants?

Save the resulting file for use in the next exercises.

Create a point group. Building a surface using a point group is a common task. Among other criteria, you may want to filter out any points with zero or negative elevations from your Topo point group.

> **Master It** Create a new point group called Topo that includes all points *except* those with elevations of zero or less. Use the DWG created in the previous exercise or start with `Master_It.dwg`.

Export points to LandXML and ASCII format. It's often necessary to export a LandXML or an ASCII file of points for stakeout or data-sharing purposes. Unless you want to export every point from your drawing, it's best to create a point group that isolates the desired point collection.

> **Master It** Create a new point group that includes all the points with a raw description of TOP. Export this point group via LandXML and to a PNEZD comma-delimited text file.

Use the DWG created in the previous exercise or start with `Master_It.dwg`.

Create a point table. Point tables provide an opportunity to list and study point properties. In addition to basic point tables that list number, elevation, description, and similar options, you can customize point table formats to include user-defined property fields.

> **Master It** Create a point table for the Topo point group using the PNEZD format table style. Use the DWG created in the previous exercise or start with `Master_It.dwg`.

Chapter 4

Surfaces

One of the most primitive elements in a 3D model of any design is the surface. As you learned in the previous chapter, once survey information is gathered and points are set with elevations, then you can proceed to turn some of that information into an intelligent surface. This chapter looks at various methods of surface creation and editing. Then it moves into discussing ways to view, analyze, and label surfaces, and explores how they interact with other parts of your project.

In this chapter, you will learn to:

- Create a preliminary surface using freely available data
- Modify and update a TIN surface
- Prepare a slope analysis
- Label surface contours and spot elevations
- Import a point cloud into a drawing and create a surface model

Understanding Surface Basics

A surface in Civil 3D is built on the basis of mathematical principles of planar geometry. Each face of a surface is based on three points within a circumcircle (a circle that passes through each of the vertices of a polygon) forming a triangle and defining a plane (Figure 4.1). Each of these triangular planes shares an edge with another, and a continuous surface is made. This methodology is typically referred to as a triangulated irregular network (TIN), as shown in Figure 4.2. On the basis of Delaunay triangulation, this means that for any given (x,y) point, there can be only one unique z value within the surface (as slope is equal to rise over run, when the run is equal to 0 the result is "undefined"). What does this mean to you? It means surfaces in Civil 3D have two major limitations:

Figure 4.1
Three points defining a plane

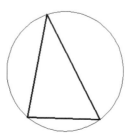

FIGURE 4.2
A triangulated irregular network (TIN)

No Thickness Operations on the basis of solid modeling are not possible. You cannot add or subtract surfaces or look for their unions as you can with a solid that has thickness in the vertical direction.

No Vertical Faces Vertical faces cannot exist in a TIN because two points on the surface cannot have the same (*x,y*) coordinate pair. At a theoretical level, this limits the ability of Civil 3D to handle true vertical surfaces, such as walls or curb structures. You must take this factor into consideration when modeling corridors, as discussed in Chapter 9, "Basic Corridors."

Beyond these basic limitations, surfaces are flexible and can describe any object's face in astonishing detail. The surfaces can range in size from a few square feet to square miles and generally process quickly.

There are two main categories of surfaces in Civil 3D: standard surfaces and volume surfaces. A standard surface is based on a single set of points, whereas a volume surface builds a surface by measuring vertical distances between surfaces. Each of these surfaces can also be a grid or TIN surface. The grid version is still a TIN upon calculation of planar faces, but the data points are arranged in a regularly spaced grid of information. The TIN surface definition is made from randomly located points that may or may not follow any pattern to their location.

Creating Surfaces

Before you can analyze a surface, you have to make one. To the land development company today, this can mean pulling information from a large number of sources, including Internet sources, old drawings, and fieldwork. Working with each requires some level of knowledge about the reliability of the information and how to handle it in Civil 3D. In this section, we'll look at how you can obtain data from a couple of free sources and bring it into your drawing, create new surfaces, and make a volume surface.

Before creating surfaces, you need to know a bit about the components that can be used as part of a surface definition:

LandXML Files   These typically come from an outside source or are exported from another project. LandXML has become the *lingua franca* of the land development industry. These files include information about points and triangulation, making replication of the original surface a snap.

TIN Files   Typically, a TIN file will come from a land development project on which you or a peer has worked. These files contain the baseline TIN information from the original surface and can be used to replicate it easily.

DEM Files   Digital Elevation Model (DEM) files are the standard format files from governmental agencies and GIS systems. These files are typically very large in scale but can be great for planning purposes.

Point Files   Point files work well when you're working with large data sets where the points themselves don't necessarily contain extra information. Examples include laser scanning or aerial surveys.

Point Groups   Civil 3D point groups or survey point groups can be used to build a surface from their respective members and maintain the link between the membership in the point group and being part of the surface. In other words, if a point is removed from a group used in the creation of a surface, it is also removed from the surface.

Boundaries   Boundaries are closed polylines that determine the visibility of the TIN inside the polyline. Outer boundaries are often used to eliminate stray triangulation, whereas others are used to indicate areas that could perhaps not be surveyed, such as a building pad.

Breaklines   Breaklines are used for creating hard-coded triangulation paths, even when those paths violate the Delaunay algorithms for normal TIN creation. They can describe anything from the top of a ridge to the flowline of a curb section. A TIN line may not cross the path of a breakline.

Drawing Objects   AutoCAD objects that have an insertion point at an elevation (e.g., text or blocks) can be used to populate a surface with points. It's important to remember that the objects themselves are not connected to the surface in any way.

Edits   Any manipulation after the surface is completed, such as adding or removing triangles or changing the datum, will be part of the edit history. These changes can be viewed in the properties of a surface and can be toggled on and off individually to make reviewing changes simple.

Working with all these elements, you can model and render almost anything you'd find in the world—and many things you wouldn't. In the next section, you start building some surfaces.

Free Surface Information

You can find almost anything on the Internet, including information about your project site that probably includes level information you can use to build a surface. For most users, free surface information can be gathered from government entities or Google Earth. You look at both in this section.

Surfaces from Government Digital Elevation Models

One of the most common forms of free data is the Digital Elevation Model (DEM). These files have been used by the U.S. Department of the Interior's United States Geological Survey (USGS) for years and are commonly produced by government organizations for their GIS systems. The DEM format can be read directly by Civil 3D, but the USGS typically distributes the data in a complex format called Spatial Data Transfer Standard (SDTS). The files can be converted using a freely available program named sdts2dem. This DOS-based program converts the files from the SDTS format to the DEM format you need. Once you are in possession of a DEM file, creating a surface from it is relatively simple, as you'll see in this exercise:

1. Start a new blank drawing from the _AutoCAD Civil 3D (Imperial) NCS template that ships with Civil 3D.

2. Switch to the Settings tab of Toolspace, right-click the drawing name, and select Edit Drawing Settings. Set the coordinate system as shown in Figure 4.3 via the Drawing Settings dialog and click OK. The coordinate system of the DEM file that you will import will be set to match the coordinate system of the drawing.

FIGURE 4.3
Civil 3D Imperial coordinate settings for DEM import

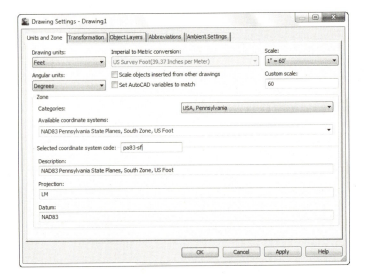

3. In Prospector, right-click the Surfaces collection and select the Create Surface option. The Create Surface dialog appears.

4. Accept the options in the dialog, and click OK to create the surface. This surface is added as Surface 1 to the Surfaces collection.

5. Expand the Surfaces ➢ Surface 1 ➢ Definition branch.

6. Right-click DEM Files and select the Add option (see Figure 4.4). The Add DEM File dialog appears.

FIGURE 4.4
Adding DEM data to a surface

7. Navigate to the Stewartstown_PA.DEM file and click Open. (Remember, all data and drawing files for this book can be downloaded from www.sybex.com/go/masteringcivil3d2012.)

8. Set the values in the Add DEM File dialog as shown in Figure 4.5 and click OK. This translates the DEM's coordinate system to the drawing's coordinate system.

FIGURE 4.5
Setting the Stewartstown_PA.DEM file properties

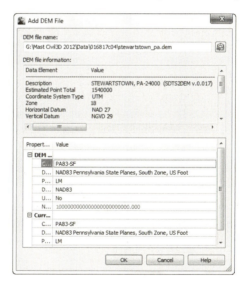

9. Right-click Surface 1 in Prospector and select Zoom To to bring the surface into view, and then right-click the surface in your drawing and select Surface Properties. The Surface Properties dialog appears.

10. On the Information tab, change the Name field entry to **Stewartstown PA**.

11. Change the Surface Style drop-down list to Border Only, and then click OK to dismiss the Surface Properties dialog.

Once you have the DEM data imported, you can pause over any portion of the surface and see that Civil 3D is providing feedback through a tooltip. This surface can be used for preliminary planning purposes but isn't accurate enough for construction purposes. The main drawback to DEM data is the sheer bulk of the surface size and point count. The Stewartstown PA DEM file you just imported contains 1.4 million points and covers more than 55 square miles. This much data can be overwhelming, and it covers an area much larger than the typical site. You'll look at some data reduction methods later in this chapter.

In addition to making a DEM a part of a TIN surface, you can build a surface directly from the DEM (select the Surfaces branch, right-click, and choose Create Surface From DEM). The drawback to this approach is that no coordinate transformation is possible. Because one of the real benefits of using georectified data is pulling in information from differing coordinate systems, we're skipping this method to focus on the more flexible method shown here.

SURFACES FROM GOOGLE EARTH

Civil 3D also includes an importing function that brings in surface and image information directly from Google Earth. The DEMs used by Google Earth were collected over a 10-day span in February 2000 by the space shuttle Endeavor. The data, known as SRTM (Shuttle Radar Topography Mission) data, is typically not updated on a large scale. Ground control was not used during the collection of data, and the mission sought to achieve a vertical accuracy of just 16 meters. Because most freely available DEMs have been gathered by digitizing USGS Quadrangles (also known as QUADs), you can generally assume that SRTM data is the best freely available information out there. In this exercise, you'll import a Google Earth location as a Civil 3D surface:

1. Download the latest version of Google Earth from `http://earth.google.com` and install it.

> **GOOGLE EARTH AND VERSIONS**
>
> This data and exercise were tested with Version 6.0.1.2032. Due to a programming change on the Google Earth side, some later versions are picky about the amount of data Civil 3D pulls. Depending on the version installed on your machine, you might want to search the Web for information regarding Civil 3D and Google Earth interactions.

2. Launch Google Earth and get connected.

3. From the main menu, choose File ➢ Open.

4. Navigate to the `Data` directory and select the `Concord Commons.kmz` file to restore a view of a site in Felton, Pennsylvania.

5. In Civil 3D, create a new drawing from the _AutoCAD Civil 3D (Imperial) NCS template and set the coordinate system as you did in the previous exercise.

6. Change to the Insert tab on the Ribbon.

CREATING SURFACES | 109

7. On the Import panel, select Google Earth ➢ Google Earth Image and Surface.
8. Press ↵ to accept the coordinate system as shown and the Surface Creation dialog will appear.
9. Accept the defaults in the Surface Creation dialog and click OK to dismiss the dialog.
10. From the main menu, choose View ➢ Zoom ➢ Extents to see something like Figure 4.6.
11. Close Google Earth.
12. Save the drawing as GE_Surface.dwg.

FIGURE 4.6
Completed Google Earth surface import

The interesting thing about surfaces built from Google Earth is that their accuracy is zoom-level dependent. This means that the tighter you are zoomed into a site in Google Earth, the better the surface you derive from that picture. Because of this dependence, you should attempt to zoom in as tightly as possible on the area of interest when using Google Earth for preliminary surface information.

DRAPING AN IMAGE

Now that you have an imported Google Earth Image and Surface, let's check out a tool that will let you drape an image (such as the image that comes from Google Earth) onto the surface.

1. If it's not already open, open GE_Surface.dwg. Select the surface.
2. From the Tin Surface tab and Surface Tools, select the Drape Image tool. The Drape Image dialog box opens, as shown below.

3. Examine the settings and press Ok to accept the defaults.
4. Select the contours and right click. Select Object Viewer from the right-click menu.
5. In the Object Viewer, select Realistic from the Visual Styles drop-down list and SW Isometric from the View Control drop-down list.
6. Your drawing should look like the image shown below.

You can see how this tool will graphically display the surface based on the contour data.

So, after doing all of this, we're going to tell you that didn't have to do that for Google Earth, since it automatically drapes. But the procedure would be the same for any image that has been brought into your drawing that contains coordinate data.

Surface Approximations

In this section, you'll work with elevated polylines. Later in this chapter, you'll work with a large point cloud delivered as a text file. These polylines are quite common, and historically it can be difficult making an acceptable surface from them.

SURFACES FROM POLYLINE INFORMATION

One common complaint about converting a drawing full of contours at elevation into a working digital surface is that the resulting contours don't accurately reflect the original data. Civil 3D includes a series of surface algorithms that work very well at matching the resulting surface to the original contour data. You'll look at those surface edits in this series of exercises.

1. Open the SurfaceFromPolylines.dwg file. Note that the contours in this file are composed of polylines.
2. In Prospector, right-click the Surfaces branch, and select the Create Surface option. The Create Surface dialog appears.
3. Leave the Type field as TIN Surface but change the Name value to **EG-Polylines**.
4. Change the description to something appropriate.
5. Change the Style drop-down list to Contours 1' and 5' (Background) and click OK to close the dialog.
6. Expand the Surfaces ➢ EG-Polylines ➢ Definition branches.
7. Right-click Contours and select the Add option. The Add Contour Data dialog appears.
8. Set the options as shown in Figure 4.7 and click OK. (You will return to the Minimize Flat Areas By options in a bit.)
9. Enter **ALL** at the command line to select all the entities in the drawing. You can dismiss Panorama if it appears and covers your screen.

FIGURE 4.7
The Add Contour Data dialog

The contour data has some tight curves and flat spots where the basic contouring algorithms simply fail. Zoom into any portion of the site, and you can see these areas by looking for the blue and cyan original contours not matching the new Civil 3D–generated contour. You'll fix that now:

1. Expand the Definition branch and right-click Edits. Select the Minimize Flat Areas option to open the Minimize Flat Areas dialog. Note that the dialog has the same options found in that portion of your original Add Contour Data dialog. You just did it as two steps to illustrate the power of these changes!

2. Click OK.

Now the contours displayed more closely match the original contour information. There might be a few instances where gaps exist between old and new contour lines, but in a cursory analysis, none was off by more than 0.4" in the horizontal direction—not bad when you're dealing with almost a square mile of contour information. You'll see how this was done in this quick exercise:

1. Zoom into an area with a dense contour spacing and select the surface.

2. Click the Surface Properties button on the Modify panel.

3. Change the Surface Style field to Contours With Points and click OK to see a drawing similar to Figure 4.8.

FIGURE 4.8
Surface data points and derived data points

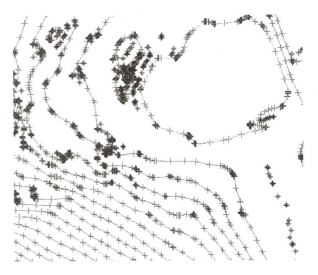

In Figure 4.8, you're seeing the points the TIN is derived from, with some styling applied to help you understand the creation source of the points. Each point in red is a point picked up from the contour data itself. The magenta points are all added data on the basis of the Minimize Flat Areas edits. These points make it possible for the Civil 3D surface to match almost exactly the input contour data.

Surfaces from Points or Text Files

Besides receiving polylines, it is common for a surveying company to also send a simple text file with points. This isn't an ideal situation because you have no information about breaklines or other surface features, but it is better than nothing or using a Google Earth–derived surface. Because you have the same aerial surface described as a series of points, you'll add them to a surface in this exercise:

1. Make a folder on your computer called C:\Mastering.
2. Place the Concord Commons.txt file in your newly created folder. Doing so ensures that future exercises will function properly.
3. Create a new drawing using the _AutoCAD Civil 3D (Imperial) NCS Template.
4. Change the Coordinates to NAD83 Pennsylvania, South Zone, US Foot (PA83-SF).
5. From the Create Ground Data panel on the Home tab, choose Surfaces ➢ Create Surface. The Create Surface dialog appears.
6. Change the Name value to **Points from Text**, and click OK to close the dialog.
7. In Prospector, expand the Surfaces ➢ Points From Text ➢ Definition branches.
8. Right-click Point Files and select the Add option. The Add Point File dialog shown in Figure 4.9 appears.

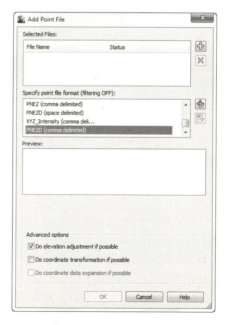

Figure 4.9
Adding a point file to the surface definition

9. Set the Format field to PNEZD (Comma Delimited). Make sure it is PNEZD, and not PENZD or your results will be off completely.

10. Click the Browse button shown in Figure 4.9. The Select Source File dialog opens.

11. Navigate to the previously created C:\Mastering folder, and select the Concord Commons.txt file. Click OK.

12. Click OK to exit the Add Point File dialog and build the surface. Panorama will appear, but you can dismiss it.

13. Right-click Points From Text Surface in Prospector and select the Zoom To option to view the new surface created.

In both the polyline and point file examples, you're making surfaces from the best information available. When you're doing preliminary work or large-scale planning, these types of surfaces are great. For more accurate and design-based surfaces, you typically have to get into field-surveyed information. We'll look at that a little later.

> **PROSPECTOR AND THE RIBBON**
>
> In previous exercises, you created new surfaces with commands from Prospector. Now we will turn our attention to the Ribbon and do the same thing. That way, you can see that there is typically more than one way to accomplish the same goal.

Surface from GIS Data

You may run into a situation where an outside firm uses GIS, or perhaps your firm is also using GIS data. Civil 3D understands GIS and can work with the data given. In this section, we'll show you how to import GIS data pertaining to surfaces.

1. Start a new drawing by using the _AutoCAD Civil 3D (Imperial) NCS template. For Metric users, use the _AutoCAD Civil 3D (Metric) NCS template. Set the coordinate system to NAD83 Georgia State Planes, West Zone, US Foot (GA83-WF).

2. In the Create Ground Data panel of the Home tab, select Surfaces ➢ Create Surface from GIS Data. The Import Gis Data Wizard - Object Options screen appears. Enter the information as shown in Figure 4.10 and click the Next button. The Import Gis wizard - Connect To Data screen appears.

3. You are importing a SHP file, so change Data Source Type to SHP.

4. Click the ellipsis next to SHP Path. Locate the contours2008.shp file (which you'll find at www.sybex.com/go/masteringcivil3d2012). Many thanks to Kevin Clark for providing this dataset for our use. The path is now populated with the location of the SHP file, as shown in Figure 4.11.

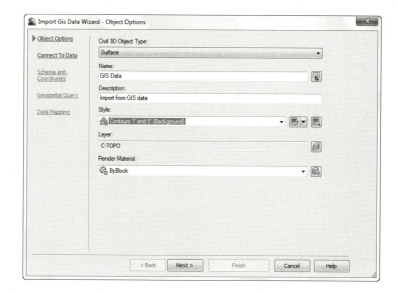

FIGURE 4.10
The Import Gis Data Wizard - Object Options screen

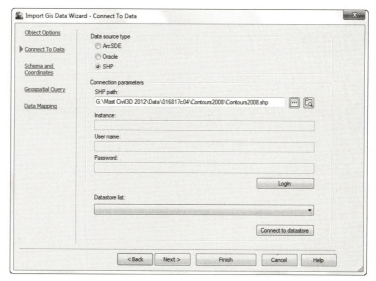

FIGURE 4.11
The Import Gis Data Wizard - Connect To Data screen

5. Click the Login button. The Import Gis Data Wizard - Schema And Coordinates screen now appears (Figure 4.12).

   You will notice that the name of the file appears as well as the coordinate system in which the SHP was created. In this case, the NAD83 Georgia State Plane, West Zone, US Foot matches what you set the drawing up with.

6. Click the box next to Feature Class and click Next.

FIGURE 4.12
The Import Gis Data Wizard - Schema And Coordinates screen

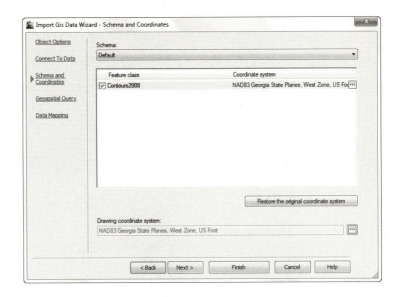

7. On the Import Gis Data Wizard - Geospatial Query screen, look at the settings but do not make any changes (Figure 4.13). Click Next.

8. On the Import Gis Data Wizard - Data Mapping screen, click the drop-down next to Civil 3D Property for the GIS field Elevation, as shown in Figure 4.14. Click the Finish button.

9. Dismiss the Panorama and zoom extents to see the imported image from the SHP file (Figure 4.15).

FIGURE 4.13
The Import Gis Data Wizard - Geospatial Query screen

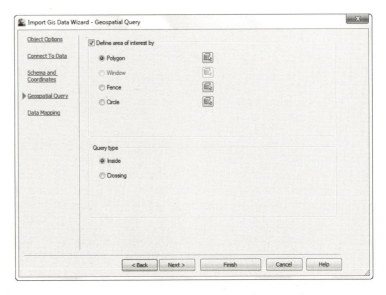

FIGURE 4.14
The Import Gis Data Wizard - Data Mapping screen

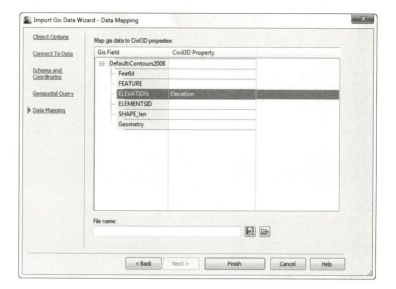

FIGURE 4.15
The finished Imported GIS contours

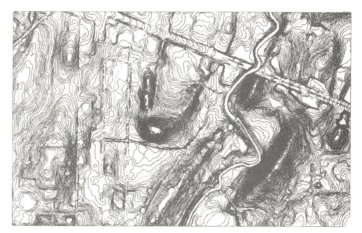

Notice that the surface name and description are the same as what you specified in step 2.

This is just another avenue for getting drawings from other sources into Civil 3D. This topic will be discussed more in depth in Chapter 17, "Interoperability."

Simply adding surface information to a TIN definition isn't enough. To get beyond the basics, you need to look at the edits and other types of information that can be part of a surface.

Refining and Editing Surfaces

Once a basic surface is built, and, in some cases, even before it is built, you can do some cleanup and modification to the TIN construction that make it much more usable and realistic. Some of these edits include limiting the input data, tweaking the triangulation, adding in breakline information, or hiding areas from view. In this section, we explore a number of ways of refining surfaces to end up with the best possible model from which to build.

Surface Properties

The most basic steps you can perform in making a better model are right in the Surface Properties dialog. The surface object contains information about the build and edit operations, along with some values used in surface calculations. These values can be used to tweak your surface to a semi-acceptable state before more manual operations are needed.

In this exercise, you'll go through a couple of the basic surface-building controls that are available. You'll do them one at a time in order to measure their effects on the final surface display.

1. Open the `SurfaceProperties.dwg` file. This is the Points From Text drawing that you worked on earlier but has the polylines frozen.
2. Expand the Surfaces branch.
3. Right-click EG and select Surface Properties. The Surface Properties dialog appears.
4. Select the Definition tab. Note the list at the bottom of the dialog.
5. Under the Definition Options at the top of the dialog, expand the Build option.

The Build options of the Definition tab allow you to tweak the way the triangulation occurs. The basic options are listed here:

Copy Deleted Dependent Objects   When you select Yes and an object that is part of the surface definition (such as the polylines you used in your aerial surface, for instance) is deleted, the information derived from that object is copied into the surface definition. Setting this option to True in the EG Surface properties will let you erase the polylines from the drawing file while still maintaining the surface information.

Exclude Elevations Less Than   Setting this to Yes puts a floor on the surface. Any point that would be built into the surface but that is lower than the floor is ignored. In the EG surface, there are calculated boundary points with zero elevations, causing real problems that can be solved with this simple click. The floor elevation is controlled by the user.

Exclude Elevations Greater Than   The idea is the same as with the preceding option, but a ceiling value is used.

Use Maximum Triangle Length   This setting attempts to limit the number of narrow "sliver" triangles that typically border a site. By not drawing any triangle with a length greater than the user input value, you can greatly refine the TIN.

Convert Proximity Breaklines To Standard   Toggling this to Yes will create breaklines out of the lines and entities used as proximity breaklines. We'll look at this more later.

Allow Crossing Breaklines   This option determines what Civil 3D should do if two breaklines in a surface definition cross each other. An (x,y) coordinate pair cannot have two z values, so some decision must be made about crossing breaklines. If you set this to Yes, you can then select whether to use the elevation from the first or the second breakline or to average these elevations.

If you look at the surface, you might not notice a blob area as shown in the top panel of Figure 4.16. When you zoom in closer to inspect the surface, it appears that there are a series of blown shots, causing the elevation to dip to zero, as shown in the bottom panel of Figure 4.16. In this next portion of the exercise, you'll limit the build options to make that blown surface "disappear":

1. Set the Exclude Elevations Less Than value to Yes.

2. Set the Value to **200** and click OK to exit the dialog. Elevations less than 200" will be excluded.

3. A warning message will appear. Civil 3D is simply warning you that your surface definition has changed. Click Rebuild The Surface to rebuild the surface. When it's done, it should look like Figure 4.17.

Figure 4.16
Overall EG surface (top). EG Surface showing blown points (bottom).

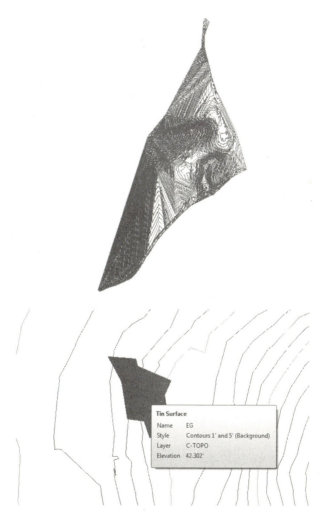

FIGURE 4.17
EG surface after ignoring low elevations

Although this surface is better than the original, there are still huge areas being contoured that probably shouldn't be. By changing the style to review the surface, you can see where you still have some issues:

1. Open the Surface Properties dialog again, and switch to the Information tab.
2. Change the Surface Style field to Contours And Triangles.
3. Click Apply. Doing so makes the changes without exiting the dialog.
4. Drag the dialog to the side so you can see the site. On the outer edges of the site, you can see some long triangles formed in areas where there was no survey taken but the surface decided to connect the triangles anyway (Figure 4.18, left).
5. Switch to the Definition tab.
6. Expand the Build option.
7. Set the Use Maximum Triangle Length value to Yes.
8. In the Maximum Triangle Length value field, type **300"**.
9. Click OK to apply and exit the dialog.
10. Click Rebuild The Surface to update and dismiss the warning message to see the revised surface (Figure 4.18, right).

FIGURE 4.18
EG surface before Maximum Triangle (left) and after (right)

The value is a bit high, but it is a good practice to start with a high value and work down to avoid losing any pertinent data. Setting this value to 225" will result in a surface that is acceptable because it doesn't lose a lot of important points. Beyond this, you'll need to look at making some edits to the definition itself instead of modifying the build options.

Surface Additions

Beyond the simple changes to the way the surface is built, you can look at modifying the pieces that make up the surface. With your drawing so far, you have merely been building from points. Although this is OK for small surfaces, you need to go further with this surface. In this section, you'll add a few breaklines and a border and finally perform some manual edits to your site.

> **YOU CAN'T ALWAYS GET WHAT YOU WANT**
>
> But sometimes you get what you need. Autodesk has included the ability to reorder the build operations on the Surface Definition tab. If you look at the lower left of the Surface Properties Definition tab shown here, you'll find that there are arrows to the left of the list box showing all the data, edits, and changes you've made to the surface.
>
> As Civil 3D builds a surface, it processes this data and information from top to bottom—in this case, adding points, and then the breaklines, and then a boundary, and so on. If a later operation modifies one of these additions or edits, the later operation takes priority. To change the processing order, select an operation, and then use the arrows at the left to push it up or down within the process. One common example of this is to place a boundary as the last operation to ensure accurate triangulation. You'll look at boundaries in the next section.
>
>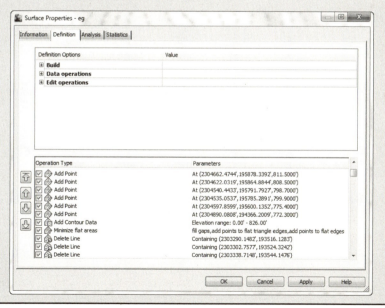

Adding Breakline Information

Breaklines can come from any number of sources. They can be approximated on the basis of aerial photos of the site that help define surface features, or they can be directly input from field book files and the Civil 3D survey functionality. Five types of breaklines are available for use:

Standard Breaklines Built on the basis of 3D lines, feature lines, or polylines, standard breaklines typically connect points already included but can contain their own elevation data. Simple-use cases for connecting the dots include linework from a survey or drawing a building pad to ensure that a flat area is included in the surface. Feature lines and 3D polylines are often used as the mechanism for grading design and include their own vertical information. An example might be the description of a parking lot area or a drainage swale behind a building.

Proximity Breaklines These breaklines allow you to force triangulation without picking precise points. They will not add vertical information to the surface.

Wall Breaklines Wall breaklines define walls in surfaces. Because of the limitation of true vertical surfaces, a wall breakline will let you approximate a wall without having to create an offset. They are defined on the basis of an elevation at a vertex, and then an elevation difference at each vertex.

From File You can select this option if a text file contains breakline information. This file can be the output of another program and can be used to modify the surface without creating additional drawing objects.

Nondestructive Breaklines This type of breakline is designed to maintain the integrity of the original surface while updating triangulation.

In most cases, you'll build your surfaces from standard, proximity, and wall breaklines. In this example, you'll add in some breaklines that describe road and surface features:

1. Open the `SurfaceBreaklines.dwg` file. This drawing has been modified and stylized per company standards. For more information on styles, see Chapter 19, "Styles."

2. Thaw the _Polylines-Road layer, if necessary. The roads are the color red.

3. Right-click on one of the red polylines and select Similar. All the red polylines are now highlighted.

4. In Prospector, expand the Surfaces ➤ EG ➤ Definition branches.

5. Right-click Breaklines and select the Add option. The Add Breaklines dialog appears.

6. Enter a description if you wish and the settings as shown in Figure 4.19. Click OK to accept the settings and close the dialog Dismiss Panorama.

7. If necessary, thaw the _Polylines-Surface layer. The Surface polylines are the color green.

8. Repeat steps 3–5 but change the description to Surface Polylines Dismiss Panorama.

9. Press ↵ to complete the command.

The surface changes reflect the breaklines added. You can see that the original border has expanded to include the breakline data. We want to clean that up so we will rectify that problem with the next exercise. On sites with more extreme grade breaks, such as those that might follow a channel or a site grading, breaklines are invaluable in building the correct surface.

FIGURE 4.19
The Add Breaklines dialog

> **CROSSING BREAKLINES**
>
> Invariably, you will see Panorama pop up with a message about crossing breaklines. In general, Civil 3D does not like breaklines that cross themselves.
>
> The Resolve Crossing Breaklines tool will let you examine those situations.
>
> 1. Click on the surface.
> 2. From the Tin Surface tab and Analyze panel, select the Resolve Crossing Breaklines tool.
> 3. At the `Please specify the types of breakline you want to find or [survey-Database/Figure/Surface]:` prompt, type **S↵ to select the surface option.**
> 4. The Crossing Breaklines tab on Panorama shows you the crossing breaklines and you can decide how you want to resolve them using Use Higher Elevation, Use Lower Elevation, Use Average Elevation, or Use Specified Elevation, as shown below. As you click on each breakline and press resolve, it disappears from the conflict list.

ADDING A SURFACE BORDER

In the previous exercise, you fixed some breakline issues. However, in the data presented, the bigger issue is still the number of inappropriate triangles that are being drawn along the edge of the site. It is often a good idea to leave these triangles untouched during the initial build of a surface, because they serve as pointers to topographical data (such as monumentation, control, utility information, and so on) that may otherwise go unnoticed without a visit to the site. This is a common problem that can be solved by using a surface border. You can sketch in a polyline to approximate a border, but the Extract Objects From Surface utility gives you the ability to use the surface itself as a starting point.

The Extract Objects From Surface utility allows you to re-create any displayed surface element as an independent AutoCAD entity. This entity can be the contours, grid, 3D faces, and so forth. In this exercise, you'll extract the existing surface boundary as a starting point for creating a more refined boundary that will limit triangulation:

1. Open the SurfacePoints.dwg file.

2. Select Extents from the Navigate panel on the View tab to view the whole surface on screen.

3. Select the surface and click Extract Objects on the Surface Tools panel to open the Extract Objects From Surface dialog.

4. Deselect the Major Contour and Minor Contour options, as shown in Figure 4.20.

FIGURE 4.20
Extracting the border from the surface object

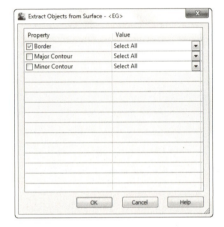

5. Click OK to finish the process. Press Esc to deselect the surface.

6. Pick the blue border line, and notice from the grips displayed that you are no longer selecting the surface but a 3D polyline.

   This polyline will form the basis for your final surface boundary. By extracting the polyline from the existing surface, you save a lot of time playing connect the dots along the points that are valid. Next, you'll refine this polyline and add it to the surface as a boundary.

7. In Prospector, right-click Point Groups and select the Properties option. The Point Groups dialog appears.

8. Move _All Points to the top of the list using the up and down arrows on the right.

9. Click OK to display all of your points on the screen.

10. Working your way around the site, grip-edit the polyline you created in step 6 to include some of the roadway. On a large site, you can see that this is a time-consuming process but worth the effort to clean up the site nicely (Figure 4.21). When complete, your polyline might look something like Figure 4.22.

Just as with breaklines, there are multiple types of surface boundaries:

Outer Boundaries   Use this type to define the outer edge of the shown boundary. When the Non-destructive Breakline option is used, the points outside the boundary are still included in the calculations; then additional points are created along the boundary line where it intersects with the triangles it crosses. This boundary trims the surface for display but does not exclude the points outside the boundary. You'll want to have your outer boundary among the last operations in your surface-building process.

Hide Boundaries   Use this type to punch a hole in the surface display for tasks like building footprints or a wetlands area that are not to be touched by design. Hidden surface areas are *not* deleted but merely not displayed.

FIGURE 4.21
Using the grips to adjust the border

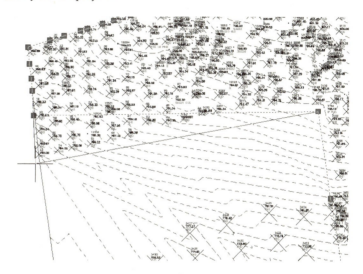

FIGURE 4.22
Revised surface border polyline

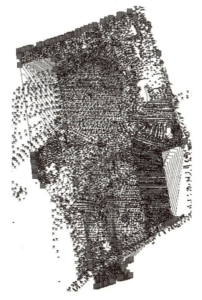

Show Boundaries Use this type to show the surface inside a hide boundary, essentially creating a donut effect in the surface display.

Data Clip Boundaries Data clip boundaries place limits on data that will be considered part of the surface from that point going *forward*. This type is different from an outer boundary in that the data clip boundary will keep the data from ever being built into the surface as opposed to limiting it after the build. Using data clip boundaries is handy when you are attempting to build Civil 3D surfaces from large data sources such as Light Detection and Ranging (LIDAR) or DEM files. Because they limit data being placed into the surface definition, you'll want to have data clips among the first operations in your surface.

The addition of every boundary is considered a separate part of the building operations. This means that the order in which the boundaries are applied controls their final appearance. For example, a show boundary selected before a hide boundary will be overridden by that hide operation. To finish the exercise, you'll add the outer boundary twice, once as a nondestructive breakline and once with a standard breakline, and observe the difference.

1. In Prospector, expand the Surfaces branch.
2. Right-click EG and select the Surface Properties option. The Surface Properties dialog appears.
3. Change Surface Style to 1" and 5" TIN Editing (Ex.).
4. In Prospector, expand the Surfaces ➢ EG ➢ Definition branches.
5. Right-click Boundaries and select the Add option. The Add Boundaries dialog opens.
6. Enter a name if you like and check the Non-destructive Breakline option.
7. Pick the polyline and notice the immediate change.
8. Zoom in on the northwest portion of your site, as shown in Figure 4.23.

FIGURE 4.23
A nondestructive border in action

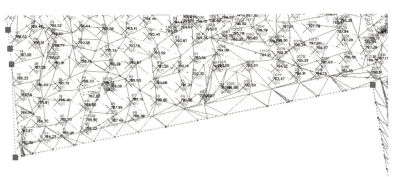

Notice how the triangulation appears to include lines to nowhere. This is the nature of the nondestructive breakline. The points you attempted to exclude from the surface are still being included in the calculation; they are just excluded from the display. This isn't the result you were after, so let's fix it now:

1. In Prospector, expand the Surfaces ➢ EG ➢ Definition branches and select Boundaries.

2. A listing of the boundaries appears in the preview area.

3. Right-click the border you just created and select Delete, as shown in Figure 4.24. Click OK in the warning dialog that tells you the selected definition items will be permanently removed from the surface.

FIGURE 4.24
Deleting a surface boundary

4. In Prospector, expand the Surfaces branch and right-click EG. Select the Rebuild option to return to the prior version of the surface.

5. Right-click Boundaries and select the Add option again. The Add Boundaries dialog appears.

6. This time, leave the Non-destructive Breakline option deselected and click OK.

7. Pick the border polyline on your screen. Notice that no triangles intersect your boundary now where it does not connect points.

8. On the main menu, choose View ➢ Zoom ➢ Extents to see the result of the border addition.

In spite of adding breaklines and a border, you still have some areas that need further correction or changes.

SURFACE CROPPING

Surface cropping is useful when working with a small portion of a much larger surface. A cropped area within a surface becomes a separate surface object (a new surface) to be managed and manipulated on a much smaller scale. In this example, you will create a cropped surface and add it to an existing drawing:

1. Open the SurfaceCrop.dwg file.

2. Select the surface and expand the Surface Tools drop-down panel on the Tin Surface: EG tab. Select Create Cropped Surface. The Create Cropped Surface dialog is displayed, as shown in Figure 4.25.

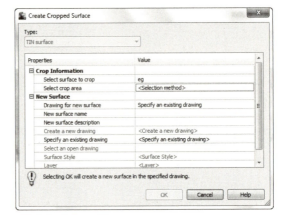

FIGURE 4.25
The Create Cropped Surface dialog

3. Select the ellipsis next to Select Crop Area and type **O**↵. Object is now displayed in the Select Crop Area Value. Select the red rectangle on the surface and press ↵.

4. Select the rectangle when prompted for a point. The rectangle is now highlighted.

5. For Drawing For New Surface, select Create A New Drawing from the drop-down, as shown in Figure 4.26.

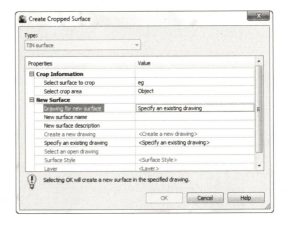

FIGURE 4.26
New surface selection

6. In the Value column for New Surface Name, enter **Cropped Area**. You can also type a description for the new surface if desired.

7. Click the ellipsis next to Create A New Drawing. The Select Template dialog opens. Select a template and click Open.

8. The Value column for Create A New Drawing is populated with the next open drawing name, such as `Drawing2.dwg`.

9. You can select a surface style and a layer for the surface to be created on. Your dialog should look like Figure 4.27.

10. Click OK to complete the command and create a cropped surface in the existing drawing.

11. Open the drawing `Surface Cropping - Final.dwg` to reveal this new surface, as shown in Figure 4.27.

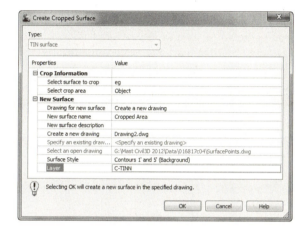

Figure 4.27
The completed Create Cropped Surface dialog

> **Rendering with Materials?**
>
> A Civil 3D surface must display triangles for rendering materials to be calculated and shown. You will inevitably forget this; it's just one of those frustrating anomalies in the program.

Manual Surface Edits

In your surface, you have a few "finger" surface areas where the surveyors went out along narrow paths from the main area of topographic data. The nature of TIN surfaces is to connect dots, and so these fingers often wind up as webbed areas of surface information that's not accurate or pertinent. A number of manual edits can be performed on a surface. These edit options are part of the definition of the surface and include the following:

Add Line Connects two points where a triangle did not exist before. This option essentially adds a breakline to the surface, so adding a breakline would generally be a better solution.

Delete Line Removes the connection between two points. This option is used frequently to clean up the edge of a surface or to remove internal data where a surface should have no triangulation at all. This can be an area such as a building pad or water surface.

Swap Edge Changes the direction of the triangulation methodology. For any four points, there are two solutions to the internal triangulation, and the Swap Edge option alternates from one solution to the other.

Add Point Allows for the manual addition of surface data. This function is often used to add a peak to a digitized set of contours that might have a flat spot at the top of a hill or mountain.

Delete Point Allows for the manual removal of a data point from the surface definition. Generally, it's better to fix the source of the bad data, but this option can be a fix if the original data is not editable (in the case of a LandXML file, for example).

Modify Point and Move Point Variations on the same idea. Modify Point moves a surface point in the z direction, whereas a Move Point is limited to horizontal movement. In both cases, only the TIN point is modified, not the original data input.

Minimize Flat Areas Performs the edits you saw earlier in this chapter to add supplemental information to the TIN and to create a more accurate surface, forcing triangulation to work in the z direction instead of creating flat planes.

Raise/Lower Surface A simple arithmetic operation that moves the surface in the z direction. This option is useful for testing rough grading schemes for balancing dirt or for adjusting entire surfaces after a new benchmark has been observed.

Smooth Surface Presents a pair of methods for supplementing the surface TIN data. Both work by extrapolating more information from the current TIN data, but they are distinctly different in their methodology:

> **Natural Neighbor Interpolation (NNI)** Adds points to a surface on the basis of the weighted average of nearby points. This data generally works well to refine contouring that is sharply angular because of limited information or long TIN connections. NNI works only within the bounds of a surface; it cannot extend beyond the original data.
>
> **Kriging** Adds points to a surface based on one of five distinct algorithms to predict the elevations at additional surface points. These algorithms create a trending for the surface beyond the known information and can therefore be used to extend a surface beyond even the available data. Kriging is very volatile, and you should understand the full methodology before applying this information to your surface. Kriging is frequently used in subsurface exploration industries such as mining, where surface (or strata) information is difficult to come by and the distance between points can be higher than desired.

Paste Surface Pulls in the TIN information from the selected surface and replaces the TIN information in the host surface with this new information. This option is helpful in creating composite surfaces that reflect both the original ground and the design intent. We'll look at pasting in Chapter 16, "Plan and Production," in the discussion on grading.

Simplify Surface Allows you to reduce the amount of TIN data being processed via one of two methods: Edge Contraction, wherein Civil 3D tries to collapse two points connected by a line to one point, or Point Removal, which removes selected surface points based on algorithms designed to reduce data points that are similar.

Manual editing should always be the last step in updating a surface. Fixing the surface is a poor substitution for fixing the underlying data the TIN is built from, but in some cases, it is the quickest and easiest way to make a more accurate surface.

Point and Triangle Editing

In this section, you'll remove triangles manually, and then finish your surface by correcting what appears to be a blown survey shot.

1. Open the SurfaceEdits.dwg file.
2. In Prospector, expand the Surfaces ➢ EG ➢ Definition branches.
3. Right-click Edits and select the Delete Line option.
4. Enter **C** as the command line to enter a crossing selection mode.
5. Start at the lower right of the pick area shown in Figure 4.28, and move to the upper-left corner as shown. Right-click or press ↵ to finish the selection.
6. Repeat this process, removing triangles until your site resembles Figure 4.29.
7. Zoom to the portion of your site with the red circle, and you'll notice a collection of contours that seems out of place.
8. Change Surface Style to Contours And Points.

FIGURE 4.28
Crossing the window selection to delete TIN lines

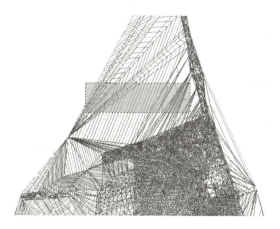

FIGURE 4.29
Surface after removal of extraneous triangles

9. Right-click Edits again and select the Delete Point option.
10. Zoom in on the area very close. You will find a series of red + markers in the area with close contours. These are blown shots and the contours are simply obeying the point elevation. In this case it is 0.
11. Delete the three markers, and notice the immediate change in the contouring.

Surface Smoothing

One common complaint about computer-generated contours is that they're simply too precise. The level of calculations in setting elevations on the basis of linear interpolation along a triangle leg makes it possible for contour lines to be overly exact, ignoring contour line trends in place of small anomalies of point information. Under the eye of a board drafter, these small anomalies were averaged out, and contours were created with smooth flowing lines.

While you can apply object-level smoothing as part of the contouring process, this process smoothes the end result but not the underlying data. In this section, you'll use the NNI smoothing algorithm to reduce surface anomalies and create a more visually pleasing contour set:

1. Open the SurfaceSmoothing.dwg file. The area to be smoothed is shown in Figure 4.30.
2. In Prospector, expand the Surfaces ➢ EG ➢ Definition branches.
3. Right-click Edits and select the Smooth Surface option. The Smooth Surface dialog opens.
4. Expand the Smoothing Methods branch, and verify that Natural Neighbor Interpolation is the Select Method value.
5. Expand the Point Interpolation/Extrapolation branch, and click in the Select Output Region value field. Click the ellipsis button.
6. Select the rectangle drawn on screen, and press ↵ to return to the Smooth Surface dialog.
7. Enter **20** for the Grid X-Spacing and Grid Y-Spacing values, and then press ↵. Note that Civil 3D will tell you how many points you are adding to the surface immediately below this input area by the value given in the Number Of Output Points field. It's grayed out, but it does change on the basis of your input values.
8. Click OK and the surface will be smoothed, as shown in Figure 4.31.

FIGURE 4.30
Area of surface to be smoothed

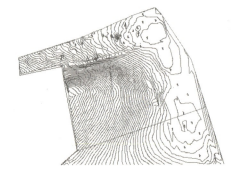

FIGURE 4.31
Using NNI to smooth the surface

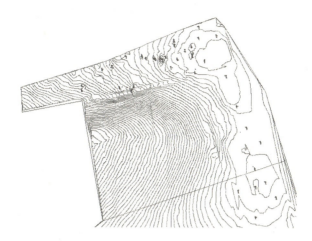

Note that we said the *surface* will be smoothed—not the contours. To see the difference, change Surface Style to Contours And Points to display your image as shown in Figure 4.32.

Note all of the points with a circle cross symbol. These points are all new, created by the NNI surface-smoothing operation. The points are part of your surface, and the contours reflect the updated surface information.

FIGURE 4.32
Points added via NNI surface smoothing

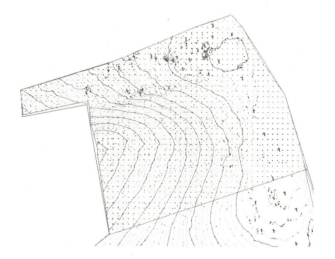

Surface Simplifying

Because of the increasing use in land development projects of GIS and other data-heavy inputs, it's critical that Civil 3D users know how to simplify the surfaces produced from these sources. In this exercise, you'll simplify the surface created from a drawing earlier in this chapter.

1. Open the SurfaceSimplifying.dwg file. For reference, the surface statistics for the EG-GIS surface are shown in Figure 4.33.

2. In Prospector, expand Surfaces ➢ EG-GIS ➢ Definition.

3. Right-click Edits and select Simplify Surface to launch the Simplify Surface wizard.
4. Select the Point Removal radio button, as shown in Figure 4.34, and click Next to move to the Region Options screen.

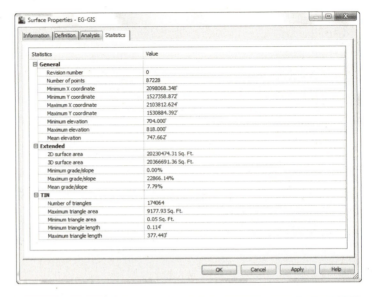

FIGURE 4.33
EG-GIS surface statistics before simplification

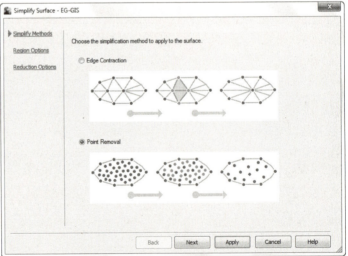

FIGURE 4.34
The Simplify Surface screen

5. Leave the Region Option set to Use Existing Surface Border. There are also options for selecting areas with a window or polygon, as well as selecting based on an existing entity. Click Next to move to the Reduction Options screen.
6. Set Percentage Of Points To Remove to 20 percent and then deselect the Maximum Change In Elevation option. This value is the maximum change allowed between the surface elevation at any point before or after the simplify process has run.

7. Click Apply. The program will process this calculation and display a Total Points Removed number, as shown in Figure 4.35. You can adjust the slider or toggle on the Maximum Change In Elevation button to experiment with different values.

8. Click Finish to dismiss the wizard and fully commit to the Simplify edit.

FIGURE 4.35
Reduction Options screen in the Simplify Surface wizard

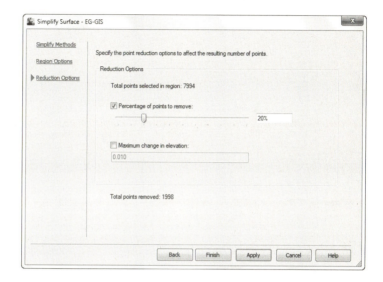

A quick visit to the Surface Properties Statistics tab shows that the number of points has been reduced. On something like an aerial topography or DEM, reducing the point count probably will not reduce the usability of the surface, but this simple 10 percent point reduction actually takes almost 20 percent off the file size. Remember, you can always remove the edit or deselect the operation on the Definition tab of the Surface Properties dialog.

The creation of a surface is merely the starting point. Once you have a TIN to work with, you have a number of ways to view the data using analysis tools and varying styles. Styles will be covered in Chapter 19.

LEVEL OF DETAIL

When you have a surface such as the SurfaceSimplifying.dwg that you have just worked with, actual manipulation of the surface can get cumbersome due to the sheer volume of data that is there. This is further evidenced when you want to rotate the view.

Luckily, Civil 3D 2012 now has a new feature called Level of Detail. This can be found on the View tab and Views panel.

It is a toggle that is either on or off. When it is invoked, you will see your surface with fewer contours when you are zoomed out. However when you zoom in, the contours are back to normal. The figure on the left below shows the surface with the Level of Detail invoked while zoomed out, and the figure to the right shows the same surface with Level of Detail invoked and zoomed in.

Surface Styling and Analysis

Once a surface is created, you can display information in a number of ways. The most common so far has been contours and triangles, but those are the basics. By using varying styles, you can show a large amount of data with one single surface. Not only can you do simple tasks such as adjust the contour interval, but Civil 3D can apply a number of analysis tools to any surface:

Contours   Allows the user to specify a more specific color scheme or linetype as opposed to the typical minor-major scheme. Commonly used in cut-fill maps to color negative colors one way, positive contours another, and the balance or zero contours yet another color.

Elevations   Creates bands of color to differentiate various elevations. You can use this tool to create a simple weighted distribution to help in development of marketing materials, hard-coded elevations to differentiate floodplain and other elevation-driven site concerns, or ranges to help a designer understand the earthwork involved in creating a finished surface.

Direction Analysis   Draws arrows showing the normal direction of the surface face. This tool is typically used for aspect analysis, helping site planners review the way a site slopes with regard to cardinal directions and the sun.

Slopes Analysis   Colors the face of each triangle on the basis of the assigned slope values. Although a distributed method is the normal setup, a common use is to check site slopes for compliance with Americans with Disabilities Act (ADA) requirements or other site slope limitations, including vertical faces (where slopes are abnormally high).

Slope Arrows   Displays the same information as a slope analysis, but instead of coloring the entire face of the TIN, this option places an arrow pointing in the downhill direction and colors that arrow on the basis of the specified slope ranges.

User-Defined Contours   Refers to contours that typically fall outside the normal intervals. These user-defined contours are useful to draw lines on a surface that are especially relevant but don't fall on one of the standard levels. A typical use is to show the normal pool elevation on a site containing a pond or lake.

Contouring Basics

Contouring is the standard surface representation on which land development plans are built. But in earlier programs such as Autodesk's Land Development, changing the contouring interval was akin to pulling teeth. With the use of styles in Civil 3D, you can have any number of styles prebuilt to allow you to quickly and painlessly change how contours are displayed. In this example, you'll copy an existing surface contouring style and modify the interval to a setting more suitable for commercial site design review:

1. Open the `SurfaceContours.dwg` file. This surface is currently displayed with a 5" minor contour and 25" major contour.

2. Select the surface by picking any contour or the boundary, and then click the Surface Properties button on the Modify panel. The Surface Properties dialog appears.

3. On the Information tab, click the down arrow next to the Style Editor button. Select the Copy Current Selection option to display the Surface Style dialog.

4. On the Information tab, change the Name field to **Contours 0.25" and 1"** and remove the description in place.

5. Switch to the Contours tab and expand the Contour Intervals property, as shown in Figure 4.36.

FIGURE 4.36
The expanded Contour Intervals setting

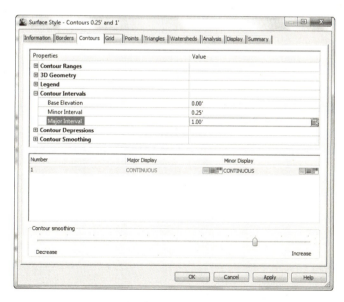

6. Change the Minor Interval value to **0.25"** and press ↵. The Major Interval value will jump to 1.25", maintaining the ratio that was previously in place.

7. Change the Major Interval value to **1.0"** and press ↵.

8. Expand the Contour Smoothing property (you may have to scroll down). Select a Smooth Contours value of True, which activates the Contour Smoothing slider bar near the bottom. Don't change this Smoothing value, but keep in mind that this gives you a level of control over how much Civil 3D modifies the contours it draws.

9. Click OK to close this dialog and then click OK again to close the Surface Properties dialog.

> **SURFACE VS. CONTOUR SMOOTHING**
>
> Remember, contour smoothing is *not* surface smoothing. Contour smoothing applies smoothing at the individual contour level but not at the surface level. If you want to make your surface contouring look fluid, you should be smoothing the surface.

The surface should be rendered faster than you can read this sentence even with the incredibly tight contour interval you've selected. This style doesn't make much sense on a site like this one, but it can be used effectively on something like a commercial site or highway entrance ramp where the low surface slope values make 1" contours close to meaningless in terms of seeing what is going on with the surface.

You skipped over one portion of the surface contours that many people consider a great benefit of using Civil 3D: depression contours. If this option is turned on via the Contours tab, ticks will be added to the downhill side of any closed contours leading to a low point. This is a stylistic option, and usage varies widely.

Now let's look at a few of the other options and areas you ignored in creating this style. There are some interesting changes from many other Civil 3D objects, as you can see in the component listing in Figure 4.37.

FIGURE 4.37
Listing of surface style components in the Plan direction

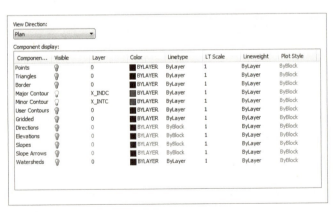

Under the Component Type column, Points, Triangles, Border, Major Contour, Minor Contour, User Contours, and Gridded are standard components and are controlled like any other object component. The plan (aka 2D), model (aka 3D), and section views are independent, and surfaces are one of the objects where different plan and model views are common. The Directions, Elevations, Slopes, and Slope Arrows components are unique to surface styles. Note that the Layer, Color, and Linetype fields are grayed out for these components. Each of these components has its own special coloring schemes, which we'll look at in the next section.

ELEVATION BANDING

Displaying surface information as bands of color is one of the most common display methods for engineers looking to make a high-impact view of the site. Elevations are a critical part of the site design process, and understanding how a site flows in terms of elevation is an important part of making the best design. Elevation analysis typically falls into two categories: showing bands of information on the basis of pure distribution of linear scales or showing a lesser number of bands to show some critical information about the site. In this first exercise, you'll use a pretty standard style to illustrate elevation distribution along with a prebuilt color scheme that works well for presentations:

1. Open the `SurfaceAnalysis.dwg` file.
2. Pick the surface on your screen.
3. Select Surface Properties from the Surface tab on the Modify panel. The Surface Properties dialog appears.
4. On the Information tab, change the Surface Style field to Elevation Banding (3D).
5. Switch to the Analysis tab.

6. Click the Run Analysis arrow in the middle of the dialog to populate the Range Details area.
7. Click OK to close the Surface Properties dialog.
8. From the View control, select SW Isometric.
9. Zoom in if necessary to get a better view.
10. From the Visual Style control, select the Conceptual option to see a semi-rendered view that should look something like Figure 4.38.

FIGURE 4.38
Conceptual view of the site with the Elevation Banding style

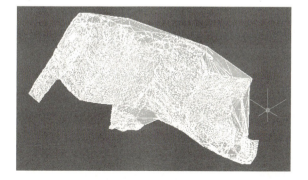

> **AUTOCAD VISUAL STYLES**
>
> The triangles seen are part of the view style and can be modified via the Visual Styles Manager. Turning the Edge mode off will leave you with a nicely gradated view of your site. You can edit the visual style by clicking the View control at the upper left of your drawing screen. You can also select View ➢ Visual Styles ➢ Visual Style Manager on the main menu.
>
>

You'll use a 2D elevation to clearly illustrate portions of the site that cannot be developed. In this exercise, you'll manually tweak the colors and elevation ranges on the basis of design constraints from outside the program:

1. Click the View control and select Top.

2. Click the Visual Style control and select 2D Wireframe.

   This site has a limitation placed in that no development can go below the elevation of 790. Your analysis will show you the areas that are below 790, a buffer zone to 791, and then everything above that.

3. Select the surface, right-click, and choose the Surface Properties option. The Surface Properties dialog appears.

4. On the Information tab, change the Surface Style field to Zoning.

5. Double-clicking in the Minimum and Maximum Elevations fields allows for direct editing. Double-clicking the Color Swatch field allows for manual picking. Modify your surface properties to match Figure 4.39. (The colors are red, yellow, and green from top to bottom, respectively.)

6. Click OK to exit the dialog.

Understanding surfaces from a vertical direction is helpful, but many times, the slopes are just as important. In the next section, you'll take a look at using the slope analysis tools in Civil 3D.

Figure 4.39
The Surface Properties dialog after manual editing

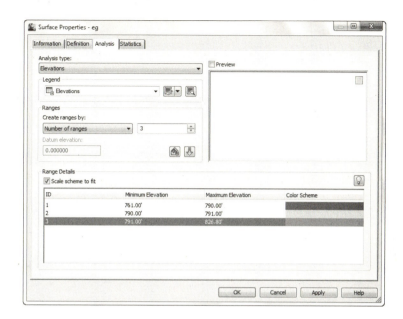

Slopes and Slope Arrows

Beyond the bands of color that show elevation differences in your models, you also have tools that display slope information about your surfaces. This analysis can be useful in checking for drainage concerns, meeting accessibility requirements, or adhering to zoning constraints. Slope is typically shown as areas of color as the elevations were or as colored arrows that indicate the downhill direction and slope. In this exercise, you'll look at a proposed site grading surface and run the two slope analysis tools:

1. Open the `SurfaceSlopes.dwg` file.
2. Select the surface in the drawing and click the Surface Properties button on the Modify panel. The Surface Properties dialog appears.
3. On the Information tab, change the Surface Style field to Slope Banding (2D).
4. Change to the Analysis tab.
5. Change the Analysis Type drop-down list to Slopes.
6. Change the Number field in the Ranges area to **5** and click the Run Analysis button. The Range Details area will populate.
7. Click OK to close the dialog. Your screen should look like Figure 4.40.

The colors are nice to look at, but they don't mean much, and slopes don't have any inherent information that can be portrayed by color association. To make more sense of this analysis, add a table:

1. Select the surface again and select the Add Legend button on the Labels & Tables panel.
2. Type **S** at the command line to select Slopes, and press ↵.
3. Press ↵ again to accept the default value of a Dynamic legend.
4. Pick a point on screen to draw the legend, as shown in Figure 4.41.

FIGURE 4.40
Slope color banding analysis

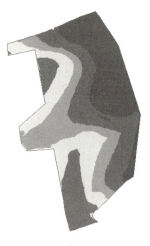

FIGURE 4.41
The Slopes legend table

Slopes Table				
Number	Minimum Slope	Maximum Slope	Area	Color
1	0.00%	3.42%	241637.72	■
2	3.42%	6.78%	398913.29	■
3	6.78%	9.34%	341321.14	■
4	9.34%	14.04%	268978.90	■
5	14.04%	Vertical	121982.26	■

By including a legend, you can make sense of the information presented in this view. Because you know what the slopes are, you can also see which way they go.

1. Select the surface and click the Surface Properties button on the Modify panel. The Surface Properties dialog appears.
2. On the Information tab, choose Slope Arrows from the Surface Style drop-down list.
3. Change to the Analysis tab.

4. Choose Slope Arrows from the Analysis Type drop-down list.

5. Change the Number field in the Ranges area to **5** and click the Run Analysis button.

6. Click OK to close the dialog.

The benefit of arrows is in looking for "birdbath" areas that will collect water. These arrows can also verify that inlets are in the right location, as shown in Figure 4.42. Look for arrows pointing to the proposed drainage locations and you'll have a simple design-verification tool.

Figure 4.42
Slope arrows pointing to a proposed inlet location

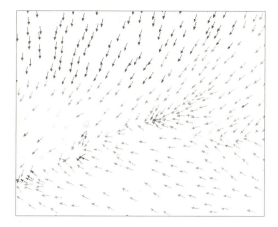

With these simple analysis tools, you can show a client the areas of their site that meet their constraints. Visually strong and simple to produce, this is the kind of information that a 3D model makes available. Beyond the basic information that can be represented in a single surface, Civil 3D also contains a number of tools for comparing surfaces. You'll compare this existing ground surface to a proposed grading plan in the Comparing Surfaces section.

Visibility Checker

New to Civil 3D 2012 are some tools that allow you to see issues that might occur even before dirt is moved. The Zone of Visual Influence tool allows you to explore what-if scenarios. In this example, a 40' tower has been proposed for the site. A concerned neighbor wants to make sure that it won't obstruct their scenic view. We want to check it from the proposed surface:

1. Open VisibilityCheck.dwg.

2. Click on the proposed surface. This is the one with the blue solid contours.

3. In the Tin Surface tab and Analyze panel, select Visibility Check➢Zone of Visual Influence.

4. Using the Intersection osnap, select the center of the proposed tower located on the lower portion of the site.

5. Type **40↵ to set the tower height to 40'**.

6. Select an intersection of the cyan color house located at the upper right of the site.

The drawing now has bands of color. The green color indicates that the object is completely visible, the yellow color indicates that the object is partially visible, and the red color indicates that the object is not visible.

So in our example, the homeowner on the upper right will be happy to know that the proposed 40′ tower will not appear in their view.

The next example will use the Point to Point tool.

1. Using the same drawing as before, zoom to the intersection of Syrah Way and Cabernet Court where you will see a car. We want to check the sight distance.

2. If it is not already selected, click on the Composite surface.

3. From the Tin Surface tab and Analyze panel, select Visibility Check ➢ Point to Point.

4. At the Specify height of eye: prompt, type **3.5**↵. This sets the height of a driver's eye while sitting in a typical car.

5. At the Specify location of eye: prompt, click where the driver would normally be seated in the vehicle.

6. The next prompt (`Specify height of target:`), is asking what the view height is that a driver can see while seated in the vehicle. Type **6**↵.

7. Click along the path where oncoming cars would be seen. An arrow is drawn on the screen. If the arrow is green, it means that the view is unobstructed. If the arrow is red, it indicates that the view is obstructed and the command prompt will tell you where the obstruction occurs.

Unfortunately, these Visual tools are not dynamic; if you change the profile, you will need to re-run the visual tools. Perhaps next release.

Comparing Surfaces

Earthwork is a major part of almost every land development project. The money involved with earthmoving is a large part of the budget, and for this reason, minimizing this impact is a critical part of the final design. Civil 3D contains a number of surface analysis tools designed to help in this effort, and you'll look at them in this section. First, a simple comparison provides feedback about the volumetric difference, and then a more detailed approach enables you to perform an analysis on this difference.

For years, civil engineers have performed earthwork using a section methodology. Sections were taken at some interval, and a plot was made of both the original surface and the proposed surface. Comparing adjacent sections and multiplying by the distance between them yields an end-area method of volumes that is generally considered acceptable. The main problem with this methodology is that it ignores the surfaces in the areas between sections. These areas could include areas of major change, introducing some level of error. In spite of this limitation, this method worked well with hand calculations, trading some accuracy for ease and speed.

With the advent of full-surface modeling, more precise methods became available. By analyzing both the existing and proposed surfaces, a volume calculation can be performed that is as good as the two surfaces. At every TIN vertex in both surfaces, a distance is measured vertically to the other surface. These delta amounts can then be used to create a third volume surface representing the difference between the surfaces. Civil 3D uses this methodology to perform its calculations, but the end-area method can still be used if desired.

Simple Volumes

When performing rough analysis, the total volume is the most important part. Once an acceptable volume has been created, more refined analysis and comparison can be performed. In this exercise, you'll compare two surfaces to pull a basic volume number and then modify the proposed grade to illustrate how quickly changes can be reviewed:

1. Open the SurfaceVolumes.dwg file.

2. Change to the Analyze tab, and then click the Volumes button on the Volumes And Materials panel to display Panorama open to the Composite Volumes tab, as shown in Figure 4.43.

FIGURE 4.43
The Composite Volumes tab in Panorama

3. Click the Create New Volume Entry button on the far left, as indicated in Figure 4.43, to create a new volume entry.

4. Click the <select surface> field under the Base Surface heading and select EG.

5. Click the <select surface> field under the Comparison Surface heading and select FG. Civil 3D will calculate the volume (Figure 4.44). Note that you can apply a cut or fill factor by typing directly into the cells for these values.

6. Without closing Panorama, move to Prospector, and expand the Surfaces ➢ FG ➢ Definition branches.

FIGURE 4.44
Composite volume calculated

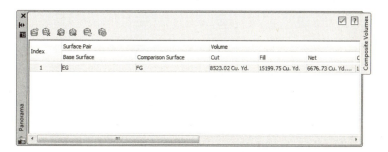

> **DON'T TOUCH THAT CLOSE BUTTON!**
>
> This utility's calculations will disappear if you close Panorama. You can export this information to XML or copy and paste it into another document if you require a record.

7. Right-click Edits and select the Raise/Lower Surface option.
8. Enter **–0.25** at the command line to drop the site 3″.
9. In Panorama, click the Recompute Volumes button (as shown in Figure 4.45) to update the calculations.

FIGURE 4.45
Recomputing the composite volume

10. Right-click the Edits list to remove the lowering edit and select Delete.
11. Return to Panorama and recompute to return to the original volume calculation.

The original design was quite good in terms of cut and fill, so in the next section you will look at a more detailed analysis of the earthwork by using a TIN volume surface.

Volume Surfaces

Using the volume utility for initial design checking is helpful, but quite often, contractors and other outside users want to see more information about the grading and earthwork for their own uses. This requirement typically falls into two categories: a cut-fill analysis showing colors or contours or a grid of cut-fill tick marks.

Color cut-fill maps are helpful when reviewing your site for the locations of movement. Some sites have areas of better material or can have areas where the cost of cut is prohibitive (such as rock). In this exercise, you'll use three of the surface analysis methods to look at the areas for cut-fill on your site:

1. Open the SurfaceVolumes.dwg file if it is not already open.
2. In Prospector, right-click the Surfaces branch and select Create Surface. The Create Surface dialog appears.
3. Change the Type field to TIN Volume Surface.
4. Expand the Information property, and change the Name to **Volume**.
5. In the Style value field, click the ellipsis button to open the Select Surface Style dialog. Select Elevation Banding (2D) and click OK.
6. Expand the Volume Surfaces property, and click in the Base Surface value field. Click the ellipsis button to open the Select Base Surface dialog. Select EG and click OK.
7. Click in the Comparison Surface value field. Click the ellipsis button to open the Select Comparison Surface dialog. Select FG and click OK. Your screen should look like Figure 4.46. Note that you can apply cut and fill factors to your calculations by filling them in here.

Figure 4.46
Creating a volume surface

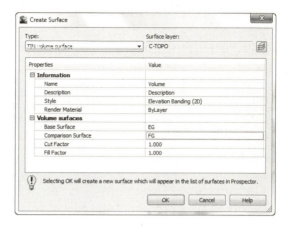

8. Click OK to complete the surface creation.

   This new volume surface appears in Prospector's Surfaces collection, but notice that the icon is slightly different, showing two surfaces stacked on each other. The color mapping currently shown is just a default set, though, and does not indicate much.

9. Right-click Volume in the Surfaces branch of the Prospector and select the Surface Properties option. The Surface Properties dialog appears.

10. Switch to the Statistics tab and expand the Volume branch.

    The value shown for the Net Volume (Unadjusted) is what was calculated in the Surface Volume utility in the previous exercise. This information can be cut and pasted into other programs for saving or other analysis if needed.

11. Switch to the Analysis tab.

12. Change the Number field in the Ranges area to **3**, and click the Run Analysis arrow.

13. Change the values in the cells by double-clicking and editing to match Figure 4.47. The colors that would be recommended are red, cyan, and green, where red indicates the worst case cut, green represents the worst case fill, and cyan represents a balance.

14. Click OK to close the dialog.

Figure 4.47
Elevation analysis settings for earthworks

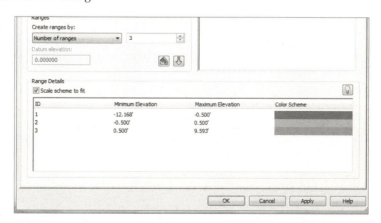

The volume surface now indicates areas of cut, fill, and areas near balancing, similar to Figure 4.48. If you leave a small range near the balance line, you can more clearly see the areas that are being left nearly undisturbed.

FIGURE 4.48
Completed elevation analysis

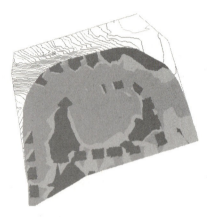

To show where large amounts of cut or fill could incur additional cost (such as compaction or excavation protection), you would simply modify the analysis range as required.

The Elevation Banding surface is great for onscreen analysis, but the color fills make it hard to plot or use in many applications. In this next exercise, you use the Contour Analysis tool to prepare cut-fill contours in these same colors:

1. Right-click Volume in the Surfaces collection of Prospector and select the Surface Properties option to open the Surface Properties dialog again.

2. On the Analysis tab, set the Analysis Type field to Contours.

3. Change the Number field in the Ranges area to **3**.

4. Click the Run Analysis button.

5. Change the ranges as shown in Figure 5.44. The contour colors are shades of red for cut, a yellow for the balance line, and shades of green for the fill areas. Click the small button to display the AutoCAD Select Color dialog.

6. Switch to the Information tab on the Surface Properties dialog, and change the Surface Style to Contours 1" and 5" (Design).

7. Click the down arrow next to the Style field and select the Copy Current Selection option. The Surface Style Editor appears.

8. On the Information tab, change the Name field to **Contours 1" and 5" (Earthworks)**.

9. Switch to the Contours tab.

10. Expand the Contour Ranges branch.

11. Change the value of the Use Color Scheme property to True. It's safe to ignore the values here because you hard-coded the values in your surface properties.

12. Click OK to close the Surface Style Editor and click OK again to close the Surface Properties dialog.

The volume surface can now be labeled using the surface-labeling functions, which you'll look at in the "Labeling the Surface" section.

To expand on the Elevation Banding surface, we can now automate the process. Say we needed to show the contractor a drawing showing elevation changes in our volume surface, but at 2′ intervals.

The old way would have been to assign elevations, and manually change the minimum and maximum elevations and the associated colors.

New to Civil 3D 2012, we now have the capability to set either a range interval or range interval with datum. Let's take a look at how this works:

1. While still in `SurfaceVolumes.dwg`, click on the volume surface and select Surface Properties from the Modify panel.
2. Click on the Analysis tab and under the Create Ranges By drop-down, you now have the Range Interval and Range Interval with Datum. Select the Range Interval with Datum option.
3. Set the interval to 2, for 2′ contour range intervals.
4. Click on the Run Analysis button and click OK.

The surface is now colorized with a range of colors representing the 2′ contour intervals for the elevations.

> **ANOTHER VOLUME OPTION—BOUNDED VOLUMES**
>
> The Bounded Volumes tool is very useful for checking volumes for a smaller area. In this example, the developer wants to know the volume of a single lot in order to start developing.
>
> 1. Open the BoundedVolumes.dwg file. This is the overall drawing, but has a thick polyline around the boundaries of Lot 4.
> 2. From the Modify tab and Ground Data panel, select Surface.
> 3. From the Surface tab and Extended Analyze tab, select Bounded Volumes.
> 4. Press ↵ and select the Volume surface.
> 5. Select the thick polyline around the perimeter of Lot 4.
> 6. Press ↵ to end the command.
> 7. Press F2 to see the volume report for Lot 4.
>
> You can see that the volume for Lot 4 is a Net cut of 535 cubic yards.
>
> You can also use this tool if you wish to set a datum elevation for a non-volume surface, such as establishing a grading pad for a building.

Labeling the Surface

Once you've created the surface model, it is time to communicate the model's information in various formats. This includes labeling contours, creating legends for the analysis you've created, or adding spot labels. These exercises work through these main labeling requirements and building styles for each.

Contour Labeling

The most common requirement is to place labels on surface-generated contours. In Land Desktop, this was one of the last steps because a change to a surface required erasing and replacing all the labels. Once labels have been placed, their styles can be modified.

PLACING CONTOUR LABELS

Contour labels in Civil 3D are created by special lines that understand their relationship with the surface. Everywhere one of these lines crosses a contour line, a label is applied. This label's appearance is based on the style applied and can be a major, minor, or user-defined contour label. Each label can have styles selected independently, so using some AutoCAD selection techniques can be crucial to maintaining uniformity across a surface. In this exercise, you'll add labels to your surface and explore the interaction of contour label lines and the labels themselves.

1. Open the `SurfaceLabeling.dwg` file.
2. Select the surface in the drawing to display the Tin Surface tab. On the Labels & Tables panel, select Add Labels ➢ Contour – Single.
3. Pick any spot on a green major contour to add a label.
4. On the Labels & Tables panel, select Add Labels ➢ Contour – Multiple.
5. Pick a point on the north and then a second point to the south, crossing a number of contours in the process. Press ↵ to end the picking.
6. On the Labels & Tables panel, select Add Labels ➢ Contour – Multiple – At Interval.
7. Pick a point near the middle right of the site and a second point across the site to the east.
8. Enter **200** at the command line for an interval value.

You've now labeled your site in three ways to get contour labels in a number of different locations. You will need additional labels in the northeast and southwest to complete the labeling, because you did not cross these contour objects with your contour label line. You can add more labels by clicking Add, but you can also use the labels created already to fill in these missing areas. By modifying the contour line labels, you can manipulate the label locations and add new labels. In this exercise, you'll fill in the labeling to the northeast:

1. Zoom to the northeast portion of the site, and notice that some of the contours are labeled only along the boundary or not at all, as shown in Figure 4.49.
2. Zoom in to any contour label placed using the Contour – Single button, and pick the text. Three grips will appear. The original contour label lines are quite apparent, but in reality, every label has a hidden label line beneath it.
3. Grab the northernmost grip and drag across an adjacent contour, as shown in Figure 4.50. New labels will appear everywhere your dragged line now crosses a contour.
4. Drop the grip somewhere to create labels as desired.

By using the created label lines instead of adding new ones, you'll find it easier to manage the layout of your labels.

FIGURE 4.49
Contour labels applied

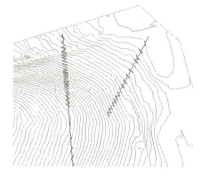

FIGURE 4.50
Grip-editing a contour label line

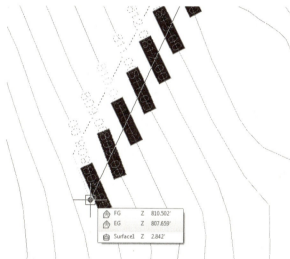

Surface Point Labels

In every site, there are points that fall off the contour line but are critical. In an existing surface, this can be the low point in a pond or a driveway that has to be matched. When you're working with commercial sites, the spot grade is the most common review element. One of the most time-consuming issues in land development is the preparation of grading plans with hundreds of individual spot grades. Every time a site grading scheme changes, these are typically updated manually, leaving lots of opportunities for error.

With Civil 3D's surface modeling, spot labels are dynamic and react to changes in the underlying surface. By using surface labels instead of points or text callouts, you can generate a grading plan early on in the design process and begin the process of creating sheets. In this section, you'll label surface slopes in a couple of ways, create a single spot label for critical information, and conclude by creating a grid of labels similar to many estimation software packages.

LABELING SLOPES

Beyond the specific grade at any single point, most grading plans use slope labels to indicate some level of trend across a site or drainage area. Civil 3D can generate the following two slope labels:

- One-point slope labels indicate the slope of an underlying surface triangle. These work well when the surface has large triangles, typically in pad or mass grading areas.

- Two-point slope labels indicate the slope trend on the basis of two points selected and their locations on the surface. A two-point slope label works by dividing the surface elevation distance between the points by the planar distance between the pick points. This works well in existing ground surface models to indicate a general slope direction but can be deceiving in that it does not consider the terrain between the points.

In this exercise, you'll apply both types of slope labels, and then look at a minor style modification that is commonly requested:

1. Open the SurfaceSlopeLabeling.dwg file.
2. Select the surface to display the Tin Surface tab. On the Labels & Tables panel, select Add Labels ➢ Slope.
3. At the command line, press ↵ to select a one-point label style.
4. Zoom in on the circle drawn on the western portion of the site and use a Center snap to place a label at its center, as shown in Figure 4.51.
5. Press Esc or ↵ to exit the command.
6. Select the surface to display the Tin Surface tab. On the Labels & Tables panel, select Add Labels ➢ Slope.
7. At the command line, press T to switch to a two-point label style.
8. Pan to the southwest portion of the site, and use an Endpoint snap to pick the northern end of the line shown in Figure 4.52.
9. Use an Endpoint snap to select the other end of the line to complete the label, and press Esc or ↵ to exit the command.

This second label indicates the average slope of the property. By using a two-point label, you get a better understanding of the trend, as opposed to a specific point.

FIGURE 4.51
A one-point slope label

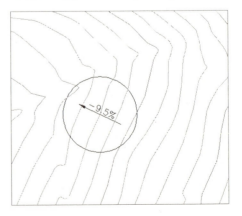

Figure 4.52
First point in a two-point slope label

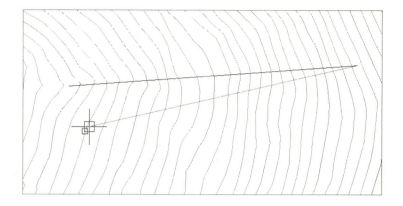

Critical Points

A typical grading plan is a sea of critical points that drive the site topography. In the past, much of this labeling and point work was done by creating coordinate geometry (COGO) points and simply displaying their properties. Although this is effective, it has two distinct disadvantages. First, these points are not reflective of the design but part of the design. This makes the sheet creation a part of the grading process, not a parallel process. Second, the addition of COGO points to any drawing and project when they're not truly needed just weighs down the design model. Point management is a mentally intensive task, and anything that can limit extraneous data is worth investigating.

Surface labels react dynamically to the surface and to the point of insertion. Moving any of these labels would update the information to reflect the surface underneath. This relationship makes it possible for one user to place labels on a grading plan while the final surface is still in flux. A change in the proposed surface is reflected in an update from the project, and an updated sheet can be on the plotter in minutes.

Surface Grid Labels

Sometimes, more than a few points are requested. Estimation software typically creates a grid of point labels that can be easily reviewed or passed to a contractor for fieldwork. In this exercise, you'll use the volume surface you generated earlier in this chapter to create a set of surface labels that reflect this requirement:

1. Open the SurfaceVolumeGridLabels.dwg file.
2. Change to the Annotate tab, and select Labels ➢ Surface ➢ Spot Elevations On Grid.
3. Press Enter and select Volume and click OK.
4. Pick a point in the southwest of the surface to set a base point for the grid.
5. Press ↵ to set the grid rotation to 0.
6. Enter 25 at the command line to set the x spacing.
7. Enter 25 at the command line to set the y spacing.
8. Click to the northeast of the surface to set the area for the labels.

9. Verify the preview box contains the Volume surface and press ↵ at the command line to continue.

10. Wait a few moments as Civil 3D generates all the labels just specified. Your drawing should look similar to Figure 4.53.

Labeling the grid is imprecise at best. Grid labeling ignores anything that might happen between the grid points, but it presents the surface data in a familiar way for engineers and contractors. By using the tools available and the underlying surface model, you can present information from one source in an almost infinite number of ways.

FIGURE 4.53
Volume surface with grid labels

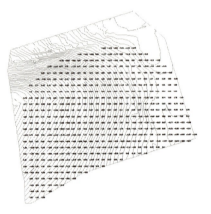

Point Cloud Surfaces

A point cloud is a huge bunch of 3D points, usually collected by laser scanner or LiDAR (Light Detection and Ranging). In a geographic information system, or GIS, point clouds are often used as a source for a digital elevation model (DEM). The technology has gotten less expensive and more accurate over the last few years, allowing LiDAR to quickly take over from traditional methods of collecting photogrammetry data.

Let Point Clouds Reign!

Point clouds in many formats can be imported to Civil 3D. The most common format is the LAS file. This binary format is a public format and at minimum contains X, Y, and Z data. LAS data can also include the following:

- Coordinate system of scanned area.
- Color; true color on RGB format.
- Intensity; LiDAR depends on lasers bouncing off objects and back to the scanning device. The intensity refers to the strength of the returned information. This value directly relates to the material from which the laser is bouncing. For example, concrete will have a stronger intensity than grass.
- Classification; this will be a number, most frequently between 0 and 9, that categorizes points based on material (i.e., ground, vegetation, or buildings).

For more information on the LAS standard, visit www.asprs.org.

Civil 3D can import a point cloud and use it in several ways. For instance, a laser scan of a bridge can be imported and placed for reference when designing a road through an existing abutment. In the example that follows, we will convert LiDAR data into a Civil 3D surface. It is important to note that point clouds often contain millions of points and require a beefy computer (and a little patience on your part) to process.

Importing a Point Cloud

A typical point cloud contains millions of points. These large files are kept external to Civil 3D in a point cloud database. After the LAS has been imported, the data is passed to three files: PRMD, IATI, and ISD. The ISD contains the points themselves and is the only file needed by CAD if the point cloud would need to be recreated or used in base AutoCAD. By default these files get created in the same directory as the DWG but can be changed when importing the information (Figure 4.54).

If the point cloud you are working with contains coordinate system information (as all the examples in this book do), the software will automatically convert the point cloud to the units and coordinate system of the drawing. For the exercises in this chapter, it does not matter whether you choose the Metric template or the Imperial template.

Civil 3D may take a long time to process these files, and you must ensure you have sufficient disk space to store them, as you will find in the following exercise:

1. Start a new file by using the default Civil 3D template of your choice. Save the file before proceeding as **Point Cloud.dwg**.

2. Select the Point Clouds collection on the Prospector tab of Toolspace and right-click.

3. Select Create Point Cloud from the menu to launch the Create Point Cloud wizard shown in Figure 4.54.

Figure 4.54
Create Point Cloud wizard

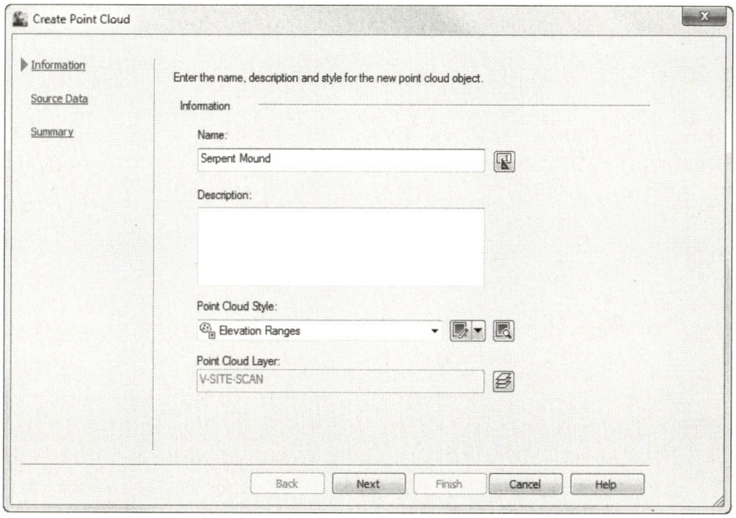

4. Set the name of the Point Cloud to **Serpent Mound**. Set Point Cloud Style to Elevation Ranges, as shown in Figure 4.54, and click the Next button. The Source Data page is displayed, as shown in Figure 4.55.

FIGURE 4.55
The Source Data page

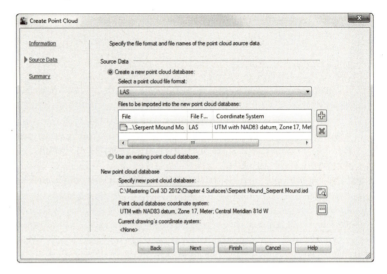

5. Using the white plus sign, select `Serpent Mound.las`. This is a large (90 megabytes) file containing roughly 1.4 million points and may require a minute or two to process. Click the Next button to display the summary page, as shown in Figure 4.56.

FIGURE 4.56
The summary page

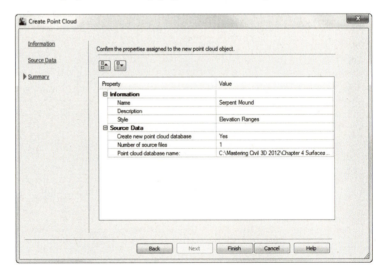

6. Accept the defaults as shown in Figure 4.56 and click Finish to process the point cloud. If the New Point Cloud Database – Processing In Background dialog appears, click Close to dismiss it. The point cloud database is processed in the background. When complete, a portion of a bounding box outlining a portion of the point cloud is displayed in the center of the screen. You must leave this drawing open to complete the next exercise.

Working with Point Clouds

Once the point cloud is visible in your drawing, you'll want to follow a few rules of thumb to prevent performance problems. The key-in POINTCLOUDDENSITY controls what percentage of the full point cloud displays on the screen at once. You can also access this value using a slider bar in the context Ribbon. However, it is easier to hit the percentage you want on the first try if you use the key-in. The lower this value, the fewer points are visible; hence the easier it will be to navigate your drawing. The POINTCLOUDDENSITY value does not have any effect on the number of points used when generating a surface model.

When you are changing view directions on a point cloud, we recommend that you use preset views and named views to flip around the object. The orbit commands should not be used, as they are a surefire way to max out your computer's RAM.

If you used the default template, your surface will be located on the V-SITE-SCAN layer. We suggest that you freeze the layer if you do not need to see the point cloud. Use Freeze instead of Off for layer management so the point cloud is not accounted for during pan, zoom, and regen operations (this is true for all AutoCAD objects, but it makes a huge difference when working with point clouds).

Creating a Point Cloud Surface

By specifying either an entire point cloud or a small region of a point cloud, you can create a new TIN surface in your drawing. Any changes to the point cloud object will render the surface definition out of date. In the following exercise, a new TIN surface is created from the point cloud previously imported:

1. Continue using the `Point Cloud.dwg` file.

2. Select the bounding box representing the point cloud to display the Point Cloud context tab as shown in Figure 4.57.

Figure 4.57
The Point Cloud context menu

3. Select the Add Points To Surface command, as shown in Figure 4.57, to display the Add Points To Surface wizard, as shown in Figure 4.58. Name the surface **Serpent Mound South**. Leave the style set to the default.

4. Click Next and the Region Options page is displayed, as shown in Figure 4.59. Choose the Window radio button, and click Define Region In Drawing.

FIGURE 4.58
The Add Points To Surface wizard

FIGURE 4.59
The Region Options page

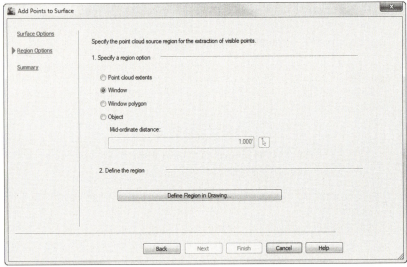

5. Define the region by creating a window around the southern half of the point cloud. Click Next to see the summary page and click Finish. Your results will look similar to those in Figure 4.60.

FIGURE 4.60
Click Finish on the summary page.

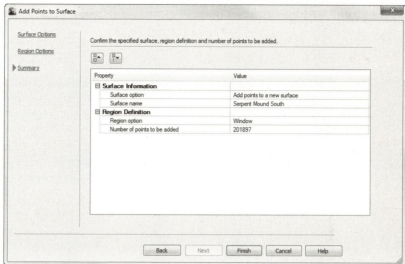

The Bottom Line

Create a preliminary surface using freely available data. Almost every land development project involves a surface at some point. During the planning stages, freely available data can give you a good feel for the lay of the land, allowing design exploration before money is spent on fieldwork or aerial topography. Imprecise at best, this free data should never be used as a replacement for final design topography, but it's a great starting point.

   Master It Create a new drawing from the Civil 3D Extended template and bring in a Google Earth surface for your home or office location. Be sure to set a proper coordinate system to get this surface in the right place.

Modify and update a TIN surface. TIN surface creation is mathematically precise, but sometimes the assumptions behind the equations leave something to be desired. By using the editing tools built into Civil 3D, you can create a more realistic surface model.

   Master It Modify your Google Earth surface to show only an area immediately around your home or office. Create an irregular-shaped boundary and apply it to the Google Earth surface.

Prepare a slope analysis. Surface analysis tools allow users to view more than contours and triangles in Civil 3D. Engineers working with nontechnical team members can create strong meaningful analysis displays to convey important site information using the built-in analysis methods in Civil 3D.

   Master It Create an Elevation Banding analysis of your home or office surface and insert a legend to help clarify the image.

Label surface contours and spot elevations. Showing a stack of contours is useless without context. Using the automated labeling tools in Civil 3D, you can create dynamic labels that update and reflect changes to your surface as your design evolves.

Master It Label the contours on your Google Earth surface at 1″ and 5″ (Design).

Import a point cloud into a drawing and create a surface model. As point cloud data becomes more common and replaces other large-scale data-collection methods, the ability to use this data in Civil 3D is key. Intensity helps postprocessing software determine the ground cover type. While Civil 3D can't do postprocessing, you can see the intensity as part of the point cloud style.

Master It Import an LAS format point cloud `Denver.las` into the Civil 3D template of your choice. As you create the point cloud file, set the style to Scaled Color Intensity - Blue. Use a portion of the file to create a Civil 3D surface model.

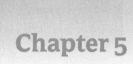

Chapter 5

Parcels

Land development projects often involve the subdivision of large pieces of land into smaller lots. Even if your projects don't directly involve subdivisions, you're often required to show the legal boundaries of your site and the adjoining sites.

In previous CAD systems, a few tools were available for parcel management. You could create AutoCAD entities, such as lines and arcs, to represent the lot boundaries and then create a closed polyline to assist in determining the parcel area. You could also create static text labels for area, bearing, and distance. Even if you took advantage of some of the parcel management tools in Land Desktop, the most minor change to the project, such as a road widening or a horizontal alignment adjustment, required days of editing, adjustment, and relabeling.

Civil 3D parcels give you a dynamic way to create, edit, manage, and annotate these legal land divisions. If you edit a parcel segment to make a lot larger, all of the affected labels will update—including areas, bearings, distances, curve information, and table information.

In this chapter, you will learn to:

- Create a boundary parcel from objects
- Create a right-of-way parcel using the right-of-way tool
- Create subdivision lots automatically by layout
- Add multiple parcel-segment labels

Creating and Managing Sites

In Civil 3D, a *site* is a collection of parcels, alignments, grading objects, and feature lines that share a common topology. In other words, Civil 3D objects that are in the same site are related to, as well as interact with, one another. The objects that react to one another are called *site geometry* objects.

Best Practices for Site Topology Interaction

At first glance, it may seem the only uses for parcels are subdivision lots, and therefore you may think you need only one site for your drawing.

However, once you begin working with parcels, you'll find features like dynamic area labels to be useful for delineating and analyzing soil boundaries; paving, open space, and wetlands areas; and any other region enclosed with a boundary. The automatic layer enforcement of parcel object styles adds to the appeal of using parcels. Using additional types of parcels will require you to come up with a site management strategy to keep everything straight.

It's important to understand how site geometry objects react to one another. Figure 5.1 shows a typical parcel that might represent a property boundary.

FIGURE 5.1
A typical property boundary

When an alignment is drawn and placed on the same site as the property boundary, the parcel splits into two parcels, as shown in Figure 5.2.

FIGURE 5.2
An alignment that crosses a parcel divides the parcel in two if the alignment and parcel exist on the same site.

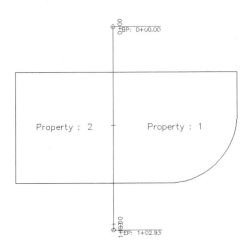

You must plan ahead to create meaningful sites based on interactions between the desired objects. For example, if you want a road centerline, a road right-of-way (ROW) parcel, and the lots in a subdivision to react to one another, they need to be in the same site (see Figure 5.3).

The alignment (or road centerline), ROW parcel, and lots all relate to one another. A change in the centerline of the road should prompt a change in the ROW parcel and the subdivision lots.

If you'd like to avoid the interaction between site geometry objects, place them in different sites. Figure 5.4 shows an alignment that has been placed in a different site from the boundary parcel. Notice that the alignment doesn't split the boundary parcel.

CREATING AND MANAGING SITES | 163

FIGURE 5.3
Alignments, ROW parcels, open space parcels, and subdivision lots react to one another when drawn on the same site.

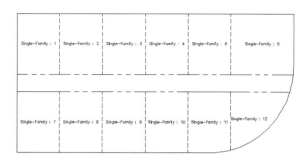

FIGURE 5.4
An alignment that crosses a parcel won't interact with the parcel if they exist on different sites.

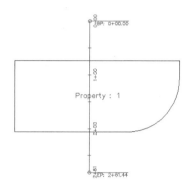

It's important that only objects that are intended to react to each other be placed in the same site. For example, in Figure 5.5 you can see parcels representing both subdivision lots and soil boundaries. Because it wouldn't be meaningful for a soil boundary parcel segment to interrupt the area or react to a subdivision lot parcel, the subdivision lot parcels have been placed in a Subdivision Lots site, and the soil boundaries have been placed in a Soil Boundaries site.

FIGURE 5.5
Parcels can be used for subdivision lots and soil boundaries as long as they're kept in separate sites.

If you didn't realize the importance of site topology, you might create both your subdivision lot parcels and your soil boundary parcels in the same site and find that your drawing looks similar to Figure 5.6. This figure shows the soil boundary segments dividing and interacting with subdivision lot parcel segments, which doesn't make any sense.

FIGURE 5.6
Subdivision lots and soil boundaries react inappropriately when placed in the same site.

Another way to avoid site geometry problems is to do site-specific tasks in different drawings and use a combination of external references and data references to share information.

For example, you could have an existing base drawing that housed the soil boundaries site, XRefed into a subdivision plat drawing that housed the subdivision lots site instead of separating the two drawings onto two different sites.

You should consider keeping your legal site plan in its own drawing. Because of the interactive and dynamic nature of Civil 3D parcels, it might be easy to accidentally grab a parcel segment when you meant to grab a manhole, and unintentionally edit a portion of your plat.

You'll see other workflow examples and drawing divisions later in this chapter, as well as in Chapter 17, "Interoperability."

If you decide to have sites in the same drawing, here are some sites you may want to create. These suggestions are meant to be used as a starting point. Use them to help find a combination of sites that works for your projects:

Roads and Lots This site could contain road centerlines, ROW, platted subdivision lots, open space, adjoining parcels, utility lots, and other aspects of the final legal site plan.

Grading Feature lines and grading objects are considered part of site geometry. If you're using these tools, you must make at least one site for them. You may even find it useful to have several grading sites.

Easements If you'd like to use parcels to manage, analyze, and annotate your easements, you may consider creating a separate site for easements.

Stormwater Management If you'd like to use parcels to manage, analyze, and annotate your stormwater subcatchment boundaries, you may consider creating a separate site for stormwater management.

As you learn new ways to take advantage of alignments, parcels, and grading objects, you may find additional sites that you'd like to create at the beginning of a new project.

> **What about the "Siteless" Alignment?**
>
> The previous section mentioned that alignments are considered site geometry objects. Civil 3D 2008 introduced the concept of the "siteless" alignment: an alignment that is placed in the <none> site. An alignment that is created in the <none> site doesn't react with other site geometry objects or with other alignments created in the <none> site.
>
> However, you can still create alignments in traditional sites, if you desire, and they will react to other site geometry objects. This may be desirable if you want your road centerline alignment to bisect a ROW parcel, for example.
>
> You'll likely find that best practices for most alignments are to place them in the <none> site. For example, if road centerlines, road transition alignments, swale centerlines, and pipe network alignments are placed in the <none> site, you'll save yourself quite a bit of site geometry management.
>
> It is important to note that although <none> sites cannot be seen or selected in a drawing, they still exist in the drawing database. For example, if you've used the <none> site option 12 times, you'll have 12 uniquely numbered <none> site definitions in the drawing database.
>
> See Chapter 6, "Alignments," for more information about alignments and sites.

Creating a New Site

You can create a new site in Prospector. You'll find the process easier if you brainstorm potentially needed sites at the beginning of your project and create those sites right away—or, better yet, save them as part of your standard Civil 3D template. You can always add or delete sites later in the project.

The Sites collection is stored in Prospector, along with the other Civil 3D objects in your drawing.

The following exercise will lead you through creating a new site that you can use for creating subdivision lots:

1. Open the `CreateSite.dwg` file, which you can download from www.sybex.com/go/masteringcivil3d2012. Note that the drawing contains alignments and a boundary parcel, as shown in Figure 5.7.
2. Locate the Sites collection on the Prospector tab of Toolspace.
3. Right-click the Sites collection, and select New to open the Site Properties dialog.
4. On the Information tab of the Site Properties dialog, enter **Subdivision Lots** for the name of your site.
5. Confirm that the settings on the 3D Geometry tab match what is shown on Figure 5.8.

FIGURE 5.7
The Create Site drawing contains alignments and soil boundary parcels.

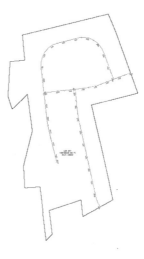

FIGURE 5.8
Confirm the settings on the 3D Geometry tab.

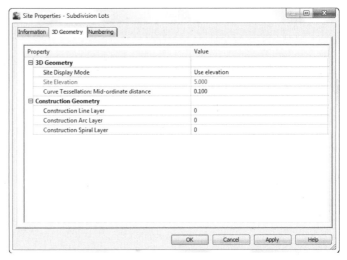

6. Confirm that the settings on the Numbering tab match Figure 5.9. Everything should be set to 1. Click OK.

7. Locate the Sites collection on the Prospector tab of Toolspace, and note that your Subdivision Lots site appears on the list.

You can repeat the process for all the sites you anticipate needing over the course of the project.

Figure 5.9
Confirm the settings on the Numbering tab.

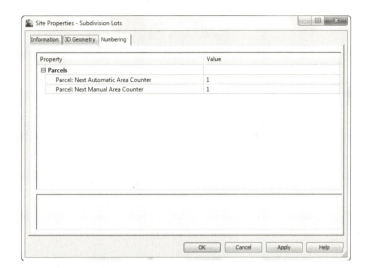

Creating a Boundary Parcel

The Create Parcel From Objects tool allows you to create parcels by choosing AutoCAD entities in your drawing or in an XRefed drawing. In a typical workflow, it's common to encounter a boundary created by AutoCAD entities, such as polylines, lines, and arcs.

When you're using AutoCAD geometry to create parcels, it's important that the geometry be created carefully and meet certain requirements. The AutoCAD geometry must be lines, arcs, polylines, 3D polylines, or polygons. It can't include blocks, ellipses, circles, or other entities. Civil 3D may allow you to pick objects with an elevation other than zero, but you'll find you get better results if you flatten the objects so all objects have an elevation of zero. Sometimes the geometry appears sound when elevation is applied, but you may notice this isn't the case once the objects are flattened. Flattening all objects before creating parcels can help you prevent frustration when creating parcels.

This exercise will teach you how to create a parcel from Civil 3D objects:

1. Open the CreateBoundaryParcel.dwg file, which you can download from www.sybex.com/go/masteringcivil3d2012. This drawing has several alignments, which were created on the Subdivision Lots site, and some AutoCAD linework representing a boundary. In addition, a parcel was created when the alignments formed closed areas in the Subdivision Lots site.

2. On the Home tab, select Parcel ➢ Choose Create Parcel From Objects on the Create Design panel.

3. At the Select lines, arcs, or polylines to convert into parcels or [Xref]: prompt, pick the red polyline that represents the site boundary. Press ↵.

4. The Create Parcels – From Objects dialog appears. Select Lot (Prop); Property; and Name Square Foot & Acres from the drop-down menus in the Site, Parcel Style, and Area Label Style selection boxes, respectively. Leave everything else set to the defaults. Click OK to dismiss the dialog.

The boundary polyline forms parcel segments that react with the alignments. Area labels are placed at the newly created parcel centroids, as shown in Figure 5.10.

FIGURE 5.10
The boundary parcel segments, alignments, and area labels

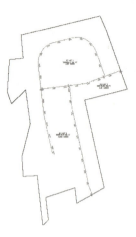

Creating a Wetlands Parcel

Although you may never have thought of things like wetlands areas, easements, and stormwater-management facilities as parcels in the past, you can take advantage of the parcel tools to assist in labeling, stylizing, and analyzing these features for your plans.

This exercise will teach you how to create a parcel representing wetlands using the transparent commands and Draw Tangent-Tangent With No Curves tool from the Parcel Layout Tools toolbox:

1. Open the `WetlandsParcel.dwg` file, which you can download from this book's web page. Note that this drawing has several alignments, parcels, and a series of points that represent a wetlands delineation.

2. Choose Parcel ➢ Parcel Creation Tools on the Create Design panel. The Parcel Layout Tools toolbar appears.

3. Click the Draw Tangent-Tangent With No Curves tool on the Parcel Layout Tools toolbar. The Create Parcels – Layout dialog appears.

4. In the dialog, select Subdivision Lots, Property, and Name Square Foot & Acres from the drop-down menus in the Site, Parcel Style, and Area Label Style selection boxes, respectively. Keep the default settings for all other options. Click OK.

5. Make sure that the Node running osnap is set. At the `Specify start point:` prompt, pick point 1. Continue picking the wetlands points in numerical order. Repeat for the other wetlands. Your drawing should look similar to Figure 5.11.

6. It's usually easier to change the appearance of the parcel and its area label after the parcel has been created. Change the style of the parcel by picking the parcel area label and choosing Parcel Properties from the Modify panel. The Parcel Properties dialog appears.

FIGURE 5.11
The wetlands defined on the site

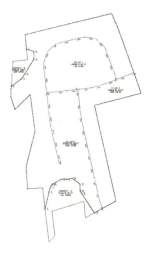

7. Select Wetlands from the drop-down menu in the Object Style selection box on the Information tab, and then click OK to dismiss the dialog. The parcel segments turn green, and a swamp hatch pattern appears inside the parcel to match the Wetlands style.

8. To change the style of the parcel area label, first select the Wetlands parcel area label, and then right-click and select Edit Area Selection Label Style. The Parcel Area Label Style dialog appears.

9. Select the Wetlands Area Label style from the drop-down menu in the Parcel Area Label Style selection box. Click OK to dismiss the dialog. A label appears, labeling the wetlands as shown in Figure 5.12. Later sections in this chapter will discuss parcel style and parcel area label style in more detail.

FIGURE 5.12
The Wetlands parcel with the appropriate label styles applied

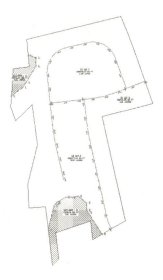

Creating a Right-of-Way Parcel

The Create ROW tool creates ROW parcels on either side of an alignment based on your specifications. The Create ROW tool can be used only when alignments are placed on the same site as the boundary parcel.

The resulting ROW parcel will look similar to Figure 5.13.

Options for the Create ROW tool include offset distance from alignment, fillet or chamfer cleanup at parcel boundaries, and alignment intersections. Figure 5.14 shows an example of chamfered cleanup at alignment intersections.

FIGURE 5.13
The resulting parcels after application of the Create ROW tool

FIGURE 5.14
A ROW with chamfer cleanup at alignment intersections

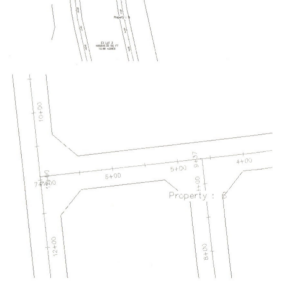

> **MAKE SURE YOUR GEOMETRY IS POSSIBLE**
>
> Make sure you provide parameters that are possible. If the program can't achieve your filleting requirements at any one intersection, a ROW parcel won't be created. For example, if you specify a 25′ filleting radius but the roads come together at a tight angle that would only allow a 15′ radius, then a ROW parcel won't be created.

Once the ROW parcel is created, it's no different from any other parcel. For example, it doesn't maintain a dynamic relationship with the alignment that created it. A change to the alignment will require the ROW parcel to be edited or, more likely, re-created.

This exercise will teach you how to use the Create ROW tool to automatically place a ROW parcel for each alignment on your site:

1. Open the `CreateROWParcel.dwg` file, which you can download from this book's web page. Note that this drawing has some alignments on the same site as the boundary parcel, resulting in several smaller parcels between the alignments and boundary.

2. Choose Parcel ➢ Create Right Of Way on the Create Design panel.

3. At the `Select parcels:` prompt, pick Ex. Lot 1, Ex. Lot 2, and Ex. Lot 3 on the screen. Press ↵ to stop picking parcels. The Create Right Of Way dialog appears, as shown in Figure 5.15.

4. Expand the Create Parcel Right of Way parameter, and enter **25'** as the value for Offset From Alignment.

5. Expand the Cleanup At Parcel Boundaries parameter. Enter **25'** as the value for Fillet Radius At Parcel Boundary Intersections. Select Fillet from the drop-down menu in the Cleanup Method selection box.

6. Expand the Cleanup at Alignment Intersections parameter. Enter **35'** as the value for Fillet Radius At Alignment Intersections. Select Fillet from the drop-down menu in the Cleanup Method selection box.

7. Click OK to dismiss the dialog and create the ROW parcels. Your drawing should look similar to Figure 5.16.

FIGURE 5.15
The Create Right Of Way dialog

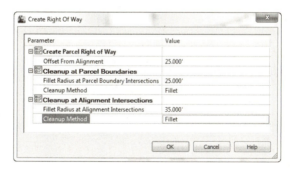

FIGURE 5.16
The completed ROW parcels

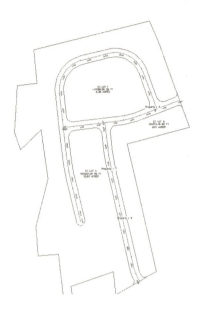

Creating a Cul-de-sac

If you look at your drawing, on the lower-left side of your right of way it looks incomplete. We will add a cul-de-sac here. Rather than go through all the mechanics of a cul-de-sac, a block has been created for you and you will insert it and turn it into a parcel. This section also introduces some editing tools.

1. Insert the `CulDeSacBlock.dwg` file using the settings shown in Figure 5.17.

FIGURE 5.17
Inserting the cul-de-sac block settings

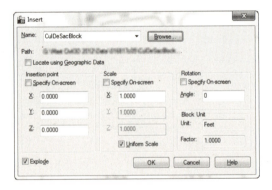

2. From the Home tab and Create Design panel, expand Parcel ➢ Create Parcel From Objects.

3. At the `Select lines, arcs, or polylines to convert into parcels or [Xref]:` prompt, draw a window around all of the cul-de-sac objects and press ↵.

CREATING A CUL-DE-SAC | 173

4. In the Create Parcels – From Objects dialog, make sure Site is set to Subdivision Lots, Parcel Style is set to Property, Area Label Style is set to Name Square Foot & Acres, and Erase Existing Entities is checked. Click OK. Your drawing should look like Figure 5.18.

FIGURE 5.18
The cul-de-sac turned into a parcel

The cul-de-sac is now a parcel, but there are some extra lines that need to be taken care of. Let's see how to clean this up a bit.

5. Select Parcel Creation Tools from the Home tab and Create Design Panel Parcel drop-down list. The Parcel Layout Tools palette opens.

6. Select the Delete Sub-entity tool. At the `Select subentity to remove:` prompt, select the right of way that interferes with the cul-de-sac, as shown in Figure 5.19.

7. Your cul-de-sac is complete, as shown on Figure 5.20.

FIGURE 5.19
Delete these portions.

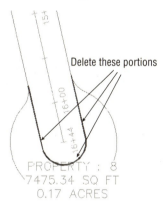

FIGURE 5.20
The finished cul-de-sac

Creating Subdivision Lot Parcels Using Precise Sizing Tools

The precise sizing tools allow you to create parcels to your exact specifications. You'll find these tools most useful when you have your roadways established and understand your lot-depth requirements. These tools provide automatic, semiautomatic, and freeform ways to control frontage, parcel area, and segment direction.

Attached Parcel Segments

Parcel segments created with the precise sizing tools are called *attached segments*. Attached parcel segments have a start point that is attached to a frontage segment and an endpoint that is defined by the next parcel segment they encounter. Attached segments can be identified by their distinctive diamond-shaped grip at their start point and no grip at their endpoint (see Figure 5.21).

In other words, you establish their start point and their direction, but they seek another parcel segment to establish their endpoint. Figure 5.22 shows a series of attached parcel segments. You can tell the difference between their start and endpoints because the start points have the diamond-shaped grips.

FIGURE 5.21
An attached parcel segment

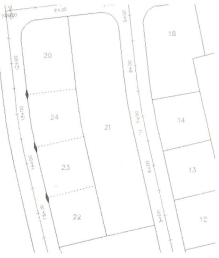

FIGURE 5.22
A series of attached parcel segments, with their endpoint at the front lot line

You can drag the diamond-shaped grip along the frontage to a new location, and the parcel segment will maintain its angle from the frontage. If the rear lot line is moved or erased, the attached parcel segments find a new endpoint (see Figure 5.23) at the next available parcel segment.

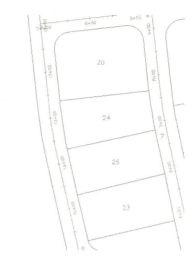

FIGURE 5.23
The endpoints of attached parcel segments extend to the next available parcel segment if the initial parcel segment is erased.

Precise Sizing Settings

The precise sizing tools consist of the Slide Angle, Slide Direction, and Swing Line tools (see Figure 5.24).

The Parcel Layout Tools toolbar can be expanded so that you can establish settings for each of the precise sizing tools (see Figure 5.25). Each of these settings is discussed in detail in the following sections.

NEW PARCEL SIZING

When you create new parcels, the tools respect your default area and minimum frontage (measured from either a ROW or a building setback line). The program always uses these numbers as a minimum; it bases the actual lot size on a combination of the geometry constraints (lot depth, frontage curves, and so on) and the additional settings that follow. Keep in mind that the numbers you establish under the New Parcel Sizing option must make geometric sense. For example, if you'd like a series of 7,500-square-foot lots that have 100' of frontage, you must make sure that your rear parcel segment allows for at least 75' of depth; otherwise, you may wind up with much larger frontage values than you desire or a situation where the software can't return a meaningful result.

FIGURE 5.24
The precise sizing tools on the Parcel Layout Tools toolbar

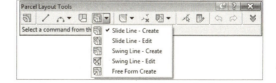

FIGURE 5.25
The settings on the Parcel Layout Tools toolbar

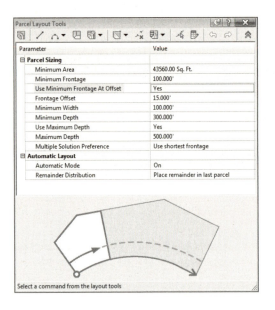

Automatic Layout

Automatic Layout has two parameters when the list is expanded: Automatic Mode and Remainder Distribution. The Automatic Mode parameter can have the following values:

On Automatically follows your settings and puts in all the parcels, without prompting you to confirm each one.

Off Allows you to confirm each parcel as it's created. In other words, this option provides you with a way to semiautomatically create parcels.

The Remainder Distribution parameter tells Civil 3D how you'd like "extra" land handled. This parameter has the following options:

Create Parcel From Remainder Makes a last parcel with the leftovers once the tool has made as many parcels as it can to your specifications on the basis of the settings in this dialog. This parcel is usually smaller than the other parcels.

Place Remainder In Last Parcel Adds the leftover area to the last parcel once the tool has made as many parcels as it can to your specifications on the basis of the settings in this dialog.

Redistribute Remainder Takes the leftover area and pushes it back through the default-sized parcels once the tool has made as many parcels as it can to your specifications on the basis of the settings in this dialog. The resulting lots aren't always evenly sized because of differences in geometry around curves and other variables, but the leftover area is absorbed.

There aren't any rules per se in a typical subdivision workflow. Typically the goal is to create as many parcels as possible within the limits of available land. To that end, you'll use a combination of AutoCAD tools and Civil 3D tools to divide and conquer the particular tract of land with which you are working.

Slide Line – Create Tool

The Slide Line – Create tool creates an attached parcel segment based on an angle from frontage. You may find this tool most useful when your jurisdiction requires a uniform lot-line angle from the right of way.

This exercise will lead you through using the Slide Line – Create tool to create a series of subdivision lots:

1. Open the CreateSubdivisionLots.dwg file, which you can download from this book's web page. Note that this drawing has several alignments on the same site as the boundary parcel, resulting in several smaller parcels between the alignments and boundary.

2. Choose Parcel ➤ Parcel Creation Tools on the Create Design panel. The Parcel Layout Tools toolbar appears.

3. Expand the toolbar by clicking the Expand The Toolbar button.

4. Change the value of the following parameters by clicking in the Value column and typing in the new values if they aren't already set. Notice how the preview window changes to accommodate your preferences:

    - Default Area: **7500.00 Sq. Ft.**
    - Minimum Frontage: **75.000'**
    - Use Minimum Frontage At Offset: **yes**
    - Frontage Offset: **25.000'**
    - Minimum Width: **75.000'**
    - Minimum Depth: **50.000'**
    - Use Maximum Depth: **no**
    - Maximum Depth: **500.000'**
    - Multiple Solution Preference: **Use shortest frontage**

5. Change the following parameters by clicking in the Value column and selecting the appropriate option from the drop-down menu, if they aren't already set:

    - Automatic Mode: **on**
    - Remainder Distribution: **Redistribute remainder**

6. Click the Slide Line – Create tool. The Create Parcels – Layout dialog appears.

7. Select Subdivision Lots, Lot (Prop), and Name Square Foot & Acres from the drop-down menus in the Site, Parcel Style, and Area Label Style selection boxes, respectively. Leave the rest of the options set to the default. Click OK to dismiss the dialog.

8. At the Select parcel to be subdivided or [Pick]: prompt, type **P** and press ↵. Pick a point on the screen inside Property 1.

9. At the Select start point on frontage: prompt, use your Endpoint osnap to pick the point of curvature along the ROW parcel segment for Property 1 (see Figure 5.26).

FIGURE 5.26
Pick the point of curvature along the ROW parcel segment.

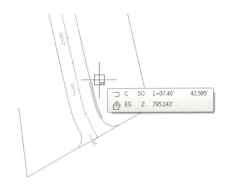

10. The parcel jig appears. Move your mouse slowly along the ROW parcel segment, and notice that the parcel jig follows the parcel segment. At the `Select end point on frontage:` prompt, use your Endpoint osnap to pick the point of curvature along the ROW parcel segment for Property 1 (see Figure 5.27).

11. At the `Specify angle or [Bearing/aZimuth]:` prompt, enter **90**↵. Notice the preview (see Figure 5.28).

FIGURE 5.27
Allow the parcel-creation jig to follow the parcel segment, and then pick the point of curvature along the ROW parcel segment.

FIGURE 5.28
A preview of the results of the automatic parcel layout

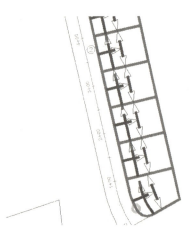

12. At the `Accept result? [Yes/No] <Yes>:` prompt, press ↵ to accept the default Yes.

13. At the `Select parcel to be subdivided or [Pick]:` prompt, press ↵, and then type **X** and press ↵ to exit the command.

14. Your drawing should look similar to Figure 5.29. Note that Property 1 still exists among the newly defined parcels and has kept its original parcel style and area label style. Note the parcels at the north end that just don't look right. We'll address that a little later in this chapter.

FIGURE 5.29
The automatically created lots

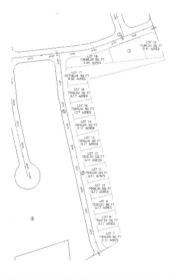

> **CURVES AND THE FRONTAGE OFFSET**
>
> In most cases, the frontage along a building setback is graphically represented as a straight line drawn tangent or parallel to, and behind, the setback. When you specify a minimum width along a frontage offset (the building setback line) in the Parcel Layout Tools dialog, and when the lot frontage is curved, the distance you enter is measured along the curve. In most cases, this result may be insignificant, but in a large development, the error could be the defining factor in your decision to add or subtract a parcel from the development.

You may find this tool most useful when you're re-creating existing lots or when you'd like to create a series of parallel lot lines with a known bearing.

Swing Line – Create Tool

The Swing Line – Create tool creates a "backward" attached parcel segment where the diamond-shaped grip appears not at the frontage but at a different location that you specify. The tool respects your minimum frontage, and it adjusts the frontage larger if necessary in order to respect your default area.

The Swing Line – Create tool is semiautomatic because it requires your input of the swing point location.

You may find this tool most useful around a cul-de-sac or in odd-shaped corners where you must hold frontage but have a lot of flexibility in the rear of the lot.

Using the Free Form Create Tool

A site plan is more than just single-family lots. Areas are usually dedicated for open space, stormwater-management facilities, parks, and public utility lots. The Free Form Create tool can be useful when you're creating these types of parcels. This tool, like the precise sizing tools, creates an attached parcel segment with the special diamond-shaped grips.

> The lot numbers were designed by the authors for the exercises. Your lot numbers may vary from those shown in the exercises.

In the following exercise, you'll use the Free Form Create tool to create a new parcel:

1. Open the CreateFreeForm.dwg file. Note that this drawing contains a series of subdivision lots.

2. Pan over to Lot 17. You can see the lot line that was drawn automatically in the previous exercise that obviously will not work (Figure 5.30).

FIGURE 5.30
Delete the highlighted parcel line.

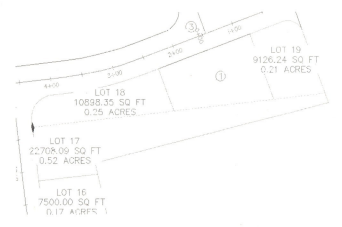

3. Delete the parcel line highlighted in Figure 5.30. The parcels readjust but Lot 20 is now much larger than needed. Let's add a line using the Free Form Create tool.

4. Select Parcel ➢ Parcel Creation Tools on the Create Design panel. Select the Free Form Create tool. The Create Parcels – Layout dialog appears.

5. Select Subdivision Lots, Lot (Prop), and Name Square Foot & Acres from the drop-down menus in the Site, Parcel Style, and Area Label Style selection boxes, respectively. Keep the default values for the remaining options. Click OK to dismiss the dialog.

6. Slide the Free Form Create attachment point around the Lot: 20 area. At the `Select attachment point:` prompt, use your Endpoint osnap to pick the endpoint, as shown in Figure 5.31.

FIGURE 5.31
Use the Free Form Create tool to select an attachment point.

7. At the `Specify lot line direction:` prompt, press ↵ to specify a perpendicular lot line direction.

8. A new parcel segment is created from your Open Space Limit point, perpendicular to the ROW parcel segment, as shown in Figure 5.32.

FIGURE 5.32
Attach the parcel segment to the marker point provided.

9. Note that a new lot parcel has formed.

10. Press ↵ to exit the Free Form Create command. Enter **X**, and then press ↵ to exit the toolbar.

11. Pick the new parcel segment so that you see its diamond-shaped grip. Grab the grip, and slide the segment along the ROW parcel segment (see Figure 5.33).

FIGURE 5.33
Sliding an attached parcel segment

12. Notice that when you place the parcel segment at a new location the segment endpoint snaps back to the rear parcel segment. This is typical behavior for an attached parcel segment.

Editing Parcels by Deleting Parcel Segments

One of the most powerful aspects of Civil 3D parcels is the ability to perform many iterations of a site plan design. Typically, this design process involves creating a series of parcels and then deleting them to make room for iteration with different parameters, or deleting certain segments to make room for easements, public utility lots, and more.

You can delete parcel segments using the AutoCAD Erase tool as shown in the previous exercise, or the Delete Sub-Entity tool on the Parcel Layout Tools toolbar.

> **BREAK THE UNDO HABIT**
>
> You'll find that parcels behave better if you use one of the segment-deletion methods described in this section to erase improperly placed parcels rather than using the Undo command.

It's important to understand the difference between these two methods. The AutoCAD Erase tool behaves as follows:

◆ If the parcel segment was originally created from a polyline (or similar parcel layout tools, such as the Tangent-Tangent With No Curves tool), the AutoCAD Erase tool erases the entire segment (see Figure 5.34).

◆ If the parcel segment was originally created from a line or arc (or similar parcel layout tools, such as the precise sizing tools), then AutoCAD Erase erases the entire length of the original line or arc (see Figure 5.35).

The Delete Sub-Entity tool acts more like the AutoCAD Trim tool. The Delete Sub-Entity tool only erases the parcel segments between parcel vertices. For example, if Lot 33, as shown in Figure 5.36, must be absorbed into Lot 37 to create a public utility lot with dual road access, you'd want to only erase the segment at the rear of Lot 33 and not the entire segment shown previously in Figure 5.35.

FIGURE 5.34
The segments indicated by the blue grips will be erased after using the AutoCAD Erase tool.

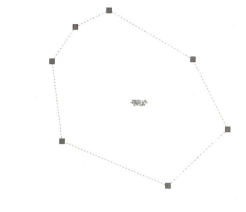

FIGURE 5.35
The AutoCAD Erase tool will erase the entire segment indicated by the blue grips.

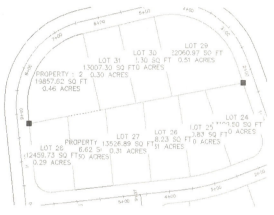

FIGURE 5.36
Use the Delete Sub-Entity tool to erase the rear parcel segment for Lot 33.

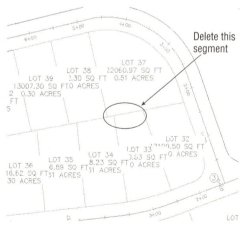

Selecting the Lot 33 label and then clicking Parcel Layout Tools on the Modify panel brings up the Parcel Layout Tools toolbar. Selecting the Delete Sub-Entity tool allows you to pick only the small rear parcel segment for Lot 33. Figure 5.37 shows the result of this deletion.

FIGURE 5.37
The rear lot line for Lot 33 was erased using the Delete Sub-Entity tool, thus creating a larger Lot 37.

The following exercise will lead you through deleting a series of parcel segments using both the AutoCAD Erase tool and the Delete Sub-Entity tool:

1. Open the DeleteSegments.dwg file. Note that this drawing contains a series of subdivision lots, along with a wetlands boundary.

2. Let's say you just received word that there was a mistake with the wetlands delineation and you need to erase the entire wetlands area. Use the AutoCAD Erase tool to erase the parcel segments that define the wetlands parcel and then freeze the wetlands points. Note that the entire parcel disappears in one shot, because it was created with the Tangent-Tangent With No Curves tool (which behaves similarly to creating a polyline).

3. Now the developer wants to maximize the number of lots instead of having large lots. Erase the lot lines as shown in Figure 5.38.

4. Re-create the lots by using the Slide Line – Create tool as discussed earlier. Use the default values. Your lots should look similar to Figure 5.39.

FIGURE 5.38
The lots to be deleted

Figure 5.39
The re-created lots

5. Next, you discover that Lots 29 and 30 need to be removed and absorbed into Property: 4. From the Parcel Layout Tools, click the Delete Sub-Entity tool.

6. At the `Select subentity to remove:` prompt, pick the line between Lots 29 and 30 and then the rear lot line of the newly combined Lot 29 and Property: 4. Press ↵ to exit the command, enter **X**, and then press ↵ to exit the Parcel Layout Tools toolbar.

The resulting parcel is displayed as shown in Figure 5.40.

Figure 5.40
The parcel after erasing the rear lot line

Best Practices for Parcel Creation

Now that you have an understanding of how objects in a site interact and you've had some practice creating and editing parcels in a variety of ways, we'll take a deeper look at how parcels must be constructed to achieve topology stability, predictable labeling, and desired parcel interaction.

Forming Parcels from Segments

In the earlier sections of this chapter, you saw that parcels are created only when parcel segments form a closed area (see Figure 5.41).

FIGURE 5.41
A parcel is created when parcel segments form a closed area.

Parcels must always close. Whether you draw AutoCAD lines and use the Create Parcel From Objects menu command or use the parcel segment creation tools, a parcel won't form until there is an enclosed polygon. Figure 5.42 shows four parcel segments that don't close; therefore, no parcel has been formed.

FIGURE 5.42
No parcel will be formed if parcel segments don't completely enclose an area.

There are times in surveying and engineering when parcels of land don't necessarily close when created from legal descriptions. In this case, you must work with your surveyor to perform an adjustment or find some other solution to create a closed polygon.

You also saw that even though parcels can't be erased, if you erase the appropriate parcel segments, the area contained within a parcel is assimilated into neighboring parcels.

Parcels Reacting to Site Objects

Parcels require only one parcel segment to divide them from their neighbor (see Figure 5.43). This behavior eliminates the need for duplicate segments between parcels, and duplicate segments must be avoided.

FIGURE 5.43
Two parcels, with one parcel segment between them

As you saw in the section on site interaction, parcels understand their relationships to one another. When you create a single parcel segment between two subdivision lots, you have the ability to move one line and affect two parcels. Figure 5.44 shows the moved parcels from Figure 5.43 once the parcel segment between them has been shifted to the left. Note that both areas change in response.

FIGURE 5.44
Moving one parcel segment affects the area of two parcels.

A mistake that many people new to Civil 3D make is to create parcels from closed polylines, which results in a duplicate segment between parcels. Figure 5.45 shows two parcels created from two closed polylines. These two parcels may appear identical to the two seen in the previous example, because they were both created from a closed polyline rectangle; however, the segment between them is actually two segments.

FIGURE 5.45
Adjacent parcels created from closed polylines create overlapping or duplicate segments.

The duplicate segment becomes apparent when you attempt to grip-edit the parcel segments. Moving one vertex from the common lot line, as seen in Figure 5.46, reveals the second segment. Also note that a sliver parcel is formed. Duplicate site geometry objects and sliver parcels make it difficult for Civil 3D to solve the site topology and can cause drawing stability problems and unexpected parcel behavior. You must avoid this situation at all costs. Creating a subdivision plat of parcels this way almost guarantees that your labeling won't perform properly and could potentially lead to data loss and drawing corruption.

FIGURE 5.46
Duplicate segments become apparent when they're grip-edited and a sliver parcel is formed.

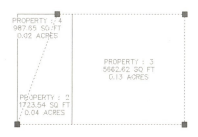

> **MIGRATING PARCELS FROM LAND DESKTOP**
>
> The Land Desktop Parcel Manager essentially created Land Desktop parcels from closed polylines. If you migrate Land Desktop parcels into Civil 3D, your resulting Civil 3D parcels will behave poorly and will almost universally result in drawing corruption.

Parcels form to fill the space contained by the original outer boundary. You should always begin a parcel-division project with an outer boundary of some sort (see Figure 5.47).

FIGURE 5.47
An outer boundary parcel

You can then add road centerline alignments to the site, which divides the outer boundary as shown in Figure 5.48.

FIGURE 5.48
Alignments added to the same site as the boundary parcel divide the boundary parcel.

It's important to note that the boundary parcel no longer exists intact. As you subdivide this site, parcel 1 is continually reallocated with every division. As road ROW and subdivision lots are formed from parcel segments, more parcels are created. Every bit of space that was contained in the original outer boundary is accounted for in the mesh of newly formed parcels (see Figure 5.49).

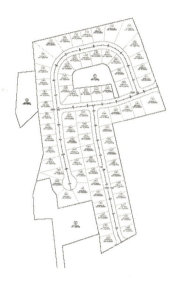

Figure 5.49
The total area of parcels contained within the original boundary sums to equal the original boundary area.

From now on, you'll consider ROW, wetlands, parkland, and open space areas as parcels, even if you didn't before. You can make custom label styles to annotate these parcels however you like, including a "no show" or none label.

 Real World Scenario

If I Can't Use Closed Polylines, How Do I Create My Parcels?

How do you create your parcels if parcels must always close, but you aren't supposed to use closed polylines to create them?

In the earlier exercises in this chapter, you learned several techniques for creating parcels. These techniques included using AutoCAD objects and a variety of parcel layout tools. Here's a summary of some of the best practices for creating parcels:

- Create closed polylines for boundaries and islands, and then use the Create Parcel From Objects menu command. Closed polylines are suitable foundation geometry in cases where they won't be subject to possible duplicate segments. The following graphic shows a boundary parcel and a designated open space parcel that were both created from closed polylines. Other examples of island parcels include isolated wetlands, ponds, or similar features that don't share a common segment with the boundary parcel.

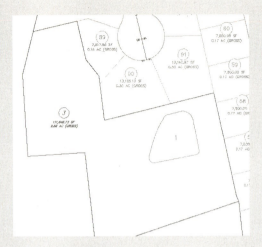

- Create trimmed/extended polylines for internal features, and then use the Create Parcel From Objects menu command. For internal features such as easements, buffers, open space, or wetlands that share a segment with the outer boundary, draw a polyline that intersects the outer boundary, but be careful not to trace over any segments of the outer boundary. Use the Create Parcel From Objects menu command to convert the polyline into a parcel segment. The following graphic shows an easement.

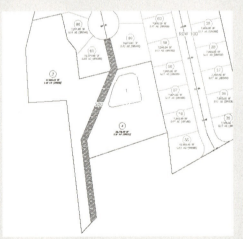

- Use the Create ROW tool or create trimmed/extended polylines, and then use the Create Parcel From Objects menu command for ROW segments. In a previous exercise, you used the Create ROW tool. You also learned that even though this tool can be useful, it can't create cul-de-sacs or changes in ROW width. In cases where you need a more intricate ROW parcel, use the AutoCAD Offset tool to offset your alignment. The resulting offsets are polylines. Use circles, arcs, fillet, trim, extend, and other tools to create a joined polyline to use as foundation geometry for your ROW parcel. Use the Create Parcel From Objects menu command to convert this linework into parcel segments for a cul-de-sac, as shown here.

- Create trimmed or extended polylines for rear lot lines, and then use the Create Parcel From Objects menu command. The precise sizing tools tend to work best when given a rear lot line as a target endpoint. Create this rear target by offsetting your ROW parcel to your desired lot depth. The resulting offset is a polyline. Use Trim, Extend, and other tools to create a joined polyline to use as foundation geometry for your rear lot parcel segment. Surveyors often prefer to lay out straight-line segments rather than curves, so for the final rear lot line cleanup, create AutoCAD lines across the back of each lot and then use Create Parcel From Objects to turn those lines into parcel segments, as shown here.

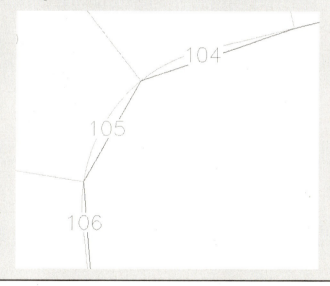

This final cleanup is best saved for the very end of the project. Parcel iterations and refinement work much better with a continuous rear lot line.

Constructing Parcel Segments with the Appropriate Vertices

Parcel segments should have natural vertices only where necessary and split-created vertices at all other intersections. A natural vertex, or point of intersection (PI), can be identified by picking a line, polyline, or parcel segment and noting the location of the grips (see Figure 5.50).

FIGURE 5.50
Natural vertices on a parcel segment

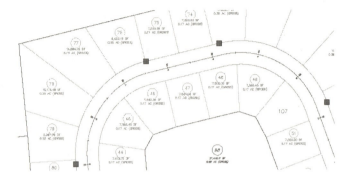

A split-created vertex occurs when two parcel segments touch or cross each other. Note that in Figure 5.51, the parcel segment doesn't show a grip even where each individual lot line touches the ROW parcel.

FIGURE 5.51
Split-created vertices on a parcel segment

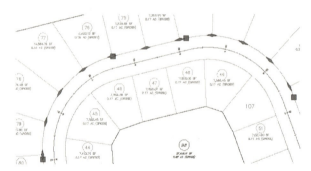

It's desirable to have as few natural vertices as possible. In the example shown previously in Figure 5.50, the ROW frontage line can be expressed as a single bearing and length from the end of the arc through the beginning of the next arc, as opposed to having several smaller line segments.

If the foundation geometry was drawn with a natural vertex at each lot line intersection, then the resulting parcel segment won't label properly and may cause complications with editing and other functions. This subject will be discussed in more detail later in the section "Labeling Spanning Segments," later in this chapter.

Parcel segments must not overlap. Overlapping segments create redundant vertices, sliver parcels, and other problems that complicate editing parcel segments and labeling as shown in Figure 5.52. Figure 5.53 shows a segment created to form parcel 3 that overlaps the rear parcel segment for the entire block. This segment should be edited to remove the redundant parcel segment across the rear of Parcel 90 to ensure good parcel topology.

FIGURE 5.52
Avoid creating overlapping parcel segments.

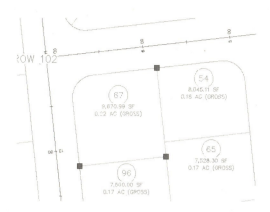

Parcel segments must not overhang. Spanning labels are designed to overlook the location of intersection formed (or T-shaped) split-created vertices. However, these labels won't span a crossing formed (X- or + [plus]-shaped) split-created vertex. Even a very small parcel segment overhang will prevent a spanning label from working and may even affect the area computation for adjacent parcels. The overhanging segment in Figure 5.53 would prevent a label from returning the full spanning length of the ROW segment it crosses.

FIGURE 5.53
Overhanging segment

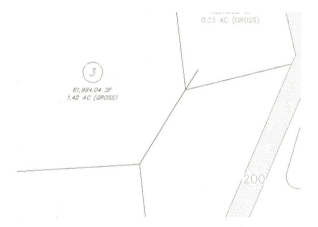

Labeling Parcel Areas

A parcel area label is placed at the parcel centroid by default, and it refers to the parcel in its entirety. When asked to pick a parcel, you pick the area label. An area label doesn't necessarily have to include the actual area of the parcel.

Area labels can be customized to suit your fancy. Figure 5.54 shows a variety of customized area labels.

Area labels often include the parcel name or number. You can rename or renumber parcels using Renumber/Rename from the Modify panel after selecting a parcel.

FIGURE 5.54
Sample area labels

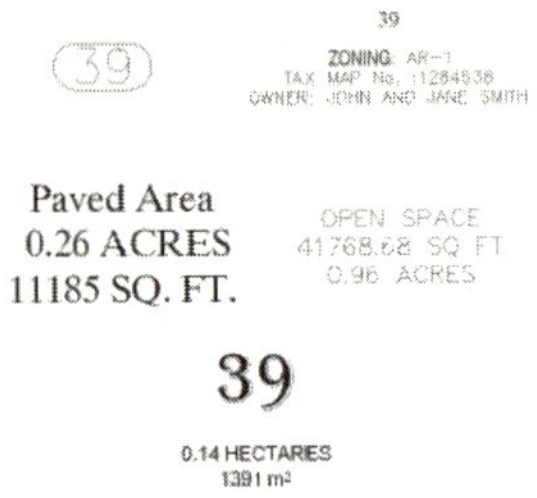

The following exercise will teach you how to renumber a series of parcels:

1. Open ChangeAreaLabel.dwg. Note that this drawing contains many subdivision lot parcels.

2. Select Lot 25 and select Renumber/Rename from the Modify panel. The Renumber/Rename Parcels dialog appears.

3. In the Renumber/Rename Parcels dialog, make sure Subdivision Lots is selected from the drop-down menu in the Site selection box. Change the value of the Starting Number selection box to **1**. Click OK.

4. At the Specify start point or [Polylines/Site]: prompt, pick a point on the screen anywhere inside the Lot 25 parcel, which will become your new Lot 1 parcel at the end of the command.

5. At the End point or [Undo]: prompt, pick a point on the screen anywhere inside the Lot 20 parcel, almost as if you were drawing a line; then, pick a point anywhere inside Lot 20 (be sure not to cross other parcel lines); and finally, pick a point inside Lot 23. Press ↵ to stop choosing parcels. Press ↵ again to end the command.

Note that your parcels have been renumbered from 1 through 15. Repeat the exercise with other parcels in the drawing for additional practice if desired.

The next exercise will lead you through one method of changing an area label using the Edit Parcel Properties dialog:

1. Continue working in the ChangeAreaLabel.dwg file.

2. Select parcel Lot 1 and select Multiple Parcel Properties from the Modify panel. At the Specify start point or [Polylines/All/Site]: prompt, pick a point on the screen anywhere inside the Single-Family: 1 parcel.

3. At the End point or [Undo]: prompt, pick a point on the screen anywhere inside Lot 2, almost as if you were drawing a line. Press ↵ to stop choosing parcels. Press ↵ again to open the Edit Parcel Properties dialog (see Figure 5.55).

4. In the Area Selection Label Styles portion of the Edit Parcel Properties dialog, use the drop-down menu to choose the Parcel Number area label style.

5. Click the Apply To All Parcels button.

FIGURE 5.55
The Edit Parcel Properties dialog

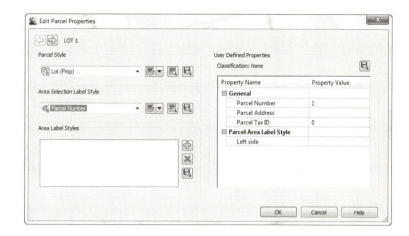

6. Click Yes in the dialog displaying the question "Apply the area selection label style to the 2 selected parcels?"

7. Click OK to exit the Edit Parcel Properties dialog.

The two parcels now have parcel area labels that call out numbers only. Note that you could also use this interface to add a second area label to certain parcels if required.

This final exercise will show you how to use Prospector to change a group of parcel area labels at the same time:

1. Continue working in the ChangeAreaLabel.dwg file.

2. In Prospector, expand the Sites ➢ Subdivision Lots ➢ Parcels collection.

3. In the Preview pane, click the Name column to sort the Parcels collection by name.

4. Hold down the Shift key, and click each Single-Family parcel to select them all. Release the Shift key, and your parcels should remain selected.

5. Slide over to the Area Label Style column. Right-click the column header and select Edit (see Figure 5.56).

FIGURE 5.56
Right-click the Area Label Style column header and select Edit.

6. In the Select Label Style dialog, select Parcel Number from the drop-down menu in the Label Style selection box. Click OK to dismiss the dialog.

7. The drawing will process for a moment. Once the processing is finished, minimize Prospector and inspect your parcels. All the Single-Family parcels should now have the Parcel Number area label style.

> ### WHAT IF THE AREA LABEL NEEDS TO BE SPLIT ONTO TWO LAYERS?
>
> You may have a few different types of plans that show parcels. Because it would be awkward to have to change the parcel area label style before you plot each sheet, it would be best to find a way make a second label on a second layer so that you can freeze the area component in sheets or viewports when it isn't needed. Here's an example where the square footage has been placed on a different layer so it can be frozen in certain viewports:
>
>
>
> You can accomplish this by creating a second parcel area label that calls out the area only:
>
> 1. Change to the Annotate tab. From the Labels & Tables panel, select Add Labels ➢ Parcel ➢ Add Parcel Labels.
> 2. Select Area from the drop-down menu in the Label Type selection box, and then select an area style label that will be the second area label.
> 3. Click Add, and then pick your parcel on screen.
>
> You'll find a second parcel area label to be a little more automatic when you place it (it already knows what parcel to reference).
>
> You can also use the Edit Parcel Properties dialog, as shown in the "Editing Parcels by Deleting Parcel Segments" section earlier in the chapter, to add a second label.

Labeling Parcel Segments

Although parcels are used for much more than just subdivision lots, most parcels you create will probably be used for concept plans, record plats, and other legal subdivision plans. These plans, such as the one shown in Figure 5.57, almost always require segment labels for bearing, distance, direction, crow's feet, and more.

Figure 5.57
A fully labeled site plan

Labeling Multiple Parcel Segments

The following exercise will teach you how to add labels to multiple parcel segments:

1. Open the `SegmentLabels.dwg` file, which you can download from this book's web page. Note that this drawing contains many subdivision lot parcels.

2. Switch to the Annotate tab, and select Add Labels from the Labels & Tables panel on the Annotate tab.

3. In the Add Labels dialog, select Parcel, Multiple Segment, Bearing Over Distance, and Delta Over Length And Radius from the drop-down menus in the Feature, Label Type, Line Label Style, and Curve Label Style selection boxes, respectively, as shown in Figure 5.58.

4. Click Add.

5. At the `Select parcel to be labeled by clicking on area label:` prompt, pick the area label for Parcel 1.

6. At the `Label direction [CLockwise/COunterclockwise]<CLockwise>:` prompt, press ↵ to accept the default and again to exit the command.

7. Each parcel segment for Parcel 1 should now be labeled. Continue picking Parcels 2 through 15 in the same manner. Note that segments are never given a duplicate label, even along shared lot lines.

8. Press ↵ to exit the command.

FIGURE 5.58
The Add Labels dialog

The following exercise will show you how to edit and delete parcel segment labels:

1. Continue working in the `SegmentLabels.dwg` file.
2. Zoom in on the label along the frontage of Parcel 8 (see Figure 5.59).
3. Select the label. You'll know your label has been picked when you see a diamond-shaped grip at the label midpoint (see Figure 5.60).

FIGURE 5.59
The label along the frontage of Parcel 8

FIGURE 5.60
A diamond-shaped grip appears when the label has been picked.

4. Once your label is picked, right-click over the label to bring up the context menu.

5. Select Flip Label from the context menu. The label flips so that the bearing component is on top of the line and the distance component is underneath the line.

6. Select the label again, right-click, and select Reverse Label. The label reverses so that the bearing now reads NW instead of SE.

7. Repeat steps 3 through 6 for several other segment labels, and note their reactions.

8. Select any label. Once the label is picked, execute the AutoCAD Erase tool or press the Delete key. Note that the label disappears.

Labeling Spanning Segments

Spanning labels are used where you need a label that spans the overall length of an outside segment, such as the example in Figure 5.61.

FIGURE 5.61
A spanning label

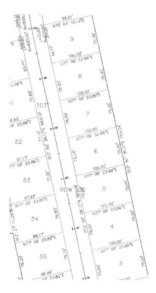

Spanning labels require that you use the appropriate vertices as discussed in detail in a previous section. Spanning labels have the following requirements:

- Spanning labels can only span across split-created vertices. Natural vertices will interrupt a spanning length.

- Spanning label styles must be composed to span the outside segment.

- Spanning label styles must be composed to attach the desired spanning components (such as length and direction arrow) on the outside segment (as shown previously in Figure 5.61), with perhaps a small offset.

Once you've confirmed that your geometry is sound and your label is properly composed, you're set to span. The following exercise will teach you how to add spanning labels to single-parcel segments:

1. Continue working in the SegmentLabels.dwg file.
2. Zoom in on the outer parcel segment that runs from Parcel 1 through Parcel 10.
3. Change to the Annotate tab and select Add Labels ➢ Parcel ➢ Add Parcel Labels from the Labels & Tables panel.
4. In the Add Labels dialog, select Single Segment, (Span) Bearing And Distance With Crow's Feet, and Delta Over Length And Radius from the drop-down menus in the Label Type, Line Label Style, and Curve Label Style selection boxes, respectively.
5. Click Add.
6. At the Select label location: prompt, pick somewhere near the middle of the outer parcel segment that runs from Parcel 1 through Parcel 10.

   A label that spans the full length between natural vertices appears (see Figure 5.61).

> **FLIP IT, REVERSE IT**
>
> If your spanning label doesn't seem to work on your first try and you've followed all the spanning label guidelines, try flipping your label to the other side of the parcel segment, reversing the label, or using a combination of both flipping and reversing.

Adding Curve Tags to Prepare for Table Creation

Surveyors and engineers often make segment tables to simplify plan labeling, produce reports, and facilitate stakeout. Civil 3D parcels provide tools for creating dynamic line and curve tables, as well as a combination of line and curve tables.

Parcel segments must be labeled before they can be used to create a table. They can be labeled with any type of label, but you'll likely find it to be best practice to create a tag-only style for segments that will be placed in a table.

The following exercise will teach you how to replace curve labels with tag-only labels, and then renumber the tags:

1. Continue working in the SegmentLabels.dwg file. Note that the labels along tight curves, such as the cul-de-sac, would be better represented as curve tags.
2. Change to the Annotate tab. Select Add Labels ➢ Parcel ➢ Add Parcel Labels from the Labels & Tables panel.

3. In the Add Labels dialog, select Replace Multiple Segment, Bearing Over Distance, and Spanning Curve Tag Only from the drop-down menus in the Label Type, Line Label Style, and Prop selection boxes, respectively.

4. At the `Select parcel to be labeled by clicking on area label or [CLockwise/ COunterclockwise]<CLockwise>:` prompt, pick the area label for Parcel 1. Note that the line labels for Parcel 1 are reset and the curve labels convert to tags.

5. Repeat step 4 for Parcels 2 through 15. Press ↵ to exit the command.

Now that each curve label has been replaced with a tag, it's desirable to have the tag numbers be sequential. The following exercise will show you how to renumber tags:

1. Continue working in the `SegmentLabels.dwg` file.

2. Zoom into the curve on the upper-left side of Parcel 11 (see Figure 5.62). Your curve may have a different number from the figure.

FIGURE 5.62
Curve tags on Parcel 11

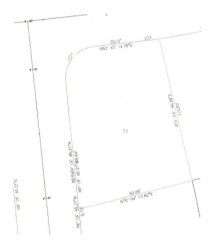

Renumber Tags

3. Select a parcel and select Renumber Tags from the Labels & Tables panel.

4. At the `Select label to renumber tag or [Settings]:` prompt, type **S**, and then press ↵.

5. The Table Tag Numbering dialog appears (see Figure 5.63). Change the value in the Curves Starting Number selection box to **1**. Click OK.

6. Click each curve tag in the drawing at the `Select label to renumber tag or [Settings]:` prompt. The command line may say `Current tag number is being used, press return to skip to next available or [Create duplicate]`, in which case you should press ↵ to skip the used number. When you're finished, press ↵ to exit the command.

FIGURE 5.63
The Table Tag Numbering dialog

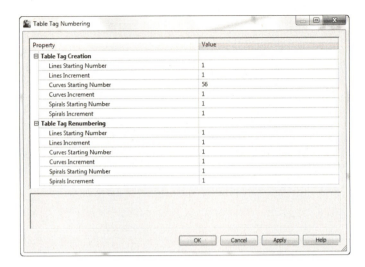

Creating a Table for Parcel Segments

The following exercise demonstrates how to create a table from curve tags:

1. Continue working in the SegmentLabels.dwg file. You should have several curves labeled with the Prop label.

2. Select a parcel and choose Add Curve under the Add Tables from the Labels & Tables panel.

3. In the Table Creation dialog, select Length Radius & Delta from the drop-down menu in the Table Style selection box. In the Selection area of the dialog, select the Apply check box for the Parcel Curve: Prop entry under Label Style Name. Keep the default values for the remaining options. The dialog should look like Figure 5.64. Click OK.

FIGURE 5.64
The Table Creation dialog

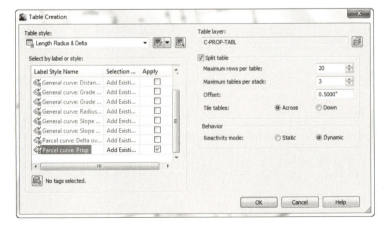

4. At the `Select upper left corner:` prompt, pick a location in your drawing for the table. A curve table appears, as shown in Figure 5.65.

FIGURE 5.65
A curve table

Curve #	Length	Radius	Delta	Chord Direction	Chord Length
C27	57.21	35.00	93.65	N71° 55' 15"W	51.05
C26	32.07	150.00	12.25	N18° 58' 08"W	32.01
C26	32.55	975.00	1.91	N11° 53' 13"W	32.55
C25	78.30	975.00	4.60	N6° 37' 48"W	78.28
C23	39.27	25.00	90.00	N38° 40' 15"E	35.36
C24	17.25	525.00	1.88	N82° 43' 47"E	17.25
C22	74.38	525.00	8.12	N77° 43' 48"E	74.31
C21	35.72	525.00	3.90	N71° 43' 21"E	35.71
C20	54.98	35.00	90.00	S65° 13' 34"E	49.50
C43	39.27	25.00	90.00	S24° 46' 25"W	35.36
C44	17.70	475.00	2.14	S70° 50' 26"W	17.70
C45	78.26	150.00	29.89	S25° 10' 21"E	77.37
C46	112.50	150.00	42.97	S71° 36' 17"E	109.88

The Bottom Line

Create a boundary parcel from objects. The first step to any parceling project is to create an outer boundary for the site.

Master It Open the `MasteringParcels.dwg` file, which you can download from www.sybex.com/go/masteringcivil3d2012. Convert the polyline in the drawing to a parcel.

Create a right-of-way parcel using the right-of-way tool. For many projects, the ROW parcel serves as frontage for subdivision parcels. For straightforward sites, the automatic Create ROW tool provides a quick way to create this parcel. Cul-de-sacs serve as a terminal point for a cluster of parcels.

Master It Continue working in the `Mastering Parcels.dwg` file. Create a ROW parcel that is offset by 25' on either side of the road centerline with 25' fillets at the parcel boundary. Then add the circles representing the cul-de-sac as a parcel.

Create subdivision lots automatically by layout. The biggest challenge when creating a subdivision plan is optimizing the number of lots. The precise sizing parcel tools provide a means to automate this process.

Master It Continue working in the `Mastering Parcels.dwg` file. Create a series of lots with a minimum of 8,000 square feet and 75' frontage.

Add multiple parcel-segment labels. Every subdivision plat must be appropriately labeled. You can quickly label parcels with their bearings, distances, direction, and more using the segment labeling tools.

Master It Continue working in the `MasteringParcels.dwg` file. Place Bearing Over Distance labels on every parcel line segment and Delta Over Length And Radius labels on every parcel curve segment using the Multiple Segment Labeling tool.

Chapter 6

Alignments

The world is 3D, but almost every design starts as a concept: a flat line on a flat piece of paper. Cutting a way through the trees, the hills, and the forests, you can design around a basic layout to get some idea of horizontal placement. This horizontal placement is the alignment and drives much of the design. This chapter shows you how alignments can be created, how they interact with the rest of the design, how to edit and analyze them, and finally, how they work with the overall project.

In this chapter, you will learn to:

- Create an alignment from a polyline
- Create a reverse curve that never loses tangency
- Replace a component of an alignment with another component type
- Create alignment tables

Alignment Concepts

Before you can efficiently work with alignments, you must understand two major concepts: the interaction of alignments and sites, and the idea of geometry that is fixed, floating, or free.

Alignments and Sites

Prior to Civil 3D 2008, alignments were always a part of a site and interacted with the topology contained in that site. This interaction led to the pickle analogy: alignments are like pickles in a mason jar. You don't put pickles and peppers in the same jar unless you want hot pickles, and you don't put lots and alignments in the same site unless you want subdivided lots.

Civil 3D now has two ways of handling alignments in terms of sites: They can be contained in a site as before, or they can be independent of a site.

Both the alignments contained in a site and independent of a site can be used to cut profiles or control corridors, but only the alignments contained in a site will react with and create parcels as a member of a site topology.

Unless you have good reason for them to interact (as in the case of an intersection), it makes sense to create alignments outside of any site object. They can be moved later if necessary. For the purpose of the exercises in this chapter, you won't place any alignments in a site.

Alignment Entities

Civil 3D recognizes four types of alignments: centerline alignments, offset alignments, curb return alignments, and miscellaneous alignments. Each alignment type can consist of three types of entities or segments: lines, arcs, and spirals. These segments control the horizontal alignment of your design. Their relationship to one another is described by the following terminology:

- Fixed segments are fixed in space (see Figure 6.1). They're defined by connecting points in the coordinate plane and are independent of the segments that occur either before or after them in the alignment. Fixed segments may be created as tangent to other components, but their independence from those objects lets you move them out of tangency during editing operations. This feature can be helpful when you're trying to match existing field conditions.

FIGURE 6.1
Alignment fixed segments

- Floating segments float in space but are attached to a point in the plane and to some segment to which they maintain tangency (see Figure 6.2). Floating segments work well in situations where you have a critical point but the other points of the horizontal alignment are flexible.

FIGURE 6.2
Alignment floating segments

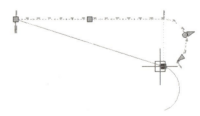

- Free segments are functions of the entities that come before and after them in the alignment structure (see Figure 6.3). Unlike fixed or floating segments, a free segment must have segments that come before and after it. Free segments maintain tangency to the segments that come before and after them and move as required to make that happen. Although some geometry constraints can be put in place, these constraints can be edited and are user dependent.

FIGURE 6.3
Alignment free segments

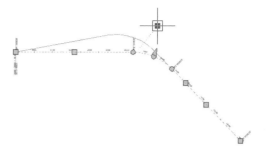

During the exercises in this chapter, you'll use a mix of these entity types to understand them better. Autodesk has also published a drawing called Playground2 that you can find by searching on the Web. This drawing contains examples of most of the types of entities that you can create.

Creating an Alignment

Alignments in Civil 3D can be created from AutoCAD objects (lines, arcs, or polylines) or by layout. This section looks at both ways to create an alignment and discusses the advantages and disadvantages of each. The exercise will use the street layout shown in Figure 6.4 as well as the different methods to achieve your designs.

FIGURE 6.4
Proposed street layout

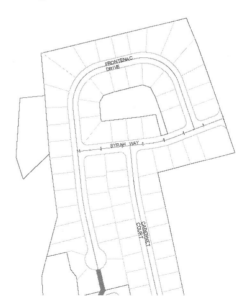

Creating from a Line, Arc, or Polyline

Most designers have used either polylines or lines and arcs to generate the horizontal control of their projects. It's common for surveyors to generate polylines to describe the center of a right of way or for an environmental engineer to draw a polyline to show where a new channel should be constructed. These team members may or may not have Civil 3D, so they use their familiar friends—the line, arc, and polyline—to describe their design intent.

Although these objects are good at showing where something should go, they don't have much data behind them. To make full use of these objects, you should convert them to native Civil 3D alignments that can then be shared and used for myriad purposes. Once an alignment has been created from a polyline, offsets can be created to represent rights of way, building lines, and so on. In this exercise, you'll convert a polyline to an alignment and create offsets:

1. Open the `AlignmentsFromPolylines.dwg` file. You can download this file from `www.sybex.com/go/masteringcivil3d2012`. You see the red polylines representing the center of roads, right of ways, and parcels.

2. Change to the Home tab and choose Alignment ➢ Create Alignment From Objects.

3. Pick the polyline labeled Syrah Way, shown previously in Figure 6.4, and press ↵. Press ↵ again to accept the default direction; the Create Alignment From Objects dialog appears.

4. Change the Name field to **Syrah Way**, and select the Centerline type, as shown in Figure 6.5.

5. Accept the other settings, and click OK.

FIGURE 6.5
The settings used to create the Syrah Way alignment

THE CREATE ALIGNMENT FROM OBJECTS DIALOG

In the Create Alignment From Objects dialog (Figure 6.5), there are many settings that you can use to create an alignment:

Name   This is the alignment name. No alignment name can be duplicated in a drawing.

Type   The alignment types can be thought of as places for objects that are alike. They can react differently depending on which type is selected.

> **Centerline**   Used mainly for centers of roads, streams, swales, etc. It places this type of alignment in the Alignments ➤ Centerline Alignments collection.
>
> **Offset**   Used for offset alignments. The difference between this and the centerline alignment is that you have the option in Alignment properties to set Offset parameters, such as naming a parent alignment and offset values. It places this alignment in the Alignments ➤ Offset Alignments collection.
>
> **Curb Return**   Used for curb returns. The difference between this and the offset alignment is that instead of offset, you have the option in Alignment properties to set Curb Return parameters, such as setting two parent alignments and offsets. It places this type of alignment in the Alignments ➤ Curb Return Alignments collection.
>
> **Miscellaneous**   This is a stripped-down alignment type that only contains Information, Stationing, and Masking tabs. It places this alignment type in the Alignments ➤ Miscellaneous Alignments collection.

The General tab contains the following options:

Description   You can be verbose here to describe your alignment.

Starting Station   Setting this with a number, either positive or negative, will be the starting stationing for the alignment. This is handy if you need to start your alignment to coincide with existing stationing, or if you wish to have your 0+00 stationing at an intersection of a road.

Site   A place to keep Civil 3D objects that you want to interact with each other. As previously mentioned, all of our alignments are put on the <None> site.

Alignment Style   You can set different styles to visually show your alignment. For more on styles, refer to Chapter 19, "Styles."

Alignment Layer   Overrides the layer that is specified in Settings for alignments.

Alignment Label Set   As with the Alignment style, you can visually select your labels.

Conversion Options   Depending on your selections, these will add curves or erase the original entities.

The Design Criteria tab contains:

Starting Design Speed   Specify the design speed of the road by typing in a value or by using design criteria such as the American Association of State Highway and Transportation Officials (AASHTO)'s 2001 design manual. You can also set rules, or expressions for minimum radius, lengths, deflections, and spiral checks. The design constraints and check sets will be covered later in this chapter.

You've created your first alignment and attached stationing and geometry point labels. It is common to create offset alignments from a centerline alignment to begin to model rights of way. In the following exercise, you'll create offset alignments and mask them where you don't want them to be seen:

1. Change to the Home tab and choose Alignment ➢ Create Offset Alignments.
2. Pick the Syrah Way alignment to open the Create Offset Alignments dialog shown in Figure 6.6.

FIGURE 6.6
The Create Offset Alignments dialog

3. Change the Incremental offset on the left to 25′. Change the Incremental offset on the right to 25′. Click OK to accept the rest of the defaults, as shown in Figure 6.6.
4. Select the offset alignment just created along the northerly right of way of Syrah Way.
5. Choose Alignment Properties from the Modify panel to open the Alignment Properties dialog.

6. Change to the Masking tab and click the Add Masking Region button to open the dialog shown in Figure 6.7. Type **0+95.00′** for the first station and **1+95.00′** for the second station when prompted. Click OK. Notice that the alignment is now masked at the intersection of Syrah Way and Frontenac Drive at the east end.

FIGURE 6.7
Creating an alignment mask

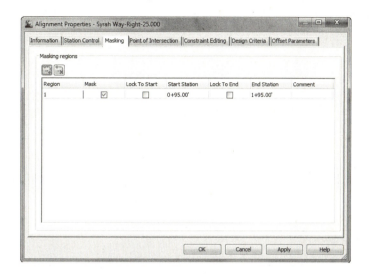

7. Repeat the process for the rest of the intersections, starting at the end of the arc on both right-of-way alignments on Syrah Way.

Note that when you selected the beginning and end of the offset alignment, the Lock To Start and Lock To End boxes were checked automatically.

Offset alignments are simple to create, and they are dynamically linked to a centerline alignment. To test this, grip the centerline alignment, select the endpoint grip, and stretch the alignment to the west. Notice the change.

> **OFFSET GRIPS AND MORE**
>
> Offset alignments have two special grips: the arrow and the plus sign. The arrow is used to change the offset value, and the plus sign is used to create a transition, called a widening, such as a turning lane. The Create Widening command can be found in the Alignment drop-down on the Create Design panel of the Home tab. Widening criteria can also be found in the Create Offset Alignments dialog.
>
> Even offset alignments with widening remain dynamic to their host alignment. Offset alignment objects can be found in Prospector in the Alignments collection.

You created an alignment from polylines and two offsets. It's ready for use in corridors, in profiling, or for any number of other uses.

Creating by Layout

Now that you've made an alignment from polylines, let's look at the other creation option: Create By Layout. You'll use the same street layout (Figure 6.4) that was provided by a planner, but instead of converting from polylines, you'll trace the alignments. Although this seems like duplicate work, it will pay dividends in the relationships created between segments:

1. Open the `AlignmentsbyLayout.dwg` file. You see the lines and arcs used to create the center of roads. The arc sections are the color green and the line sections are the color red. Everything else has been removed for clarity.

2. Change to the Home tab and choose Alignment ➢ Alignment Creation Tools from the Create Design panel. The Create Alignment – Layout dialog appears, as shown in Figure 6.8.

FIGURE 6.8
Create Alignment – Layout dialog

3. Change the Name field to **Cabernet Court** if it is not already set, and then click OK to accept the other settings as shown in Figure 6.8. The Alignment Layout Tools toolbar appears (Figure 6.9).

FIGURE 6.9
The Alignment Layout Tools toolbar

4. Click the down arrow next to the Draw Tangent-Tangent Without Curve tool at the far left, and select the Tangent-Tangent (With Curves) option (see Figure 6.10). The tool places a curve automatically; you'll adjust the curve, watching the tangents extend as needed.

FIGURE 6.10
The Tangent-Tangent (With Curves) tool

5. Using the circles as guides, pick the southernmost center of the circle of Cabernet Court using a center snap.

6. Continue to pick the center of the other three circles to finish creating this alignment. Press ↵ to end the command. Your drawing should look similar to Figure 6.11.

FIGURE 6.11
Completing the Cabernet Court alignment

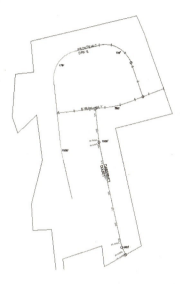

7. Click the red X button at the upper-right on the toolbar to close it.

Zoom in on the lower arc. Notice that it follows closely with the desired arc radius. Now zoom in on the upper arc, and notice that it doesn't match the arc the planner put in for you to follow. That's OK—you'll fix it in a few minutes. It bears repeating that in dealing with Civil 3D objects, it's good to get something in place and *then* refine. With Land Desktop or other packages, you didn't want to define the object until it was fully designed. In Civil 3D, you design and then refine.

The alignment you just made is one of the most basic. Let's move on to some of the others and use a few of the other tools to complete your initial layout. In this exercise, you build the alignment at the north end of the site, but this time you use a floating curve to make sure the two segments you create maintain their relationship:

1. Change to the Home tab and choose Alignment ➢ Alignment Creation Tools from the Create Design panel. The Create Alignment – Layout dialog appears.

2. In the Create Alignment – Layout dialog, do the following:
   ◆ Change the Name field to **Frontenac Drive**.
   ◆ Set the Alignment Style field to Layout.
   ◆ Set the Alignment Label Set field to Major And Minor Only.

3. Click OK, and the Alignment Layout Tools toolbar appears.

4. Select the Draw Fixed Line – Two Points tool.

5. Pick the two points circled in Figure 6.12, using Endpoint snaps and working south to north to draw the fixed line. When you've finished, the command line will state `Specify start point:`.

FIGURE 6.12
Pick the two points circled on Frontenac Drive.

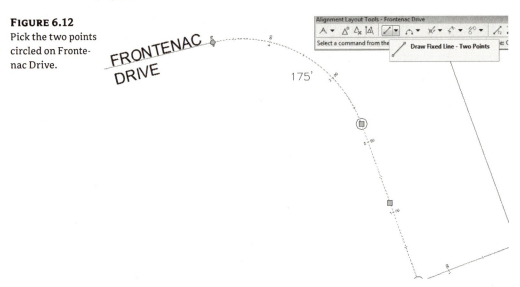

6. Click the down arrow next to the Add Fixed Curve (Three Points) tool on the toolbar, and select More Floating Curves ➢ Floating Curve (From Entity End, Through Point), as shown in Figure 6.13.

FIGURE 6.13
Selecting the Floating Curve tool

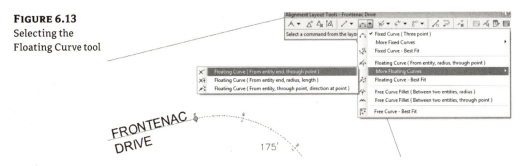

7. Ctrl+click the fixed-line segment you drew in steps 4 and 5. A blue rubber band should appear, indicating that the alignment of the curve segment is being floated off the endpoint of the fixed segment (see Figure 6.14). The Ctrl+click is required to make the pick activate the alignment segment and not the polyline entity.

FIGURE 6.14
Adding a floating curve to the Frontenac Drive alignment

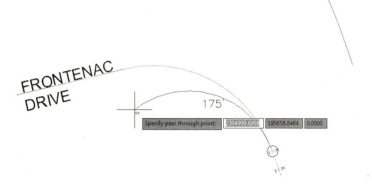

8. Pick the endpoint of the arc of the Frontenac Drive polyline arc segment.
9. Right-click or press ↵ to exit the command.
10. Close the toolbar to return to Civil 3D.

Pick the Frontenac Drive alignment and then pick the grip on the upper end of the line and pull it away from its location. Notice that the line and the arc move in sync, and tangency is maintained (see Figure 6.15).

FIGURE 6.15
Floating curves maintain their tangency.

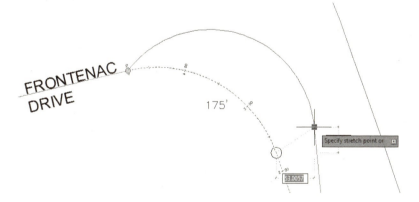

Best Fit Alignments from Lines and Curves

The Create Best Fit Alignment command can use AutoCAD blocks, entities, points, COGO points, or Feature lines. You can also simply click on the screen. The Line and Curve drop-down menus on the Alignment Layout toolbar include options for Floating and Fixed Lines by Best Fit, as well as Best Fit curves in all three flavors: Fixed, Float, and Free. These options are helpful when you're doing rehab work or other jobs where some form of the data already exists. It is similar to what we covered in Chapter 1 in the "Best Fit Entities" section. Let's see how it works with alignments:

1. Open the `AlignmentsBestFit.dwg` file, which you can download from this book's web page.

2. From the Home tab and Create Design panel, choose Alignment ➢ Alignment Creation Tools. The Create Alignment – Layout dialog appears.

3. Enter **Best Fit Lines** in the Name field. Leave the rest at their defaults and click OK. The Alignment Layout Tools – Best Fit Lines palette opens.

4. Click the down arrow next to the Draw Fixed Line – Two Points tool on the toolbar, and select the Fixed Line – Best Fit option, as shown in Figure 6.16. The Tangent By Best Fit dialog opens (Figure 6.17).

Figure 6.16
Selecting Fixed Line – Best Fit

Figure 6.17
The Tangent By Best Fit dialog

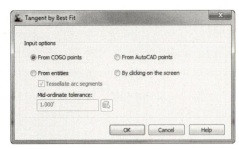

Here, you can choose various methods to create a best fit line alignment: From COGO points, From Entities, From AutoCAD points, or by clicking on the screen.

5. Pick the By Clicking On The Screen radio button and click OK.

6. With the running endpoint osnap, click on all the endpoints of the line. As you progress, you see a red dashed line being formed. In your selections, this line looks at all the endpoints selected in order to create the best fit line alignment (Figure 6.18). When you get to the last endpoint, press ↵ to open the Regression Data window.

Figure 6.18
The best fit line being formed

7. In the Regression Data window, you can choose to exclude endpoints, or force them to be a pass-through endpoint by checking the boxes. As you do, notice the changes that occur on your best fit line alignment (Figure 6.19).

FIGURE 6.19
Regression Data window

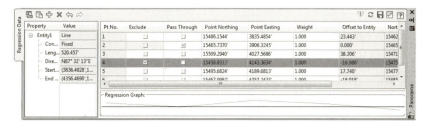

8. Click the green check mark on the upper right-hand side of the Regression Data window to accept and dismiss the window. The alignment is complete.

The best fit alignment by curve works the same way as the Best Fit Line option, as shown in Figure 6.20.

FIGURE 6.20
The best fit curve

Reverse Curve Creation

Next, let's look at a more complicated alignment construction—building a reverse curve connecting two curves:

1. Open the AlignmentReverse.dwg file.
2. Change to the Home tab and choose Alignment ➢ Alignment Creation Tools from the Create Design panel. The Create Alignment – Layout dialog appears.

3. In the Create Alignment – Layout dialog, do the following:
   - Change the Name field to **Reverse**.
   - Set the Alignment Style field to Layout.
   - Set the Alignment Label Set field to Major Minor And Geometry Points.
4. Click OK, and the Alignment Layout Tools toolbar appears.
5. Start by drawing a fixed line from the north end of the western portion to its endpoint using the same Draw Fixed Line (Two Points) tool as before.
6. Use the Floating Curve (From Entity, Radius, Through Point) tool to draw a curve from the end of this segment.
7. At the Specify radius <200>: prompt, enter **500**.
8. At the Is curve solution angle [Greaterthan180/Lessthan180] <Lessthan180>: prompt, press ↵.
9. Click on the other end of the arc.
10. The command continues. Select the arc.
11. The second arc radius is 400′ and it is less than 180.
12. The command now prompts: Is curve compound or reverse to curve before? [Compound/Reverse] <Compound>:. Type **R**↵ to specify that it is a reverse curve.
13. Click the endpoint of the lower arc and then continue to finish the alignment by selecting the two-point line (Figure 6.21).

FIGURE 6.21
Segment layout for the Reverse Curve alignment

The alignment now contains a perfect reverse curve. Move any of the pieces, and you'll see the other segments react to maintain the relationships shown in Figure 6.22. This flexibility in design isn't possible with the converted polylines you used previously. Additionally, the flexibility of the Civil 3D tools allows you to explore an alternative solution (the reverse curve) as opposed to the basic solution. Flexibility is one of Civil 3D's strengths.

FIGURE 6.22
Curve relationships during a grip-edit

You've completed your initial layout. There are some issues with curve sizes, and the reverse curve may not be acceptable to the designer, but you'll look at those changes later in the section "Component Level Editing."

Creating with Design Constraints and Check Sets

Starting in Civil 3D 2009, users have the ability to create and use design constraints and design check sets during the process of aligning and creating design profiles. Typically, these constraints check for things like curve radius, length of tangents, and so on. Design constraints use information from AASHTO or other design manuals to set curve requirements. Check sets allow users to create their own criteria to match local requirements, such as subdivision or county road design. First, you'll make one quick set of design checks:

1. Open the CreatingChecks.dwg file.

2. Change to the Settings tab in Toolspace, and expand the Alignment ➢ Design Checks menu branch.

3. Right-click the Line Folder, and select New to display the New Design Check dialog.

4. Change the name to **Subdivision Tangent**.

5. Click the Insert Property drop-down menu, and select Length.

6. Click the greater-than/equals symbol (>=), and then type **100** in the Expression field as shown. When complete, your dialog should look like Figure 6.23. Click OK to close the dialog.

7. Right-click the Curve folder, and create the Subdivision Curve design check, as shown in Figure 6.24.

FIGURE 6.23
The completed Subdivision Tangent design check. The result of a design check is true or false; in this case, it tells you whether the alignment line segment is equal to or longer than 100′.

FIGURE 6.24
The completed Subdivision Curve design check. The >= indicates that an acceptable curve is equal to or greater than 200′. Without the equal portion, a curve would require a radius of 200.01′ to pass the check.

8. Right-click the Design Check Sets folder, and select New to display the Alignment Design Check Set dialog.

9. Change the name to **Subdivision Streets**, and then switch to the Design Checks tab.

10. Click the Add button to add the Subdivision Tangent check to the set.

11. Choose Curve from the Type drop-down list, and click Add again to complete the set as shown in Figure 6.25.

FIGURE 6.25
The completed Subdivision Streets design check set

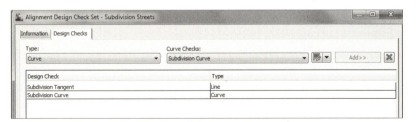

CREATING AN ALIGNMENT | 221

Once you've created a number of design checks and design check sets, you can apply them as needed during the design and layout stage of your projects.

> **DESIGN CHECKS VERSUS DESIGN CRITERIA**
>
> In typical fashion, the language used for this feature isn't clear. What's the difference? A design check uses basic properties such as radius, length, grade, and so on to check a particular portion of an alignment or profile. These constraints are generally dictated by a governing agency based on the type of road involved. Design criteria use speed and related values from design manuals such as AASHTO to establish these geometry constraints.

Think of having a suite of check sets, with different sets for each city and type of street or each county, or for design speed. In the next exercise, you'll see the results of your Subdivision Streets check set in action:

1. Open the `CheckingAlignments.dwg` file.

2. Change to the Home tab and choose Alignment ➢ Alignment Creation Tools from the Create Design tab. The Create Alignment – Layout dialog appears.

3. Click the Design Criteria tab, and set the design speed to **30 mi/h**, as shown in Figure 6.26. Set the check boxes as shown in the figure. (Note that the Use Criteria-Based Design check box must be selected to activate the other two.) Click OK to close the dialog.

FIGURE 6.26
Setting up design checks during the creation of alignments

4. Select the Tangent-Tangent With Curves option on the left side of the Alignment Layout toolbar. The curve radius on the northernmost part of Frontenac Drive is 175′ and is left in place to illustrate the design check failure indicators.

5. Connect the center points of the circles on the screen to create the alignment shown in Figure 6.27. Notice the exclamation-point symbol, which indicates that a design check has been violated.

Figure 6.27
Completed alignment layout

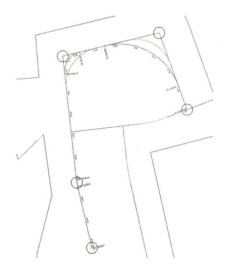

Now that you know how to create an alignment that doesn't pass the design checks, let's look at different ways of modifying alignment geometry. As you correct and fix alignments that violate the assigned design checks, those symbols will disappear.

Editing Alignment Geometry

The general power of Civil 3D lies in its flexibility. The documentation process is tied directly to the objects involved, so making edits to those objects doesn't create hours of work in updating the documentation. With alignments, there are three major ways to edit the object's horizontal geometry without modifying the underlying construction:

Graphical   Select the object, and use the various grips to move critical points. This method works well for realignment, but precise editing for things like a radius or direction can be difficult without construction elements.

Tabular   Use Panorama to view all the alignment segments and their properties; type in values to make changes. This approach works well for modifying lengths or radius values, but setting a tangent perpendicular to a screen element or placing a control point in a specific location is better done graphically.

Segment   Use the Alignment Layout Parameters dialog to view the properties of an individual piece of the alignment. This method makes it easy to modify one piece of an alignment that is complicated and that consists of numerous segments, whereas picking the correct field in a Panorama view can be difficult.

In addition to these methods, you can use the Alignment Layout Tools toolbar to make edits that involve removing components or adding to the underlying component count. The following sections look at the three simple edits and then explain how to remove components from and add them to an alignment without redefining it.

Grip-Editing

You already used graphical editing techniques when you created alignments from polylines, but those techniques can also be used with considerably more precision than shown previously. The alignment object has a number of grips that reveal important information about the elements' creation (see Figure 6.28).

Figure 6.28
Alignment grips

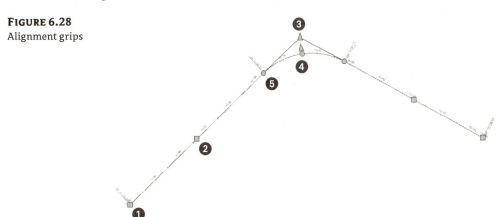

You can use the grips in Figure 6.28 to do the following actions:

◆ The square grip at the beginning of the alignment, Grip 1, indicates a segment point that can be moved at will. This grip doesn't attach to any other components.

◆ The square grip in the middle of the tangents, Grip 2, allows the element to be translated. Other components attempt to hold their respective relationships, but moving the grip to a location that would break the alignment isn't allowed.

◆ The triangular grip at the intersection of tangents, Grip 3, indicates a Point of Intersection (PI) relationship. The curve shown is a function of these two tangents and is free to move on the basis of incoming and outgoing tangents, while still holding a radius.

◆ The triangular grip near the middle of the curve, Grip 4, lets the user modify the radius directly. The tangents must be maintained, so any selection that would break the alignment geometry isn't allowed.

◆ Circular grips on the end of the curve, Grip 5, allow the radius of the curve to be indirectly changed by changing the point of the Point of Curvature (PC) of the alignment. You make this change by changing the curve length, which in effect changes the radius.

CERT OBJECTIVE

In the following exercise, you'll use grip-edits to make one of your alignments match the planner's intent more closely:

1. Open the `EditingAlignments.dwg` file.

2. Expand the Alignments branch in Prospector, right-click Frontenac Drive, and select Zoom To.

3. Zoom in on one of the curves in the middle of the alignment. This curve was inserted using the default settings and doesn't match the guiding polyline well.

4. Select the alignment to activate the grips.

5. Select the triangular grip that appears near the PI, as shown in Figure 6.29, and use your scroll wheel to zoom out.

FIGURE 6.29
Grip-editing the Frontenac Drive curve

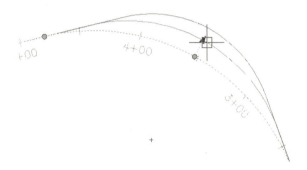

6. Select the circular grip shown in Figure 6.29, and use a Nearest snap to place it on the magenta polyline. Doing so changes the radius without changing the PI.

Your alignment now follows the planned layout. With no knowledge of the curve properties or other driving information, you've quickly reproduced the design's intent.

Tabular Design

When you're designing on the basis of governing requirements, one of the most important elements is meeting curve radius requirements. It's easy to work along an alignment in a tabular view, verifying that the design meets the criteria. In this exercise, you'll verify that your curves are suitable for the design:

1. If necessary, open the `EditingAlignments.dwg` file.

2. Zoom to the Frontenac Drive alignment, and select it in the drawing window to activate the grips.

3. Select Geometry Editor from the Modify panel. The Alignment Layout Tools toolbar opens.

4. Select the Alignment Grid View tool.

5. Panorama appears, with all the elements of the alignment listed along the left. You can use the scroll bar along the bottom to review the properties of the alignment if necessary. Note that the columns can be resized as well as toggled off by right-clicking the column headers. The segment selected in the alignment grid view is also highlighted in the model, which can also make identifying the segment easier.

> **CREATING AND SAVING CUSTOM PANORAMA VIEWS**
>
> If you right-click a column heading and select Customize near the bottom of the menu, you're presented with a Customize Columns dialog. This dialog allows you to set up any number of column views, such as Road Design or Stakeout, that show different columns. These views can be saved, allowing you to switch between views easily. This feature is a great change from previous versions where the column view changes weren't held or saved between viewings.

6. Notice that you can click in the Radius field to change it. You will learn more about constraints later this chapter.

7. Click the check box to dismiss Panorama, and then close the toolbar.

Panorama allows for quick and easy review of designs and for precise data entry, if required. Grip-editing is commonly used to place the line and curve of an alignment in an approximate working location, but then you use the tabular view in Panorama to make the values more reasonable—for example, to change a radius of 292.56 to 300.00.

Component-Level Editing

Once an alignment gets more complicated, the tabular view in Panorama can be hard to navigate, and deciphering which element is which can be difficult. In this case, reviewing individual elements by picking them on screen can be easier:

1. Continue with the EditingAlignments.dwg file.

2. Zoom to Frontenac Drive, and select it to activate the grips.

3. Select Geometry Editor from the Modify panel. The Alignment Layout Tools toolbar appears.

4. Select the Sub-Entity Editor tool to open the Alignment Layout Parameters dialog.

5. Select the Pick Sub-Entity tool on the Alignment Layout Tools toolbar.

6. Pick the first line on the northeast corner of the site to display its properties in the Alignment Layout Parameters dialog (see Figure 6.30). Since this alignment was created

from objects, all line segments are not constrained—that is, they do not hold any relationship between the other segments that make up the alignment.

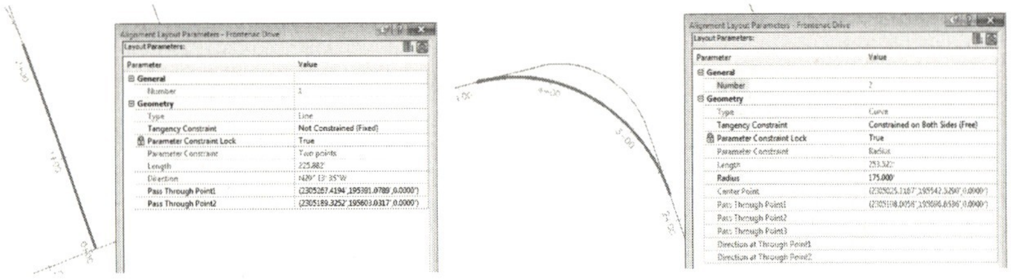

FIGURE 6.30
The Alignment Layout Parameters dialogs for the first line (on the left) and the first curve (on the right) on Frontenac Drive

7. Zoom in, and pick the first curve. Notice that the Radius Tangency now reports that it is constrained on both sides (Free).

8. Change the value in the Radius field to **300**, and watch the screen update. This value is too far from the original design intent to be a valid alternative.

9. Change the value in the Radius field to **100**, and again watch the update. This value is closer to the design and is acceptable.

10. Close the Alignment Layout Parameters dialog and the Alignment Layout Tools toolbar.

By using the Alignment Layout Parameters dialog, you can concisely review all the individual parameters of a component. In each of the editing methods discussed so far, you've modified the elements that were already in place. We will look at how to change the makeup of the alignment itself, not just the values driving it. But now, let's look at some of the constraints and understand how they work.

Understanding Alignment Constraints

In the previous exercise you were exposed to constraints. The various constraints will help keep geometry together to maintain tangency or to maintain the radius.

When an alignment is created from objects, the lines are not constrained (Fixed). This is the same if you select the Draw Two Lines – Fixed from the Alignment Layout Tools toolbar. The curves are all constrained on both sides (Free). When you grip-edit a line or arc, the lines maintain tangency to the arc, but the arc loses its original radius (Figure 6.31).

You may have noticed in Panorama the Tangency Constraint field. What you may not have noticed is that you can click on any segment and change constraints (Figure 6.32). You can also change the constraints in the Sub-entity Editor.

1. Select the Frontenac Drive alignment. From the Modify panel, select Geometry Editor.

2. In the Alignment Layout Tools toolbar, select the Alignment Grid View.

3. Click in the Tangency Constraint field for the first curve and change to Not Constrained (Fixed).

4. Now grip-edit the curve and notice how it does not maintain any tangency or radius (Figure 6.33).

FIGURE 6.31
Gripping on an alignment with curves constrained on both sides

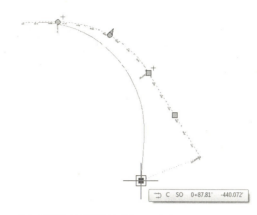

FIGURE 6.32
The tangency constraints in Panorama

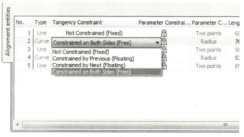

FIGURE 6.33
Gripping on an alignment with curves using Not Constrained

5. Click in the Tangency Constraint field for the first curve and change to Constrained By Previous (Floating).

6. Grip-edit the curve again and notice that the curve maintains its tangency with the previous line, but does not for the following line (Figure 6.34).

7. Click in the Tangency Constraint field for the first curve and change to Constrained By Next (Floating).

8. Grip-edit the curve. See that the curve now maintains its tangency with the following line, but not the previous line (Figure 6.35).

Figure 6.34
Gripping on an alignment with curves using Constrained By Previous

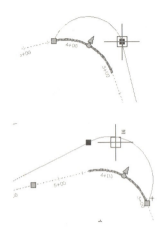

Figure 6.35
Gripping on an alignment with curves using Constrained By Next

When you click on an alignment that contains curves and select alignment properties, some additional options become available. In the Point Of Intersection tab, you can select whether you want to visually show points of intersection by a change in alignment direction or by individual curves and curve group. The default is to not display any implied points of intersection (Figure 6.36).

Figure 6.36
The Point Of Intersection tab

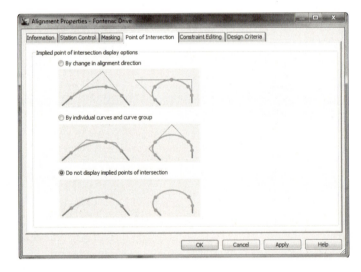

In the Constraint Editing tab, you can select if you wish to always perform any implied tangency constraint swapping and whether to lock all parameter constraints (Figure 6.37).

Figure 6.37
The Constraint Editing tab

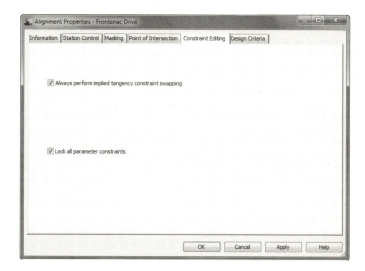

Changing Alignment Components

One of the most common changes is adding a curve where there was none before or changing the makeup of the curves and tangents already in place in an alignment. Other design changes can include swapping out curves for tangents or adding a second curve to smooth a transition area.

 Real World Scenario

Sometimes the Planner Is Right

It turns out that your perfect reverse curve isn't allowed by the current ordinances for subdivision design! In this example, you'll go back to the design the planner gave you, and place a tangent between the curves:

1. Open the `AlignmentReverseEdit.dwg` file.
2. Select the ReverseCurve alignment to activate the grips.
3. Select Geometry Editor from the Modify panel. The Alignment Layout Tools toolbar appears.
4. Select the Delete Sub-Entity tool.
5. Pick the two curves to remove them. Note that the last tangent is still part of the alignment—it just isn't connected.
6. Select the Free Curve Fillet (Between Two Entities, Through Point) option.
7. Ctrl+click the line on the south, and then using the midpoint osnap, select the midpoint of the very small line that originally connected the two arcs.

8. Repeat the Free Curve Fillet (Between Two Entities, Through Point) option. Select the arc just drawn, and then select the upper line endpoint.
9. Use the Delete Sub-Entity tool to delete the small line. Your reverse curve is complete.

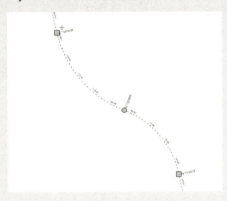

So far in this chapter, you've created and modified the horizontal alignments, adjusted them on screen to look like what your planner delivered, and tweaked the design using a number of different methods. Now let's look beyond the lines and arcs and get into the design properties of the alignment.

Alignments as Objects

Beyond the simple nature of lines and arcs, alignments represent other things such as highways, streams, sidewalks, or even flight patterns. All these items have properties that help define them, and many of these properties can also be part of your alignments. In addition to obvious properties like names and descriptions, you can include functionality such as superelevation, station equations, reference points, and station control. This section will look at other properties that can be associated with an alignment and how to edit them.

Renaming Objects

The default naming convention for alignments is flexible (and configurable) but not descriptive. In previous sections, you ignored the descriptions and left the default names in place, but now let's modify them. In addition, you'll learn the easy way to change the object style and how to add a description.

Most of an alignment's basic properties can be modified in Prospector. In this exercise, you'll change the name in a couple of ways:

1. Open the AlignmentProperties.dwg file, and make sure Prospector is open.
2. Expand the Alignments/Centerline Alignments collection, and note that Alignment – (1) through Alignment – (3) are listed as members.
3. Click the Centerline Alignments branch, and the individual alignments appear in a preview area (see Figure 6.38).

FIGURE 6.38
The Alignments collection listed in the preview area of Prospector

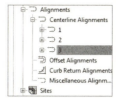

4. Down in the grid view section, click in the Name field for Alignment – (3), and pause briefly before clicking again. The text highlights for editing.

5. Change the name to **Cabernet Court**, and press ↵. The field updates, as does Prospector.

6. Click in the Description field, and enter a description. Press ↵.

7. Click the Style field, and the Select Label Style dialog appears. Select Layout from the drop-down menu, and click OK to dismiss. The screen updates.

That's one method. The next is to use the AutoCAD Object Properties Manager (OPM) palette:

1. Open the OPM palette by using the Ctrl+1 keyboard shortcut or some other method.

2. Select Alignment – (1) in the drawing (the horizontal alignment). The OPM looks like Figure 6.39.

3. Click in the Name field, and change the name to **Syrah Way**.

4. Click in the Description field, and enter a description. Click OK.

5. Press Esc on your keyboard to deselect all objects, and close the OPM dialog if you'd like.

FIGURE 6.39
Alignment – (1) in the AutoCAD Object Properties Manager palette

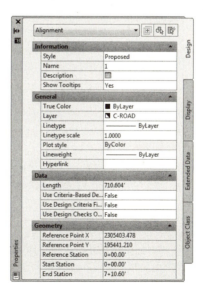

The final method involves getting into the Alignment Properties dialog, your access point to information beyond the basics:

1. In the main Prospector window, right-click Alignment – (2), and select Properties. This is the alignment at the top. The Alignment Properties dialog for Alignment – (2) opens.
2. Change to the Information tab if it isn't selected.
3. Change the name to **Frontenac Drive**, and enter a description in the Description field.
4. Set Object Style to Existing.
5. Click Apply. Notice that the dialog header updates immediately, as does the display style in the drawing.
6. Click OK to exit the dialog.

Now that you've updated your alignments, let's make them all the same style for ease of viewing. The best way to do this is in the Prospector preview window:

1. Pick the Alignments ➤ Centerline Alignments branch, and highlight one of the alignments in the preview area.
2. Press Ctrl+A to select them all, or pick the top and then Shift+click the bottom item. The idea is to pick *all* of the alignments.
3. Right-click the Style column header and select Edit (see Figure 6.40).

FIGURE 6.40
Editing alignment styles en masse via Prospector

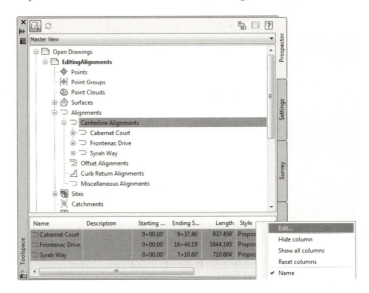

4. Select Layout from the drop-down list in the Select Label Style dialog that appears, and click OK. Notice that all alignments pick up this style. Although the dialog is named Edit Label Style, you are editing the object style.

> **DON'T FORGET THIS TECHNIQUE**
>
> This technique works on every object that displays in the List Style preview: parcels, pipes, corridors, assemblies, and so on. It can be painfully tedious to change a large number of objects from one style to another using any other method.

The alignments now look the same, and they all have a name and description. Let's look beyond these basics at the other properties you can modify and update.

The Right Station

At the end of the process, every alignment has stationing applied to help locate design information. This stationing often starts at zero, but it can also tie to an existing object and may start at some arbitrary value. Stationing can also be fixed in both directions, requiring station equations that help translate between two disparate points that are the basis for the stationing in the drawing.

One common problem is an alignment that was drawn in the wrong direction. Thankfully, Civil 3D has a quick edit command to fix that:

1. Open the `AlignmentProperties.dwg` file, and make sure Prospector is open.
2. Pick Syrah Way and choose Reverse Direction from the Modify drop-down panel.
3. A warning message appears, reminding you of the consequences of such a change. Click OK to dismiss it.
4. The stationing reverses, with 0+00 now at the west end of the street.

This technique allows you to reverse an alignment almost instantly. The warning that appears is critical, though! When an alignment is reversed, the information that was derived from its original direction may not translate correctly, if at all. One prime example of this is design profiles: they don't reverse themselves when the alignment is reversed, and this can lead to serious design issues if you aren't paying attention.

Beyond reversing, it's common for alignments to not start with zero. For example, the Cabernet Court alignment may be a continuation of an existing street, and it makes sense to make the starting station for this alignment the end station from the existing street. In this exercise, you'll set the beginning station:

1. Select the Cabernet Court alignment.
2. Select Alignment Properties from the Modify panel.
3. Switch to the Station Control tab. This tab controls the base stationing and lets you create station equations.
4. Enter **315.62** in the Station field in the Reference Point section (see Figure 6.41), and click Apply.
5. Dismiss the warning message that appears, and click Apply again. The Station Information section in the top right updates. These options can't be edited but provide a convenient way for you to review the alignment's length and station values.

FIGURE 6.41
Setting a new starting station on the Cabernet Court alignment

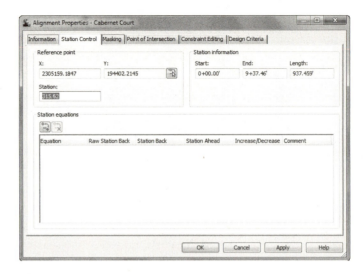

In addition to changing the value for the start of the alignment, you could use the Pick Reference Point button, as shown earlier in Figure 6.41, to select another point as the stationing reference point.

Station equations can occur multiple times along an alignment. They typically come into play when plans must match existing conditions or when the stationing has to match other plans, but the lengths in the new alignment would make that impossible without some translation. In this exercise, you'll add a station equation about halfway down Cabernet Court for illustrative purposes:

1. On the Station Control tab of the Alignment Properties dialog, click the Add Station Equations button.

2. Change the Station value to **725**. (Again, you're going for illustration, not reality!)

3. Click Apply, and notice the change in the Station Information section (see Figure 6.42).

FIGURE 6.42
Cabernet Court station equation in place

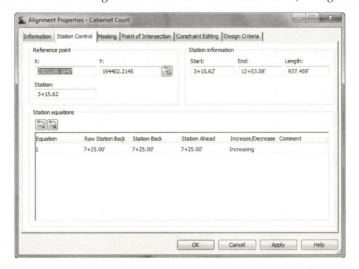

Click OK to close the dialog, and review the stationing that has been applied to the alignment.

Stationing constantly changes as alignments are modified during the initial stages of a development or as late design changes are pushed back into the plans. With the flexibility shown here, you can reduce the time you spend dealing with minor changes that seem to ripple across an entire plan set.

Assigning Design Speeds

One driving part of transportation design is the design speed. Civil 3D considers the design speed a property of the alignment, which can be used in labels or calculations as needed. In this exercise, you'll add a series of design speeds to Frontenac Drive. Later in the chapter, you'll label these sections of the road:

1. Bring up the Alignment Properties dialog for Frontenac Drive to the north using any of the methods discussed.

2. Switch to the Design Criteria tab.

3. Click in the Design Speed field for Number 1, and enter **30**. This speed is typical for a subdivision street.

4. Click the Add Design Speed button.

5. Click in the Start Station field for Number 2. A small Pick On Screen button appears to the right of the Start Station value, as shown in Figure 6.43.

Figure 6.43
Setting the design speed for a Start Station field

6. Click the Pick On Screen button, and then use a snap to pick the PC on the southwest portion of the site, near station 2 + 25.

7. Enter a value of **20** in the Design Speed field for Number 2.

8. Click the Pick On Screen button again to add one more design speed portion, and either pick or enter **9+00** to select the end of the tangent.

9. Enter a value of **30** for this design speed. When complete, the tab should look like Figure 6.44.

In a subdivision, these values can be inserted for labeling purposes. In a highway design, they can be used to drive the superelevation calculations that are critical to a working design. Chapter 11, "Superelevations," looks at this subject.

Figure 6.44
The design speeds assigned to Frontenac Drive

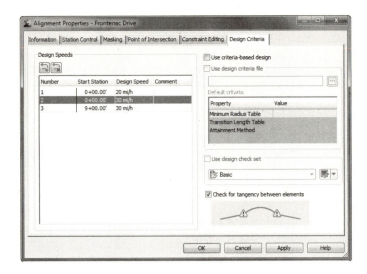

Labeling Alignments

Labeling in Civil 3D is one of the program's strengths, but it's also an easy place to get lost. There are myriad options for every type of labeling situation under the sun, and keeping them straight can be difficult. In this section, you'll begin by building label styles for stationing along an alignment, culminating in a label set. Then, you'll create styles for station and offset labels, using reference text to describe alignment intersections. Finally, you'll add a street-name label that makes it easier to keep track of things.

The Power of Label Sets

When you think about it, any number of items can be labeled on an alignment, before getting into any of the adjoining objects. These include major and minor stations, geometry points, design speeds, and profile information. Each of these objects can have its own style. Keeping track of all these individual labeling styles and options would be burdensome and uniformity would be difficult, so Civil 3D features the concept of *label sets*.

A label set lets you build up the labeling options for an alignment, picking styles for the labels of interest, or even multiple labels on a point of interest, and then save them as a set. These sets are available during the creation and labeling process, making the application of individual labels less tedious. Out of the box, a number of sets are available, primarily designed for combinations of major and minor station styles along with geometry information.

You'll learn how to create individual label styles in Chapter 19. We will use the out-of-the-box styles over the next couple of exercises and then pull them together with a label set. At the end of this section, you'll apply your new label set to the alignments.

Major Station

Major station labels typically include a tick mark and a station callout.

Geometry Points

Geometry points reflect the PC, PT, and other points along the alignment that define the geometric properties. The existing label style doesn't reflect a plan-readable format, so you'll copy it and make a minor change in this exercise.

In this exercise, you'll apply a label set to all of your alignments and then see how an individual label can be changed from the set:

1. Open the `AlignmentLabels.dwg` file.

2. Select the Cabernet Court alignment on the screen. Cabernet Court is the southern alignment.

3. Right-click, and select Edit Alignment Labels to display the Alignment Labels dialog shown in Figure 6.45.

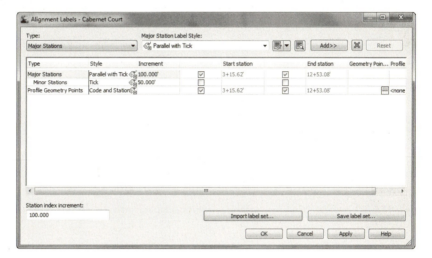

FIGURE 6.45
The Alignment Labels dialog for Cabernet Court

4. Click the Import Label Set button near the bottom of this dialog.

5. In the Select Style Set drop-down list, select the Paving Label Set and click OK.

6. The Style field for the alignment labels populates with the option you selected.

7. Click OK to dismiss the dialog.

8. Repeat this process across the rest of the alignments.

9. When you've finished, zoom in on any of the major station labels.

10. Hold down the Ctrl key, and select the label. Notice that a single label is selected, not the label set group.

11. Right-click and select Label Properties.

12. The Label Properties dialog appears; click another label style from the drop-down list.
13. Change the Major Station Label Style value to Parallel with Tick, and change the Flip Label value to True.
14. Press Esc to deselect the label item and exit this dialog.

By using alignment label sets, you'll find it easy to standardize the appearance of labeling and stationing across alignments. Building label sets can take some time, but it's one of the easiest, most effective ways to enforce standards.

> **CTRL+CLICK? WHAT IS THAT ABOUT?**
>
> Prior to AutoCAD Civil 3D 2008, clicking an individual label picked the label and the alignment. Because labels are part of a label set object now, Ctrl+click is the *only* way to access the Flip Label and Reverse Label functions!

Station Offset Labeling

Beyond labeling an alignment's basic stationing and geometry points, you may want to label points of interest in reference to the alignment. Station offset labeling is designed to do just that. In addition to labeling the alignment's properties, you can include references to other object types in your station offset labels. The objects available for referencing are as follows:

- Alignments
- COGO points
- Parcels
- Profiles
- Surfaces

In this exercise, you'll use an alignment reference to create a label suitable for labeling the intersection of two alignments. It will pick up the stationing information from both:

1. Go to the Annotate tab of the Ribbon. From the Add Labels drop-down, select Alignment ➢ Add Alignment Labels. The Add Labels dialog appears.
2. In the Label Type drop-down list, select Station Offset.
3. In the Station Offset Label Style drop-down, select Alignment Intersection.
4. Leave the Marker Style field alone, but remember that you could use any of these styles to mark the selected point.
5. Click the Add button.
6. Pick the Syrah Way alignment.

ALIGNMENTS AS OBJECTS | 239

7. Snap to the intersection of Syrah Way and Frontenac Drive. It is the one on the right.

8. Enter **0** for the offset amount, and press ↵.

9. The command line prompts you to `Select Alignment for Label Style Component Intersecting Alignment`. Pick the Frontenac Drive alignment.

10. Click the Add button again, and repeat the process at the other two alignment intersections.

There are two things to note in this process: first, you click Add between adding labels because Civil 3D otherwise assumes you want to use the same reference object for every instance of the label; second, the labels are sitting right on the point of interest. Drag them to a convenient location, and you're set to go. When you do this, your label should look something like the one shown in Figure 6.46.

FIGURE 6.46
The Alignment Intersection label style in use

Using station offset labels and their reference object ability, you can label most site plans quickly with information that dynamically updates. Because of the flexibility of labels in terms of style, you can create "design labels" that are used to aid in modeling yet never plot and aren't seen in the final deliverables. This is ideal in corridor modeling, as discussed in Chapter 10, "Advanced Corridors, Intersections, and Roundabouts."

SEGMENT LABELING

Every land development professional has a story about the developer who named an entire subdivision after their kids, grandkids, dogs, golf buddies, favorite bars, and so on. As these plans work their way through reviewing agencies, there are inevitably changes, and the tedium of changing a street name on 45 pages of construction documents can't be described.

Thankfully, as you've already seen in the station offset label, you can access the properties of the alignment to generate a label. In this exercise, you'll use that same set of properties to create street-name labels that are applied and always up-to-date:

1. Open the `AlignmentSegments.dwg` file.

2. Change to the Annotate tab and choose the Add Labels button on the Labels & Tables panel to display the Add Labels dialog shown in Figure 6.47.

FIGURE 6.47
The Add Labels dialog

3. In the Label Type field, select Single Segment from the drop-down. Then, in the Line Label Style field, select the Street Names style. Note that curves and spirals have their own set of styles. If you click a curve during the labeling process, you won't get a street name—you'll get something else.

4. Click the Add button.

5. Pick various line segments around the drawing. Each street is labeled with the appropriate name. You may need to flip the label so that it does not interfere with the station labels.

6. Click the Close button to close the Add Labels dialog.

Days of work averted! The object properties of an alignment can be invaluable in documenting your design. Creating a collection of styles for all the various components and types takes time but pays you back in hours of work saved on every job.

Alignment Tables

There isn't always room to label alignment objects directly on top of them. Sometimes doing so doesn't make sense, or a reviewing agency wants to see a table showing the radius of every curve in the design. Documentation requirements are endlessly amazing in their disparity and seeming randomness. Beyond labels that can be applied directly to alignment objects, you can also create tables to meet your requirements and get plans out the door.

You can create four types of tables:

- Lines
- Curves
- Spirals
- Segments

Each of these is self-explanatory except perhaps the Segments table. That table generates a mix of all the lines, curves, and spirals that make up an alignment, essentially re-creating the alignment in a tabular format. In this section, you'll generate a new line table and draw the segment table that ships with the product.

All the tables work in a similar fashion. Go to the Annotate tab of the Ribbon. From the Add Tables drop-down, select Alignment, and then pick a table type that is relevant to your work. The Table Creation dialog appears (see Figure 6.48).

FIGURE 6.48
The Table Creation dialog

You can select a table style from the drop-down list or create a new one. Select a table layer by clicking the blue arrow. The selection area determines how the table is populated. All the label-style names for the selected type of component are presented, with a check box to the right of each one. Applying one of these styles enables the Selection Rule, which has the following two options:

Add Existing Any label using this style that currently exists in the drawing is converted to a tag format, substituting a key number such as L1 or C27, and added to the table. Any labels using this style created in the future will *not* be added to the table.

Add Existing and New Any label using this style that currently exists in the drawing is converted to a tag format and added to the table. In addition, any labels using this style created in the future will be added to the table.

To the right of the Select section is the Split Table section, which determines how the table is stacked up in model space once it's populated. You can modify these values after a table is generated, so it's often easier to leave them alone during the creation process.

Finally, the Behavior section provides two selections for the Reactivity Mode: Static and Dynamic. These selections determine how the table reacts to changes in the driving geometry. In some cases in surveying, this disconnect is used as a safeguard to the platted data, but in general, the point of a 3D model is to have live labels that dynamically react to changes in the object.

Before you draw any tables, you need to apply labels so the tables will have data to populate. In this exercise, you'll throw some labels on your alignments, and then you'll move on to drawing tables in the next sections:

1. Open the `AlignmentSegments.dwg` file.

2. Change to the Annotate tab and choose Add Labels from the Labels & Tables panel to open the Add Labels dialog.

3. In the Feature field, select Alignment, and then in the Label Type field, select Multiple Segment from the drop-down list. With this option, you'll click each alignment one time, and every subcomponent will be labeled with the style selected here.

4. Verify that the Line Label Style field (not the General Line Label Style) is set to Bearing Over Distance. You won't be left with these labels—you just want them for selecting elements later.

5. Click Add, and select all three alignments.

6. Click Close to close the Add Labels dialog.

Now that you've got labels to play with, let's build some tables.

CREATING A LINE TABLE

Most line tables are simple: a line tag, a bearing, and a distance. You'll also see how Civil 3D can translate units without having to change anything at the drawing level:

1. Choose the Add Tables drop-down arrow from the Labels & Tables panel, and choose Alignment ➢ Add Line to open the Table Creation dialog.

2. Set the dialog options as shown in Figure 6.49, and click OK.

FIGURE 6.49
Creating a line table

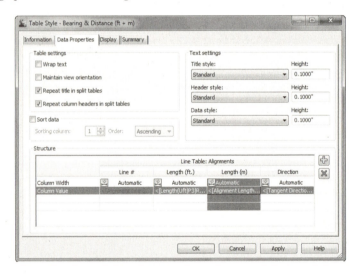

3. Pick a point on screen, and the table will generate.

Pan back to your drawing, and you'll notice that the line labels have turned into tags on the line segments. After you've made one table, the rest are similar. Be patient as you create tables—a lot of values must be tweaked to make them look just right. By drawing one on screen and then editing the style, you can quickly achieve the results you're after.

An Alignment Segment Table

An individual segment table allows a reviewer to see all the components of an alignment. In this exercise, you'll draw the segment table for Frontenac Drive:

1. Choose the Add Tables drop-down arrow from the Labels & Tables panel, and choose Alignment ≻ Add Segment to open the Table Creation dialog.

2. In the Select Alignment field, choose the Frontenac Drive alignment from the drop-down list, and click OK.

3. Pick a point on the screen, and the table will be drawn.

The Bottom Line

Create an alignment from a polyline. Creating alignments based on polylines is a traditional method of building engineering models. With Civil 3D's built-in tools for conversion, correction, and alignment reversal, it's easy to use the linework prepared by others to start your design model. These alignments lack the intelligence of crafted alignments, however, and you should use them sparingly.

Master It Open the `MasteringAlignments-Objects.dwg` file, and create alignments from the linework found there.

Create a reverse curve that never loses tangency. Using the alignment layout tools, you can build intelligence into the objects you design. One of the most common errors introduced to engineering designs is curves and lines that aren't tangent, requiring expensive revisions and resubmittals. The free, floating, and fixed components can make smart alignments in a large number of combinations available to solve almost any design problem.

Master It Open the `MasteringAlignments-Reverse.dwg` file, and create an alignment from the linework on the right. Create a reverse curve with both radii equal to 200 and with a pass-through point at the intersection of the two arcs.

Replace a component of an alignment with another component type. One of the goals in using a dynamic modeling solution is to find better solutions, not just a solution. In the layout of alignments, this can mean changing components out along the design path, or changing the way they're defined. Civil 3D's ability to modify alignments' geometric construction without destroying the object or forcing a new definition lets you experiment without destroying the data already based on an alignment.

Master It Convert the reverse curve indicated in the `MasteringAlignments-Rcurve.dwg` file to a floating arc that is constrained by the following segment. Then change the radius of the curves to 150′.

Create alignment tables. Sometimes there is just too much information that is displayed on a drawing, and to make it clearer, tables are used to show bearings and distances for lines, curves, and segments. With their dynamic nature, these tables are kept up-to-date with any changes.

Master It From the `Mastering Alignments-Table.dwg`, make a line table, curve table, and segment table. Use whichever style you want to accomplish this.

Chapter 7

Profiles and Profile Views

Profile information is the backbone of vertical design. Civil 3D takes advantage of sampled data, design data, and external input files to create profiles for a number of uses. Even the most basic designs require profiles. In this chapter we'll look at creation tools, editing profiles, and display styles, and you'll learn about ways to get your labels just so. Profile views are a different subject and will be covered in more detail in the next chapter.

In this chapter, you will learn to:

- Sample a surface profile with offset samples
- Lay out a design profile on the basis of a table of data
- Add and modify individual components in a design profile
- Apply a standard label set

The Elevation Element

The whole point of a three-dimensional model is to include the elevation element that's been missing for years. But to get there, designers and engineers still depend on a flat 2D representation of the vertical dimension as shown in a profile view (see Figure 7.1).

FIGURE 7.1
A typical profile view of the surface elevation along an alignment

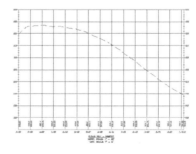

A profile is nothing more than a series of data pairs in a station, elevation format. There are basic curve and tangent components, but these are purely the mathematical basis for the paired data sets. In Civil 3D, you can generate profile information in one of the following three ways:

- Sampling from a surface involves taking vertical information from a surface object every time the sampled alignment crosses a TIN line of the surface.

- Using a layout to create a profile allows you to input design information, setting critical station and elevation points, calculating curves to connect linear segments, and typically working within requirements laid out by a reviewing agency.

- Creating from a file lets you point to a specially formatted text file to pull in the station and elevation pairs. Doing so can be helpful in dealing with other analysis packages or spreadsheet tabular data.

This section looks at all three methods of creating profiles.

Surface Sampling

Working with surface information is the most elemental method of creating a profile. This information can represent a simple existing or proposed surface, a river flood elevation, or any number of other surface-derived data sets. Within Civil 3D, surfaces can also be sampled at offsets, as you'll see in the next series of exercises. Follow these steps:

1. Open the `ProfileSampling.dwg` file shown in Figure 7.2. (Remember, all data files can be downloaded from this book's web page at www.sybex.com/go/masteringcivil3d2012.)

Figure 7.2
The drawing you'll use for this exercise

2. Select Create Surface Profile from the Profile drop-down on the Create Design panel to open the Create Profile From Surface dialog, as shown in Figure 7.3.

FIGURE 7.3
The Create Profile From Surface dialog

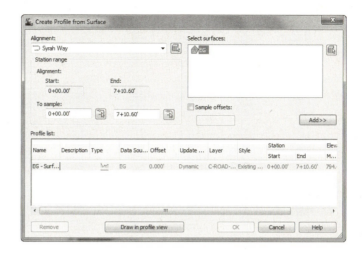

This dialog has a number of important features, so take a moment to see how it breaks down:

- The upper-left quadrant is dedicated to information about the alignment. You can select the alignment from a drop-down list, or you can click the Pick On Screen icon. The Station Range area is automatically set to run from the beginning to the end of the alignment, but you can control it manually by entering the station ranges in the To Sample text boxes or by using the station pick icons.

- The upper-right quadrant controls the selection of the surface and the offsets. You can select a surface from the list, or you can click the Pick On Screen icon. Beneath the Select Surfaces list box is a Sample Offsets check box. The offsets aren't applied in the left and right direction uniformly. You must enter a negative value to sample to the left of the alignment or a positive value to sample to the right. In all cases, the profile isn't generated until you click the Add button. You can add multiple offsets here—for example: –50, 25,–10,–5,0,5,10,25,50

- The Profile List box displays all profiles associated with the alignment currently selected in the Alignment drop-down menu. This area is generally static (it won't change), but you can modify the Update Mode, Layer, and Style columns by clicking the appropriate cells in this table. You can stretch and rearrange the columns to customize the view.

3. Select the Syrah Way option from the Alignment drop-down menu if it isn't already selected.

4. In the Select Surfaces list area, select EG.

5. Select the Sample Offsets check box to make the entry box active, and enter **–25,25**.

6. Click Add again. The dialog should look like Figure 7.4.

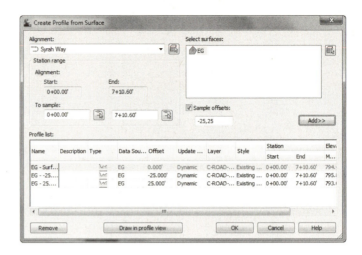

Figure 7.4
The Create Profile From Surface dialog showing the profiles sampled on surface EG

7. In the Profile List area, select the cell in the Style column that corresponds to the –25.000' value in the Offset column (see Figure 7.4) to activate the Pick Profile Style dialog.

8. Select the Left Sample Profile option, and click OK. The style changes from Existing Ground Profile to Left Sample Profile in the table.

9. Select the cell in the Style column that corresponds to the 25.000' value in the Offset column.

10. Select the Right Sample Profile option, and click OK. The dialog should look like Figure 7.5.

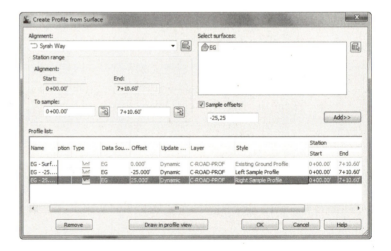

Figure 7.5
The Create Profile From Surface dialog with styles assigned on the basis of the Offset value

11. Click OK to dismiss this dialog.
12. The Events tab in Panorama appears, telling you that you've sampled data or if an error in the sampling needs to be fixed. Click the green check mark or the X to dismiss Panorama.

Profiles are dependent on the alignment they're derived from, so they're stored as profile branches under their parent alignment on the Prospector tab, as shown in Figure 7.6.

FIGURE 7.6
Alignment profiles on the Prospector tab

> **AREN'T YOU GOING TO DRAW THE PROFILE VIEW?**
>
> Under normal conditions, you would click the Draw In Profile View button to go through the process of creating the grid, labels, and other components that are part of a completed profile view. You'll skip that step for now because you're focusing on the profiles themselves.

By maintaining the profiles under the alignments, it becomes simpler to review what has been sampled and modified for each alignment. Note that the profiles are dynamic and continuously update, as you'll see in this exercise:

1. Open the `DynamicProfiles.dwg` file.
2. On the Viewport Controls located at the upper left of the drawing screen, select Viewport Configurations ➢ Two: Horizontal.
3. Click in the top viewport to activate it.
4. On the Prospector tab, expand the Alignments branch to view the alignment types, expand the Centerline Alignments branch, and right-click Syrah Way. Select the Zoom To option, as shown in Figure 7.7.
5. Click in the bottom viewport to activate it.
6. Expand the Alignments ➢ Centerline Alignments ➢ Syrah Way ➢ Profile Views branches.
7. Right-click Syrah Way, and select Zoom To. Your screen should now look similar to Figure 7.8.

FIGURE 7.7
The Zoom To option on Syrah Way

FIGURE 7.8
Splitting the screen for plan and profile editing

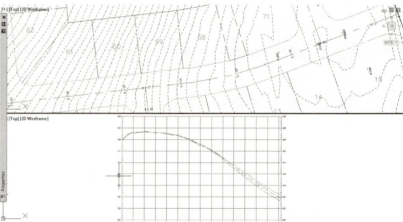

8. Click in the top viewport.

9. Zoom out so you can see more of the plan view.

10. Pick the alignment to activate the grips, and stretch a beginning grip to lengthen the alignment, as shown in Figure 7.9.

11. Click to complete the edit. The alignment profile (the red lines) automatically adjusts to reflect the change in the starting point of the alignment. Note that the offset profiles move dynamically as well.

By maintaining the relationships between the alignment, the surface, the sampled information, and the offsets, Civil 3D creates a much more dynamic feedback system for designers. This system can be useful when you're analyzing a situation with a number of possible solutions, where the surface information will be a deciding factor in the final location of the alignment. Once you've selected a location, you can use this profile view to create a vertical design, as you'll see in the next section.

FIGURE 7.9
Grip-editing the alignment

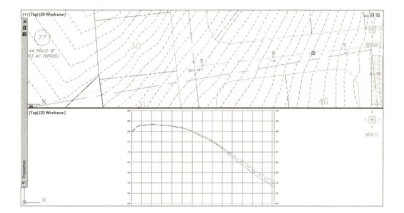

> **LEFT TO RIGHT AND RIGHT TO LEFT?**
>
> You may have noticed that the alignment for Syrah Way is drawn right to left but the profile shows it left to right. It is often desirable to have the plan and profile go in the same direction. There are a number of things you can do to fix this:
>
> ◆ Rotate the plan view 180 degrees. If you have your labeling all set to be plan-readable, it will follow along nicely.
>
> ◆ Consider changing the profile to read from right to left:
>   ◆ Right-click the profile view (grid) and select Profile View Properties.
>   ◆ On the Information tab, change Object Style to Full Profile - RIGHT TO LEFT.
>   ◆ Click OK.

To learn more about styles, refer to Chapter 19, "Styles."

Layout Profiles

Working with sampled surface information is dynamic, and the improvement over previous generations of Autodesk Civil design software is profound. Moving into the design stage, you'll see how these improvements continue as you look at the nature of creating design profiles. By working with layout profiles as a collection of components that understand their relationships with each other as opposed to independent finite elements, you can continue to use the program as a design tool instead of just a drafting tool.

You can create layout profiles in two basic ways:

◆ PVI-based layouts are the most common, using tangents between points of vertical intersection (PVIs) and then applying curve parameters to connect them. PVI-based editing allows editing in a more conventional tabular format.

◆ Entity-based layouts operate like horizontal alignments in the use of free, floating, and fixed entities. The PVI points are derived from pass-through points and other parameters that are used to create the entities. Entity-based editing allows for the selection of individual entities and editing in an individual component dialog.

You'll work with both methods in the next series of exercises to illustrate a variety of creation and editing techniques. First, you'll focus on the initial layout, and then you'll edit the various layouts.

LAYOUT BY PVI

PVI layout is the most common methodology in transportation design. Using long tangents that connect PVIs by derived parabolic curves is a method most engineers are familiar with, and it's the method you'll use in the first exercise:

1. Open the `LayoutProfiles-PVI.dwg` file.

2. On the Home tab's Create Design panel, select Profiles ➢ Profile Creation Tools.

3. Pick the Syrah Way profile view by clicking one of the grid lines. The Create Profile dialog appears, as shown in Figure 7.10.

FIGURE 7.10
The Create Profile dialog

4. Make the changes as shown in Figure 7.10. Click OK. The Profile Layout Tools toolbar appears. Notice that the toolbar is *modeless*, meaning it stays open even if you do other AutoCAD operations such as Pan or Zoom.

At this point, you're ready to begin laying out your vertical design. Before you do, however, note that this exercise skips two options. The first is the idea of profile label sets. You'll explore them in a later section of this chapter. The other option is Criteria-Based Design on the Design tab. Criteria-based design operates in profiles similar to Chapter 6, "Alignments," in that the software compares the design speed to a selected design table (typically AASHTO 2001 in the North America releases) and sets minimum values for curve K values. This can be helpful when you're laying out long highway design projects, but most site and subdivision designers have other criteria to design against.

5. On the toolbar, click the arrow by the Draw Tangents tool on the far left.

6. Select the Curve Settings option. The Vertical Curve Settings dialog opens, as shown in Figure 7.11.

7. The Select Curve Type drop-down menu should be set to Parabolic, and the Length values in both the Crest Curves and Sag Curves areas should be 150.000′, as shown in Figure 7.11. Selecting a Circular or Asymmetric curve type activates the other options in this dialog.

FIGURE 7.11
The Vertical Curve Settings dialog

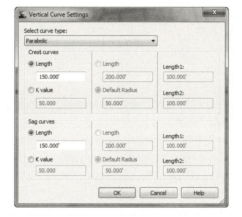

TO K OR NOT TO K

You don't have to choose. Realizing that users need to be able to design using both, Civil 3D 2012 lets you modify your design based on what's important. You can enter a K value to see the required length, and then enter a nice even length that satisfies the K. The choice is up to you.

8. Click OK to close the Vertical Curve Settings dialog.

9. On the Profile Layout Tools toolbar, click the arrow next to the Draw Tangents Without Curves tool again. This time, select the Draw Tangents With Curves option.

10. Use a Center osnap to pick the center of the circle at far right in the profile view. A jig line, which will be your layout profile, appears. Remember, the profile is reversed so that station 0+00 is lined up with the plan. Therefore, Station 0+00 is on the right of the profile.

11. Continue working your way across the profile view, picking the center of each circle with a Center osnap.

12. Right-click or press ↵ after you select the center of the last circle; your drawing should look like Figure 7.12.

FIGURE 7.12
A completed layout profile with labels

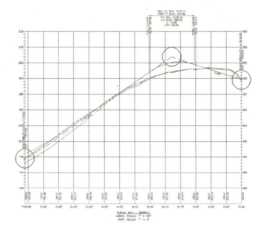

> **WHAT IS A JIG?**
>
> A *jig* is a temporary line shown onscreen to help you locate your pick point. Jigs work in a similar way to osnaps in that they give feedback during command use but then disappear when the selection is complete. Civil 3D uses jigs to help you locate information on the screen for alignments, profiles, profile views, and a number of other places where a little feedback goes a long way.

The layout profile is labeled with the Complete Label Set you selected in the Create Profile dialog. As you'd expect, this labeling and the layout profile are dynamic. If you click and then zoom in on this profile line, not the labels or the profile view, you'll see something like Figure 7.13.

FIGURE 7.13
The types of grips on a layout profile

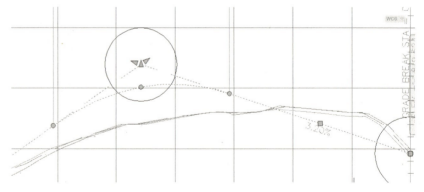

The PVI-based layout profiles include the following unique grips:

- The red triangle at the PVI point is the PVI grip. Moving this alters the inbound and outbound tangents, but the curve remains in place with the same design parameters of length and type.

- The triangular grips on either side of the PVI are sliding PVI grips. Selecting and moving either moves the PVI, but movement is limited to along the tangent of the selected grip. The curve length isn't affected by moving these grips.

- The circular grips near the PVI and at each end of the curve are curve grips. Moving any of these grips makes the curve longer or shorter without adjusting the inbound or outbound tangents or the PVI point.

Although this simple pick-and-go methodology works for preliminary layout, it lacks a certain amount of control typically required for final design. For that, you'll use another method of creating PVIs:

1. Open the `LayoutProfiles-Transparent.dwg` file. Make sure the Transparent Commands toolbar (Figure 7.14) is displayed somewhere on your screen.

FIGURE 7.14
The Transparent Commands toolbar

2. On the Home tab's Create Design panel, select Profiles ➤ Profile Creation Tools.

3. Pick a grid line on the Cabernet Court profile view to display the Create Profile – Draw New dialog.

4. Change the name to **Cabernet Court FG** and then click OK.

5. On the Profile Layout Tools toolbar, click the drop-down menu by the Draw Tangents Without Curves tool, and select the Draw Tangents With Curves tool, as in the previous exercise. Use a Center osnap to snap to the center of the circle at the left edge of the profile view.

6. On the Transparent Commands toolbar, select the Profile Station Elevation command.

7. Pick a grid line on the profile view. If you move your cursor within the profile grid area, a vertical red line, or *jig*, appears; it moves up and down and from side to side. Notice the tooltips.

8. Enter **175**↵ at the command line for the station value. If you move your cursor within the profile grid area, a horizontal and vertical jig appears (see Figure 7.15), but it can only move vertically along station 175. If you've turned on Dynamic Input, the text in the white box shows the elevation along the 175 jig, and the text in the black box shows the horizontal and vertical location of the cursor.

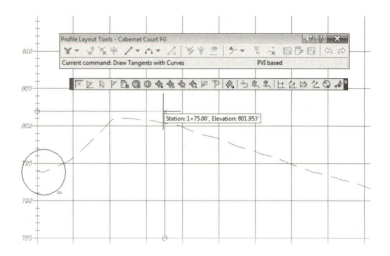

FIGURE 7.15
A jig appears when you use the Profile Station Elevation transparent command.

9. Enter **802↵** at the command line to set the elevation for the second PVI.

10. Press Esc only once. The Profile Station Elevation command is no longer active, but the Draw Tangents With Curves tool that you previously selected on the Profile Layout Tools toolbar continues to be active.

11. On the Transparent Commands toolbar, select the Profile Grade Station command.

12. Enter **−4↵** at the command line for the profile grade.

13. Enter **525↵** for the station value at the command line. Press Esc only once to deactivate the Profile Grade Station command.

14. On the Transparent Commands toolbar, select the Profile Grade Length command.

15. Enter **9.78↵** at the command line for the profile grade.

16. Enter **225↵** for the profile grade length.

17. Press Esc only once to deactivate the Profile Grade Length command and to continue using the Draw Tangent With Curves tool.

18. Use a Center osnap to select the center of the circle along the far-right side of the profile view.

19. Press ↵ to complete the profile. Your profile should look like Figure 7.16.

Using PVIs to define tangents and fitting curves between them is the most common approach to create a layout profile, but you'll look at an entity-based design in the next section.

THE ELEVATION ELEMENT | 257

LAYOUT BY ENTITY

Working with the concepts of fixed, floating, and free entities as you did in Chapter 6, you'll lay out a design profile in this exercise:

1. Open the `LayoutProfiles-Entity.dwg` file.
2. On the Home tab's Create Design panel, select Profiles ➤ Profile Creation Tools.
3. Pick a grid line on the Syrah Way profile view to display the Create Profile dialog.
4. Change the name to **Syrah Way FG**, click OK, and open the Profile Layout Tools toolbar.

FIGURE 7.16
Using the Transparent Commands toolbar to create a layout profile

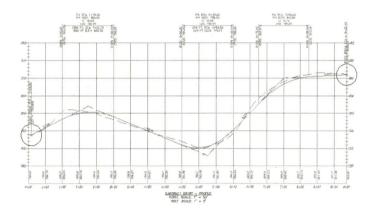

RIGHT TO LEFT

Remember that as we use the Syrah Way profile it has been created from right to left.

5. Click the arrow by the Draw Fixed Tangent By Two Points tool, and select the Fixed Tangent (Two Points).
6. Using a Center osnap, pick the circle at the right edge of the profile view. A rubber-banding line appears.
7. Using a Center osnap, pick the circle located at approximately station 1+50. A tangent is drawn between these two circles.
8. Using a Center osnap, pick the circle located at approximately station 3+00. Another rubber-banding line appears.
9. Using a Center osnap, pick the circle located at the left edge of the profile view. A second tangent is drawn. Right-click to exit the Fixed Tangent (Two Points) command; your drawing should look like Figure 7.17.

258 | **CHAPTER 7** PROFILES AND PROFILE VIEWS

FIGURE 7.17
Layout profile with two tangents drawn

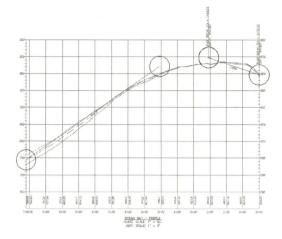

10. Click the arrow next to the Draw Fixed Parabola By Three Points tool on the Profile Layout Tools toolbar.

11. Choose the More Fixed Vertical Curves ➢ Fixed Vertical Curve (Entity End, Through Point) option.

12. Pick the right tangent to attach the fixed vertical curve. Remember to pick the tangent line and not the end circle. A rubber-banding line appears.

13. Using a Center osnap, select the circle located at approximately station 3+00.

14. Right-click to exit the Fixed Vertical Curve (Entity End, Through Point) command. Your drawing should look like Figure 7.18.

FIGURE 7.18
Completed curve from the entity end

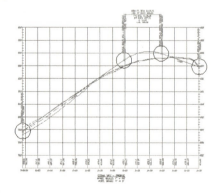

With the entity-creation method, grip editing works in a similar way to other layout methods. You'll look at more editing methods after trying the final creation method.

THE BEST FIT PROFILE

You've surveyed along a centerline, and you need to closely approximate the tangents and vertical curves as they were originally designed and constructed.

A best fit profile can be created from AutoCAD blocks, AutoCAD 3D polylines, AutoCAD points, COGO points, a surface profile, or feature lines. The most common option is the surface profile. The tool attempts to run a complex algorithm to determine the best fit profile path, including both tangents and vertical curves. However, the only best fit option available for determining vertical curves is the maximum curve radius. The maximum curve radius is a formula rarely used in the design of parabolic curves, the most common curves found in roadway design. Unlike the Best Fit tool for lines and curves as seen in Chapter 1, "The Basics of Civil 3D," this tool has no options for selecting or deselecting points, making the tool somewhat cumbersome.

The Create Best Fit Profile tool is found on the Profile drop-down on the Create Design panel on the Home tab.

In many cases, the tools you choose to use in a given scenario, as promising as the name may sound, may actually prove to be more of a hindrance than they are helpful. The Best Fit Profile tool may be one of these cases. The tool attempts to perform an analysis on all the sampled existing ground profile points in a profile view, the options are limited, and in the end, you may find a good pair of eyes and graphical analysis using other profile creation tools to be less time consuming. It's important to understand all the tools at your disposal, but you are not required to use each one in a production environment. If results don't meet your needs and you've spent considerable time with a single tool, find another tool, or tools, that accomplish the same task. Remember, at the end of the day, you still have sheets to cut and plans to get out the door. You're limited only by your selection of tools and your creativity.

CREATING A PROFILE FROM A FILE

Working with profile information in Civil 3D is nice, but it's not the only place where you can create or manipulate this sort of information. Many programs or analysis packages generate profile information. One common case is the plotting of a hydraulic grade line against a stormwater network profile of the pipes. When information comes from outside the Civil 3D program, it's often output in myriad formats. If you convert this format to the required format for Civil 3D, the profile information can be input directly.

Civil 3D has a specific format. Each line is a PVI definition (station and elevation). Curve information is an optional third bit of data on any line. Here's one example:

```
0    550.76
127.5  552.24
200.8  554   100
256.8  557.78  50
310.75  561
```

In this example, the third and fourth lines include the curve length as the optional third piece of information. The only inconvenience of using this input method is that the information in Civil 3D doesn't directly reference the text file. Once the profile data is imported, no dynamic relationship exists with the text file.

In this exercise, you'll import a small text file to see how the function works:

1. Open the `ProfilefromFile.dwg` file.
2. On the Home tab's Create Design panel, select Profiles ➢ Create Profile From File.
3. Select the `ProfileFromFile.txt` file, and click Open. The Create Profile dialog appears.
4. In the dialog, choose Frontenac Drive for the Name, Design Profile from the Profile Style, and Complete Label Set from the Profile Label Set drop-down menu.
5. Click OK. Your drawing should look like Figure 7.19.

FIGURE 7.19
A completed profile created from a file

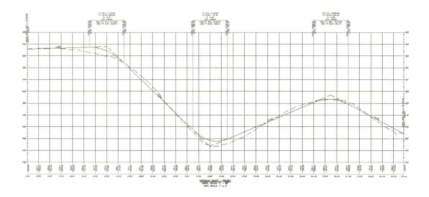

Now that you've tried the three main ways of creating profile information, you'll edit a profile in the next section.

Profile Layout Tools

Although we have touched on many of the available tools in the Profile Layout Tools palette, there are still many that we have not.

Draw Tangents   Profile layout point to point. No curves.

Draw Tangents With Curves   Profile layout point to point with curve type and length determined from Curve Settings.

Curve Settings   Sets the type of curve (Parabolic, Circular, Asymmetric), K factor, and length.

Insert PVI   Inserts a PVI at the specified location.

Delete PVI   Deletes a PVI at the specified location.

Move PVI   Moves a PVI at the specified location.

Convert AutoCAD Line And Spline   Takes a singular line/spline and converts it into a profile object.

Insert PVI Tabular   Allows you to enter PVI station and elevation information in a table-like dialog.

Raise/Lower PVI   Allows you to raise or lower a profile at a specified distance and either the entire profile or a specified station range.

Copy Profile   Copies a profile.

PVI or Entity-Based   Display profile layout tools based on either PVI or Entity.

Select PVI   Opens the Profile Layout Parameters dialog for the selected PVI.

Delete Entity   Deletes a profile subobject curve and turns it into tangents.

Profile Layout Parameters   Opens the Profile Layout Parameters dialog.

Profile Grid Box   Opens the Entities Panorama showing all segments of the profile.

Editing Profiles

The methods just reviewed let you quickly create profiles. You saw how sampled profiles reflect changes in the parent alignment and how some grips are available on layout profiles, and you also imported a text file that could easily be modified. In all these cases, the editing methods left something to be desired, from either a precision or a dynamic relationship viewpoint.

This section looks at profile editing methods. The most basic is a more precise grip-editing methodology, which you'll learn about first. Then you'll see how to modify the PVI-based layout profile, how to change out the components that make up a layout profile, and how to use editing functions that don't fit into a nice category.

GRIP PROFILE EDITING

Once a layout is in place, sometimes a simple grip edit will suffice. But for precision editing, you can use a combination of the grips and the tools on the Transparent Commands toolbar, as in this short exercise:

1. Open the `GripEditingProfiles.dwg` file.

2. Zoom in, and pick the layout profile for Frontenac Drive (the blue line) to activate its grips.

3. Pick the triangular grip pointing upward on the left vertical curve that is around Sta. 8+00 to begin a grip stretch of the PVI, as shown in Figure 7.20.

FIGURE 7.20
Grip-editing a PVI

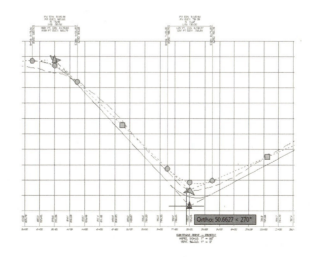

4. On the Transparent Commands toolbar, select the Profile Station Elevation command.

5. Pick a grid line on the profile view.

6. Enter **801↵** at the command line to set the profile station.

7. Enter **781↵** to set the profile elevation.

8. Press Esc to deselect the layout profile object you selected in step 2, and regenerate your view to complete the changes, as shown in Figure 7.21.

FIGURE 7.21
Completed grip edit using the transparent commands for precision

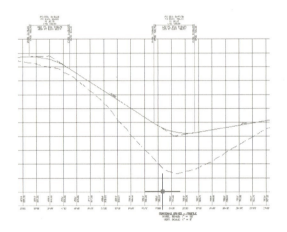

The grips can go from quick-and-dirty editing tools to precise editing tools when you use them in conjunction with the transparent commands in the profile view. They lack the ability to precisely control a curve length, though, so you'll look at editing a curve next.

PARAMETER AND PANORAMA PROFILE EDITING

Beyond the simple grip edits, but before changing out the components of a typical profile, you can modify the values that drive an individual component. In this exercise, you'll use the Profile Layout Parameter dialog and the Panorama palette set to modify the curve properties on your design profile:

1. Open the `ParameterEditingProfiles.dwg` file.

2. Pick the layout profile for Frontenac Drive (the blue line) to activate its grips.

3. Select Geometry Editor from the Modify Profile panel.

4. On the Profile Layout Tools toolbar, click the Profile Layout Parameters tool to open the Profile Layout Parameters dialog.

5. Click the Select PVI tool, and zoom in to click near the PVI near station 13+25 to populate the Profile Layout Parameters dialog (Figure 7.22).

FIGURE 7.22
The Profile Layout Parameters dialog

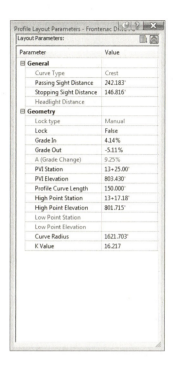

Values that can be edited are in black; the rest are mathematically derived and can be of some design value but can't be directly modified. The two buttons at the top of the dialog adjust how much information is displayed.

6. Change the Profile Curve Length in the Profile Layout Parameters dialog to **250.000′** (see Figure 7.23).

FIGURE 7.23
Direct editing of the curve layout parameters

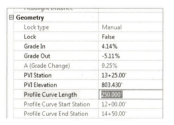

7. Close the Profile Layout Parameters dialog by clicking the red X in the upper-right corner. Then, press the Esc key to deactivate the Select PVI tool.

8. On the Profile Layout Tools toolbar, click the Profile Grid View tool to activate the Profile Entities tab in Panorama. Panorama allows you to view all the profile components at once, in a compact form.

9. Scroll right in Panorama until you see the Profile Curve Length column.

4. On the Profile Layout Tools toolbar, click the Insert PVI tool.

the tangents on either side of this new PVI is affected, and the profile is adjusted accord-

6. On the Profile Layout Tools toolbar, click the Delete Entity tool. Notice the grips disappear and the labels update.

7. Zoom in, and pick the curve entity near station 2+50 to delete it. Then right-click to update the display.

8. On the Profile Layout Tools toolbar, click the arrow by the Draw Fixed Vertical Curve

11. Repeat steps 6–10, but use the PVI near 5+50.

12. Right-click to exit the command and update the profile display.

FIGURE 7.26
The completed editing of the curve using component-level editing

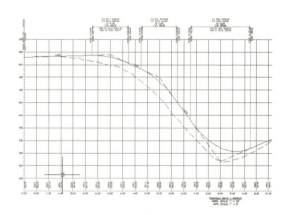

Editing profiles using any of these methods gives you precise control over the creation and layout of your vertical design. In addition to these tools, some of the tools on the Profile Layout Tools toolbar are worth investigating and somewhat defy these categories. You'll look at them next.

OTHER PROFILE EDITS

Some handy tools exist on the Profile Layout Tools toolbar for performing specific actions. These tools aren't normally used during the preliminary design stage, but they come into play as you're working to create a final design for grading or corridor design. They include raising or lowering a whole layout in one shot, as well as copying profiles. Try this exercise:

1. Open the `OtherProfileEdits.dwg` file.
2. Zoom to the Cabernet Court Profile if it not already showing.
3. Pick the Cabernet Court FG profile to activate its grips. It is the blue line.
4. Select Geometry Editor from the Modify Profile panel. The Profile Layout Tools toolbar appears.

5. Click the Raise/Lower PVIs tool. The Raise/Lower PVI Elevation dialog shown here appears:

6. Set Elevation Change to **–3.000′**. Click the Station Range radio button, and set the Start value to 0+50′ and the End value to 9+00.00′. Click OK.
7. Click the Copy Profile tool to display the Copy Profile Data dialog.

8. Click OK to create a new layout profile directly on top of Layout 1. Note that after picking the newly created layout profile, the Profile Layout Tools toolbar now references this newly created Layout (2) profile.
9. Use the Raise/Lower tool to drop Layout (2) by 0.5′ to simulate an edge-of-pavement design. Your drawing should look similar to this:

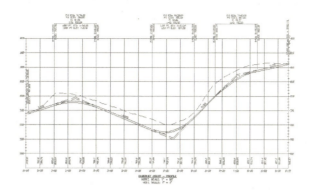

Using the layout and editing tools in these sections, you should be able to design and draw a combination of profile information presented to you as a Civil 3D user.

Matching Profile Elevations

Up to now, you have learned how to use some of the available tools for modifying profiles. Now let's try to tie it all together in our subdivision.

If you take a look at Figure 7.27, you will see that Syrah Way actually has three road intersections, which are circled: two for Frontenac Drive and one for Cabernet Court. We need to know the elevation for the intersecting point between these intersections. First, let's look at the station where the intersections occur:

Figure 7.27
Syrah Way and the intersecting roads

1. Open the RoadsMatchProfiles.dwg file.

2. From the Annotate tab and Labels & Tables panel, select Add Labels. The Add Labels dialog opens.

3. Change the feature type to Alignment, label type to Station Offset - Fixed Point, and station offset label style type to Plan View Alignment Station Check, as shown in Figure 7.28. Click Add.

4. At the `Select alignment:` prompt, click on the alignment for Syrah Way. This sets Syrah Way as the main alignment that will be used.

5. At the `Select point:` prompt, use an endpoint osnap and use Frontenac Drive Station 0+00 as the point.

6. At the `Select alignment for label style component Reference Alignment:` prompt, select anywhere on Frontenac Drive.

7. Right-click to end the command.

FIGURE 7.28
Syrah Way and the intersecting roads

A label showing the station of Syrah Way and Frontenac Drive intersection appears. You will need this information for the next step, but first rerun the Plan View Alignment Station Check label for the next two intersections. Your drawing should look similar to Figure 7.29.

FIGURE 7.29
Syrah Way and the intersecting roads

Now that you have established the stations at which the intersections occur, let's put the intersecting stations into practice on the profiles:

1. Continuing with the same drawing, set your viewports to Two: Horizontal so you can see the Syrah Way plan view on top and the profile view on the bottom.

2. From the Annotate tab and Labels & Tables panel, select Add Labels. The Add Labels dialog opens.

3. Change the feature type to Profile View, label type to Station Elevation, and station elevation label style to Station And Elevation Reference Alignment RIGHT; then click Add.

4. At the `Select a profile view:` prompt, select Syrah Way.

5. At the `Specify station:` prompt, refer to the plan view and type **145** as the first label indicates. Remember, the Syrah Way profile goes from right to left.

6. At the `Specify elevation:` prompt, pick anywhere on your drawing. For this case, the location is not important as we will address that in a bit.

7. You are now prompted to select alignment for the label style component Alignment Ref:. Select the top viewport and choose Frontenac Drive.

8. Press Esc. Your label should look like Figure 7.30.

FIGURE 7.30
The alignment intersection label

You can see that everything is populated except the elevation, which is the most important piece of information needed.

9. Select the newly created label. In the Properties General area, click in the Profile1 Object. Click the Selector icon.

10. At the `Select an object <or press Enter key to select from list>:` prompt, press ↵, select the Syrah Way FG object, and click OK.

The Elevation portion of the label is now populated with the elevation, which is 824.50. This is the starting elevation of Frontenac Drive in order to make both roads meet.

Repeat steps 1–10 for the other two elevation labels. If you wish to move the label, we recommend that you turn Ortho mode on so you keep the same station. Your profile should look like Figure 7.31.

Now you have all the pieces you need. Modify the profiles using any of the previously mentioned methods so that the beginning of Frontenac Drive matches elevation 824.50 and the end of Cabernet Court matches elevation of 811.36. Further, you can lock these elevations since they are critical by selecting the lock icon in the Profile Entities Panorama.

FIGURE 7.31
The Alignment Intersection label updated

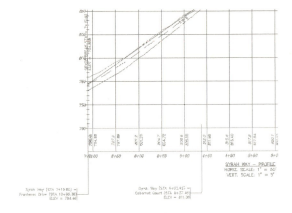

Profile Display

No matter how profiles are created, they need to be shown and labeled to make the information more understandable. In this section, you'll learn about options for the profile labels.

> **STYLES: WHERE TO LOOK**
>
> The exercises in this chapter have many different styles created to show variety. You'll learn more about styles in Chapter 19, "Styles." It's okay to take a peek ahead once in a while.

Profile Labels

It's important to remember that the profile and the profile view aren't the same thing. The labels discussed in this section are those that relate directly to the profile. This usually means station-based labels, individual tangent and curve labels, or grade breaks.

APPLYING LABELS

As with alignments, you apply labels as a group of objects separate from the profile. In this exercise, you'll learn how to add labels along a profile object:

1. Open the `ApplyingProfileLabels.dwg` file.

2. Pick the layout profile (the blue line) to activate the profile object.

3. Select Edit Profile Labels from the Labels panel to display the Profile Labels dialog (see Figure 7.32).

   Selecting the type of label from the Type drop-down menu changes the Style drop-down menu to include styles that are available for that label type. Next to the Style drop-down menu are the usual Style Edit/Copy button and a preview button. Once you've selected a style from the Style drop-down menu, clicking the Add button places it on the profile. The middle portion of this dialog displays information about the labels that are being applied to the profile selected; you'll look at that in a moment.

FIGURE 7.32
An empty Profile Labels dialog

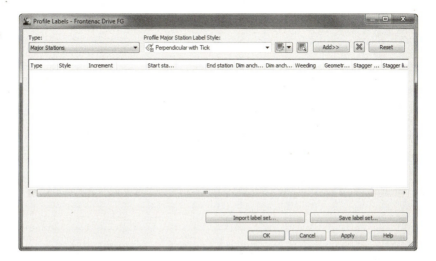

4. Choose the Major Stations option from the Type drop-down menu. The name of the second drop-down menu changes to Profile Major Station Label Style to reflect this option. Verify that Perpendicular With Tick is selected in this menu.

5. Click Add to apply this label to the profile.

6. Choose Horizontal Geometry Points from the Type drop-down menu.

7. The name of the Style drop-down menu changes to Profile Horizontal Geometry Point. Select the Standard option, and click Add again to display the Geometry Points dialog shown in Figure 7.33. This dialog lets you apply different label styles to different geometry points if necessary.

8. Deselect the Alignment Beginning and Alignment End rows, as shown in Figure 7.33, and click OK to close the dialog.

9. Click the Apply button. Drag the dialog out of the way to view the changes to the profile (see Figure 7.34).

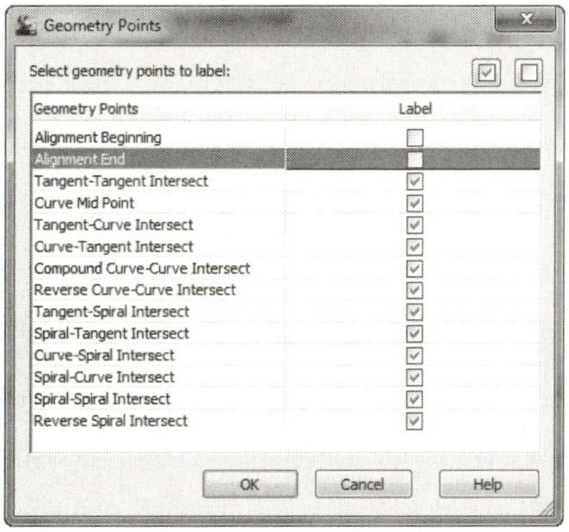

FIGURE 7.33
The Geometry Points dialog appears when you apply labels to horizontal geometry points.

FIGURE 7.34
Labels applied to major stations and alignment geometry points

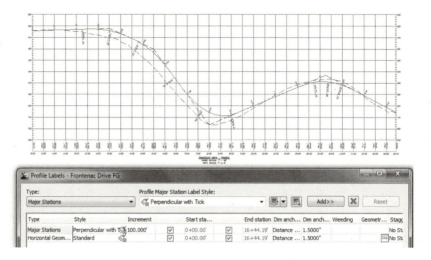

10. In the middle of the Profile Labels dialog, change the Increment value in the Major Stations row to **50**, as shown in Figure 7.35. This modifies the labeling increment only, not the grid or other values.

FIGURE 7.35
Modifying the major station labeling increment

11. Click OK to close the Profile Labels dialog.

As you can see, applying labels one at a time could turn into a tedious task. After you learn about the types of labels available, you'll revisit this dialog and look at the two buttons at the bottom for dealing with label sets.

Station Labels

Labeling along the profile at major, minor, and alignment geometry points lets you insert labels similar to a horizontal alignment. In this exercise, you'll modify a style to reflect a plan-readable approach and remove the stationing from the first and last points along the profile:

1. Pick one of the station labels previously created and select Edit Label Group from the Modify panel to display the Profile Labels dialog.

2. Deselect the Start Station and End Station check boxes for the Major Stations label.

3. Change the value for the Start Station to **100** and the value of the End Station to **803**, as shown in Figure 7.36.

FIGURE 7.36
Modifying the values of the starting and ending stations for the major labels

4. Click the icon next to Major Stations in the Style field to display the Pick Label Style dialog.

5. Select the Perpendicular With Tick Orientation.

6. Click OK to close the Profile Labels dialog. Instead of each station label being oriented so that it's perpendicular to the profile at the station, all station labels are now oriented vertically along the top of the profile at the station.

By controlling the frequency, starting and ending station, and label style, you can create labels for stationing or for conveying profile information along a layout profile.

Line Labels

Line labels in profiles are typically used to convey the slope or length of a tangent segment. In this exercise, you'll add a length and slope to the layout profile:

1. Pick the layout profile and select Edit Profile Labels from the Labels panel to display the Profile Labels dialog.

2. Change the Type field to the Lines option. The name of the Style drop-down menu changes to Profile Tangent Label Style. Select the Length and Percent Grade option.

3. Click the Add button, and then click OK to exit the dialog. The profile view should look like Figure 7.37.

FIGURE 7.37
A new line label applied to the layout profile

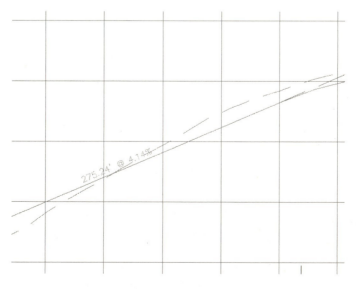

> **WHERE IS THAT DISTANCE BEING MEASURED?**
>
> The *tangent slope length* is the distance along the horizontal geometry between vertical curves. This value doesn't include the tangent extensions. There are a number of ways to label this length; be sure to look in the Text Component Editor if you want a different measurement.

CURVE LABELS

Vertical curve labels are one of the most confusing aspects of profile labeling. Many people become overwhelmed rapidly because there's so much that can be labeled and there are so many ways to get all the right information in the right place. In this quick exercise, you'll look at some of the special label anchor points that are unique to curve labels and how they can be helpful:

1. Pick the layout profile and select Edit Profile Labels from the Labels panel to display the Profile Labels dialog.

2. Choose the Crest Curves option from the Type drop-down menu. The name of the Style drop-down menu changes. Select the Crest and Sag option.

3. Click the Add button to apply the label.

4. Choose the Sag Curves option from the Type drop-down menu. The name of the Style drop-down menu changes to Profile Sag Curve Label Style. Click Add to apply the same label style.

5. Click OK to close the dialog; your profile should look like Figure 7.38.

FIGURE 7.38
Curve labels applied with default values

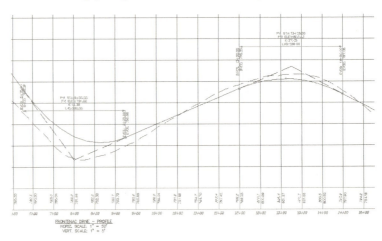

Most labels are applied directly on top of the object being referenced. Because typical curve labels contain a large amount of information, putting the label right on the object can yield undesired results. In the following exercise, you'll modify the label settings to review the options available for curve labels:

1. Pick the layout profile and select Edit Profile Labels from the Labels panel to display the Profile Labels dialog.

2. Scroll to the right in the middle of the dialog, and locate the Dim Anchor Opt column.

3. Change the Dim Anchor Opt value for the Sag Curves to Distance Below.

4. Change the Dim Anchor Val to **2'**.

5. Change the Dim Anchor Val value for the Crest Curves to **2'** as well. Your dialog should look like the one shown in Figure 7.39.

6. Click OK to close the dialog.

FIGURE 7.39
Curve labels with distance values inserted

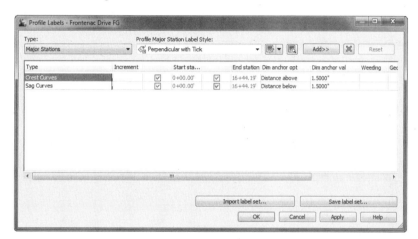

The labels can also be grip-modified to move higher or lower as needed, but you'll try one more option in the following exercise:

1. Pick the layout profile and select Edit Profile Labels from the Labels panel to display the Profile Labels dialog.

2. Scroll to the right, and change both Dim Anchor Opt values for the Crest and Sag Curves to Graph View Top.

3. Change the Dim Anchor Val for both curves to **−0.5'**, and click OK to close the dialog. Your drawing should look like Figure 7.40.

FIGURE 7.40
Curve labels anchored to the top of the graph

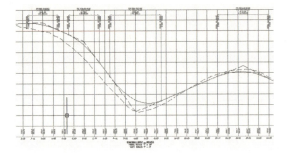

By using the top or bottom of the graph as the anchor point, you can apply consistent and easy labeling to the curve, regardless of the curve location or size.

GRADE BREAKS

The last label style typically involved in a profile is a grade-break label at PVI points that don't fall inside a vertical curve, such as the beginning or end of the layout profile. Additional uses include things like water-level profiling, where vertical curves aren't part of the profile information or existing surface labeling. In this exercise, you'll add a grade-break label and look at another option for controlling how often labels are applied to profile data:

1. Open the `Grade Break Profile Labels.dwg` file.

2. Pick the red surface profile, and then select Edit Profile Labels from the Labels panel to display the Profile Labels dialog.

3. Choose Grade Breaks from the Type drop-down menu. The name of the Style drop-down menu changes. Select the Station Over Elevation style and click the Add button.

4. Click Apply, and drag the dialog out of the way to review the change. It should appear as in Figure 7.41.

FIGURE 7.41
Grade-break labels on a sampled surface

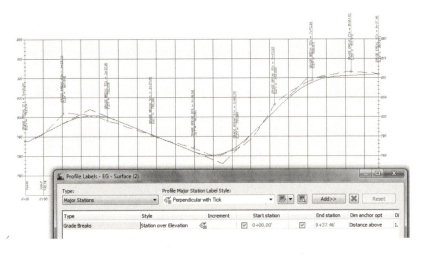

A sampled surface profile has grade breaks every time the alignment crosses a surface TIN line. Why wasn't your view coated with labels?

5. Scroll to the right, and change the Weeding value to **50′**.

6. Click OK to dismiss the dialog. Your profile should look like the one in Figure 7.42.

FIGURE 7.42
Grade-break labels with a 50′ Weeding value

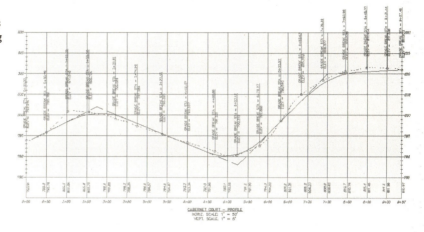

Remember just as with almost every other label in Civil 3D, you can Ctrl+click a label to select a single one for deletion.

Weeding lets you control how frequently grade-break labels are applied. This makes it possible to label dense profiles, such as a surface sampling, without being overwhelmed or cluttering the view beyond usefulness.

As you've seen, there are many ways to apply labeling to profiles, and applying these labels to each profile individually could be tedious. In the next section, you'll build a label set to make this process more efficient.

STAGGERING PROFILES

When you are using Grade Breaks, there is an option in the Profile Labels dialog that will allow you to stagger the labels. The options as shown are Stagger Both Sides, No Staggering, Stagger To Right, and Stagger To Left.

If you do not see all of the grade breaks, the Weeding value needs to be adjusted. In this example, the value is set to a default of 100′:

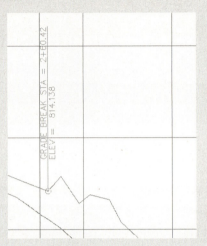

Setting a lower value, for example 10′, will show more grade breaks and you can see the staggered label effect.

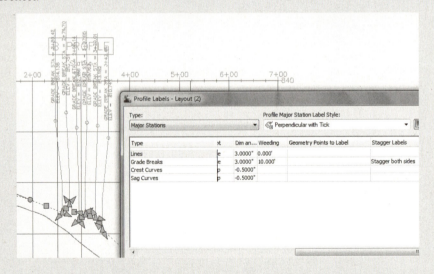

In addition, you can set the stagger line lengths by entering values in the Stagger Line 1 Height and Stagger Line 2 Height text boxes.

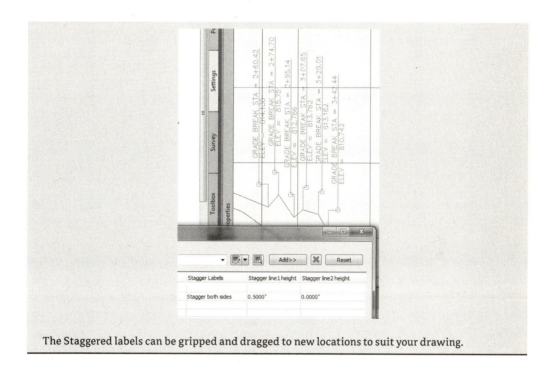

The Staggered labels can be gripped and dragged to new locations to suit your drawing.

PROFILE LABEL SETS

Applying labels to both crest and sag curves, tangents, grade breaks, and geometry with the label style selection and various options can be monotonous. Thankfully, Civil 3D gives you the ability to use label sets, as in alignments, to make the process quick and easy. In this exercise, you'll apply a label set, make a few changes, and export a new label set that can be shared with team members or imported to the Civil 3D template. Follow these steps:

1. Delete the grade-break labels from the previous exercise.
2. Pick the layout profile, and then select Edit Profile Labels from the Labels panel to display the Profile Labels dialog.
3. Click the Import Label Set button near the bottom of the dialog to display the Select Style Set dialog.
4. Select the Standard option from the drop-down menu, and click OK.
5. Click OK again to close the Profile Labels dialog and see the profile view, as in Figure 7.43.

FIGURE 7.43
Profile with the Standard label set applied

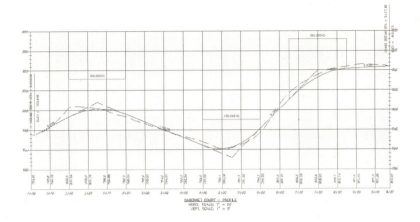

6. Pick the layout profile and select Edit Profile Labels from the Labels panel to display the Profile Labels dialog.

7. Click Import Label Set to display the Select Style Set dialog.

8. Select the Complete Label Set option from the drop-down menu, and click OK.

9. Double-click the icon in the Style cell for the Lines label. The Pick Label Style dialog opens. Select the Length and Percent Grade option from the drop-down menu, and click OK.

10. Double-click the icon in the Style cell for the Crest label. The Pick Label Style dialog opens. Select the Crest and Sag option from the drop-down menu. Repeat this step for the Sag Curve label and click OK.

11. Click the Apply button, and drag the dialog out of the way, as shown in Figure 7.44, to see the changes reflected.

FIGURE 7.44
Establishing the Road Profile Labels set

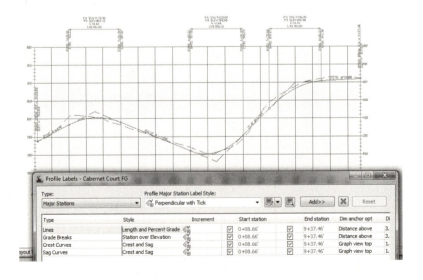

12. Click the Save Label Set button to open the Profile Label Set dialog and create a new profile label set.

13. On the Information tab, change the name to **Road Profile Labels**. Click OK to close the Profile Label Set dialog.

14. Click OK to close the Profile Labels dialog.

15. On the Settings tab of Toolspace, select Profile ➤ Label Styles ➤ Label Sets. Note that the Road Profile Labels set is now available for sharing or importing to other profile label dialogs.

> **SOMETIMES YOU DON'T WANT TO SET EVERYTHING**
>
> Resist the urge to modify the beginning or ending station values in a label set. If you save a specific value, that value will be applied when the label set is imported. For example, if you set a station label to end at 15+00 because the alignment is 15+15 long, that label will stop at 15+00, even if the target profile is 5,000′ long!

Label sets are the best way to apply profile labeling uniformly. When you're working with a well-developed set of styles and label sets, it's quick and easy to go from sketched profile layout to plan-ready output.

Profile Views

Working with vertical data is an integral part of building the Civil 3D model. Once profile information has been created in any number of ways, displaying it to make sense is another task. It can't be stated enough that profiles and profile views are not the same thing in Civil 3D. The profile view is the method that Civil 3D uses to display profile data. A single profile can be shown in an infinite number of views, with different grids, exaggeration factors, labels, or linetypes. In this part of the chapter, you'll look at the various methods available for creating profile views.

Creating Profile Views During Sampling

The easiest way to create a profile view is to draw it as an extended part of the surface sampling procedure. In this brief exercise, you'll sample a surface and then create the view in one series of steps:

1. Open the ProfileViews.dwg file. (Remember, all data files can be downloaded from this book's web page.)

2. Change to the Home tab and select Profile ➤ Create Surface Profile from the Create Design panel to display the Create Profile From Surface dialog.

3. In the Alignment text box, select Syrah Way. In the Select Surface list box, select the EG surface. Click the Add button.

4. Click the Draw In Profile View button to move into the Create Profile View wizard, shown in Figure 7.45.

282 | **CHAPTER 7** PROFILES AND PROFILE VIEWS

Profile views are created with the help of a wizard. The wizard offers the advantage of stepping through all the options involved in creating a view or simply accepting the command settings and creating the profile view quickly and simply.

5. Verify that Syrah Way is selected in the Alignment drop-down list and Profile View is selected in the Profile View Style, and click Next.

6. Verify that the Station Range area has the Automatic option selected and click Next.

7. Verify that the Profile View Height field has the Automatic option selected and click Next.

8. Click the Create Profile View button in the Profile Display Options window.

9. Pick a point on screen somewhere to the right of the site and surface to draw the profile view, as shown in Figure 7.46.

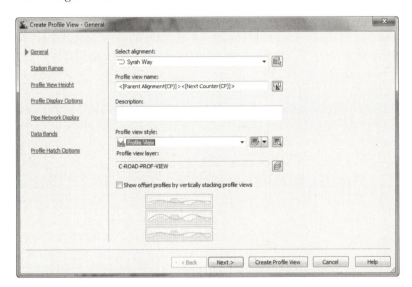

FIGURE 7.45
The Create Profile View wizard

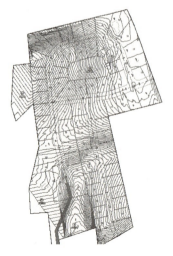

FIGURE 7.46
The completed profile view for Syrah Way

By combining the profile sampling step with the creation of the profile view, you have avoided one more trip to the menus. This is the most common method of creating a profile view, but we'll look at a manual creation in the next section.

Creating Profile Views Manually

Once an alignment has profile information associated with it, any number of profile views might be needed to display the proper information in the right format. To create a second, third, or tenth profile view once the sampling is done, you must use a manual creation method. In this exercise, you'll create a profile view manually for an alignment that already has a surface-sampled profile associated with it:

1. Open the `ProfileViews.dwg` file if you have not already done so.

2. Change to the Home tab and select Profile View ➢ Create Profile View from the Profile & Section Views panel.

3. In the Select Alignment text box, select Cabernet Court from the drop-down list. The profile was already sampled from the surface.

4. In the Profile View Style drop-down list, select the Full Grid style.

5. Click the Create Profile View button and pick a point on screen to draw the profile view, as shown in Figure 7.47.

FIGURE 7.47
The completed profile view of Cabernet Court

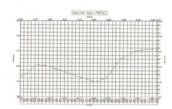

Using these two creation methods, you've made simple views, but you look at a longer alignment in the next exercise, and some more of the options available in the Create Profile View wizard.

Splitting Views

Dividing up the data shown in a profile view can be time consuming. Civil 3D's Profile View wizard is used for simple profile view creation, but the wizard can also be used to create manually limited profile views, staggered (or stepped) profile views, multiple profile views with gaps between the views, and stacked profiles (aka three-line profiles). You'll look at these variations on profile view creation in this section.

Creating Manually Limited Profile Views

Continuous profile views like you made in the first two exercises work well for design purposes, but they are often unusable for plotting or exhibiting purposes. In this exercise, you'll sample a surface, and then use the wizard to create a manually limited profile view. This variation will allow you to control how long and how high each profile view will be, thereby making the views easier to plot or use for other purposes.

1. Open the `ProfileViewsSplit.dwg` file.
2. Change to the Home tab and select Profile ➢ Create Surface Profile from the Create Design panel to display the Create Profile From Surface dialog.
3. In the Alignment text box, select Frontenac Drive; in the Select Surface list box, select the EG surface and click the Add button.
4. Click the Draw In Profile View button to enter the wizard.
5. In the Profile View Style drop-down list, select the Full Grid style and click Next.
6. In the Station Range area, select the User Specified Range radio button. Enter **0** for the Start station and **8+00** for the End station, as shown in Figure 7.48. Notice the preview picture shows a clipped portion of the total profile. Click Next.
7. In the Profile View Height area, select the User Specified radio button. Set the Minimum height to **781** and the Maximum height to **820**.
8. Click the Create Profile View button and pick a point on screen to draw the profile view. Your screen should look similar to Figure 7.49.

Creating Staggered Profile Views

When large variations occur in profile height, the graph must often be split just to keep from wasting much of the page with empty grid lines. In this exercise, you use the wizard to create a staggered, or stepped, view:

1. Open the `ProfileViewsStaggered.dwg` file.
2. Change to the Home tab and select Profile ➢ Create Surface Profile from the Create Design panel to display the Create Profile From Surface dialog.
3. In the Alignment text box, select Frontenac Drive; in the Select Surface list box, select the EG surface and click the Add button.
4. Click the Draw In Profile View button to enter the wizard. Click Next to move to the Station Range options.
5. Make sure that Station Range is set to Automatic to allow the view to show the full length. Click Next.
6. In the Profile View Height field, select the User Specified option and set the values to **785.00′** and **825.00′** as shown in Figure 7.50.

FIGURE 7.48
The start and end stations for the user-specified profile view

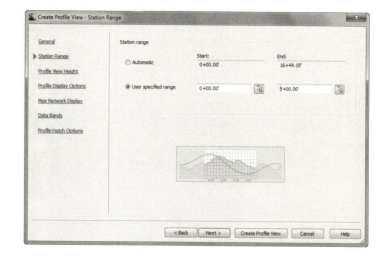

FIGURE 7.49
Applying user-specified station and height values to a profile view

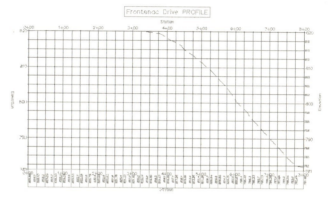

FIGURE 7.50
Split Profile View settings

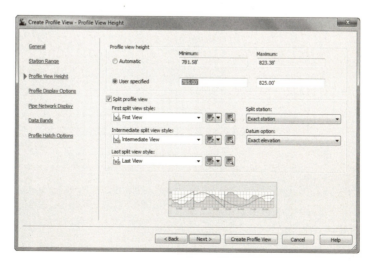

7. Check the Split Profile View options and set the view styles, as shown in Figure 7.50.

8. Click the Create Profile View button and pick a point on screen to draw the staggered display, as shown in Figure 7.51.

FIGURE 7.51
A staggered (stepped) profile view created via the wizard

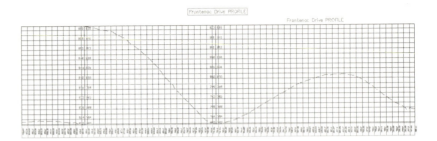

The profile view is split into views according to the settings that were selected in the Create Profile View wizard in step 7. The first section shows the profile from 0 to the station where the elevation change of the profile exceeds the limit for height. The next section displays the same and so forth for the rest of the profile. Each of these sections is part of the same profile view and can be adjusted via the Profile View Properties dialog.

CREATING GAPPED PROFILE VIEWS

Profile views must often be limited in length and height to fit a given sheet size. Gapped views are a way to show the entire length and height of the profile, by breaking the profile into different sections with "gaps" or spaces between each view.

When you are using the Plan and Production Tools (covered in Chapter 16, "Plan and Production"), the gapped profile views are automatically created.

In this exercise, you use a variation of the Create Profile View wizard to create gapped views automatically:

1. Open the `ProfileViewsStaggered.dwg` file if you haven't already. *If you did the previous exercise, skip steps 2 and 3.*

2. Change to the Home tab and select Profile ➢ Create Surface Profile from the Create Design panel to display the Create Profile From Surface dialog.

3. In the Alignment text box, select Frontenac Drive; in the Select Surface list box, select the EG surface and click the Add button. Click OK to exit the dialog.

4. Change to the Home tab and select Profile View ➢ Create Multiple Profile Views to display the Create Multiple Profile Views wizard (see Figure 7.52).

5. In the Select Alignment drop-down list, select Frontenac Drive, and in the Profile View Style drop-down list, select the Full Grid option, as shown in Figure 7.52. Click Next.

FIGURE 7.52
The Create Multiple Profile Views wizard

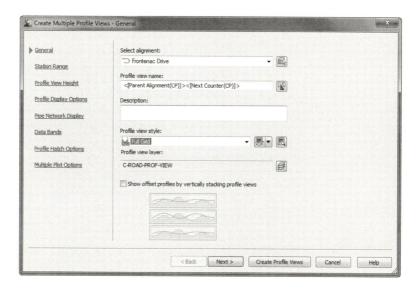

6. On the Station Range page, make sure the Automatic option is selected. This page is also where you set the length of each view. Click Next.

7. On the Profile View Height page, make sure the Automatic option is selected. Note that you could use the Split Profile View options from the previous exercise here as well. Click Next.

8. On the Profile Display Options page, scroll across until you get to Labels. Change the style to _No Labels.

9. Click the Multiple Plot Options text on the left side of the page to jump to that step in the wizard. This step controls whether the gapped profile views will be arranged in a column, row, or a grid. The Frontenac Drive alignment is fairly short, so the gapped views will be aligned in a row. However, it could be prudent with longer alignments to stack the profile views in a column or a compact grid, thereby saving screen space.

10. Click the Create Profile Views button and pick a point on screen to create a view similar to Figure 7.53.

The gapped profile views are the two profile views on the bottom of the screen and, just like the staggered profile view, show the entire alignment from start to finish. Unlike the staggered view, however, the gapped view is separated by a "gap" into two views. In addition, the gapped views are independent of each other so they have their own styles, properties, and labeling associated with them, making them useful when you don't want a view to show information that is not needed on a particular section. This is also the primary way to create divided profile views for sheet production.

Note that when using the Create Multiple Views option, every profile view is the full length as defined in the wizard, even if the alignment is not that long.

FIGURE 7.53
The staggered and gapped profile views of the Frontenac alignment

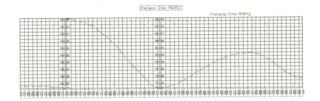
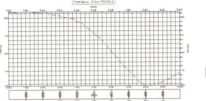
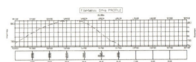

CREATING STACKED PROFILE VIEWS

In some parts of the United States, a three-line profile view is a common requirement. In this situation, the centerline is displayed in a central profile view, with left and right offsets shown in profile views above and below the centerline profile view. These are then typically used to show top-of-curb design profiles in addition to the centerline design. In this exercise, you look at how the Create Profile View wizard makes generating these views a simple process:

1. Open the StackedProfiles.dwg file. This drawing has sampled profiles for the Cabernet Court alignment at center as well as left and right offsets.

2. Change to the Home tab and select Profile View ➢ Create Profile View to display the Create Profile View wizard.

3. Select Cabernet Court from the Select Alignment drop-down list.

4. Check the Show Offset Profiles By Vertically Stacking Profile Views option on the General page of the wizard.

5. Click the Stacked Profile text on the left side of the wizard (it's a hyperlink) to jump to the Stacked Profile step; the page looks like Figure 7.54.

6. Set the style for each view as shown in Figure 7.54 and click Next.

7. Toggle the Draw option for the first profile (EG – Surface (13)) as shown in Figure 7.55. Note that Middle View - [1] is currently selected in the Select Stacked View To Specify Options For list box. Also note that your EG Surface numbers may not be the same as the example.

FIGURE 7.54
Setting up stacked profile views

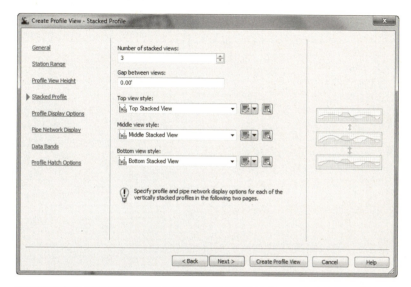

FIGURE 7.55
Setting the stacked view options for each view

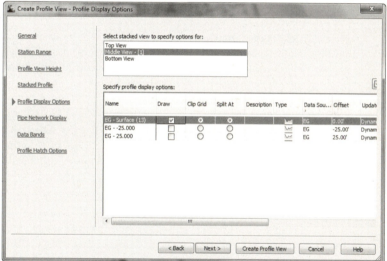

8. Click Top View in the Select Stacked View To Specify Options For list box, and then toggle on the EG - - 25.000 profile (the second profile listed). Note the two hyphen (-) symbols because of the naming template that ships in Civil 3D. This is the left-hand offset.

9. Click Bottom View in the Select Stacked View To Specify Options For list box, and then toggle on the EG - 25.000 profile (the third profile listed). This is the right-hand offset.

10. Click the Create Profile View button, and snap to the center of the circle to the southeast of the site as a pick point. Your result should look similar to Figure 7.56.

FIGURE 7.56
Completed stacked profiles

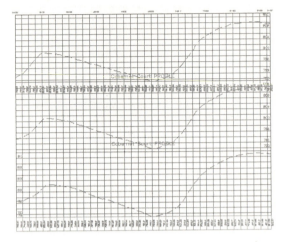

The styles for each profile can be adjusted, as can the styles for each profile view. The stacking here simply automates a process that many users found tedious. Once you have created layout profiles, they can also be added to these views by editing the profile view properties.

Profile Utilities

One common requirement is to compare profile data for objects that are aligned similarly but not parallel. Another is the ability to project objects from a plan view into a profile view. The abilities to superimpose profiles and project objects are both discussed in this section.

Superimposing Profiles

In a profile view, a profile is sometimes superimposed to show one profile adjacent to another (e.g., a ditch adjacent to a road centerline). In this brief exercise, you'll superimpose one of your street designs onto the other to see how they compare over a certain portion of their length:

1. Open the `SuperimposeProfiles.dwg` file. This drawing has two profile views created, one with a layout profile.

2. Change to the Home tab and select Profile ➢ Create Superimposed Profile. Civil 3D will prompt you to select a source profile.

3. Zoom into the Syrah Way profile view and pick the cyan layout profile. Civil 3D will prompt you to select a destination profile view for display.

4. Pick the Parker Place profile view to display the Superimpose Profile Options dialog shown in Figure 7.57.

Figure 7.57
The Superimpose Profile Options dialog

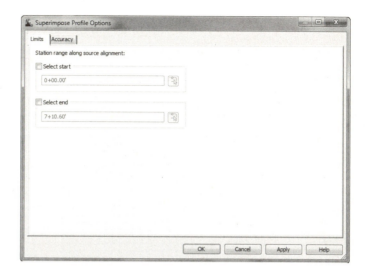

5. Click OK to dismiss the dialog, accepting the default settings.
6. Zoom in on the right side of the Frontenac Drive profile view to see the superimposed data, as shown in Figure 7.58.

Figure 7.58
The Syrah Way layout profile superimposed on the Frontenac Drive profile view

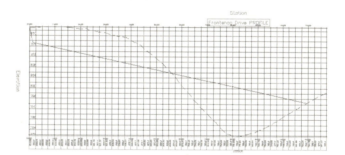

Note that the vertical curve in the Syrah Way layout profile has been approximated on the Frontenac Drive profile view, using a series of PVIs. Superimposing works by projecting a line from the target alignment (Frontenac Drive) to an intersection with the other source alignment (Syrah Way).

The target alignment is queried for an elevation at the intersecting station and a PVI is added to the superimposed profile. Note that this superimposed profile is still dynamic! A change in the Syrah Way layout profile will be reflected on the Frontenac Drive profile view.

Object Projection

Some AutoCAD and some AutoCAD Civil 3D objects can be projected from a plan view into a profile view. The list of available AutoCAD objects includes points, blocks, 3D solids, and 3D polylines. The list of available AutoCAD Civil 3D objects includes COGO points, feature lines, and survey figures. In the following exercise, you'll project a 3D object into a profile view:

1. Open the `ObjectProjection.dwg` file.

2. Change to the Home tab and select Profile View ➢ Project Objects To Profile View from the Profile & Section Views panel. Select the Fire Hydrant object located in the center of the circle and press ↵. Civil 3D will prompt you to select a profile view.

3. On the Syrah Way profile, select a grid line. The Project Objects To Profile View dialog opens.

4. Select the EG Elevation Option and verify that the other options match those in Figure 7.59. Click OK to dismiss the dialog, and review your results as shown in Figure 7.60.

Figure 7.59
A completed Project Objects To Profile View dialog

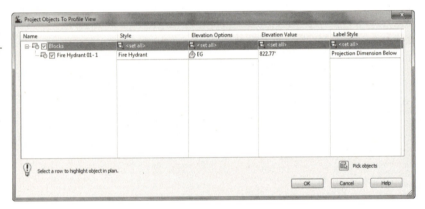

Figure 7.60
The COGO point object projected into a profile view

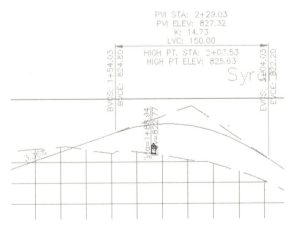

We actually wanted the Fire Hydrant to show on the Proposed surface. No problem. Follow these steps:

1. Click on the Fire Hydrant in profile view.

2. Right-click and select Projection Object Properties. The Projection View Properties dialog opens.

3. In the Projection View Properties dialog, click the <set all> area by the Elevation Options and select Syrah Way FG, as shown in Figure 7.61. Click OK.

4. The Fire Hydrant is now reprojected to the Syrah Way Finished Grade (FG), as shown in Figure 7.62.

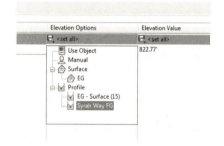

FIGURE 7.61
Select the Syrah Way FG elevation.

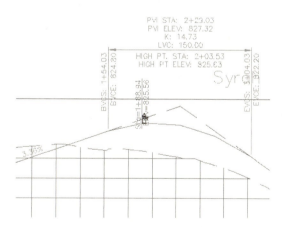

FIGURE 7.62
The Fire Hydrant reprojected onto the Syrah Way FG

Once an object has been projected into a profile view, the Profile View Properties dialog will display a new Projections tab. Projected objects will remain dynamically linked with respect to their plan placement. Because profile views and section views are similar in nature, objects can be projected into section views in the same fashion.

Editing Profile Views

Once profile views have been created, things get interesting. The number of modifications to the view itself that can be applied, even before editing the styles, makes profile views one of the most flexible pieces of the Civil 3D package. In this series of exercises, you'll look at a number of changes that can be applied to any profile view in place.

Profile View Properties

Picking a profile view and selecting Profile View Properties from the Modify View panel yields the dialog shown in Figure 7.63. The properties of a profile include the style applied, station and elevation limits, the number of profiles displayed, and the bands associated with the profile view. If a pipe network is displayed, a tab labeled Pipe Networks will appear.

FIGURE 7.63
Typical Profile View Properties dialog

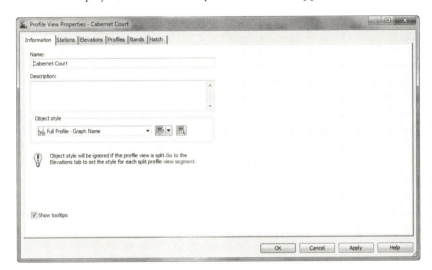

ADJUSTING THE PROFILE VIEW STATION LIMITS

In spite of the wizard, there are often times when a profile view needs to be manually adjusted. For example, the most common change is to limit the length or height (or both) of the alignment that is being shown so it fits on a specific size of paper or viewport. You can make some of these changes during the initial creation of a profile view (as shown in a previous exercise), but you can also make changes after the profile view has been created.

One way to do this is to use the Profile View Properties dialog to make changes to the profile view. The profile view is a Civil 3D object, so it has properties and styles that can be adjusted through this dialog to make the profile view look like you need it to.

1. Open the `ProfileViewProperties.dwg` file.

2. Zoom to the Cabernet Court profile view.

3. Pick a grid line, and select Profile View Properties from the Modify View panel.

4. On the Stations tab, click the User Specified Range radio button, and set the value of the End station to 7+00, as shown in Figure 7.64.

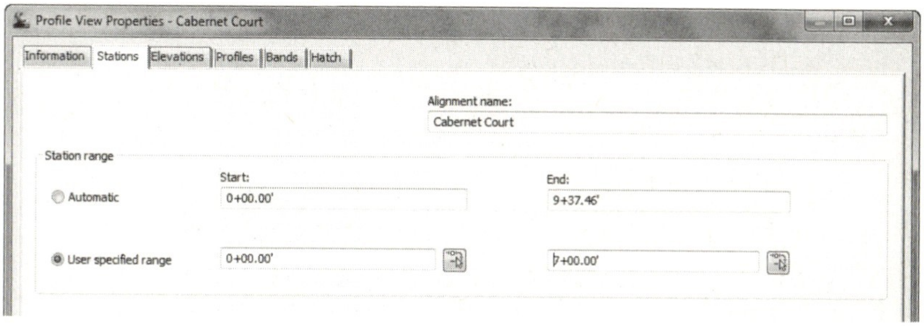

FIGURE 7.64
Adjusting the end station values for Cabernet Court

5. Click OK to close the dialog. The profile view will now reflect the updated end station value.

   One of the nice things about Civil 3D is that copies of a profile view retain the properties of that view, making a gapped view easy to create manually if they were not created with the wizard.

6. Enter **Copy**↵ on the command line. Pick the Cabernet Court profile view you just modified.

7. Press F8 on your keyboard to toggle on Ortho mode, and then press ↵.

8. Pick a base point and move the crosshairs to the right. When the crosshairs reach a point where the two profile views do not overlap, pick that as your second point, and press ↵ to end the COPY command.

9. Pick the copy just created and select Profile View Properties from the Modify View panel. The Profile View Properties dialog appears.

10. On the Stations tab, change the stations again. This time, set the Start field to **7+00** and the End field to **9+37.46**. The total length of the alignment will now be displayed on the two profile views, with a gap between the two views at station 7+00. Click OK, and your drawing will look like Figure 7.65.

FIGURE 7.65
A manually created gap between profile views

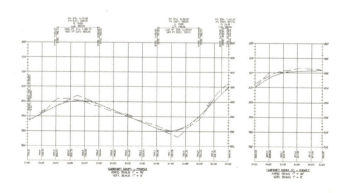

In addition to creating gapped profile views by changing the profile properties, you can show phase limits by applying a different style to the profile in the second view.

Adjusting the Profile View Elevations

Another common issue is the need to control the height of the profile view. Civil 3D automatically sets the datum and the top elevation of profile views on the basis of the data to be displayed. In most cases this is adequate, but in others, this simply creates a view too large for the space allocated on the sheet or wastes a large amount of that space.

1. Open the `ProfileViewProperties.dwg` file if you have not already done so.
2. Pan or zoom to the Syrah Way profile view.
3. Select Profile View Properties from the Modify View panel. The Profile View Properties dialog appears.
4. Change to the Elevations tab.
5. In the Elevation Range section, check the User Specified Height radio button and enter the minimum and maximum heights, as shown in Figure 7.66.

Figure 7.66
Modifying the height of the profile view

6. Click OK to close the dialog. The profile view of Syrah Way should reflect the updated elevations, as shown in Figure 7.67.

Figure 7.67
The updated profile view with the heights manually adjusted

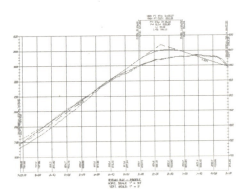

The Elevations tab can also be used to split the profile view and create the staggered view that you previously created with the wizard:

1. Pick the Frontenac Drive profile view, right-click a grid line, and select the Profile View Properties option to open the Profile View Properties dialog.
2. Switch to the Elevations tab. In the Elevations Range area, click the User Specified Height radio button.

3. Check the Split Profile View option.

4. Notice that the Height field is now active. Set the height to **36**.

5. In the Split Profile View Data section, click the plus (+) icon twice and complete the dialog as shown in Figure 7.68.

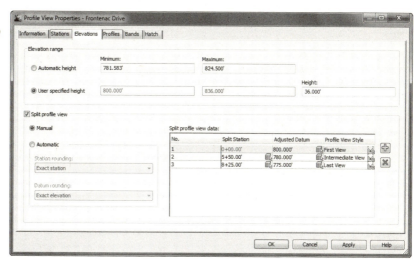

FIGURE 7.68
The Elevations tab

6. Click OK to exit the dialog. The profile view should look like Figure 7.69.

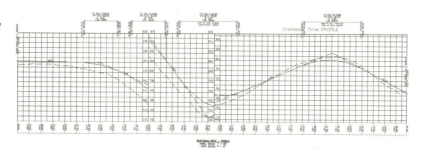

FIGURE 7.69
A split profile view for the Frontenac Drive alignment

Automatically creating split views is a good starting point, but you'll often have to tweak them as you've done here. The selection of the proper profile view styles is an important part of the Split Profile View process. We'll look at styles in Chapter 19.

Profile Display Options

Civil 3D allows the creation of literally hundreds of profiles for any given alignment. Doing so makes it easy to evaluate multiple design solutions, but it can also mean that profile views get very crowded. In this exercise, you'll look at some profile display options that allow the toggling of various profiles within a profile view:

1. Open the `ProfileViewProperties1.dwg` file.
2. Pick the Cabernet Court profile view, and select Profile View Properties from the Modify View panel. The Profile View Properties dialog appears.
3. Switch to the Profiles tab.
4. Uncheck the Draw option in the EG Surface row and click the Apply button.
5. Drag your dialog out of the way and your profile view should look similar to Figure 7.70.

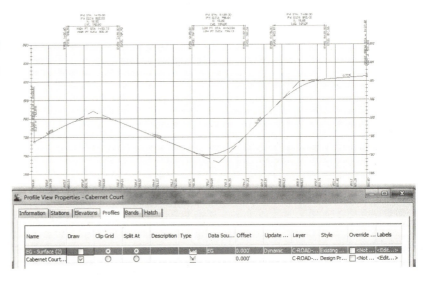

FIGURE 7.70
The Cabernet Court profile view with the Draw option toggled off

Toggling off the Draw option for the EG surface has created a profile view style in which a profile of the existing ground surface will not be drawn on the profile view.

The sampled profile from the EG surface still exists under the Cabernet Court alignment; it simply isn't shown in the current profile view. Now that you've modified a number of styles, let's look at another option that is available on the Profile View Properties dialog: bands.

PROFILE VIEW BANDS

Data bands are horizontal elements that display additional information about the profile or alignment that is referenced in a profile view. Bands can be applied to both the top and bottom of a profile view, and there are six different band types:

Profile Data Bands Display information about the selected profile. This information can include simple elements such as elevation, or more complicated information such as the cut-fill between two profiles at the given station.

Vertical Geometry Bands Create an iconic view of the elements making up a profile. Typically used in reference to a design profile, vertical data bands make it easy for a designer to see where vertical curves are located along the alignment.

Horizontal Geometry bands Create a simplified view of the horizontal alignment elements, giving the designer or reviewer information about line, curve, and spiral segments and their relative location to the profile data being displayed.

Superelevation Bands Display the various options for Superelevation values at the critical points along the alignment.

Sectional Data Bands Can display information about the sample line locations, the distance between them, and other sectional-related information.

Pipe Data Bands Can show specific information about each pipe or structure being shown in the profile view.

In this exercise, you'll add bands to give feedback on the EG and layout profiles, as well as horizontal and vertical geometry:

1. Open the ProfileViewBands.dwg file.

2. Zoom out and down to pick the Frontenac Drive profile view, and select Profile View Properties from the Modify View panel. The Profile View Properties dialog appears.

3. Click the Bands tab, as shown in Figure 7.71.

FIGURE 7.71
The Bands tab of the Profile View Properties dialog

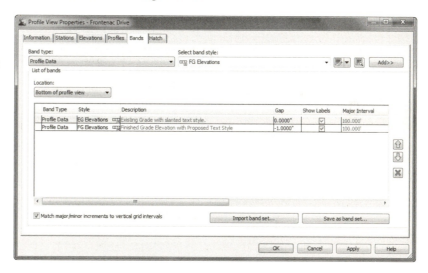

4. Verify that the Band Type drop-down is set to Profile Data and set the Select Band Style drop-down to Elevations And Stations. Click the Add button to display the Geometry Points To Label In Band dialog shown in Figure 7.72.

5. Click OK to close the dialog and return to the Profile View Properties dialog. The Profile Data band should now be listed in the middle of the dialog.

6. Set the Location drop-down to Top Of Profile View.

7. Change the Band Type drop-down to the Horizontal Geometry option and choose the Geometry option from the Select Band Style drop-down. Click Add. The Horizontal Geometry band will now be added to the table in the List Of Bands section.

FIGURE 7.72
The Geometry Points To Label In Band dialog

8. Change the Band Type drop-down to Vertical Geometry. Do not change the Select Band Style field from its current selection (Geometry). Click Add. The Vertical Geometry band will also be added to the table in the List Of Bands section.

9. Click OK to exit the dialog. Your profile view should look like Figure 7.73.

FIGURE 7.73
Applying bands to a profile view

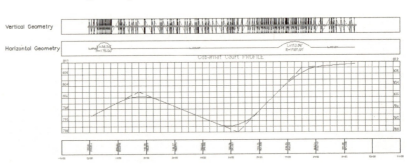

However, there are obviously problems with the bands. The Vertical Geometry band is a mess and is located above the title of the profile view, whereas the Horizontal Geometry band actually overwrites the title. In addition, the elevation information has two numbers with different rounding applied. In this exercise, you'll fix those issues:

1. Pick the Cabernet Court profile view, and then select Profile View Properties from the Modify View panel. The Profile View Properties dialog appears.

2. Switch to the Bands tab.

3. Verify that the Location drop-down in the List Of Bands section is set to Bottom Of Profile View.

4. Verify that the "Match major/minor increments to vertical grid intervals" option at the bottom of the screen is selected. Checking this option ensures that the major/minor intervals of the profile data band match the major/minor profile view style's major/minor grid spacing.

5. The Profile Data band is the only band currently listed in the table in the List Of Bands section. Scroll right in the Profile Data row and notice the two columns labeled Profile 1 and Profile 2. Change the value of Profile 2 to Cabernet Court FG, as shown in Figure 7.74.

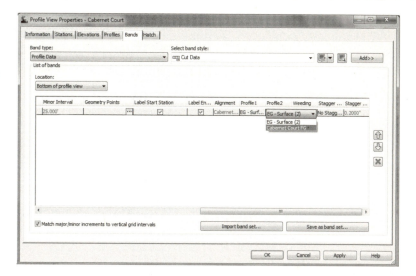

FIGURE 7.74
Setting the profile view bands to reference the Cabernet Court FG profile

6. Change the Location drop-down to Top Of Profile View.

7. The Horizontal Geometry and the Vertical Geometry bands are now listed in the table as well. Scroll to the right again, and set the value of Profile 1 in the Vertical Geometry band to Cabernet Court FG.

8. Scroll back to the left and set the Gap value for the Horizontal Geometry band to 1.5". This value controls the distance from one band to the next or to the edge of the profile view itself.

9. Click OK to close the dialog. Your profile view should now look like Figure 7.75.

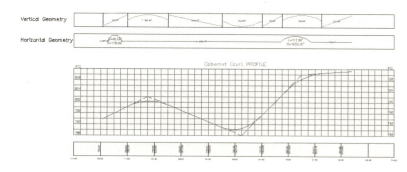

FIGURE 7.75
Completed profile view with the Bands set appropriately

Bands use the Profile 1 and Profile 2 designation as part of their style construction. By changing the profile referenced as Profile 1 or 2, you change the values that are calculated and displayed (e.g., existing versus proposed elevations). These bands are just more items that are driven by styles, which you will learn more about in Chapter 19.

Profile View Hatch

Many times it is necessary to shade cut/fill areas in a profile view. The settings on the Hatch tab are used to specify upper and lower cut/fill boundary limits for associated profiles (see Figure 7.76). Shape styles from the General Multipurpose Styles collection found on the Settings tab of Toolspace can also be selected here. These settings include the following:

Cut Area Click this button to add hatching to a profile view in areas of the cut.

Fill Area Click this button to add hatching to a profile view in areas of the fill.

Multiple Boundaries Click this button to add hatching to a profile view in areas of a cut/fill where the area must be averaged between two existing profiles (for example, finished ground at the centerline vs. the left and right top of a curb).

From Criteria Click this button to import Quantity Takeoff Criteria.

Figure 7.77 shows a cut and fill hatched profile.

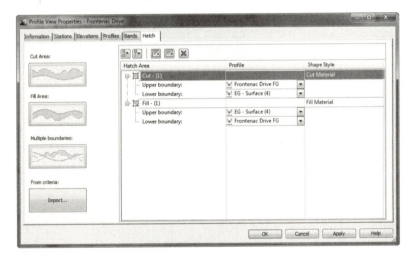

Figure 7.76
Shape style selection on the Hatch tab of the Profile View Properties dialog

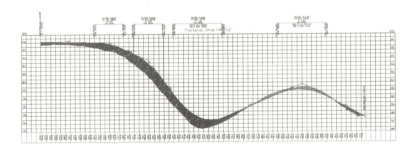

Figure 7.77
The Frontenac Drive Profile shown with cut and fill shading

Mastering Profiles and Profile Views

One of the most difficult concepts to master in AutoCAD Civil 3D is the notion of which settings control which display property. Although the following two rules may sound overly simplistic, they are easily forgotten in times of frustration:

- Every object has a label and a style.
- Every label has a style.

Furthermore, if you can remember that there is a distinct difference between a profile object and a profile view object you place it in, you'll be well on your way to mastering profiles and profile views. When in doubt, select an object, right-click, and pay attention to the Civil 3D commands available.

Profile View Style Selection

Selection of a profile view style is straightforward, but because of the large number of settings in play with a profile view style, the changes can be dramatic. In the following quick exercise, you'll change the style and see how much a profile view can change in appearance:

1. Open the ProfileViewsModify.dwg file. Remember that all files can be downloaded from www.sybex.com/go/masteringcivil3d2012.

2. Pick a grid line in the Syrah Way profile view and right-click. Notice the available commands.

3. Select the Profile View Properties option. The Profile View Properties dialog opens.

4. On the Information tab, change Object Style to Major Grids and click OK to arrive at Figure 7.78.

Figure 7.78
The Syrah Way profile view with the Major Grids style applied

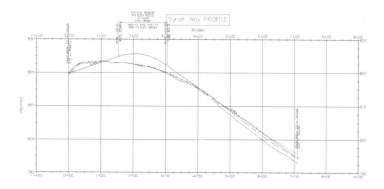

A profile view style includes information such as labeling on the axis, vertical scale factors, grid clipping, and component coloring. Using various styles lets you make changes to the view to meet requirements without changing any of the design information. Changing the style is a straightforward exercise.

Labeling Styles

Now that the profile is labeled, the profile view grid spacing is set, and the titles all look good, it's time to add some specific callouts and detail information. Civil 3D uses profile view labels and bands for annotating.

VIEW ANNOTATION

Profile view annotations label individual points in a profile view, but they are not tied to a specific profile object. These labels can be used to label a single point or the depth between two points in a profile. We say "depth" because the label recognizes the vertical exaggeration of the profile view and applies the scaling factor to label the correct depth. Profile view labels can be either station elevation or depth labels. In this exercise, you'll use both:

1. Open the `ProfileViewLabels.dwg` file.
2. Zoom in on the Syrah Way profile view.
3. Switch to the Annotate tab and select Add Labels from the Labels & Tables panel. The Add Labels dialog opens.
4. In the Feature drop-down, select Profile View; in the Label Type drop-down, verify that Station Elevation is selected; and in the Station Elevation Label Style drop-down, make sure that Station And Elevation is selected.
5. Click the Add button.
6. Click a grid line in the Syrah Way profile view. Zoom in on the right side so that you can see the point where the EG and layout profiles cross over.
7. Pick this profile crossover point visually, and then pick the same point to set the elevation and press ↵. Your label should look like Figure 7.79.

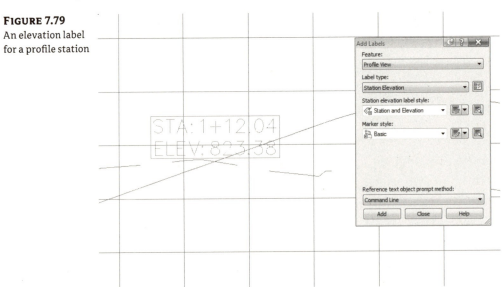

FIGURE 7.79
An elevation label for a profile station

8. In the Add Labels dialog, change both Label Type and Depth Label Style to the Depth option. Click the Add button.

9. Click a grid line on the Syrah Way profile view.

10. Pick a point along the layout profile and then pick a point along the EG profile and press ↵. The depth between the two profiles will be measured, as shown in Figure 7.80.

11. Close the Add Labels dialog.

Figure 7.80
A depth label applied to the Syrah Way profile view

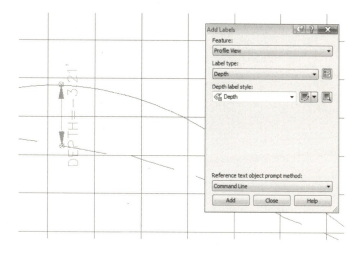

Why Don't Snaps Work?

There's no good answer to this question. For a number of releases now, users have been asking for the ability to simply snap to the intersection of two profiles. We mention this because you'll try to snap and wonder if you've lost your mind. You haven't—it just doesn't work. Maybe next year? If you are after a solution (it's not elegant), you can draw lines on top of the profile.

Depth labels can be handy in earthworks situations where cut and fill become critical, and individual spot labels are important to understanding points of interest, but most design documentation is accomplished with labels placed along the profile view axes in the form of data bands. The next section describes these band sets.

Band Sets

Band sets are simply collections of bands, much like the profile label sets or alignment label sets. In this exercise, you'll save a band set, and then apply it to a second profile view:

1. Open the ProfileViewBandSets.dwg file.

2. Pick the Syrah Way profile view, and select Profile View Properties from the Modify View panel. The Profile View Properties dialog opens.

306 | **CHAPTER 7** PROFILES AND PROFILE VIEWS

3. Switch to the Bands tab.
4. Click the Save As Band Set button to display the Band Set dialog in Figure 7.81.

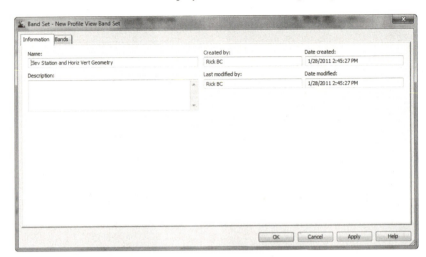

FIGURE 7.81
The Information tab for the Band Set dialog

5. In the Name field, enter **Elev Station and Horiz Vert Geometry**.
6. Click OK to close the Band Set dialog.
7. In the Profiles tab, click the Clip Grid radio button next to Frontenac Drive FG.
8. Click OK to close the Profile View Properties dialog.
9. Pick the Frontenac Drive profile view, and select Profile View Properties from the Modify View panel.
10. Switch to the Bands tab.
11. Click the Import Band Set button, and the Band Set dialog opens.
12. Select the Elev Station and Horiz Vert Geometry option from the drop-down list and click OK.
13. Select Top Of Profile View from the Location drop-down list.
14. Scroll over on the Vertical Geometry and set Profile1 to Frontenac Drive FG.
15. Click OK to exit the Profile View Properties dialog.

16. From the Home tab and Modify panel, select Match Properties. You can also type **MA↵**.
17. At the `Select source object:` prompt, select the profile grid on Syrah Way.
18. At the `Select destination object(s) or [Settings]:` prompt, select the profile grid on Frontenac Drive.
19. The grid for Frontenac Drive now matches the Syrah Way profile. Your profile view should look like Figure 7.82.

Figure 7.82
Completed profile view after importing the band set and matching properties

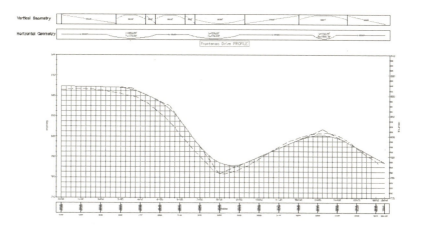

Your Frontenac Drive profile view now looks like the Syrah Way profile view. Band sets allow you to create uniform labeling and callout information across a variety of profile views. By using a band set, you can apply myriad settings and styles that you've assigned to a single profile view to a number of profile views. The simplicity of enforcing standard profile view labels and styles makes using profiles and profile views simpler than ever.

> **BRING OUT THE BAND**
>
> Once design speeds have been assigned to an alignment, and superelevation has been calculated, you'll find the Superelevation View command on the Modify panel of the Alignment context menu. Although not the focus of this chapter, superelevation views behave much like profile views, and you can access their properties via the right-click menu after selecting a view.

The Bottom Line

Sample a surface profile with offset samples.   Using surface data to create dynamic sampled profiles is an important advantage of working with a three-dimensional model. Quick viewing of various surface slices and grip-editing alignments makes for an effective preliminary planning tool. Combined with offset data to meet review agency requirements, profiles are robust design tools in Civil 3D.

Master It   Open the `MasteringProfile.dwg` file and sample the ground surface along Alignment A, along with offset values at 15′ left and 25′ right of the alignment.

Lay out a design profile on the basis of a table of data.   Many programs and designers work by creating pairs of station and elevation data. The tools built into Civil 3D let you input this data precisely and quickly.

Master It In the `Mastering Profiles.dwg` file, create a layout profile on Alignment B with the following information:

Station	PVI Elevation	Curve Length
0+00	812.76	
1+45	818.59	250'
5+22	794.48	

Add and modify individual components in a design profile. The ability to delete, modify, and edit the individual components of a design profile while maintaining the relationships is an important concept in the 3D modeling world. Tweaking the design allows you to pursue a better solution, not just a working solution.

Master It In the `MasteringProfile.dwg` file, on profile B insert a PVI at Sta 3+52, Elevation 812. Modify the curve so that it is 100' and then, add a 175' parabolic vertical curve at the newly created point.

Apply a standard label set. Standardization of appearance is one of the major benefits of using Civil 3D styles in labeling. By applying label sets, you can quickly create plot-ready profile views that have the required information for review.

Master It In the `MasteringProfile.dwg` file, apply the Road Profile Labels set to all layout profiles.

Chapter 8

Assemblies and Subassemblies

Roads, ditches, trenches, and berms usually follow a predictable pattern known as a *typical section*. Assemblies are how you tell Civil 3D what these typical sections look like. Assemblies are made up of smaller components called *subassemblies*. For example, a typical road section assembly contains subassemblies such as lanes, sidewalks, and curbs.

In this chapter, the focus will be on understanding where these assemblies come from and how to build and manage them.

In this chapter, you will learn to:

- Create a typical road assembly with lanes, curbs, gutters, and sidewalks
- Edit an assembly
- Add daylighting to a typical road assembly

Subassemblies

A *subassembly* is a building block of a typical section, known as an *assembly*. Examples of subassemblies include lanes, curbs, sidewalks, channels, trenches, daylighting, and any other component required to complete a typical corridor section.

An extensive catalog of subassemblies has been created for use in Civil 3D. More than 100 subassemblies are available in the standard catalogs, and each subassembly has a list of adjustable *parameters*. There are also about a dozen generic links you can use to further refine your most complex assembly needs. From ponds and berms to swales and roads, the design possibilities are almost infinite.

You will add subassemblies to a design by clicking on them from the subassembly tool palette, as you'll see later in this chapter. By default, Civil 3D has several tool palettes created for corridor modeling.

You can access these tool palettes by changing to the Home tab and clicking the Tool Palettes button on the Palettes panel. When Civil 3D is installed, you have an initial set of the most commonly used assemblies and subassemblies ready to go, as shown in Figure 8.1. However, there are more subassemblies available to you in the Corridor Modeling Catalog.

> **GETTING TO THE TOOL PALETTES**
>
> The exercises in this chapter depend heavily on the use of the Tool Palette feature of AutoCAD when pulling together assemblies from subassemblies. To avoid some redundancy, we will omit the initial step of opening the tool palette in every single exercise. If it's open, leave it open; if it's closed, open it. In case you need a reminder, the easiest way to open the Tool Palette feature is Ctrl+3.

FIGURE 8.1
The Civil 3D subassemblies tool palette

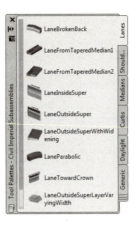

The Corridor Modeling Catalog

If the default set of subassemblies is not adequate for your design situation, check the Corridor Modeling Catalog for one that will work.

ACCESSING THE CORRIDOR MODELING CATALOG

The Corridor Modeling Catalog is installed by default on your local hard drive. Change to the Modify tab and then choose Assembly from the Design panel to open the Assembly contextual tab.

On the Assembly contextual tab, click the Catalog button on the Launch Pad panel to open a content browser interface. Choose either the Metric or Imperial catalog to explore the entire collection of subassemblies available in each category (see Figure 8.2).

FIGURE 8.2
The front page of the Corridor Modeling Catalog

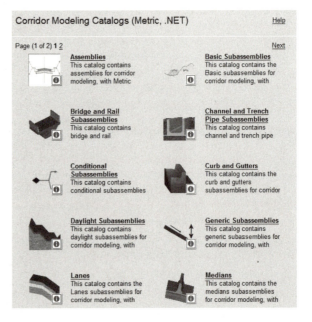

Accessing Subassembly Help

Later, this chapter will point out other shortcuts to access the extensive subassembly documentation. You can get quick access to information by right-clicking any subassembly entry on the Corridor Modeling Catalog page and selecting the Help option (see Figure 8.3).

Figure 8.3
Accessing the Help file through the Corridor Modeling Catalog

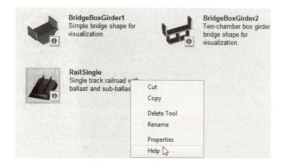

The Subassembly Reference in the Help file provides a detailed breakdown of each subassembly, examples for its use, its parameters, a coding diagram, and more. While you're searching the catalog for the right parts to use, you'll find the Subassembly Reference infinitely useful.

Adding Subassemblies to a Tool Palette

If you'd like to add additional subassemblies to your tool palettes, you can use the i-drop to grab subassemblies from the catalog and drop them onto a tool palette. To use the i-drop, click the small blue *i* next to any subassembly, and continue to hold down your left mouse button until you're over the desired tool palette. Release the button, and your subassembly should appear on the tool palette (see Figure 8.4).

Figure 8.4
Using the i-drop to add a subassembly to a tool palette

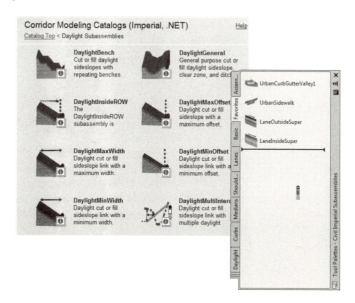

Building Assemblies

You build an assembly from the Home tab by choosing Assembly ➤ Create Assembly from the Create Design panel. The result is the main assembly baseline. This is the point on the assembly that gets "hooked in" to your design alignment and profile. A typical assembly baseline is shown in Figure 8.5.

FIGURE 8.5
Creating an assembly (left); an assembly baseline (right)

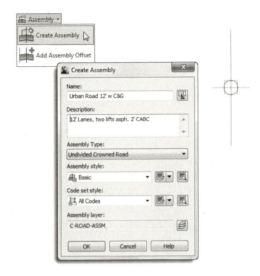

When an assembly is created, you have the option of telling Civil 3D what type of assembly this will be:

- Undivided Crowned Road
- Undivided Planar Road
- Divided Crowned Road
- Divided Planar Road
- Other

These categories will help the software determine the axis of rotation options in superelevation, if needed.

The Pre-Cooked Assemblies

On the Assemblies tab of the tool palette, you will find a selection of predefined, completed assemblies (Figure 8.6). These are a great starting point for beginners who are looking for examples of how subassemblies are put together. There are examples of simple roadway sections as well as more advanced items, such as intersection and roundabout examples. To use one, click on the desired assembly, then click to place it in your drawing.

FIGURE 8.6
Predefined assemblies

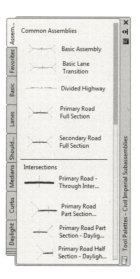

Creating a Typical Road Assembly

The process for building an assembly requires the use of the Tool Palette feature and the AutoCAD Properties palette, both of which can be docked. You'll quickly learn how to best orient these palettes with your limited screen real estate. If you run dual monitors, you may find it useful to place both of these palettes on your second monitor.

The exercise that follows builds a typical assembly using LaneOutsideSuper, UrbanCurbGutterGeneral, UrbanSidewalk, and DaylightMaxOffset subassemblies (see Figure 8.7).

FIGURE 8.7
A typical road assembly

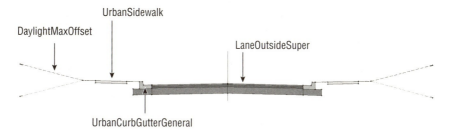

Let's have a more detailed look at each component you'll use in the following exercise. A quick peek into the subassembly Help will give you a breakdown of attachment options; input parameters; target parameters; output parameters; behavior; layout mode operation; and the point, link, and shape codes.

The LaneOutsideSuper Subassembly The LaneOutsideSuper subassembly is the best all-purpose subassembly for lanes. It can superelevate if needed, and allows for up to four layers of materials. The default width of 12′ (3.6m) can be adjusted in the parameters or can be used with an offset alignment to control its width. (See Figure 8.8.)

FIGURE 8.8
The LaneOutsideSuper subassembly

The UrbanCurbGutterGeneral Subassembly The UrbanCurbGutterGeneral subassembly (Figure 8.9) is another standard component that creates an attached curb and gutter. Looking into the subassembly Help, you'll see a diagram of the BasicCurbandGutter with callouts for its seven parameters: side; insertion point; gutter width and slope; and curb height, width, and depth. You can adjust these parameters to match many standard curb-and-gutter configurations.

FIGURE 8.9
The UrbanCurbGutterGeneral subassembly

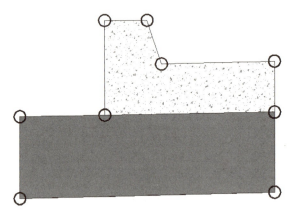

The UrbanSidewalk Subassembly The UrbanSidewalk subassembly (Figure 8.10) creates a sidewalk and terrace buffer strips. The Help file lists the following five parameters for the UrbanSidewalk subassembly: side, width, depth, buffer width 1, and buffer width 2. These parameters let you adjust the sidewalk width, material depth, and buffer widths to match your design specification. You can adjust the slope for the entire length of the subassembly as a unit, but not the individual segments.

The UrbanSidewalk subassembly can return quantities of concrete (or other sidewalk construction material) but not gravel bedding or other advanced material layers.

FIGURE 8.10
The UrbanSidewalk subassembly

The DaylightMaxOffset Subassembly The DaylightMaxOffset subassembly (Figure 8.11) is a nice "starter" for creating simple, single-slope daylight instructions for your corridor. The maximum offset in the parameters is measured from the baseline (in our example, the centerline of the road). The slope will attempt a default of 4:1, but will adjust if it needs to in order to keep inside your specified maximum offset.

Figure 8.11
The Daylight-MaxOffset Subassembly

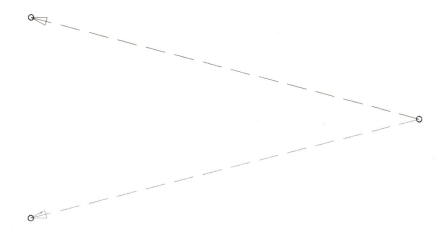

> ### What's With the Funny Names?
>
> You'll notice that all subassemblies have names with no spaces. This is because of the underlying .NET coding that makes up a subassembly. When you place one of these in your project it will retain the name from the tool palette and place a number after it to ensure unique names for all subassemblies in your drawing. Later in this chapter, you'll see how to rename them to something more user-friendly!

In the following exercise, you'll build a typical road assembly using these subassemblies. Follow these steps:

1. Create a new drawing from the Civil 3D template of your choice. Be sure your Civil 3D tool palette is showing the subassembly set appropriate for your drawing units (or you may end up with monster 12 meter lanes!). Change the active tool palette by right-clicking the edge of the tool palette, as shown in Figure 8.12.

2. Change to the Home tab and choose Assembly ➢ Create Assembly from the Create Design panel. The Create Assembly dialog opens.

3. Enter **Urban 14′ Single-Lane** (or Urban 4.5m Single-Lane) in the Name text box. Set the Assembly Type to Undivided Crowned Road. Make sure the Assembly Style text box is set to Basic and the Code Set Style text box is set to All Codes. Click OK.

4. Pick a location in your drawing for the assembly—somewhere in the center of your screen is fine.

5. Locate the Lanes tab on the tool palette. Position the palette on your screen so that you can clearly see the assembly baseline.

316 | **CHAPTER 8** ASSEMBLIES AND SUBASSEMBLIES

6. Click the LaneOutsideSuper button on the tool palette. The AutoCAD Properties palette appears. Position the palette on your screen so that you can clearly see both the assembly baseline and the Lanes tool palette.

7. Locate the Advanced-Parameters section on the Design tab of the AutoCAD Properties palette (Figure 8.13). This section lists the LaneOutsideSuper parameters. Make sure the Side parameter says Right, set the Crown Point on Inside to Yes, and change the Width parameter to 14′ (4.5 m). Your Properties palette should resemble Figure 8.13.

FIGURE 8.12
Right-click the edge of the tool palette to change assembly sets if needed.

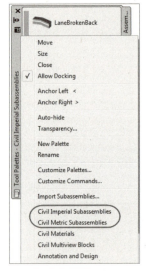

FIGURE 8.13
The Advanced Parameters inside the Properties palette

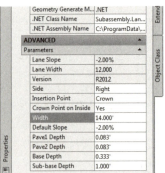

8. The command line states `Select marker point within assembly or [RETURN for Detached]`: Click anywhere on the red assembly marker to place the first lane.

9. Switch to the Curbs tab in the subassemblies palette.

10. Click the UrbanCurbGutterGeneral button on the tool palette. The Advanced section of the AutoCAD Properties palette's Design tab lists the UrbanCurbGutterGeneral parameters. Verify the Side parameter is set to Right; you will accept the parameter defaults, so no changes are needed.

11. The command line states Select marker point within assembly or [RETURN for Detached]:. Click the circular point marker located at the top right of the LaneOutsideSuper subassembly. This marker represents the top-right edge of pavement (see Figure 8.14).

FIGURE 8.14
The Urban-CurbGutterGeneral subassembly placed on the LaneOutsideSuper subassembly

IF YOU GOOF...

Often, the first instinct when a subassembly is misplaced is to Undo or erase the wayward piece. However, if you have spent lots of time diligently tweaking parameters, there is a way to fix things without redoing the subassembly. Select the errant subassembly and use the Move To Assembly option in the context tab. Use this instead of the base-AutoCAD Move tool to get the best results. Using regular AutoCAD Move may cause unexpected results in the corridor.

If you placed everything correctly but forgot to change a parameter or two, there's an easy fix for that, too. Cancel out of any active subassembly placement and select the subassembly you wish to change. Use good-old AutoCAD properties to quickly access the parameters.

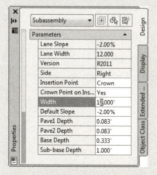

Most subassembly parameters can be changed from base-AutoCAD properties. For more heavy-duty modifications (such as side or renaming), you will want to get into the Subassembly Properties discussed later in this chapter.

318 | **CHAPTER 8** ASSEMBLIES AND SUBASSEMBLIES

12. In the Curbs tab, click the UrbanSidewalk button on the tool palette. In the Advanced section of the Design tab on the AutoCAD Properties palette, verify the Side parameter is set to Right, the Sidewalk Width parameter to 5′ (1.5 m), and the Inside Boulevard Width and Outside Boulevard Width parameters to 2′ (0.7 m).

13. The command line states `Select marker point within assembly or [RETURN for Detached]:`. Click the circular point marker on the UrbanCurbGutterGeneral subassembly that represents the top rear of the curb to attach the UrbanSidewalk subassembly (see Figure 8.15).

FIGURE 8.15
The BasicSidewalk subassembly placed on the UrbanCurbGutterGeneral subassembly

14. Switch to the Daylight tab on the Civil subassemblies palette. Select the DaylightMaxOffset subassembly. In the Advanced-Parameters area, set the Max Offset from Baseline to 50′ (17 m).

15. The command line states `Select marker point within assembly or [RETURN for Detached]:`. Select the circular marker on the outermost point of the sidewalk subassembly. Your drawing should now resemble Figure 8.16.

FIGURE 8.16
The complete right side of the assembly with DaylightMaxOffset

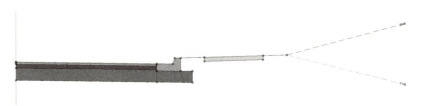

16. To complete the left side, you will use the Mirror Subassemblies command, as shown in Figure 8.17. Select all four subassemblies to the right of the baseline.

FIGURE 8.17
The context tab with subassembly modification tools

17. The context tab will show a variety of tools, including Mirror Subassemblies (Figure 8.17). Select Mirror Subassemblies, and then click the assembly baseline. Your assembly should now resemble Figure 8.7 from earlier in the chapter.

You have now completed a typical road assembly. Save your drawing if you'd like to use it in a future exercise.

Getting the Most from Subassembly Help

Each subassembly is capable of accomplishing different tasks in your design. There is no way to tell just by looking at the icon all the acrobatics that an assembly can do. For a detailed rundown of each parameter, and what can be done with a subassembly, you will need to pop into the help files.

Before we go into the help files, let's dig deeper into the anatomy of a subassembly. A subassembly is made up of three basic parts: *links*, *marker points*, and *shapes*, as shown in Figure 8.18. Each piece plays a role in your design and is used for different purposes at each stage of the design process.

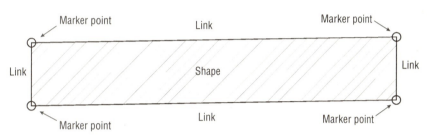

FIGURE 8.18
Schematic showing parts of an assembly

Links

Links are the linear components to your assembly. A link usually represents the top or bottom of a material but can also be used as a spacer between subassemblies.

Links can have codes assigned to them that Civil 3D uses to build the design. Think of these codes as nicknames. In the example assembly you created in the previous exercise, you used a sidewalk subassembly. On the sidewalk the topmost link has the codes Top and Sidewalk (Figure 8.19a); on the lane subassembly the topmost codes are Top and Pave (Figure 8.19b).

Links will be your primary source of data when creating proposed surfaces from your corridors.

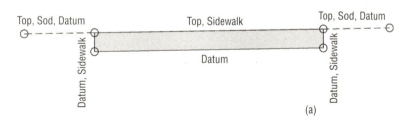
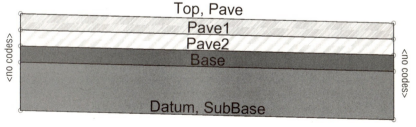

FIGURE 8.19
Link codes on the UrbanSidewalk Subassembly (a) and link codes on the LaneOutsideSuper Subassembly (b)

MARKER POINTS

Marker points are located at the endpoints of every link and usually are represented by the circles you see on the subassemblies as shown in Figures 8.20a and 8.20b. As you experienced in the previous exercise, the markers are used in assembly creation to "click" subassemblies together.

When a corridor is created, the marker points show real might. Markers are used as "hooks" to attach to alignments and/or profiles, known as *targets*, as discussed in Chapter 9, "Basic Corridors." Markers are the starting point for *feature lines* generated by the corridor, which are used for a variety of purposes that we will discuss in the upcoming chapters.

FIGURE 8.20
Marker point codes on the UrbanSidewalk subassembly (a) and marker point codes on the LaneOutsideSuper subassembly (b)

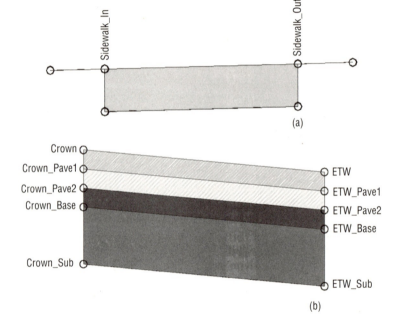

SHAPES

Shapes are the areas inside a closed formation of links. For example, Figure 8.21a and Figure 8.21b show different subassemblies with shape codes labeled. Shapes are used in end-area material quantity calculations. At the time an assembly is created, you do not need to consider what material these shapes represent. After your corridor is complete, you will specify what materials the codes represent upon computing materials.

Jumping into Help

Subassembly Help is extremely—well, helpful! There are many doors into the help files, including from the Corridor Modeling Catalog as you saw earlier. Right-click on any subassembly in the tool palette and select Help as shown in Figure 8.22.

FIGURE 8.21
Shape codes on the UrbanSidewalk Subassembly (a) and shape codes on the LaneOutsideSuper subassembly (b)

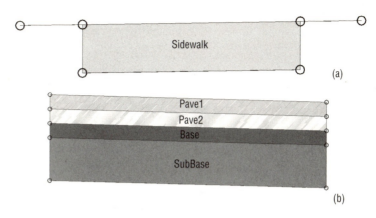

FIGURE 8.22
Getting to the subassembly help file for UrbanCurbGutterGeneral

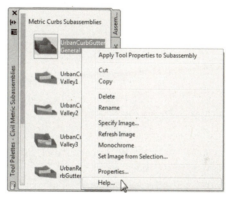

HELP 101

When you access Subassembly Help, it will take you to the help file specific to the subassembly you are working with. At the top, you will see a diagram showing the location of the numeric parameters.

For most subassemblies, the default attachment point will be the topmost-inside marker point. The help file will tell you if this differs for the subassembly you are looking at (Figure 8.23). Scroll further down to see detailed explanation of every parameter.

TARGET PARAMETERS AND OUTPUT PARAMETERS

Target parameters are a listing of what targets can be set for a subassembly. In other words, if a marker point can be "hooked" to something (i.e., an offset alignment), the target parameters will tell you what to look for in the corridor targets and the type of object it can hook. The help file will also tell you whether the target is optional or required. We will get into target parameters and setting targets in Chapter 9.

Output parameters are values calculated on corridor build, such as the cross-slope of a lane. In several subassemblies, there is an advanced option called Parameter Reference that can use output parameters instead of a constant as a slope.

FIGURE 8.23
The top portion of Subassembly Help shown with subassembly parameters

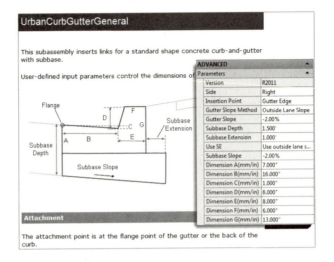

READING A CODING DIAGRAM

The coding diagram gives a list of all the codes used on the subassembly you are working with. Every named marker point, link, and shape is listed here. Not all subassembly components have explicit names, such as L10 shown in Figure 8.24. If the marker point, shape, or link is not included in the table, it is considered uncoded.

FIGURE 8.24
Coding diagram and name table for UrbanCurbGutter-General

Point / Link	Code	Description
P1	Flange	Flange point of the gutter
P2	Flowline_Gutter	Gutter flowline point
P3	TopCurb	Top-of-curb
P4	BackCurb	Back-of-curb
L1 – L3	Top, Curb	Finish grade on the curb and gutter
L7	Subbase Datum	
S1	Curb	Curb-and-gutter concrete area
S2	Subbase	

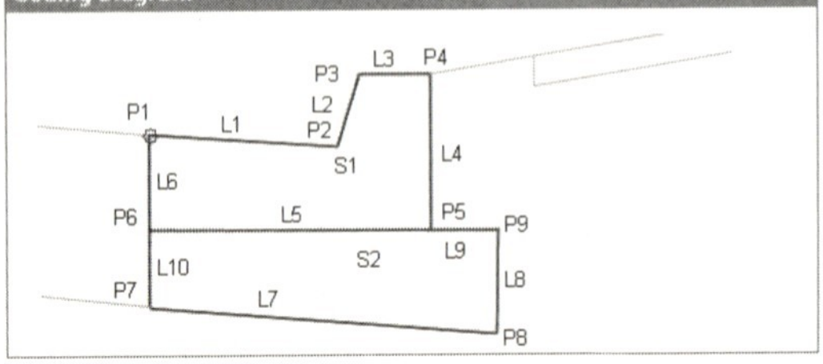

Commonly Used Subassemblies

Once you gain some skills in building assemblies, you can explore the Corridor Modeling Catalog to find subassemblies that have more advanced parameters so that you can get more out of your corridor model. For example, if you must produce detailed schedules of road materials such as asphalt, coarse gravel, fine gravel, subgrade material, and so on, the catalog includes lane subassemblies that allow you to specify those thicknesses for automatic volume reports.

The following section includes some examples of different components you can use in a typical road assembly. Many more alternatives are available in the Corridor Modeling Catalog. The Help file provides a complete breakdown of each subassembly in the catalog; you'll find this useful as you search for your perfect subassembly.

Each of these subassemblies can be added to an assembly using exactly the same process specified in the first exercise in this chapter. Choose your alternative subassembly instead of the basic parts specified in the exercise, and adjust the parameters accordingly.

Common Lane Subassemblies

Although the LaneOutsideSuper subassembly is suitable for many roads, you may need different road lanes for your locality or design situation.

LaneInsideSuper On a two-lane, undivided highway there is practically no difference between the LaneOutsideSuper and LaneInsideSuper. However, when a divided highway with superelevation is created, LaneInsideSuper is needed to control the superelevation rate of the lanes closest to the median, as shown in Figure 8.25.

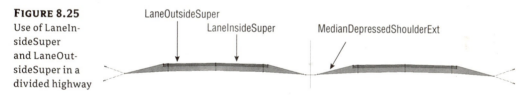

Figure 8.25
Use of LaneInsideSuper and LaneOutsideSuper in a divided highway

LaneParabolic The LaneParabolic subassembly (Figure 8.26) is used for road sections that require a parabolic lane in contrast to the linear grade of LaneOutsideSuper. The LaneParabolic subassembly also adds options for two pavement depths and a base depth. This is useful in jurisdictions that require two lifts of asphalt and granular sub-base material. Taking advantage of these additional parameters gives you an opportunity to build corridor models that can return more detailed quantity takeoffs and volume calculations.

Figure 8.26
The LaneParabolic subassembly and parameters

Note that the LaneParabolic subassembly doesn't have a Side parameter. The parabolic nature of the component results in a single attachment point that would typically be the assembly centerline marker. Keep in mind that subassemblies are AutoCAD objects and, therefore, can be moved to the right or to the left (as well as up and down) when specifications require design profiles along the top back of curb as opposed to the centerline of a travel way.

LaneBrokenBack If your design calls for multiple lanes, and those lanes must each have a unique slope, investigate the LaneBrokenBack subassembly (Figure 8.27). This subassembly provides parameters to change the road-crown location and specify the width and slope for each lane. Like LaneParabolic, the LaneBrokenBack subassembly provides parameters for additional material thicknesses.

FIGURE 8.27
The LaneBroken-Back subassembly and parameters

The LaneBrokenBack subassembly, like LaneOutsideSuper, allows for the use of target alignments and profiles to guide the subassembly horizontally and/or vertically.

COMMON SHOULDER AND CURB SUBASSEMBLIES

There are many types of curbs, and the UrbanCurbGutterGeneral subassembly can't model them all. Sometimes you may need a mountable curb, or perhaps you need a shoulder instead. In those cases, the Corridor Modeling Catalog provides many alternatives:

UrbanCurbGutterValley (1, 2, or 3) The UrbanCurbGutterValley subassemblies are great if you need mountable curbs (Figure 8.28). UrbanCurbGutterValley 1, 2, and 3 vary slightly in how they handle the sub-base slope. UrbanCurbGutterValley 1 also differs because it comes to a point instead of offering a width at the top of curb.

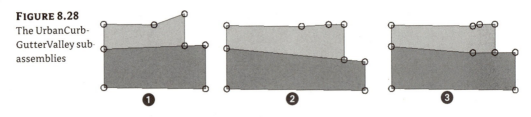

FIGURE 8.28
The UrbanCurbGutterValley subassemblies

BasicShoulder BasicShoulder (see Figure 8.29) is another simple yet effective subassembly for use with road sections that require a shoulder. The predefined shape for this subassembly is Pave1, which is good if you are planning to treat this as a paved shoulder and quantify the material with the Pave1 from a lane.

FIGURE 8.29
The BasicShoulder subassembly and parameters

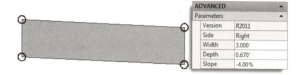

ShoulderExtendSubbase and ShoulderExtendAll   Shoulders that can work with your lanes in a superelevation situation, as these two do, are extremely helpful. These two subassemblies (shown in Figures 8.30a and 8.30b) will "play nice" with your breakover removal settings, as you will see in Chapter 11, "Superelevation."

FIGURE 8.30
ShoulderExtend-Subbase subassembly (a) and ShoulderExtendAll subassembly (b)

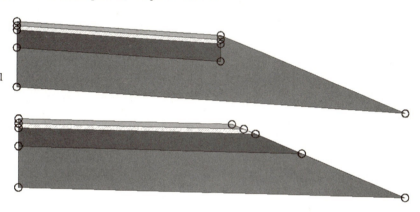

Editing an Assembly

As you saw earlier in this chapter, good-old AutoCAD properties are an option for changing subassembly parameters for one or more subassemblies of the same type. However, there are a handful of settings that can only be controlled in the Civil 3D Subassembly Properties. For example, the side (left or right) is an item must be changed in the Subassembly Properties.

Editing a Single Subassembly

Once your assembly is created, you can edit individual subassemblies as follows:

1. Pick the subassembly you'd like to edit. This will bring up the context tab.
2. Select the Subassembly Properties option from the Modify Subassembly panel.
3. The Subassembly Properties dialog appears. Click the Subassembly Help button at the bottom right of the dialog if you want to access the help page that gives detailed information about the use of this particular subassembly.
4. Switch to the Parameters tab to access the same parameters you saw in the AutoCAD Properties palette when you first placed the subassembly.
5. Click inside any field on the Parameters tab to make changes.

Editing the Entire Assembly

Sometimes it's more efficient to edit all the subassemblies in an assembly at once. To do so, pick the assembly baseline marker, or any subassembly that is connected to the assembly you'd like to edit. This time, select the Assembly Properties option from the Modify Assembly panel.

Renaming the Assembly

The Information tab on the Assembly Properties dialog gives you an opportunity to rename your assembly and provide an optional description. It is good practice to be consistent and detailed in your assembly names (for example, **Divided 4-Lane 12′ w Paved Shoulder**). With informative assembly names, you will eliminate much of the guesswork when it comes to building corridors.

Changing Parameters

The Construction tab on the Assembly Properties dialog houses each subassembly and its parameters. You can change the parameters for individual subassemblies by selecting the subassembly on the left side of the Construction tab and changing the desired parameter on the right side.

Renaming Groups and Subassemblies

Note that the left side of the Construction tab displays a list of groups. Under each group is a list of the subassemblies in use in your assembly. A new group is formed every time a subassembly is connected directly to the assembly marker.

For example, in Figure 8.31 you see Group - (1). The first subassembly under Group - (1) is LaneOutsideSuper - (37). If you dig into its parameters on the right side of the dialog, you'll learn that this lane is attached to the right side of the assembly marker, a UrbanCurbGutterGeneral is attached to right side of the LaneOutsideSuper, and a UrbanSidewalk is attached to the right side of the UrbanCurbGutterGeneral. The next group, Group - (3), is identical but attached to the left side of the assembly marker.

FIGURE 8.31
The Construction tab shows the default group and subassembly naming.

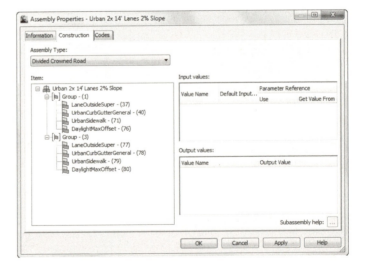

The automatic naming conventions are somewhat cryptic, and it is convenient not to have to dig into the subassembly parameters to determine which side of the assembly a certain group is on. Later, when you're making complex corridors, you'll be provided with a list of subassemblies to choose from; it's certainly easier to figure out which LaneOutsideSuper you need to choose when your choice is `Lane - Right` as opposed to `LaneOutsideSuper - (37)`. Therefore, it's in your best interest to rename your subassemblies once you've built your assembly.

You can rename both groups and subassemblies on the Construction tab of the Assembly Properties dialog by double-clicking on the group or subassembly you wish to rename.

There is no official best practice for renaming your groups and subassemblies, but you may find it useful to designate what type of subassembly it is, what side of the assembly it falls on, and other distinguishing features (see Figure 8.32). For example, if a lane is to be designated as a transition lane or a generic link used as a ditch foreslope, it would be useful to name them descriptively.

FIGURE 8.32
The Construction tab showing renamed groups and subassemblies

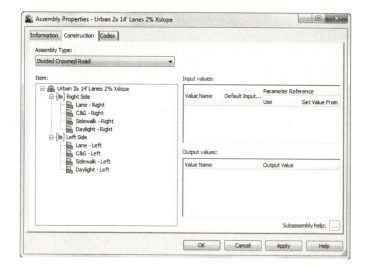

Creating Assemblies for Nonroad Uses

There are many uses for assemblies and their resulting corridor models aside from road sections. The Corridor Modeling Catalog also includes components for retaining walls, rail sections, bridges, channels, pipe trenches, and much more. In Chapter 9, you'll use a channel assembly and a pipe-trench assembly to build corridor models. Let's investigate how those assemblies are put together by building a channel assembly for a stream section:

1. Create a new drawing from the DWT of your choice, or continue working in your drawing from the first exercise in this chapter.

2. Change to the Home tab and choose Assembly ➢ Create Assembly from the Create Design panel. The Create Assembly dialog opens.

3. Enter **Channel** in the Name text box. Confirm that the Assembly Style text box is set to Basic and that Code Set Style is set to All Codes. Click OK.

4. Specify a location in your drawing for the assembly. Somewhere in the center of your screen where you have room to work is fine.

5. Locate the Trench Pipes tab on the tool palette. Position the palette on your screen so that you can clearly see the assembly baseline.

6. Click the Channel button on the tool palette. The AutoCAD Properties palette appears.

7. Locate the Advanced section of the Design tab on the AutoCAD Properties palette. You'll place the channel with its default parameters and make adjustments through the Assembly Properties dialog, so don't change anything for now. Note that there is no Side parameter. This subassembly will be centered on the assembly marker.

8. The command line states `Select marker point within assembly or [RETURN for Detached]:`. Pick the assembly center-point marker, and a channel is placed on the assembly (see Figure 8.33).

FIGURE 8.33
The Channel subassembly placed on the assembly center point marker

9. Press Esc to leave the assembly-creation command and dismiss the palette.

10. Select the center assembly baseline marker and select Assembly Properties from the Modify Assembly panel.

11. The Assembly Properties dialog appears. Switch to the Construction tab.

12. Select the Channel entry on the left side of the dialog (under the Group). Click the Subassembly Help button located at bottom right in the dialog's Construction tab.

13. The Subassembly Reference portion of the AutoCAD Civil 3D 2012 Help file appears. Familiarize yourself with the diagram and input parameters for the Channel subassembly. Especially note the Attachment Point, Bottom Width, Depth, and Sideslope parameters. The attachment point indicates where your baseline alignment and profile will be applied.

14. Minimize the Help file.

15. To match the engineer's specified design, you need a stream section 6′ (2 m) deep with a 3′ (1 m)-wide bottom, 1:1 sideslopes, and no backslopes. Change the following parameters in the Assembly Properties dialog:

    ◆ Bottom Width: 3′ (1 m)
    ◆ Depth: 6′ (2 m)
    ◆ Left and Right Backslope Width: 0′ (0 m)
    ◆ Sideslope: 1:1

16. Click OK, and confirm that your completed assembly looks like Figure 8.34.

Figure 8.34
A completed channel assembly

 Real World Scenario

A Pipe Trench Assembly

Projects that include piping, such as sanitary sewers, storm drainage, gas pipelines, or similar structures, almost always include trenching. The trench must be carefully prepared to ensure the safety of the workers placing the pipe, as well as provide structural stability for the pipe in the form of bedding and compacted fill.

The corridor is an ideal tool for modeling pipe trenching. With the appropriate assembly combined with a pipe-run alignment and profile, you can not only design a pipe trench but also use cross-section tools to generate section views, materials tables, and quantity takeoffs. The resulting corridor model can also be used to create a surface for additional analysis.

The following exercise will lead you through building a pipe trench corridor based on an alignment and profile that follow a pipe run, and a typical trench assembly:

1. Create a new drawing using the DWT of your choice, or continue working in your drawing from the previous exercise.
2. Change to the Home tab and choose Assembly ➢ Create Assembly from the Create Design panel. The Create Assembly dialog opens.

3. Enter **Pipe Trench** in the Name text box to change the assembly's name. Confirm that the Assembly Style text box is set to Basic and Code Set Style is set to All Codes. Click OK.

4. Pick a location in your drawing for the assembly. Somewhere in the center of your screen where you have room to work is fine.

5. Locate the Trench Pipes tab on the tool palette. Position the palette on your screen so that you can clearly see the assembly baseline.

6. Click the TrenchPipe1 button on the tool palette. The AutoCAD Properties palette appears. Position the AutoCAD Properties palette on your screen so that you can clearly see both the assembly baseline and the tool palette.

7. Locate the Advanced section of the Design tab on the AutoCAD Properties palette. This section lists the TrenchPipe1 parameters. You'll place TrenchPipe1 with its default parameters and make adjustments through the Assembly Properties dialog, so don't change anything for now. Note that there is no Side parameter. This subassembly will be placed centered on the assembly marker.

8. The command line states `Select marker point within assembly or [RETURN for Detached]:`. Pick the assembly center-point marker. A TrenchPipe1 subassembly is placed on the assembly as shown here:

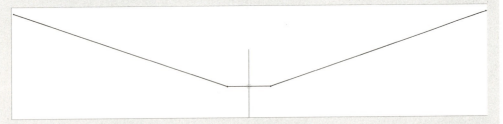

9. Press Esc to leave the assembly-creation command and dismiss the AutoCAD Properties palette.

10. Select the assembly marker and select Assembly Properties from the Modify Assembly panel.

11. The Assembly Properties dialog appears. Switch to the Construction tab.

12. Select the TrenchPipe1 assembly entry on the left side of the dialog. Click the Subassembly Help button located at the bottom right.

13. The Subassembly Reference portion of the AutoCAD Civil 3D 2012 Help file appears. Familiarize yourself with the diagram and input parameters for the TrenchPipe1 subassembly. In this case, the profile grade line will attach to a profile drawn to represent the pipe invert. Because the trench will be excavated deeper than the pipe invert to accommodate gravel bedding, you'll use the bedding depth parameter in a moment. Also note under the Target Parameters that this subassembly needs a surface target to determine where the sideslopes terminate.

14. Minimize the Help file.

15. To match the engineer's specified design, the pipe trench should be 3′ (1 m) deep and 4′ (1.3 m) wide with 2:1 sideslopes and 1′ (0.3 m) of gravel bedding. Change the following parameters in the Assembly Properties dialog:

    ◆ Width: 4′ (1.3 m)
    ◆ Sideslope: 2:1
    ◆ Bedding Depth: 1′ (0.3m)
    ◆ Offset To Bottom: 3′ (1 m)

16. Click OK.

17. Confirm that your completed assembly looks like the graphic shown here, and save your drawing.

This assembly will be used to build a pipe-trench corridor in Chapter 9.

Specialized Subassemblies

Despite the more than 100 subassemblies available in the Corridor Modeling Catalog, sometimes you may not find the perfect component. Perhaps none of the channel assemblies exactly meet your design specifications, and you'd like to make a more customized assembly, or neither of the sidewalk subassemblies allows for the proper boulevard slopes. Maybe you'd like to try to do some preliminary lot grading using your corridor, or mark a certain point on your subassembly so that you can extract important features easily.

You can handle most of these situations by using subassemblies from the Generic Subassembly Catalog (see Figure 8.35). These simple and flexible components can be used to build almost anything, although they lack the coded intelligence of some of the more intricate assemblies (such as knowing if they're paved, grass, or similar, and understanding things like sub-base depth, and so on).

Using Generic Links

Let's look at two examples where you might take advantage of generic links.

The first example involves the typical road section you built in the first exercise in this chapter. You saw that UrbanSidewalk doesn't allow for differing cross-slopes. If you need a 3′ (1 m)-wide terrace with a 3 percent slope, and then a 5′ (1.5 m) sidewalk with a 2 percent slope, followed by another buffer strip that is 6′ (2 m) wide with a slope of 5 percent, you can use generic links to assist in the construction of the proper assembly:

1. Open the `Subassembly Practice.dwg` file (which you can download from this book's web page).
2. In Prospector, locate and expand the assemblies group. Right-click on Sidewalk With Generic Links and select Zoom To.
3. Locate the Generic tab on the tool palette. Position the palette on your screen so that you can clearly see the assembly baseline.
4. Click the LinkWidthandSlope subassembly, and the AutoCAD Properties dialog appears. Position the dialog on your screen so that you can clearly see both the assembly baseline and the tool palette.
5. Scroll down to the Advanced Parameters section of Properties and change the parameters as follows to create the first buffer strip:
   ◆ Side: Right
   ◆ Width: 3′ (1 m)
   ◆ Slope: 3%
6. The command line states `Select marker point within assembly or [RETURN for Detached]:`. Select the circular point marker on the right UrbanCurbGutterGeneral subassembly, which represents the top back of the curb. A Link subassembly appears, as shown in Figure 8.36.

FIGURE 8.35
The Generic Subassembly tool palette

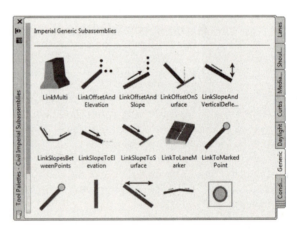

FIGURE 8.36
The Generic Link subassembly (parameters shown for illustration)

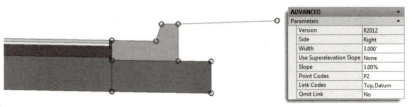

7. Switch to the Curbs tab of the tool palette. Click the UrbanSidewalk button.

8. In the Advanced Parameters area of Properties, change the parameters as follows to create the sidewalk:

   - Side: Right
   - Sidewalk Width: 5' (1.5 m)
   - Slope: 2%
   - Inside Boulevard Width: 0'
   - Outside Boulevard Width: 0'

9. The command line states Select marker point within assembly or [RETURN for Detached]:. Select the circular point marker on the right LinkWidthandSlope subassembly. An UrbanSidewalk subassembly appears, as shown in Figure 8.37.

Figure 8.37
The UrbanSidewalk subassembly (parameters shown for illustration)

10. Switch to the Generic tab of the tool palette. Click the LinkWidthandSlope button, and the AutoCAD Properties palette appears. Position the palette on your screen so that you can still see the assembly baseline and the tool palette.

11. In the Advanced Parameters area of Properties, change the parameters as follows to create the second buffer strip:

    - Side: Right
    - Width: 6' (2 m)
    - Slope: 5%

    Your drawing should now look like Figure 8.38.

Figure 8.38
The sidewalk and terrace strips (parameters shown for illustration)

12. Select the two generic links and the sidewalk assembly.

13. Select Mirror Subassemblies from the Modify Subassembly panel. The command line displays Select marker point within assembly:.

14. Select the marker point on the left back of curb. The completed assembly should look like Figure 8.39.

FIGURE 8.39
The completed assembly

15. Save the drawing if you'd like to use it in a future exercise.

You've now created a custom sidewalk terrace for a typical road.

Daylighting with Generic Links

The second example involves the channel section you built earlier in this chapter. This exercise will lead you through using the LinkSlopetoSurface generic subassembly, which will provide a surface target to the Channel assembly that will seek the target assembly at a 25 percent slope. For more information about surface targets, see Chapters 9 and 10. Follow these steps:

1. Continue working in, or open, the `Subassembly Practice.dwg` file (which you can download from this book's web page). You do not need to have the previous exercise completed to continue.

2. In Prospector, locate and expand the assemblies group. Right-click on Channel With Link To Surface and select Zoom To.

3. Locate the Generic tab on the tool palette.

4. Click the LinkSlopetoSurface button.

5. In the Advanced Parameters area of Properties, change the parameters as follows to create a surface target link:
   - Side: Right
   - Slope: 25%

6. The command line states `Select marker point within assembly or [RETURN for Detached]:`. Click the circular point marker at the upper right on the Channel subassembly that is farthest away. A surface target link appears (see Figure 8.40).

7. To complete the left side of the assembly, repeat steps 4 through 6, and change the Side parameter to the Left option.

8. The completed assembly should look like Figure 8.41.

Adding a surface link to a Channel assembly provides a surface target for the assembly. Now that you've added the LinkSlopetoSurface, you can specify your existing ground as the surface target, and the subassembly will grade between the top of the bank and the surface for you. You can achieve additional flexibility for connecting to existing ground with the more complex Daylight subassemblies, as discussed in the next section.

Figure 8.40
LinkSlopetoSurface subassembly (parameters shown for illustration)

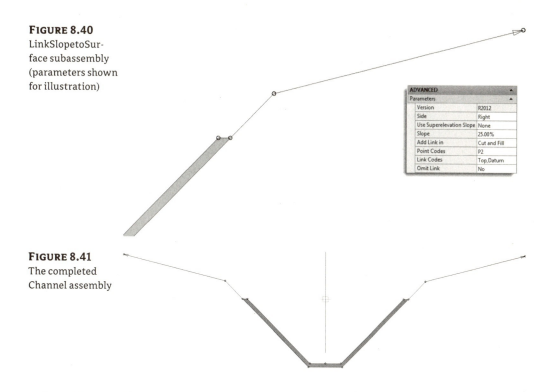

Figure 8.41
The completed Channel assembly

Working with Daylight Subassemblies

In previous examples we worked with daylight subassemblies, but it is now time to take a closer look at what they can do for us.

A daylight subassembly contains the instructions to Civil 3D as to how it is to extend a link to a target surface. The instructions might include a ditch or berm before looking for existing ground. Others provide a straight shot but with contingencies for certain design conditions. Figure 8.42 shows the many options you have for adding a daylight to an assembly.

Figure 8.42
Daylight subassemblies in the Corridor tool palette

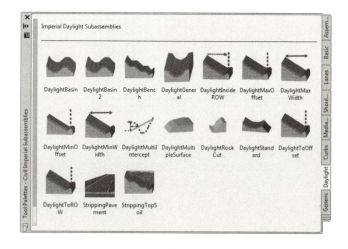

In the following exercise, you'll use the DaylightInsideROW subassembly. This subassembly contains parameters for specifying the maximum distance from the centerline or offset alignments. If the 4:1 slope hits the surface inside the ROW, no adjustment is made to the slope. If 4:1 causes the daylight to hit outside of the ROW, the slope adjusts to stay inside the specified location:

1. Continue working in, or open the `Subassembly Practice.dwg` file (which you can download from this book's web page). You do not need to have the previous exercise completed to continue.

2. In Prospector, locate and expand the assemblies group. Right-click on Add Daylight Subassembly and select Zoom To.

3. Locate the Daylight tab on the tool palette.

4. Click the DaylightInsideROW button on the tool palette.

5. In the Advanced Parameters area of Properties, change the following parameters to create the daylight as required:

    ROW Offset from Baseline: 33' (10 m)

    Leave all other parameters at their default values.

6. The command line states `Select marker point within assembly or [RETURN for Detached]:`. Select the circular point marker on the farthest right link.

7. Stay in the subassembly placement, but change the following parameter before placing the left side:

    ROW Offset from Baseline: -33' (-10 m)

    Notice there is no Left or Right parameter. The negative value in the ROW Offset From Baseline parameter is what tells Civil 3D the daylight is to the left. You can now dismiss the Properties palette.

8. Pick the DaylightInsideROW subassembly, and then choose Subassembly Properties from the Modify Subassembly panel.

9. Switch to the Parameters tab in the Subassembly Properties dialog.

10. Click the Subassembly Help button in the lower-right corner. The Subassembly Reference opens in a new window. Familiarize yourself with the options for the DaylightInsideROW subassembly, especially noting the optional parameters for a lined material, a mandatory daylight surface target, and an optional ROW offset target that can be used to override the ROW offset specified in the parameters.

11. Minimize the Subassembly Reference window. The completed assembly should look like Figure 8.43.

When to Ignore Daylight Input Parameters

The first time you attempt to use many Daylight subassemblies, you may become overwhelmed by the sheer number of parameters.

ADVANCED			
Parameters		Fill 1 Slope	Horizontal
Version	R2012	Fill 2 Width	0.000'
Side	Left	Fill 2 Slope	Horizontal
Daylight Link	Include Daylight link	Fill 3 Width	0.000'
Cut Test Point Link	3	Fill 3 Slope	Horizontal
Cut 1 Width	0.000'	Flat Fill Slope	6.00:1
Cut 1 Slope	Horizontal	Flat Fill Max Height	5.000'
Cut 2 Width	0.000'	Medium Fill Slope	4.00:1
Cut 2 Slope	Horizontal	Medium Fill Max Height	10.000'
Cut 3 Width	0.000'	Steep Fill Slope	2.00:1
Cut 3 Slope	Horizontal	Guardrail Width	2.000'
Cut 4 Width	0.000'	Guardrail Slope	-2.00%
Cut 4 Slope	Horizontal	Include Guardrail	Omit Guardrail
Cut 5 Width	0.000'	Width to Post	1.000
Cut 5 Slope	Horizontal	Rounding Option	None
Cut 6 Width	0.000'	Rounding By	Length
Cut 6 Slope	Horizontal	Rounding Parameter	1.500'
Cut 7 Width	0.000'	Rounding Tessellation	6
Cut 7 Slope	Horizontal	Place Lined Material	None
Cut 8 Width	0.000'	Slope Limit 1	1.00:1
Cut 8 Slope	Horizontal	Material 1 Thickness	1.000'
Flat Cut Slope	6.00:1	Material 1 Name	Rip Rap
Flat Cut Max Height	5.000'	Slope Limit 2	2.00:1
Medium Cut Slope	4.00:1	Material 2 Thickness	0.500'
Medium Cut Max Height	10.000'	Material 2 Name	Rip Rap
Steep Cut Slope	2.00:1	Slope Limit 3	4.00:1
Fill 1 Width	0.000'	Material 3 Thickness	0.333'
		Material 3 Name	Seeded Grass

The good news is that many of these parameters are unnecessary for most uses. For example, many Daylight subassemblies, such as DaylightGeneral (shown above), include multiple cut-and-fill widths for complicated cases where the design may call for test scenarios. If your design doesn't require this level of detail, leave those parameters set to zero.

Some Daylight subassemblies include guardrail options. If your situation doesn't require a guardrail, leave the default parameter set to the Omit Guardrail option and ignore it from then on. Another common, confusing parameter is Place Lined Material, which can be used for riprap or erosion-control matting. If your design doesn't require this much detail, ensure that this parameter is set to None, and ignore the thickness, name, and slope parameters that follow.

If you're ever in doubt about which parameters can be omitted, investigate the Help file for that subassembly.

FIGURE 8.43
An assembly with the daylight subassembly attached to each side (parameters shown for illustration)

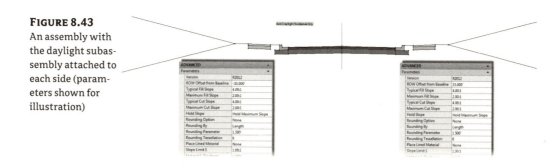

ALTERNATIVE DAYLIGHT SUBASSEMBLIES

At least a dozen Daylight subassemblies are available, varying from a simple cut-fill parameter to a more complicated benching or basin design. Your engineering requirements may dictate something more challenging than the exercise in this section. Here are some alternative Daylight subassemblies and the situations where you might use them. For more information on any of these subassemblies and the many other daylighting choices, see the AutoCAD Civil 3D 2012 Subassembly Reference in the help file.

DaylightToROW The DaylightToROW subassembly (Figure 8.44a) differs slightly from the DaylightInsideROW (Figure 8.44b). DaylightToROW constantly adjusts the slope to stay a certain distance away from your ROW, as specified by the Offset Adjustment input parameter. For example, you can have a ROW alignment specified, but use this subassembly to tell Civil 3D to always stay 3′ inside the ROW line.

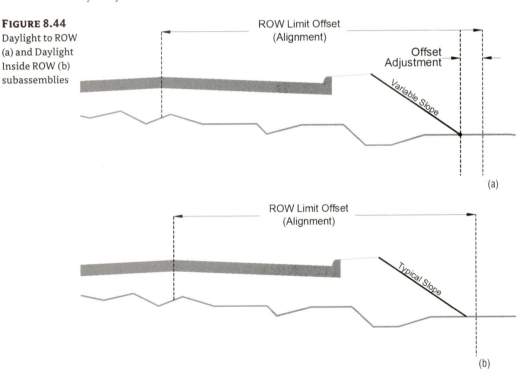

FIGURE 8.44
Daylight to ROW (a) and Daylight Inside ROW (b) subassemblies

BasicSideSlopeCutDitch In addition to including cut-and-fill parameters, the BasicSideSlopeCutDitch subassembly (see Figure 8.45) creates a ditch in a cut condition. This is most useful for road sections that require a roadside ditch through cut sections but omit it when passing through areas of fill. If your corridor model is revised in a way that changes the location of cut-and-fill boundaries, the ditch will automatically adjust.

Several subassemblies display the "LayoutMode" text on the design assembly. This will not display on the completed corridor.

FIGURE 8.45
The BasicSide-SlopeCutDitch subassembly in layout mode

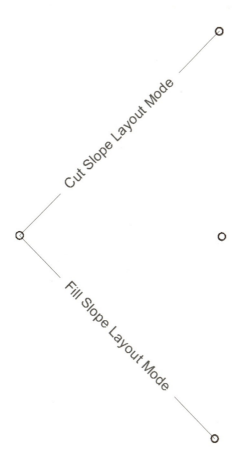

DaylightBasin Many engineers must design berms to contain roadside swales when the road design is in the fill condition. The process for determining where these berms are required is often tedious. The DaylightBasin subassembly (see Figure 8.46) provides a tool for automatically creating these "false berms." The subassembly contains parameters for the specification of a basin (which can be easily adapted to most roadside ditch cross sections as well) and parameters for containment berms that appear only when the subassembly runs into areas of roadside cut.

FIGURE 8.46
The DaylightBasin subassembly

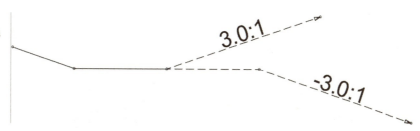

Advanced Assemblies

As you get to know Civil 3D better, you will want it to do more for you. With the tools you are given and your own creativity and problem-solving skills, Civil 3D can create some serious designs. Offset assemblies and marked point assemblies are weapons in your arsenal of awesome.

Offset Assemblies

Offset assemblies are an advanced option when you want to model a coordinating structure whose design is related to the main assembly. An example of where an offset assembly would be helpful is a main road adjacent to a meandering bike path. The bike path generally follows the main road, but its alignment is not always parallel and the profile might be altogether different. Figure 8.47 shows what the assembly for a bike path to the left of a road would look like.

To use an offset assembly, go to Home ➢ Assembly ➢ Add Assembly Offset. You will be prompted to select the main assembly and place the offset in the graphic. The location of the offset assembly in relation to the main assembly will have no effect on the final design.

Once the offset assembly is placed, the construction of an offset is identical to any other assembly. We will use an example of an assembly with an offset in Chapter 10, "Advanced Corridors, Intersections and Roundabouts."

FIGURE 8.47
An example of an assembly with an offset to the left representing a bike path

Marked Points and Friends

The marked point assembly is a small but powerful subassembly found in the Generic palette. Consisting only of a single marker, you can place this on an assembly to flag a location. You can use the marked point by itself to generate a feature line where no marker currently exists, say in the midpoint of a lane link. Where marked points really shine is when used with one of the subassemblies designed to look for a marked point.

When using a marked point, name it right away, and make note of that name for using it with its "friends" (Figure 8.48).

FIGURE 8.48
Name the marked point in the Advanced Parameters.

Linking to a Marked Point

In the example shown in Figure 8.49, a LinkToMarkedPoint2 subassembly is placed on the right side of the bike path pavement. The LinkToMarkedPoint2 subassembly has been created to look for the marked point on the left side of the sidewalk buffer.

Figure 8.49
Add the name of the marked point before you place it on the assembly.

In the Advanced Parameters of the subassembly, place the name of the marked point. Add the name of the marked point before you place it on the assembly. Be sure to tell your subassembly the name of the marked point before you place it on your assembly.

At this stage, the geometry for a subassembly using a marked point is not known. The final geometry will be determined when you plug it into a corridor. All subassemblies that use the marked point will appear with the "Layout Mode" placeholder. Subassemblies designed to look for a marked point include:

- LinkToMarkedPoint
- LinkToMarkedPoint2
- LinkSlopesBetweenPoints
- MedianRaisedWithCrown
- MedianRaisedConstantSlope
- MedianDepressed
- Channel
- ChannelParabolicBottom
- OverlayParabolic
- UrbanReplaceCurbGutter (1 and 2)

> **Making Sure Your Marked Point Processes**
>
> Always place the marked point before the links that use it to avoid having to reorder subassemblies in the Construction tab of Assembly Properties. If the marked point is listed below the subassembly that needs it in the Construction tab, Civil 3D will not process it.

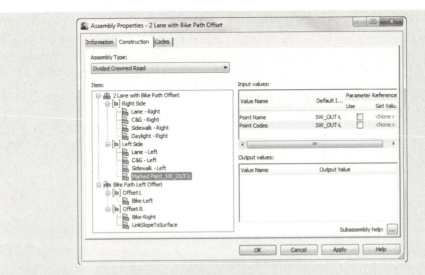

To reorder subassemblies in this dialog, right-click the subassembly and select Move Up or Move Down as needed.

Assembling Your Assemblies

The more geometry changes that occur throughout your corridor, the more assemblies you will have. Civil 3D offers several tools to keep your assemblies organized and available for future use.

Storing a Customized Subassembly on a Tool Palette

Customizing subassemblies and creating assemblies are both simple tasks. However, you'll save time in future projects if you store these assemblies for later use.

A typical jurisdiction usually has a finite number of allowable lane widths, curb types, and other components. It would be extremely beneficial to have the right subassemblies with the parameters already available on your tool palette.

The following exercise will lead you through storing a customized subassembly on a tool palette:

1. Open the `Storing Subassemblies and Assemblies.dwg` file (which you can download from this book's web page).

2. Be sure your Subassemblies tool palette is displayed.

3. Right-click in the Tool Palette area, and select New Palette to create a new tool palette. Enter **My Road Parts** in the Name text box.

4. Select the sidewalk sub-base from the C&G 12′ Lanes Sidewalk 4:1 Daylight assembly. You'll know it's selected when you can see it highlighted and the grip appears at the assembly baseline.

5. Click on the dashed portion of the subassembly (i.e., a subassembly marker, *not* the grip point) and drag the assembly into the tool palette. It may take you several tries to get the click-and-drag timing correct, but it will work. You'll know it is working when the cursor appears with a plus sign in the tool palette.

6. When you release the mouse button, an entry appears on your tool palette for GenericPavementStructure. Right-click this entry, and select the Properties option. The Tool Properties dialog appears (see Figure 8.50).

FIGURE 8.50
The Tool Properties dialog

7. Replace GenericPavementStructure with **Sidewalk Subbase** in the Name text box. You can also change the image, description, and other parameters in this dialog. Click OK.

8. Try this process for several lanes and curbs in the drawing, if desired. The resulting tool palette looks similar to Figure 8.51.

FIGURE 8.51
A tool palette with three customized subassemblies

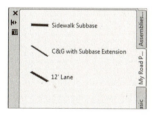

Note the tool palette entries for each subassembly point to the location of the `Subassembly.NET` directory, and not to this drawing. If you share this tool palette, make sure the subassembly directory is either identical or accessible to the person with whom you're sharing.

Storing a Completed Assembly on a Tool Palette

In addition to storing individual subassemblies on a tool palette, it's often useful to warehouse entire completed assemblies. Many jurisdictions have several standard road cross sections; once each standard assembly has been built, you can save time on future similar projects by pulling in a prebuilt assembly.

The process for storing an assembly on a tool palette is nearly identical to the process of storing a subassembly. Simply select the assembly baseline, hover your cursor over the assembly baseline, left-click, and drag to a palette of your choosing.

It's usually a good idea to create a library drawing in a shared network location for common completed assemblies and to create all assemblies in that drawing before dragging them onto the tool palette. By using this approach, you'll be able to test your assemblies for validity before they are rolled into production.

> **ORGANIZING ASSEMBLIES WITHIN A DRAWING**
>
> Civil 3D 2012 has new features that help you keep a drawing with many assemblies organized. In Prospector, you will see your listing of assemblies and an entry for Unassigned Subassemblies.
>
> Unassigned Subassemblies are orphaned parts that are not attached to any main assembly. They may be left over from some assembly customization or they may just be a mistake. In either case, you will want to clean these out. Right-click on the Unassigned Assemblies collection and select Erase All Unreferenced Assemblies.
>
>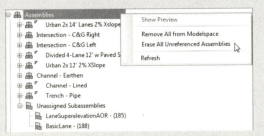
>
> You can also choose to remove the display of the assemblies from modelspace by right-clicking on the Assemblies collection. This hides the display of the assembly but retains its definition in the drawing.
>
> You can still use a hidden assembly in a corridor. If you need it visible again for editing purposes, right-click on the assembly and select Insert To Modelspace.
>
> The niftiest part of this new method of organizing assemblies is that they can now be part of your Civil 3D template without having them visible.

The Bottom Line

Create a typical road assembly with lanes, curbs, gutters, and sidewalks. Most corridors are built to model roads. The most common assembly used in these road corridors is some variation of a typical road section consisting of lanes, curb, gutter, and sidewalk.

> **Master It** Create a new drawing from the DWT of your choice. Build a symmetrical assembly using LaneInsideSuper, UrbanCurbGutterValley1, and LinkWidthAndSlope for terrace and buffer strips adjacent to the UrbanSidewalk. Use widths and slopes of your choosing.

Edit an assembly. Once an assembly has been created, it can be easily edited to reflect a design change. Often, at the beginning of a project, you won't know the final lane width. You can build your assembly and corridor model with one lane width and then later change the width and rebuild the model immediately.

> **Master It** Working in the same drawing, edit the width of each LaneInsideSuper to 14′ (4.3 m), and change the cross slope of each LaneInsideSuper to −3.08%.

Add daylighting to a typical road assembly. Often, the most difficult part of a designer's job is figuring out how to grade the area between the last engineered structure point in the cross section (such as the back of a sidewalk) and existing ground. An extensive catalog of daylighting subassemblies can assist you with this task.

> **Master It** Working in the same drawing, add the DaylightMinWidth subassembly to both sides of your typical road assembly. Establish a minimum width between the outermost subassembly and the daylight offset of 10′ (3 m).

Chapter 9

Basic Corridors

The corridor object is a three-dimensional road model that combines the horizontal geometry of an alignment, the vertical geometry of a profile, and the cross-sectional geometry of an assembly.

Corridors range from extremely simple roads to complicated highways and interchanges. This chapter focuses on building several simple corridors that can be used to model and design roads, channels, and trenches.

In this chapter, you will learn to:

- Build a single baseline corridor from an alignment, profile, and assembly
- Use targets to add lane widening
- Create a corridor surface
- Add an automatic boundary to a corridor surface

Understanding Corridors

In its simplest form, a corridor is a three-dimensional combination of an alignment, a profile, and an assembly (see Figure 9.1).

FIGURE 9.1
A corridor shown in 3D view

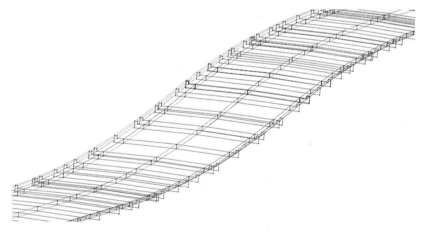

You can also build corridors with additional combinations of alignments, profiles, and assemblies to make complicated intersections, interchanges, and branching streams (see Figure 9.2).

The horizontal properties of the alignment, the vertical properties of the profile, and the cross-sectional properties of the assembly are merged together to form a dynamic model that can be used to build surfaces, sample cross sections, and much more.

Most commonly, corridors are thought of as being used to model roads, but they can also be adapted to model berms, streams (see Figure 9.3), lagoons, trails, and even parking lots.

FIGURE 9.2
An intersection modeled with a corridor

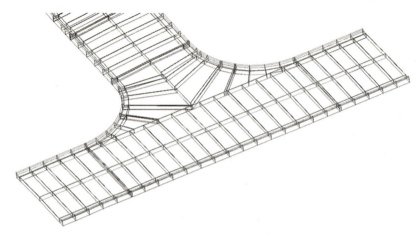

FIGURE 9.3
A stream modeled with a corridor

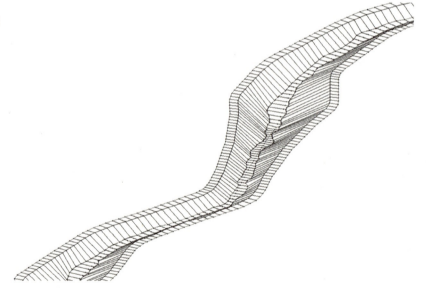

Creating a Simple Corridor

Corridors are made up of several components. *Marker points* and *links* are coded into the subassemblies that comprise the assembly, as you saw in Chapter 8, "Assemblies and Subassemblies." *Feature lines* glue the points and links together along the baseline.

CREATING A SIMPLE CORRIDOR | **349**

First, let's look at some important terminology you will want to become familiar with before proceeding. *Baseline*, *regions*, *assemblies*, *frequency*, and *targets* are all parts of a corridor that you will encounter even on your first design.

Baseline

The first ingredient in any corridor is an alignment. This alignment is referred to as a *baseline*. Every baseline alignment needs a corresponding design profile.

In this chapter, the baseline will correspond to the alignment and profile representing the crown of a road. However, this is not always the case, as we will explore in more depth in Chapter 10, "Advanced Corridors." As your designs become more detailed, you will have corridors with multiple baselines.

Regions

When the geometry along a baseline changes enough to warrant a new assembly, a new *region* is needed. Regions specify the station range where a specific assembly is applied to the design. There may be many regions along a baseline to accommodate design geometry.

Regions must not overlap each other and must progress in ascending station values along the baseline.

ASSEMBLIES

You took a long, hard look at assemblies in Chapter 8. Figure 9.4 is a quick refresher on some of the terminology you encountered.

Remember, before you build a corridor it is a good idea to have your initial assemblies created. Don't forget to give your subassemblies recognizable names (Figure 9.5).

FIGURE 9.4
You will use these assembly parts to build a corridor.

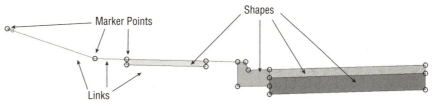

FIGURE 9.5
Make corridor creation easier by giving recognizable names to subassemblies.

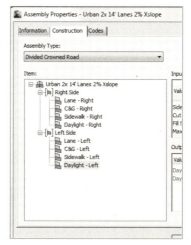

Frequency

Frequency refers to how often the assembly is applied to the corridor design. You can set the frequency distance for the corridor as a whole, but in most cases, you should apply it at the region level.

The frequency value will vary depending on the situation. The default frequencies are 25′ in Imperial units and a chunky 20 m for metric units. Civil 3D will place frequency lines at special stations such as superelevation key stations, horizontal design stations, and profile design stations. Users can create their own frequency stations for things like driveways or culvert crossings. Table 9.1 shows some typical frequencies for common design situations.

TABLE 9.1: Frequency guideline

DESIGN SITUATION	TYPICAL FREQUENCY
Civil 3D Default	25′ (20 m)
Intersection Curb Return	5′ (2 m)
Alignment Curve	10′ (3 m)

Corridor Feature Lines

After a corridor is created, you will see a series of lines running parallel (or mostly parallel) to the alignment; these lines are called *corridor feature lines*. Corridor feature lines, sometimes referred to simply as feature lines, are the result of Civil 3D playing connect-the-dots with marker points between the frequency lines.

Everywhere the assembly contains a named marker point, Civil 3D creates a feature line with the same name.

These special feature lines are the third dimension that takes a corridor from being simply a collection of cross sections to being a model with meaningful flow (see Figure 9.6).

Later on in this chapter, we will take a closer look at corridor feature lines.

Create Simple Corridor vs. Create Corridor

On the Home tab, in the Create Design panel, the Corridor command has two options: Create Simple Corridor and Create Corridor. The only difference in the commands is what Civil 3D asks you to do along the way.

> Create Simple Corridor has a step for naming the corridor right away, and presents you with the Target Mapping dialog before continuing with one initial region. Create Corridor assumes you've had your coffee for the day; you need to remember to name the corridor and set targets on your own. Create Corridor also allows you to add multiple regions on the first build.
>
> At the end of each process, the result is identical: a corridor object to which you can add regions and baselines, and manipulate to your heart's content.

FIGURE 9.6
The anatomy of a corridor

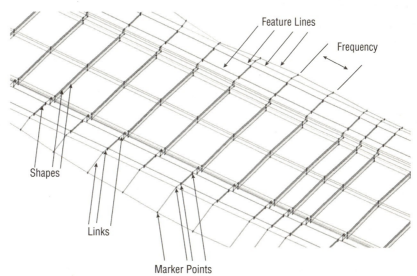

This exercise gives you hands-on experience in building a corridor model from an alignment, a profile, and an assembly:

1. Open the `Simple Corridor.dwg` file, which you can download from www.sybex.com/go/masteringcivil3d2012. Note that the drawing has an alignment, a profile view with two profiles, and an assembly, as well as an existing ground surface.

2. Change to the Home tab and select Corridor ➢ Create Simple Corridor from the Create Design panel. The Create Simple Corridor dialog opens.

3. In the Name text box, name the corridor **Cabernet Court Corridor**. Keep the default values for Corridor Style and Corridor Layer (see Figure 9.7).

4. Click OK to dismiss the dialog and continue.

5. At the `Select baseline alignment <or press enter key to select from list>:` prompt, press ↵ and select the Cabernet Court alignment from the list. Click OK after highlighting Cabernet Court.

6. At the `Select a profile <or press enter key to select from list>:` prompt, pick the Finished Ground profile (the blue profile with labels) in the drawing. Alternatively, you could press ↵ and select your profile from a list.

7. At the Select an assembly <or press enter key to select from list>: prompt, pick the vertical line of the assembly in the drawing. Alternatively, you could press ↵ and select your assembly from a list.

8. You are now shown the Target Mapping dialog for this corridor (Figure 9.8). Click the field that says <Click here to set all>. Highlight EG. Click OK to set the surface; then click OK to complete the corridor creation process.

FIGURE 9.7
Change the corridor name to something recognizable.

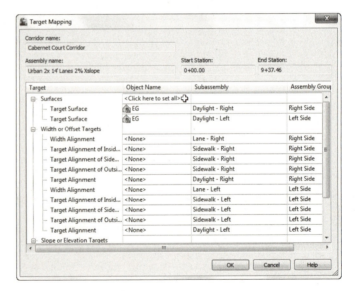

FIGURE 9.8
Target Mapping dialog

9. You will receive several messages in Panorama, which read 0+00.00': Intersection with target could not be computed, Intersection Point doesn't exist, or something similar. You will rectify this issue in the following steps. Dismiss the Panorama. Your corridor should look like Figure 9.9.

Now, let's try to figure out why we got those messages back in Panorama. In plan view, everything looks fine. However, a look at the corridor in the Object Viewer tells a different story (see Figure 9.10).

CREATING A SIMPLE CORRIDOR | **353**

FIGURE 9.9
The nearly completed corridor

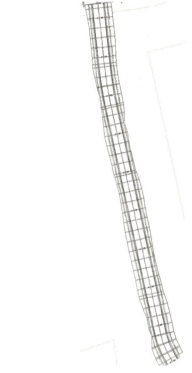

FIGURE 9.10
A "waterfall" at station 0+00

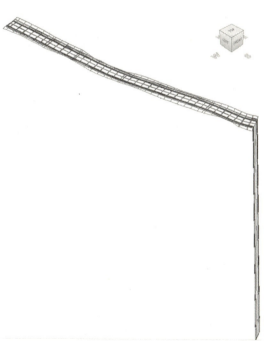

Corridor Properties

10. (Optional) If you'd like to view the corridor in the Object Viewer, select one of the corridor lines. Select Object Viewer from the context tab. Use the View Control drop-down from the top of the Object Viewer to show SW Isometric. After you have examined the corridor, dismiss the Object Viewer.

11. Select the corridor and click Corridor Properties from the context tab. In the Corridor Properties dialog, switch to the Parameters tab.

    Notice in Figure 9.11 that the start station for the region is 0+00, which is coming from the alignment. However, the design profile begins at 0+08.6607. The difference in these values is causing the waterfall. To resolve this, tell the corridor not to start processing the region until the start of the design profile. Note that even though the display value shows only two decimal places, the corridor examines up to eight decimal places of precision. To make sure you are starting the region in the correct location, round up at the thousandths decimal place.

FIGURE 9.11
Corridor Properties dialog, Parameters tab

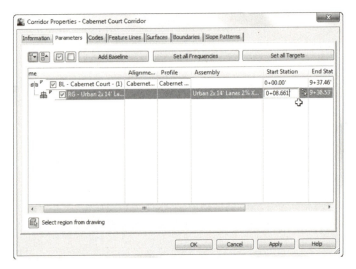

12. Click the Start Station field for the region and type **8.661**. Station notation is not needed. Click OK to allow the corridor to reprocess. No errors will appear in Panorama.

A quick look in the Object Viewer should reveal that the waterfall is gone.

Rebuilding Your Corridor

A corridor is a *dynamic* model—which means that if you modify any of the objects used to create the corridor, the corridor must be updated to reflect those changes. For example, if you make a change to the Finished Ground profile, the corridor needs to be rebuilt to reflect the new design. The same principle applies to changes to alignments, assemblies, target surfaces, and any other corridor ingredients or parameters.

You can rebuild the corridor by selecting the corridor object graphically and choosing Rebuild from the context tab. You can also access the Rebuild command by right-clicking on the corridor name from Prospector, as shown in Figure 9.12.

FIGURE 9.12
Right-click the corridor name in the Corridor collection in Prospector to rebuild it.

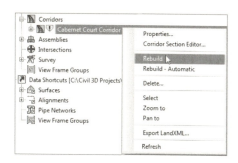

We recommend that you leave the option for Rebuild Automatic unchecked. Every time a change is made that affects your corridor, the corridor will go through the rebuilding process, during which you cannot work. The larger or more complex the corridor, the longer the rebuild process will take.

Troubleshooting Corridor Problems

Your corridor may not be perfect on your first iteration of the design. Get comfortable getting into and working with the Parameters tab of the Corridor Properties dialog. This tab will be your first stop to examine what might be amiss in your corridor model.

Whether you are building your first corridor or your five hundredth, odds are good that you will run into one of the following common issues:

Problem   Your corridor seems to fall off a cliff, meaning the beginning or ending station of your corridor drops down to zero.

Typical Cause   Your design profile starts and/or ends at a different station value than your alignment.

Fix   This is exactly what we ran into in the first exercise. The corridor takes the initial station range from your alignment. However, most designers don't tie into existing ground at the exact alignment start and end stations, so we need to adjust the corridor stations accordingly.

If you need to check the exact start and end stations of the design profile, the best place to do so is the Profile Data tab in Profile Properties dialog (as shown in Figure 9.13). Edit your corridor region to begin and end at the design profile station. To ensure you are within the design profile range, round up a smidgeon at the beginning (as we did in the exercise) and round down at the end.

Alternately, you can use the station picker to select the first and last corridor region stations by snapping to the frequency lines that correspond to your profile geometry.

Problem   Your corridor seems to take longer to build and has irregular frequency stations. Also, your daylighting may not extend out to where you expect it (see Figure 9.14).

Typical Cause   You accidentally chose the Existing Ground profile instead of the Finished Ground profile for your baseline profile. Most corridors are set up to place a frequency line at every vertical geometry point, and a profile, such as the Existing Ground profile, has many more vertical geometry points than a layout profile (in this case, the Finished Ground profile). These additional points on the Existing Ground profile are the cause of the unexpected sample lines and flags that something is wrong.

Figure 9.13
Check your Profile Properties dialog to examine the station range of the design profile.

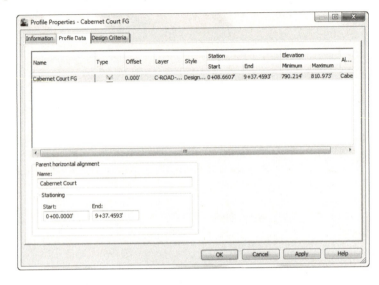

Figure 9.14
An example of unexpected corridor frequency

Fix Always use care to choose the correct profile. Either physically pick the profile on screen or make sure your naming conventions clearly define your finished grade as finished grade. If your corridor is already built, pick your corridor, right-click, and choose Corridor Properties. On the Parameters tab of the Corridor Properties dialog, change the baseline profile from Existing Ground to Finished Ground. Figure 9.15 shows the Parameters tab with Finished Ground properly listed as the baseline profile.

FIGURE 9.15
Setting the design profile in the Parameters tab of the Corridor Properties dialog

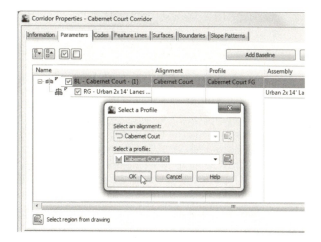

Adding a surface target throws another variable into the mix. Here is a list of some of the most typical problems new users face and how to solve them:

Problem Your corridor doesn't show daylighting even though you have a Daylight subassembly on your assembly. You may get a Target Object Not Found or a similar error message in Event Viewer.

Typical Cause You forgot to set the surface target when you created your surface.

Fix If your corridor is already built, pick your corridor, right-click, and select Corridor Properties. On the Parameters tab of the Corridor Properties dialog, click Set All Targets. The Target Mapping dialog opens, and its first entry is Surfaces. Click in the Object Name column field. This will prompt you to choose a surface for the Daylight subassembly to target.

Problem Your corridor seems to be missing patches of daylighting. You may also get an error message in Event Viewer.

Typical Cause Your target surface doesn't fully extend the full length of your corridor or your target surface is too narrow at certain locations.

Fix Add more data to your target surface so that it is large enough to accommodate daylighting down the full length of the corridor. If this is not possible, omit daylighting through those specific stations, and once your corridor is built, do hand grading using feature lines or grading objects. You can also investigate other subassemblies such as Link Offset To Elevation that will meet your design intention without requiring a surface target.

Problem Your corridor daylighting falls short of a tie-in to the existing ground surface. You may get an error message in Event Viewer, such as No Intersection With Link Found.

Typical Cause Your Daylight subassembly parameters are too restrictive to grade all the way to your target surface. The Daylight link cannot find the target surface within the grade, width, or other parameters you've set in the subassembly properties.

Fix Revisit your Daylight subassembly settings to give the program a wider offset or steeper grade. If your settings cannot be adjusted, you'll have to adjust your horizontal and/or vertical design to properly grade.

Corridor Feature Lines

Corridor feature lines are first drawn connecting the same point codes. For example, a feature line will work its way down the corridor and connect all the TopCurb points. If there are TopCurb points on the entire length of your corridor, then the feature line does not have any decisions to make. If your corridor changes from having a curb to having a grassed buffer or ditch, the feature line needs to figure out where to go next.

The Feature Lines tab of the Corridor Properties dialog has a drop-down menu called Branching (see Figure 9.16), with two options—Inward and Outward. Inward branching forces the feature line to connect to the next point it finds toward the baseline. Outward branching forces the feature line to connect to the next point it finds away from the baseline.

FIGURE 9.16
The Feature Lines tab of the Corridor Properties dialog

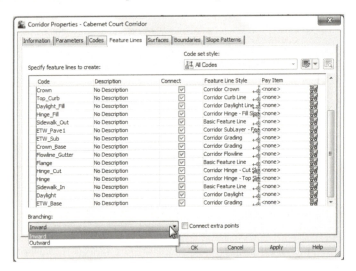

As mentioned earlier, a feature line will only connect the same point codes by default. However, the Feature Lines tab of the Corridor Properties dialog allows you to eliminate certain feature lines on the basis of the point code. For example, if for some reason you did not want your TopCurb points connected with a feature line, you could toggle that feature line off.

When a corridor is built, the feature line that is created is a similar type of object to the feature line you will use in Chapter 15, "Grading," for grading purposes. The difference is that corridor feature lines are locked in the corridor object. There is not much we can do with them until they are extracted.

Corridor feature lines can be extracted from a corridor to produce alignments, profiles, and grading feature lines. Once an alignment or profile has been extracted, its geometry can be adjusted independently. In this case, the profile or alignment no longer retains a link to the corridor from which it originated. Once a profile or alignment has been extracted, you can use it as a target back in the corridor that formed it.

In the case of extracting a feature line, we have the choice of keeping the line dynamically linked to the corridor geometry or making it an entirely separate entity. If the feature line is linked, it can be used for grading and cannot be manipulated in any way other than modifying the corridor (i.e., it cannot be grip-edited). If the feature line is not linked, it can be used for grading or as a target in a corridor. The nonlinked feature line can be grip-edited and is completely divorced from the originating corridor. Both types of feature lines can be used as breaklines in surface models.

In the following exercise, a corridor feature line is extracted to produce an alignment and profile:

1. Open the Extract Feature Line.dwg file.

2. Change to the Home tab and select Alignment ➤ Create Alignment From Corridor from the Create Design panel.

3. Select the right edge of the corridor near the note in the drawing (the note is for reference only). This opens the Select A Feature Line dialog shown in Figure 9.17.

FIGURE 9.17
Selecting the Daylight feature line to be extracted as an alignment

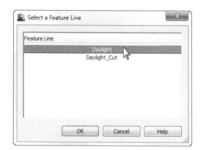

4. Select the Daylight feature line, as shown in Figure 9.17.

5. Click OK to close the Select A Feature Line dialog. The Create Alignment From Objects dialog is displayed, as shown in Figure 9.18.

6. Change your options to match those shown in Figure 9.18. Notice that the Create Profile option has been selected.

7. Click OK to dismiss the Create Alignment From Objects dialog. The Create Profile – Draw New dialog opens, as shown in Figure 9.19.

FIGURE 9.18
The completed Create Alignments From Objects dialog

FIGURE 9.19
The Create Profile – Draw New dialog

8. Change your options to match those shown in Figure 9.19.

9. Click OK to dismiss the dialog and then press ↵ to exit the command.

10. Review the alignment and profile as shown on the Prospector tab of Toolspace (see Figure 9.20).

FIGURE 9.20
A completed alignment and profile shown on the Prospector tab of Toolspace

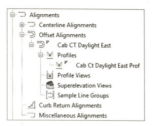

11. Next, you will extract a corridor feature line and keep it dynamically linked to the corridor. From the Home tab, in the Create Design panel, select Feature Line From Corridor.

12. Select the feature line labeled in the drawing as Hinge. Verify Hinge when prompted to select a feature line in the Select A Feature Line dialog. Click OK.

13. Select the Name check box and name the feature line **Hinge East**. Set this and all other settings to match Figure 9.21.

FIGURE 9.21
The Create Feature Line From Corridor dialog with the option to create a dynamic link turned on

14. Select the newly created feature line. Note that it can be selected but does not have grips. Take a look at the context tab and note which options are grayed out and which are not.

Understanding Targets

Every subassembly is programmed to have certain capabilities, but it needs some guidance from the user to figure out exactly how to do what is wanted. For example, a daylight subassembly contains instructions for how it is to extend to a surface model, but it is up to you to tell it which surface model to work with. Many of the lane subassemblies can be used to hook into an offset alignment and/or profile, but need your help in determining which alignment and/or profile is desired.

These instructions to the subassembly are called *targets*. When you built the corridor in the first exercise, Civil 3D prompted you to add targets. In that case, you chose a surface target and continued. However, surface targets are just the tip of the iceberg.

Using Target Alignments and Profiles

So far, all of the corridor examples you looked at have a constant cross section. In the next section, we'll take a look at what happens when a portion of your corridor needs to transition to a wider section and then transition back to normal.

Many subassemblies have been programmed to allow for not only a baseline attachment point but also additional attachment points on target alignments and/or profiles. Be sure to check the subassembly help file to make sure the subassembly you are using will accept targets if you need them. In Chapter 8, you learned that you can right-click on any subassembly to enter the help file. For instance, the BasicLane subassembly will show None in the Target Parameters area of the help file, but LaneOutsideSuper lists Width and Outside Elevation.

Think of a subassembly as a rubber band that is attached both to the baseline of the corridor (such as the road centerline) and the target alignment. As the target alignment, such as a lane widening, gets further from the baseline, the rubber band is stretched wider. As that target alignment transitions back toward the baseline, the rubber band changes to reflect a narrower cross section.

Figure 9.22 shows what happens to a cross section when various targets are set for the left edge of traveled way for a lane subassembly. Figure 9.22a shows the assembly as it was originally placed in the drawing. The width from the original assembly is 14' with a cross slope of 2.00 percent to the edge of pavement. Figure 9.22b shows how the geometry changes if an alignment target is set for the edge of pavement. Notice that because there is no profile specified to change the elevation, the 2.00 percent cross slope is held and the lane width is the only geometry that changes. Figure 9.22c shows that if both an alignment and profile are specified for an edge of pavement alignment, the design cross slope and width both change. Lastly, you see the assembly with just a profile target assigned to the edge of pavement in Figure 9.22d. In this case, the width stays at 14' but the elevation of the edge of pavement is dictated by the profile.

The following exercise shows you how to set targets using alignments and profiles for a corridor lane widening:

1. Open the Target Practice.dwg file. Note that this drawing has an incomplete corridor. You will be using the Target Mapping dialog to add the alignments and profiles that are created for you in this drawing.

2. Freeze the corridor using the Layer Freeze command from the Layers panel on the Home tab. Note that the Widening EOP alignment represents the edge of pavement (EOP) for a street parking zone (Figure 9.23).

3. Click the Layer Previous icon in the Layers panel to bring the corridor layer back.

4. Pan over to the cross-section view to the right in the drawing. Note the current cross slope and edge of pavement (EOP) offsets.

5. Pan to the profiles. The dark blue lines represent the EOP profile in the respective views. The light pink profile is the centerline profile superimposed for reference.

6. Pan back over to the corridor location. Select the corridor and click Corridor Properties from the context tab.

7. On the Parameters tab, scroll to the right until the ellipsis button for region RG - Urban 2x 14′ Lanes 2% Xslope - (3) target column is visible. Click the ellipsis button, as shown in Figure 9.24.

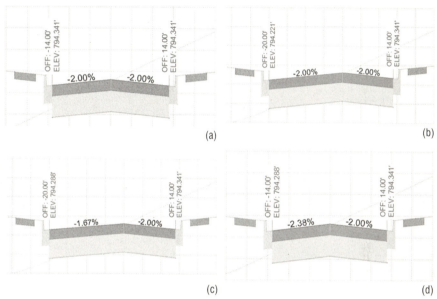

FIGURE 9.22
How geometry changes with a target on the left: Original assembly geometry (a), assembly with width alignment target only (b), assembly with both alignment and profile target (c), and assembly with only profile target set (d)

FIGURE 9.23
Alignments in plan view that will be used in the exercise (rotated for illustration purposes)

FIGURE 9.24
Click the ellipsis button to set targets for the region.

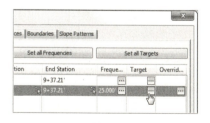

8. The Target Mapping dialog opens. Set the target surface to EG by clicking Click Here To Set All.

9. Click the Object Name field to the right of Width Alignment for Lane-Right. Highlight Cab Ct EOP-R alignment near the top of the Set Width Or Offset Target dialog. Click Add. Be sure the Cab Ct EOP-R appears in the listing near the bottom of the dialog, as shown in Figure 9.25. Click OK.

FIGURE 9.25
Highlight the alignment used as the width target and click Add.

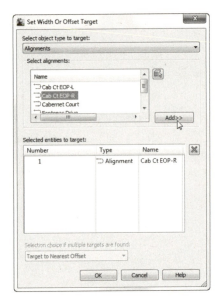

10. Repeat step 9 for the Lane-Left width alignment. The alignment will be Cab Ct EOP-L.

11. Back in the Target Mapping dialog, scroll down to the Slope Or Elevation Targets area. Select the Object Name field next to Outside Elevation Profile for Lane-Right.

12. In the Set Slope Or Elevation Target dialog, select Cab Ct EOP-R from the Alignment drop-down.

13. Highlight Cab Ct EOP-R Prof from the listing and click Add. Verify that the profile has been added to the selected target listing (as shown in Figure 9.26) and click OK.

14. Repeat steps 12 and 13 for the Left EOP profile target. The profile name to be used is Cab Ct EOP-L Prof.

15. The Target Mapping dialog should now look like Figure 9.27. Click OK to dismiss the dialog.

16. Click OK to dismiss the Corridor Properties dialog and allow the corridor to rebuild. The corridor should now be following the EOP alignments and resemble Figure 9.28.

17. Pan over to reexamine the cross section. The section should now show that the targets, rather than the original assembly geometry, control the offset and elevations.

FIGURE 9.26
Set the alignment, and then highlight the profile used as the elevation target and click Add.

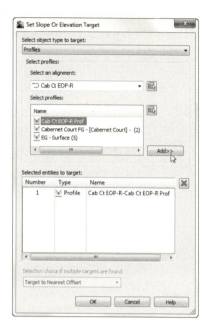

FIGURE 9.27
Targets set for surface, width, and elevation for left and right sides

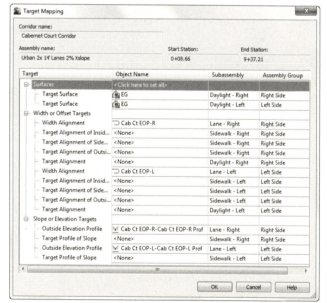

FIGURE 9.28
The completed exercise in plan view (rotated for illustration purposes)

Corridor Properties Modifications Made Easy

It can be tedious to modify alignments, profiles, starting stations, ending stations, frequencies, and targets along a corridor. Each of these items can be modified in the item view at the bottom (or side, depending on your orientation) of Prospector as shown here:

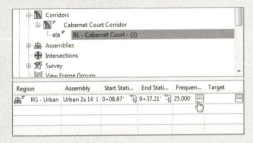

In the following quick exercise, you will use Toolspace to access and modify Frequency values for the Cabernet Court region:

1. Continue working in the `Target Practice.dwg` from the previous exercise. If you did not complete this exercise, you can use `Target Practice - Complete.dwg`.
2. In Prospector, expand the Cabernet Court Corridor and highlight BL - Cabernet Court - (3).
3. Select the Frequency ellipsis button from the Item view, as shown here:

4. Set the Frequency Along Curves value to 5'.
5. Click the Plus sign to add a driveway station. At the `Specify Station along Alignment:` prompt, type **3+30.85**. Press ↵ to accept the value, and then press ↵ again to return to the Frequency dialog box.
6. Click OK. The corridor will rebuild and the additional frequency lines should be visible in the plan.

Editing Sections

Once your corridor is built, chances are you will want to examine the corridor in section view, tweak a station here or there, and check for problems. For a station-by-station look at a corridor, pick the corridor and in the context tab, go to the Modify Corridor Sections panel and select Section Editor (see Figure 9.29).

Once you are in the Section Editor, you are in a purely data-driven view. That means that this is a live, editable section of the corridor, and is not for plotting purposes (Figure 9.30).

FIGURE 9.29
Some of the many tools available on the Corridor context tab, including Section Editor

FIGURE 9.30
The Corridor design in the Section Editor

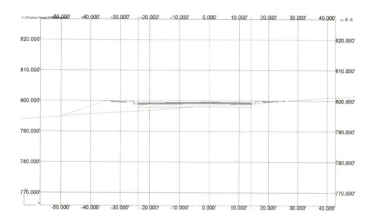

The Station Selection panel on the Section Editor context tab allows you to move forward and backward through your corridor to see what each section looks like.

If you wish to edit a section, you may do so geometrically in the graphic or through the Parameter Editor palette, but not both. To graphically edit a link, hold down the Ctrl key on the keyboard while selecting the item. This will activate grips you can use to relocate or stretch the link (Figure 9.31).

If you would rather use the Parameter Editor to make changes that are more precise to your section, select the Parameter Editor button from the Section Editor tab (Figure 9.32).

Look through the listing of subassemblies and their current values. When you find a value you wish to edit, click on the field to override the current value, as shown in Figure 9.33.

The changes you make can be applied to just the section you are viewing or to a range of stations in the region you are working in (use the Apply To A Station Range button in the Corridor edit tools to do so).

To exit the Section Editor, click the Close button on the Close panel.

FIGURE 9.31
A daylight link ready for grip-editing in the Section Editor

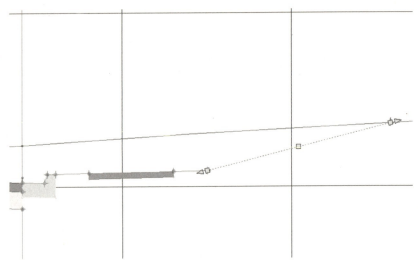

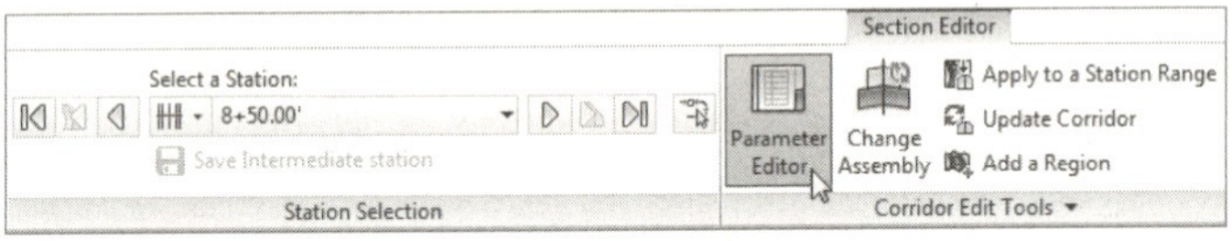

FIGURE 9.32
Some of the many tools from the Section Editor tab, including Parameter Editor

FIGURE 9.33
The corridor Parameter Editor

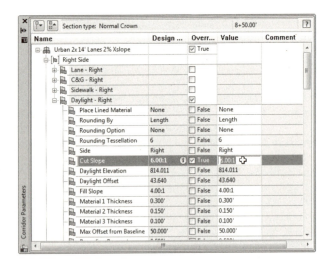

ACK! STUCK IN BIZZARRO COORDINATE SYSTEM

If you accidentally exit your drawing without exiting the Section Editor, you will return to a drawing in a rotated coordinate system. It may be difficult to see your design, but don't panic.

At the command line type **Plan.↵ W.↵**. You will see your project, but there is still another step.

In the View tab, set the current UCS to World, as shown in the following graphic. The UCS icon will return to normal and you can continue working.

Creating a Corridor Surface

A corridor provides the raw material for surface creation. Just as you would use points and breaklines to make a surface, a corridor surface uses corridor points as point data and uses feature lines and links like breaklines.

The Corridor Surface

Civil 3D does not automatically build a corridor surface when you build a corridor. From examining subassemblies, assemblies, and the simple corridors you built in the previous exercises, you have probably noticed that there are many "layers" of points, links, and feature lines. Some represent the very top of the finished ground of your road design, some represent subsurface gravel or concrete thicknesses, and some represent subgrade, among other possibilities. You can choose to build a surface from any one of these layers or from all of them. Figure 9.34 shows an example of a TIN surface built from the links that are all coded Top, which would represent final finished ground.

FIGURE 9.34
A surface built from Top code links

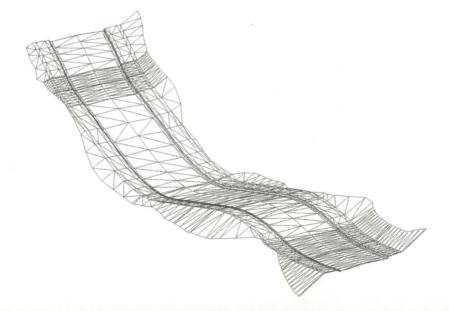

When you first create a surface from a corridor, the surface is dependent on the corridor object. This means that if you change something that affects your corridor and then rebuild the corridor, the surface will also update.

A corridor surface shows up as a true surface under the Surfaces branch in Prospector. After you create the initial corridor surface, you can create a static export of the surface by changing to the Home tab and choosing Surfaces ➤ Create Surface From Corridor on the Create Ground Data panel. A detached surface will not react to corridor changes and can be used to archive a version of your surface.

Creation Fundamentals

You create corridor surfaces on the Surfaces tab in the Corridor Properties palette using the following two steps (which are examined in detail later in this section):

Click Create A Corridor Surface to add a surface entry. Then choose data to add, and click the + sign.

You can choose to create your corridor surface on the basis of links, feature lines, or a combination of both.

Creating a Surface from Link Data

Most of the time, you will build your corridor surface from links. As discussed earlier, each link is coded with a name such as Top, Pave, Datum, and so on. Choosing to build a surface from Top links will create a surface that triangulates between the points at the link vertices that represent the final finished grade.

The most commonly built link-based surfaces are Top and Datum; however, you can build a surface from any link code in your corridor. Figure 9.35 shows a schematic of how links are used to form the most common surfaces.

> #### Overhang Correction for Whacked-out Datum Surfaces
>
> A common situation with assemblies is a material that juts out past another material, such as curb sub-base. Triangulated surfaces cannot contain caves or perfectly vertical faces, so Civil 3D needs to go around the material in a logical manner. What Civil 3D sees as logical and what you actually want from the surface may not agree, as shown here:
>
>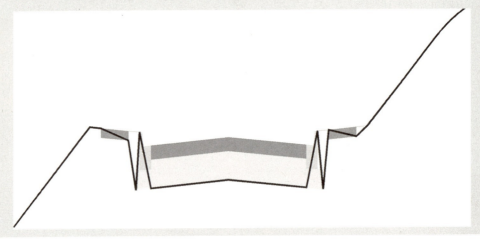

If you have a datum surface doing an unexpected zig-zag, change the Overhang Correction setting to Bottom Links. This setting is in the Corridor Properties – Surfaces tab. After your corridor rebuilds, the result will resemble Figure 9.35b.

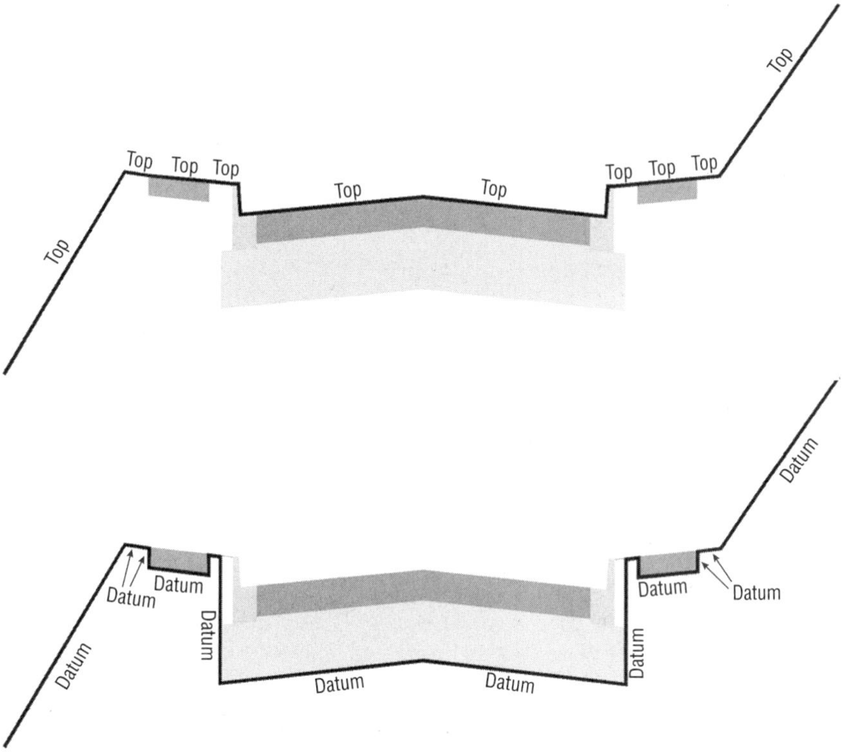

FIGURE 9.35
Schematic of Top links connecting to form a surface (a) and schematic of Datum links connecting to form a surface (b)

When building a surface from links, you have the option of checking a box in the Add As Breakline column. Checking this box will add the actual link lines themselves as additional breaklines to the surface. In most cases, especially in intersection design, checking this box forces better triangulation.

Creating a Surface from Feature Lines

There might be cases where you would like to build a simple surface from your corridor—for example, by using just the crown and edge-of-travel way. If you build a surface from feature lines only or a combination of links and feature lines, you have more control over what Civil 3D uses as breaklines for the surface.

If you added all the topmost corridor feature lines to your surface entry and built a surface, you would get a very similar result as if you had added the Top link codes.

CREATING A SURFACE FROM BOTH LINK DATA AND FEATURE LINES

A link-based surface can be improved by the addition of feature lines. A link-based surface does not automatically include the corridor feature lines, but instead uses the link vertex points to create triangulation. Therefore, the addition of feature lines ensures that triangulation occurs where desired. This is especially important for intersection design, curves, and other corridor surfaces where triangulation around tight corners is critical. Figure 9.36 shows the Surfaces tab of the Corridor Properties dialog where a Top link surface will be improved by the addition of Back_Curb, ETW, and Top_Curb feature lines.

FIGURE 9.36
The Surfaces tab indicates that the surface will be built from Top links as well as from several feature lines.

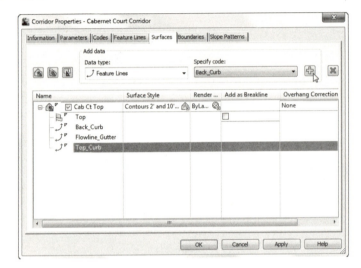

If you are having trouble with triangulation or contours not behaving as expected, experiment with adding a few feature lines to your corridor surface definition.

OTHER SURFACE TASKS

You can do several other tasks on the Surfaces tab. You can set a Surface Style, assign a meaningful name, and provide a description for your surface. Alternatively, you can do all those things once the surface appears in the drawing and through Prospector.

Adding a Surface Boundary

Surface boundaries are critical to any surface, but especially so for corridor surfaces. Tools that automatically and interactively add surface boundaries, using the corridor intelligence, are available. Figure 9.37 shows a corridor surface before the addition of a boundary.

You can create corridor surface boundaries using the Boundaries tab of the Corridor Properties dialog. Figure 9.38 shows a corridor surface after the application of an automatic boundary. Notice how the extraneous contours have been eliminated along the line of intersection between the existing ground and the proposed ground (the daylight line), thereby creating a much more accurate surface.

FIGURE 9.37
A corridor surface before the addition of a boundary

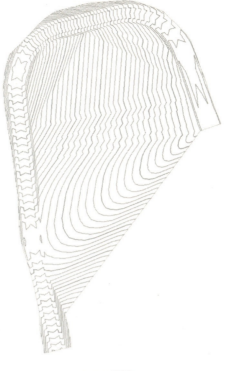

FIGURE 9.38
A corridor surface after the addition of an automatic daylight boundary

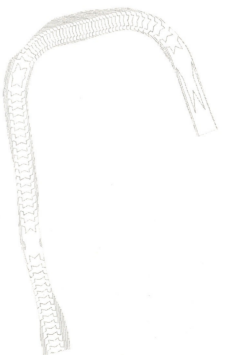

BOUNDARY TYPES

There are several tools to assist in corridor surface boundary creation. They can be automatic, semiautomatic, or manual in nature depending on the complexity of the corridor.

You access these options on the Boundaries tab of the Corridor Properties dialog by right-clicking the name of your surface entry, as shown in Figure 9.39.

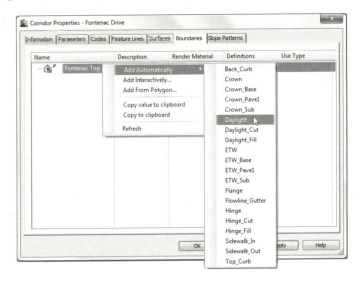

FIGURE 9.39
Corridor Surface Boundary options for a corridor containing a single baseline

The following corridor boundary methods are listed in order of desirability. Corridor Extents As Outer Boundary is the most user-friendly, whereas Add From Polygon is fast but needs constant updating as it is not dynamically linked to the corridor.

Corridor Extents As Outer Boundary   This option is available only when you have a corridor with multiple baselines. With this selection, Civil 3D will shrink-wrap the corridor, taking into account intersections and various daylight options on different alignments. Corridor Extents As Outer Boundary will probably be your most-used boundary option.

Add Automatically   The Add Automatically boundary tool allows you to pick a feature line code to use as your corridor boundary. This tool is available only for single-baseline corridors. Because this tool is the most automatic and easiest to apply, you will use it almost every time you build a single-baseline corridor.

Add Interactively   The Add Interactively boundary tool allows you to work your way around a multibaseline corridor and choose which corridor feature lines you would like to use as part of the boundary definition.

This is the better option than Add From Polygon if Add Automatically and Corridor Extents are not available. It takes a bit of patience to trace the corridor, but the result is a dynamically linked boundary that changes if the corridor changes.

Add From Polygon   The Add From Polygon tool allows you to choose a closed polyline or polygon in your drawing that you would like to add as a boundary for your corridor surface. This method is quick, but the resulting boundary is not dynamic to your design.

The next exercise leads you through creating a corridor surface with an automatic boundary:

1. Open the `Corridor Surface.dwg` file. Note that there is a completed corridor in this drawing.
2. Pick the Frontenac Drive corridor and select Corridor Properties from the Modify panel. The Corridor Properties dialog opens.
3. Switch to the Surfaces tab.

4. Click the Create A Corridor Surface button on the far left side of the dialog. You should now have a surface entry in the bottom half of the dialog.
5. Click the surface entry under the Name column and change the default name of your surface to **Frontenac-Top Surface**.
6. Confirm that Links has been selected from the drop-down menu in the Data Type selection box and that Top has been selected from the drop-down menu in the Specify Code selection box. Click the + button to add Top Links to the Surface Definition.
7. Click OK to leave this dialog, and examine your surface. The road surface should look fine; however, because you have not yet added a boundary to this surface, undesirable triangulation is occurring outside your corridor area.
8. Expand the Surfaces branch in Prospector. Note that you now have a corridor surface listed.
9. Pick the corridor and select Corridor Properties from the Modify panel. The Corridor Properties dialog opens. (If the Corridor Properties button is not available on the Modify panel, you may have accidentally chosen the corridor surface.)
10. In the Corridor Properties dialog, switch to the Boundaries tab.
11. Right-click Frontenac – Top Surface in the listing. Hover over the Add Automatically flyout, and select Daylight as the feature line that will define the outer boundary of the surface.
12. Confirm that the Use Type column says Outside Boundary to ensure that the boundary definition will be used to define the desired extreme outer limits of the surface.
13. Click OK to dismiss the dialog. Examine your surface, and note that the triangulation terminates at the Daylight point all along the corridor model.
14. Experiment with making changes to your finished grade profile, assembly, or alignment geometry and rebuilding both your corridor and finished ground surface.

> **REBUILD: LEAVE IT ON OR OFF?**
>
> Upon rebuilding your corridor, your surface will need to be updated. Typically, the best practice is to leave Rebuild – Automatic off for corridors and keep Rebuild – Automatic on for surfaces. This practice is usually OK for your corridor-dependent surfaces. The surface will only want to rebuild when the corridor is rebuilt. For very large corridors, this may become a bit of a memory lag, so try it both ways to see what you like best.

COMMON SURFACE CREATION PROBLEMS

Some common problems encountered when creating surfaces are as follows:

Problem Your corridor surface does not appear or seems to be empty.

Typical Cause You might have created the surface entry but no data.

Fix Open the Corridor Properties dialog and switch to the Surfaces tab. Select an entry from the drop-down menus in the Data Type and Specify Code selection boxes, and click the + sign. Make sure your dialog shows both a surface entry and a data type, as shown in Figure 9.40.

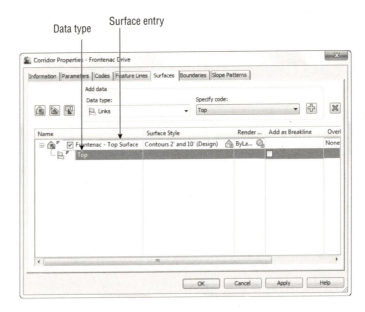

FIGURE 9.40
A surface cannot be created without both a surface entry and a data type.

Problem Your corridor surface does not seem to respect its boundary after a change to the assembly or surface-building data type (in other words, you switched from link data to feature lines).

Typical Cause Automatic and interactive boundary definitions are dependent on the codes used in your corridor. If you remove or change the codes used in your corridor, the boundary needs to be redefined.

Fix Open the Corridor Properties dialog and switch to the Boundaries tab. Erase any boundary definitions that are no longer valid (if any). Redefine your boundaries.

Problem Your corridor surface seems to have gaps at PCs and PTs near curb returns.

Typical Cause You may have encountered an error in rounding at these locations, and you may have inadvertently created gaps in your corridor. This is commonly the result of building a corridor using two-dimensional linework as a guide, but some segments of that linework do not touch.

Fix Be sure your corridor region definitions produce no gaps. You might consider using the PEDIT command to join lines and curves representing corridor elements that will need to be modeled later. You might also consider setting a COGO point at these locations (PCs, PTs, and so on) and using the Node object snap instead of the Endpoint object snap to select the same location each time you are required to do so.

Performing a Volume Calculation

One of the most powerful aspects of Civil 3D is having instant feedback on your design iterations. Once you create a preliminary road corridor, you can immediately compare a corridor surface to existing ground and get a good understanding of the earthwork magnitude. When you make an adjustment to the finished grade profile and then rebuild your corridor, you can see the effect that this change had on your earthwork within a minute or two, if not sooner.

Even though volumes were covered in detail in Chapter 4, "Surfaces," it is worth revisiting the subject here in the context of corridors.

This exercise uses a TIN-to-TIN composite volume calculation; average end area and other section-based volume calculations are covered in Chapter 12, "Cross Sections and Mass Haul."

1. Open the Corridor Surface Volume.dwg file. Note that this drawing has a completed corridor, a top, and datum corridor surfaces.

2. Change to the Analyze tab and choose Volumes ➢ Volumes from the Volumes And Materials panel.

3. The Composite Volume palette in Panorama appears.

4. Click the Create New Volume Entry button toward the top left of the Volume palette. A Volume entry with an Index of 1 should appear in the palette.

5. Click inside the cell in the Base Surface column and select EG.

6. Click inside the cell in the Comparison Surface column and select Frontenac Drive – Datum.

7. A Cut/Fill breakdown should appear in the remaining columns, as shown in Figure 9.41. Make a note of these numbers.

Figure 9.41
Panorama showing an example of a volume entry and the cut/fill results

8. Leave Panorama open on your screen (make it smaller, if desired), and pan over to the proposed profile for Frontenac Drive.

9. Select the Finished Ground profile and adjust it by grip editing any PVI.
10. Select the corridor, right-click, and choose Rebuild Corridor. Notice that the corridor changes, and therefore the corridor surface changes as well.
11. Click Recompute Volumes in Panorama and note the new values for cut and fill.
12. Repeat steps 9 through 11 and see if you can get the Net cut-fill within 1,000 cubic yards.

A common volume problem is as follows:

Problem   Your volume number does not update.

   Typical Cause   You might have forgotten to rebuild the corridor, rebuild the corridor surface, or click Recompute Volumes.

   Fix   Check Prospector to see if either your corridor or corridor surface is out of date. First, rebuild the corridor, and then rebuild the corridor surface.

Non-Road Corridors

Corridors are not just for roads. Once you have the basics from this chapter down, your ingenuity can take hold.

Corridors can be used for far more than just road designs. You explore some more advanced corridor models in Chapter 10, but there are plenty of simple, single-baseline applications for alternative corridors such as channels, berms, streams, retaining walls, and more. You can take advantage of several specialized subassemblies or build your own custom assembly using a combination of generic links. Figure 9.42 shows an example of a stream corridor.

One of the subassemblies discussed in Chapter 8 is the Channel subassembly. The following exercise shows you how to apply this subassembly to design a simple stream:

1. Open the `Corridor Stream.dwg` file. Note that there is an alignment that represents a stream centerline, a profile that represents the stream normal water line, and an assembly created using the Channel and LinkSlopetoSurface subassemblies.

2. Change to the Home tab and choose Corridor ➢ Create Simple Corridor on the Create Design panel. The Create Simple Corridor dialog opens.

3. Name your corridor **Stream Channel**. Click OK.

4. Follow the prompts, and pick the Stream CL alignment, the Stream NWL profile, and the Project Stream assembly. The Target Mapping dialog will appear after all objects have been picked.

5. In the Target Mapping dialog, choose the Existing Ground surface for all surface targets. Keep the default values for the additional targets. Click OK to dismiss the dialog.

6. The stream corridor will build and will look similar to Figure 9.43. Select the corridor and choose Corridor Section Editor from the Modify panel. Navigate through the stream cross sections. When you are finished viewing the sections, dismiss the dialog by clicking the X on the Close panel.

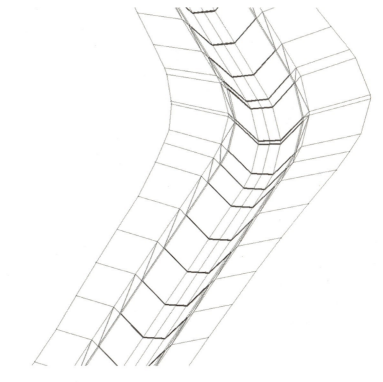

FIGURE 9.42
A simple stream corridor viewed in 3D built from the Channel subassembly and a generic link subassembly

FIGURE 9.43
The completed stream corridor

This corridor can be used to build a surface for a TIN-to-TIN volume calculation or can be used to create sections and generate material quantities, cross-sectional views, and anything else that can be done with a more traditional road corridor.

Real World Scenario

CREATING A PIPE TRENCH CORRIDOR

Another use for a corridor is a pipe trench. A pipe trench corridor is useful for determining quantities of excavated material, limits of disturbance, trench-safety specifications, and more. This graphic shows a completed pipe trench corridor:

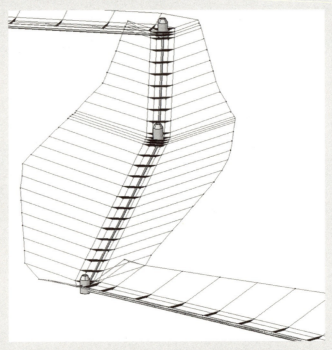

One of the subassemblies discussed in Chapter 8 is the TrenchPipe1 subassembly. The following exercise leads you through applying this subassembly to a pipe trench corridor:

1. Open the `Corridor Pipe Trench.dwg` file. Note that there is a pipe network, with a corresponding alignment, profile view, and pipe trench assembly. Also note that there is a profile drawn that corresponds with the inverts of the pipe network.

2. Change to the Home tab and choose Corridor ➢ Create Simple Corridor. Name the corridor **Pipe Trench**. Click OK.

3. Follow the prompts, press ↵, and pick the Pipe Centerline alignment from the list, the Bottom of Pipe profile, and the Pipe Trench assembly as your corridor components. Once these selections are made, the Target Mapping dialog appears.

4. In the Target Mapping dialog, choose Existing Ground as the target surface. Click OK.

5. The corridor will build. Select the corridor and choose Corridor Section Editor from the Modify panel. Browse the cross sections through the trench.

6. When you are finished viewing the sections, dismiss the dialog by clicking the X on the Close panel.

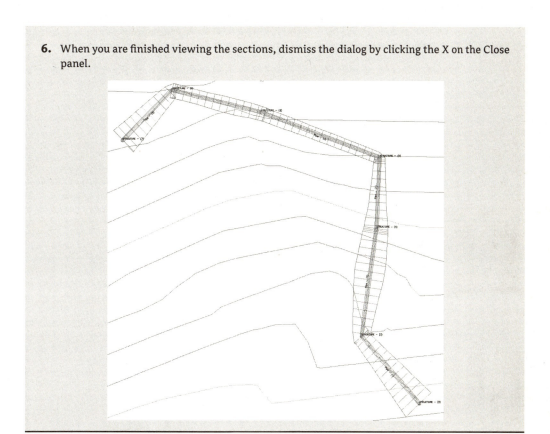

The Bottom Line

Build a single-baseline corridor from an alignment, profile, and assembly. Corridors are created from the combination of alignments, profiles, and assemblies. Although corridors can be used to model many things, most corridors are used for road design.

Master It Open the `Mastering Corridors.dwg` file. Build a corridor on the basis of the Project Road alignment, the Project Road Finished Ground profile, and the Project Typical Road Assembly.

Use targets to add lane widening. Targets are an essential design tool used to manipulate the geometry of the road.

Master It Open `Mastering Corridor Widening`. Set the Lane - L to target the USH 10 Left 16 alignment and set Lane - R to target the USH 10 Right alignment.

Create a corridor surface. The corridor model can be used to build a surface. This corridor surface can then be analyzed and annotated to produce finished road plans.

Master It Continue working in the `Mastering Corridors.dwg` file or open `Mastering Corridors - Step 1 Complete`. Create a corridor surface from Top links.

Add an automatic boundary to a corridor surface. Surfaces can be improved with the addition of a boundary. Single-baseline corridors can take advantage of automatic boundary creation.

Master It Continue working in the `Mastering Corridors.dwg` file or open `Mastering Corridors - Step 2 Complete`. Use the Automatic Boundary Creation tool to add a boundary using the Daylight code.

Chapter 10

Advanced Corridors, Intersections, and Roundabouts

In Chapter 9, "Basic Corridors," you built several simple corridors and began to see the dynamic power of the corridor model. The focus of that chapter was to get things started, but it's unrealistic to think that a project would have only one road in the middle of nowhere with no intersections, no adjustments, and no complications. You may be having trouble visualizing how you'll build a corridor to tackle your more complex design projects, such as the one pictured in Figure 10.1.

FIGURE 10.1
A corridor model for the example subdivision

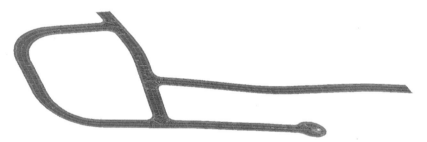

This chapter focuses on taking your corridor-modeling skills to a new level by introducing more tools to your corridor-building toolbox, such as intersecting roads, cul-de-sacs, advanced techniques, and troubleshooting.

This chapter assumes that you've worked through the examples in the alignments, profiles, profile view, assemblies, and basic corridor chapters. Without a strong knowledge of the foundation skills, many of the tasks in this chapter will be difficult.

In this chapter, you will learn to:

- Create corridors with noncenterline baselines
- Add alignment and profile targets to a region for a cul-de-sac
- Use the Interactive Boundary tool to add a boundary to the corridor surface

Getting Creative with Corridor Models

New users often ask more experienced users to teach them how to design an intersection (or a cul-de-sac, or a site, or anything) using Civil 3D. By the end of this chapter, you'll understand why this request is not only unrealistic, but probably impossible. There are as many ways to design an intersection as there are intersections in the world.

The best you can do is to learn how the corridor tools can be applied to a few typical scenarios. But don't take this chapter as gospel: Use the skills you learn here to create a foundation for your own models in your own design situations. Users often dismiss an intersection from being applicable to their situation because it doesn't include a turn lane or perhaps their intersection comes together at an odd angle. This is unfortunate, because the same fundamental tools can be adapted to accommodate additional design constraints.

Another example you may consider is adapting the corridor model for use in a parking lot or in a commercial site. As you saw in Chapter 9 with stream and pipe-trench corridors, the corridor model is not a road-only tool. It can be used for ponds, berms, curbs, and gutters, and much more.

Civil 3D in general, and the corridor model specifically, won't be truly useful to you unless you can see them as limitless, flexible models that you control to your design constraints. Build something; try something. If it doesn't work, look back through the chapter for more ideas and keep refining, improving, and learning.

Using Alignment and Profile Targets to Model a Roadside Swale

The previous chapter included an example where the road lane used an alignment and profile target to add a variable width and elevation to the edge of traveled way.

This chapter deals with a roadside swale that follows a variable horizontal alignment, as well as a vertical profile that doesn't follow the centerline of the road. This happens frequently when existing culvert crossings must be met or when you have different slope requirements for the roadside swale.

Corridor Utilities

To create an alignment and profile for the swale, you'll take advantage of some of the corridor utilities found in the Launch Pad panel (Figure 10.2) by selecting the corridor.

FIGURE 10.2
A bounty of corridor utilities on the Launch Pad panel

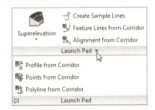

The utilities on this panel are as follows:

Superelevation This button will jump you to the alignment superelevation parameters. See Chapter 11, "Superelevation," for a detailed look at how Civil 3D creates banked curves for your design speed.

Create Sample Lines Corridors and sample lines are both linked to alignments. Civil 3D gives you a shortcut to the sample line creation tool. Chapter 12, "Cross Sections and Mass Haul," explores the creation and uses of sample lines.

Feature Lines From Corridor   This utility extracts a grading feature line from a corridor feature line. This grading feature line can remain dynamic to the corridor, or it can be a static extraction. Typically, this extracted feature line will be used as a foundation for some feature-line grading or projection grading. If you choose to extract a dynamic feature line, it can't be used as a corridor target due to possible circular references.

Alignments From Corridor   This utility creates an alignment that follows the horizontal path of a corridor feature line. You can use this alignment to create target alignments, profile views, special labeling, or anything else for which a traditional alignment could be used. Extracted alignments are not dynamic to the corridor.

Profile From Corridor   This utility creates a profile that follows the vertical path of a corridor feature line. This profile appears in Prospector under the baseline alignment and is drawn on any profile view that is associated with that baseline alignment. This profile is typically used to extract edge of pavement (EOP) or swale profiles for a finished profile view sheet or as a target profile for additional corridor design, as you'll see in this section's exercise. Extracted profiles aren't dynamic to the corridor.

Points From Corridor   This utility creates Civil 3D points that are based on corridor point codes. You select which point codes to use, as well as a range of corridor stations. A Civil 3D point is placed at every point-code location in that range. These points are a static extraction and don't update if the corridor is edited. For example, if you extract COGO points from your corridor and then revise your baseline profile and rebuild your corridor, your COGO points won't update to match the new corridor elevations.

Polyline From Corridor   This utility extracts a 3D polyline from a corridor feature line. The extracted 3D polyline isn't dynamic to the corridor. You can use this polyline as is or flatten it to create road linework.

The following exercise takes you through revising a model from a symmetrical corridor with roadside swales to a corridor with a transitioning roadside swale centerline. You'll also take advantage of some of the static extraction corridor utilities discussed in this section:

1. Open the `Corridor Swale.dwg` file, which you can download from www.sybex.com/go/masteringcivil3d2012.

   Note that the drawing contains a symmetrical corridor (see Figure 10.3), which was built using an assembly that includes two roadside swales. This drawing has been split into two modelspace views so you can see what is going on in plan and in profile at the same time.

2. Select the corridor, and choose the Alignment From Corridor tool from the Launch Pad panel. When prompted, pick the corridor feature line that represents the swale on the left side of the road centerline, as noted in the drawing, to create an alignment.

   Note that if you select one of the corridor feature lines to activate the Corridors tab, that feature line will be created by default. You can pick one of the frequency lines to avoid this potential pitfall!

3. In the Create Alignment From Objects dialog (Figure 10.4), name the alignment **Swale Left**. Keep the default values for the Style and Label options as shown in Figure 10.4. Be sure to uncheck Create Profile. Click OK to finish. Press Esc to exit the command.

FIGURE 10.3
The initial corridor with symmetrical roadside swales

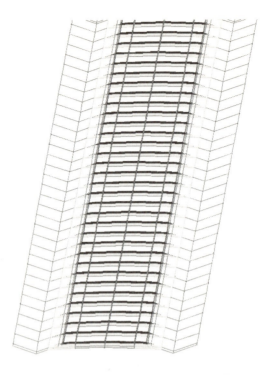

FIGURE 10.4
The Create Alignment From Objects dialog box appears when you're extracting a feature line.

You should now have a new alignment with labels in the plan view, as shown in Figure 10.5.

FIGURE 10.5
The extracted alignment

Next, you will extract the profile in a separate step. By performing these actions in two steps, you ensure that the profile will be associated with the main alignment, rather than with the offset alignment you created in the previous step.

4. Select the Corridor and select Profile From Corridor from the Launch Pad panel of the contextual Ribbon tab.

5. Select the same Swale Left feature line as you did in the previous step. (You may need to use the DRAWORDER command to access the feature line hiding behind the new alignment.) When the feature line is selected, you will see the Create Profile dialog, as shown in Figure 10.6. Name the profile **Swale Left Profile** and click OK.

FIGURE 10.6
Create Profile dialog

6. You should now see the profile you just created in the bottom viewport. The profile represents the vertical path of the feature line representing the swale centerline, as shown in Figure 10.7.

7. Pan over to your newly extracted swale centerline alignment. Using the alignment geometry tools and transparent commands you learned in previous chapters, add a PI at Sta. 46+00, with an offset of 15' to the left (Hint: -15) in plan.

8. In the Profile view, grip-edit the swale centerline profile from 1182.93' in elevation to elevation 1170', as shown in Figure 10.8. Press Esc to exit the command.

FIGURE 10.7
A portion of the resulting extracted profile

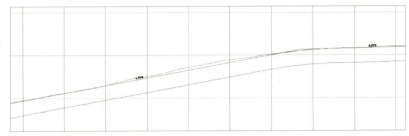

FIGURE 10.8
Stretch a PVI to provide an exaggerated low spot.

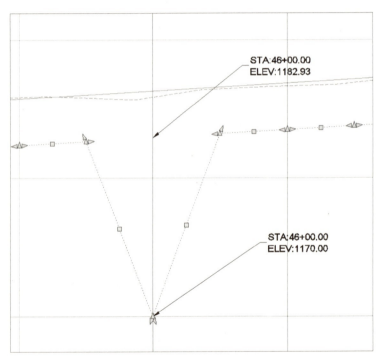

9. Select the corridor and select Corridor Properties from the Modify panel to open the Corridor Properties dialog. Switch to the Parameters tab.

10. In the region, scroll over to the ellipsis button to open the Target Mapping dialog.
11. In the Width or Offset Target area (Figure 10.9), set the Foreslope - L subassembly to target the Swale Left Target Alignment.
12. In the Slope or Elevation Targets area, set the Foreslope - L subassembly to target the Swale Left Profile. Figure 10.9 shows the Target Mapping dialog with the alignments and profiles appropriately mapped.

13. Click OK to dismiss the Target Mapping dialog. Click OK again to dismiss the Corridor Properties dialog and rebuild the corridor. When viewed in 3D, the corridor should now look like Figure 10.10.

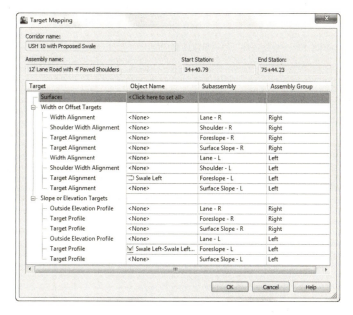

FIGURE 10.9
Set the Foreslope - L subassembly to follow the Swale Left alignment and profile.

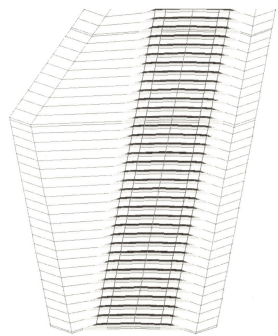

FIGURE 10.10
The adjusted corridor

Note that the corridor has been adjusted to reflect the new target alignment and profile. Also note that you may want to increase the sampling frequency. You can view the sections using the View/Edit Corridor Section tools. You can also view the corridor in 3D by picking it, right-clicking, and choosing Object Viewer. Use the 3D Orbit tools to change your view of the corridor.

Multiregion Baselines

A question many people ask when working with corridors is, "At what point do I need another region?" The answer is simple: If you need a different assembly, you need a different region.

In the following example, you will step through adding an additional region to an existing baseline:

1. Open the drawing `Multi-Region Corridor.dwg`.

2. Select the corridor that exists in the drawing and select Corridor Properties from the contextual Ribbon tab.

3. On the Parameters tab, notice there is a baseline containing a single region. Right-click on the region and select Split Region, as shown in Figure 10.11.

FIGURE 10.11
Right-click to access Region Creation tools.

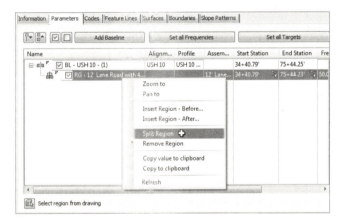

4. At the command line, enter **4000**↵ to create a split at 40+00. Create a second split at 55+00 by entering **5500**↵. Press ↵ again to return to the Corridor Parameters dialog. (If you see a warning dialog referring to 0+00 being outside of the station limits, click OK and continue.)

5. You should have three regions at this step, as shown in Figure 10.12. Change the assembly for the middle region to use the 12' Lane Road w Guardrail assembly.

FIGURE 10.12
New regions with different assemblies applied

6. Scroll over and click the Target ellipsis button. Notice that the Targets have reset to <none> for this region. Set the Targets as shown in Figure 10.13. Click OK when Target Mapping is complete. Then click OK to build the corridor.

7. You should now see additional feature lines in the plan view representing the top of the guardrail. If you choose to examine the corridor in the Object Viewer, it will resemble Figure 10.14.

FIGURE 10.13
Targets for the middle region

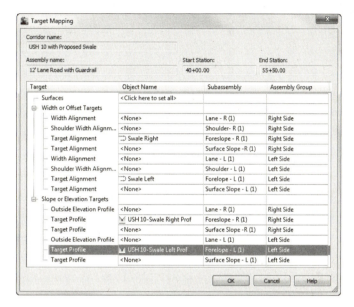

FIGURE 10.14
The completed corridor with guardrails

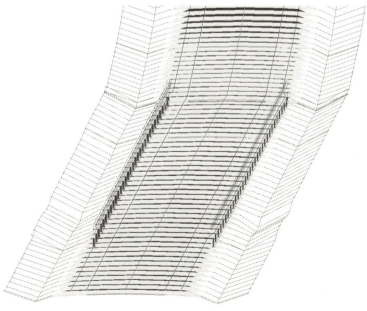

Modeling a Cul-de-sac

Even if you never plan to build one in real life, understanding what is going on in a cul-de-sac corridor model will set you on the right path for building more complex models. If you truly understand the principles explained in the section that follows, then expanding your repertoire to include intersections and roundabouts will become much easier.

Using Multiple Baselines

Up to this point, every corridor we've examined has had a single baseline. In our examples, the centerline alignment has been the main driving force behind the corridor design. In this section, the training wheels are coming off! You've seen the last of single baseline corridors in this book.

A cul-de-sac by itself can be modeled in two baselines, as shown in Figure 10.15. The procedures that follow will work for most cul-de-sacs, symmetrical, asymmetrical, and hammerhead styles. You will need the centerline alignment and design profile, as well as an edge of pavement alignment and design profile.

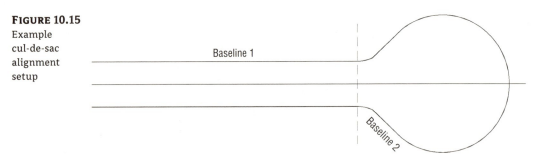

FIGURE 10.15
Example cul-de-sac alignment setup

In the section leading to the cul-de-sac bulb, the corridor can be modeled using the tools you learned in Chapter 9. From station 0+00 to 1+50, the centerline of the road is the baseline, with an assembly using two lanes and with the crown as the baseline marker. However, once the curvature of the bulb starts at 1+50.00, the centerline is no longer an acceptable baseline.

Assemblies are always applied perpendicular to a baseline. To get correct pavement grades in the bulb, you'll need to hop over to the edge of pavement alignment for a baseline. This second baseline will use a partial assembly that is built with the main assembly marker at the outside edge of pavement (Figure 10.16). The crown of the road will stretch to meet the centerline alignment and profile as its targets.

FIGURE 10.16
Assembly used for designing off the edge of pavement

It helps to think of the assembly as radiating away from the baseline, from the assembly base outward, toward a target (Figure 10.17). Because the assemblies are applied to the baseline in a perpendicular manner, using the edge of pavement for a baseline in curved areas (such as cul-de-sac bulbs or curb returns) will result in a smooth, properly graded pavement surface.

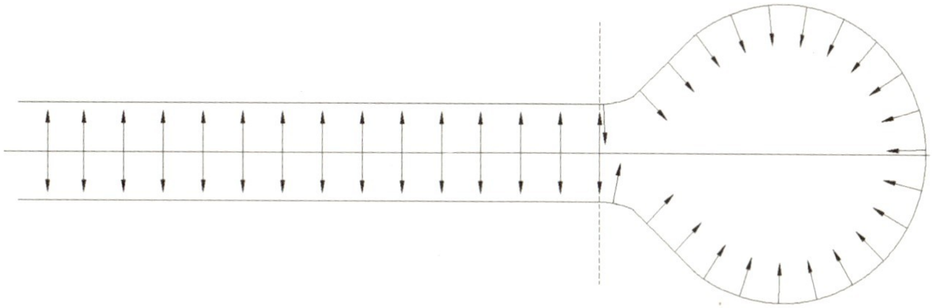

FIGURE 10.17
It helps to think of the assemblies radiating away from the baseline toward the targets.

Establishing EOP Design Profiles

One of the most challenging parts of any "fancy" corridor (i.e., cul-de-sac, intersection, or roundabout) is establishing design profiles for noncenterline alignments. You must have design profiles for every alignment used as a target—there's no way around it—but it does not have to be a painful process to obtain them.

Using a simple, preliminary corridor and the profile creation tools you learned in Chapter 7, "Profiles and Profile Views," establishing an edge of pavement (EOP) profile can go quickly.

In the exercise that follows, you will work through the steps of creating an EOP profile:

1. Open the Cul-de-sac_Profile.dwg file, which you can download from this book's web page.

   This drawing contains several alignments and an existing surface that is turned off.

2. From the Home tab, click Corridor ➤ Create Simple Corridor. Name the corridor **PRELIM**. Click OK.

3. Click the centerline near the cul-de-sac bulb, Frontenac Drive. For the profile, use Frontenac Drive - FG. For assembly, use PRELIM.

4. No targets are needed in this preliminary corridor, so click OK to complete the corridor.

5. Select the new corridor and open Corridor Properties. On the Surfaces tab, click the leftmost button to start a corridor surface. With Data Type set to Links and Code set to Top, click Add.

6. On the Boundaries tab, right-click the PRELIM Surface, click Add Automatically, and use ETW as the outer boundary. Click OK.

7. You should now see contours in your drawing representing the 2 percent crossfall from the centerline. Select the red EOP alignment for the cul-de-sac. In the contextual Ribbon tab, click Surface Profile.

8. In the Create Profile From Surface dialog, highlight PRELIM Surface and click Add. Click the Draw In Profile View button.

9. Leave all the defaults in the Create Profile View dialog and click Create Profile View. Click in the graphic to the right of the surface to place the view.

   The profile you are seeing will have gaps in the middle that need grading information. However, the bulb portion of the profile only needs to exist between 16+87.00 and 19+42.89. They are the PC and PT stations for the EOP alignment in plan.

10. Select the profile view and click Profile Creation Tools on the contextual Ribbon tab. Click OK to accept the defaults in the Create Profile dialog.

11. In the Profile Layout toolbar, select Draw Tangents. Snap to the intersection of the preliminary profile and the grid line at 16+50. Next, snap to the endpoint where the preliminary surface trails off at station 17+05.88 (elevation 792.74). The next VPI will be the peak of the preliminary profile in the center of the view (station 18+14.94, elevation 790.35).

    Snap to the endpoint where the preliminary profile picks up again (station 19+24.01, elevation 792.74). Finally, snap to the grade break at station 19+78.23, elevation 794.68.

12. Press ↵ to complete the command.

    You now have a proposed profile that is acceptable to use in the cul-de-sac corridor.

Putting the Pieces Together

You have all the pieces in place to perform the first iterations of this cul-de-sac design.

The following exercise will walk you through the steps to put the cul-de-sac together. You will complete several steps and let the corridor build to observe what is happening at each stage. This exercise will also encourage you to get comfortable using Corridor Properties to make design modifications.

1. Open the `Cul-de-sac_Design.dwg` file, which you can download from this book's web page. This drawing contains the cul-de-sac centerline alignment and profile, the EOP alignment and profile, and the assemblies needed to complete the process. The PRELIM corridor layer is frozen.

2. From the Home tab, on the Create Design panel select Corridor ➢ Create Corridor. Select the Frontenac Drive centerline alignment, the Frontenac Drive FG profile, and the Urban 2×14′ Lanes 2% Xslope assembly.

3. In the Create Corridor dialog, name the corridor **Cul-de-Sac**.

4. Click the Pick Station button for the start station of the first region. Snap to the start of the EOP alignment on the west side, as shown in Figure 10.18 (left). This will result in a start station of 11+42.86 in the Create Corridor dialog.

5. Click the Pick Station icon for the end station of the first region. Snap to the PT station on the east side of the cul-de-sac bulb, as shown in Figure 10.18 (right). This will result in the region end station of 15+35.41.

FIGURE 10.18
Setting the start station (left) and end station (right) for the centerline region

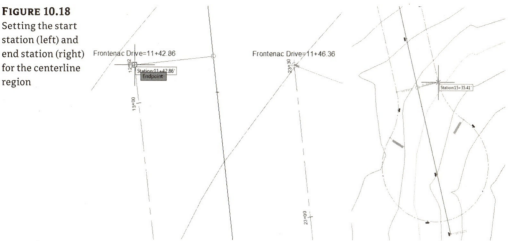

6. Scroll over and click the Target ellipsis. Set the surface target to EG. Click OK to dismiss the Target Mapping dialog. Click OK to let the corridor build.

   Examine the corridor you just created. It should start south of the intersection with the Syrah Way alignment and end before the curvy part of the cul-de-sac bulb.

7. Select the corridor and click Corridor Properties. If necessary, correct any station problems with the first region.

8. In the Parameters tab of Corridor Properties, click Add Baseline. Select the Cul-de-Sac EOP as the alignment. Click OK.

9. In the Profile Column, click <Click here...>. Select the Cul-de-Sac EOP - FG profile (this is the profile that you learned how to develop in the previous exercise).

10. Right-click on the newly added baseline and select Add Region, as shown in Figure 10.19. Select the Curb Right assembly and click OK.

11. Click the Pick Station button for the Curb Right region start station. Select the PC station of the corridor bulb. This will result in a start station value of 16+87.00. Click the Pick Station icon for the end station of the region. Select the PT station of the corridor bulb. This will result in an end station of 19+42.89. Click OK and let the corridor build once again.

    Have a look at the corridor in its current state (see Figure 10.20). Even if you are not exactly sure what you are looking at, you should at least see a few things are amiss with the corridor so far.

FIGURE 10.19
Right-click the baseline to add a region.

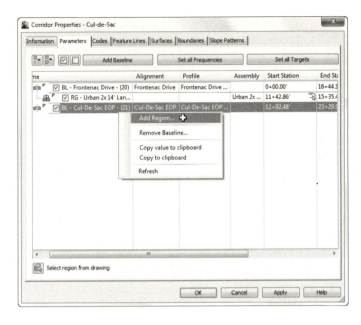

FIGURE 10.20
The cul-de-sac corridor several steps away from completion

There are three things that need to be modified before the cul-de-sac is complete. The lane needs to extend to the center of the cul-de-sac bulb and the daylight surface needs to be set, both of which can be corrected in the Target Mapping area. Finally, you will want to increase the frequency around the curvy area to get a smoother, more precise design.

In the next steps, you will correct these issues and complete the cul-de-sac.

12. Select the corridor and return to Corridor Properties (last time in this exercise, I promise). Scroll over, and click the ellipsis to enter the Target Mapping dialog for the region named RG - Curb Right - (36). (The number following the region name may vary.)

13. Set Target Surface to EG. Set Width Alignment for Lane - L to Frontenac Drive. Set Outside Elevation Profile to Frontenac Drive FG for Lane - L. Click OK.

14. Click the frequency ellipsis next to the current value of 25'. Set the frequency for both tangents and curves to 5'. Click OK to dismiss the Frequency dialog.

15. Click OK and let the corridor build one last time. The completed corridor will look like Figure 10.21.

FIGURE 10.21
The completed cul-de-sac corridor. Gorgeous!

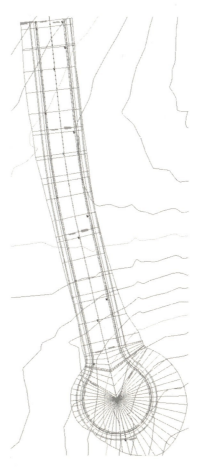

Troubleshooting Your Cul-de-Sac

People make several common mistakes when modeling their first few cul-de-sacs:

Your cul-de-sac appears with a large gap in the center. If your curb line seems to be modeling correctly but your lanes are leaving a large empty area in the middle (see Figure 10.22), chances are pretty good that you forgot to assign targets or perhaps assigned the incorrect targets.

FIGURE 10.22
A cul-de-sac without targets

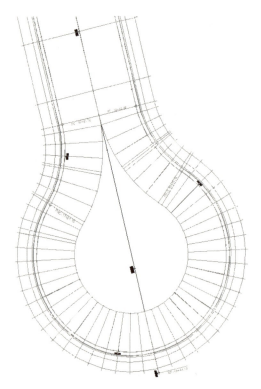

Fix this problem by opening the Target Mapping dialog for your region and checking to make sure you assigned the road centerline alignment and FG profile for your transition lane. If you have a more advanced lane subassembly, you may have accidentally set the targets for another subassembly somewhere in your corridor instead of the lane for the cul-de-sac transition, especially if you have poor subassembly-naming conventions. Poor naming conventions become especially confusing if you used the Map All Targets button.

Your cul-de-sac appears to be backward. Occasionally, you may find that your lanes wind up on the wrong side of the EOP alignment, as shown in Figure 10.23. The direction of your alignment will dictate whether the lane will be on the left or right side. In the example from the previous exercise, the alignment was running counterclockwise around the cul-de-sac bulb; therefore, the lane was on the left side of the assembly.

FIGURE 10.23
A cul-de-sac with the lanes modeled on the wrong side without targets (left) and with targets (right)

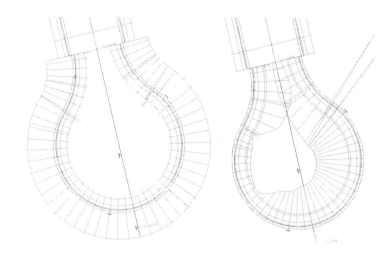

You can fix this problem by changing the assembly to the correct side.

Your cul-de-sac drops down to 0. A common problem when you first begin modeling cul-de-sacs, intersections, and other corridor components is that one end of your baseline drops down to 0. You probably won't notice the problem in plan view, but once you build your surface (see Figure 10.24, left) or rotate your corridor in 3D (see Figure 10.24, right), you'll see it. This problem will always occur if your region station range extends beyond the proposed profile.

FIGURE 10.24
(Right) Contours indicating that the corridor surface drops down to 0, and (left) a corridor viewed in 3D showing a drop down to 0

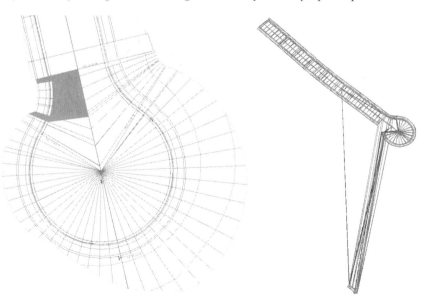

The fix for this is the same as what you saw in the first exercise in Chapter 9—that is, make sure your region station range jives with the design profile length. You may need to extend the design profile in some cases, but usually the station range will do the trick.

Your cul-de-sac seems flat. When you're first learning the concept of targets, it's easy to mix up baseline alignments and target alignments. In the beginning, you may accidentally choose your EOP alignment as a target instead of the road centerline. If this happens, your cul-de-sac will look similar to Figure 10.25.

FIGURE 10.25
A flat cul-de-sac with the wrong lane target set

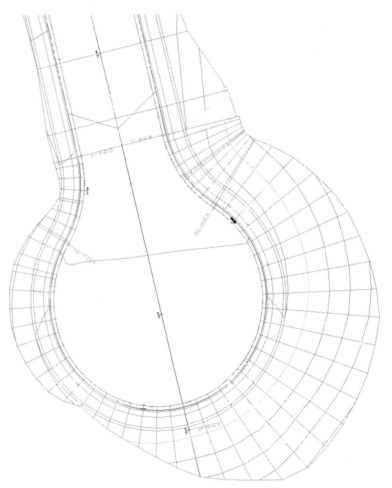

You can fix this problem by opening the Target Mapping dialog for this region and making sure the target alignment is set to the road centerline and the target profile is set to the road centerline FG profile.

Intersections: The Next Step Up

Ask yourself the following question: "Did I understand why we did what we did to build the cul-de-sac in the previous section?" If the answer is "Not really," you may want to review a few topics before proceeding. If the answer is, "Yeah, mostly," then you are ready to move to the next level of corridor complexity: intersections.

The steps that follow apply to all intersections, regardless of whether it is a T-shaped intersection, a four-way intersection, perfectly perpendicular, or skewed at an angle.

CERT OBJECTIVE

Corridor modeling is an iterative process. The more advanced your model, the more iterations it may take to get to the correct design. You will often not know the final design parameters until you see how the model relates to existing conditions or ties into other pieces of the design. Get comfortable jumping in and out of corridor properties and identifying regions within your corridor.

Plan what alignments, profiles, and assemblies you'll need to create the right combination of baselines, regions, and targets to model an intersection that will interact the way you want. It helps to create a simple sketch, as shown in Figure 10.26.

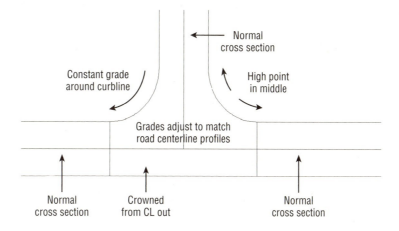

Figure 10.26
Plan your intersection model in sketch form.

Figure 10.27 shows a sketch of required baselines. As you saw in the previous example, *baselines* are the horizontal and vertical foundation of a corridor. Each baseline consists of an alignment and its corresponding finished ground (FG) profile. You may never have thought of edge of pavement (EOP) in terms of profiles, but after building a few intersections, thinking that way will become second nature. The Intersection tool on the Create Design panel of the Home tab will create EOP baselines as curb return alignments for you, but it will rely on your input for curb return radii.

Figure 10.28 breaks each baseline into regions where a different assembly or different target will be applied. Once the intersection has been created, target mapping as well as other particulars can be modified as needed.

FIGURE 10.27
Required baselines for modeling a typical intersection

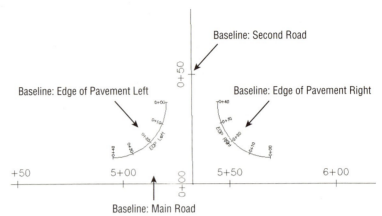

FIGURE 10.28
Required regions for modeling an intersection created by the Intersection tool

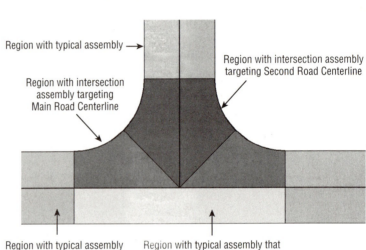

Using the Intersection Wizard

All the work of setting baselines, creating regions, setting targets, and applying the correct frequencies can be done manually for an intersection. However, Civil 3D contains an automated Intersection tool that can handle many types of intersections.

On the basis of the schematic you drew of your intersection, your main road will need several assemblies to reflect the different road cross sections. Figure 10.29 shows the full range of potential assemblies you may need in an intersection and the design situations in which they may arise.

FIGURE 10.29
Various assembly schematics and applications

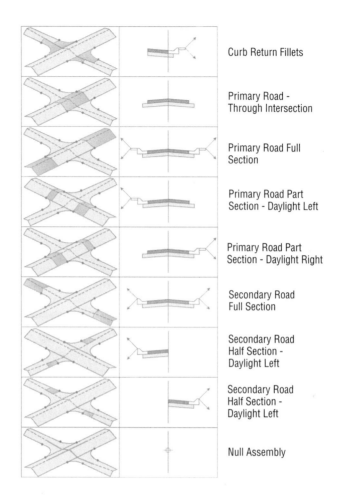

ASSEMBLY SETS

When you are ready to create an intersection, you do not need to have all the special assemblies created. On the Corridor Regions page of the Create Intersection wizard, you will see a list of the assemblies Civil 3D plans to use. If the assemblies are not already part of the drawing, they will get pulled in automatically when you click Create Intersection.

The default intersection assemblies are general and may not work for your design situation. You will want to create and save an assembly set of your own.

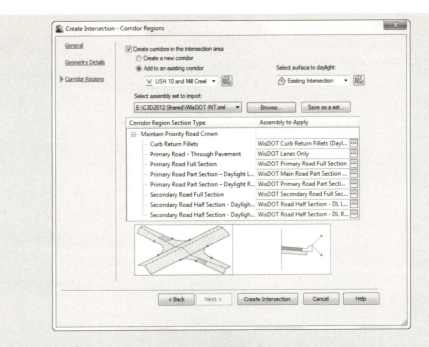

In a file that contains all of your desired assemblies, work through the Create Intersection wizard to get to the Corridor Regions page. Click the ellipsis to select the appropriate assembly for each Corridor Region Section Type. Once the listing is complete, click the Save As A Set button.

Civil 3D creates an XML file that stores the listing of the assemblies. It also creates a copy of each assembly as a separate DWG file. Save the set in a network shared location so your office colleagues can use the set as well.

The next time an intersection is created, you can use the assembly set by clicking Browse and selecting the XML file. Civil 3D will pull in your assemblies, saving lots of time!

This exercise will take you through building a typical peer-road intersection using the Create Intersection wizard:

1. Open the Intersection.dwg file, which you can download from this book's web page. The drawing contains two centerline alignments (USH 10 and Mill Creek Drive) that are part of the same corridor.

2. Select the corridor in plan view to activate the contextual tab. From the Modify Corridor panel, click Corridor Properties. Switch to the Parameters tab.

   There are already two baselines with three regions each in the corridor. These are the regions for modeling the areas outside of the intersection. The regions that use the Null Assembly are empty placeholders to prevent feature lines from crossing through the intersection. This way, there is an empty spot for the corridor design to reside.

3. After making note of the current state of the corridor, click OK to close the Corridor Properties.

4. From the Home tab, choose the Intersections ➤ Create Intersection tool on the Create Design panel. Using the Intersection object snap, choose the intersection of the two existing alignments.

5. When prompted to select the main road alignment, click the USH 10 alignment that runs vertically in the project.

6. The Create Intersection – General dialog will appear (Figure 10.30). Name the intersection **USH 10 and Mill Creek Drive**. Set the Intersection corridor type to Primary Road Crown Maintained, as shown in Figure 10.30, and click Next.

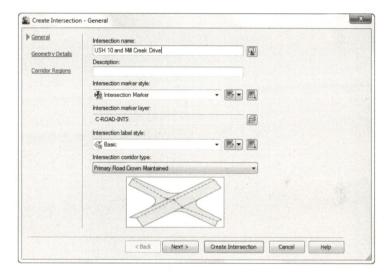

FIGURE 10.30
The Create Intersection – General dialog

7. In the Geometry Details page (Figure 10.31), verify that USH 10 is the primary road by looking at the Priority listing. If USH 10 is not at the top of the list, use the arrow buttons on the right side.

8. Click the Offset Parameters button.

   ◆ Set the offset values for both the left and right sides to 24′ for USH 10.

   ◆ Set the left and right offset values for Mill Creek Drive to 18′.

   ◆ Select the check box Create New Offsets From Start To End Of Centerlines.

   ◆ At this step, the screen should resemble Figure 10.32. Click OK to close the Offset Parameters dialog.

FIGURE 10.31
The Geometry Details page of the Create Intersection wizard

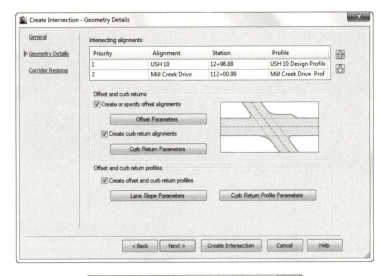

FIGURE 10.32
Intersection offset parameters

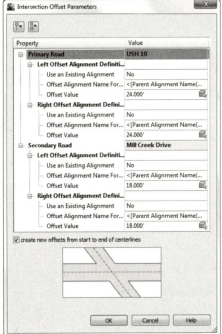

9. Click the Curb Return Parameters to enter the Curb Return Parameters dialog.

   ◆ For all four quadrants of the intersection, place a check mark next to Widen Turn Lane For Incoming Road and Widen Turn Lane For Outgoing Road.

- Click the Next button at the top of the dialog to move from quadrant to quadrant. As shown in Figure 10.33, a temporary glyph will help you determine which quadrant you are currently modifying.

- For all four curb returns, click the check marks for Widen Turn Lane For Incoming Road and Widen Turn Lane For Outgoing Road. Leave turn lane values at their defaults for each quadrant.

- When you reach the last quadrant, you will see that the Next button is grayed out. This means that you have successfully worked through all four curb returns.

- Click OK to return to wizard's Geometry Details page.

FIGURE 10.33
Adding lane widening to the SW - Quadrant of the intersection

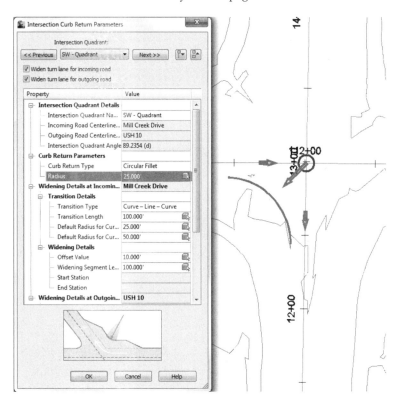

Some locales require that lane slopes flatten out to a 1 percent cross-slope in an intersection. If this is the case for you, you can change the lane slope parameters in the Intersection Lane Slope Parameters dialog (Figure 10.34).

Civil 3D is performing the task of generating the curb return profile. The profile will be at least as long as the rounded curb plus the turn lanes that are added in this exercise. If you wish to have Civil 3D generate even more than the length needed, you can specify that in the Intersection Curb Return Profile Parameters area (Figure 10.35).

FIGURE 10.34
Lane slope parameters control the cross-slope in the intersection.

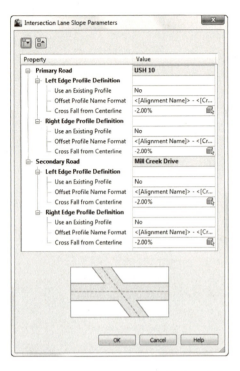

FIGURE 10.35
Curb Return Profile Parameters extend the Civil 3D–generated profile beyond the curb returns by the value specified.

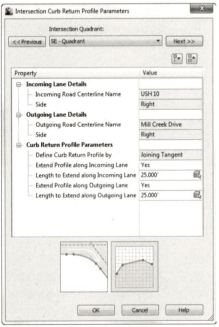

10. We will be keeping all default settings in both the Lane Slope Parameters area (Figure 10.34) and the Curb Return Parameters area (Figure 10.35). Click Next to continue to the Corridor Regions page (Figure 10.36).

   The Corridor Regions page is where you control which assemblies are used for the different design locations around the intersection. Clicking each entry in the Corridor Region Section Type list will give you a clear picture of which assemblies you should use and where (as shown at the bottom of Figure 10.36). If your assemblies have the same names as the default assemblies, as is the case in this example, they will be pulled from the current drawing. Alternately, you can click the ellipsis to select any assembly from the drawing.

   If you don't have all the necessary assemblies at this point, you can still create your intersection. Civil 3D will pull in the default set of assemblies. You can always modify these assemblies after they are brought in.

11. Set the Create Corridor In Intersection option to Add To An Existing Corridor. Click through the Corridor Region Section Type list to view the schematic preview for each. Do not make any changes to the assembly listing. Click Create Intersection.

12. After a few moments of processing, you will see a corridor appear at the intersection of the roads. Use the REGEN command if you do not see the frequency lines. Your corridor should now resemble Figure 10.37.

   Notice there are still some gaps where our initial corridor regions do not meet up with our corridor. In the next step, you will use grips to extend the regions to meet the intersection portion of the corridor.

FIGURE 10.36
The Corridor Regions page drives the assemblies used in the intersection.

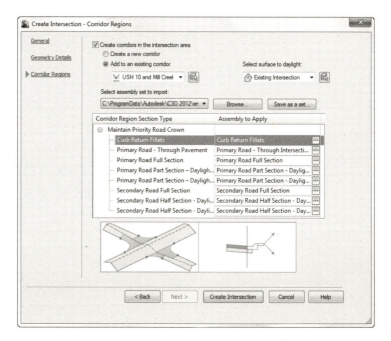

FIGURE 10.37
The nearly completed intersection

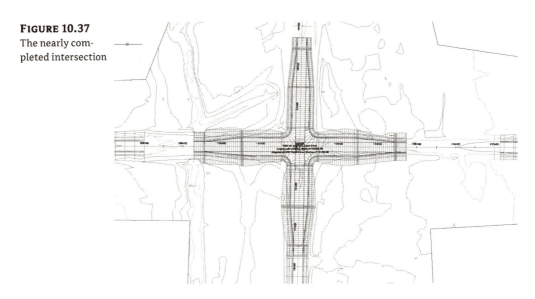

13. Select the corridor. Use the diamond-shaped grips to pull the regions in toward the intersection, as shown in Figure 10.38a. If you are having difficulty doing so or are having computer performance problems, you can also adjust the region stations in the Parameters tab of the Corridor Properties dialog (Figure 10.38b).

   Notice how the region highlighted in the Parameters tab lights up graphically as well. This will help you determine which region to edit, even in the largest of corridors.

   To move from region to region without searching through the somewhat daunting list, use the Select Region From Drawing button.

   When the corridor is complete, it will look like Figure 10.39.

FIGURE 10.38
Change the regions leading up to the intersections using grips (a) or in Corridor Properties (b)

FIGURE 10.38
(continued)

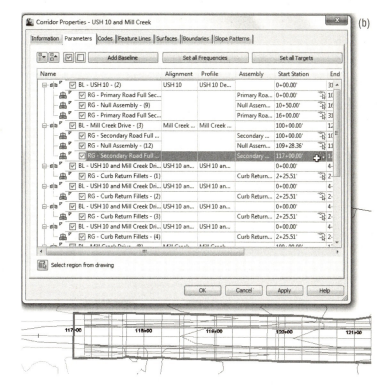

(b)

FIGURE 10.39
The completed intersection

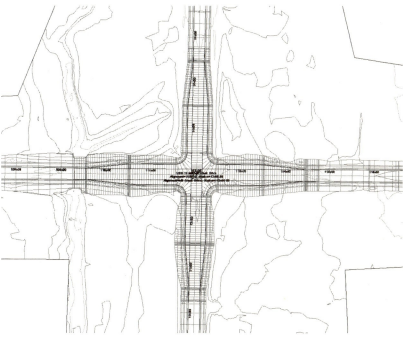

Manually Modeling an Intersection

In some cases, you may find it necessary to model an intersection manually. Five-way intersections and intersections containing superelevated curves can't be created with the automated tools alone. You need to understand what the intersection tool is doing behind the scenes before you can be a true corridor guru. The next example will take you through an overview of the manual steps.

At the point where the example begins, a few pieces are already in place: the centerline alignments and profiles, EOP alignments and profiles, and the full road assemblies. The corridor for the intersection is started for you, but it only contains one baseline. For simplicity's sake, daylight subassemblies have been omitted from this example.

In this example, you'll add a baseline for an intersecting road to your corridor:

1. Open the `Manual Intersection.dwg` file, which you can download from this book's web page.

2. Open the Corridor Properties dialog for Intersection1 (Cab-Syrah), and switch to the Parameters tab.

3. Click Add Baseline. The Pick Horizontal Alignment dialog opens.

4. Pick Syrah Way alignment. Click OK to dismiss the dialog.

5. Click in the Profile field on the Parameters tab of the Corridor Properties dialog. The Select A Profile dialog opens.

6. Pick the Syrah Way FG. Click OK to dismiss the dialog.

   You now have a new baseline, but the region still needs to be added.

7. In the Corridors Properties dialog, right-click BL - Syrah Way and select Add Region.

8. In the Pick an Assembly dialog, select Full Road Section. Click OK to dismiss the dialog.

9. Expand BL - Syrah Way by clicking the small + sign, and see the new region you just created.

10. Click OK to dismiss the Corridor Properties dialog; your corridor will automatically rebuild. Your corridor should now look like Figure 10.40.

FIGURE 10.40
The Syrah Way baseline added

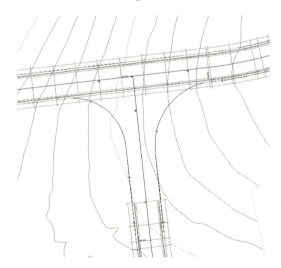

Notice that the region for the second baseline extends all the way through the intersection. You must now split the Syrah region to accommodate the curb returns.

11. Open the Corridor Properties dialog and switch to the Parameters tab. Using the techniques covered in the previous exercise, right-click to split the region at 3+50.00 and 5+50.00.

12. Switch the middle subassembly to Daylight Right. Your corridor parameters will now match Figure 10.41.

13. Click OK to update the corridor. You will see that the way has been cleared for the curb assemblies to be applied as shown in Figure 10.42.

FIGURE 10.41
Split the Syrah Way baseline into 3 regions to accommodate curb returns

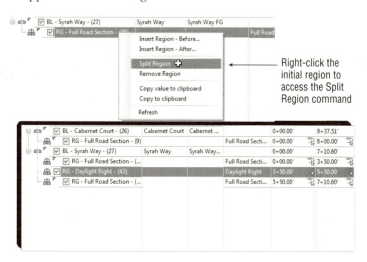

FIGURE 10.42
The intersection corridor...so far

Creating an Assembly for the Intersection

You built several assemblies in Chapter 8, "Assemblies and Subassemblies," but most of them were based on the paradigm of using the assembly marker along a centerline. This next exercise leads you through building an assembly that attaches at the EOP.

This exercise uses the LaneOutsideSuper subassembly (see Figure 10.43), but you're by no means limited to this subassembly in practice.

FIGURE 10.43
An assembly for an intersection curb return

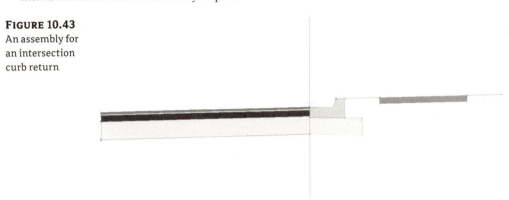

You can also use the LaneInsideSuper and LaneTowardCrown subassemblies, as well as any other lane with an alignment and profile target.

To create an assembly suitable for use on the filleted alignments of the intersection, follow these steps:

1. Open the `Manual Intersection_Assembly.dwg` file (which you can download from this book's web page), or continue working in your drawing from the previous exercise.

2. From Prospector, expand the Assemblies group and locate Curb Return Fillets. Right-click on this assembly and select Zoom To. The assembly base has been placed in the drawing for you but does not contain any subassemblies.

3. Open your tool palette, and switch to the Lanes palette.

4. Click the LaneOutsideSuper subassembly. On the Properties palette, change the side to Left.

5. Click on the main baseline assembly to add the LaneOutsideSuper subassembly to the left side of the assembly.

6. Switch to the Curbs tab of the tool palette. Select UrbanCurbGutterGeneral. Change the side to Right and leave all other values at their defaults. Click to add the curb and gutter to the right side of the assembly.

7. Also on the Curbs tab of the tool palette, add an UrbanSidewalk assembly. Set the boulevard widths to 2' for both inside and outside. Click to place the assembly.

   You could combine steps 6 and 7 by using the Copy To Assembly option after selecting the curb and gutter along with the sidewalk from the Daylight Right Assembly.

Your assembly should now look like Figure 10.43.

> **TAKE THE PEBBLE FROM MY HAND, GRASSHOPPER**
>
> If you are a little worried about the downward slope of the lane in the subassembly you just created, you needn't be. Remember that the slope and length of the lane will be controlled by the alignment and profile target in the corridor. The geometry that you see in the assembly is purely preliminary. If this concept makes sense to you, you are ready for bigger and better corridors!
>
> And of course, if you'd rather see that slope go +2% in the subassembly, you can change it in the good old AutoCAD properties, but keep in mind it won't make one bit of difference to the corridor.

Adding Baselines, Regions, and Targets for the Intersections

The Intersection assembly attaches to alignments created along the EOP. Because a baseline requires both horizontal and vertical information, you also need to make sure that every EOP alignment has a corresponding finished ground (FG) profile.

It may seem awkward at first to create alignment and profiles for things like the EOP, but after some practice you'll start to see things differently. If you've been designing intersections using 3D polylines or feature lines, think of the profile as a vertical representation of a feature line and the profile grid view as the feature-line elevation editor. If you've designed intersections by setting points, think of alignment PIs and profile PVIs as points. If you need a low point midway through the EOP, as indicated in the sketch at the beginning of this section, you'll add a PVI with the appropriate elevation to the EOP FG profile.

Here are some things to keep in mind:

Site Geometry Make sure any alignments you create are either *siteless* (placed on the <none> site) or placed on another appropriate site.

Naming Conventions Instead of allowing your alignments and profiles to be named Alignment-1, Alignment-2, and so on, give each one a meaningful name that will help you keep them straight so that you can identify them on a list and locate them in plan. The same rule applies for the FG profiles. If the alignment is named EOP Right, then name the profile EOP Right FG or something similar. Explore alignment- and profile-name templates to see if you can help yourself by automating the naming.

Organization Figure out a way to keep your profile views organized. Perhaps line them up next to the appropriate road centerline, or use some other convention. If you just stick them anywhere, you'll have a difficult time navigating and finding what you need.

> **USING MULTIPLE TARGETS**
>
> Civil 3D allows you to select more than one item in Width Or Offset Target and Slope Or Elevation Target areas. You can use any number of targets. You can even mix and match the types of targets you use, such as alignments with polylines, survey figures, and feature lines.

In the example that follows, you need two alignment targets and two profile targets in each curb region. To help Civil 3D figure out what you want, you need to verify that it will use the target that it finds *first* as it radiates out from the baseline. In most cases, you can click the Target To Nearest Offset option found in the target selection dialogs.

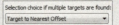

In the following graphic, the arrows represent the assemblies seeking out a target. From station 1+60.60 to 2+19.56, the Syrah Way alignment is closer to the baseline; therefore, Syrah Way's geometry is used as the target. From station 2+19.56 to 3+28.46, Cabernet Court's alignment is closer and is used as the target.

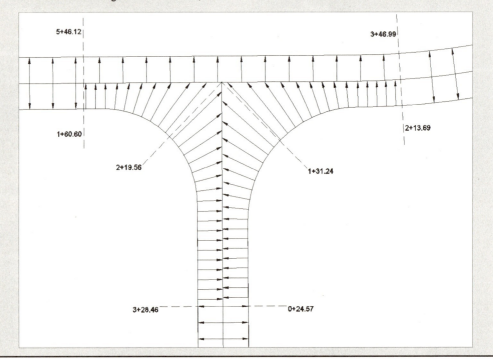

You'll finalize the corridor intersection in the next exercise. Along the way, you'll examine the corridor in various stages of completion so that you'll understand what each step accomplishes. In practice, you'll likely continue working until you build the entire model. Follow these steps:

1. Open the `Manual Intersection_Targets.dwg` file (which you can download from this book's web page), or continue working in your drawing from the previous exercise.

2. (Optional) Zoom to the area where the profile views are located. Notice that existing and proposed ground profiles are created for both EOP alignments (see Figure 10.44). The step of developing the design profiles has been completed for you.

3. Select the corridor. Open the Corridor Properties dialog and switch to the Parameters tab.

4. Click Add Baseline. The Pick Horizontal Alignment dialog opens.

5. Select the Cab-Syrah EOP W alignment. Click OK to dismiss the dialog.

6. Click in the Profile field on the Parameters tab of the Corridor Properties dialog.

7. In the Select A Profile dialog, select Cab-Syrah EOP W- FG. Click OK to dismiss the dialog.

8. Right-click BL - Cab-Syrah EOP W and select Add Region.

9. In the Pick An Assembly dialog, select Curb Return Fillets. Click OK to dismiss the dialog.

10. Expand the baseline, and see the new region you just created. The Parameters tab of the Corridor Properties dialog will look like Figure 10.44.

FIGURE 10.44
The Parameters tab with a new baseline and region

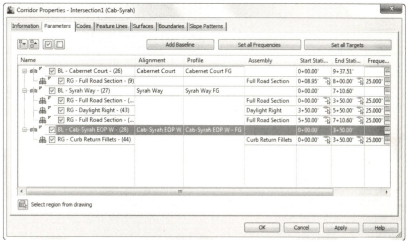

11. Click OK to dismiss the Corridor Properties dialog and automatically rebuild your corridor. Your corridor should now look similar to Figure 10.45.

    What is missing from the corridor? We need to rein in the station range, to set targets, and to add more frequency lines along the curves. Oh, and eventually we'll want to do this all to the east side of the road as well.

12. Select the corridor in plan view, click Corridor Properties from the contextual tab, and switch to the Parameters tab. To set the correct stations for the regions, use the station picker and snap to where we stopped the main assembly along Syrah Way. This will be station 1+60.60. Set the end station using the station picker and snapping to where the normal region ends on Cabernet Court. This will be station 3+28.46.

13. Next, set the frequency by clicking the ellipsis button. Set both tangents and curves to 5'. Click OK.

FIGURE 10.45
Your corridor after applying the Curb Return Fillets assembly

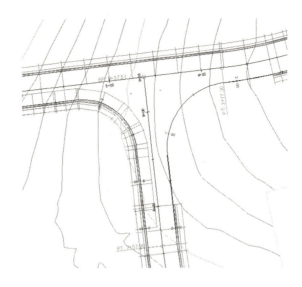

14. Click the ellipsis for targets.

    ◆ Click the Width Alignment target for LaneOutsideSuper. Click the Select From Drawing button. Click both Syrah Way and Cabernet Court centerline alignments.

    ◆ Press ↵ when complete.

    ◆ In the Set Width Or Offset Target dialog, click Add. Both alignments will appear in the listing, as shown in Figure 10.46.

    ◆ Click OK when complete.

FIGURE 10.46
In the Set Width Or Offset Target dialog, use the Select From Drawing button to pick both Syrah Way and Cabernet Court centerlines.

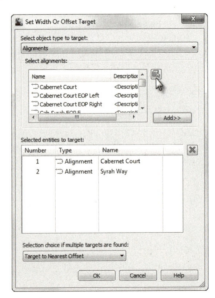

15. Click the Outside Elevation Profile target for LaneOutsideSuper.

    ◆ From the Select and Alignment drop-down, select Cabernet Court from the list.

    ◆ Highlight the Cabernet Court FG profile and click Add.

    ◆ Change the selected alignment to Syrah Way.

    ◆ Highlight the Syrah Way FG profile and again click Add. Both profiles will appear in the listing, as shown in Figure 10.47.

    ◆ Click OK.

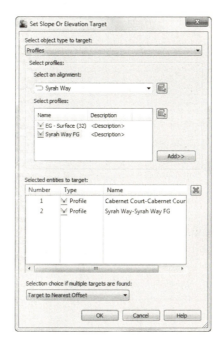

FIGURE 10.47
In the Set Slope Or Elevation Target dialog, pick alignment first, and then add the FG profile.

16. Click OK to dismiss target mapping. Click OK to let the corridor build. Your intersection should now look like Figure 10.48.

17. Repeat steps 4–17 for the Cab-Syrah EOP E. Here are some hints to get you going: The region should go from station 0+24.57 to 2+13.69. The frequency should be 5′ for both tangents and curves. The target mapping will be identical to the region you created in steps 15–16.

18. Let the corridor build one last time and admire your work. The completed intersection should look like Figure 10.49.

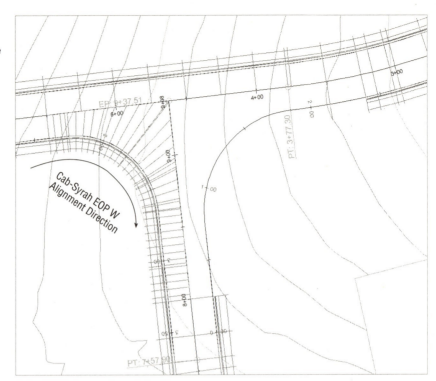

FIGURE 10.48
Three baselines completed—one to go

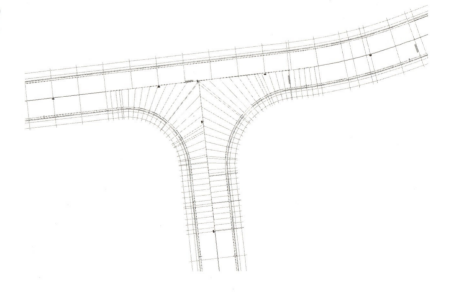

FIGURE 10.49
The completed intersection

Troubleshooting Your Intersection

The best way to learn how to build advanced corridor components is to go ahead and build them, make mistakes, and try again. This section provides some guidelines on how to "read" your intersection to identify what steps you may have missed.

Your lanes appear to be backward. Occasionally, you may find that your lanes wind up on the wrong side of the EOP alignment, as in Figure 10.50. The most common cause is that your assembly is backwards from what is needed based on your alignment direction.

FIGURE 10.50
An intersection with the lanes modeled on the wrong side

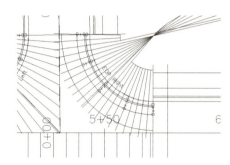

Fix this problem by editing your subassembly to swap the lane to the other side of the assembly. If the assembly is used in another region that is correct, just make a new assembly that is the mirror reverse of the other assembly and apply the new one to the alignment.

In this exercise, you used an intersection assembly with the transition lane on the left side, so you made sure the left intersection EOP alignment ran clockwise and the right intersection EOP alignment ran counterclockwise. It's also easy to reverse an alignment and quickly rebuild the corridor if you catch the mistake after the corridor is built.

Your intersection drops down to zero. A common problem when modeling corridors is the cliff effect, where a portion of your corridor drops down to zero. You probably won't notice in plan view, but if you rotate your corridor in 3D using the Object Viewer (see Figure 10.51), you'll see the problem. The most common cause for this phenomenon is incorrect region stationing.

FIGURE 10.51
A corridor viewed in 3D, showing a drop down to zero

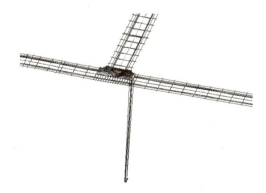

Fix this problem by making sure your baseline profile exists where you need it and make note of the station range. Set your station range in the corridor to be within the correct range.

Your lanes extend too far in some directions.   There are several variations on this problem, but they all appear similar to Figure 10.52. All or some of your lanes extend too far down a target alignment, or they may cross one another, and so on.

FIGURE 10.52
The intersection lanes extend too far down the main road alignment.

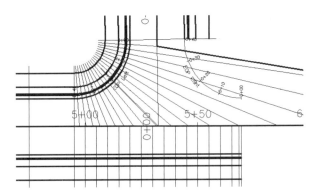

This occurs when a target alignment and profile have been omitted for one or more regions. In the case of Figure 10.52, the EOP Left baseline region was only set to one alignment. In an intersection, you need two targets in a corner region to model the road correctly.

Your lanes don't extend far enough.   If your intersection or portions of your intersection look like Figure 10.53, you neglected to set the correct target alignment and profile.

FIGURE 10.53
Intersection lanes don't extend out far enough.

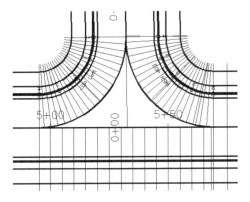

You can fix this problem by opening the Target Mapping dialog for the appropriate regions and double-checking that you assigned targets to the right subassembly. It's also common to accidentally set the target for the wrong subassembly if you use Map All Targets or if you have poor naming conventions for your subassemblies.

Checking and Fine-Tuning the Corridor Model

Recall that the EOP profiles are developed by creating a preliminary corridor surface model. To create the EOP elevations, you filled in the gaps by creating the EOP design profile where the preliminary surface stops and picks back up again on the adjacent road (or in the case of the cul-de-sac, the opposite side of the street).

You will want to check the elevations of the corridor model against the profiles you created. The preliminary surface model was a best guess for elevations; now we need to reconcile our corridor geometry with those best-guess profiles. Figure 10.54 shows a common elevation problem that arises in the first iteration of intersection design.

FIGURE 10.54
A potential elevation problem in the intersection

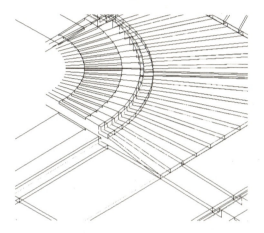

The first step in correcting this is creating a corridor surface model out of Top links. The surface will show us exactly how Civil 3D is interpreting our design, elevation-wise.

Before beginning this exercise, review the basic corridor surface building sections in Chapter 9. When you think you are ready, follow these steps:

1. Open the `Manual Intersection_Surface.dwg` file (which you can download from this book's web page), or continue working in your drawing from the previous exercise.

2. Select the corridor in plan view, click Corridor Properties from the contextual tab, and switch to the Surfaces tab.

3. Click the Create A Corridor Surface button. A Corridor Surface entry appears. Rename the surface to **Corridor Surface** by clicking on the surface name.

4. Ensure that Links is selected as Data Type and that Top appears under Specify Code. Click the + button. An entry for Top appears under the Corridor Surface entry.

5. Switch to the Boundaries tab. Right-click on Corridor Surface and select Corridor Extents as the outer boundary.

6. Click OK, and your corridor will automatically rebuild.

7. Freeze the corridor to get a better look at the corridor surface. You should have a corridor surface that appears similar to Figure 10.55.

8. Thaw any frozen layers.

FIGURE 10.55
A first-draft corridor surface

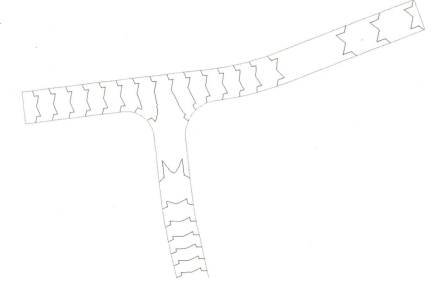

The next exercise will lead you through placing some design labels to assist in perfecting your model and then show you how to easily edit your EOP FG profiles to match the design intent on the basis of the draft corridor surface built in the previous section.

Don't forget a few basics that will help you navigate this exercise easier. If you wish to see the drawing without the corridor, then freeze the corridor layer, which is C-ROAD-CORR-INTR. You can change the view of your surface around by changing the active style or by freezing C-TOPO. Also, remember that draw order may be an issue. You will have corridor feature lines overlapping alignments, so it may be best to use the Send To Back option on the corridor.

1. Open the `Manual Intersection_Labels.dwg` file (which you can download from this book's web page), or continue working in your drawing from the previous exercise.

2. For the next few steps, either freeze the corridor layer or send your corridor display order to back.

3. Switch to the Annotate tab on the Ribbon. Click the Add Labels button. The Add Labels dialog box will appear.

4. Set Feature to Alignment and Label Type to Station Offset. Set Station Offset Label Style to Intersection Centerline Label, as shown in Figure 10.56. Set Marker Style to Basic Circle With Cross. Click Add.

FIGURE 10.56
Add alignment labels to identify potential problem areas.

5. You composed this label to reference two alignments and two profiles. The command line will read `Select Alignment:`. Click Syrah Way.

6. The command line will read `Specify station along alignment:`. Use object snaps to specify the intersection of Syrah Way and Cabernet Court.

7. At the `Specify station offset:` prompt, type **0** (zero) and press ↵.

8. At the `Select profile for label style component Profile 1:` prompt, right-click to bring up a list of profiles and choose Syrah Way FG. Click OK to dismiss the dialog.

9. At the `Select alignment for label style component Alignment 2:` prompt, pick the Cabernet Court alignment.

10. At the `Select profile for label style component Profile 2:` prompt, right-click to open a list of profiles, and choose Cabernet Court FG. Click OK to dismiss the profile listing. Press Esc to exit the labeling command.

11. Thaw the C-ROAD-CORR-INTR layer if it's frozen.

12. Pick the label, and use the square-shaped grip to drag the label somewhere out of the way. Your label should look like Figure 10.57. Keep the Add Labels dialog open for the next part of the exercise.

    Note that the crown elevations of the Cabernet Court FG and Syrah Way FG are both equal to 811.204′, which is great! If they differed at all, you'd have some adjustments to make on the FG profiles where the roads meet.

    The next part of the exercise guides you through the process of adding a label to help determine what elevations should be assigned to the start and end stations of the EOP Right and EOP Left alignments.

13. In the Add Labels dialog, select Station Offset for Label Type, Intersection EOP Label for Station Offset Label Style, and Basic Circle With Cross for Marker Style. Click Add.

FIGURE 10.57
The intersection centerline design label

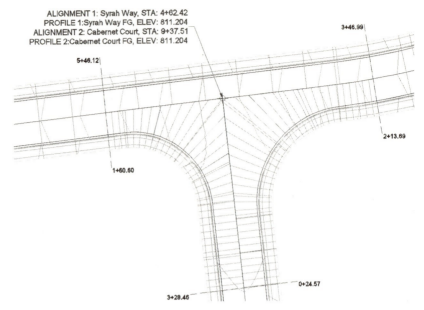

14. You composed this label for the EOP that references a surface elevation and the corresponding profile elevation for comparison. At the `Select Alignment:` prompt, pick the Cab-Syrah EOP W alignment.

15. At the `Specify Station:` prompt, use your Endpoint osnap to pick the region start station of the EOP Right alignment (this is labeled in the exercise drawing for your guidance).

16. At the `Specify Station Offset:` prompt, type **0** (zero) and press ↵.

17. At the `Select surface for label style component Corridor Surface:` prompt, right-click and from the list of surfaces, select Corridor Surface.

18. At the `Select Profile for label style component Proposed Profile Elevation:` prompt, press ↵, and pick Cab-Syrah EOP W - FG.

19. Click again at the end of the Cab-Syrah EOP W region. Because you have already specified the surface and profile, the label will pop right in.

20. Press Esc to finish the command but keep the Add Labels dialog open.

21. Select the labels. Click and drag the square grip and add leaders to make them easier to read. Your labels should look like Figure 10.58.

22. Repeat the previous steps to provide labels for start and end region stations of the Cab-Syrah EOP E alignment.

23. Once all the labels have been placed, confirm that your corridor looks like Figure 10.59.

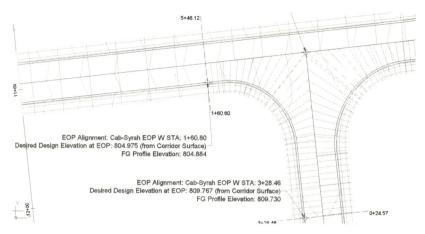

FIGURE 10.58
The intersection profile/surface comparison label

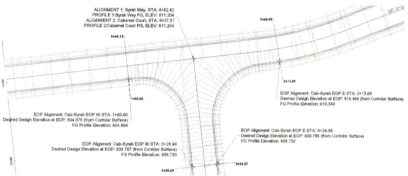

FIGURE 10.59
The corridor with all labels placed

As you can observe from the labels, the profiles and the surface model are darn close. In fact, if the 0.1' bust doesn't bother you, you can skip the next steps. If you want your design to be as perfect as it can be, carry on:

1. Open the `Manual Intersection_Adjustment.dwg` file (which you can download from this book's web page), or continue working in your drawing from the previous exercise.

2. If you continue working in the previous drawing, you will want to split your model space into two views for easier working. Change to the View tab and Set Viewports ➢ Two: Vertical. Press ↵ at the command line to specify a vertical split. Your screen should look similar to Figure 10.60.

3. Pick the Cab-Syrah EOP W - FG profile (the blue line in the profile view). From the contextual Ribbon, click Geometry Editor.

4. On the Profile Layout Tools toolbar, click Insert PVIs Tabular.

5. In the Insert PVIs dialog box, set the vertical curve type to None. Type **160.6** for the station. (Station notation is not needed.) Type **804.975** for the elevation. Click OK.

6. Select the corridor and choose Rebuild Corridor from the contextual tab. Your first curb return label will now report a perfect match between the Corridor Surface elevation and the FG Profile elevation (Figure 10.61).

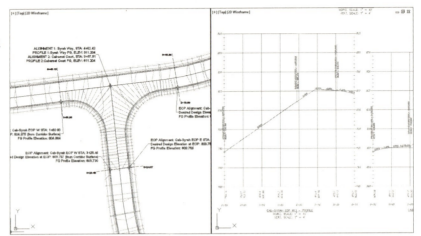

FIGURE 10.60
Use a split screen to see the plan and profile simultaneously.

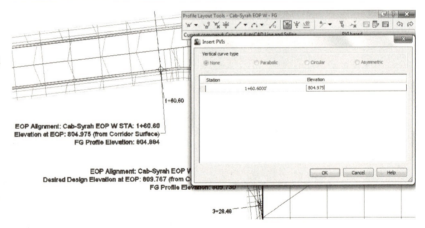

FIGURE 10.61
Add PVI stations and elevations to match the desired FG elevations.

7. Repeat steps 3 through 5 for station 3+28.46 on Cab-Syrah EOP W - FG and station 2+13.69 on Cab-Syrah EOP E - FG (station 0+24.56 does not need this adjustment since it already happens to match perfectly).

8. Pick your corridor. Right-click and choose Object Viewer.

9. Navigate through the Object Viewer to confirm that your corridor model is now appropriately tied together at the EOP. Note that any changes in the centerline won't automatically update your EOP alignments. Your corridor should look like Figure 10.62.

Figure 10.62
The properly modeled intersection

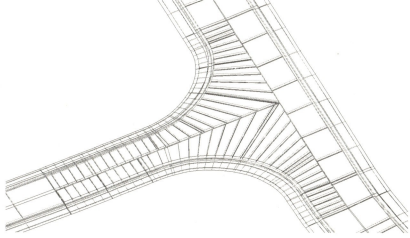

10. Exit the Object Viewer showing just the corridor and look at your surface. Pick the corridor surface, and use the Object Viewer to study the TIN in the intersection area.

Study the contours in the intersection area. You are not quite done perfecting the surface model, but you are getting close. In the next section you will learn about tools you can use to make the corridor surface model even more precise.

Refining a Corridor Surface

Once your model makes more sense, you can build a better corridor surface. At any time you can continue to edit, refine, and optimize your corridor as you gain new information—and because all the corridor elements are connected and labeled, it will take only a few minutes for any edits to be reflected in the corridor surface.

In this exercise, you'll use links as breaklines and add feature lines. You may wish to review the section "Creating a Corridor Surface" in Chapter 9 that explains corridor surface models in more detail. When you're building a surface from links, you have the option of selecting a check box in the Add As Breakline column. Doing so adds the actual link lines as additional breaklines to the surface. In most cases, especially intersection design, selecting this check box forces better triangulation.

In the next exercise, you'll add a few meaningful corridor feature lines to the surface to force triangulation along important features like Edge Of Travel Way and Top Of Curb.

The more appropriate data you add to the surface definition, the better your surface (and therefore your contours) will look—right from the beginning, which means fewer edits and less grading by hand. Follow these steps:

1. Open the `Manual Intersection_Refining.dwg` file (which you can download from this book's web page), or continue working in your drawing from the previous exercise.

2. Pick one of the contours from the corridor surface in the drawing. Click Surface Properties from the contextual Ribbon tab. The Surface Properties dialog opens.

3. Change Surface Style Of Corridor Surface to No Display so you don't accidentally pick it when choosing a corridor boundary. Click OK to dismiss the dialog.

4. Select the corridor, and choose Corridor Properties from the contextual Ribbon tab. Switch to the Surfaces tab. In the row for the Top, under Corridor Surface, click the check box in the Add As Breakline column if it is not already checked.

5. Select Feature Lines from the drop-down menu in the Data Type selection box.

6. Use the drop-down menu in the Specify Code selection box and the + button to add the Back_Curb, ETW, Flange, Flowline_Gutter, and Top_Curb feature lines to the Corridor Surface, as shown in Figure 10.63.

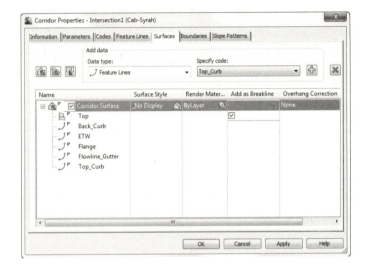

FIGURE 10.63
Adding feature lines to the corridor surface

7. Click OK to dismiss the dialog, and your corridor will automatically rebuild, along with your corridor surface.

8. Select your corridor surface under the Surfaces branch in Prospector. Right-click and choose Surface Properties. The Surface Properties dialog opens. Change Surface Style to Contours 1′ and 5′ (Prop), and click OK. Click OK to dismiss the Surface Properties dialog.

You can make additional, optional edits to the TIN if you're still unhappy with your road surface. Edits that may prove useful include increasing the corridor frequency in select regions, adjusting the start and end stations in a region, or using surface-editing commands to swap edges or delete points.

 Real World Scenario

WHEN AUTOMATIC SURFACE BOUNDARIES ARE NOT AVAILABLE

You will come to a point in the corridor modeling where you need to add a corridor surface boundary, but the best options (Add Automatically ➢ Daylight or Corridor Extents As Outer Boundary) are not available. There will also be situations where Corridor Extents As Outer Boundary does not go where you want it to go. In those situations, you will need to use the Add Interactively tool.

Add Interactively allows you to direct Civil 3D to use the correct feature line as the boundary. You will trace your desired outer bounds with a temporary graphic called a *jig*.

1. Open the Concord Commons Corridor.dwg file, which you can download from the book's web page. You will see a surface in this file that needs some serious reining in.
2. Select the corridor, and click Corridor Properties on the contextual Ribbon. Select the Boundaries tab.
3. Right-click on the Concord Commons Corridor Top surface and select Corridor Extents as the outer boundary. Click OK.

   Pan around the drawing. The contours are tidier, but the surface inside the loop formed by Frontenac Drive has not been cleaned up. In the next steps of the exercise, you will clean up the surface using the Add Interactively tool.
4. Set your AutoCAD object snaps to use only Endpoint and Nearest. These will be the most helpful osnaps to steer Civil 3D in the right direction.
5. Select the corridor and open the Corridor Properties dialog. On the Boundaries tab, right-click Corridor Boundary(1) and select Remove Boundary.
6. Right-click Concord Commons Corridor Top and select Add Interactively.
7. Start on the right side and click the Sidewalk_Out feature line. Trace it with your cursor. When the jig is going to the correct location, click to commit the boundary. The following image shows what the jig will look like:

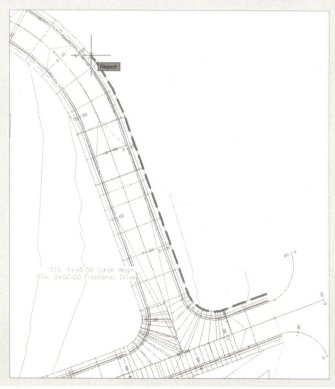

As you move around, you will see the jig tend to move unexpectedly at region boundaries. Steer the jig where you want it to go by clicking and tracing. If you made a mistake, type **U** for undo at the command line. When you reach a point where you need to cross a street, use the endpoint snap to jump the jig across, as shown here:

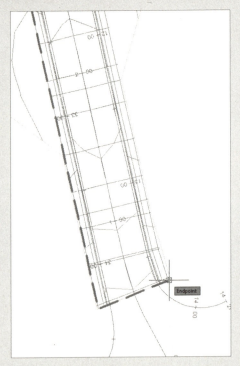

If you click a location where there is more than one feature line, or you are zoomed out to where your pickbox encompasses two lines, you will see a dialog like this one asking you to pick the line you are after:

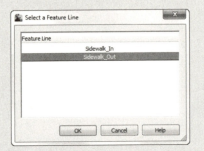

8. After you have finally come back around to where you started the boundary, type **C** (for close) at the command line and press ↵.
9. Right-click on the name of the surface again. Select Add Interactively. This time, repeat the process for the inner boundary, again using the Sidewalk_Out feature line as the boundary.

10. When you complete the inner loop and return to the Corridor Properties dialog box, set the boundary type to Hide Boundary, as shown here.

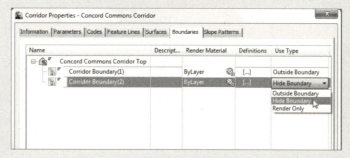

Yes, this process can be tedious. However, your work will pay off when changes are made to your design. These boundaries are dynamic to the design, unlike a polyline boundary.

Using an Assembly Offset

In Chapter 9, you completed a road-widening example with a simple lane transition. Earlier in this chapter, you worked with a roadside-ditch transition, intersections, and cul-de-sacs. These are just a few of the techniques for adjusting your corridor to accommodate a widening, narrowing, interchange, or similar circumstances. There is no single method for building a corridor model; every method discussed so far can be combined in a variety of ways to build a model that reflects your design intent.

Another tool in your corridor-building arsenal is the assembly offset. In Chapter 8, you had your first glimpse of an offset assembly, but in the example that follows you will have a chance to use one for a bike path design.

Notice in Figure 10.64 how the frequency lines in the corridor are running perpendicular to the main alignment. The bike path is an alignment that is not a constant offset through the length of the corridor. In this scenario, the cross section of the bike path itself is skewed. This could prove problematic when computing end area volumes for the bike path pavement. This is the result of using an assembly where all of the design is based off one main baseline assembly.

Figure 10.64
A bike bath modeled with a traditional assembly

There are several advantages to using an offset assembly instead. The offset assembly requires its own alignment and profile for design. In the corridor that results, a secondary set of frequency lines is generated perpendicular to the offset alignment, as shown in Figure 10.65. Additionally, you can use a marked point assembly to model the ditch between the bike path and the main road.

FIGURE 10.65
Modeling a bike path with an assembly offset

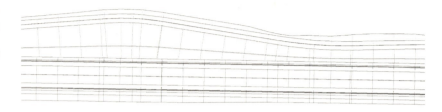

There are many uses of offset assemblies besides bike paths. Typical examples of when you'll use an assembly offset include transitioning ditches, widening roads, creating traffic-calming lanes, and introducing interchanges. The assembly in Figure 10.66, for example, includes two assembly offsets. The assembly could be used for transitioning roadside swales, similar to the first exercise in this chapter.

FIGURE 10.66
An assembly with two offsets representing roadside swale centerlines

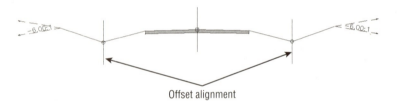

Even though each offset requires its own profile, it isn't always necessary to use Profile ➢ Profile Creation Tools from the Create Design panel and design a profile from scratch. Use some creativity to figure out additional methods to achieve your design intent. Extracting a profile from a corridor, sampling a profile from a first-draft surface, importing a profile from another source, copying a main-road profile and moving it to different elevation, and superimposing profiles are all valid methods for creating profiles for targeting and assembly offsets.

In this exercise, you will model a bike path with an assembly offset:

1. Open the `Highway 10 with Bike Path_Assembly.dwg` file, which you can download from this book's web page.

2. Zoom to the area of the drawing where the assemblies are located. You'll see an incomplete assembly called 12' Lane Road With GR and Bikepath.

3. Click on the main assembly marker. From the contextual Ribbon tab, click Add Assembly Offset in the Modify Assembly panel.

4. At the Specify offset location: prompt, click to the left of the 12′ Lane Road With GR And Bikepath assembly, leaving enough room for the bike lane and ditch. Your result should look like Figure 10.67.

FIGURE 10.67
12′ Lane Road With GR And Bikepath assembly, so far

5. Open the Civil Imperial Subassembly tool palette. Switch to the Basic tab. Select the BasicLane subassembly.
6. In the Advanced Parameters in the Properties dialog box, set Side to Right and Width to 5′. Click to place the subassembly on the offset assembly.
7. Switch the subassembly side to the left and place it on the offset assembly as well. Your assembly should now look like Figure 10.68.

FIGURE 10.68
12′ Lane Road With GR And Bikepath assembly with the BasicLane as a bike path

Next you will use a MarkPoint assembly to set the stage for building a ditch between the bike path and the main road.

8. Switch to the Generic tab in the subassemblies tool palette. Click the MarkPoint subassembly.
9. Change the Point Name to **BIKE** (use all capital letters).
10. Click on the outermost point of the left shoulder subassembly. Your marker will look like Figure 10.69.

FIGURE 10.69
A close-up of the MarkPoint subassembly

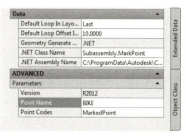 

11. From the Subassemblies tool palette, click the LinkSlopesBetweenPoints subassembly. Set Marked Point Name to **BIKE** (use all capital letters again). Set Ditch Width to 0.5′ and click on the right side of the bike path. The offset assembly will now look like Figure 10.70.

FIGURE 10.70
The LinkSlopes-
BetweenPoints
subassembly in
layout mode

12. Add a LinkSlopeToSurface generic link subassembly. Set the Side to Left and Slope to 25%. Place the subassembly on the left side of the bike path. The completed assembly will look like Figure 10.71.

FIGURE 10.71
The completed
assembly with
offset

Next, you will create a corridor using this new assembly:

1. Open the `Highway 10 with Bikepath_Corridor.dwg` file (which you can download from this book's web page), or continue working in your drawing from the previous exercise.

2. From the Home tab, select Corridor ➢ Create Simple Corridor. Name the corridor **Bike Path** and click OK.

3. For the baseline alignment, right-click and select USH 10. Click OK.

4. For Profile, right-click and select USH 10 Roadway CL Prof. Click OK.

5. For the assembly, select 12′ Lane Road With GR And Bikepath.

6. The only targets needed are the surface targets. Click the <Click here to set all> link and choose Existing Intersection. Click OK.

7. Dismiss the error that comes up in Panorama—you will rectify that issue in the next step. Pan over to the corridor. Notice that the alignment to the west, which represents the bike path centerline, is currently being ignored by the corridor.

8. Select the corridor and open the Corridor Properties dialog. Switch to the Parameters tab. Notice Offset - (1) is not associated with an alignment.

9. Click the alignment field for Offset - (1), select Bike Path, and click OK. Click the profile field for Offset - (1), select Bike Path FG, and click OK.

10. The Bike Path alignment is slightly shorter than the main USH 10 alignment, which explains why you were getting target errors in Panorama. To correct the corridor errors, set the start station for the Offset - (1) region to 0+25.00. Set the end station to 40+00. Click OK to rebuild the corridor. Your completed corridor will resemble the example back in Figure 10.65.

11. Select the corridor by clicking on one of the frequency lines anywhere near the middle of the alignment. Click Section Editor. You may want to change your annotation scale to 1"=1' to get an unobstructed view of your masterpiece.

12. You'll see only your mainline assembly in the Section Editor initially. To see the offset baseline, select the Offset - (1) baseline from the drop-down menu, as shown in Figure 10.72.

FIGURE 10.72
Inside the corridor Section Editor

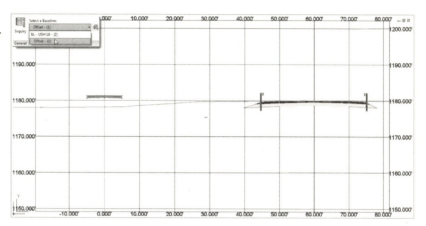

THE TROUBLE WITH BOWTIES

In your adventures with corridors, chances are pretty good that you'll create an overlapping link or two. These overlapping links are known affectionately as *bowties*. A mild example can be seen in the following graphic:

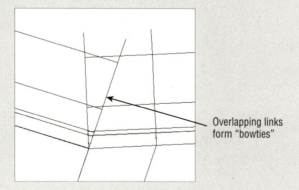

Overlapping links form "bowties"

An even more pronounced example can be seen in the following river corridor image:

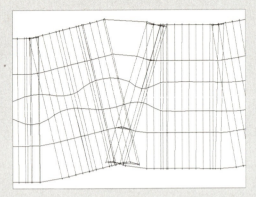

Bowties are problematic for several reasons. In essence, the corridor model has created two or more points at the same x and y locations with a different z, making it difficult to build surfaces, extract feature lines, create a boundary, and apply code-set styles that render or hatch.

When your corridor surface is created, the TIN has to make some assumptions about crossing breaklines that can lead to strange triangulation and incoherent contours, such as in the following graphic:

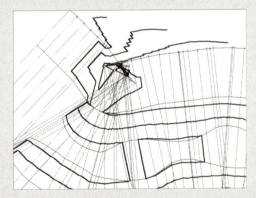

When you create a corridor that produces bowties, the corridor won't behave as expected. Choosing Corridors ➤ Utilities to extract polylines or feature lines from overlapping corridor areas yields an entity that is difficult to use for additional grading or manipulation because of extraneous, overlapping, and invalid vertices. If the corridor contains many overlaps, you may have trouble even executing the extraction tools. The same concept applies to extracted alignments, profiles, and COGO points.

If you try to add an automatic or interactive boundary to your corridor surface, either you'll get an error or the boundary jig will stop following the feature line altogether, making it impossible to create an interactive boundary.

To prevent these problems, the best plan is to try to avoid link overlap. Be sure your baseline, offset, and target alignments don't have redundant or PI locations that are spaced excessively close.

If you initially build a corridor with simple transitions that produce a lot of overlap, try using an assembly offset and an alignment besides your centerline as a baseline. Another technique is to split your assembly into several smaller assemblies and to use your target assemblies as baselines, similar to using an assembly offset. This method was used to improve the river corridor shown in the previous graphics. The following graphic shows the two assemblies that were created to attach at the top of bank alignments instead of the river centerline:

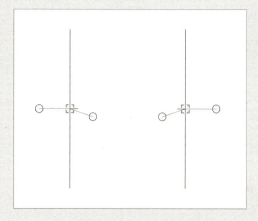

The resulting corridor is shown in the following graphic:

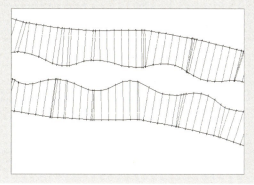

The TIN connected the points across the flat bottom and modeled the corridor perfectly, as you can see in the following image:

Another method for eliminating bowties is to notice the area where they seem to occur and then adjust the regions. If your daylight links are overlapping, perhaps you can create an assembly that doesn't include daylighting and create a region to apply that new assembly.

If overlap can't be avoided in your corridor, don't panic. If your overlaps are minimal, you should still be able to extract a polyline or feature line—just be sure to weed vertices and clean up the extracted entity before using it for projection grading. You can create a boundary for your corridor surface by drawing a regular polyline around your corridor and adding it as a boundary to the corridor surface under the Surfaces branch in Prospector. The surface-editing tools, such as Swap Edge, Delete Line, and Delete Point, can also prove useful for the final cleanup and contour improvement of your final corridor surface.

As you gain more experience building corridors, you'll be able to prevent or fix most overlap situations, and you'll also gain an understanding of when they aren't having a detrimental effect on the quality of your corridor model and resulting surface.

Using a Feature Line as a Width and Elevation Target

You've gained some hands-on experience using alignments and profiles as targets for swale, intersection, and cul-de-sac design. Civil 3D adds options for corridor targets beyond alignments and profiles. You can use grading feature lines, survey figures, or polylines to drive horizontal and/or vertical aspects of your corridor model.

> **DYNAMIC FEATURE LINES CANNOT BE USED AS TARGETS**
>
> It's important to note that dynamic feature lines extracted using the Feature Lines From Corridor tool can't be used as targets. The possibility of circular references would be too difficult for the program to anticipate and resolve.

Imagine using an existing polyline that represents a curb for your lane-widening projects without duplicating it as an alignment, or grabbing a survey figure to assist with modeling an existing road for a rehabilitation project. The next exercise will lead you through an example where a lot-grading feature line is integrated with a corridor model:

1. Open the `Feature Line Target.dwg` file, which you can download from this book's web page. This drawing includes a corridor as well as a yellow feature line that runs through a few lots.

2. Zoom to the area of the drawing where the assemblies are located. There is an assembly that includes a LinkWidthAndSlope subassembly attached to the sidewalk on the left side. You'll be using the yellow feature line as a target for this subassembly.

3. Zoom to the corridor. Select the corridor, and choose Corridor Properties from the contextual Ribbon tab.

4. Switch to the Parameters tab in the Corridor Properties dialog. Click the Set All Targets button. The Target Mapping dialog appears.

5. Click the `<None>` field next to Target Alignment for the LinkWidthAndSlope subassembly. The Set Width Or Offset Target dialog appears.

6. Choose Feature Lines, Survey Figures And Polylines from the Select Object Type To Target drop-down list (see Figure 10.73).

FIGURE 10.73
The Select Object Type To Target drop-down list as seen in both the Set Width Or Offset Target and Set Slope Or Elevation Target dialogs

7. Click the Select From Drawing button. When you see the command-line prompt `Select feature lines, survey figures or polylines to target:`, select the yellow feature line and then press ↵. The Set Width Or Offset Target dialog reappears, with an entry in the Selected Entries To Target area. Click OK to return to the Target Mapping dialog.

If you stopped at this point, the horizontal location of the feature line would guide the LinkWidthAndSlope assembly, and the vertical information would be driven by the slope set in the subassembly properties. Although this has its applications, most of the time you'll want the feature line elevations to direct the vertical information. The next few steps will teach you how to dynamically apply the vertical information from the feature line to the corridor model.

8. Click the <None> field next to Target Profile for the LinkWidthAndSlope subassembly. The Set Slope Or Elevation Target dialog appears.

9. Make sure Feature Lines, Survey Figures And Polylines is selected in the Select Object Type To Target drop-down.

10. Click the Select From Drawing button. When you see the command-line prompt `Select feature lines, survey figures or polylines to target:`, select the yellow feature line and then press ↵. The Set Slope Or Elevation Target dialog reappears, with an entry in the Selected Entries To Target area. Click OK to return to the Target Mapping dialog.

11. Click OK to return to the Corridor Properties dialog.

12. Click OK to exit the Corridor Properties dialog. The corridor will rebuild to reflect the new target information and should look similar to Figure 10.74.

FIGURE 10.74
The corridor now uses the grading feature line as a width and elevation target.

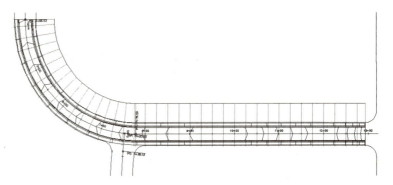

Once you've linked the corridor to this feature line, any edits to this feature line will be incorporated into the corridor model. You can establish this feature line at the beginning of the project and then make horizontal edits and elevation changes to perfect your design. The next few steps will lead you through making some changes to this feature line and then rebuilding the corridor to see the adjustments.

13. Select the corridor, right-click, and choose Display Order ➤ Send To Back. Doing so sends the corridor model behind the target feature line.

14. Switch on your Ortho setting. Use the AutoCAD Move command to move the feature line back approximately 10–20 feet, as shown in Figure 10.75. Note that you could also edit individual vertices or use any of the horizontal feature-line editing tools available in the Grading menu and on the Feature Line toolbar.

Figure 10.75
Move the feature line approximately 10–20 feet deeper into the lot.

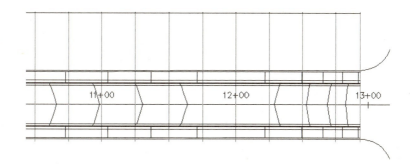

15. Select the feature line. Right-click and choose Raise/Lower from the context menu.

16. When you see the command-line prompt `Specify elevation difference <1.00>:`, type **5** and then press ↵. Each vertex of the feature line rises 5′ vertically. Note that you could also edit individual vertices or use any of the vertical feature-line editing tools available in the Grading menu and on the Feature Line toolbar.

17. Select the corridor. Right-click and choose Rebuild Corridor. The corridor will rebuild to reflect the changes to the target feature line and should look like Figure 10.76.

Figure 10.76
The completed corridor model

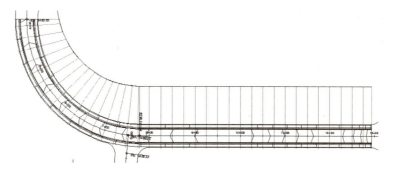

Edits to targets—whether they're feature lines, alignments, profiles, or other Civil 3D objects—drive changes to the corridor model, which in turn drives changes to any corridor surfaces, sections, section views, associated labels, and other objects that are dependent on the corridor model. Brainstorm ways that you can take advantage of this dynamic connection, such as making a corridor surface and then a quick profile or two, so that you can see your iterations of the feature line in immediate action as you work through your design.

Roundabouts: The Mount Everest of Corridors

If you really understand what went on earlier in this chapter, you are almost ready for roundabout design. You may want to wait to tackle your first roundabout until after reading Chapter 15, "Grading."

The same concepts apply to a roundabout as for a standard road junction, but you will have several more regions, baselines, and corresponding profiles.

This section will help you prepare files for roundabout design. A roundabout is best done in several corridors:

- Preliminary corridor for circulatory road
- Main corridor with approaches and circulatory road
- Curb island corridors (optional)

Drainage First

Based on your existing ground surface, determine the general direction that you want water to flow away from the center of the roundabout. Use grading tools and a feature line to create the general drainage direction of the roundabout.

Chapter 15 will go into much more depth on creating grading. You will certainly want to have an understanding of grading basics before you tackle a roundabout.

Create a feature line that represents the highest elevations. In the example shown in Figure 10.77, the feature line slopes downward and acts as a ridge to separate water flow. The grading tools are then used to create grading objects and a corresponding surface model called "Roundabout Grading."

FIGURE 10.77
Feature lines and grading create a preliminary surface to ensure proper drainage through the roundabout.

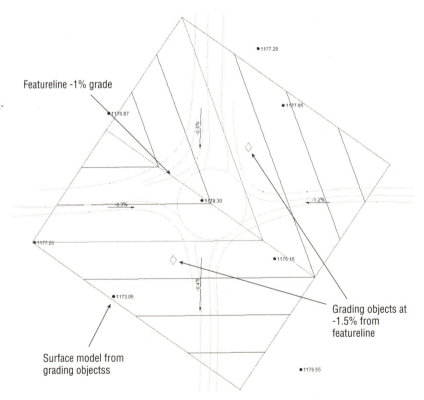

The Roundabout Grading surface will be the basis for our profile elevations through the rest of the design process.

Roundabout Alignments

Roundabouts need alignments to guide the design for the same reasons that an intersection needs them. Alignments will be baselines and targets for the approaches and rotary. Create alignments manually with the tools you learned earlier in this chapter, or start with the handy roundabout layout tool.

The roundabout layout tools create horizontal data based on the location of the center of the roundabout and the approach alignments.

In the exercise that follows, you will create the Civil 3D alignments needed to create a roundabout:

1. Open the Roundabout Layout.dwg file, which you can download from this book's web page.

2. From the Home tab, in the Create Design panel, select Intersections Roundabout ➢ Create Roundabout, as shown in Figure 10.78.

FIGURE 10.78
Access the Create Roundabout tool from the Home tab.

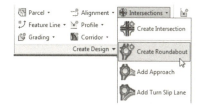

3. At the Specify roundabout centerpoint: prompt, use your intersection osnap to select the point where the approaches meet.

4. At the Select approach road: prompt, select all four alignments leading into the roundabout. Press ↵ when you are done.

   You now see the Create Roundabout – Circulatory Road screen, shown in Figure 10.79.

5. From the Predefined Parameters To Import drop-down, choose R=75. This will be the radius from the center of the roundabout to the outermost circular edge of pavement. Set Alignment Layer to C-ROAD and Alignment Label Set to Major And Minor Only. Click Next.

   Now, you'll design the approach road exit and entry geometry. The options in the Create Roundabout – Approach Roads screen (see Figure 10.80) can be set independently for each approach, or you can click Apply To All, which will set the geometry for all four approaches.

FIGURE 10.79
The first roundabout layout screen for designing the main circulatory road

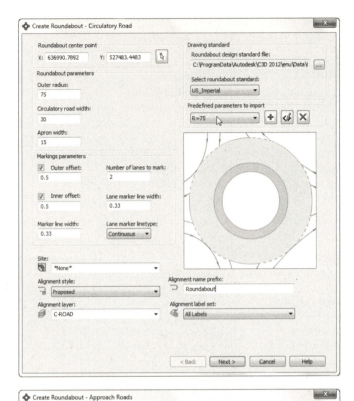

FIGURE 10.80
Approach road widths at entry and exit

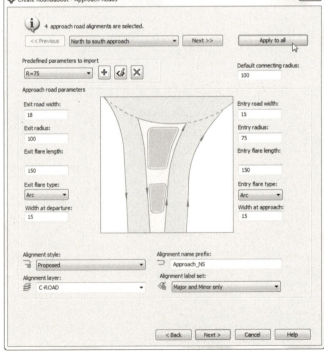

6. Set Predefined Parameters To Import to R=75. For Alignment Label Set, choose Major And Minor Only. Then click the Apply To All button and click Next.

7. In the Create Roundabout – Islands screen (see Figure 10.81), again set Parameters To Import to R=75. Click Apply To All and then click Next.

FIGURE 10.81
Roundabout Islands parameters

The final screen of the Create Roundabout wizard deals with pavement markings and signage. Notice that you can specify your own blocks for the signs that will be placed in this process.

Everything created in this last step is an AutoCAD polyline or block. The polylines have a global width set to indicate pavement marking thicknesses. These thicknesses are set in the Markings And Signs screen (Figure 10.82).

8. Leave all defaults in the Create Roundabout – Markings And Signs screen (Figure 10.82). Click Finish, and your roundabout should resemble Figure 10.83.

Finally, you will add a turn lane in the NW quadrant of the roundabout. When you're creating slip turn lanes, remember that the turn radius must be large enough to fillet the exit and entry roads without overlapping the other alignments.

When selecting the approach entry and exit alignments, you need to click the shorter approach alignments created by Civil 3D rather than the original approach road. For this reason, the exercise has you select inside the islands, just to be sure.

FIGURE 10.82
Pavement markings galore!

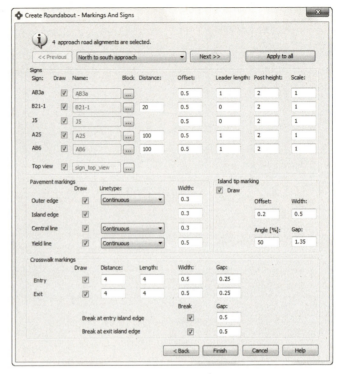

FIGURE 10.83
Completed roundabout alignment layout

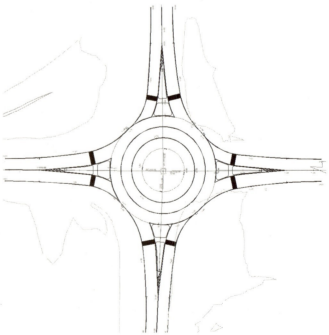

9. From the Home tab, in the Create Design panel select Intersections Add Turn Slip Lane. When prompted to select the entry approach, select the north approach alignment inside the curb island. When prompted for the exit approach, select the west approach alignment inside the curb island, as shown in Figure 10.84.

FIGURE 10.84
Entry and exit approach alignments for the slip lane

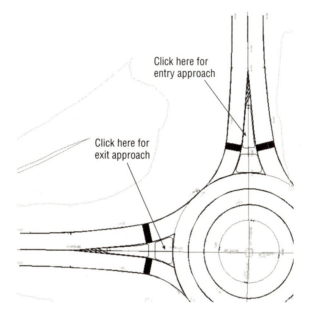

10. In the Draw Slip Lane screen shown in Figure 10.85, set the lane width to **14** and the radius to **150**. For the alignment layer, choose C-ROAD and for Alignment Label Set, choose Major And Minor Only. Click OK.

FIGURE 10.85
Adding a slip lane

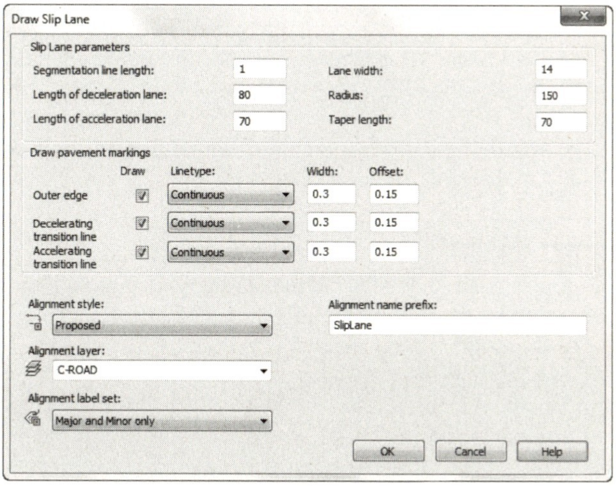

Your roundabout will now look like Figure 10.86. The warning symbol is an indicator that the deceleration lane is not tangent to the outer alignment curve, which is normal in this situation.

FIGURE 10.86
The completed alignment layout with a slip lane

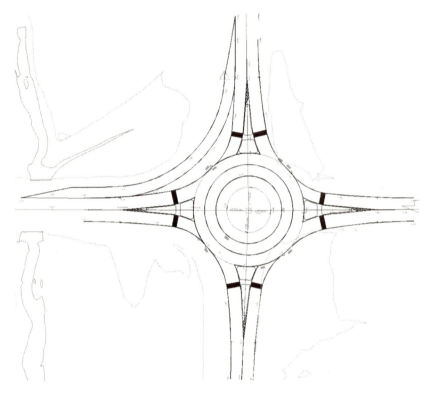

You now have all the alignments you need to start your roundabout design. At this point, you can add geometry to the alignments and modify what Civil 3D has created for you. The pavement markings and signs will only stay dynamic to the roundabout if you modify the initial approaches.

The horizontal layout is complete, but the roundabout design is far from done. No vertical data has been created; that is up to you.

Center Design

All profiles need to meet at the elevations inside the traveled way in the circular pavement area of the roundabout. Therefore, the main circle design comes first.

The example you are seeing has been modified slightly from the default layout created by Civil 3D. The main circular alignment is located at the centerline of the circular portion of the roundabout. You can use any of the circular alignments created by Civil 3D as the basis for this step, as long as your assembly works with the design. Remember to make note of which direction the alignment goes, to ensure the assembly you create is not backwards.

Extract profiles for the main circle design from the Existing Ground and Roundabout Grading surfaces, as shown in Figure 10.87.

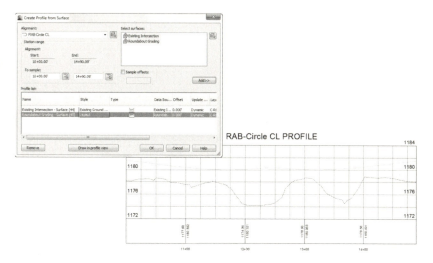

FIGURE 10.87
Extract surface profiles around the main circular alignment.

The assembly you create for this preliminary design will also be used in the main design. Decide which alignment will be used as the circular design basis, and create an assembly based on your alignment location and desired geometry, as shown in Figure 10.88.

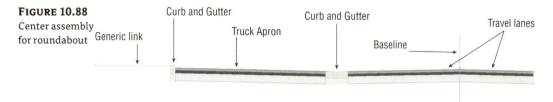

FIGURE 10.88
Center assembly for roundabout

When the assembly is complete, you will apply the assembly to the first of several corridors you create. Create a corridor called **Preliminary Roundabout Center**. Use the main circle alignment (called RAB-Center CL in this example) as the baseline and use the Roundabout Grading profile as the corridor profile. There are no targets or frequencies to set in this corridor.

Create a Top link surface from the Preliminary Roundabout Center. Any approach alignments that touch this surface need to tie into this surface elevation. At this point the roundabout will resemble Figure 10.89.

Profiles for All

You have all the preliminary surfaces in place, and you have all the alignments you need, so it is time to extract profiles from your various surfaces.

Extract surface profiles for all the approach alignments by sampling the existing ground, drainage surface, and preliminary corridor surface.

These profiles will look something like Figure 10.90.

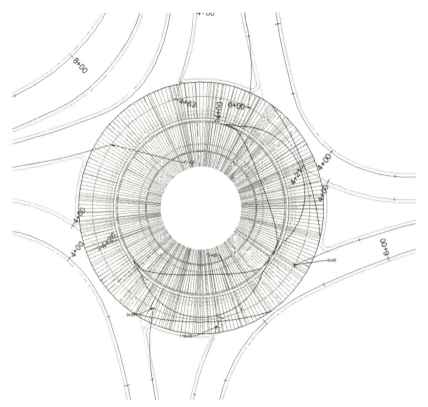

FIGURE 10.89
Preliminary center corridor and surface

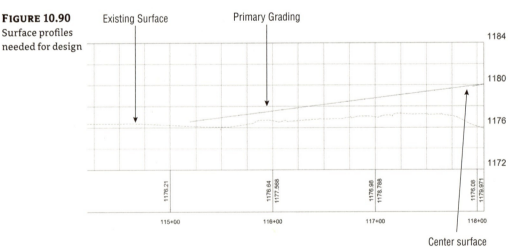

FIGURE 10.90
Surface profiles needed for design

When you create your design, you will see all your design considerations in the profile views. No matter how you decide to tie into existing ground and slope upward toward the surface, your design needs to tie into the preliminary center surface, as shown in Figure 10.91.

FIGURE 10.91
Design profiles must tie into the center.

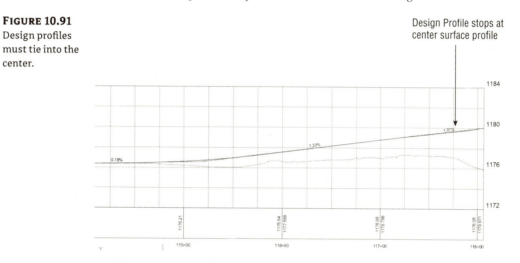

Use techniques you learned earlier in this chapter to assist you. Labels are an especially valuable tool for roundabouts. Keep your profile views organized, as you will have at least three for each approach. If you have a slip turn lane, you will have a profile for that as well.

Tie It All Together

Stretch your legs and go for more coffee. It is time to put this thing together into a completed corridor. When you model the corridor initially, ignore curb islands—you will add them as individual corridors in a later step.

A simple roundabout can be completed using as few as three assemblies. In our example, however, the slip turn lane necessitates a total of four assemblies. In addition to the RAB-Center assembly we examined in Figure 10.88, you will need three more assemblies, as shown in Figure 10.92.

FIGURE 10.92
Assemblies needed for the main roundabout corridor

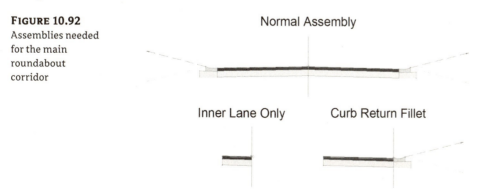

These assemblies will be tied to the EOP alignments and profiles as baselines, similar to a traditional intersection. Each quadrant of the roundabout will target at least two alignments and profiles, as shown in Figure 10.93.

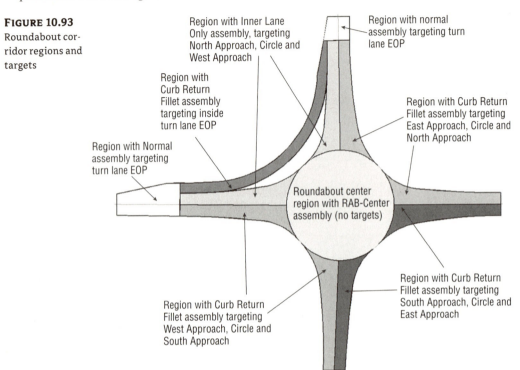

FIGURE 10.93
Roundabout corridor regions and targets

Keep in mind the direction of your alignments. If you build the corridor in stages, check the corridor periodically to make sure it is building correctly.

Create a corridor surface from your completed roundabout lanes. You will likely have to use the Add Interactively tool to add the boundary correctly.

Finishing Touches

The median islands are the last parts to go on the corridor. You can create them using simple grading objects, but since this is a book about mastering skills and this is a chapter about corridors, you should examine the dynamic way.

Create a simple assembly containing the curb and gutter for the curb islands. This will be the assembly that you use with the median corridors (Figure 10.94).

Each median island will need its own alignment. Take note of the direction of the alignments to make sure they are compatible with the curb island assembly. If necessary, change the direction of the alignments using the Reverse Direction tool in the Modify panel of the contextual Ribbon tab. Figure 10.95 shows the bypass island and north island with directions.

FIGURE 10.94
Simple assembly containing just the curb and gutter on the median islands

FIGURE 10.95
Several median island alignments with direction shown

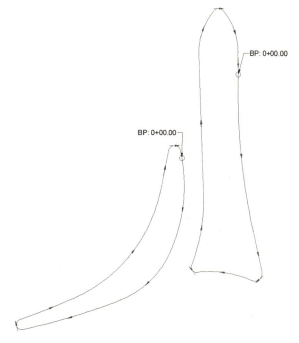

The good news is that the elevation data for the medians is already complete. Your main corridor's surface will act as the profile for each individual median. This also means that once the curb return corridors are created, they will be dynamic to the main corridor. Once these little

456 | **CHAPTER 10** ADVANCED CORRIDORS, INTERSECTIONS, AND ROUNDABOUTS

corridors are created and surface model information has been obtained, you can set them to Rebuild – Automatic and forget all about them.

Extract a profile for all the curb return alignments from the Top link surface model from the main roundabout corridor, as shown in Figure 10.96. You do not need to see this profile in a view, so you can click OK to extract.

When the design roundabout corridors are complete and surfaces are made, the next step is to merge the surfaces together. Create a final surface model and paste the main roundabout design in first. After the main corridor is pasted in, paste the smaller median corridor surfaces (see Figure 10.97).

FIGURE 10.96
Extract the profile for medians from your main roundabout surface.

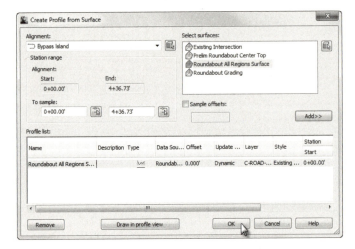

FIGURE 10.97
The completed roundabout

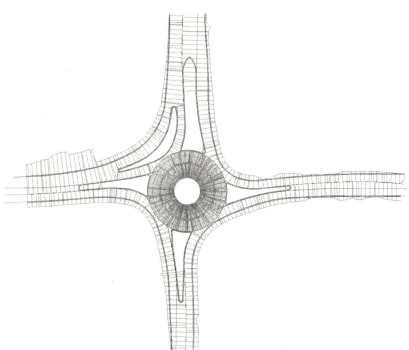

Your next step is to use the corridor fine-tuning techniques you learned earlier in this chapter to ensure your grades are correct and the design is correct. To see an example of a completed corridor using these steps, take a look at `Completed Roundabout Corridor example.dwg`, which you can download from www.sybex.com/go/masteringcivil3d2012.

The Bottom Line

Create corridors with noncenterline baselines. Although for simple corridors you may think of a baseline as a road centerline, other elements of a road design can be used as a baseline. In the case of a cul-de-sac, the EOP, the top of curb, or any other appropriate feature can be converted to an alignment and profile and used as a baseline.

> **Master It** Open the `Mastering Advanced Corridors.dwg` file, which you can download from www.sybex.com/go/masteringcivil3d2012. Add the cul-de-sac alignment and profile to the corridor as a baseline. Create a region under this baseline that applies the Typical Intersection assembly.

Add alignment and profile targets to a region for a cul-de-sac. Adding a baseline isn't always enough. Some corridor models require the use of targets. In the case of a cul-de-sac, the lane elevations are often driven by the cul-de-sac centerline alignment and profile.

> **Master It** Continue working in the `Mastering Advanced Corridors.dwg` file. Add the Second Road alignment and Second Road FG profile as targets to the cul-de-sac region. Adjust Assembly Application Frequency to 5′, and make sure the corridor samples are profile PVIs.

Use the Interactive Boundary tool to add a boundary to the corridor surface. Every good surface needs a boundary to prevent bad triangulation. Bad triangulation creates inaccurate and unsightly contours. Civil 3D provides several tools for creating corridor surface boundaries, including an Interactive Boundary tool.

> **Master It** Continue working in the `Mastering Advanced Corridors.dwg` file. Create an interactive corridor surface boundary for the entire corridor model.

Chapter 11

Superelevation

Whether you call it banking, cant, camber, or cross-slope in your native land, Civil 3D's superelevation commands will serve your needs. These tools have matured greatly since the early releases, and Civil 3D 2012 has added even more functionality.

Before you can understand the superelevation tools, you should have a good grasp of alignments, assemblies, and corridors. You may want to refer to the corresponding chapters for a refresher on those topics before attempting to put superelevation into practice.

In this chapter, you will learn to:

- Add superelevation to an alignment
- Create a superelevation view

Getting Ready for Super

Several pieces need to be in place before superelevation can be applied to the design. You will need a design criteria file appropriate for your region, design speeds applied to an alignment, and an assembly that is capable of superelevating.

There are a lot of abbreviations and terminology thrown around when it comes to superelevation, so let's take a look at that first.

When an assembly is applied without superelevation, the geometry comes directly from the original design, as shown in Figure 11.1.

FIGURE 11.1
An example assembly without superelevation

As the assembly begins its entrance into a curve with superelevation applied to it, it will first start to lose its normal crown. Figure 11.2 shows the same assembly at the End Normal Crown (ENC) station.

FIGURE 11.2
At the End Normal Crown (ENC) station, the default lane slope starts to change.

When the assembly has one side flattened out, as shown in Figure 11.3, this is called Level Crown (LC).

When the assembly straightens out into a plane that matches the inside lane, this is called Reverse Crown (RC), as shown in Figure 11.4.

If accommodations have been made for shoulder slope rollover and breakover removal, the shoulder will shift as well. Figure 11.5a shows where the superelevated lane becomes steep enough to cause an outside rollover problem with the outside shoulder. The shoulder begins to adjust to increase the safety of the road. Figure 11.5b shows the change in the lower shoulder.

FIGURE 11.3
Level Crown entering a left-hand turn

FIGURE 11.4
Reverse Crown (RC)

FIGURE 11.5
Begin Shoulder Rollover (BSR) (left) and Low Shoulder Match (LSM) (right)

Finally, the lane gets to its maximum slope at the Begin Full Super (BFS) station. As you can see in Figure 11.6, all the geometry adjustment has taken place.

FIGURE 11.6
Begin Full Super (BFS)

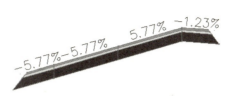

On the way out of the curve, you will see the geometry transitioning back to its original design. It will pass through End Full Super (EFS), Low Shoulder Match (LSM), Reverse Crown (RC), Level Crown (LC), Begin Normal Crown (BNC), and finally back to Begin Normal Shoulder (BNS).

Now that you are familiar with the terminology and abbreviations Civil 3D uses, let's get started on some design.

Design Criteria Files

Having the correct design criteria file in place is the first step to applying superelevation to your corridor. These XML-based files contain instructions to the software on when to flag your design for geometry problems both horizontally and vertically. Design criteria files are the brains behind how your road behaves when superelevation is applied to the design.

Several design criteria files are supplied with Civil 3D upon installation. The out-of-the-box standards include AASHTO 2001 and AASHTO 2004 for both metric and US units. Several of the country kits include design criteria files for your locality if you are outside of the United States. If country or state kits do not exist for your situation, you can create your own, user-defined files.

To create your own design criteria, select any alignment and click the Design Criteria Editor icon from the context tab. It is easiest to modify an existing table in your desired units, rather than starting from an empty file. Be sure to click the Save As icon before making any changes.

User-Defined Criteria Files

If you create design criteria files for your region or design scenario, you will need to be able to share the file with anyone who will be working with your design.

Inside your organization, the best way to handle the design criteria file is to move it to a shared network location. If you are feeling squeamish that a less experienced person may accidentally modify the file, you can set the XML file to read-only.

The default location for design standards is:

    C:\ProgramData\Autodesk\C3D 2012\enu\Data\Corridor Design Standards\

Once you re-path to the design criteria file for an alignment, the location will be saved with the alignment.

If you are collaborating with someone without direct access to your design criteria files, you will need to send that person the XML file with your DWG.

Inside the Design Criteria Editor (Figure 11.7), you will see three headings: Units, Alignments, and Profiles. The Units page tells Civil 3D what type values it will be using in the file. The Alignments page is used for checking design, creating superelevation, and widening outside curves. The Profiles page provides tabular data for minimum K value used to check vertical design.

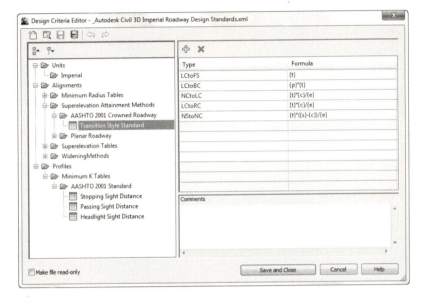

FIGURE 11.7 Inside the Design Criteria Editor

This chapter will focus on the alignment-related tables in the file. Civil 3D will graphically flag alignments when the design speed specified in the alignment properties has a radius less than the value specified in the Minimum Radius Table. The Minimum Radius Tables from AASHTO use superelevation rates in the table names, but this does not lock you into that rate for applying superelevation to the corridor. In other words, just because you use a more conservative value in your radius check, that doesn't mean you can't superelevate at a steeper rate. The tables are independent of one another.

Also in the Alignments page you will find the superelevation attainment equations. These equations determine the distance between superelevation critical stations. Familiarize yourself with the terminology and locations represented by these stations, as shown in Figure 11.8.

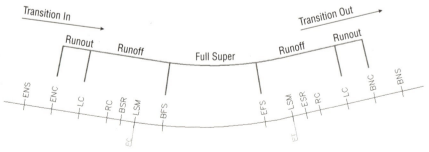

FIGURE 11.8 Superelevation critical stations and regions calculated by Civil 3D

In the following exercise, you will modify an example design criteria file and save it:

1. Open the `Highway 10 Criteria.dwg` file, which you can download from this book's web page at www.sybex.com/go/masteringcivil3d2012.

2. Select the Highway 10 alignment that already exists in the drawing. From the Modify panel on the context tab, select Design Criteria Editor.

3. Click the Open button at the top of the dialog box. Browse to the `MasteringExample.xml` file.

4. Expand the Alignments category. Expand the Superelevation Tables area. There is only one superelevation table in this example for 4% maximum slope. Right-click on Superelevation Tables and select New Superelevation Table.

5. Double-click the new table and rename it **Example 6% Super**.

6. Right-click on Example 6% Super and select New SuperelevationTypeByTable. Expand the new section. Right-click SuperelevationTypeByTable and select New Superelevation Design Speed, as shown in Figure 11.9.

FIGURE 11.9
Adding a design speed to the design criteria file

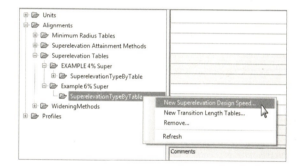

7. Civil 3D will place a new design speed with a default of 10mph in the listing. Double-click the Design Speed 10 and change the design speed to 30.

8. Highlight your example design speed. The right side of the dialog box will have an empty table containing columns for Radius and Superelevation Rate. Click the first field in the Radius column to start entering data. Add several radius and superelevation values, as shown in Figure 11.10.

9. Add a note in the Comment field that reads **Example Data Only**.

10. Click the Save And Close button at the bottom of the editor. Click Save Changes and exit when prompted.

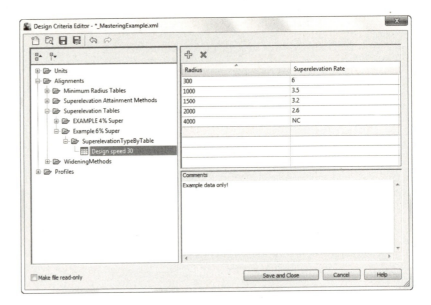

FIGURE 11.10
Adding example data using the Design Criteria Editor

Ready Your Alignment

Superelevation stations are connected to alignment curves. The design speed from the alignment properties is needed at each curve to specify which superelevation rate tables to use from the design criteria. The design speed has an effect on the distance between superelevation critical stations and the cross-slope used when the road is at full-super.

It is a good idea to get your alignment geometry and design speed locations finalized before attaching superelevation. If a change is made to your alignment, the superelevation stationing will be marked as out of date.

Super Assemblies

As a general rule, if the lane subassembly has the word "super" somewhere in its name, it will respond to superelevation. If you want to verify that the lane you are choosing will behave the way you want it to in a superelevation situation, you can right-click it from the tool palette and access the subassembly help.

As long as you stay away from the Basic tab, all of the shoulder and curb subassemblies have parameters you can set to dictate how the assembly is to behave when an adjacent lane superelevates.

Most subassemblies that are capable of superelevating are intended for use where the pivot point for the cross section is at the center crown of the road. When the pivot point is at the center of the road, the baseline profile dictates the final elevation of the crown of the road. Figure 11.4 shows an example of a two-lane highway (a) and a four-lane divided highway (b) that are designed to be used with the superelevation tools. The examples in Figure 11.11 show assemblies that will superelevate using the center crown of the road as a pivot point.

FIGURE 11.11
Two-lane road ready for super (a); four-lane road used in super (b)

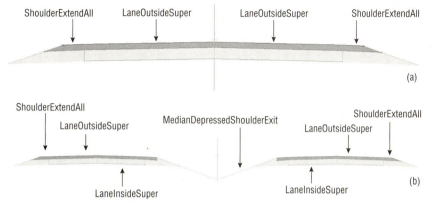

Axis of Rotation Support

The newest addition to the subassembly family is the LaneSuperelevationAOR. This axis of rotation (AOR) subassembly can be used when the centerline of the road is not the pivot point for superelevation. The "flag" symbols (as shown in Figure 11.12) indicate potential pivot points on the assembly.

FIGURE 11.12
AOR subassemblies used on an undivided, crowned roadway

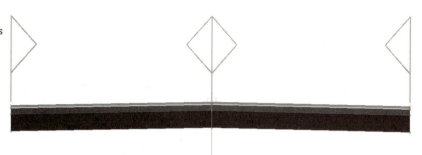

The flag symbols on LaneSuperelevationAOR indicate where the lane can be pinned down and used as a pivot point. When the axis of rotation is not the centerline of the road, the lane geometry is used to determine the change in elevation that will occur as a result.

> **Axis of Rotation (AOR)**
>
> The following is a list of the limitations and other factors to be aware of if you decide to take these new assemblies for a spin:
>
> ◆ Don't use an offset assembly with an axis of rotation superelevation assembly. Doing so can throw off the superelevation calculation.
>
> ◆ When working with curbs, medians, and shoulders, keep an eye out for the parameter SE AOR Unsupported. If this parameter is set to 1, it means that the subassembly will not adjust for superelevations other than at the center. None of the curb and gutter subassemblies will adjust for breakover or rollover, but most of the shoulder assemblies do. This is a "hard-coded" parameter that cannot be changed by the end user.

Real World Scenario

THE TIPPING POINT

In some design situations that use superelevation, the crown of the road is not the ideal pivot point. In the following example, you will create a four-lane divided highway that pivots inside the curve rather than at the crowns during superelevation.

1. Open the file AOR Assembly.dwg.
2. Zoom into the assembly that was started for you in this drawing. Note that a generic link has been placed on each side as a spacer for the subassemblies that you will be adding in the next steps.
3. Open your subassembly tool palette and find the Lanes tab. Select the LaneSuperelevationAOR subassembly.
4. In the Advanced Parameters of the properties, set Side to Right and Use Superelevation to Right Lane Outside.
5. Place the subassembly on the right side, using the generic link marker as its "hanger."

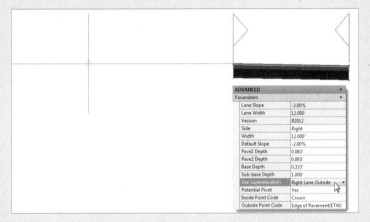

6. Change the Advanced Parameters so that Side is set to Left and Use Superelevation is set to Right Lane Inside. Place this subassembly in the drawing to form the lane as shown here.

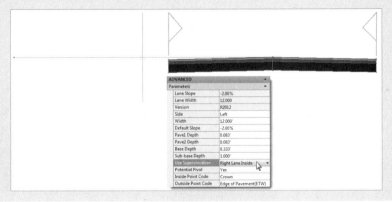

7. Repeat the process for the left side of the assembly. Use the following graphic to help set the correct sides and superelevation parameters. Note that the inside lane must be set to Left Lane Inside and the outside lane must be set to Left Lane Outside.

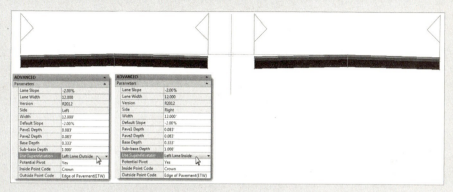

8. Select the MarkPoint subassembly from the Generic tab. Name the Marker **MEDIAN** and place it on the inside left edge of pavement.
9. On the Medians tab, select MedianDepressed. Be sure the Marked Point Name is set to MEDIAN.

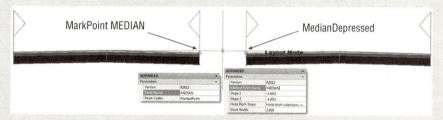

10. As the final step in building this assembly, select both of the original generic link spacers and set the Omit Link property to Yes. If time permits, add the ShoulderExtendAll subassembly to the outermost edges.

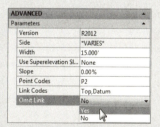

11. Select the alignment that runs through the project. On the context tab, click Superelevation ➢ Calculate/Edit Superelevation. Click the Calculate Superelevation Now option when notified that no data exists.
12. Set the Roadway Type to Divided Crown With Median. Set the Pivot method to Inside Of Curve. Set the Median treatment to Distorted Median. Click Next.

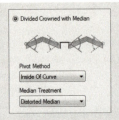

13. On the Lanes page, set the normal lane width to **12'** and the normal lane slope to **-2.00%** for a Symmetric Roadway. Click Next.
14. On the Shoulder Control page, leave the settings at their defaults and click Next.
15. On the Attainment page, set the active criteria file to `Autodesk Civil 3D Imperial (2004) Roadway Design Standards.xml`. Set the Superelevation rate table to AASHTO 2004 Customary eMax 6%, using the 4 Lane Transition Length table. Toggle on Automatically Resolve Overlap. Click Finish.

16. In Prospector, select the Corridor, right-click, and select Properties. Right-click on BL-Route 66 and select Add Region. Select the Route 66 Main assembly and click OK. Click OK to let the corridor build.
17. Select the corridor in plan and enter the Section Editor to examine your corridor. You should observe that superelevation is occurring based on our specified pivot point of inside the curves.

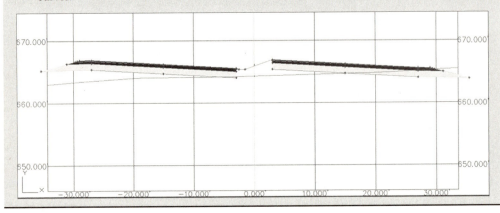

Applying Superelevation to the Design

In releases of Civil 3D prior to 2011, superelevation parameters were tucked in with alignment properties. Today, superelevation is primarily tied to alignment curves, but Civil 3D also takes into account other factors such as curve station locations and assembly geometry.

Start with the Alignment

To begin applying superelevation to the design, select your alignment.

1. Open the Highway 10 Super.dwg file, which you can download from this book's web page.

2. Select the Highway 10 alignment that already exists in the drawing. From the context tab, select Superelevation ➢ Calculate/Edit Superelevation.

3. When prompted by the dialog box, click Calculate Superelevation Now.

4. On the Calculate Superelevation - Roadway Type screen, select the Undivided Crowned radio button. From the Pivot Method drop-down, choose Center, as shown in Figure 11.13. Click Next.

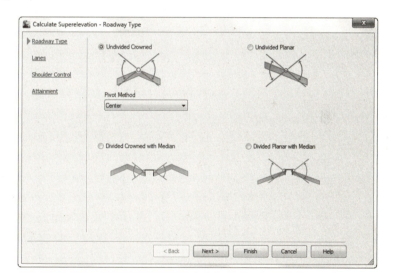

Figure 11.13
Roadway type specification for superelevation

5. On the Lanes screen, verify that the Symmetric Road check box is selected. Set Normal Lane Width to **12′**. Set Normal Lane Slope to **-2.00%**, as shown in Figure 11.14, and click Next.

6. On the Shoulder Control screen, make sure that Calculate is selected. Set Normal Shoulder Width to **6′**. Set Normal Shoulder Slope to **-5.00%**.

7. For Shoulder Slope Treatment, set the Low Side option to Breakover Removal. Set the High Side option to Default Slopes. Set the maximum rollover to **8.00%**. Your Shoulder Control screen should look like Figure 11.15. Click Next.

FIGURE 11.14
Lane information

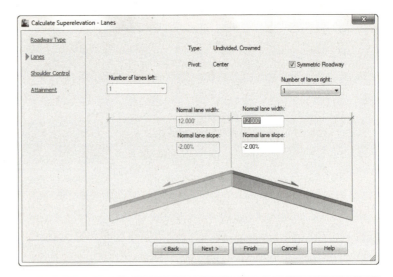

FIGURE 11.15
Shoulder Control and Breakover Removal parameters

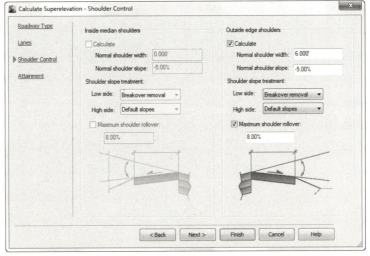

8. On the Attainment screen, verify that the design criteria file is set to AASHTO 2004 Standard. Set the superelevation rate table to AASTO 2004 US Customary eMax 4%.

9. Set the transition length table to 2 Lane. Set the Attainment method to AASHTO 2004 Crowned Roadway. Leave all other settings at their defaults, as shown in Figure 11.16. Click Finish.

10. You should now see the superelevation table appear with the data resulting from the wizard. Examine your alignment, as you should now have labels showing the superelevation critical stations created by the wizard.

As you click in the table, you will see helpful glyphs showing you which superelevation station and corresponding curve you are editing, as shown in Figure 11.17.

FIGURE 11.16
Finalizing the superelevation on the Attainment screen

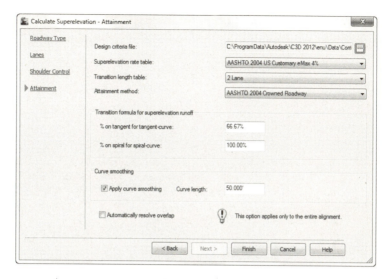

FIGURE 11.17
Superelevation table with glyphs in the graphic

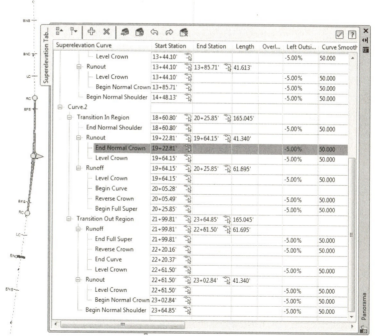

Transition Station Overlap

It is not uncommon to have overlap warnings in your superelevation table. You should resolve the transition station overlap before you continue your design.

Overlap occurs when there is not enough room between curves to fully transition out of one curve and back into the next. Transition station overlap will always occur when a reverse curve or compound curve exists in your alignment. As you can see in Figure 11.18, Curve 1 does not complete its transition out until station 16+82.56, but according to the attainment calculations, Curve 2 will begin affecting the shoulder starting at station 15+35.24.

FIGURE 11.18
Superelevation table showing overlap between two curves

	Station 1	Station 2	Station 3	Length		Overlap	Left Outside
⊟ Transition Out Region	15+26.92'	16+82.56'	155.643'				
⊟ Runoff	15+26.92'	15+78.53'	51.612'				
— End Full Super	15+26.92'					-5.00%	50.000
— Reverse Crown	15+36.92'				⚠	-5.00%	50.000
— End Curve	15+44.12'				⚠		
— Level Crown	15+78.53'				⚠	-5.00%	50.000
⊟ Runout	15+78.53'	16+20.14'	41.613'				
— Level Crown	15+78.53'				⚠	-5.00%	50.000
— Begin Normal Crown	16+20.14'				⚠	-5.00%	50.000
— Begin Normal Shoulder	16+82.56'				⚠	-5.00%	50.000
⊟ Curve.2							
⊟ Transition In Region	15+35.24'	17+00.29'	165.045'				
End Normal Shoulder	15+35.24'				⚠	-5.00%	50.000
⊟ Runout	15+97.25'	16+38.59'	41.340'				
— End Normal Crown	15+97.25'				⚠	-5.00%	50.000
— Level Crown	16+38.59'				⚠	-5.00%	50.000
⊟ Runoff	16+38.59'	17+00.29'	61.695'				
— Level Crown	16+38.59'				⚠	-5.00%	50.000
— Begin Curve	16+79.73'				⚠		
— Reverse Crown	16+79.93'				⚠	-5.00%	50.000
— Begin Full Super	17+00.29'					-5.00%	50.000

You have several options for fixing superelevation overlap. You can choose to have Civil 3D rectify the overlap for you, or you can manually modify the stations in the table. You can also change the stationing for superelevation by modifying the superelevation view, which we will discuss later in this chapter.

To have Civil 3D clear the overlap for you, click the warning symbol that appears in the Superelevation Tabular Editor. Civil 3D resolves overlap by omitting noncritical stations and/or by compressing the transition length between certain stations. In the case of a reverse curve, Civil 3D will pivot the road from full-super to full-super, without transitioning back to normal crown. Be sure to verify that the software has made the update that meets the requirements of your locale.

PRESERVING YOUR DELICATE MANUAL EDITS

You may spend hours getting your superelevation stations to work out correctly. However, it just takes one wrong button click to blow away all your time-consuming edits to the tabular input. To save you from yourself (or the click-happy intern), back up your superelevation tables by exporting them to a file. You'll find the Export Superelevation Data button at the top of the tabular input.

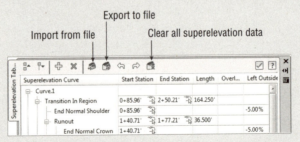

When you need to reimport the superelevation data, clear any data and click Import From File.

Superelevation Views

Superelevation views are a graphic representation of the roadway superelevation. Grip edits to the graphical view will also edit the superelevation stations. The view itself is not intended for plotting. The superelevation view plots station against lane slope to form a sort of profile of the left and right edge of pavement.

At first glance, the superelevation view may seem difficult to read, but with a little explanation it can shed a lot of light on what is going on with your lane slopes. Figure 11.19 shows the superelevation diagram for a two-lane road that contains a reverse curve. The superelevation graphic plots the station value against the percent cross-slope of each edge of pavement. The upper line shows the behavior of the right edge of the pavement, and the lower line shows the left edge of the pavement.

FIGURE 11.19
The superelevation view

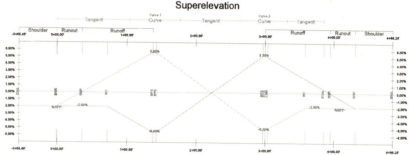

Where no superelevation is applied, the graph data is flat. As the assembly twists into position during superelevation, the distances between the lines becomes greater as the right edge slopes up to a maximum superelevation of 5.6%. You can also see in Figure 11.19 how Civil 3D has resolved overlap between the two curves. Neither line goes back to the default position at -2.00% until the end of the alignment. The crisscross in the center indicates that the roadway transitions directly from one max super to the other.

Place the view into Civil 3D by selecting the alignment and clicking Superelevation ➢ Create Superelevation view. To simplify the view and make it easier to edit, you can choose which lanes to show or hide at the bottom of the view, as shown in Figure 11.20. You should also change the colors for left and right edge of pavement by clicking the ByBlock option and changing it to the color of your choice.

To modify superelevation data using the graphic, click on it to activate the numerous grips. Figure 11.21 shows a close-up of some of the grips you will see.

The diamond-shaped grips can be slid in one axis to modify stationing (the horizontally oriented grips) or slope (the vertically oriented grips). The rectangular grip can be moved to reduce the maximum lane slope when it is in a full-super state.

FIGURE 11.20
Creating a super-elevation view

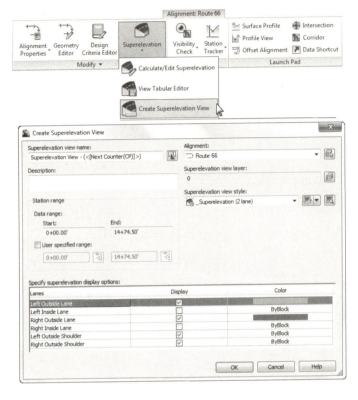

FIGURE 11.21
Grip-editing is another method for fine-tuning your superelevation stationing.

SUPERELEVATION: IT'S NOT JUST FOR HIGHWAYS

Don't get trapped into thinking that just because you don't do highway design, you can't use some of the same tools. Some firms use superelevation to help design waterslides. The fun in learning Civil 3D is finding new ways to use the tools intended to solve one problem to address another problem entirely. You'll know you've mastered the software when you find yourself using the tools in this manner.

The Bottom Line

Apply superelevation to an alignment. Civil 3D has very convenient and flexible tools that will apply safe, correct superelevation to an alignment curve.

> **Master It** Open `Master Super.dwg` file, which you can download from `www.sybex.com/go/masteringcivil3d2012`. Set the design speed of the road to 20 miles per hour and apply superelevation to the entire length of the alignment. Use AASHTO 2004 Design Criteria with an eMax of 6%.

Create a superelevation view. Superelevation views are a great place to get a handle on what is going on in your roadway design. You can visually check the geometry as well as make changes to the design.

> **Master It** Continue working in the drawing from the previous exercise. Create a superelevation view for the alignment.

Chapter 12

Cross Sections and Mass Haul

Cross sections are used in Civil 3D to allow the user to have a graphic confirmation of design intent, as well as to calculate the quantities of materials used in a design. Sections must have at least two types of Civil 3D objects to be created: an alignment and a surface. Other objects, such as pipes, structures, and corridor components, can be sampled in a sample line group, which is used to create the graphical section that is displayed in a section view. These section views and sections remain dynamic throughout the design process, reflecting any changes made to the sampled information. This process reduces potential errors in materials reports, keeping often costly mistakes from happening during the construction process.

In this chapter, you will learn to:

- Create sample lines
- Create section views
- Define materials
- Generate volume reports

The Corridor

Before you create sample lines, you often start with a corridor. The corridor allows you to display the materials being used, as well as show the new surface with cut-and-fill areas. In this chapter, the corridor is a relatively short roadway (937') designed for a residential subdivision (see Figure 12.1).

This corridor has both a top surface and a datum surface created for inclusion in the sample line group, as shown in Figure 12.2. Creating surfaces from the different links and feature lines in a corridor allows you to use sections to calculate volumes between those surfaces. These volumes are calculated by specifying which surfaces to compare when you create a materials list.

> **CREATING THE BEST POSSIBLE SURFACE FOR SAMPLING**
>
> Note that you can create corridor surfaces in two different ways—from links and from feature lines. Links will provide you with a total surface along the width of a corridor, such as the top of pavement, top of base, and top of sub-base. Feature lines require selecting a few more objects to add into a corridor surface to accurately create the surface.

FIGURE 12.1
The Concord Commons corridor showing Cabernet Court road

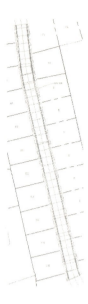

FIGURE 12.2
The Corridor Properties dialog

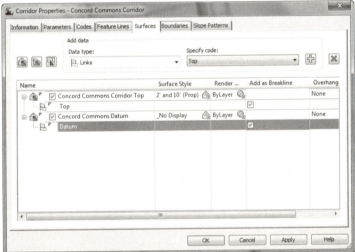

When you create your sample line group, you will have the option to sample any surface in your drawing, including corridor surfaces, the corridor assembly itself, and any pipes in your drawing. The sections are then sampled along the alignment with the left and right widths specified and at the intervals specified. Once the sample lines are created, you can then choose to create section views or to define materials.

Lining Up for Samples

Sample lines are the engine underneath both sections and materials and are held in a collection called sample line groups. One alignment can have multiple sample line groups, but a sample line group can sample only one alignment. Sample lines typically consist of two components: the sample lines and the sample line labels, as shown in Figure 12.3.

FIGURE 12.3
Sample lines consist of the lines and their labels.

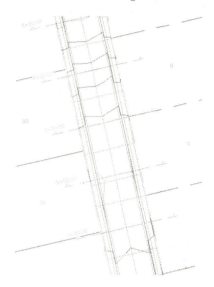

If you pick a sample line, you will see it has three different types of grips, as shown in Figure 12.4. The diamond grip on the alignment allows you to move the sample line along the alignment. The triangular grip on the end of the sample line allows you to move the sample line along an extension of the line, making it either longer or shorter. The square grip on the end of the sample line allows you to not only move the sample line in or out, but also move it in any direction on the XY-plane.

FIGURE 12.4
The three types of grips on a sample line

To create a sample line group, change to the Home tab and choose Sample Lines from the Profile & Section Views panel. After selecting the appropriate alignment, the Create Sample Line Group dialog shown in Figure 12.5 will display. This dialog prompts you to name the sample line group, apply a sample line and label style, and choose the objects in your drawing that you would like to sample. Every object that is available will be displayed in this box, with an area to set the section style, whether to sample the data, what layer each sampled item would be applied to, and a setting to specify whether the data should be static or dynamic. For example, you would typically select your existing ground (EG) surface to be sampled, displayed with an EG style, and be static. Your finished grade (FG) surface would also be sampled, but would be displayed with an FG style and be dynamic.

Once the sample data has been selected, the Sample Line Tools toolbar will appear, as shown in Figure 12.6. This toolbar is context-sensitive and is displayed only when you are creating sample lines.

FIGURE 12.5
The Create Sample Line Group dialog

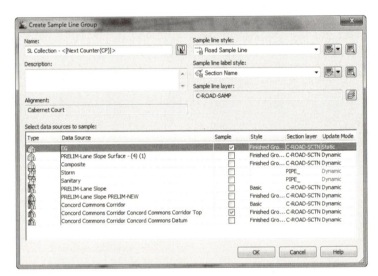

FIGURE 12.6
The Sample Line Tools toolbar

Once you have completed the Sample Line Creation process, close the toolbar, and the command ends. Because most of the information is already set for you in this toolbar, the Sample Line Creation Methods button is the only one that is really needed. This gives you the following five options for creating sample lines:

- By Range Of Stations
- At A Station
- From Corridor Stations
- Pick Points On Screen
- Select Existing Polylines

In Civil 3D, these options are listed in order from most used to least used. Because the most common method of creating sample lines is from one station to another at set intervals, the By Range Of Stations option is first. You can use At A Station to create one sample line at a specific station. From Corridor Stations allows you to insert a sample line at each corridor assembly insertion. Pick Points On Screen allows you to pick any two points to define a sample line. This option can be useful in special situations, such as sampling a pipe on a skew. The last option, Select Existing Polylines, lets you define sample lines from existing polylines.

> **A Warning About Using Polylines to Define Sample Lines**
>
> Be careful when picking existing polylines to define sample lines. Any osnaps used during polyline creation can throw off the Z-values of the section, sometimes giving undesirable results.

To define sample lines, you need to specify a few settings. Figure 12.7 shows the settings that need to be defined in the Create Sample Lines – By Station Range dialog. Right Swath Width is the width from the alignment that you sample. Most of the time this distance is greater than the ROW distance. You also select Sampling Increments, and choose whether to include special stations, such as horizontal geometry (PC, PT, and so on), vertical geometry (PVC, high point, low point, and so on), and superelevation critical stations.

Figure 12.7
Sample line settings

Creating Sample Lines along a Corridor

Before creating cross sections, you must sample the information that will be displayed. You do so by creating sample lines, which are part of a sample line group. Only one alignment can be sampled per sample line group. When creating sample lines, you will have to determine the frequency of your sections and the objects that you want included in the section views. In the following exercise, you create sample lines for Cabernet Court:

1. Open the sections1.dwg file, which you can download from www.sybex.com/go/masteringcivil3d2012.

2. Change to the Home tab and choose Sample Lines from the Profile & Section Views panel.

3. Press ↵ to display the Select Alignment dialog.

4. Select the Cabernet Court alignment and click OK. You can also pick the alignment in the drawing. The Create Sample Line Group dialog opens.

5. Adjust your settings to match those shown in Figure 12.8 and click OK.

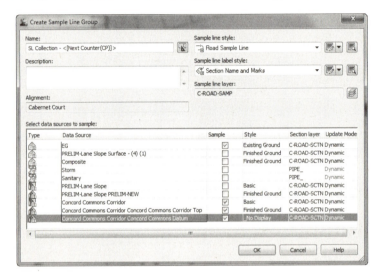

FIGURE 12.8
Use these settings in the Create Sample Line Group dialog.

6. On the Sample Line Tools toolbar, click the Sample Line Creation Methods drop-down arrow and then select By Range Of Stations. Observe the settings but do not change anything.

7. Click OK, and press ↵ to end the command.

8. If you receive a Panorama view telling you that your corridor is out of date and may require rebuilding, dismiss it.

The sample lines can be edited by using the grips as just described or by selecting a sample line and choosing Edit Sample Line from the Modify panel. This command displays both the Sample Line Tools toolbar and the Edit Sample Line dialog, allowing you to pick your alignment. Once the alignment is picked, the sample lines can be edited individually or as a group. The Edit Sample Lines dialog allows you to pick a sample line and edit the information on an individual basis, but it is much more efficient to edit all the sample lines at the same time.

Editing the Swath Width of a Sample Line Group

There may come a time when you will need to show information outside the limits of your section views or not show as much information. To edit the width of a section view, you will have to change the swath width of a sample line group. These sample lines can be edited manually on an individual basis, or you can edit the entire group at once. In this exercise, you edit the widths of an entire sample line group:

1. Continue working on `sections1.dwg`, or you can open the `sections2.dwg` file.

2. Select a sample line and choose Edit Sample Line from the Modify panel. The Sample Line Tools toolbar and the Edit Sample Line dialog appear.

3. Click the down arrow for the sample line editing tools on the toolbar and then click the Edit Swath Widths For Group button. The Edit Sample Line Widths dialog appears.

4. Type **100** in both the Left and Right Swath Width text boxes. Click OK.

5. Press ↵ to end the command.

6. If you receive a Panorama view telling you that your corridor is out of date and may require rebuilding, dismiss it.

7. Examine your sample lines, noting the wider sample lines.

Creating the Views

Once the sample line group is created, it is time to create views. Views can be created in three ways: single view, all views, or all views-by-page. Creating views-by-page allows you to lay out your cross sections on a sheet-by-sheet basis. This is accomplished by creating a page setup for your proposed sheet size and defining that page setup in your sheet style. You can arrange the section views by either rows or columns and specify the space between each consecutive section view. This even allows you to put a predefined grid on your cross-section sheet. Although the setup is quite tedious, the payoff is incredible if you plot many sheets' worth of cross sections at a time. Figure 12.9 shows a layout containing section views arranged to plot by page.

Figure 12.9
Section views arranged to plot by page

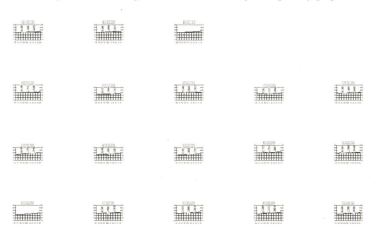

Section views are nothing more than a window showing the section. The view contains horizontal and vertical grids, tick marks for axis annotation, the axis annotation itself, and a title. Views can also be configured to show horizontal geometry, such as the centerline of the section, edges of pavement, and right-of-way. Tables displaying quantities or volumes can also be shown for individual sections.

Creating a Single-Section View

There are occasions when all section views are not needed. In these situations, a single-section view can be created. In this exercise, you create a single-section view of station 5 + 50 from the sample lines created in the previous exercise:

1. Continue with sections1.dwg, or you can open the sections3.dwg file.

2. Change to the Home tab and choose Section Views ➢ Create Section View from the Profile & Section Views panel. The Create Section View Wizard appears.

3. Make sure the settings in the Create Section View Wizard match the settings shown in the following figures. The first screen of the wizard is the General page, shown in Figure 12.10. It allows you to select the alignment, sample line group name, desired sample line or station, and the section view style. You can navigate from one page to another by either clicking Next at the bottom of the screen or clicking the links on the left side of the screen.

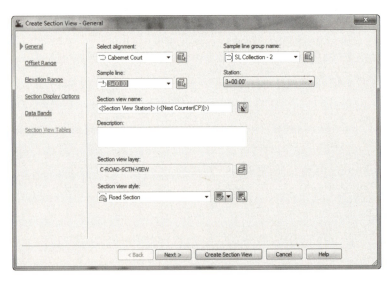

FIGURE 12.10
The General page of the Create Section View Wizard

The second page is where you specify the offset range, as shown in Figure 12.11. You can make the section views the same width by entering a user-specified offset or specify that the width of the section view be controlled by the width of the sample line.

FIGURE 12.11
The Offset Range page of the Create Section View Wizard

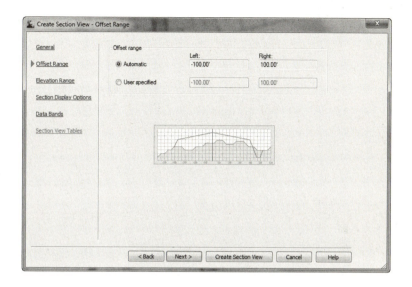

The third page, shown in Figure 12.12, is where you choose the elevation range. You can specify that the height of the section view be set automatically based on the depth of cut or fill, or make it a consistent height by clicking the User Specified radio button and entering an elevation range.

FIGURE 12.12
The Elevation Range page of the Create Section View Wizard

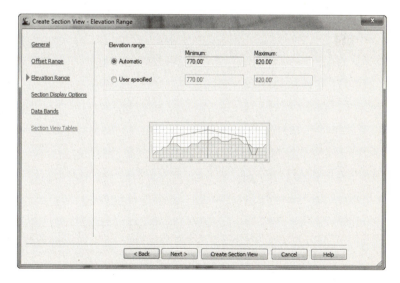

The fourth page contains the section display options, as shown in Figure 12.13. On this page, you can pick which sections to draw in the view, indicate if and how you want the grid clipped in the section view, and specify section view labels and section styles.

FIGURE 12.13
The Section Display Options page of the Create Section View Wizard

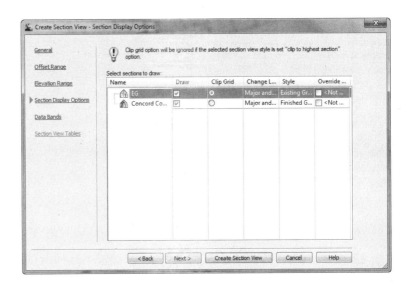

The fifth page, shown in Figure 12.14, lets you specify the data band options. Here, you can select band sets to add to the section view, pick the location of the band, and choose the surfaces to be referenced in the bands.

FIGURE 12.14
The Data Bands page of the Create Section View Wizard

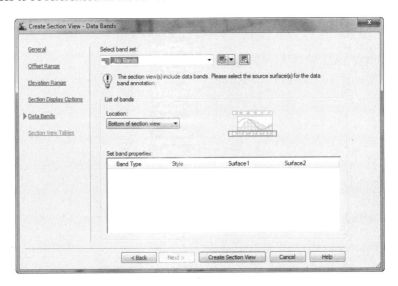

The sixth and last page, shown in Figure 12.15, is where you set up the section view tables. Note that this screen will be available only if you have already computed materials for the sample line group. On this page, you can select the type of table and the table style, and select the position of the table relative to the section view. Notice the graphic on the right side of the window that illustrates the current settings.

FIGURE 12.15
The Section View Tables page of the Create Section View Wizard

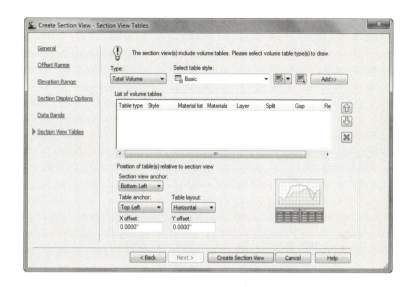

4. Select Create Section View.
5. Pick any point in the drawing area to place your section view.
6. Examine your section view. The display should match Figure 12.16.
7. Close the drawing without saving.

FIGURE 12.16
The finished section view

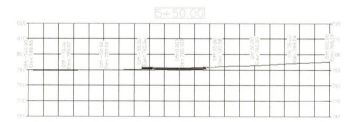

Section View Object Projection

Civil 3D has the ability to project AutoCAD points, blocks, 3D solids, 3D polylines, AutoCAD Civil 3D COGO points, feature lines, and survey figures into section and profile views. Each of the objects listed can be projected to a section view and labeled appropriately. See Chapter 7, "Profiles and Profile Views," to learn more. You access the command by changing to the Home tab's Profile And Section Views panel and selecting Section Views ➢ Project Objects To Section.

It's a Material World

Once alignments are sampled, volumes can be calculated from the sampled surface or from the corridor section shape. These volumes are calculated in a materials list and can be displayed as a label on each section view or in an overall volume table, as shown in Figure 12.17.

The volumes can also be displayed in an XML report, as shown in Figure 12.18.

FIGURE 12.17
A total volume table inserted into the drawing

Total Volume Table

Station	Fill Area	Cut Area	Fill Volume	Cut Volume	Cumulative Fill Vol	Cumulative Cut Vol
0+50.00	0.00	0.00	0.00	0.00	0.00	0.00
1+00.00	0.00	160.06	0.00	147.06	0.00	147.06
1+50.00	6.18	88.59	5.72	230.23	5.72	377.29
2+00.00	17.28	34.69	21.72	114.15	27.45	491.44
2+50.00	27.93	14.43	41.86	45.49	69.31	536.93
3+00.00	18.82	21.75	43.29	33.51	112.60	570.43
3+50.00	11.34	47.35	27.93	63.99	140.53	634.42
4+00.00	5.93	69.10	15.99	107.83	156.52	742.25
4+50.00	0.91	94.56	6.34	151.54	162.86	893.78
5+00.00	6.11	54.52	6.50	138.03	169.36	1031.82
5+50.00	9.57	44.74	14.52	91.91	183.88	1123.73
6+00.00	15.79	24.02	23.49	63.67	207.37	1187.40
6+50.00	2.93	98.57	17.33	113.51	224.70	1300.91
7+00.00	0.00	114.41	2.71	197.20	227.41	1498.11
7+50.00	2.97	86.53	2.86	184.58	230.27	1682.69
8+00.00	3.86	89.09	6.32	162.61	236.59	1845.30
8+50.00	4.02	103.89	7.30	178.69	243.89	2023.98
9+00.00	0.00	0.00	3.73	96.20	247.62	2120.18

FIGURE 12.18
A total volume XML report shown in Microsoft Internet Explorer

Volume Report

Project: C:\Users\Rick BC\appdata\local\temp\sections4-saved_1_1_0002.sv$

Alignment: Cabernet Court
Sample Line Group: SL Collection - 3
Start Sta: 0+50.000
End Sta: 9+00.000

Station	Cut Area (Sq.ft.)	Cut Volume (Cu.yd.)	Reusable Volume (Cu.yd.)	Fill Area (Sq.ft.)	Fill Volume (Cu.yd.)	Cum. Cut Vol. (Cu.yd.)	Cum. Reusable Vol. (Cu.yd.)	Cum. Fill Vol. (Cu.yd.)	Cum. Net Vol. (Cu.yd.)
0+50.000	0.00	0.00	0.00	0.00	0.00	0.00	0.00	0.00	0.00
1+00.000	160.06	147.06	147.06	0.00	0.00	147.06	147.06	0.00	147.06
1+50.000	88.59	230.23	230.23	6.18	5.72	377.29	377.29	5.72	371.56
2+00.000	34.69	114.15	114.15	17.28	21.72	491.44	491.44	27.45	463.99
2+50.000	14.43	45.49	45.49	27.93	41.86	536.93	536.93	69.31	467.62
3+00.000	21.75	33.51	33.51	18.82	43.29	570.43	570.43	112.60	457.83
3+50.000	47.35	63.99	63.99	11.34	27.93	634.42	634.42	140.53	493.89
4+00.000	69.10	107.83	107.83	5.93	15.99	742.25	742.25	156.52	585.73
4+50.000	94.56	151.54	151.54	0.91	6.34	893.78	893.78	162.86	730.92
5+00.000	54.52	138.03	138.03	6.11	6.50	1031.82	1031.82	169.36	862.45
5+50.000	44.74	91.91	91.91	9.57	14.52	1123.73	1123.73	183.88	939.84
6+00.000	24.02	63.67	63.67	15.79	23.49	1187.40	1187.40	207.37	980.03
6+50.000	98.57	113.51	113.51	2.93	17.33	1300.91	1300.91	224.70	1076.21

IT'S A MATERIAL WORLD | 489

Once a materials list is created, it can be edited to include more materials or to make modifications to the existing materials. For example, soil expansion (fluff or swell) and shrinkage factors can be entered to make the volumes more accurately match the true field conditions. This can make cost estimates more accurate, which can result in fewer surprises during the construction phase of any given project.

Creating a Materials List

Materials can be created from surfaces or from corridor shapes. Surfaces are great for earthwork because you can add cut or fill factors to the materials, whereas corridor shapes are great for determining quantities of asphalt or concrete. In this exercise, you practice calculating earthwork quantities for the Cabernet Court corridor:

1. Continue working in `sections1.dwg`, or you can open the `sections4.dwg` file.

2. Change to the Analyze tab and choose Compute Materials from the Volumes And Materials panel. The Select A Sample Line Group dialog appears.

3. In the Select Alignment field, verify that the alignment is set to Cabernet Court.

4. Click OK. The Compute Materials dialog appears.

5. Select Earthworks from the drop-down menu in the Quantity Takeoff Criteria drop-down box.

6. Click the Object Name cell for the Existing Ground surface, and select EG from the drop-down menu.

7. Click the Object Name cell for the Datum surface, and select Concord Commons Corridor Concord Commons Datum (this is the name of the corridor followed by the name of the surface) from the drop-down menu.

8. Verify that your settings match those shown in Figure 12.19.

9. Click OK.

10. Save the drawing.

FIGURE 12.19
The settings for the Compute Materials dialog

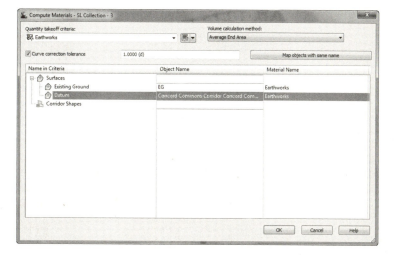

Creating a Volume Table in the Drawing

In the preceding exercise, materials were created that represent the total dirt to be moved or used in the sample line group. In the next exercise, you insert a table into the drawing so you can inspect the volumes:

1. Continue using the sections4.dwg file (or sections1.dwg if you're working in that).

2. Change to the Analyze tab and choose Total Volume Table from the Volumes And Materials panel. The Create Total Volume Table appears.

3. Verify that your settings match those shown in Figure 12.20. Pay close attention and make sure that Reactivity Mode at the bottom of the dialog is set to Dynamic. This will cause the table to update if any changes are made to the sample line collection.

FIGURE 12.20
The Create Total Volume Table dialog settings

4. Click OK.

5. Pick a point in the drawing to place the volume table. The table indicates a Cumulative Fill Volume of 247.62 cubic yards and a Cumulative Cut Volume of 2,120.18 cubic yards, as shown in Figure 12.21. This means you will have to haul off just over 1,870 cubic yards of dirt, or about five tri-axle truckloads from the site during road construction. This might not seem like a lot, but we have only analyzed one roadway and the actual lot grading can increase that number.

6. Save the drawing.

Figure 12.21
The total volume table

Station	Fill Area	Cut Area	Fill Volume	Cut Volume	Cumulative Fill Vol	Cumulative Cut Vol
0+50.00	0.00	0.00	0.00	0.00	0.00	0.00
1+00.00	0.00	160.06	0.00	147.06	0.00	147.06
1+50.00	6.18	58.59	5.72	230.23	5.72	377.29
2+00.00	17.28	34.69	21.72	114.15	27.45	491.44
2+50.00	27.93	14.43	41.86	45.49	69.31	536.93
3+00.00	18.82	21.75	43.29	33.51	112.60	570.43
3+50.00	11.34	47.35	27.93	63.99	140.53	634.42
4+00.00	5.93	69.10	15.99	107.83	156.52	742.25
4+50.00	0.91	94.56	6.34	151.54	162.86	893.78
5+00.00	6.11	54.52	6.50	138.03	169.36	1031.82
5+50.00	9.57	44.74	14.52	91.91	183.88	1123.73
6+00.00	15.79	24.02	23.49	63.67	207.37	1187.40
6+50.00	2.93	98.57	17.33	113.51	224.70	1300.91
7+00.00	0.00	114.41	2.71	197.20	227.41	1498.11
7+50.00	2.97	86.53	2.86	184.58	230.27	1682.69
8+00.00	3.86	89.09	6.32	162.61	236.59	1845.30
8+50.00	4.02	103.89	7.30	178.69	243.89	2023.98
9+00.00	0.00	0.00	3.73	96.20	247.62	2120.18

Adding Soil Factors to a Materials List

Because this design obviously has an excessive amount of fill, the materials need to be modified to bring them closer in line with true field numbers. For this exercise, the shrinkage factor will be assumed to be 0.80 and 1.20 for the expansion factor (20 percent shrink and swell). These numbers are arbitrary—numbers used during an actual design will be based on soil type and conditions. In addition to these numbers (which Civil 3D represents as Cut Factor for swell and Fill Factor for shrinkage), you can specify a Refill Factor value. This value specifies how much cut can be reused for fill. For this exercise, assume a Refill Factor value of 1.00:

1. Continue using drawing `sections4.dwg`.

2. Change to the Analyze tab and choose Compute Materials from the Volumes And Materials panel. The Select A Sample Line Group dialog appears.

3. Select the Cabernet Court alignment and the SL Collection – 3 sample line group, if not already selected.

4. Click OK. The Edit Material List dialog appears.

5. Enter a Cut Factor of **1.20** and a Fill Factor of **0.80**, and verify that all other settings are the same as in Figure 12.22. Click OK.

6. Examine the Total Volume table again. Notice that the new Cumulative Fill Volume is 198.09 cubic yards and the new Cumulative Cut Volume is 2,544.22 cubic yards.

7. Save the drawing.

CAN YOU HAVE ACCURATE VOLUME NUMBERS WITHOUT SECTIONS?

Yes, you can. Civil 3D has the capability to add cut and fill factors to both volume surfaces during creation and in the surface volumes panorama.

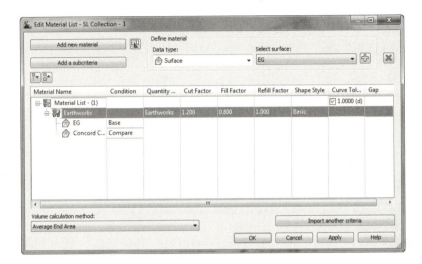

FIGURE 12.22
The Edit Material List dialog

Volumes can also be created in a format that can be printed and put into a project documentation folder. This is accomplished by creating a volume report, which is populated through LandXML. This report will open and display in your browser, but you can convert it to Word or Excel format with a simple copy and paste from the XML report. The XML report style sheet can be edited. The following is a sample of the default code in the style sheet:

```
- <CrossSect name="1+00.00" number="2"
 sta="100" staEq="100" areaCut="160.05542428531"
areaUsable="160.05542428531" areaFill="0"
volumeCut="176.470710087465"
volumeUsable="176.470710087465"
volumeFill="0" cumVolumeCut="176.470710087465"
cumVolumeUsable="176.470710087465" cumVolumeFill="0"
massHaul="176.470710087465">
- <MaterialCrossSects>
- <MaterialCrossSect name="Earthworks(Cut)"
area="160.05542428531" volume="176.470710087465"
cumVolume="176.470710087465">
- <MaterialCrossSectEnvelop area="160.05542428531">
  <CrossSectPnt OE="-25.000000, 798.700342" />
  <CrossSectPnt OE="-25.000000, 799.308899" />
  <CrossSectPnt OE="-18.636164, 799.554025" />
  <CrossSectPnt OE="-12.983610, 800.034982" />
```

```
<CrossSectPnt OE="-7.134326, 800.364341" />
<CrossSectPnt OE="0.000000, 800.542347" />
<CrossSectPnt OE="11.591210, 800.831555" />
<CrossSectPnt OE="12.106582, 800.844414" />
<CrossSectPnt OE="25.000000, 801.248116" />
<CrossSectPnt OE="24.000000, 798.477101" />
<CrossSectPnt OE="20.000000, 798.064101" />
<CrossSectPnt OE="18.000000, 798.357101" />
<CrossSectPnt OE="16.000000, 796.383767" />
<CrossSectPnt OE="0.000000, 796.803185" />
<CrossSectPnt OE="-16.000000, 796.587009" />
<CrossSectPnt OE="-18.000000, 798.560342" />
<CrossSectPnt OE="-20.000000, 798.600342" />
<CrossSectPnt OE="-24.000000, 798.347342" />
<CrossSectPnt OE="-25.000000, 798.700342" />
</MaterialCrossSectEnvelop>
</MaterialCrossSect>
```

Generating a Volume Report

Volume reports can be included on a drawing but normally aren't because of liability issues. However, it is often necessary to know what these volumes are and have some record of them. Civil 3D provides you with a way to create a report that is suitable for printing or for transferring to a word processing or spreadsheet program. In this exercise, you'll create a volume report for the Concord Commons corridor:

1. Open the sections5.dwg file.

2. Change to the Analyze tab and choose Volume Report from the Volumes And Materials panel. The Report Quantities dialog appears.

3. Verify that Material List – (1) is selected in the dialog and click OK.

4. You may get a warning message that says "Scripts are usually safe. Do you want to allow scripts to run?" Click Yes.

5. Note the cut-and-fill volumes and compare them to your volume table in the drawing. Close the report when you are done viewing it.

6. Close the drawing without saving.

A Little More Sampling

Although it is good practice to create your section views as one of the last steps in your project, occasionally data is added that needs to be shown in a section view after the section views are created. To accomplish this, you need to add that data to the sample line group. You can do so by navigating to the sample line group and editing its properties. The Sections tab shown in Figure 12.23 shows the Sample More Sources button you click to add more data to the sample line group.

FIGURE 12.23
The Sections tab of the Sample Line Group Properties dialog

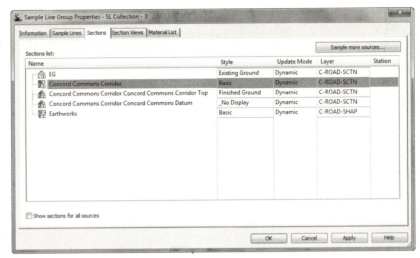

Once you click Sample More Sources, the Section Sources dialog opens. This dialog allows you to either add more sample sources to or delete sample sources from the sample line group. In this example, you have a sanitary sewer network that was added to the project. Because you need to show the locations of the sanitary pipes with respect to the designed road, you need to add the sanitary sewer network to the sample line group. To do so, simply select the sanitary sewer network and click Add, as shown in Figure 12.24.

FIGURE 12.24
Adding a sanitary sewer network to the sample line group

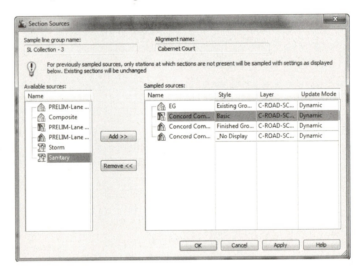

In prior releases of Civil 3D, it was difficult to add more information to a section view once the sample lines were created. Quite often, it was easier to delete the sample line group and create a new one from scratch, including the information that you wanted to show. In this exercise, you'll add a pipe network to a sample line group and inspect the existing section views to ensure that the pipe network was added correctly:

1. Open the sections6.dwg file.

2. In Prospector, expand the Alignments ➢ Centerline Alignments ➢ Cabernet Court ➢ Sample Line Groups ➢ SL Collection – 3 branches, as shown in Figure 12.25.

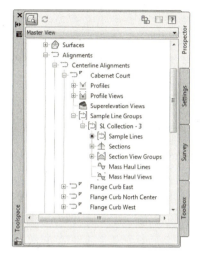

FIGURE 12.25
The location of the sample line group, under the Alignments branch

3. Right-click SL Collection – 3 and select Properties. The Sample Line Group Properties dialog appears.

4. Switch to the Sections tab if necessary.

5. Click Sample More Sources. The Section Sources dialog appears.

6. Select the Sanitary Network from the drop-down menu in the Available Sources selection box.

7. Click Add.

8. Click OK to dismiss the Section Sources dialog.

9. Click OK again to dismiss the Sample Line Group Properties dialog.

10. Close the drawing without saving.

Annotating the Sections

Now that the views are created, you can add annotation to further explain design intent. Labels can be added through the code set style or by adding labels to the section view itself. These section labels can be used to label the section offset, elevation, or slope. You can see an example in Figure 12.26.

FIGURE 12.26
Adding labels to a section view will help explain design intent.

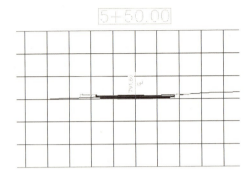

> ### 🌐 Real World Scenario
>
> **ADDING LABELS TO THE SECTION VIEW**
>
> There is often a need for labels to show what the various graphics mean in a section view. To show exact elevations, labels are much more efficient than scaling information from the grid. In this exercise, you'll label the section view elevation at the designed centerline of the road:
>
> 1. Open the sections7.dwg file.
> 2. Change to the Annotate tab and choose Add Labels from the Labels & Tables panel. The Add Labels dialog appears.
> 3. Select Section View from the drop-down menu in the Feature selection box.
> 4. Select Offset Elevation from the drop-down menu in the Label Type selection box.
> 5. Select CL Elevation from the drop-down menu in the Label Style selection box.
> 6. Click Add.
> 7. Pick a grid on the section view to select the view.
> 8. Pick the top centerline of the corridor in the view. It may be easier to pick using the Intersection osnap.
> 9. Click Close in the Add Labels dialog. You now have labels reflecting the section view elevation at the centerline.
> 10. Close the drawing without saving.

Mass Haul

Now that you have an understanding of cross sections and you can compute how much cut-fill is required at each cross section, let's take a look at the "big picture"—that is, Mass Haul.

Mass Haul is basically the art of moving dirt around with the least amount of dirt brought into the site to satisfy grading requirements (borrowed) or excess dirt hauled out of the site (waste).

Many contractors like to see a Mass Haul diagram so they can visualize the overall road and what dirt they need to move around. The goal is to achieve zero balance—that is, no cut or fill. Figure 12.27 shows Syrah Way depicted in a Mass Haul diagram.

FIGURE 12.27
Syrah Way Mass Haul diagram

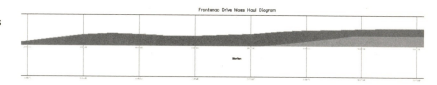

Taking a Closer Look at the Mass Haul Diagram

Figure 12.27 showed a typical Mass Haul diagram, but what did it all mean? Let's take a closer look at our example and explore some common terminology used.

The Mass Haul diagram is based on Elevation 0 (zero). When the line appears above the zero elevation point, it is showing Net Cut values. When the line appears below the zero elevation point, it is showing Net Fill values, as you can see in Figure 12.28.

As the Mass Haul diagram continues, it shows the cumulative effect of Net Cut and Fill for the alignment. When the Net Cut and Net Fill converge at the zero elevation, the alignment at that point is balanced (Figure 12.29).

FIGURE 12.28
The Elevation, Net Cut, and Net Fill on the Mass Haul diagram

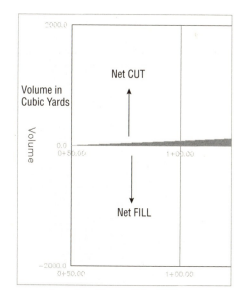

FIGURE 12.29
The Cumulative earthwork for the alignment

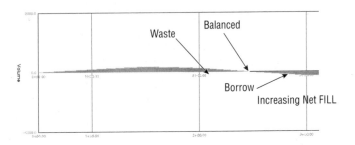

Here is some of the nomenclature you might encounter:

Balanced The state where the cumulative cut and fill volumes are equal.

Origin Point The beginning of the Mass Haul diagram, typically at Station 0 + 00, but can vary depending on your stationing.

Borrow A negative value typically at the end of the Mass Haul diagram that indicates Fill material that will need to be brought into the site.

Waste A positive value typically at the end of the Mass Haul diagram that indicates Cut material that will need to be hauled out of the site.

Free Haul Earthwork that has been contractually agreed upon to be moved by a contractor. This typically involves a contracted distance.

Over Haul Earthwork that has is not contractually agreed upon to be moved. This excess can be used for borrow pits or waste piles.

Create a Mass Haul Diagram

Now, let's put it all together and build a Mass Haul diagram in Civil 3D for Frontenac Drive and you'll see how easy it is:

1. Open the `MassHaul.dwg` file. Remember, you can download all the data files from this book's web page.

2. From the Analyze tab and Volumes And Materials panel, select Mass Haul. The Create Mass Haul Diagram dialog opens.

3. Make changes as shown in Figure 12.30 and click Next.

4. On the Mass Haul Display Options area (Figure 12.31), explore the options and click Next.

5. In the Balancing Options area, make the changes as shown in Figure 12.32 and click Create Diagram.

6. Find a clear spot on your drawing to place the Mass Haul diagram. The diagram should look similar to Figure 12.33.

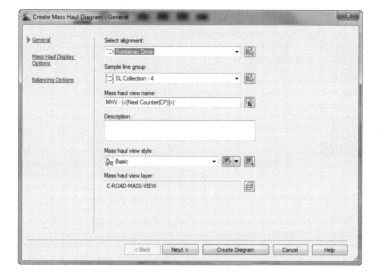

FIGURE 12.30
The Create Mass Haul Diagram dialog, General options

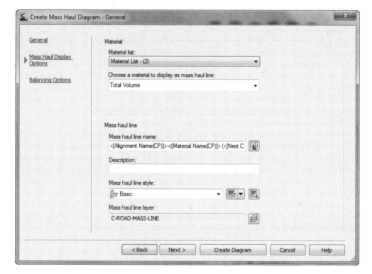

FIGURE 12.31
The Create Mass Haul Diagram options, Mass Haul Display Options page

FIGURE 12.32
The Create Mass Haul diagram, Balancing Options page

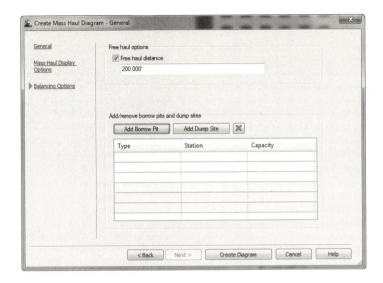

FIGURE 12.33
The completed Mass Haul diagram

The Create Mass Haul Diagram Dialog Explained

Now that we have created a Mass Haul diagram, let's look into each of the wizard screens.

THE GENERAL AREA

This screen contains general options for the Mass Haul diagram. Refer to Figure 12.30:

Select Alignment   Select the alignment that has been previously sampled. You can select from the drop-down menu, or by using the drawing selector icon to the right of the alignment.

Sample Line Group   This is the predefined sample group that you wish to use for the Mass Haul diagram. Remember that an alignment can have multiple sample line groups. You can select the sample line group by using the drop-down menu or by using the drawing selector icon to the right of the sample line group.

Mass Haul View Name   You can name the Mass Haul diagram by typing in a name, or by leaving it at the default. The name will appear as MHV – 1 (or whatever the next logical number is).

Description   Enter a verbose description of the Mass Haul diagram here.

Mass Haul View Style   You can set predefined looks of Mass Haul diagrams or create your own. For more on styles, refer to Chapter 19, "Styles."

Mass Haul View Layer   You can set a specific layer for the Mass Haul diagram.

Mass Haul Display Options

This section shows more options for visual display. Refer to Figure 12.31:

Material List   This is the material list that was predefined when you used the Compute Materials dialog.

Choose a Material to Display as the Mass Haul Line.   You can choose Total Volume, Total Cut Volume, Total Fill Volume, Total Unusable Volume, Ground Removed, and Ground Fill.

Mass Haul Line Name   You can give a unique name for the Mass haul line:

Description   Enter a verbose description for the Mass haul line.

Mass Haul Line Style   You can set predefined looks of Mass Haul lines or create your own.

Mass Haul Line Layer   You can set a specific layer for the Mass Haul line.

Balancing Options

In this section, you set options such as free haul options and adding borrow pits and/or dump sites. Refer to Figure 12.32.

Free Haul Options   By checking this box, you set the agreed-upon maximum distance that earth will be hauled. When you check the box, you enter a distance in the box beneath it.

Add/Remove Borrow Pits and Dump Sites   You can enter (by station and cubic yard) the area(s) you wish to place a borrow pit and/or dump site.

Editing a Mass Haul Diagram

When you create a Mass Haul diagram, you can easily modify parameters and get instant feedback on your diagram. Follow these steps to see how:

1. Continuing in the `MassHaul.dwg`, click on the Mass Haul Object (not the grid).

2. From the Modify panel, select Balancing Options. The Mass Haul Line Properties dialog opens. It looks similar to part of the Balancing Options section shown earlier.

   If you look at your Mass Haul diagram (Figure 12.34), you can see that nearly all the earthwork for Frontenac Drive involves Net Cut, which means hauling away dirt. There are measures that can be taken, such as looking at the finished ground and seeing if the profile can be modified to alleviate all of this earth being moved. If you can't modify the profile, then the Balancing Options come into play.

FIGURE 12.34
Net Cut shown on Frontenac Drive

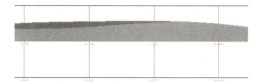

3. The Free Haul Distance is presently set for 200'. Change it to 500' and then click Apply.

4. Drag the dialog away from the screen to see the changes. The blue color indicates the area of free haul and the reddish color indicates the Overhaul area as shown in Figure 12.35.

FIGURE 12.35
The Mass Haul diagram with (a) setting free haul distance to 200' and (b) setting free haul distance to 500'

By changing the free haul distance, you can see the amount of earth that will be moved as part of the contract (free) versus that which is not free.

We can further tweak this amount to cut down on the net cut values, by adding a dump site.

5. Click the Add Dump Site button and make changes shown in Figure 12.36. Click the Apply button and move the dialog out of the way to see the results (Figure 12.37).

6. You can continue to tweak the station location and the capacity values. Remember, what we are after is a balanced site.

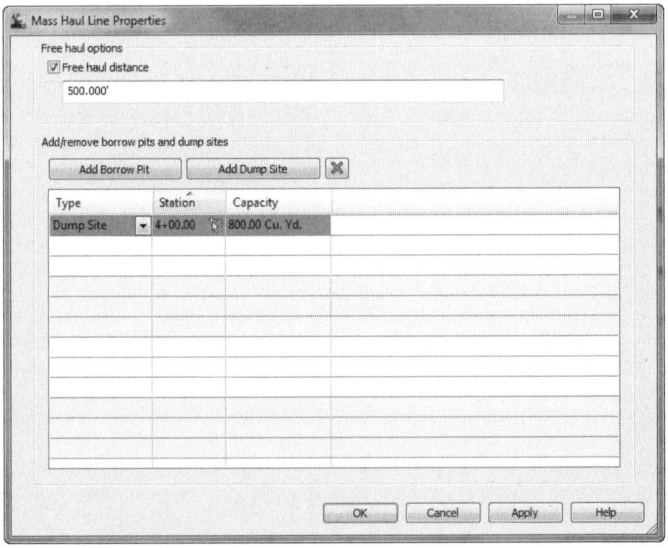

FIGURE 12.36
The Mass Haul Line Properties dialog: adding a Dump Site.

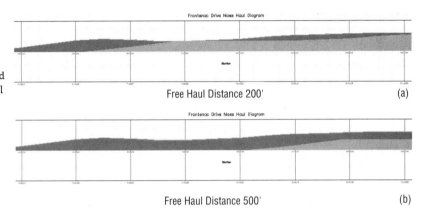

FIGURE 12.37
Frontenac Drive Mass Haul diagram with the added dump site

The Bottom Line

Create sample lines. Before any section views can be displayed, sections must be created from sample lines.

Master It Open `sections1.dwg` and create sample lines along the alignment every 50′.

Create section views. Just as profiles can only be shown in profile views, sections require section views to display. Section views can be plotted individually or all at once. You can even set them up to be broken up into sheets.

Master It In the previous drawing, you created sample lines. In that same drawing, create section views for all the sample lines.

Define materials. Materials are required to be defined before any quantities can be displayed. You learned that materials can be defined from surfaces or from corridor shapes. Corridors must exist for shape selection and surfaces must already be created for comparison in materials lists.

Master It Using `sections4.dwg`, create a materials list that compares EG with Cabernet Court Top Road Surface.

Generate volume reports. Volume reports give you numbers that can be used for cost estimating on any given project. Typically, construction companies calculate their own quantities, but developers often want to know approximate volumes for budgeting purposes.

Master It Continue using `sections4.dwg`. Use the materials list created earlier to generate a volume report. Create an XML report and a table that can be displayed on the drawing.

Chapter 13

Pipe Networks and Part Builder

Before you can begin modeling and designing your pipe network, consider what components you'll need in your design. And what about creating a nonstandard structure or pipe? Part Builder can handle that job. Once you understand the parts used to design and construct pipe networks, it's time to assemble those parts into a system or network.

In this chapter, you will learn to:

- Create a pipe network by layout
- Create an alignment from network parts and draw parts in profile view
- Label a pipe network in plan and profile
- Create a dynamic pipe table

Planning a Typical Pipe Network

Before you begin designing a pipe network, it's important to brainstorm all of the parts you'll need to construct the network, how these objects will be represented in plan and profile, and the behavior of these parts (which you'll specify using the slope, cover, rim, and sump parameters). Once you have a list of the elements you need, you can locate the appropriate parts in the part catalog, build the appropriate rule sets, create the proper styles to match your CAD standard, and tie it all together in a parts list.

Let's look at a typical sanitary sewer design. You typically start by going through the sewer specifications for the jurisdiction in which you're working. There is usually a published list of allowable pipe materials, manhole details, slope parameters, and cover guidelines. Perhaps you have concrete and PVC pipe manufacturer catalogs that contain pages of details for different manholes, pipes, and junction boxes. There is usually a recommended symbology for your submitted drawings—and, of course, you have your own in-house CAD standards. Assemble this information, and make sure you address these issues:

- Recommended structures, including materials and dimensions (be sure to attach detail sheets)
- Structure behavior, such as required sump, drops across structures, and surface adjustment
- Structure symbology
- Recommended pipes, including materials and dimensions (again, be sure to attach detail sheets)

- Pipe behavior, such as cover requirements; minimum, maximum, and recommended slopes; velocity restrictions; and so on
- Pipe symbology

The following is an example of a completed checklist:

Sanitary Sewers in Sample County

Recommended Structures: Standard concentric manhole, small-diameter cleanout.

Structure Behavior: All structures have 1.5' (0.46 m) sump, rims, a 0.10' (0.03 m) invert drop across all structures. All structures designed at finished road grade.

Structure Symbology: Manholes are shown in plan view as a circle with an S inside and a diameter that corresponds to the actual diameter of the manhole. Cleanouts are shown as a solid, filled circle with a diameter that corresponds to the actual diameter of the cleanout. (See Figure 13.1.)

Manholes are shown in profile view with a coned top and rectangular bottom. Cleanouts are shown as a rectangle (see Figure 13.2).

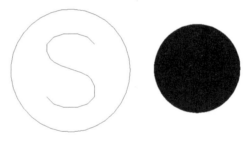

FIGURE 13.1
Sanitary sewer manhole in plan view (left) and a cleanout in plan view (right)

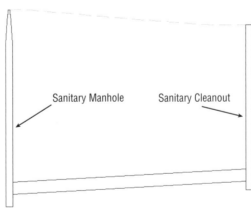

FIGURE 13.2
Profile view of a sanitary sewer manhole (left) and a cleanout (right)

Recommended Pipes: 8" (20.32 cm), 10" (25.4 cm), and 12" (30.48 cm) PVC pipe per manufacturer specifications.

Pipe Behavior: Pipes must have cover of 4′ (1.22 m) to the top of the pipe; the maximum slope for all pipes is 10 percent, although minimum slopes may be adjusted to optimize velocity as follows:

Sewer Size	Minimum Slope
8″ (20.32 cm)	0.40%
10″ (25.4 cm)	0.28%
12″ (30.48 cm)	0.22%

Pipe Symbology: In plan view, pipes are shown with a CENTER2 linetype line that has a thickness corresponding to the inner diameter of the pipe. In profile view, pipes show both inner and outer walls, with a hatch between the walls to highlight the wall thickness (see Figure 13.3).

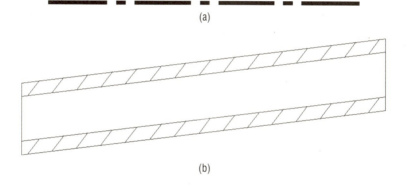

Figure 13.3
Sanitary pipe in plan view (a) and in profile view (b)

The Part Catalog

Once you know what parts you require, you need to investigate the part catalog to make sure these parts (or a reasonable approximation) are available.

The part catalog is a collection of two domains that contain two catalogs each. Structures are considered one domain, and Pipes are the second. The Structures domain consists of a Metric Structures catalog and a US Imperial Structures catalog; the Pipes domain consists of a Metric Pipes catalog and a US Imperial Pipes catalog.

Although you can access the parts from the catalogs while creating your parts lists in Civil 3D, you can't examine or explore the catalogs easily while in the Civil 3D interface. It's useful to understand where these catalogs reside and how they work.

The part catalog is installed locally by default at:

`C:\ProgramData\Autodesk\C3D 2012\enu\Pipes Catalog\`

Note that all paths in this chapter are the Windows 7 install paths. If you're running Civil 3D on any other Windows operating systems, please check the Civil 3D Users Guide for information on the `Pipes Catalog` folder install location. Also, note that the `ProgramData` folder is hidden by default.

If you can't locate the `Pipes Catalog` folder, it may be because your network administrator installed the catalogs at a network location when Civil 3D was deployed.

The Structures Domain

To learn more about how a catalog is organized, let's explore the `US Imperial Structures` folder.

The first file of interest in the `US Imperial Structures` folder is an HTML document called US Imperial Structures (see Figure 13.4).

FIGURE 13.4
The US Imperial Structures HTML document

Double-click this file. Internet Explorer opens with a window so you can explore the US Imperial Structures catalog. A tree with different structure types on the left is under the Catalog tab. Expanding the tree allows you to explore the types of structures that are available. You may have to allow ActiveX controls to view the file.

Structures that fall into the same type have behavioral properties in common but may vary in shape and proportion. The four structure types in the default catalogs are as follows:

- Inlet-Outlets
- Junction Structures With Frames
- Junction Structures Without Frames
- Simple Shapes

Under each structure type are several *shapes*. The shape spells out the details of how the structure is shaped and proportioned, and it shows what happens to each dimension when the size increases. If you drill into the Junction Structures With Frames type and highlight the AeccStructConcentricCylinder_Imperial shape, you can see this in action (see Figure 13.5).

In this view, you can see the available sizes of the concentric cylindrical structure, such as 48″ (121.92 cm), 60″ (152.40 cm), 72″ (182.88 cm), and 96″ (243.84 cm), but you may not edit them. Editing must be completed in the Part Builder, as discussed later in this chapter. Follow the table to see what happens to the cone height, wall thickness, floor thickness, and frame diameter and height as the diameter dimension increases. The structure may get bigger or smaller, but its basic form remains the same and its behavior is predictable.

Explore the other structures of the Junction Structures With Frames type, and note why they're all considered to be the same type. Their shape changes, but their fundamental behavior and intent are similar. For example, each has a frame, each has a similar size range, and each is used as a junction for pipes.

FIGURE 13.5
A closer look at the structure type and shape

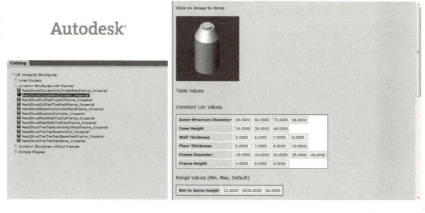

Let's return to the example of a sanitary sewer network. You need a standard concentric manhole and a simple, small-diameter cleanout, and most likely you have specification sheets from the concrete products company handy. You know that you'll need to have a least a couple of junction structures, and based on the required structures' spec sheets, you know you need a junction structure with frame. Your manhole most closely resembles the concentric cylindrical structure, and the 48″ (121.92 cm) and 60″ (152.40 cm) match the allowable sizes. For the cleanout, the cylindrical slab top structure is the appropriate shape and behavior, but you need a 6″ size. The smallest size available by default is 15″ (38.10 cm). Make notes on your checklist to add a part size for a 6″ (15.24 cm) cylindrical slab top structure. (You'll take care of that in the next section.)

Something to keep in mind as you're searching for the appropriate structures to meet your standard is that the part you choose doesn't necessarily have to be a perfect match for your specified standard detail. The important things to look for are general shape, insertion behavior, and key dimensions.

Ask yourself the following questions:

- Will I be able to orient this structure in an appropriate way (is it round, rectangular, concentric, or eccentric)?

- Will I be able to label the insertion point (rim elevation) correctly?

- Will this structure look the way I need it to in the plan, profile, and section views? Does it look the same viewed from every angle, or would some views show a wider/narrower structure?

- Can I adjust my structure style in plan and profile views to display the structure the way I want to see it?

If you can find a standard shape that is conducive to all of these, it's worth your time to try it out before resorting to building custom parts with Part Builder.

For example, if a certain catch basin required in your town has a unique frame and is an elongated rectangle, try one of the standard rectangle frame structures first (see Figure 13.6). You probably already use a certain type of block in your CAD drawings to represent this type of catch basin, and that block can be applied to the structure style. It's not always necessary to build a custom part for small variations in shape. The important thing is that you model and label your rims and inverts properly. You'll learn more about this in the next few sections.

FIGURE 13.6
Each catch basin can be modeled with the same rectangular slab top structure.

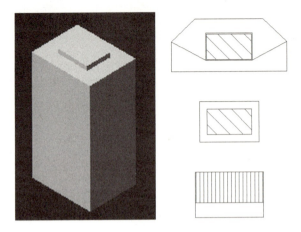

THE PIPES DOMAIN

The second part domain is named Pipes. Pipes have only one type, which is also called Pipes.

Locate the catalog HTML file under the US Imperial Pipes folder, and explore this catalog the same way you explored the US Imperial Structures catalog.

Double-click the HTML file. Internet Explorer opens with a window that allows you to view the US Imperial Pipes catalog. A tree with different pipe shapes appears on the left under the Catalog tab. Expanding the tree allows you to explore the pipe shapes that are available. Pipe shapes are broken into smaller categories by material (in the case of a circular pipe) or orientation (for an elliptical pipe).

The four pipe shapes in the default catalogs are as follows:

- Circular Pipes (Concrete, Ductile Iron, and Polyvinyl Chloride)
- Egg-Shaped Pipes (Concrete)
- Elliptical Pipes (Concrete Vertical and Concrete Horizontal)
- Rectangular Pipes (Concrete)

Drill into the Circular Pipes shape, and highlight the AeccCircularConcretePipe_Imperial material (see Figure 13.7). Let's dig into this a bit more.

In this view, you can see the different inner pipe diameters and their corresponding wall thicknesses. Explore the other pipe shapes to get a feel for what is available.

You need 8" (20.32 cm), 10" (25.40 cm), and 12" (30.48 cm) pipe for the sanitary sewer example. The spec sheet from the PVC pipe supplier indicates that the inner diameters and corresponding wall thicknesses are appropriate for your design.

Again, keep in mind as you're searching for the appropriate pipes to meet your standard that the pipe you choose doesn't necessarily have to be a perfect match for your specified standard detail. For example, your storm-drainage system might have the option to use high-density polyethylene (HDPE) pipe, and you might think you'd have to build a custom pipe type using Part Builder.

However, examine your symbology and labeling requirements a little further before jumping into building a custom part. The wall thickness of HDPE pipe is different from the same size PVC, but that may not be important for your drawing representation or labeling.

If you show HDPE as a single line or a double line representing inner diameter, wall thickness is irrelevant. You can create a custom label style to label the pipe HDPE instead of PVC.

If you show HDPE represented as its outer diameter or have particular requirements to show wall thickness in crossing and profile views, then building a custom pipe in Part Builder may be your best option (see Figure 13.8).

FIGURE 13.7
A closer look at the pipe shape and material

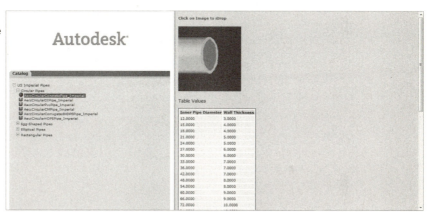

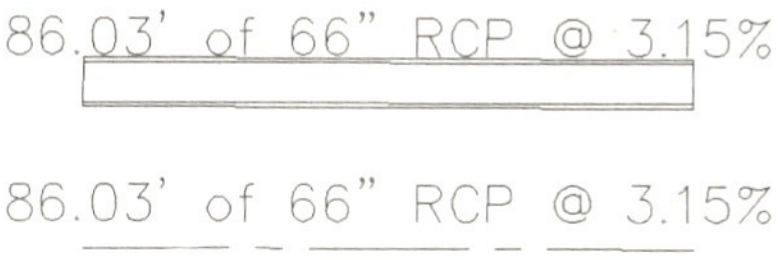

FIGURE 13.8
The top pipe must specify the outer diameter exactly. The bottom pipe, which uses a centerline to represent the inner diameter, doesn't have this requirement.

The important things to consider are general shape and wall thickness. Ask yourself these questions:

- Does this pipe approximate the shape of my specified pipe?
- Does this pipe offer the correct inner diameter choice?

- How important is wall thickness to my model, plans, profiles, and sections?
- Can I adjust my pipe style in plan and profile views to display the pipe the way I need to see it?
- Can I create a pipe label that will label this pipe the way I need it labeled?

Again, if you can find a standard pipe that is conducive to all of these, it's worth your time to try it before you resort to building custom pipes with Part Builder.

THE SUPPORTING FILES

Each part family in the catalog has three corresponding files located at:

`C:\ProgramData\Autodesk\C3D 2012\enu\Pipes Catalog\`

Two files are mandatory sources of data required for the part to function properly, and one is an optional bitmap that provides the preview image you see in the catalog browser and parts-list interface.

First, dig into the `US Imperial Pipes` folder as follows:

`C:\ProgramData\Autodesk\C3D 2012\enu\Pipes Catalog\US Imperial Pipes\Circular Pipes`

This reveals the following files for the Imperial Circular Pipe:

`partname.dwg`   This part drawing file contains the geometry that makes up the part as well as the definition of the parametric relationships.

`partname.xml`   This part XML file is created as soon as a new part file is started in Part Builder. This file contains information on the parameter sizes, such as wall thickness, width, and diameter.

`partname.bmp`   This part image file is used as a visual preview for that particular part in the HTML catalog (as you saw in the previous section). This file is optional.

You'll study some of these files in greater detail later this chapter.

Mark on your checklist which parts you'd like to use from the standard catalog, and also note the parts that you may have to customize using Part Builder.

Part Builder

Part Builder is an interface that allows you to build and modify pipe network parts. Part Builder is accessed by selecting the drop-down list under Create Design from the Home tab. At first, you may use Part Builder to add a few missing pipes or structure sizes. As you become more familiar with the environment, you may build your own custom parts from scratch.

This section is intended to be an introduction to Part Builder and a primer in some basic skills required to navigate the interface. It isn't intended to be a robust "how-to" for creating custom parts. Civil 3D includes three detailed tutorials for creating three types of custom structures. The tutorials lead you through creating a cylindrical manhole structure, a drop inlet manhole structure, and a vault structure. You can find these tutorials by going to Help ➤ Tutorials and then navigating to Part Builder Tutorials.

> **BACK UP THE PART CATALOGS**
>
> Here's a warning: before exploring Part Builder in any way, it's critical that you make a backup copy of the part catalogs. Doing so will protect you from accidentally removing or corrupting default parts as you're learning and will provide a means of restoring the original catalog.
>
> The catalog (as discussed in the previous section) can be found by default at:
>
> `C:\ProgramData\Autodesk\C3D 2012\enu\Pipes Catalog\`
>
> To make a backup, copy this entire directory and then save that copy to a safe location, such as another folder on your hard drive or network, or to a CD.
>
> We recommend that you do this and use the backup file for the exercises here. To do this, you will have to point to the new location by clicking the drop-down list in the Create Design panel and selecting Set Pipe Catalog Location. Select the icon next to catalog folder and point it to your saved location.

The parts in the Civil 3D pipe network catalogs are parametric. Parametric parts are dynamically sized according to a set of variables, or parameters. In practice, this means you can create one part and use it in multiple situations.

For example, in the case of circular pipes, if you didn't have the option of using a parametric model, you'd have to create a separate part for each diameter of pipe you wanted, even if all other aspects of the pipe remained the same. If you had 10 pipe sizes to create, that would mean 10 sets of `partname.dwg`, `partname.xml`, and `partname.bmp` files, as well as an opportunity for mistakes and a great deal of redundant editing if one aspect of the pipe needed to change.

Fortunately, you can create one parametric model that understands how the different dimensions of the pipe are related to each other and what sizes are allowable. When a pipe is placed in a drawing, you can change its size. The pipe will understand how that change in size affects all the other pipe dimensions such as wall thickness, outer diameter, and more; you don't have to sort through a long list of individual pipe definitions.

Part Builder Orientation

The Civil 3D pipe network catalogs are drawing-specific. If you're in a metric drawing, you need to make sure the catalog is mapped to metric pipes and structures, whereas if you're in an imperial drawing, you'll want the imperial. By default, the Civil 3D templates should be appropriately mapped, but it's worth the time to check. Set the catalog by changing to the Home tab and selecting the drop-down list on the Create Design panel. Verify the appropriate folder and catalog for your drawing units in the Pipe Network Catalog Settings dialog (see Figure 13.9), and you're ready to go.

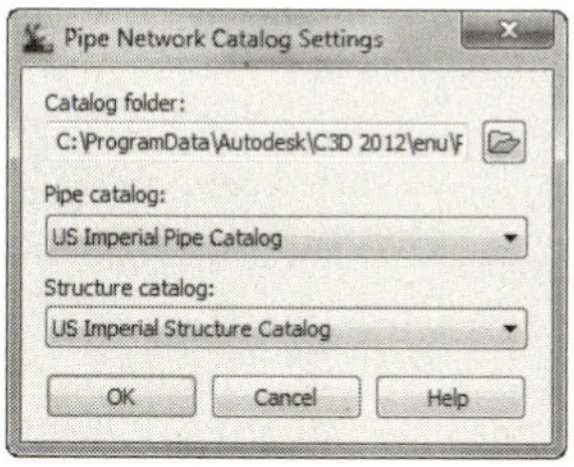

FIGURE 13.9
Choose the appropriate folder and catalog for your drawing units.

Understanding the Organization of Part Builder

The vocabulary used in the Part Builder interface is related to the vocabulary in the HTML catalog interface that you examined in the previous section, but there are several differences that are sometimes confusing.

To open Part Builder, select the drop-down list on the Create Design panel of the Home tab.

The first screen that appears when you start the Part Builder is Getting Started – Catalog Screen (see Figure 13.10).

FIGURE 13.10
The Getting Started – Catalog Screen

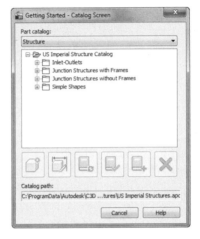

At the top of this window is a drop-down menu for selecting the pipe catalog. The choices, in this case Pipe and Structure, are based on what has been set for the drawing (either Metric or Imperial).

Below the Part Catalog input box is a listing of chapters. (In terms of Part Builder vocabulary, a pipe chapter is roughly equivalent to the catalog interface term shape.) The US Imperial Pipe Catalog has four default chapters: Circular Pipes, Egg-Shaped Pipes, Elliptical Pipes, and Rectangular Pipes. You can create new chapters for different-shaped pipes, such as Arch Pipe.

The US Imperial Structure Catalog also has four default chapters: Inlets-Outlets, Junction Structures With Frames, Junction Structures Without Frames, and Simple Shapes. You can create new chapters for custom structures. (In terms of Part Builder vocabulary, a structure chapter is roughly equivalent to the catalog interface term type.)

You can expand each chapter folder to reveal one or more part families. The US Imperial Circular Pipe Chapter has six default families (Concrete Pipe, Corrugated HDPE Pipe, Corrugated Metal Pipe, Ductile Iron Pipe, HDPE Pipe, and PVC Pipe). Pipes that reside in the same family typically have the same parametric behavior, with only differences in size.

The US Imperial Structure Catalog has four default chapters (Inlet-Outlets, Concentric Cylindrical Structure NF, Cylindrical Junction Structure NF, and Rectangular Junction Structure NF). Like pipes, structures that reside in the same family typically have the same parametric behavior, with only differences in size.

As Table 13.1 shows, a series of buttons on the Getting Started – Catalog Screen lets you perform various edits to chapters, families, and the catalog as a whole.

PART BUILDER ORIENTATION | 515

TABLE 13.1: The Part Builder Catalog Tools

ICON	FUNCTION
	The New Parametric Part button creates a new part family.
	The Modify Part Sizes button allows you to edit the parameters for a specific part family.
	The Catalog Regen button refreshes all the supporting files in the catalog when you've finished making edits to the catalog.
	The Catalog Test button validates the parts in the catalog when you've finished making edits to the catalog.
	The New Chapter button creates a new chapter.
	The Delete button deletes a part family. Use this button with caution, and remember that if you accidentally delete a part family, you can restore your backup catalog as mentioned in the beginning of this section.

EXPLORING PART FAMILIES

The best way to get oriented to the Part Builder interface is to explore one of the standard part families. In this case, click the Part Catalog drop-down list and select Pipe. Then drill down through the US Imperial Pipe Catalog ➤ Circular Pipe Chapter ➤ Concrete Pipe family by highlighting Concrete Pipe and clicking the Modify Part Sizes button. A Part Builder task pane appears with `AeccCircularConcretePipe_Imperial.dwg` on the screen, as shown in Figure 13.11.

You can close Part Builder by clicking on the X on the Part Builder palette. Click No when asked to save the changes to the Concrete Pipe. And click No when asked to save the drawing.

FIGURE 13.11
The Parametric Building environment

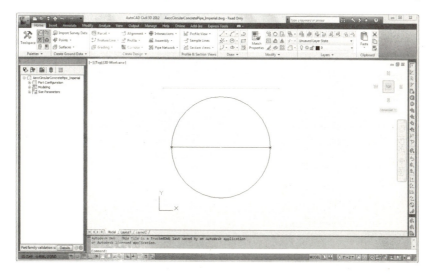

The Part Builder task pane, or Content Builder (Figure 13.12), is well documented in the Civil 3D User's Guide. Please refer to the User's Guide for detailed information about each entry in Content Builder.

FIGURE 13.12
Content Builder

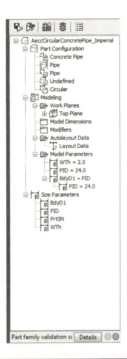

> **IS THAT ALL THERE IS TO PART BUILDER?**
>
> There is much more to Part Builder than this book can cover. As a matter of fact, a separate book could be written just on that subject. So if you are hungering for more information on Part Builder, we'd like to point you the website of Cyndy Davenport. Cyndy has been a repeat speaker at Autodesk University on Part Builder and Civil 3D. You can find her blog at http://c3dcougar.typepad.com/.

Adding a Part Size Using Part Builder

The hypothetical municipality requires a 12″ (30.48 cm) sanitary sewer cleanout. After studying the catalog, you decide that Concentric Cylindrical Structure With No Frames is the appropriate shape for your model, but the smallest inner diameter size in the catalog is 48″ (121.92 cm). The following tutorial gives you some practice in adding a structure size to the catalog—in this case, adding a 12″ (30.48 cm) structure to the US Imperial Structures catalog:

1. You can make changes to the US Imperial Structures catalog from any drawing that is mapped to that catalog, which is probably any imperial drawing you have open. For this exercise, start a new drawing from _AutoCAD Civil 3D (Imperial) NCS.dwt.

2. On the Home tab, select the drop-down list on the Create Design panel and select Part Builder.

3. Choose Structure from the drop-down list in the Part Catalog selection box.
4. Expand the Junction Structure Without Frames chapter.
5. Highlight the Concentric Cylindrical Structure NF (no frames) part family.
6. Click the Modify Part Sizes button.
7. The Part Builder interface opens `AeccStructConcentricCylinderNF_Imperial.dwg` along with the Content Builder task pane.
8. Expand the Size Parameters tree.
9. Right-click the SID (Structure Inner Diameter) parameter, and choose Edit. The Edit Part Sizes dialog appears.
10. Locate the SID column (see Figure 13.13). Double-click inside the box, and note that a drop-down menu shows the available inner diameter sizes: 48, 60, 72, and 96" (121.92, 152.40, 182.88, and 243.84 cm).

11. Locate the Edit button. Make sure you're still active in the SID column cell, and then click Edit. The Edit Values dialog appears. Click Add, and type **12** (**30.48** cm), as shown in Figure 13.14. Click OK to close the Edit Values dialog, and click OK again to close the Edit Part Sizes dialog.
12. Click the small X in the upper-right corner of the Content Builder task pane to exit Part Builder. When prompted with Save Changes To Concentric Cylindrical Structure NF, click Yes. (You could also click Save in Content Builder to save the part and remain active in the Part Builder interface.)
13. You're back in your original drawing. If you created a new parts list in any drawing that references the US Imperial Structures catalog, the 12" (30.48 cm) structure would now be available for selection.

FIGURE 13.13
Choosing a part size

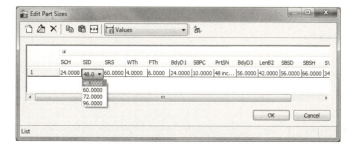

FIGURE 13.14
Add the 12" value to the Edit Values dialog

Sharing a Custom Part

You may find that you need to go beyond adding pipe and structure sizes to your catalog and build custom part families or even whole custom chapters. Perhaps instead of building them yourself, you're able to acquire them from an outside source.

The following section can be used as a reference for adding a custom part to your catalog from an outside source, as well as sharing custom parts that you've created. The key to sharing a part is to locate the three files mentioned earlier.

Adding a custom part size to your catalog requires these steps:

1. Locate the `partname.dwg`, `partname.xml`, and (optionally) `partname.bmp` files of the part you'd like to obtain.

2. Make a copy of the `partname.dwg`, `partname.xml`, and (optionally) `partname.bmp` files.

3. Insert the `partname.dwg`, `partname.xml`, and (optionally) `partname.bmp` files in the correct folder of your catalog.

4. Run the **partcatalogregen** command in Civil 3D.

Adding an Arch Pipe to Your Part Catalog

This exercise will teach you how to add a premade custom part to your catalog:

1. You can make changes to the US Imperial Pipes catalog from any drawing that is mapped to that catalog, which is probably any imperial drawing you have open. For this exercise, start a new drawing from the `_AutoCAD Civil 3D (Imperial) NCS.dwt` file.

2. Create a new folder called **Arch Pipes** in your backup directory. Out of the box, it is located here:

   `C:\ProgramData\Autodesk\C3D 2012\enu\Pipes Catalog\US Imperial Pipes\`

   This directory should now include five folders: `Arch Pipes`, `Circular Pipes`, `Egg-Shaped Pipes`, `Elliptical Pipes`, and `Rectangular Pipes`.

3. Copy the `Concrete Arch Pipe.dwg` and `Concrete Arch Pipe.xml` files into the `Arch Pipe` folder. (Note that there is no optional bitmap for this custom pipe.)

4. Return to your drawing, and enter **PARTCATALOGREGEN** in the command line. Press P to regenerate the Pipe catalog, and then press ↵ to exit the command.

   If you created a new parts list at this point in any drawing that references the US Imperial Pipes catalog, the arch pipe would be available for selection.

5. To confirm the addition of the new pipe shape to the catalog, locate the catalog HTML file at

   `C:\ProgramData\Autodesk\C3D 2012\enu\Pipes Catalog\US Imperial Pipes\Imperial Pipes.htm`

   Explore the catalog as you did in the "The Part Catalog" section earlier.

Part Rules

At the beginning of this chapter, you made notes about your hypothetical municipality having certain requirements for how structures and pipes behave—things like minimum slope, sump depths, and pipe-invert drops across structure. Depending on the type of network and the complexity of your design, there may be many different constraints on your design. Civil 3D allows you to establish structure and pipe rules that will assist in respecting these constraints during initial layout and edits. Some rules don't change the pipes or structures during layout but provide a "violation only" check that you can view in Prospector.

Rules are separated into two categories—structure rules and pipe rules—and are collected in rule sets. You can then add these rule sets to specific parts in your parts list, which you'll build at the end of this chapter.

Structure Rules

Structure rule sets are located on the Settings tab of Toolspace, under the Structure tree.

For a detailed breakdown of structure rules and how they're applied, including images and illustrations, please see the Civil 3D Users Guide. This section will serve as a reference when you're creating rules for your company standards.

Under the Structure Rule Set tree, right-click Basic and click Edit. Click the Add Rule button on the Rules tab in the Structure Rule Set dialog. The Add Rule dialog appears, which allows you to access all the various structure rules (see Figure 13.15).

FIGURE 13.15
The Add Rule dialog

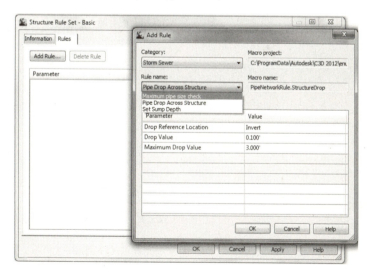

MAXIMUM PIPE SIZE CHECK

The Maximum Pipe Size Check rule (see Figure 13.16) examines all pipes connected to a structure and flags a violation in Prospector if any pipe is larger than your rule. This is a violation-only rule—it won't change your pipe size automatically.

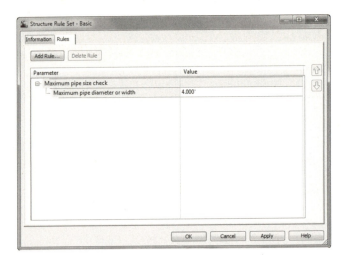

FIGURE 13.16
The Maximum Pipe Size Check rule option

PIPE DROP ACROSS STRUCTURE

The Pipe Drop Across Structure rule (see Figure 13.17) tells any connected pipes how their inverts (or, alternatively, their crowns or centerlines) must relate to one another.

When a new pipe is connected to a structure that has the Pipe Drop Across Structure rule applied, the following checks take place:

- A pipe drawn to be exiting a structure has an invert equal to or lower than the lowest pipe entering the structure.

- A pipe drawn to be entering a structure has an invert equal to or higher than the highest pipe exiting the structure.

- Any minimum specified drop distance is respected between the lowest entering pipe and the highest exiting pipe.

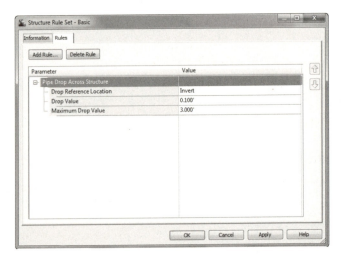

FIGURE 13.17
The Pipe Drop Across Structure rule options

In the hypothetical sanitary sewer example, you're required to maintain a 0.10′ invert drop across all structures. You'll use this rule in your structure rule set in the next exercise.

SET SUMP DEPTH

The Set Sump Depth rule (Figure 13.18) establishes a desired sump depth for structures. It's important to add a sump-depth rule to all of your structure rule sets; otherwise, Civil 3D will assume a sump that is most often undesirable and is difficult to modify once your structures have been drawn.

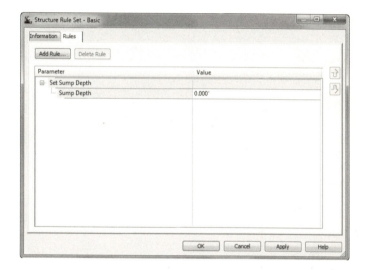

FIGURE 13.18
The Set Sump Depth rule options

In the hypothetical sanitary sewer example, all the structures have a 1.5′ sump depth. You'll use this rule in your structure rule set in the next exercise.

Pipe Rules

Pipe rule sets are located on the Settings tab of Toolspace, under the Pipe tree. For a detailed breakdown of pipe rules and how they're applied, including images and illustrations, please see the Civil 3D Users Guide.

After you right-click on a Pipe Rule Set and click Edit, you can access all the pipe rules by clicking the Add Rule button on the Rules tab of the Pipe Rule Set dialog.

COVER AND SLOPE RULE

The Cover And Slope rule (Figure 13.19) allows you to specify your desired slope range and cover range. You'll create one Cover And Slope rule for each size pipe in the hypothetical sanitary sewer example.

COVER ONLY RULE

The Cover Only rule (Figure 13.20) is designed for use with pressure-type pipe systems where slope can vary or isn't a critical factor.

FIGURE 13.19
The Cover And Slope rule options

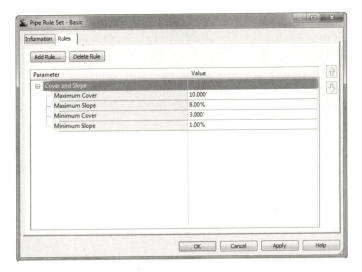

FIGURE 13.20
The Cover Only rule options

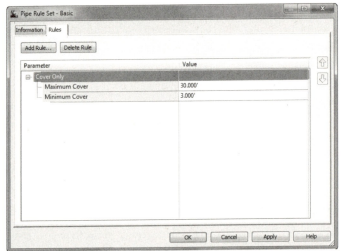

Length Check

Length Check is a violation-only rule; it won't change your pipe length size automatically. The Length Check options (see Figure 13.21) allow you to specify a minimum and maximum pipe length.

Pipe To Pipe Match Rule

The Pipe To Pipe Match rule (Figure 13.22) is also designed for use with pressure-type pipe systems where there are no true structures (only null structures), including situations where pipe is placed to break into an existing pipe. This rule determines how pipe inverts are assigned when two pipes come together, similar to the Pipe Drop Across Structure rule.

FIGURE 13.21
The Length Check rule options

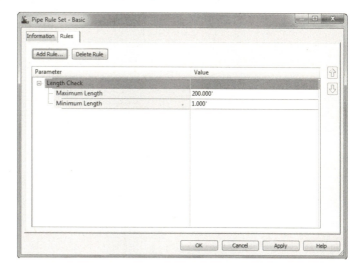

FIGURE 13.22
The Pipe To Pipe Match rule options

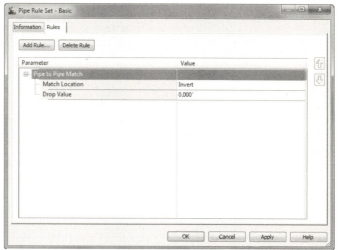

SET PIPE END LOCATION RULE

The Set Pipe End Location rule (Figure 13.23) is new to Civil 3D 2012. It addresses an issue that has been plaguing engineers for some time. There is now the capability to set where the pipe end is located on the structure. You have the option of setting the start and end locations of the pipe to Structure Center (the way it has been), Structure Inner Wall, or Structure Outer Wall.

Figure 13.23
The Set Pipe End Location rule options

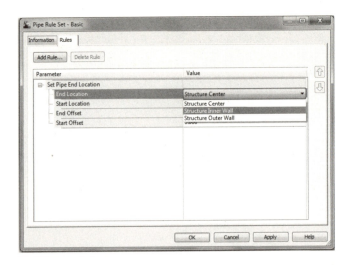

Creating Structure and Pipe Rule Sets

In this exercise, you'll create one structure rule set and three pipe rule sets for a hypothetical sanitary sewer project:

1. Continue working in your drawing from the previous exercise.
2. Locate the Structure Rule Set on the Settings tab of Toolspace under the Structure tree. Right-click the Structure rule set, and choose New.
3. On the Information Tab, enter **Sanitary Structure Rules** in the Name text box.
4. Switch to the Rules tab. Click the Add Rule button.
5. In the Add Rule dialog, choose Pipe Drop Across Structure in the Rule Name drop-down list. Click OK. (You can't change the parameters.)
6. Confirm that the parameters in the Structure Rule Set dialog are the following:

Drop Reference Location	Invert
Drop Value	0.1′
Maximum Drop Value	3′

These parameters establish a rule that will match your hypothetical municipality's standard for the drop across sanitary sewer structures.

7. Click the Add Rule button.
8. In the Add Rule dialog, choose Set Sump Depth in the Rule Name drop-down list. Click OK. (You can't change the parameters.)
9. Change the Sump Depth parameter to 1.5′ in the Structure Rule Set dialog to meet the hypothetical municipality's standard for sump in sanitary sewer structures.
10. Click OK.

11. Locate the Pipe Rule Set on the Settings tab of Toolspace under the Pipe tree. Right-click the Pipe Rule Set, and choose New.
12. On the Information tab, enter **8 Inch Sanitary Pipe Rules** for the name.
13. Switch to the Rules tab. Click Add Rule.
14. In the Add Rule dialog, choose Cover and Slope in the Rule Name drop-down list. Click OK. (You can't change the parameters.)
15. Modify the parameters to match the constraints established by your hypothetical municipality for 8" pipe, as follows:

Maximum Cover	10'
Maximum Slope	10%
Minimum Cover	4'
Minimum Slope	0.40%

16. Click OK.
17. Select the rule set you just created (8 Inch Sanitary Pipe Rules). Right-click, and choose Copy.
18. On the Information tab, enter **10 Inch Sanitary Pipe Rules** in the Name text box.
19. Modify the parameters to match the constraints established by your hypothetical municipality for a 10" pipe, as follows:

Maximum Cover	10'
Maximum Slope	10%
Minimum Cover	4'
Minimum Slope	0.28%

20. Repeat the process to create a rule set for the 12" pipe using the following parameters:

Maximum Cover	10'
Maximum Slope	10%
Minimum Cover	4'
Minimum Slope	0.22%

21. You should now have one structure rule set and three pipe rule sets.
22. Save your drawing—you'll use it in the next exercise.

Parts List

When you know what parts you need, what they need to look like, and how you want them to behave, you can standardize your needs in the form of a parts list. If you think of the part catalog, the styles in your template, and the rule sets as being your well-stocked workshop, the parts list is the toolbox that you fill with only the equipment you need to get the job done. Parts lists are stored in your standard Civil 3D template so they're at your fingertips when new jobs are created.

For example, when you're designing a sanitary sewer system, you may need only a small spectrum of PVC pipe sizes and manhole types that follow a few sets of rules and require only a style or two. You wouldn't want to have to sort through your entire collection of parts, rules, and styles every time you created a sanitary sewer network. You can make a parts list called Sanitary Sewers (or something similar) and stock it with the pipes, structures, styles, and rules you'll need to get the job done.

Similarly, depending on the type of work you do, you'll want at least a Storm Drainage parts list with concrete pipe, catch basins, storm manholes, applicable rule sets and styles, and a Water Network parts list containing PVC pipe and null structures as well as some cover-only rule sets. As you begin your first few pilot projects, you'll begin to see which parts lists are most useful, and you can continue to build them as part of your standard template.

You can create parts lists by clicking the Parts Lists object and from the Pipe Networks tab and Network Tools, choose Create Parts List from the drop-down. We will be discussing this later in this chapter.

Exploring Pipe Networks

Parts in a pipe network have relationships that follow a network paradigm. A pipe network, such as the one in Figure 13.24, can have many branches. In most cases, the pipes and structures in your network will be connected to each other; however, they don't necessarily have to be physically touching to be included in the same pipe network.

Figure 13.24
A typical Civil 3D pipe network

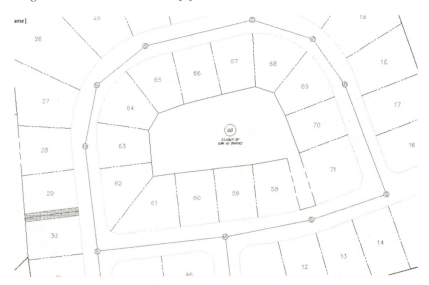

Land Desktop with Civil Design and some other civil engineering–design programs don't design piping systems using a network paradigm; instead, they use a branch-by-branch or "run" paradigm (see Figure 13.25). Although it's possible to separate your branches into their own separate pipe networks in Civil 3D, your design will have the most power and flexibility if you change your thinking from a "run-by-run" to a network paradigm.

Figure 13.25
A pipe network with a single-pipe run

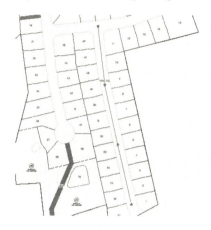

Pipe Network Object Types

Pipes are components of a pipe network that primarily represent pipes but can be used to represent any type of conduit such as culverts, gas lines, or utility cables. They can be straight or curved, and although primarily used to represent gravity systems, they can be adapted and customized to represent pressure and other types of systems such as water networks and force mains. The standard catalog has pipe shapes that are circular, elliptical, egg-shaped, and rectangular and are made of materials that include PVC, RCP, DI, and HDPE. You can use Part Builder (discussed earlier this chapter) to create your own shapes and materials if the default shapes and dimensions can't be adapted for your design.

Structures are the components of a pipe network that represent manholes, catch basins, inlets, joints, and any other type of junction between two pipes. The standard catalog includes inlets, outlets, junction structures with frames (such as manholes with lids or catch basins with grates), and junction structures without frames (such as simple cylinders and rectangles). Again, you can use Part Builder to create your own shapes and materials if needed.

Null structures are created automatically when two pipes are joined together without a structure; they act as a placeholder for a pipe endpoint. They have special properties, such as allowing pipe cleanup at pipe intersections. Most of the time, you'll create a style for them that doesn't plot or is invisible for plotting purposes.

Creating a Sanitary Sewer Network

Earlier, you prepared a parts list for a typical sanitary sewer network. This chapter will lead you through several methods for using that parts list to design, edit, and annotate a pipe network.

There are several ways to create pipe networks. You can do so using the Civil 3D pipe layout tools. Limited tools are also available for creating pipe networks from certain AutoCAD and Civil 3D objects, such as lines, polylines, alignments, and feature lines.

Creating a Pipe Network with Layout Tools

Creating a pipe network with layout tools is much like creating other Civil 3D objects, such as alignments. After naming and establishing the parameters for your pipe network, you're presented with a special toolbar that you can use to lay out pipes and structures in plan, which will also drive a vertical design.

Establishing Pipe Network Parameters

This section will give you an overview of establishing pipe network parameters. Use this section as a reference for the exercises in this chapter. When you're ready to create a pipe network, select Pipe Network ➢ Pipe Network Creation Tools from the Create Design panel on the Home tab. The Create Pipe Network dialog appears (see Figure 13.26), and you can establish your settings.

FIGURE 13.26
The Create Pipe Network dialog

Before you can create a pipe network, you must give your network a name, but more important, you need to assign a parts list for your network. As you saw earlier, the parts list provides a toolkit of pipes, structures, rules, and styles to automate the pipe network design process. It's also important to select a reference surface in this interface. This surface will be used for rim elevations and rule application.

When creating a pipe network, you're prompted for the following options:

Network Name   Choose a name for your network that is meaningful and that will help you identify it in Prospector and other locations.

Network Description   The description of your pipe network is optional. You might make a note of the date, the type of network, and any special characteristics.

Network Parts List   Choose the parts list that contains the parts, rules, and styles you want to use for this design.

Surface Name Choose the surface that will provide a basis for applying cover rules as well as provide an insertion elevation for your structures (in other words, rim elevations). You can change this surface later or for individual structures. For proposed pipe networks, this surface is usually a finished ground surface.

Alignment Name Choose an alignment that will provide station and offset information for your structures in Prospector as well as any labels that call for alignment stations and/or offset information. Because most pipe networks have several branches, it may not be meaningful for every structure in your network to reference the same alignment. Therefore, you may find it better to leave your Alignment option set to None in this dialog and set it for individual structures later using the layout tools or structure list in Prospector.

Using the Network Layout Creation Tools

After establishing your pipe network parameters in the Create Pipe Network dialog (shown previously in Figure 13.26), click OK; the Network Layout Tools toolbar appears (see Figure 13.27). No other command can be executed while the toolbar is active.

FIGURE 13.27
The Network Layout Tools toolbar

Clicking the Pipe Network Properties tool displays the Pipe Network Properties dialog, which contains the settings for the entire network. If you mistyped any of the parameters in the original Create Pipe Network dialog, you can change them here. In addition, you can set the default label styles for the pipes and structures in this pipe network.

The Pipe Network Properties dialog contains the following tabs:

Information On this tab, you can rename your network, provide a description, and choose whether you'd like to see network-specific tooltips (see Figure 13.28).

FIGURE 13.28
The Information tab of the Pipe Network Properties dialog

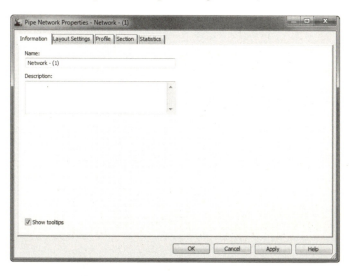

Layout Settings Here you can change the default label styles, parts list, reference surface and alignment, master object layers for plan pipes and structures, as well as name templates for your pipes and structures (see Figure 13.29).

FIGURE 13.29
The Layout Settings tab of the Pipe Network Properties dialog

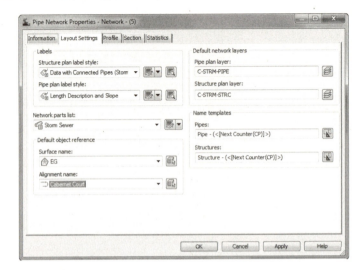

Profile On this tab, you can change the default label styles and master object layers for profile pipes and structures (see Figure 13.30).

FIGURE 13.30
The Profile tab of the Pipe Network Properties dialog

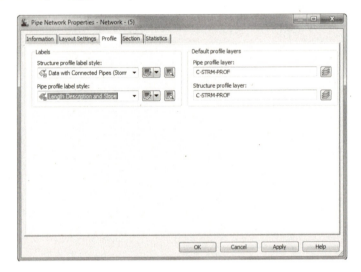

Section Here you can change the master object layers for network parts in a section (see Figure 13.31).

Statistics This tab gives you a snapshot of your pipe network information, such as elevation information, pipe and structure quantities, and references in use (see Figure 13.32).

Figure 13.31
The Section tab of the Pipe Network Properties dialog

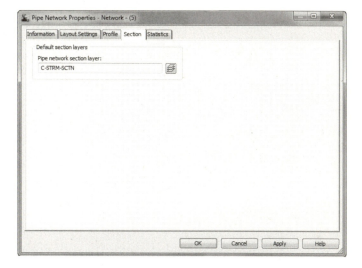

Figure 13.32
The Statistics tab of the Pipe Network Properties dialog

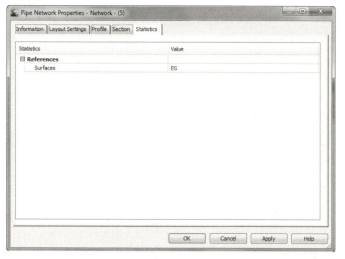

The Select Surface tool on the Network Layout Tools toolbar allows you to switch between reference surfaces while you're placing network parts. For example, if you're about to place a structure that needs to reference the existing ground surface but your network surface was set to a proposed ground surface, you can click this tool to switch to the existing ground surface.

> **Using a Composite Finished Grade Surface for Your Pipe Network**
>
> It's cumbersome to constantly switch between a patchwork of different reference surfaces while designing your pipe network. You may want to consider creating a finished grade-composite surface that includes components of your road design, finished grade, and even existing ground. You can create this finished grade-composite surface by pasting surfaces together, so it's dynamic and changes as your design evolves.

The Select Alignment tool on the Network Layout Tools toolbar lets you switch between reference alignments while you're placing network parts, similar to the Select Surface tool.

The Parts List tool allows you to switch parts lists for the pipe network.

The Structure drop-down list (see Figure 13.33a) lets you choose which structure you'd like to place next, and the Pipes drop-down list (see Figure 13.33b) allows you to choose which pipe you'd like to place next. Your choices come from the network parts list.

FIGURE 13.33
(a) The Structure drop-down list, and (b) the Pipes drop-down list

The options for the Draw Pipes and Structures category let you choose what type of parts you'd like to lay out next. You can choose Pipes and Structures, Pipes Only, or Structures Only.

Placing Parts in a Network

You place parts much as you do other Civil 3D objects or AutoCAD objects such as polylines. You can use your mouse, transparent commands, dynamic input, object snaps, and other drawing methods when laying out your pipe network.

If you choose Pipes and Structures, a structure is placed wherever you click, and the structures are joined by pipes. If you choose Structures Only, a structure is placed wherever you click, but the structures aren't joined. If you choose Pipes Only, you can connect previously placed structures. If you have Pipes Only selected and there is no structure where you click, a null structure is placed to connect your pipes.

While you're actively placing pipes and structures, you may want to connect to a previously placed part. For example, there may be a service or branch that connects into a structure along the main trunk. Begin placing the new branch. When you're ready to tie into a structure, you get a circular connection marker (shown at the top of Figure 13.34) as your cursor comes within connecting distance of that structure. If you click to place your pipe when this marker is visible, a structure-to-pipe connection is formed (shown at the bottom of Figure 13.34).

If you're placing parts and you'd like to connect to a pipe, hover over the pipe you'd like to connect to until you see a connection marker that has two square shapes. Clicking to connect to the pipe breaks the pipe in two pieces and places a structure (or null structure) at the break point.

FIGURE 13.34
The Structure Connection marker (top), and the Pipe Connection marker (bottom)

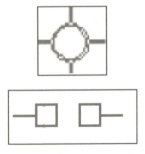

The Toggle Upslope/Downslope tool changes the flow direction of your pipes as they're placed. In Figure 13.35, Structure 9 was placed before Structure 10.

FIGURE 13.35
Using (a) the Downslope toggle and (b) the Upslope toggle to create a pipe network leg

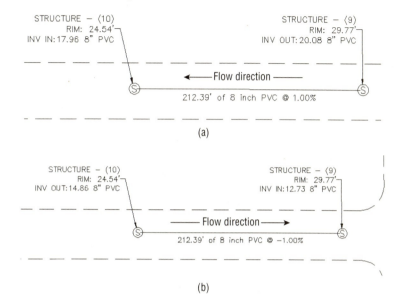

OPTIMIZING THE COVER BY STARTING UPHILL

If you're using the Cover and Slope rule for your pipe network, you'll achieve better cover optimization if you begin your design at an upstream location and work your way down to the connection point.

The Cover and Slope rule prefers to hold minimum slope over optimal cover. In practice, this means that as long as minimum cover is satisfied, the pipe will remain at minimum slope. If you start your design from the upstream location, the pipe is forced to use a higher slope to achieve minimum cover. The following graphic shows a pipe run that was created starting from the upstream location (right to left):

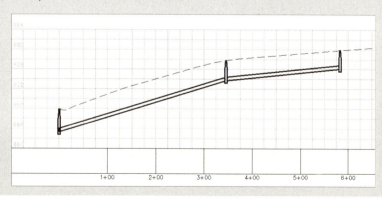

> When you start from the downhill side of your project, the Minimum Slope Rule is applied as long as minimum cover is achieved. The following graphic shows a pipe run that was created starting from the downstream location (left to right):
>
>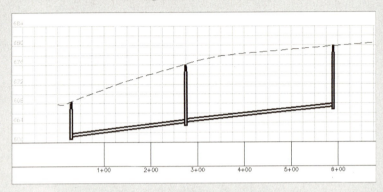
>
> Notice how the slope remains constant even as the pipe cover increases. Maximum cover is a *violation only* rule, which means it never forces a pipe to increase slope to remain within tolerance; it only provides a warning that maximum cover has been violated.

Click Delete Pipe Network Object to delete pipes or structures of your choice. AutoCAD Erase can also delete network objects, but using it requires you to leave the Network Layout Tools toolbar.

Clicking Pipe Network Vistas brings up Panorama (see Figure 13.36), where you can make tabular edits to your pipe network while the Network Layout Tools toolbar is active.

FIGURE 13.36
Pipe Network Vistas via Panorama

The Pipe Network Vistas interface is similar to what you encounter in the Pipe Networks branch of Prospector. Many people think Pipe Network Vistas isn't as user friendly as the Prospector interface because it doesn't remember your preferred column order the way Prospector does. The advantage of using Pipe Network Vistas is that you can make tabular edits without leaving the Network Layout Tools toolbar. You can edit pipe properties, such as Invert and Slope, on the Pipes tab, and you can edit structure properties, such as Rim and Sump, on the Structure tab.

Creating a Sanitary Sewer Network

This exercise will apply the concepts taught in this section and give you hands-on experience using the Network Layout Tools toolbar:

1. Open Pipes-Exercise1.dwg, which you can download from www.sybex.com/go/masteringcivil3d2012.

2. Expand the Surfaces branch in Prospector. This drawing has several surfaces which have a _No Display style applied to simplify the drawing. Expand the Alignments and Centerline Alignments branches, and notice that there are several road alignments.

3. On the Home tab's Create Design panel, select Pipe Network ➢ Pipe Network Creation Tools.

4. In the Create Pipe Network dialog (shown previously in Figure 13.26), give your network the following information:

    - Network Name: **Sanitary Sewer Network**
    - Network Description: **Sanitary Sewer Network created by** *your name* **on** *today's date*
    - Network Parts List: **Sanitary Sewer**
    - Surface Name: **Composite**
    - Alignment Name: **Cabernet Court**
    - Structure Label Style: **Data with Connected Pipes (Sanitary)**
    - Pipe Label Style: **Length Description and Slope**

5. Click OK. The Network Layout Tools toolbar appears.

6. Choose Concentric Structure 48 Dia 18 Frame 24 Cone from the drop-down list in the Structure menu and 8 Inch PVC from the Pipe list.

7. Click the Draw Pipes And Structures tool. Place two structures along Cabernet Court (the lower vertical road) somewhere between the road centerline and the right-of-way by selecting structure locations on your screen. (Doing so also places one pipe between the structures.) Note that the command line gives you additional options for placement. You can refer to the sections on pipe network creation and editing for additional methods of placement.

8. Without exiting the command, go back to the Network Layout Tools toolbar and change the Pipe drop-down from 8 Inch PVC to 10 Inch PVC and then place another structure. Notice that the diameter of the pipe between your second and third structures is 10".

9. Press ↵ to exit the command.

10. Next, you'll add a connecting length of pipe. Go back to the Network Layout Tools toolbar, and select 8 Inch PVC from the Pipe drop-down menu. Click the Draw Pipes And Structures tool button again. Add a structure off to one side of the road, and then connect to one of your structures within the road right-of-way. You'll know you're about to connect to a structure when you see the connection marker (a round, golden-colored glyph shaped like the one in Figure 13.37) appear next to your previously inserted structure.

Figure 13.37
The connection marker appears when your cursor is near the existing structure.

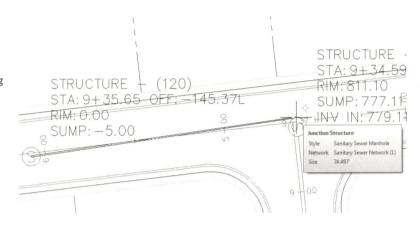

11. Press ↵ to exit the command. Observe your pipe network, including the labeling that automatically appeared as you drew the network.

12. Expand the Pipe Networks branch in Prospector in Toolspace, and locate your sanitary sewer network. Click the Pipes branch; the list of pipes appears in the preview pane. Click the Structures branch; the list of structures appears in the preview pane.

13. Experiment with tabular and graphical edits, drawing parts in profile view, and other tasks described throughout this chapter.

Creating a Storm Drainage Pipe Network from a Feature Line

If you already have an object in your drawing that represents a pipe network (such as a polyline, an alignment, or a feature line), you may be able to take advantage of the Create Pipe Network From Object command in the Pipe Network drop-down menu.

This option can be used for applications such as converting surveyed pipe runs into pipe networks and bringing forward legacy drawings that used AutoCAD linework to represent pipes. Because of some limitations described later in this section, it isn't a good idea to use this in lieu of Create Pipe Network By Layout for new designs.

It's often tempting in Civil 3D to rely on your former drawing habits and try to "convert" your AutoCAD objects into Civil 3D objects. You'll find that the effort you spend learning Create Pipe Network By Layout pays off quickly with a better-quality model and easier revisions.

The Create Pipe Network From Object option creates a pipe for every linear segment of your object and places a structure at every vertex of your object. For example, the polyline with three line segments and three arcs, shown in Figure 13.38, is converted into a pipe network containing three straight pipes, three curved pipes, and seven structures—one at the start, one at the end, and one at each vertex.

This option is most useful for creating pipe networks from long, single runs. It can't build branching networks or append objects to a pipe network. For example, if you create a pipe network from one feature line and then, a few days later, receive a second feature line to add to that pipe network, you'll have to use the pipe-network editing tools to trace your second feature line; no tool lets you add AutoCAD objects to an already-created pipe network.

Figure 13.38
(a) A polyline showing vertices, and (b) a pipe network created from the polyline

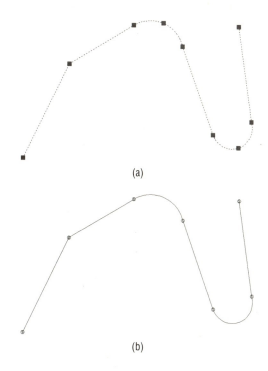

This exercise will give you hands-on experience building a pipe network from a feature line with elevations:

1. Open the `Pipes-Exercise2.dwg` file. (It's important to start with this drawing rather than use the drawing from an earlier exercise.)

2. Expand the Surfaces branch in Prospector. This drawing has several surfaces which have a _No Display style applied to simplify the drawing.

3. Expand the Alignments and Centerline Alignments branches, and notice that there are several road alignments. In the drawing, a cyan feature line runs through Syrah Way (the horizontal road) and then goes onto Frontenac Drive. This feature line represents utility information for a storm-drainage line. The elevations of this feature line correspond with centerline elevations that you'll apply to your pipe network.

4. Choose Create Pipe Network From Object from the Pipe Network drop-down.

5. At the `Select Object or [Xref]:` prompt, select the cyan feature line. You're given a preview (see Figure 13.39) of the pipe-flow direction that is based on the direction in which the feature line was originally drawn.

6. At the `Flow Direction [OK/Reverse] <Ok>:` prompt, press ↵ to choose OK. The Create Pipe Network From Object dialog appears.

FIGURE 13.39
Flow-direction preview

7. In the dialog, give your pipe network the following information:

    ◆ Network Name: **Storm Network**

    ◆ Network Description: **Storm Network created by** *your name* **on** *today's date* **from Feature Line**

    ◆ Network Parts List: **Storm Sewer**

    ◆ Pipe To Create: **12 Inch Concrete Pipe**

    ◆ Structure To Create: **15 x 15 Rect Structure 24 dia Frame**

    ◆ Surface Name: **Composite**

    ◆ Alignment Name: **<none>**

    ◆ Erase Existing Entity check box: Selected

    ◆ Use Vertex Elevations check box: Selected

    The two check boxes at the bottom of the dialog allow you to erase the existing entity and to apply the object's vertex elevations to the new pipe network.

    If you select Use Vertex Elevations, the pipe rules for your chosen parts list will be ignored. The elevations from each vertex will be applied as a center elevation of each pipe endpoint and not the invert as you may expect. For example, if you had a feature line that you created from survey shots of existing pipe inverts and used this method, your newly created pipe network would have inverts that were off by an elevation equal to the inner diameter of your newly created pipe.

8. Click OK. A pipe network is created.

If you do find that you've created a second network using this tool, you can actually merge them together. Simply select the newly created network, and from the Pipe Networks tab's Modify panel, select Merge Networks. This will let you combine all your object-created networks into a more manageable single network.

Changing Flow Direction

Choosing a pipe network object and right-clicking opens the Pipe Networks contextual tab. To change flow direction, select Change Flow Direction from the Modify panel. Change Flow Direction allows you to reverse the pipe's understanding of which direction it flows, which comes into play when you're using the Apply Rules command and when you're annotating flow direction with a pipe label-slope arrow.

Changing the flow direction of a pipe doesn't make any changes to the pipe's invert. By default, a pipe's flow direction depends on how the pipe was drawn and how the Toggle Upslope/Downslope tool was set when the pipe was drawn:

- If the toggle was set to Downslope, the pipe flow direction is set to Start To End, which means the first endpoint you placed is considered the start of flow and the second endpoint is established as the end of flow.

- If the toggle was set to Upslope when the pipe was drawn, the pipe flow direction is set to End To Start, which means the first endpoint placed is considered the end for flow purposes and the second endpoint the start.

After pipes are drawn, you can set two additional flow options—Bi-directional and By Slope—in Pipe Properties:

Start To End   A pipe label-flow arrow shows the pipe direction from the first pipe endpoint drawn to the second endpoint drawn, regardless of invert or slope.

End To Start   A pipe label-flow arrow shows the pipe direction from the second pipe endpoint drawn to the first pipe endpoint drawn, regardless of invert or slope.

Bi-directional   Typically, this is a pipe with zero slope that is used to connect two bodies that can drain into each other, such as two stormwater basins, septic tanks, or overflow vessels. The direction arrow is irrelevant in this case.

By Slope   A pipe label-flow arrow shows the pipe direction as a function of pipe slope. For example, if End A has a higher invert than End B, the pipe flows from A to B. If B is edited to have a higher invert than A, the flow direction flips to be from B to A.

Editing a Pipe Network

You can edit pipe networks in several ways:

- Using drawing layout edits such as grip, move, and rotate
- Grip-editing the pipe size
- Using vertical-movement edits using grips in profile (see the "Vertical Movement Edits Using Grips in Profile" section later in this chapter)
- Using tabular edits in the Pipe Networks branch in Prospector
- Right-clicking a network part to access tools such as Swap Part or Pipe/Structure Properties
- Returning to the Network Layout Tools toolbar by right-clicking the object and choosing Edit Network
- Selecting a network part to access the Pipe Networks contextual tab on the Ribbon

With the exception of the last option, each of these methods is explored in the following sections.

Editing Your Network in Plan View

When selected, a structure has two types of grips, shown in Figure 13.40. The first is a square grip located at the structure insertion point. You can use this grip to grab the structure and stretch/move it to a new location using the insertion point as a base point. Stretching a structure results in the movement of the structure as well as any connected pipes. You can also scroll through Stretch, Move, and Rotate by using your spacebar once you've grabbed the structure by this grip.

FIGURE 13.40
Two types of structure grips

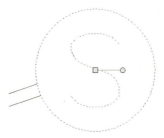

The second structure grip is a rotational grip that you can use to spin the structure about its insertion point. This is most useful for aligning eccentric structures, such as rectangular junction structures.

Also note that common AutoCAD Modify commands work with structures. You can execute the following commands normally (such as from a toolbar or keyboard macro): Move, Copy, Rotate, Align, and Mirror (Figure 13.41a). (Scale doesn't have an effect on structures.) Keep in mind that the Modify commands are applied to the structure model itself; depending on how you have your style established, it may not be clear that you've made a change. For example, if you execute the Mirror command, you can select the structures to see the results or use a 3D visual style, such as in Object Viewer, to see the modeled parts, as shown in Figure 13.41b.

FIGURE 13.41
Mirrored structures seen (a) in plan view by their style and (b) in 3D wireframe visual style using Object Viewer

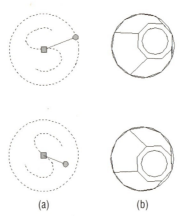

(a) (b)

You can use the AutoCAD Erase command to erase network parts. Note that erasing a network part in plan completely removes that part from the network. Once erased, the part disappears from plan, profile view, Prospector, and so on.

When selected, a pipe end has two types of grips (see Figure 13.42). The first is a square Endpoint-Location grip. Using this grip, you can change the location of the pipe end without constraint. You can move it in any direction; make it longer or shorter; and take advantage of Stretch, Move, Rotate, and Scale by using your spacebar.

FIGURE 13.42
Two types of
Pipe-End grips

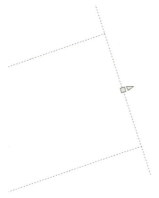

The second grip is a Pipe-Length grip. This grip lets you extend a pipe along its current bearing.

A pipe midpoint also has two types of grips (see Figure 13.43). The first is a square Location grip that lets you move the pipe using its midpoint as a base point. As before, you can take advantage of Stretch, Move, Rotate, and Scale by using your spacebar.

FIGURE 13.43
Two types of pipe
Midpoint grips

The second grip is a triangular-shaped Pipe-Diameter grip. Stretching this grip gives you a tooltip showing allowable diameters for that pipe, which are based on your parts list. Use this grip to make quick, visual changes to the pipe diameter.

Also note that common AutoCAD Modify commands work with pipes. You can execute the following commands normally (such as from a toolbar or keyboard macro): Move, Copy, Rotate, Align, Scale, and Mirror. Remember that executing one of these commands often results in the modified pipe becoming disconnected from its structures. After the completion of a Modify command, be sure to right-click your pipe and choose Connect To Part to remedy any disconnects.

You can use the AutoCAD Erase command to erase network parts. Note that erasing a network part in plan completely removes that part from the network. Once erased, the part disappears from plan view, profile view, Prospector, and so on.

Dynamic Input and Pipe Network Editing

Dynamic Input (DYN) has been in AutoCAD-based products since the 2006 release, but many Civil 3D users aren't familiar with it. For some of the more command line–intensive Civil 3D tasks, DYN isn't always useful, but for pipe network edits, it provides a visual way to interactively edit your pipes and structures.

To turn DYN on or off at any time, click DYN at the bottom of your Civil 3D window.

Structure Rotation   While DYN is active, you get a tooltip that tracks rotation angle.

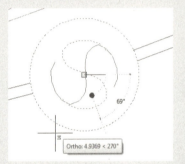

Press your down arrow key to get a pop-up menu that allows you to specify a base point followed by a rotation angle, as well as options for Copy and Undo. DYN combined with the Rotate command is beneficial when you're rotating eccentric structures for proper alignment.

Pipe Diameter   When you're using the Pipe-Diameter grip edit, DYN gives you a tooltip to assist you in choosing your desired diameter. Note that the tooltip depends on your drawing units.

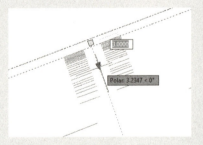

Pipe Length   This is probably the most common reason to use DYN for pipe edits. Choosing the Pipe Length grip when DYN is active shows tooltips for the pipe's current length and preview length, as well as fields for entering the desired pipe total length and pipe delta length. Use your Tab key to toggle between the Total Length and Delta Length fields. One of the benefits of using DYN in this interface is that even though you can't visually grip-edit a pipe to be shorter than its original length, you can enter a total length that is shorter than the original length. Note that the length shown and edited in the DYN interface is the 3D center-to-center length.

Pipe Endpoint Edits Similar to pipe length, pipe endpoint location edits can benefit from using DYN. The active fields give you an opportunity to input x- and y-coordinates.

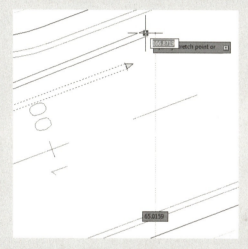

Pipe Vertical Grip Edits in Profile Using DYN in profile view lets you set exact invert, centerline, or top elevations without having to enter the Pipe Properties dialog or Prospector. Choose the appropriate grip, and note that the active DYN field is Tracking Profile Elevation. Enter your desired elevation, and your pipe will move as you specify.

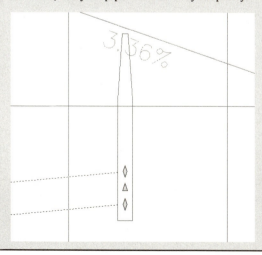

Making Tabular Edits to Your Pipe Network

Another method for editing pipe networks is in a tabular form using the Pipe Networks branch in Prospector (see Figure 13.44).

FIGURE 13.44
The Pipe Networks branch in Prospector

To edit pipes in Prospector, highlight the Pipes entry under the appropriate pipe network. For example, if you want to edit your sanitary sewer pipes, expand the Sanitary Sewers branch and select the Pipes entry. You should get a preview pane that lists the names of your pipes and some additional information in a tabular form. The same procedure can be used to list the structures in the network.

White columns can be edited in this interface. Gray columns are considered calculated values and therefore can't be edited.

You can adjust many things in this interface, but you'll find it cumbersome for some tasks. The interface is best used for the following:

Batch Changes to Styles, Render Materials, Reference Surfaces, Reference Alignments, Rule Sets, and So On   Use your Shift key to select the desired rows, and then right-click the column header of the property you'd like to change. Choose Edit, and then select the new value from the drop-down menu. If you find yourself doing this on every project for most network parts, confirm that you have the correct values set in your parts list and in the Pipe Network Properties dialog.

Batch Changes to Pipe Description   Use your Shift key to select the desired rows, and then right-click the Description column header. Choose Edit, and then type in your new description. If you find yourself doing this on every project for most network parts, check your parts list. If a certain part will always have the same description, you can add it to your parts list and prevent the extra step of changing it here.

Changing Pipe or Structure Names You can change the name of a network part by typing in the Name field. If you find yourself doing this on every project for every part, check that you're taking advantage of the Name templates in your Pipe Network Properties dialog (which can be further enforced in your Pipe Network command settings).

You can Shift-select and copy the table to your Clipboard and insert it into Microsoft Excel for sorting and further study. (This is a static capture of information; your Excel sheet won't update along with changes to the pipe network.)

This interface can be useful for changing pipe inverts, crowns, and centerline information. It's not always useful for changing the part rotation, insertion point, start point, or endpoint. It isn't as useful as many people expect because the pipe inverts don't react to each other. If Pipe A and Pipe B are connected to the same structure, and Pipe A flows into Pipe B, changing the end invert of Pipe A does *not* affect the start invert of Pipe B automatically. If you're used to creating pipe design spreadsheets in Excel using formulas that automatically drop connected pipes to ensure flow, this behavior can be frustrating.

Context Menu Edits

You can perform many edits at the individual part level by using your right-click menu.

If you realize you placed the wrong part at a certain location—for example, if you placed a catch basin where you need a drainage manhole—use the Swap Part option on the context menu (see Figure 13.45). You're given a list of all the parts from all the parts lists in your drawing.

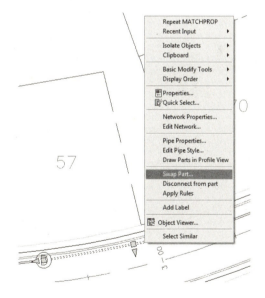

Figure 13.45
Right-clicking a network part brings up a context menu with many options, including Swap Part.

The same properties listed in Prospector can be accessed on an individual part level by right-clicking and choosing Pipe Properties or Structure Properties. A dialog like the Structure Properties dialog in Figure 13.46 opens, with several tabs that you can use to edit that particular part.

FIGURE 13.46
The Part Properties tab in the Structure Properties dialog gives you the opportunity to perform many edits and adjustments.

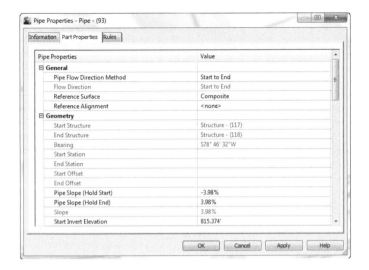

Editing with the Network Layout Tools Toolbar

You can also edit your pipe network by retrieving the Network Layout Tools toolbar. This is accomplished by selecting a pipe network object, right-clicking, and choosing Edit Network, or by changing to the Modify tab and clicking Pipe Network on the Design panel.

Once the toolbar is up, you can continue working exactly the way you did when you originally laid out your pipe network.

This exercise will give you hands-on experience in making a variety of edits to a sanitary and storm-drainage pipe network:

1. Open the EditingPipesPlan.dwg file. This drawing includes a sanitary sewer network and a storm drainage network as well as some surfaces and alignments.

2. Select the structure STM STR 2 in the drawing. Right-click and choose Swap Part. Select the 2 × 4 structure from the Rectangular Junction Structure NF. Click OK.

3. Select the newly placed catch basin so that you see the two Structure grips. Use the Rotational grip and your nearest osnap to align the catch basin as shown in Figure 13.47.

FIGURE 13.47
Rotate and move the catch basin into place along the curb.

4. Use the AutoCAD Erase command to erase Structure-(5).

5. Click DYN to turn it on, and select the labeled pipe. Use the triangular Endpoint grip and the DYN tooltip to lengthen it to a total length of 200′ (60.96 m), as shown in Figure 13.48. (Note that this is the 3D Center To Center Pipe Length.)

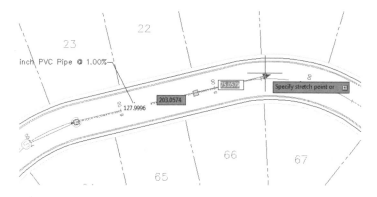

FIGURE 13.48
Lengthen the pipe to 200′ (60.96 m).

TOGGLING NUMBERS

When you grip the triangular Endpoint grip, DYN shows the amount that the pipe is being lengthened. We want to lengthen the entire pipe run. Pressing the Tab key toggles between the additional pipe lengthened versus the entire pipe length.

---

6. Select any pipe in the network. Right-click and choose Edit Network.

7. Select Draw Structures Only from the drop-down menu in the Draw Pipes And Structures selection box. Place a concentric structure at the end of the lengthened pipe discussed in step 5.

8. Select Structure-(1) in the drawing. Right-click and choose Structure Properties. Switch to the Part Properties tab. Scroll down to the Sump Depth field, and change the value to 0′. Click OK to exit the Structure Properties dialog.

9. Expand the Pipe Networks ➤ Networks ➤ Sanitary Sewer Network branches in Prospector in Toolspace, and select the Structures entry. Use the tabular interface in the preview pane area of Prospector to change the names of Structures (10) through (13) to MH1 through MH4.

I Thought I Took Care of That Sump?

In step 8, you were instructed to make the sump depth 0. So why isn't the bottom of the structure even with the pipe then? The answer lies in Part Builder (discussed earlier). Structures are created using the outermost edges, and in the case of structures this includes a 6" (15.24 cm) sump.

You can go into Part Builder to fix this by adding a zero value to the Floor Thickness list:

1. Start Part Builder.
2. Select the Concentric Cylinder part and click Edit.
3. Right-click the FTh value (FTh = Floor Thickness).
4. Highlight the FTh value and select Edit.
5. In the Edit Values dialog, click the Add button.
6. Type **0** for the new value and click OK.
7. Exit Part Builder, saving the part.
8. Back in Prospector, choose Settings ➢ Pipe Network ➢ Parts Lists ➢ Storm Sewer.
9. Expand the part you changed earlier, right-click, and choose Edit.
10. In the Floor Thickness, notice that there is now a Floor Thickness of 0.

Creating an Alignment from Network Parts

On some occasions, certain legs of a pipe network require their own stationing. Perhaps most of your pipes are shown on a road profile, but the legs that run offsite or through open space require their own profiles. Whatever the reason, it's often necessary to create an alignment from network parts. Follow these steps:

1. Open the `AlignmentFromNetworkParts.dwg` file.
2. Select the CB1 structure, which is the first structure and will be station 0 + 00 on the alignment.
3. Choose Alignment From Network on the Launch Pad panel.
4. The command line reads `Select next Network Part or [Undo]`. Select the DMH2 structure, which is the last structure on the alignment.
5. Press ↵, and a dialog appears that is almost identical to the one you see when you create an alignment from the Alignments menu. Name and stylize your alignment as appropriate. Notice the Create Profile And Profile View check box on the last line of the dialog. Leave the box selected, and click OK.
6. The Create Profile From Surface dialog appears (see Figure 13.49). This dialog is identical to the one that appears when you create a profile from a surface. Choose the Composite

surface, and click Draw In Profile View. (See Chapter 7, "Profiles and Profile Views," for further information about sampling profiles from surfaces.)

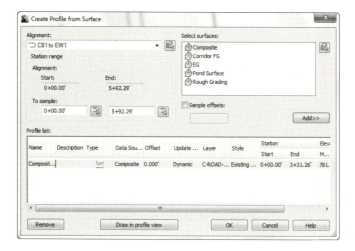

FIGURE 13.49
The Create Profile from Surface dialog

7. You see the Create Profile View Wizard (see Figure 13.50). Click the Next button in the wizard until you reach the Pipe Network Display page. You should see a list of pipes and structures in your drawing. Make sure Yes is selected for each pipe and structure in the Storm Network only. Click Create Profile View, and place the profile view to the right of the site plan.

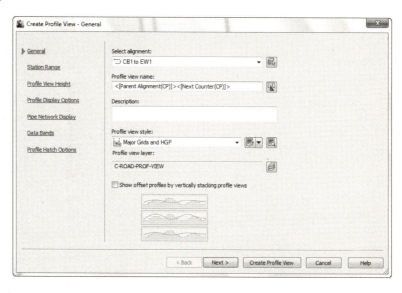

FIGURE 13.50
The Create Profile View Wizard

8. You see three structures and two pipes drawn in a profile view, which is based on the newly created alignment (see Figure 13.51).

FIGURE 13.51
Creating a profile view

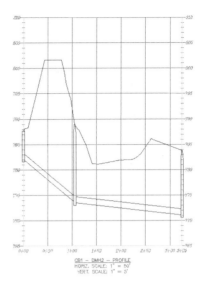

Drawing Parts in Profile View

If you've already created an alignment and profile view—for example, if you're going to show your pipes on the same profile view as your road design—select a network part, right-click, and choose Draw Parts In Profile from the Network Tools panel. When you're using this command, it's important to note that only selected parts are drawn in your chosen profile view.

If you neglect to choose specific parts that are meaningful to show in your profile view, you'll end up with a result like the one shown in Figure 13.52.

FIGURE 13.52
A profile view with inappropriate pipe network parts drawn on it

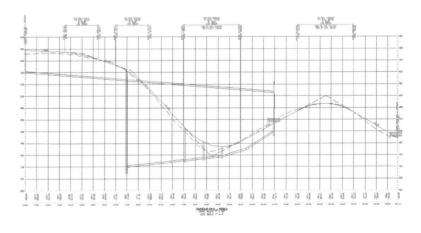

Also keep in mind when adding parts to your profile view that depending on the location of your alignment with respect to your pipes and structures, the labeled length from the model may not be the same as a pipe length you scale or measure from the profile view.

Profiles and profile views are always cut with respect to an alignment. Therefore, pipes are shown in the profile view on the basis of how they appear along that alignment or how they cross that alignment. Unless your alignment *exactly follows the centerline of your network parts*, your pipes will likely show some drafting distortion.

Let's look at Figure 13.53 as an example. This particular jurisdiction requires that all utilities be profiled along the road centerline. There's a road centerline, a Storm network that jogs across the road to connect with another catch basin.

At least two potentially confusing elements show up in your profile view. First, the distance between structures (2D Length – Center To Center) isn't the same in plan and profile (see Figure 13.54) because the storm pipe doesn't run parallel to the alignment. Because the labeling reflects the network model, all labeling is true to the 2D Length – Center To Center or any other length you specify in your label style.

FIGURE 13.53
These pipe lengths will be distorted in profile view.

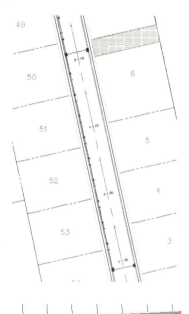

FIGURE 13.54
Pipe labels in plan view (top) and profile view (bottom)

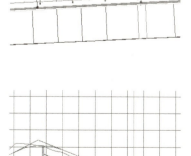

The second potential issue is that the invert of your crossing storm pipe is shown at the point where the storm pipe *crosses the alignment* and not at the point where it crosses the sanitary pipe (see Figure 13.55).

FIGURE 13.55
The invert of a crossing pipe is drawn at the location where it crosses the alignment.

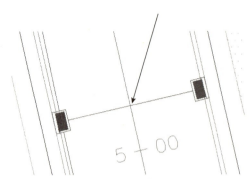

Vertical Movement Edits Using Grips in Profile

Although you can't make changes to certain part properties (such as pipe length) in profile view, pipes and structures both have special grips for changing their vertical properties in profile view.

When selected, a structure has two grips in profile view (see Figure 13.456). The first is a triangular-shaped grip representing a rim insertion point. This grip can be dragged up or down and affects the model structure-insertion point.

FIGURE 13.56
A structure has two grips in profile view.

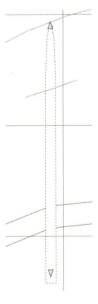

Moving this grip can affect your structure insertion point two ways, depending on how your structure properties were established:

- If your structure has Automatic Surface Adjustment set to True, grip-editing this Rim Insertion Point grip changes the surface adjustment value. If your reference surface changes, then your rim will change along with it, plus or minus that surface adjustment value.

- If your structure has the Automatic Surface Adjustment set to False, grip-editing this Rim grip modifies the insertion point of the rim. No matter what happens to your reference surface, the rim will stay locked in place.

Typically, you'll use the Rim Insertion Point grip only in cases where you don't have a surface for your rims to target to or if you know there is a desired surface adjustment value. It's tempting to make a quick change instead of making the improvements to your surface that are fundamentally necessary to get the desired rim elevation. One quick change often grows in scope. Making the necessary design changes to your target surface will keep your model dynamic and, in the long run, will make editing your rim elevations easier.

The second grip is a triangular grip located at the sump depth. This grip doesn't represent structure invert. In Civil 3D, only pipes truly have invert elevation. The structure uses the connected pipe information to determine how deep it should be. When the sump has been set at a depth of 0, the sump elevation equals the invert of the deepest connected pipe.

This grip can be dragged up or down. It affects the modeled sump depth in one of two ways, depending on how your structure properties are established:

- If your structure is set to control sump by depth, editing with the Sump grip changes the sump depth. The depth is measured from the structure insertion point. For example, if the original sump depth was 0, grip-editing the sump 0.5' (15.24 cm) lower would be the equivalent of creating a new sump rule for a 0.5' (15.24 cm) depth and applying the rule to this structure. This sump will react to hold the established depth if your reference surface changes, your connected pipe inverts change, or something else is modified that would affect the invert of the lowest connected pipe. This triangular grip is most useful in cases where most of your pipe network will follow the sump rule applied in your parts lists, but selected structures need special treatment.

- If your structure is set to control sump by elevation, adjusting the Sump grip changes that elevation. When sump is controlled by elevation, sump is treated as an absolute value that will hold regardless of the structure insertion point. For example, if you grip-edit your structure so its depth is 8.219' (2.51 m), the structure will remain at that depth regardless of what happens to the inverts of your connected pipes. The Control Sump By Elevation parameter is best used for existing structures that have surveyed information of absolute sump elevations that won't change with the addition of new connected pipes.

When selected, a pipe end has three grips in profile view (see Figure 13.57). You can grip-edit the invert, crown, and centerline elevations at the structure connection using these grips, resulting in the pipe slope changing to accommodate the new endpoint elevation.

FIGURE 13.57
Three grips for a pipe end in profile view

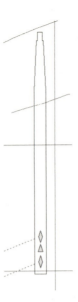

When selected, a pipe in profile view has one grip at its midpoint (see Figure 13.58). You can use this grip to move the pipe vertically while holding the slope of the pipe constant.

FIGURE 13.58
Use the Midpoint grip to move a pipe vertically.

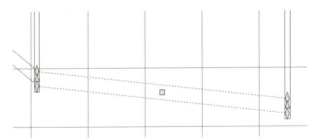

You can access pipe or structure properties by choosing a part, right-clicking, and choosing Pipe Properties.

Removing a Part from Profile View

If you have a part in profile view that you'd like to remove from the view but not delete from the pipe network entirely, you have a few options.

AutoCAD Erase can remove a part from profile view; however, that part is then removed from every profile view in which it appears. If you have only one profile view, or if you're trying to delete the pipe from every profile view, this is a good method to use.

A better way to remove parts from a particular profile view is through the profile view properties. You can access these properties by selecting the profile view, right-clicking, and choosing Profile View Properties.

The Pipe Networks tab of the Profile View Properties dialog (see Figure 13.59) provides a list of all pipes and structures that are shown in that profile view. You can deselect the check boxes next to parts you'd like to omit from this view. Also note that you can add parts to your view by deselecting the Show Only Parts Drawn In Profile View check box and making changes in the Draw column.

FIGURE 13.59
Deselect parts to omit them from a view.

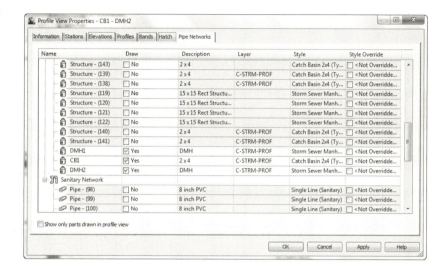

Showing Pipes That Cross the Profile View

If you have pipes that cross the parent alignment of your profile view, you can show them with a crossing style. A pipe must cross the parent alignment to be shown as a crossing in profile, and the vertical location of the pipe is shown where it crosses that alignment (see Figure 13.60).

FIGURE 13.60
A pipe crossing a profile

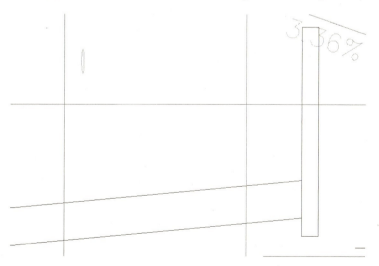

For example, if you created an alignment directly from your network parts and then created a profile view for that alignment, any crossing pipes would be shown at the elevation where they cross the main run because your alignment and pipes coincide (see Figure 13.61).

FIGURE 13.61
The invert of a storm crossing that runs along an alignment is shown in profile at the elevation where it crosses the sanitary line.

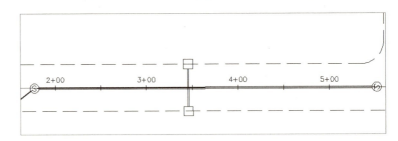

If a leg of a pipe network runs offset from the road centerline, such as that shown in Figure 13.62, but you're showing those pipes in profile along the centerline alignment, any crossing pipes are shown where they *intersect the road centerline alignment.*

FIGURE 13.62
The invert of a storm crossing that is offset from the alignment is shown in profile at the elevation where it crosses the alignment.

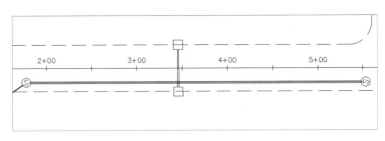

When pipes enter directly into profiled structures, they can be shown as ellipses through the Display tab of the Structure Style dialog (see Figure 13.63). See Chapter 19, "Styles," for more information about creating structure styles.

FIGURE 13.63
Pipes that cross directly into a structure can be shown as part of the structure style.

The first step to display a pipe crossing in profile is to add the pipe that crosses your alignment to your profile view by either selecting the pipe, right-clicking, and selecting Draw Parts In Profile from the Network Tools panel, or by checking the appropriate boxes on the Pipe Network tab of the Profile View Properties dialog. When the pipe is added, it's distorted when it's projected onto your profile view—in other words, it's shown as if you wanted to see the entire length of pipe in profile (see Figure 13.64).

FIGURE 13.64
The pipe crossing is distorted.

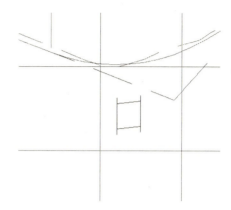

The next step is to override the pipe style *in this profile view only*. Changing the pipe style through pipe properties won't give you the desired result. You must override the style on the Pipe Networks tab of the Profile View Properties dialog (see Figure 13.65).

FIGURE 13.65
Correct the representation in the Profile View Properties dialog.

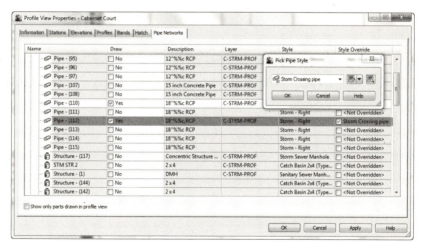

Locate the pipe you just added to your profile view, and scroll to the last column on the right (Style Override). Select the Style Override check box, and choose your pipe crossing style. Click OK. Your pipe should now appear as an ellipse.

If your pipe appears as an ellipse but suddenly seems to have disappeared in the plan and other profiles, chances are good that you didn't use the Style Override but accidentally changed the pipe style. Go back to the Profile View Properties dialog, and make the necessary adjustments; your pipes will appear as you expect.

Exploring the Tools on the Pipe Network Tab

We have already looked at some of the tools available on the Pipe Networks tabs. You can access this tab by clicking on any pipe or structure. In this section, we will look at these tools.

THE LABELS & TABLES PANEL

The Labels & Tables Panel is where you do the most annotation labeling for pipes. It can either be selected from the Panel, or if you click on a pipe network, you can choose from the contextual menu.

Add Labels   You can add labels for the selected pipe network using the suboptions such as Entire Network Plan, Entire Network Profile, Entire Network Section, Single Part Plan, Single Part Profile, Single Part Section, Spanning Pipes Plan, and Spanning Pipes Profile.

Add Tables   You can add a structure or pipe table.

Reset Labels   This will allow you to reset a label that is referenced in the drawing to match the labels that are set within the drawing.

THE GENERAL TOOLS PANEL

While not specifically for pipe networks or even Civil 3D for that matter, these tools are placed here for convenience.

Properties   Toggles the Properties palette on and off.

Object Viewer   Allows you to view the selected object or objects in 3D via a separate viewer.

Isolate Objects   Selected objects will be the only objects visible on the screen. This is useful if you are working in a tight area and do not wish to see extraneous objects, and it is much quicker than clicking to turn off or freeze layers.

These tools are available via the General Tools panel drop-down:

Select Similar   When this tool is invoked, it will select all similar type objects.

Quick Select or QSelect   Opens the Quick Select dialog, which allows custom filtering.

Draw Order Icons   Allow you to move objects either to the front or back of other objects, or above or behind a specific object.

Google Earth Timespan   Opens the Timespan For Google Earth dialog, where you can set specific times for use in Google Earth. This will allow you to view your model in Google Earth with the specified times. It is useful for sun/shade studies.

The Modify Panel

The Modify Panel is where you do editing of an existing pipe network. It can also be selected by clicking on a pipe network and selecting it from the contextual menu.

Network Properties tool   Opens the Pipe Network Properties dialog.

Pipe Properties Tool   Opens the Pipe Properties dialog. This tool only works when a pipe is the selected object.

Edit Pipe Style Tool   Opens the Pipe Style dialog. This tool only works when a pipe is the selected object. For more on pipe styles, see Chapter 19.

Structure Properties Tool   Opens the Structure Properties dialog. This tool only works when a structure is the selected object.

Edit Structure Style Tool   Opens the Structure Style dialog. This tool only works when a structure is the selected object. For more on structure styles, see Chapter 19.

Edit Pipe Network Tool   Opens the Network Layout Tools toolbar.

Connect and Disconnect Part Tools   Allow you to connect pipes to structures that may have been disconnected and to disconnect pipes from structures.

Swap Part Tool   Allows you to replace a structure or pipe type with another one from a Swap Part Size dialog.

Split and Merge Network Tools   Allow you to take an existing network and split it into two networks, or take an existing pipe network and merge it onto another pipe network.

The Modify panel drop-down list contains the following tools:

Rename Parts Tool   Opens the Rename Pipe Network Parts dialog. Here, you can rename pipes and structures, modify pipe numbering, and decide how you want to handle conflicting names or numbers.

Apply Rules Tool   Forces the set rule on the selected objects.

Change Flow Direction Tool   Changes the path of the selected objects. It is very important to have this option set correctly if you are going to use any of the analysis programs.

The Network Tools Panel

The Network Tools panel contains the following tools:

Create Parts List   Allows you to create new parts list via the Network Parts List – New Parts List dialog.

Create Full Parts List   Takes all the parts available in the parts catalog and creates a list called Full Catalog.

Edit Parts List   Opens the Parts List dialog, where you can select the network to add to an existing or new pipe network.

Set Network Catalog   Opens the Pipe Network Settings dialog, where you can select whether you wish to use Imperial or Metric parts, and also sets the location of the Network Catalog.

Part Builder   Opens the Part Builder tool.

Draw Parts In Profile View   Adds the selected pipe network objects into an existing profile.

The Analyze Panel

The Analyze Panel contains the following tools for doing various checks on a pipe network.

Interference Check Properties   Allows you to create an interference check between parts, whether or not they are on the same network. The following tools are available via the submenu:

Create Interference Check Tool   This is the same as the Interference Check Properties tool.

When you select the Interference Check Properties tool, you are prompted to select Interference and then the Interference Check Properties dialog opens. The Information tab displays basic information such as the name of the interference set, style, rendering material, and layer. The Criteria tab contains specific editable items, such as whether to use 3D proximity check, distance, and scale factors. The Statistics tab is a combined listing of the other tabs but also allows you to see the networks used for the interference check.

Interference Properties Tool   This may seem similar to the Interference Check Properties tool, even down to the Interference Check Properties dialog. It contains an Information tab (where you can name the individual interference objects) and a Statistics tab (which contains the X-Y location of the selected interference object).

Edit Interference Style Tool   Opens the Interference Style dialog, where you can change the visual parameters for the interference object. For more on interference styles, see Chapter 19.

Storm Sewers Tool   Gives you options for importing, exporting, and editing within the Storm Sewers program.

Edit In Storm And Sanitary Analysis Tool   Opens the Storm and Sanitary Analysis (SSA) program. For more on SSA, see Chapter 14, "Storm and Sewer Analysis."

The Launch Pad Panel

The Launch Pad Panel contains the following tools:

Alignment From Network   Allows you to create an alignment from Pipe Network parts.

Storm Sewers Tool   Opens the Hydroflow Storm Sewers program.

Hydrographs Tool   Opens the Hydraflow Hydrographs program.

Express Tool   Opens the Hydraflow Express program.

Adding Pipe Network Labels

Once you've designed your network, it's important to annotate the design in a pleasing way. This section focuses on pipe network–specific label components in plan and profile views (see Figure 13.66).

FIGURE 13.66
Typical pipe network labels (top) in plan view and (bottom) in profile view

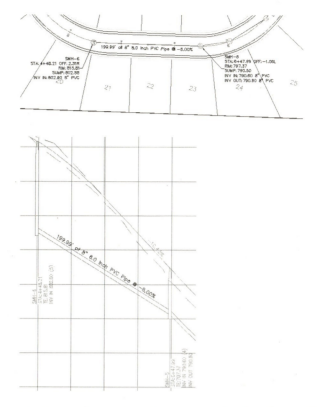

Like all Civil 3D objects, the Pipe and Structure label styles can be found in the Pipe and Structure branches of the Settings tab in Toolspace and are covered in Chapter 19.

Creating a Labeled Pipe Network Profile Including Crossings

This exercise will apply several of the concepts in this chapter to give you hands-on experience producing a pipe-network profile that includes pipes that cross the alignment:

1. Open the `Pipes-Exercise3.dwg` file. (It's important to start with this drawing rather than use the drawing from an earlier exercise.)

2. Explore the drawing. It has two pipe networks: a sanitary network that partially doesn't follow the road centerline and a small storm network that crosses the road and the sanitary network.

3. From the Modify tab, choose Pipe Network to open the Pipe Networks contextual tab. Select Alignment From Network on the Launch Pad panel.

4. At the command-line prompt, first select the sanitary structure 16, and then select structure 18 and press ↵ to create an alignment from the sanitary sewer network.

5. Choose the following options in the Create Alignment From Network Parts dialog:

    ◆ Site: <none>

    ◆ Name: **SMH16 to SMH18 Alignment**

- Description: **Sanitary Alignment for SMH16 to SMH18**
- Alignment Style: _None
- Alignment Label Set: _No Labels
- Create Profile And Profile View check box: Selected

Click OK.

6. Sample the EG and Corridor FG surfaces in the Create Profile From Surface dialog.

7. Click Draw In Profile View to open the Create Profile View dialog. Click the Next button in the Create Profile View Wizard until you reach the Pipe Network Display page. You should see a list of pipes and structures in your drawing. Make sure Yes is selected for each pipe and structure under Sanitary Sewer Network. Click Create Profile View, and place the profile view to the right of the site plan.

8. Select either a pipe or a structure to open the Pipe Network contextual tab, and on the Labels & Tables panel, select Add Labels ➢ Entire Network Profile.

9. The alignment information is missing from your structure labels because the alignment was created after the pipe network. Expand the Sanitary Sewer Network branch in Prospector in Toolspace to set your new alignment as the reference alignment for these structures. Click the Structures entry, and find the Structures list in the preview pane.

10. Select the structures 16–18 in the preview pane using Shift+click. Scroll to the right until you see the Reference Alignment column. Right-click the column header and select Edit. Choose Sanitary Network Alignment from the drop-down list.

Once your reference alignment has been changed, type **REGEN ALL** to see your structure labels.

11. Pan your drawing until you see the Storm layout located on the middle horizontal road (Syrah Way). The sanitary has already been laid out on the Syrah Way profile. We want to add just the storm where it crosses the sanitary.

12. Add the storm crossing by first choosing Pipe Network from the Design panel on the Modify tab. The Pipe Network contextual tab opens. From the Network Tools panel, select Draw Parts In Profile. Be careful to type **S** in the command line so you select just the crossing pipe and not the entire storm network. Press ↵ and then choose the profile view. Override the pipe style in this profile view only by following the procedure in the "Showing Pipes That Cross the Profile View" section earlier in this chapter.

Pipe Labels

Civil 3D makes no distinction between a plan-pipe label and a profile-pipe label. The same label can be used both places.

A pipe label is composed identically to most other labels in Civil 3D. In addition to text, line, and block, structure labels can have a flow arrow and reference text.

You can derive a large number of pipe properties from the model and use them in your label. If it's part of the design, it can probably be labeled. Pipe label styles will be covered in Chapter 19.

> **SPANNING PIPE LABELS**
>
> In addition to single-part labels, pipes shown in either plan or profile view can be labeled using the Spanning Pipes option. This feature allows you to choose more than one pipe; the length that is reported in the label is the cumulative length of all pipes you choose.
>
> Unlike Parcel spanning labels, no special label-style setting is required to use this tool. The Spanning Pipes option is on the Annotate tab. Select Add Labels ➢ Pipe Network ➢ Spanning Pipes Profile to access the command.

Structure Labels

As with pipes, Civil 3D makes no distinction between a plan-structure label and a profile-structure label. The same label can be used both places. You can learn more about structure label styles in Chapter 19.

Special Profile Attachment Points for Structure Labels

Although Civil 3D makes no distinction between plan labels and profile labels, structure labels have two special attachment points in profile view that you need to understand before you can harness their power.

Typical structure labels from the default template have limited flexibility, such as the label shown in Figure 13.67.

FIGURE 13.67
An example of a standard structure label from the default template

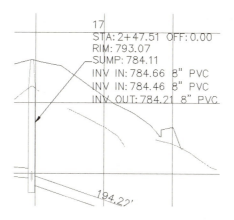

If you highlight a structure label in profile, two cyan-colored grips appear. The lower grip appears at Profile View Elevation Zero. This is the structure dimension. Right now, it's fixed at elevation zero, which makes sense on the basis of the default settings. The location of the structure dimension is based on the profile view, and it can be grip-edited and stretched vertically after placement.

The second cyan-colored grip is the structure-label location. It's set to At Middle of Structure per the default settings. The structure-label location is always tied to the top, middle, or bottom of the structure and can't be grip-edited after placement.

You learned earlier that you can grip-edit the structure dimension to override the feature settings. If you grip-edit the structure dimension, you can drag the line vertically.

If you ever need to reset or customize the placement of this label, you can select the label, right-click, and choose Label Properties.

In the Properties dialog shown in Figure 13.68, you see Dimension Anchor Option from the feature settings and a Dimension Anchor Value, which is equal to the distance you stretched the grip. If you want to shorten the line so it touches the Graph View Top, change Dimension Anchor Value to 0.

FIGURE 13.68
The AutoCAD Properties palette showing Dimension Anchor Option

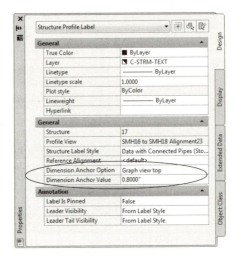

Creating an Interference Check between a Storm and Sanitary Pipe Network

In design, you must make sure pipes and structures have appropriate separation. You can perform some visual checks by rotating your model in 3D and plotting pipes in profile and section views (see Figure 13.69). Civil 3D also provides a tool called Interference Check that makes a 3D sweep of your pipe networks and lets you know if anything is too close for comfort.

The following exercise will lead you through creating a pipe network Interference Check to scan your design for potential pipe network conflicts:

1. Open the `InterferenceStart.dwg` file. The drawing includes a sanitary sewer pipe network and a storm drainage network.

2. Select a part from either network and choose Interference Check from the Analyze panel. You'll see the prompt `Select a part from the same network or different network:`. Select a part from the network not chosen.

3. The Create Interference Check dialog appears (Figure 13.70). Name the Interference Check **Exercise**, and confirm that Sanitary Sewer Network and Storm Network appear in the Network 1 and Network 2 boxes.

4. Click 3D Proximity Check Criteria, and the Criteria dialog appears (see Figure 13.71).

    You're interested in finding all network parts that are within a certain tolerance of one another, so enter **3.000′** (**0.91** m) in the Use Distance box. This setting creates a buffer to help find parts in all directions that might interfere.

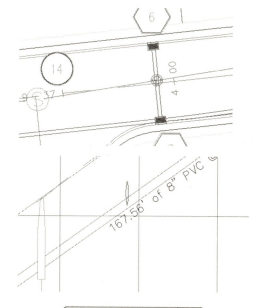

FIGURE 13.69
(Top) Two pipe networks may interfere vertically where crossings occur. (Bottom) Viewing your pipes in profile view can also help identify conflicts.

FIGURE 13.70
The Create Interference Check dialog

FIGURE 13.71
Criteria for the 3D proximity check

If you were interested only in physical, direct collisions between parts in your networks, leave the Apply 3D Proximity Check check box deselected and dismiss this dialog.

> **AN INTERFERENCE CLOUD?**
>
> Think of the Interference Check as if each part is surrounded by a 3D "cloud." When that cloud touches another part, interference is flagged. If you specify distance, such as 2' (0.61 m), your cloud is created 2' (0.61 m) in all directions around each part. When you specify scale, such as 1.5, your cloud makes each part 1.5 times larger than its actual size.

5. Click OK to exit the Criteria dialog, and click OK to run the Interference Check. You see a dialog that alerts you to two interferences. Click OK to dismiss this dialog.

6. Zoom in on the crossing of pipes of CB3 and CB6 (Storm Network) and the crossing sanitary pipe. A small marker has appeared, as shown in Figure 13.72.

7. From the View control, select SW Isometric. Zoom in on the crossing of pipes of CB3 and CB6 and the crossing sanitary pipe. The interference marker appears in 3D, as shown in Figure 13.73.

FIGURE 13.72
The interference marker in plan view

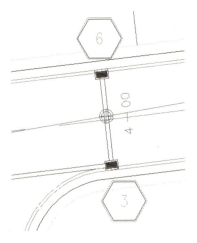

FIGURE 13.73
The interference marker in 3D

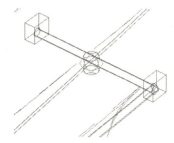

8. Using the same view navigation technique as you used in step 7, change back to a top-view orientation.

9. Drill into the Pipe Networks branch of Prospector, and you see an entry for Interference Checks. Note that each instance of interference is listed in the preview pane for further study. Making edits to your pipe network flags the interference check as "out of date." You can rerun Interference Check by right-clicking Interference Check in Prospector. You can also edit your criteria in this right-click menu.

Creating Pipe Tables

Just like with parcels and labels, labeling pipes can turn into a mess with all the labels set on the plan, such as pipes and structures (see Figure 13.74). In this section, you will explore table creation for pipes and structures.

FIGURE 13.74
Pipes and structures labeled on a plan

In the following exercise, you will create a pipe network table for the sanitary sewer structures:

1. Open the `CreatingPipeTables.dwg` file which you can download from this book's web page.

2. Click on the Sanitary Sewer Manhole 1.

3. From the Labels & Tables panel, select Add Tables ➢ Add Structure. The Structure Table Creation dialog opens (Figure 13.75).

FIGURE 13.75
The Structure Table Creation dialog

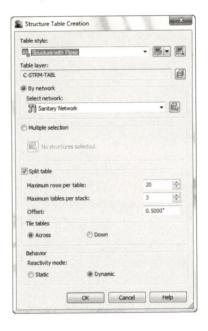

4. Click OK to accept the default settings. Place the table to the right of your plan. The table should look similar to Figure 13.76.

FIGURE 13.76
The finished structure table

STRUCTURE TABLE			
STRUCTURE NAME:	DETAILS:	PIPES IN:	PIPES OUT
12	48" RIM = 825.60 SUMP = 821.8 INV OUT = 821.91		Pipe – (111), 8" PVC INV OUT =821.91
2	48" RIM = 824.52 SUMP = 817.7 INV IN = 817.78 INV OUT = 817.78	Pipe – (101), 8" PVC INV IN =817.78	Pipe – (102), 8" PVC INV OUT =817.76
3	48" RIM = 822.81 SUMP = 815.3 INV IN = 815.37 INV OUT = 816.37	Pipe – (102), 8" PVC INV IN =815.37	Pipe – (103), 8" PVC INV OUT =815.37
1	48" RIM = 822.17 SUMP = 818.4 INV OUT = 818.48		Pipe – (101), 8" PVC INV OUT =818.45
4	48" RIM = 819.88 SUMP = 814.2 INV IN = 814.31 INV OUT = 814.31	Pipe – (103), 8" PVC INV IN =814.31	Pipe – (104), 8" PVC INV OUT =814.31
5	48" RIM = 818.20 SUMP = 809.3 INV IN = 809.44 INV OUT = 809.44	Pipe – (104), 3" PVC INV IN =809.44	Pipe – (105), 8" PVC INV OUT =809.44
	48"		

The table creation for pipes is similar to the structure table creation process:

1. Click on a Sanitary Sewer pipe.

2. From the Labels & Tables panel, select Add Tables ➢ Add Pipe. The Pipe Table Creation dialog opens (Figure 13.77).

FIGURE 13.77
The Pipe Table Creation dialog

3. Click OK to accept the default settings. Place the table to the right of the structure table. The pipe table should look similar to Figure 13.78.

FIGURE 13.78
The finished pipe table

NAME	SIZE	LENGTH	SLOPE	MATERIAL
Pipe – (101)	8"	70.05'	1.00%	PVC
Pipe – (102)	8"	240.56'	1.00%	PVC
Pipe – (103)	8"	106.00'	1.00%	PVC
Pipe – (104)	8"	93.53'	5.21%	PVC
Pipe – (105)	8"	225.24'	6.76%	PVC
Pipe – (106)	8"	100.34'	8.00%	PVC
Pipe – (107)	8"	99.68'	3.67%	PVC
Pipe – (108)	8"	122.61'	0.99%	PVC
Pipe – (109)	8"	213.96'	3.99%	PVC
Pipe – (110)	8"	225.12'	5.01%	PVC
Pipe – (111)	8"	87.48'	4.00%	PVC
Pipe – (112)	8"	167.56'	8.00%	PVC
Pipe – (113)	8"	249.06'	6.97%	PVC
Pipe – (114)	8"	340.81'	2.83%	PVC
Pipe – (115)	8"	247.31'	0.74%	PVC
Pipe – (116)	8"	194.22'	3.34%	PVC
Pipe – (117)	8"	160.24'	7.89%	PVC
Pipe – (118)	8"	250.78'	6.70%	PVC

Exploring the Table Creation Dialog

Since the structure and pipe table creation dialogs are similar, we will cover both of them in this section.

- The Table Style option allows you to select a table look or style. You can select the available styles from the arrow to the right of the table style name. You can also create new, copy, edit, or pick a table style. For more on table styles, see Chapter 19.

- The Table Layer option will set the overall layer for the table.

- With the By Network radio button selected, you can select the network to create a table from the drop-down list, or by using the pick icon, which will allow you to pick the network on the drawing.

- With the Multiple Selection radio button selected and by using the pick icon, you can select structures or pipes (depending on which table type is selected) from the drawing. You can pick pipes or structures regardless of the network.

- The Split table check box will allow you to split the table if it gets too large. You can specify the maximum number of rows per table and maximum number of tables per stack. Additionally, you can set the offset distance between the stacked tables.

- You can choose whether you want the split tables tiled across or down.

- The last option to choose defines the behavior you want for the table: Static or Dynamic. A static table will not update if any changes are made to the pipe network, such as swapping a part. Dynamic will update the table to those changes.

Changes to a Pipe Network and Pipe Tables

Let's see how a predefined table reacts with network changes by following this exercise:

1. Open the `PipeTableStaticDynamic.dwg` file. This drawing contains two identical structure tables except one is static and one is dynamic.

2. Click on the sanitary sewer pipe that is between structure 1 and structure 2.

3. From the Modify panel, select Pipe Properties.

4. In the Pipe Properties dialog, select the Part Properties tab and change Start Invert Elevation from 818.476′ (249.47 m) to 800.000′ (243.84 m). This change is extreme and won't make sense on the actual design, but is meant to show the dynamic relationship between the pipe network and the structure table.

5. Click OK and compare the table that has been set to static versus the table set to dynamic.

The Table Panel Tools

When you click on a table, the Table panel opens and has several tools available. We'll look at each in this section.

THE GENERAL TOOLS PANEL

The tools here are the same as mentioned earlier when we discussed the pipe network tools.

The Modify Panel

The Modify Panel contains the following tools:

Table Properties Tool   Opens the Table Properties dialog. You can set the table style and choose whether to split the table with all the options mentioned earlier. In addition, you can force realignment of stacks and force content updating.

Edit Table Style Tool   Opens the Table Style dialog. For more on editing table styles, see Chapter 19.

Static Mode Tool   Turns a dynamic table into a static table.

Update Content Tool   Forces an update on a table.

Realign Stacks Tool   Readjusts the table columns back to the default setting.

Add Items Tool   Adds additional pipe data to the table added after the table was created.

Remove Items Tool   Removes pipe or structure objects from a table.

Replace Items Tool   Allows you to swap pipe or structures in the table.

The Bottom Line

Create a pipe network by layout.   After you've created a parts list for your pipe network, the first step toward finalizing the design is to create a Pipe Network By Layout.

> **Master It**   Open the `MasteringPipes.dwg` file. Choose Pipe Network Creation Tools from the Pipe Network drop-down in the Create Design panel on the Home tab to create a sanitary sewer pipe network. Use the Finished Ground surface, and name only structure and pipe label styles. Don't choose an alignment at this time. Create 8" (20.32 cm) PVC pipes and concentric manholes. There are blocks in the drawing to assist you in placing manholes. Begin at the START HERE marker, and place a manhole at each marker location. You can erase the markers when you've finished.

Create an alignment from network parts and draw parts in profile view.   Once your pipe network has been created in plan view, you'll typically add the parts to a profile view on the basis of either the road centerline or the pipe centerline.

> **Master It**   Continue working in the `MasteringPipes.dwg` file. Create an alignment from your pipes so that station zero is located at the START HERE structure. Create a profile view from this alignment, and draw the pipes on the profile view.

Label a pipe network in plan and profile.   Designing your pipe network is only half of the process. Engineering plans must be properly annotated.

> **Master It**   Continue working in the `MasteringPipes.dwg` file. Add the Length Description And Slope style to profile pipes and the Data With Connected Pipes (Sanitary) style to profile structures. Add the alignment created in the previous exercise to all pipes and structures.

Create a dynamic pipe table.   It's common for municipalities and contractors to request a pipe or structure table for cost estimates or to make it easier to understand a busy plan.

> **Master It**   Continue working in the `MasteringPipes.dwg` file. Create a pipe table for all pipes in your network.

Chapter 14

Storm and Sanitary Analysis

One of the trickiest and most controversial parts of a civil designer's job is determining where water and wastewater will go in a site. Civil 3D offers a variety of tools for designing and analyzing the hydraulic and hydrology portions of your project. Hydraulic calculations for pipe design takes into account many factors that influence the end product. Hydrology is part science and part art, with a dash of voodoo. Luckily, Storm and Sanitary Analysis (SSA) is here to help you sort it all out.

SSA is a separate program that launches from the Analysis tab in Civil 3D or from the Autodesk folder off of the Windows Start menu. This is the portion of the product that can compute site runoff, pipe sizing calculations, and design pressure pipe systems.

In this chapter, you will learn to:

- Create a catchment object
- Export pipe data from Civil 3D into SSA

Getting Started on the CAD Side

When you are getting ready to perform hydraulic or hydrologic computations using SSA, there are several steps that can be completed on the CAD side of Civil 3D.

You will use Civil 3D tools to find drainage points, catchment areas, and flow paths.

Water Drop

In any hydrologic system, it is important to know the direction water will flow on a surface. The Water Drop command is an excellent tool for determining where your outfall location should be. You will want to locate preliminary flow paths before defining the drainage basin.

To use the Water Drop command, select the surface you wish to analyze. Then select the Water Drop command from the Analyze panel of the Surface tab.

In the following exercise, you will use the Water Drop command on the existing site to determine the current flow pattern:

1. Open the `WaterDrop.dwg` file, which you can download from www.sybex.com/go/masteringcivil3d2012.

2. For this exercise, you will find it helpful to turn off object snaps and polar tracking.

3. Select the surface EG by clicking on any contour or portion of the boundary.

4. From the Analyze panel of the Surface tab, click Water Drop. The Water Drop settings will appear. Verify that the Place Marker At Start Point option is set to Yes, as shown in Figure 14.1. Click OK.

FIGURE 14.1
Water Drop command options

5. Get a feel for the site drainage by clicking on various areas throughout the site. There is no wrong place to click, as long as you are inside the surface boundary.

   An X represents the place your cursor clicked. The place you click is a location on the surface where water hits the surface. You will see lines forming in the direction water flows from the area. If you see an X but no path, it could mean you are in a flat area or inside a small depression. After you click in several locations, your drawing will resemble Figure 14.2.

6. Click around as many places as you would like to examine. Press Esc when you finish.

   The water drop paths that are created are polylines without any special intelligence. The next step demonstrates how to remove them if your surface model changes or you need to clear them from the screen.

FIGURE 14.2
Surface with water drop paths

7. (Optional) Select one of the X symbols and the water drop path. Be sure your surface is not selected. Right-click and choose Select Similar. All of the flow paths and corresponding Xs will be highlighted. Remove them by pressing Delete on your keyboard.

Catchments

The first step in any runoff computation is finding the area of the watersheds (also known as drainage basins or catchments) that contribute to the flow.

A new object type has been added to Civil 3D 2012: *catchments*. Catchments work with your surface models to determine the area draining to a point you specify. They are also used to find the drainage path within the catchment to compute time of concentration (Tc).

Catchment Groups

You must create a catchment group before creating catchments. Catchment groups are a way to separate predevelopment and postdevelopment watershed areas. Similar to sites, a catchment in one catchment group will not interfere with objects in another catchment group. Catchment groups do not interact with each other, allowing catchments from differing groups to overlay each other.

Catchment Creation

In Figure 14.3 you see a catchment with all its related features. You will need a grasp of the terminology used in the software before proceeding. It is useful to think about a hypothetical raindrop hitting the catchment to grasp what the various components represent.

Figure 14.3
A catchment with several flow path segments

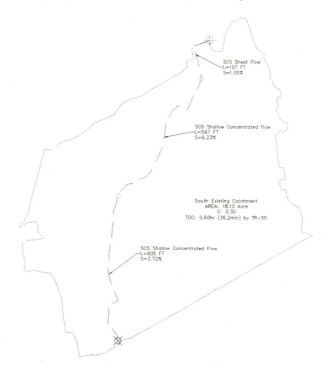

Catchment Boundary The outline represents the catchment boundary, often called the basin divide or watershed boundary. This represents the outermost points of the catchment object. Any raindrops that hit inside this boundary will flow to the same point.

Discharge Point Toward the bottom of the boundary in Figure 14.3 is the discharge point. This is the main point of analysis used for the catchment. All raindrops that hit the basin ultimately end up at this location.

Hydraulically Most Distant Point Near the top of the catchment boundary in Figure 14.3 is a marker indicating the hydraulically most distant point. A raindrop that lands at this point will take the longest to arrive at the discharge point.

Catchment Flow Path In Figure 14.3, the catchment flow path is shown as a dashed line running through the catchment. This linear path represents the course the raindrop takes on its journey from the hydraulically most distant point to the discharge point. The catchment flow path is used to determine the slope and length for computing Tc for hydrology calculations.

Flow Path Segments By default, a flow path is created with an average slope through its overall length. Using flow path segments, you can break up the flow path into smaller pieces with the average slope computed per segment. Each segment can have different computation methods associated with them for Tc computations. The computation methods are as follows:

- SCS Shallow Concentrated Flow
- SCS Sheet Flow
- SCS Channel Flow

Time of Concentration Time of concentration is the time it takes for a drop of rain to travel from the hydraulically most distant point to the discharge point. The default method for determining this value is TR-55. Civil 3D uses the flow type, slope, and user-specified Manning's roughness to determine the time. Alternately, you can input a user-defined time of concentration.

SCS, WHAT? TR, WHO?

TR-55 is short for a technical document put out in 1986 by a branch of the US Department of Agriculture. The agency prior to 1996 was called SCS (Soil Conservation Service), but is now known as NRCS (National Resources Conservation Service). Technical Release 55 "Urban Hydrology for Small Watersheds" is the seminal work behind hydrology computations in the United States.

This document, which has been added to the chapter dataset (which you can download from this book's webpage) for your reference, describes the procedures needed for determining the curve number (a.k.a. runoff coefficient), when to apply the different flow types, and how to compute Tc for each flow type.

Civil 3D incorporates the Tc computation directly in the catchment object. However, TR-55 does have some limitations that you need to keep an eye out for in Civil 3D:

- Minimum time of concentration = 0.1 hour (6 minutes).
- If you're using sheet flow, it should be your first flow segment and should not exceed 300′. Some newer literature says sheet flow should not exceed 200′. Check with your local regulating authority for specific requirements. Typically, sheet flow morphs into shallow concentrated flow before 200′.

- Each Civil 3D catchment supports one entry for curve number. However, the curve number is not used for determining the time of concentration. If you have nonhomogeneous land and/or soil types, leave the default, but be sure to compute the composite curve number in the SSA portion of the software.

- If you need to use a method other than TR-55 for Tc, you can specify User-Defined and leave the entry blank until you export to the SSA portion of the product.

Finally, keep in mind that Civil 3D catchments are not dynamic to the surface model. If the surface model changes, you will need to re-create your catchment.

Catchment Options

You can choose to create a catchment from a surface model or by converting a closed polyline. You will get the best results when defining your catchment by converting a polyline to a catchment area. Manually delineate your watershed areas using the Water Drop command you used in the previous section as a guide.

Create Catchment From Object   In most cases, this will be the method you should use for watershed creation. Use Create Catchment From Object if you have watershed data created ahead of time (i.e., from an imported GIS file or existing DWG). You will also need to use the Create Catchment From Object option for watersheds that are adjacent to your project's limits of disturbance. Even if you choose to use an existing polyline as the catchment boundary, you can still use the surface model to determine the flow path location and slope.

Create Object From Surface   Use the Create Catchment From Surface option if you have a well-formed surface model with adequate data for Civil 3D to compute the watershed area. Surface models with many flat areas, sparse data, or multiple hide boundaries will not work well for this tool. If you do not have good luck with this tool, don't be surprised. Make sure the surface model used for catchment creation is complete before defining a catchment object. The catchment is not dynamically linked to the surface model; therefore, it will not update if changes are made to the surface model. If you are using the hydrology tool on an existing ground surface (predevelopment), make sure all surveyed shots and breaklines are added before proceeding. If you are using the tools on a proposed site (postdevelopment), make sure corridor surface models, grading surface models, and any proposed elevation data are pasted together in the surface model used for analysis.

In the following exercise, you will use multiple tools to delineate several watershed areas for a predeveloped site. You will also use the site characteristics to specify Manning's roughness for SCS Sheet flow and SCS Channel flow. The site is covered in a dense grass.

1. Open the Catchment-1.dwg file, which you can download from this book's web page.

2. For this exercise you will find it helpful to turn on the Insert object snap.

3. On the Analyze tab, locate the Ground Data panel. Choose Catchments ➢ Create Catchment Group, as shown in Figure 14.4. Name the group **Pre-developed**. Click OK.

4. On the Analyze tab, locate the Ground Data panel. Choose Catchments ➢ Create Catchment From Surface.

5. When you see the prompt Specify the Discharge Point:, select the insertion point of the storm structure labeled Existing Inlet (S).

FIGURE 14.4
Accessing catchment tools from the Analyze tab

6. When the Create Catchment From Surface dialog appears:

   ◆ Name the catchment **South**.

   ◆ Set the surface to Drainage GE-Survey.

   ◆ Click the select structure icon to set the Reference Pipe network structure to Existing Inlet (S).

   ◆ Alternately, you can right-click to pick structures from a list.

   ◆ Set Catchment Label Style to Name Area And Properties.

   ◆ Set Runoff Coefficient to **0.4**. (This is the runoff coefficient used in the Rational method and will not affect our Tc value.)

   If your dialog looks like Figure 14.5, click OK.

7. After a moment, a red line representing the catchment boundary will form. Press Esc on your keyboard to dismiss the command.

FIGURE 14.5
Create Catchment From Surface settings

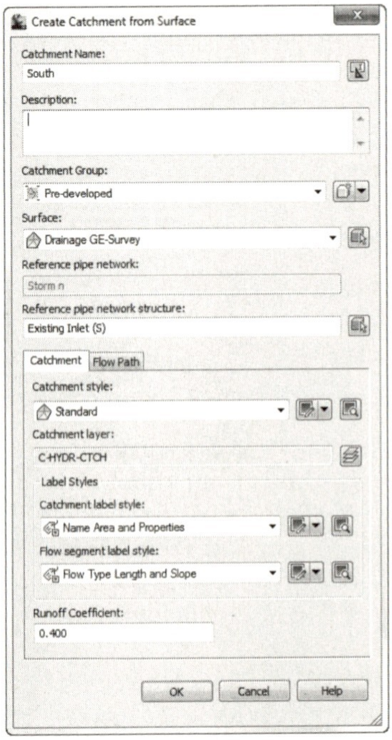

8. On the Analyze tab, locate the Ground Data panel. Choose Catchments ➢ Create Catchment From Object.

9. Select the closed polyline north of the first catchment. At the prompt `Select a polyline on the uphill end to use as a flow path (or Press Esc to skip):`, click the eastern end of the blue, dashed polyline that runs across the catchment.

10. In the Create Catchment From Object dialog that appears:

    ◆ Name the catchment **North**.

    ◆ Click the select structure icon to set the Reference Pipe network structure to Existing Inlet (N).

    ◆ Set Catchment Label Style to Name Area And Properties.

    ◆ Set Runoff Coefficient to **0.4**.

    ◆ Uncheck Erase Existing Entities. Your dialog will now look like Figure 14.6a.

    ◆ Switch to the Flow Path tab.

    ◆ Set Flow Path Slopes to From Surface.

    ◆ Set Surface to Drainage GE-Survey.

    ◆ Your Flow Path tab will now look like Figure 14.6b. Click OK. Press Esc on your keyboard to exit the command.

FIGURE 14.6
Create Catchment From Object settings (a) and the Flow Path tab (b)

(a)   (b)

Your new catchment should resemble Figure 14.7.

FIGURE 14.7
The North catchment and flow path

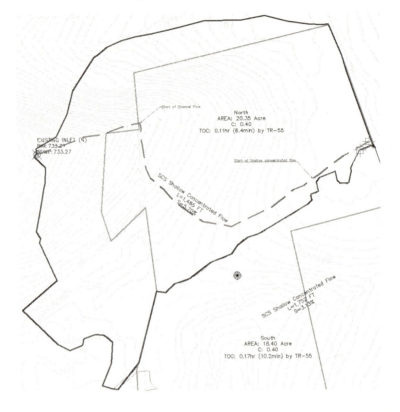

11. Select the North catchment by clicking anywhere on the red boundary or on the dashed flow path line.
12. Right-click and select Edit Flow Segments, as shown in Figure 14.8.

FIGURE 14.8
Right-click to access Edit Flow Segments.

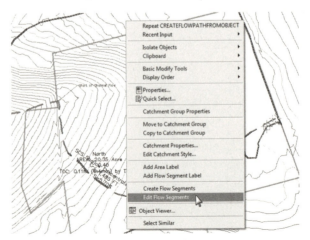

13. Panorama will appear showing the continuous path as SCS Shallow Concentrated Flow. Change Surface Type to Short Grass Pasture.

14. Click the plus sign in the upper-left corner to create a flow segment.

15. Create the new flow segment by clicking at the location on the flow line that crosses contour line 822'.

    The new segment is placed in the Panorama listing from uphill to downhill. Segment 1 is the segment you just created and Segment 2 represents everything else downhill.

16. This new segment represents sheet flow. Change the flow type for Segment 1 to SCS Sheet Flow. Note the input fields change to accommodate this type of flow.

    - Set the 2Yr-24Hr Rainfall value to **2.9"**.
    - Set the Manning's Roughness value to **0.24**.

    The rainfall information used in this step was found in the PennDOT drainage design manual, which you can check for yourself at ftp://ftp.dot.state.pa.us/public/bureaus/design/PUB584/PDMChapter07A.pdf.

    The Manning's roughness for this area is based on coefficients for long grass found in the TR-55 document.

17. Click the plus sign and add the next segment at the abrupt bend in the flow path near the site boundary. This is where our flow enters a swale. For Segment 3, change Flow Type to SCS Channel Flow.

    - Set the Manning's Roughness value to **0.05**.
    - Set the Cross-Sectional Area value to **27** square feet.
    - Set the Wetted Perimeter value to **28.2'**.

    Manning's roughness was determined based on an earthen swale from TR-55. The cross-sectional area and wetted perimeter are site-specific measurements of the swale.

    Your panorama should now look like Figure 14.9. You have used Civil 3D to compute the Tc for the north catchment.

18. Save the drawing.

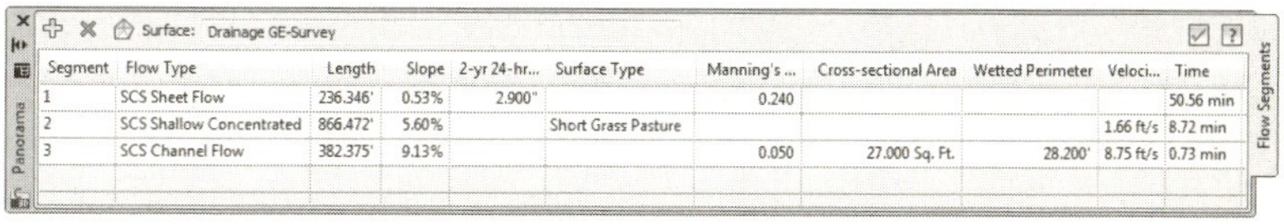

FIGURE 14.9
Three flow segments with different characteristics

Segment	Flow Type	Length	Slope	2-yr 24-hr...	Surface Type	Manning's ...	Cross-sectional Area	Wetted Perimeter	Veloci...	Time
1	SCS Sheet Flow	236.346'	0.53%	2.900"		0.240				50.56 min
2	SCS Shallow Concentrated	866.472'	5.60%		Short Grass Pasture				1.66 ft/s	8.72 min
3	SCS Channel Flow	382.375'	9.13%			0.050	27.000 Sq. Ft.	28.200'	8.75 ft/s	0.73 min

Real World Scenario

CATCHMENT AREAS TO GIS

There are many situations where you will need to work with GIS data to send data to outside sources. You can use GIS data for anything from land uses to soil types. Luckily, Civil 3D contains all of Autodesk's Map 3D program, allowing you to work with geographical data.

When catchments are exported to SSA from Civil 3D, the Tc, runoff coefficient, and area are exported as data only. However, you may be asked to provide ESRI Shape files (SHP) of your watershed areas.

To get physical geometry from Civil 3D, use parcels instead of catchments and then use GIS tools to move the data to SHP format:

1. Open the file Parcel Export.dwg, which you can download from this book's web page.
2. From the Export panel on the Output tab, click Export Civil Objects To SDF.

3. The SDF file will automatically be stored in the same folder as the DWG file you are working in. If you wish to change the location, click the ellipsis button. Click OK to confirm the coordinate system and create the SDF file.

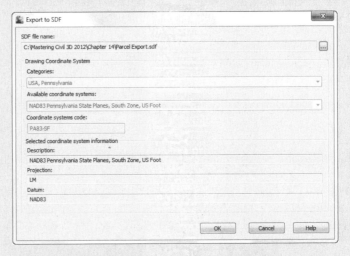

The SDF file has been created, but to use it in SSA you must convert it to an SHP file. In the next steps you will be using the Map commands to perform the conversion. You will change to the Map part of the program by switching to the Planning And Analysis Workspace.

4. Change to the Planning And Analysis Workspace by clicking the gear icon in the lower-right corner of your Civil 3D screen. On the Data panel of the Home tab, click Connect.

5. On the left side of the Data Connect screen, click Add SDF Connection. On the right side of the dialog, click the browse icon beside the Source File field to browse to the file you created in step 3. Click Connect. No additional action is needed for the SDF data.

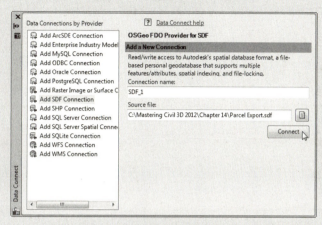

6. Next, highlight Add SHP Connection. Click the Folder icon and browse to the folder location where the project resides.

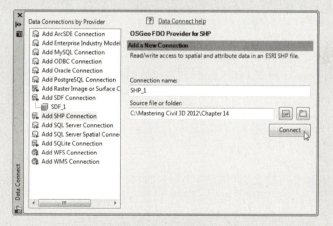

7. Click Connect. The folder you chose will be the destination for SHP files you will create in the steps that follow. No additional steps are needed in the Data Connect screen and you can click the X to close it.

8. Switch to the Create tab and locate the Feature Data Store panel. Click Bulk Copy.

9. In the From side of the screen on the left, set the Source to SDF_1. Scroll down the listing and put a check mark next to Parcels.

10. On the right side, set the To Target to SHP_1. Scroll down to where the command has automatically matched parcel information.

11. Double-click the field for Autogenerated_SDF_ID and rename it to **ID**. This step is necessary because the SHP file only supports property names containing fewer than 12 characters.

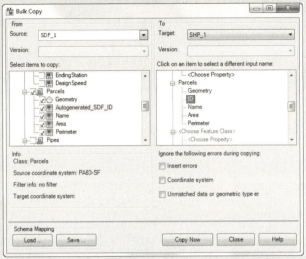

12. Click Copy Now. You will be prompted with a message warning you that the Bulk Copy operation cannot be undone. Click Continue Bulk Copy.

13. The Bulk Copy Results message should pop up, informing you that 13 objects were copied. Click OK.

14. Close the Bulk Copy dialog and close the drawing.

Exporting Pipes to SSA

When your catchments and your pipe networks are ready to analyze, switch to the Analyze tab and select Edit In Storm and Sanitary Analysis. You must have at least one pipe to export.

When you export, the pipe network information, inlets, pipes, and elevations convert over to SSA. Catchment data is shown with area, Tc, and runoff coefficient as data only. The graphic area that you define on the Civil 3D end is not needed.

When you first export a project over, SSA will ask if you would like to create a new project or open an existing project. If you are adding pipes to an existing network that has been analyzed in SSA, click the Open Existing Project radio button and browse to the location of your stored SPF file.

When you create a new project, SSA does a few grunt-work things for you. First, it converts all your junctions and pipes to the SSA format (via Hydraflow's STM file, oddly enough) and asks if you would like to save the log file. Generally, saving the log file is not necessary unless you are having difficulty importing certain objects. In most cases, you can click No to continue.

Once you are in the SSA interface, you can access your active project by clicking the Plan View tab at the top of the window. You will see your DWG file as an underlay to the project you are working on.

> **HOW DOES SSA DECIDE WHAT'S AN INLET AND WHAT'S A MANHOLE?**
>
> Buried deep in the bowels of your command settings are the Part Matchup Settings. On the Settings tab, choose Pipe Networks ➢ Commands and double-click the EditInSSA option. Once you are there, expand the Storm Sewers Migration Defaults area.
>
> Click the Part Matching Defaults field. Click the ellipsis to enter the Part Matchup Settings dialog.
>
>
>
> The Import tab handles how parts from SSA or Hydraflow behave on the way into Civil 3D. The Export tab handles how SSA or Hydraflow handles parts from Civil 3D. As you can tell from poking around here, there isn't always a perfect match for every part you use.

> There are a few quirks and limitations with the intended SSA workflow:
>
> - Pipe networks that contain parts from multiple-part families (for example, a pipe network may contain a part from the HDPE family and another part from the Concrete family) will reimport back into Civil 3D with only one part family.
>
> - To prevent SSA and Civil 3D from changing your pipe sizes, make a single part family for all circular pipes. Be sure that the "master" part family contains all possible sizes. Once it is reimported back to Civil 3D, use Swap Part to change back to the original part family type.
>
> - For advanced users, there is a super-secret XML file located in `C:\Program Files(x86)\Autodesk\SSA 2012\Samples\Part Matching`. For more information about using and editing this file, see the Part Matching help file.
>
> - Culverts are treated the same as a single pipe in SSA. Once you export a culvert into SSA, you'll need to set entrance and exit conditions.
>
> A little part tune-up will be needed after the round-trip; however, the time savings in using the export commands are still very much worth it.

Storm and Sanitary Analysis

SSA can perform a number of water- and wastewater-related tasks. In the remainder of this chapter, you will be going through several examples of what SSA can do, but you are just seeing the tip of the proverbial iceberg.

In the next section, you will focus on the workflow between Civil 3D and SSA.

In the following section, you will learn how to create subbasins (catchments) directly in SSA. You'll also see examples of the Rational method and TR-55 and learn about SSA reports.

Guided Tour of SSA

SSA is not at all like AutoCAD. It is a true modeling software that shows a schematic of your design in a plan view. On the left side of the screen you will see the data tree. Like Prospector in Civil 3D, the SSA data tree gives you access to tools and information about objects.

The easiest way to add new structures, pipes, or catchments to the project is to do so from the Elements toolbar along the top of the screen. Here you will find everything you need to build a project from scratch if you did not start out in Civil 3D.

The next stop on our tour is the Project Options dialog (Figure 14.10). Access Project Options by double-clicking on the listing in the data tree. Project Options is where you set your units, hydrology method, routing method, and force main equation. Depending on the scope of your project, you may not need to worry about every equation used. In the following example, you will use the Rational method with TR-55 Tc calculations. Since you are calculating runoff only, you do not need to set the other project options. Once you have finished entering the options shown in Figure 14.10, click OK.

Whenever you are designing in SSA, always work upstream to downstream. If your system includes a pump, you should work from highest elevation to lower elevations.

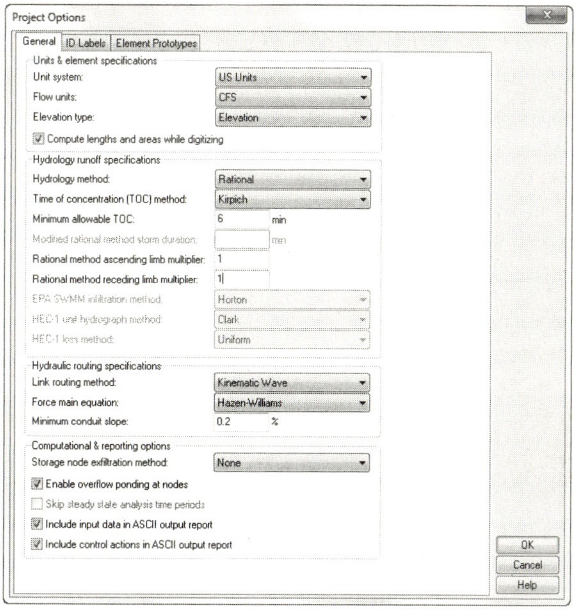

FIGURE 14.10
SSA Project Options dialog

An SSA project consists of three main object types: subbasins, nodes, and links:

Subbasins Subbasins are important for runoff computations but are not mandatory if you are inputting flows manually. Subbasins can be defined directly in SSA or brought over as data from Civil 3D. SSA can also turn an SHP file directly into a subbasin.

Nodes Nodes are the workhorses of SSA, and can take several forms:

Junction Node Nodes are where all calculations are done in SSA. If you want information along a pipe, for example, you will need to place a junction node at the location of interest. These can be used as null structures, water, or pollutant infiltration points.

Outfall Node Outfall node is the end of your system. An outfall can be where proposed sewer ties into existing, where sanitary sewer enters a treatment facility or a culvert discharge location. An outfall can be attached to only one link.

Flow Diversion Node If a node has two outgoing links, you will need to use a flow diversion. Flow diversions most commonly represent multiple outflow points on a detention basin, such as a weir or orifice. They can also represent flow diversion in a combined sewer system or anywhere an overflow check is in use.

Inlets When SSA converts structures from Civil 3D, it assumes an inlet node by default. Inlets are used in storm systems and can collect flow from a subbasin. You can also manually enter flows.

Storage Nodes Anywhere water is detained on your site, a storage node can model the scenario. Storage nodes can represent ponds, wet wells, underground storage, or a residential rain barrel.

Links Links connect nodes to each other. Links represent pipes, channels, curbs, and gutters, or culverts.

> **HYDRAFLOW, WE HAVEN'T FORGOTTEN YOU**
>
> When it comes to flexibility in storm design and routing, SSA beats the pants off of Hydraflow any day. In general, there is a lot of overlap between the two products. For basic storm sewer calculations, either product will work great. When your designs get more complex and require hydrograph and storage routing with infiltration and exfiltration, SSA has a much better solution.
>
> You can still access all the Hydraflow functions from the Analyze tab ➢ Design panel flyout.
>
>
>
> Hydraflow is still part of the picture when it comes to exporting and importing data to and from Civil 3D. Hydraflow's file format does a much better job than LandXML at keeping the data integrity of your pipes and structures.
>
>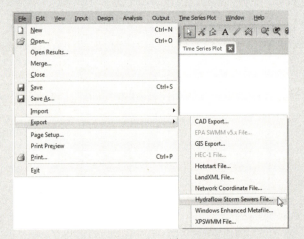
>
> When exporting from SSA, choose File ➢ Export ➢ Hydraflow Storm Sewers File for the best path back to Civil 3D.

Hydrology Methods

SSA is capable of computing rainfall runoff using any of the following methods:

Rational Method The Rational method is used to approximate the peak discharge (or flow, Q) from an impervious area (A) using the tried and true Q=ciA. This is the most common method used for sizing sewer pipes.

Imagine yourself outside in a rainstorm, staring at a single point on your driveway (assuming your driveway is impervious and on a nice, even slope). At first, the water on your driveway is coming right from the sky. After a while, water flows in from rain falling elsewhere in your neighborhood. Now imagine that the storm stops as soon as raindrops from the farthest point in your neighborhood reach the point on your driveway.

The Rational method is based on the assumption that it takes the same amount of time for those farthest raindrops to get to you as they do to run off your driveway completely.

The first step to using the Rational method is to delineate the drainage basin, as described earlier in this chapter, and assign composite runoff coefficients. SSA needs an intensity-duration-frequency (IDF) curve for the study area to compute the rainfall.

Modified Rational and DeKalb Rational   These hydrology methods are similar to the Rational method, except that they are intended to approximate the storage volume needed on small, simple detention basins. Instead of the storm duration equaling two times the time of concentration, the storm duration is longer. The longer storm duration results in a larger volume than the Rational method but results in a lower peak discharge.

The same information is needed for the Modified and DeKalb Rational methods as for the Rational method, and you must have a value in the Project Options dialog for Modified Rational Storm Duration.

Both the Rational method and the Modified Rational method are intended for small watersheds, usually less than 20 acres (8.1 hectares).

SCS TR-55 Method and TR-20 Method   More gifts from the USDA, the SCS TR-55 runoff method is a more complex method that is appropriate for study areas up to 2,000 acres (809 hectares).

For the SCS TR-55 and TR-20 methods, SSA contains tools for looking up computation variables. You will find a curve number (CN) lookup table inside the subbasin. SSA will even help you determine the composite curve number. The TR-55 hydrology method is also where you will use SSA rain gauges.

HEC-1   The US Army Corps of Engineers' Hydraulic Engineering Center has a computer program called HEC-1 for hydrology calculations. The HEC-1 method in SSA duplicates the results from the Army Corps' program. Their method is used in situations where there is an existing body of water, such as a stream. For an explanation of this method directly from the source, go to `http://www.hec.usace.army.mil/publications/TrainingDocuments/TD-15.pdf`.

EPA-SWMM   The US Environmental Protection Agency (EPA) Storm Water Management Model (SWMM) method is used for developed watershed areas and can take into account many more factors than the other analysis methods. SWMM allows for localized ponding effects, which can be entered into the subbasin physical properties area. This method also allows for the effects of existing soil moisture, evaporation, and snowmelt in the system. When using EPA-SWMM, be sure to set all the methods you intend to use for infiltration and exfiltration in the Project Options dialog.

> **A FEW HINTS FOR WORKING IN SSA**
>
> Since this is not traditional CAD, the program behaves differently than you might expect. This list contains a few hints to help you along the way:
>
> - Make sure you download and install the Civil 3D 32-bit object enabler. SSA needs the object enabler to see Civil 3D objects such as pipes, structures, and nodes. Since SSA is a 32-bit application under the hood, download the 32-bit version of the object enabler even if your computer is on a 64-bit operating system.
>
> - Unlike AutoCAD, SSA keeps you in a command until you press Esc, or click the Select Element tool from the toolbar.
>
> - Another "gotcha" are the tabs along the top of the screen. To dismiss them, use the red X that appears on the tab itself. It is a common mistake to dismiss the program accidentally by clicking the incorrect red X.
>
>
>
> - Don't forget to save! Unlike CAD, there is no auto-save.
>
> - Also, there is no Undo command in SSA. Luckily, adding and removing elements is a simple process. If something gets goofed up beyond repair, you could always close without saving.
>
> It may take a while to get used to, but SSA is a powerhouse analysis tool.

In the following example, you will work through a simplified example to get the feel for the SSA interface:

1. Launch SSA from the stand-alone icon in your computer's Start menu. This will launch SSA with an empty project.

2. In the SSA interface, choose File ➤ Import ➤ Layer Manager (DWG/DXF/TIF/more), as shown in Figure 14.11. Click the ellipsis next to the Image/CAD File field. Browse to the file SSA Intro.dwg, which you can download from this book's web page. Place a check mark next to Watermark Image and click OK.

   When you import the CAD file, make a mental note of the many file formats that can be used to import data to SSA, including EPA SWMM.

3. Click the green Add Subbasin button from the elements toolbar along the top of the window. Using this tool, click around the boundary lines that represent the example watersheds. When you are close to completing a shape, you can right-click and select Done to close the shape.

   There are no snap or curve drawing tools in SSA. Digitize as closely as possible using multiple small segments. If you make a mistake, you can right-click and select Delete Last Segment. Later on in this chapter we will look more closely at modifying vertices of a completed shape.

4. When you have completed both shapes, your graphic will resemble Figure 14.12.

FIGURE 14.11
To show the AutoCAD information as an underlay, import it as a background layer

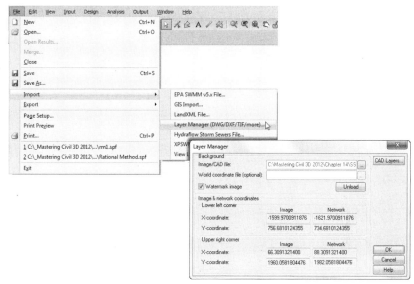

FIGURE 14.12
New subbasins in SSA

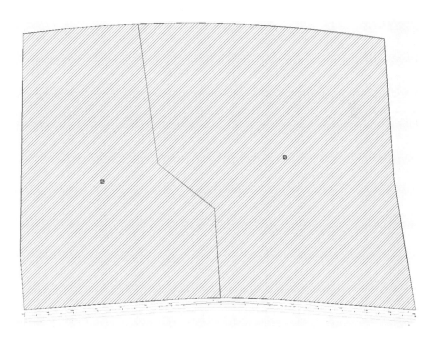

5. Switch to the Inlet tool by clicking the Inlet button at the top of the SSA window.

6. Place two inlets on the south side of the site. An X has been placed at each inlet location to help you locate the recommended positions.

When you have finished placing inlets, press Esc on your keyboard or click the arrow icon in SSA to get back to Selection mode. If you accidentally place more inlets than you wanted, you can right-click the wayward inlet while in Selection mode and select Delete from the menu.

7. Click the Outfall icon. Click to place the outfall structure in the location marked with a circle at the southeastern corner of the site.

8. Next, switch to the Conveyance Link tool. Working from left to right across the project, click the upstream inlet. Resist the temptation to click and drag. Connect the first inlet with a straight segment to the second inlet. Next, connect the second inlet to the outfall. At this point your SSA file should resemble Figure 14.13.

FIGURE 14.13
Pipes connecting inlets to each other and the outfall

9. Double-click the first conveyance link you created. You will see the Conveyance Links dialog. The easiest place to rename the pipes is in the small grid at the bottom of the box, as shown in Figure 14.14. When your cursor is in the field to rename a pipe, a dark black square will appear on the pipe in the main graphic, helping to indicate which pipe you are editing. Rename the first link to **Pipe1**. Rename the second conveyance link to **Pipe2**.

FIGURE 14.14
The Conveyance Links dialog is the main pipe and channel control tool.

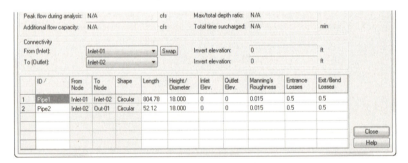

So far, you've told SSA what your drainage areas are and you've established a pipe network. Next, you need to tell SSA which subbasin drains to which inlet.

10. Switch to the selection tool by pressing Esc if it is not already active. Right-click the icon that represents the western subbasin and select Connect To, as shown in Figure 14.15.

11. Connect the subbasin to the western inlet (Inlet-01) by clicking the inlet symbol in plan view. A line will form to indicate the connection was successfully created. Repeat the process for the eastern subbasin and inlet (Inlet-02). At the end of the process your SSA screen will resemble Figure 14.16.

FIGURE 14.15
Connecting a subbasin to an inlet

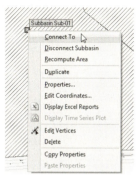

FIGURE 14.16
Subbasins assigned to inlets

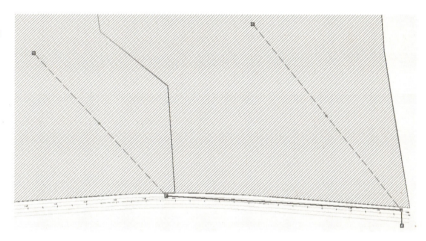

Before SSA can run any analysis on the project, you must make one last crucial connection. Both of these inlets are on grade, which means water will flow toward them, but in a big storm, some water may flow past into the next downstream inlet. The flow along the gutter of the road between inlets is called a *bypass link*.

12. Click the Add Conveyance Link button. Connect Inlet-01 to Inlet-02 again, but this time follow the curve of the road. This conveyance link represents the gutter flow of any water that doesn't make it into Inlet-01 and continues on to Inlet-02.

13. Add a conveyance link between Inlet-02 and the outfall. To visually differentiate the culvert from the bypass link, you can put a small kink in the line, as shown in Figure 14.17. When you modify link properties, you will have the opportunity to compensate for the change in length this technique causes.

14. Rename the new links to **Overland1** and **Overland2**, respectively.

All the pieces are here; next you will assign elevations to the pipes, inlets, outfalls, and curb flows.

15. Double-click any of the links to open the Conveyance Link dialog. Using the tabular input at the bottom screen, key in the values for Diameter, Inlet Elevation, Outlet Elevation, and Manning's Roughness, as shown in Figure 14.18.

16. For the overland flow links, you will use a rectangular open channel to approximate gutter flow. For both Overland1 and Overland2 links, set the width to 2′, as shown in Figure 14.19. Click Close when complete.

FIGURE 14.17
The bypass link between Inlet-02 and the outfall

FIGURE 14.18
Set pipe and gutter elevation data as shown.

FIGURE 14.19
Set the channel type and width as shown.

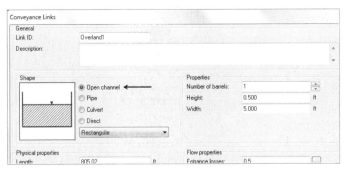

17. Next, you will set the Inlet elevations by double-clicking on Inlet-01.

- Using the Inlets dialog, set the number of inlets for both Inlet-01 and Inlet-02 to **2**, as shown toward the top of Figure 14.20.
- Set Invert Elevation and Rim Elevations, as shown at the bottom of Figure 14.20.
- For Inlet-01, set Roadway/Gutter Bypass Link to Overland1, as shown in Figure 14.20.

- For Inlet-02, set Roadway/Gutter Bypass Link to Overland2.
- Click Close when complete.

18. Double-click the Outfall location. Set the Outfall elevation to **480**, as shown in Figure 14.21. Under Boundary Condition, set Type to Free. Click Close.

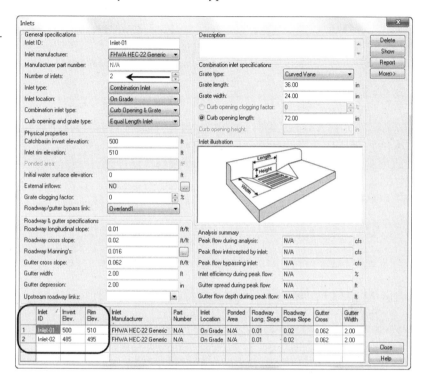

FIGURE 14.20 Setting Invert Elevation and design settings

FIGURE 14.21 Outfall elevation and settings

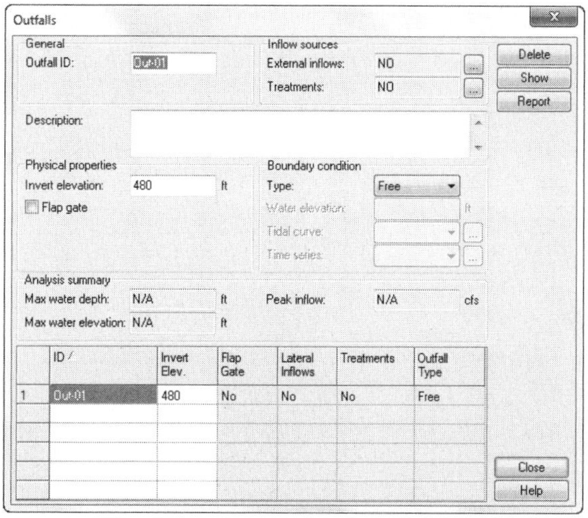

Now that all the elevations are set, you must tell SSA how much rain you expect to see in your system.

19. In the data tree on the left side of the SSA window, click IDF Curves, as shown in Figure 14.22.

FIGURE 14.22
IDF Curves are found in the data tree.

20. Click Load. Browse to the file `York Co PA.idfdb`. This is the rainfall data for our example file. After loading the file, set the ID to **York, PA, USA**, as shown in Figure 14.23. Click Close.

FIGURE 14.23
IDF curves entered in SSA

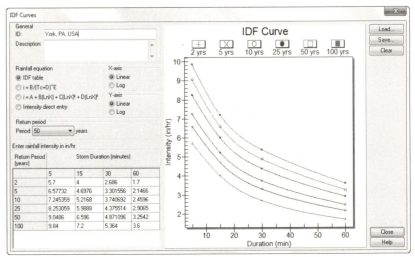

21. In the data tree on the far left, double-click Project Options. (This dialog takes a moment to open.)

   ◆ Set Hydrology Method to Rational.
   ◆ Set Time Of Concentration (TOC) Method to Kirpich.

FIGURE 14.24
Verify your settings and click OK.

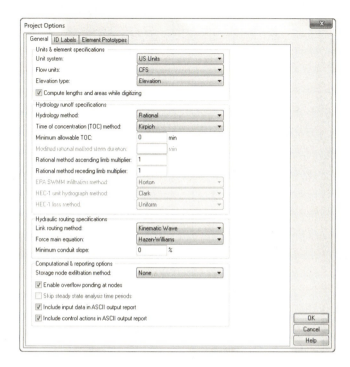

Your Project Options dialog should resemble Figure 14.24. Click OK.

22. Double-click Analysis Options from the data tree.

    ◆ On the General tab, set the End Analysis On value to **02:00:00**. Your dates will vary from what is shown in Figure 14.25, but the important thing is that the duration of the analysis is large enough to accommodate the Rational method.

    ◆ On the Storm Selection tab, set Return Period to 25 years for a single storm analysis.

    ◆ Click OK when complete.

The last step before you are ready to run the design is setting the subbasin information for use with the Kirpich method of time of concentration. The Kirpich method calculates Tc based on the average slope and area characteristics of the site.

23. Double-click the green subbasin icon located in the centroid of the western subbasin.

    ◆ For the western subbasin, set the flow length to **750**, and set the average slope to **3%**.

    ◆ Switch to the Runoff Coefficient tab and set the runoff coefficient to 0.37.

    ◆ Switch to the eastern subbasin by selecting its row at the bottom of the dialog.

    ◆ Set the eastern subbasin flow length to **700**, and set its average slope to **2.5%**.

    ◆ When your dialog resembles Figure 14.26, click Close.

FIGURE 14.25
Setting the design storm in Analysis Options

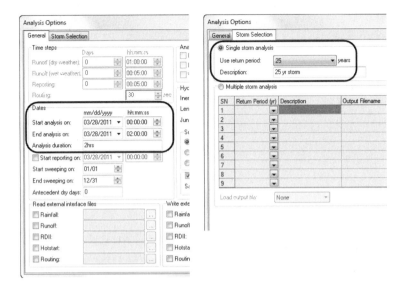

FIGURE 14.26
Final preparation of the subbasin data

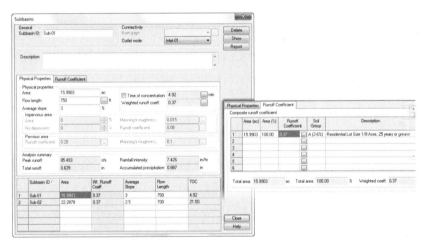

24. Click Perform Analysis. After a moment, the analysis will complete. Click OK.

    In the SSA graphic, you will see that conveyance link Overland1 is flooded. Later on in the chapter you will look at reporting capabilities to see details of this predicament.

25. Save the file as **SSA Intro.spf**.

From Civil 3D, with Love

Earlier in this chapter, you saw some of the preparations that must take place before you click the Edit In Storm And Sanitary Analysis button. Once inside SSA, a few more pieces of information need to be added before SSA can do its analysis.

In the following example, you will see what becomes of catchment objects and pipes once they are imported into SSA:

1. From Civil 3D, open the file C3DtoSSA.dwg. You can download this file from this book's web page.

2. On the Design panel of the Analysis tab, click Edit In Storm And Sanitary Analysis. Verify that there is a checkmark next to the STORM network. Click OK in the Export To Storm Sewers dialog, as shown in Figure 14.27.

FIGURE 14.27
The pipe network on its way to SSA.

3. SSA launches. Click OK to create a new project. After a moment, SSA will let you know that you have successfully imported the Hydraflow Storm Sewers file. Click No, as you do not need to save the log file.

4. At the top of the SSA window, click Plan View. Examine the inlets and catch basins that have been imported.

   Notice that an offsite outfall has been added to each inlet. This is because you don't model the overland links in Civil 3D. When inlets are first imported into SSA, the program assumes they are on grade. Water that bypasses on grade inlets needs to go somewhere, so SSA automatically throws in an outfall to take care of the excess water. In the next steps you will remove the outfalls and create the overland links.

5. Right-click on the outfall that has been added to Structure 1. Click Delete, as shown in Figure 14.28. Click OK to confirm the deletion.

6. Right-click and delete the outfall to the far right.

7. Add overland conveyance links similar to the previous exercise. Add a link between Inlet Structure 1 and Inlet Structure 2. Add a link between Inlet Structure 2 and the outfall.

8. Double-click one of the conveyance links to edit the links in tabular form. Set the pipe and link design values as shown in Figure 14.29.

9. For the overland links, change the channel type to Open Channel. Set the shape type to User-Defined, as shown in Figure 14.30.

FIGURE 14.28 Right-click and click Delete to remove the extraneous outfall.

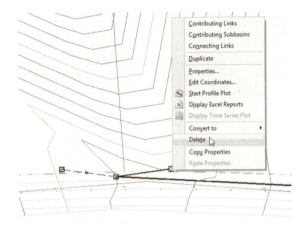

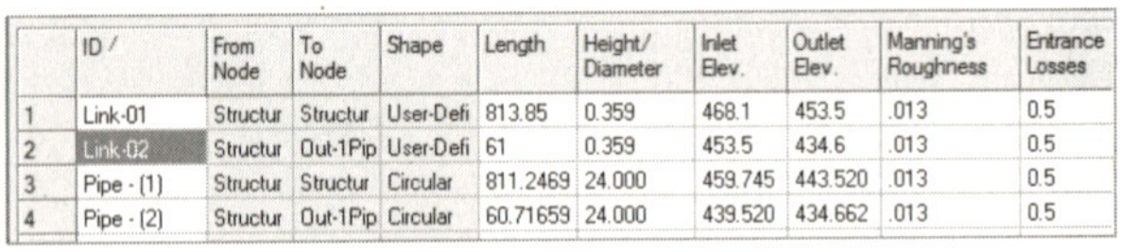

FIGURE 14.29 Design values for links and pipes

	ID	From Node	To Node	Shape	Length	Height/Diameter	Inlet Elev.	Outlet Elev.	Manning's Roughness	Entrance Losses
1	Link-01	Structur	Structur	User-Defi	813.85	0.359	468.1	453.5	.013	0.5
2	Link-02	Structur	Out-1Pip	User-Defi	61	0.359	453.5	434.6	.013	0.5
3	Pipe - (1)	Structur	Structur	Circular	811.2469	24.000	459.745	443.520	.013	0.5
4	Pipe - (2)	Structur	Out-1Pip	Circular	60.71659	24.000	439.520	434.662	.013	0.5

FIGURE 14.30 Gutter slope in cross-section; this will be the overland channel shape.

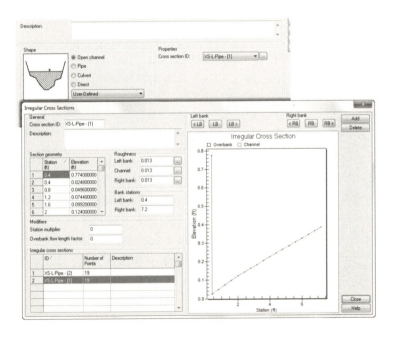

10. Click the ellipsis next to Cross-Section ID to open the Irregular Cross Sections dialog.

    The shape you see in Figure 14.30 is an exaggerated view of a curb and gutter. In the previous example, you approximated the curb and gutter channel using a rectangular open channel. In this exercise, you will use one of the cross sections created for you by the export process.

11. For the active Cross Section ID, enter **XS-L-Pipe - (1)** and then click Close. Make sure both Links 01 and 02 are using the XS-L-Pipe (1) cross section for open channel flow. Click Close when your conveyance links are complete.

12. Double-click Project Options in the data tree. Set the Hydrology method to SCS TR-55. Set the Time Of Concentration method to SCS TR-55. Set Minimum Allowable TOC to **6** minutes. At the end of this step, your Project Options dialog should look like Figure 14.31. Click OK.

FIGURE 14.31
Project Options for this example

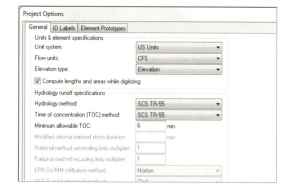

13. Back in the graphic, double-click the west subbasin. Switch to the SCS TR-55 TOC tab. You will input the following information to compute TOC for the western subbasin:

    ◆ On the Sheet Flow tab, set the Manning's roughness value to **0.17**.

    This value is based off the TR-55 listing for different land conditions. The listing of values can be found by selecting the ellipsis next to the entry field.

    ◆ Set Flow Length to **150′**.

    This value is a site-specific measurement. Remember that in TR-55 sheet flow is considered to turn into shallow concentrated flow after 200′.

    ◆ Set Slope to **5%**.

    This is another site-specific value and represents the average overall slope of the subbasin.

    ◆ Click the ellipsis next to the 2Yr-24Hr Rainfall and in the resulting dialog, choose Pennsylvania from the State drop-down and York from the County drop-down, as shown in Figure 14.32. Click OK.

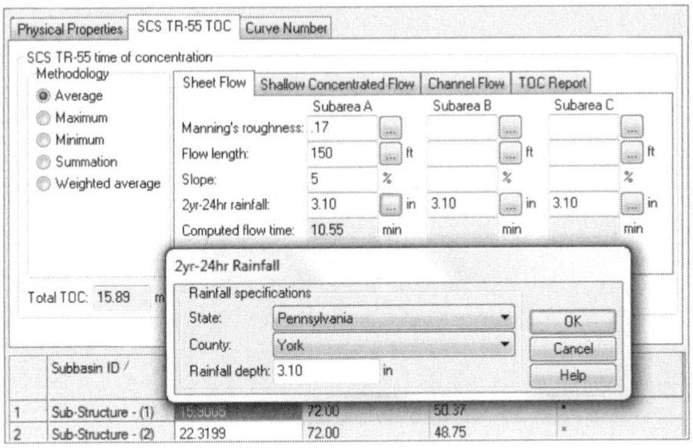

FIGURE 14.32
Sheet Flow input and rainfall location

Selecting the rainfall location at this point is helping you determine the Tc. Your actual design storm duration may vary. Later in this chapter you'll learn how rainfall works.

- Switch to the Shallow Concentrated Flow tab. Set Flow Length to **250'**.
- Set Slope to **5%**.
- Set the surface type to Short Grass Pasture.

The velocity for sheet flow will calculate based off the nomograph, which you can examine by clicking the ellipsis button next to Velocity. Your results should resemble the top of Figure 14.33.

FIGURE 14.33
Shallow concentrated flow and channel flow for TR-55

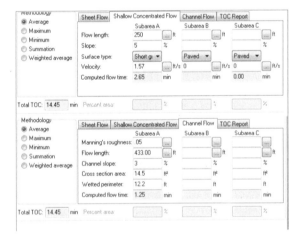

- Switch to the Channel Flow tab and set the Manning's roughness value to **0.05**.
- Set the flow length to **433'**.
- Set the channel slope to **3%**.

- Set Cross Section Area to **14.5**.
- Set Wetted Perimeter to **12.2′**.

Wetted Perimeter and Cross Section Area are site-specific values and are needed to calculate the hydraulic radius used in the Manning's roughness equation. Figure 14.34 shows a schematic of what these values mean in channel flow.

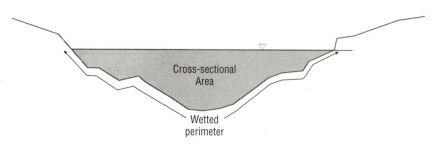

FIGURE 14.34
The Cross Section Area and Wetted Perimeter values are used to find the Tc for channel flow.

After entering the values for channel flow, your dialog will now resemble the bottom of Figure 14.34.

14. Still working with Sub-Structure - (1), switch to the Curve Number tab. Set the curve number values for the subbasin as shown in Figure 14.35.

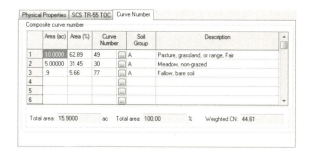

FIGURE 14.35
Use these values on the Curve Number tab to create a weighted curve number.

15. Switch to the eastern subbasin by highlighting Sub-structure- (2) in the tabular listing. The Tc parameters for this subbasin are:

    - Sheet Flow: Manning's roughness of **0.17**, Flow Length of **200′**, and Slope of **4.5%**.
    - Shallow Concentrated Flow: Flow Length of **612′**, Slope of **4%**, Surface Type set to Short Grass Pasture.
    - Channel Flow: Manning's Roughness of **0.05**, Flow Length of **305**, Channel Slope of **2%**, Cross Section Area of **12.2′**, Wetted Perimeter of **8.4′**.

16. Still working with Sub-Structure - (2), switch to the Curve Number tab and set the curve number to **45** for the entire area.

17. Click Close. Save the file as **C3DtoSSA.spf** for use in the next exercise.

Make It Rain

There are several methods for telling SSA the rainfall information for the model. A typical rainfall event in Herten, the Netherlands, is going to be vastly different from Arizona.

In the previous exercise, you set up many of the physical characteristics of the hydrologic system. You did not, however, tell SSA what type of rain to expect on the site.

Different hydrology methods require different assumptions about rainfall. In the Rational method, you used an intensity-based rainfall model (IDF Curve) because you were primarily interested in peak runoff. Because this is a TR-55 example, you will need to use rainfall information appropriate for this method.

1. Continue working in the file from the previous exercise or open SSA-Rain.spf in the Storm And Sanitary Analysis program. You can download this file from this book's web page.

2. From the top of the SSA screen, select Add Rain Gauge. Click anywhere in the graphic to place the rain gauge in the project. Press Esc on your keyboard to return to Selection mode.

3. Double-click the New Rain gauge. Rename the Rain Gauge to **Design Storm**. Set Rain Data Format Type to Intensity. Set Incremental Interval to **0:20**. At this step your Rain Gauge dialog will look like Figure 14.36.

FIGURE 14.36
The Rain Gauge dialog

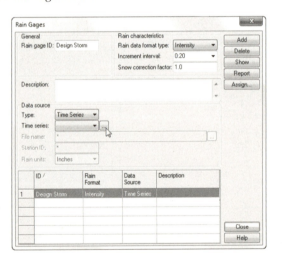

4. Click the ellipsis next to Time Series.

5. In the Time Series dialog, click Add. Rename the Time Series to **York, PA 50.**

6. Set the data type to Standard Rainfall and click the Rainfall Designer button.

7. Set Rainfall Type to Intensity. Set State to Pennsylvania and County to York. Set Return Period to **50 years.**

8. Make sure that the Unit intensity check mark is set to SCS Type II 24-hour, as shown in Figure 14.37. Click OK.

FIGURE 14.37
Rainfall Designer dialog

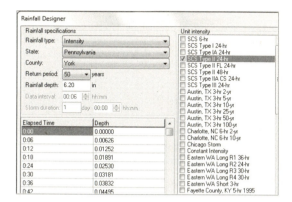

9. Once your Time Series dialog resembles figure 14.38, click Close.

FIGURE 14.38
The completed time series

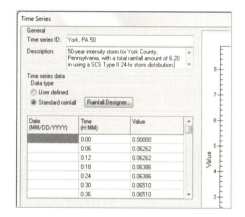

10. In the Rain Gauge dialog, click Assign. Click Yes when asked if you'd like to assign the rain gauge to all subbasins. Close the Rain Gauge dialog.

11. Double-click Inlet Structure - (1). Set Roadway Gutter Bypass Link to Link-01. Switch to Structure - (2) by selecting it from the table at the bottom of the Inlets dialog. Set Roadway Gutter Bypass Link to Link-02. Click Close.

12. Double-click Analysis Options on the data tree.

13. On the General tab, set the Analysis Duration to **1d** by changing the date in the End Analysis On field to one day after the Start Analysis On (the exact date will vary depending on when you open the file).

14. On the Storm Selection tab, set Single Storm to Use Assigned Rain Gauge. Click OK.

15. Click Perform Analysis.

Running Reports from SSA

There are many places in SSA to view the results of your work. After you've run the analysis, the link and node listing will tell you if your pipe is underdesigned or if there is water backing up in the manholes.

From virtually every dialog in SSA, clicking the Report button will create an Excel spreadsheet from the data you are observing. You will see the Report button for links, nodes, junctions, outfalls, and subbasins.

After running the analysis, you will want to see performance information and graphical representations of the relationship between the parts of your design. The Reports toolbar is at the top of the SSA screen. Figure 14.39 shows the different types of reports that you can choose.

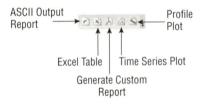

FIGURE 14. 39
The Reports toolbar

First, we'll take a look at the hydrograph that has been created for the site.

1. From SSA, open the file `SSA-Reports.spf`. You can download this file from this book's web page.

2. Click the Perform Analysis button.

3. Click the Time Series Plot button.

4. On the left side of the screen, expand the `Subbasins` folder, as shown in Figure 14.38.

5. Click the word Runoff. A plus sign will appear. Click the plus sign to expand the listing.

6. Put a check mark next to both subbasins, as shown in Figure 14.40.

7. Right-click anywhere in the graph area. Select Export Time Series Plot ➢ CAD Export, as shown in Figure 14.41.

FIGURE 14. 40
The resulting hydrograph from the TR-55 Hydrology analysis

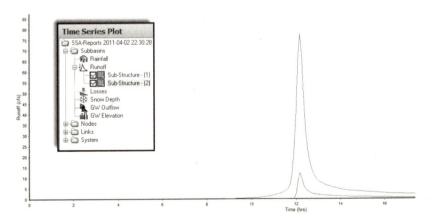

Figure 14.41
Export To CAD

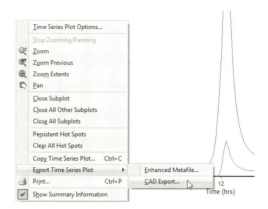

8. Save the resulting DWG to your desktop as **SSA-OUT.dwg**.

9. Remain in the SSA project for the next exercise.

A common analysis you will want to examine is the profile plot. On the left side of the SSA window you will see the Profile Plot option. Once the profile plot settings appear, you can graphically select the nodes you wish to include in your output.

In this next exercise, you will examine a profile plot for the pipes in the system.

1. From SSA, continue working in or open the file SSA-Reports.spf. You do not need to have completed the previous exercise to continue—you can download this file from the book's web page.

2. Click the Perform Analysis button. From the Reports toolbar, select Profile Plot.

3. On the left side of the screen you will see where you can specify Starting Node and Ending Node. You can also specify the link you'd like to plot by clicking on it in the graphic. Click the link Pipe - (1). The pipe will highlight in the plan view.

4. Click Plot Options. In the Plot Options dialog, place a check mark next to Maximum Flow, Maximum Velocity, and Maximum Depth. In the Other Display Specifications, place a check mark next to Maximum EGL, Critical Depth, HGL Markers, and Show Flooded Node. Your Profile Plot Options dialog should resemble Figure 14.42. Click OK.

5. Click Show Plot. Your graphic should resemble Figure 14.43.

After your break from the world of Civil 3D, it is time to export SSA modified data back to CAD. At first glance, you may be thinking that CAD Export or LandXML may be the way to go. However, those options are not the best way to maintain the integrity of your pipe and structure objects. The best option for getting your edits back into Civil 3D is to use File ➤ Export ➤ Hydraflow Storm Sewers File.

FIGURE 14.42
Use the Profile Plot Options dialog to control the display of elements in the graphic.

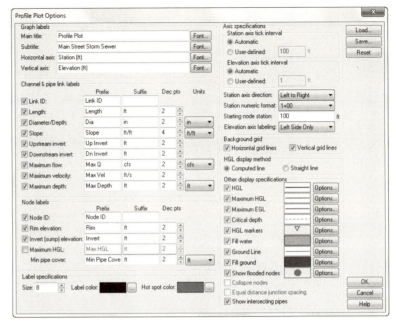

FIGURE 14.43
A sample profile plot

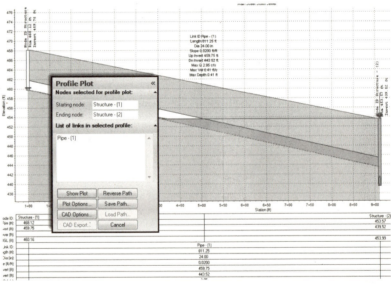

The following exercise will walk you through the steps needed to get back to Civil 3D:

1. From SSA, open the file `SSAtoC3D.spf`. You can download this file from this book's web page.

2. Choose File ➢ Export ➢ Hydraflow Storm Sewers File, as shown in Figure 14.44.

FIGURE 14.44
Export back to Civil 3D via Hydraflow

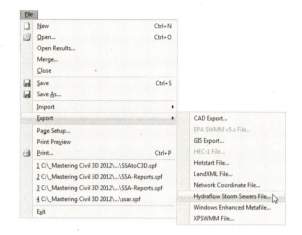

3. Save the file as **SSAtoC3D.stm** on your computer's desktop.
4. Close SSA.
5. In Civil 3D, open the file **C3DtoSSA.dwg**. From the Import panel, on the Insert tab, select Storm Sewers, as shown in Figure 14.45.
6. Browse to the STM file you exported and click Open. You will get a warning that there is already a pipe network with the same name in the file, as shown in Figure 14.46.
7. Click Update The Existing Pipe Network. After a moment, your pipes and structures should reflect any size or elevation changes that may have occurred in SSA.

FIGURE 14.45
Import the file as if it were a storm sewers file

FIGURE 14.46
Click Update The Existing Pipe Network.

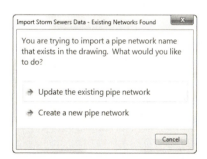

The Bottom Line

Create a catchment object. Catchment objects are the newest object type that Civil 3D can use to determine the area and time of concentration of a site. You can create a catchment from surface, but in most cases you will use the option to create from polyline.

Master It Create a catchment for predeveloped conditions on the site.

Export pipe data from Civil 3D into SSA. Civil 3D by itself cannot do pipe flow or runoff calculations. For this reason, it is important to export Civil 3D pipe networks into the Storm And Sanitary Analysis portion of the product.

Master It Verify your pipe export settings and then export to SSA.

Chapter 15

Grading

Beyond creating streets and sewers, cul-de-sacs, and inlets, much of what happens to the ground as a site is being designed must still be determined. Describing the final plan for the earthwork of a site is a crucial part of bringing the project together. This chapter examines feature lines and grading groups, which are the two primary tools of site design. These two functions work in tandem to provide the site designer with tools for completely modeling the land.

In this chapter, you will learn to:

- Convert existing linework into feature lines
- Model a simple linear grading with a feature line
- Model planar site features with grading groups

Working with Grading Feature Lines

There are two types of feature lines: corridor feature lines and grading feature lines. Corridor feature lines are discussed in Chapter 9, "Basic Corridors," and grading feature lines are the focus of this chapter. It's important to note that grading feature lines can be extracted from corridor feature lines, and you can choose whether or not to dynamically link them to the host object.

As discussed in Chapter 4, "Surfaces," terrain modeling can be defined as the manipulation of triangles created by connecting points and vertices to achieve Delaunay triangulation. In Land Desktop (LDT) and other software, this is often done with the use of native 3D polylines. In Civil 3D, the creation of the feature line object adds a level of control and complexity not available to 3D polylines. In this section, you look at the feature line, various methods of creating feature lines, some simple elevation edits, planar editing functionality, and labeling of the newly created feature lines.

Accessing Grading Feature Line Tools

The Feature Line creation tools can be accessed from the Home tab's Create Design panel, as shown in Figure 15.1.

The Feature Line editing tools can be accessed via the Design panel on the Modify tab, or via the Feature Line context tab that's available after you select an existing feature line (see Figure 15.2).

FIGURE 15.1
The Feature Line drop-down menu on the Create Design panel

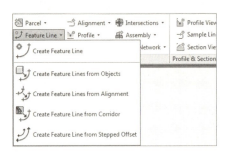

FIGURE 15.2
The Feature Line context tab accessed by selecting an existing feature line

One thing to remember when working with feature lines is that they do belong in a site. Feature lines within the same site snap to each other in the vertical direction and can cause some confusion when you're trying to build surfaces. If you're experiencing some weird elevation data along your feature line, be sure to check out the parent site. If the concept of a parent site or sites in general doesn't make much sense to you, be sure and look at Chapter 5, "Parcels," before going too much further. Sites are a major part of the way feature lines interact with each other, and many users who have problems with grading completely ignore sites as part of the equation.

The next few sections break down the various tools in detail. You'll use almost all of them in this chapter, so in each section you'll spend some time getting familiar with the available tools and the basic concepts behind them.

Creating Grading Feature Lines

There are five primary methods for creating feature lines, as shown in Figure 15.1. They generate similar results but have some key differences:

- The Create Feature Line tool allows you to create a feature line from scratch, assigning elevations as you go. These elevations can be based on direct data input at the command line, slope information, or surface elevations.

- The Create Feature Lines From Objects tool converts lines, arcs, polylines, and 3D polylines into feature lines. This process also allows elevations to be assigned from a surface or grading group.

- The Create Feature Lines From Alignment tool allows you to build a new feature line from an alignment, using a profile to assign elevations. This feature line can be dynamically tied to the alignment and the profile, making it easy to generate 3D design features based on horizontal and vertical controls.

- The Create Feature Line From Corridor command is used to export a grading feature line from a corridor feature line.

- The Create Feature Line From Stepped Offset tool is used to create a feature line from an offset and the difference in elevation from a feature line, survey figure, polyline, or 3D polyline.

You explore each of these methods in the next few exercises. In this exercise, you will be creating a swale from feature lines.

1. Open the CreatingFeatureLines.dwg file. (Remember, all data can be downloaded from www.sybex.com/go/masteringcivil3d2012.) This drawing has the outer base line of the subdivision drawn and an overland swale.

2. Change to the Home tab and select Feature Line ➢ Create Feature Line from the Create Design panel to display the Create Feature Lines dialog. We need to create a new site in which to put the swale. Just like parcels, grading objects will react with like objects in a site. So by isolating this swale, we can ensure that everything will be drawn properly before committing it to a site. Or you can always leave it on the created site and then create composite surfaces (see Chapter 4 for more information on composite surfaces).

3. Click the Create New button to the right of the site name. The Site Properties dialog opens. Enter **Swale** for the name of this site and click OK. The Create Feature Lines dialog should now look like Figure 15.3.

FIGURE 15.3
The Create Feature Lines dialog

4. Click OK. The command prompt now reads Specify start point:.

5. Use an Endpoint osnap to pick the line on the right (closest to the red line).

6. Enter S↵ at the command line to use a surface to set elevation information. In the Select Surface dialog, use the drop-down to choose EG and click OK.

7. Press ↵ to accept the elevation offered.

8. Use an Endpoint osnap to pick the end of the left side of the line.

9. Enter **-5.2**↵ to set the grade between points.

10. Press ↵ to exit the command. Your screen should look like Figure 15.4.

FIGURE 15.4
Setting the grade between points

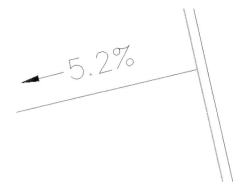

Looking back at Figure 15.3, note that there is an unused option for assigning a Name value to each feature line. Using names does make it easier to pick feature line objects if you decide to use them for building a corridor object. The Name option is available for each method of creating feature line objects, but will generally be ignored for this chapter except for one exercise covering the renaming tools available.

This method of creating a feature line connecting a few points seems pretty tedious to most users. In the next exercise, you convert an existing polyline to a feature line and set its elevations on the fly:

1. Open the `CreatingFeatureLines.dwg` file if it isn't open already.

2. Change to the Home tab and choose Feature Line ➢ Create Feature Lines From Objects from the Create Design panel.

3. Pick the red closed polyline representing the limits of grading.

4. Right-click and select Enter, or press ↵. The Create Feature Lines dialog will appear. Put the polyline on a site called Rough Grading. This will be the beginning of our overall grading plan.

There are some differences in the Create Feature Lines dialog from that shown on Figure 15.3. Notably, the Conversion Options near the bottom of the dialog are now active, so take a look at the options presented:

- Erase Existing Entities removes the object onscreen and replaces it with a feature line object. This avoids the creation of duplicate linework, but could be harmful if you wanted your linework for planimetric purposes.

- Assign Elevations lets you set the feature line point of intersection (PI) elevations from a surface or grading group, essentially draping the feature line on the selected object.

- Weed Points decreases the number of nodes along the object. This option is handy when you're converting digitized information into feature lines.

5. Check the Assign Elevations box. The Erase Existing Entities option is on by default, and you will not select the Weed Points option.

6. Click OK to display the Assign Elevations dialog. Here you can select a surface to pull elevation data from, or assign a single elevation to, all PIs.

7. Make sure that the EG surface is selected. Check the Insert Intermediate Grade Break Points option. This inserts a PI at every point along the feature line where it crosses an underlying TIN line.

8. Click OK.

9. Pick the site outline (which is now the color green), and the grips will look like Figure 15.5.

FIGURE 15.5
Conversion to a feature line object

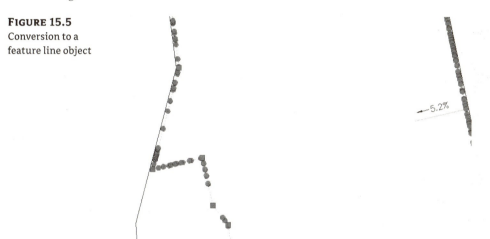

Note that in Figure 15.5 there are two types of grips: circular and square. Feature lines offer feedback via the grip shape. A square feature line grip indicates a full PI. This node can be moved in the x, y, and z directions, manipulating both the horizontal and vertical design. Circular grips are elevation points only. In this case, the elevation points are located at the intersection of the original polyline and the TIN lines existing in the underlying surface. Elevation points can only be slid along a given feature line segment, adjusting the vertical design, but cannot be moved in a horizontal plane. This combination of PIs and elevation points makes it easy to set up a long element with numerous changes in design grade that will maintain its linear design intent if the endpoints are moved.

Both of the methods used so far assume static elevation assignments for the feature line. They're editable, but are not physically related to other objects in the drawing. This is generally acceptable, but sometimes it's necessary to have a feature line that is dynamically related to an object. For grading purposes, it is often ideal to create a horizontal representation of a vertical profile along an alignment. Rather than build a corridor model as discussed in Chapters 9 and 10, a dynamic feature line can be extracted from a profile along an alignment, offset both horizontally and vertically, and used for grading. In the following example, a dynamic feature line is extracted from an alignment. Elevations for the vertices of the feature line are extracted from a profile, and finally, offset using the Create Feature Line From Stepped Offset tool to represent a swale.

1. Open the `SteppedOffset.dwg` file.

2. Change to the Home tab and select Feature Line ➢ Create Feature Lines From Alignment in the Create Design panel.

3. Select the Cabernet Court alignment at the lower portion of the site (the vertical alignment) to display the Create Feature Line From Alignment dialog. Make changes as shown in Figure 15.6. Note the Create Dynamic Link To The Alignment option near the bottom.

FIGURE 15.6
The Create Feature Line From Alignment dialog

4. Deselect the option to weed points, as shown in Figure 15.6. Click OK.

5. Change to the Home tab and select Feature Line ➢ Create Feature Line From Stepped Offset.

6. Enter **25** (**7.62** m) at the command line and press ↵.

7. Select the dynamic feature line along the alignment.

8. Select a point to the right of the alignment when you see the prompt `Specify side to offset or [Multiple]:`.

9. The command prompt will offer you several choices for specifying elevations or grades to be used along the feature lines. If necessary, type the appropriate letters until

the command prompt reads Specify elevation difference or [Grade/Slope/Elevation/Variable]:.

10. Enter **0** as the elevation difference at the command line and press ↵. This simply offsets the line from the road centerline to the right-of-way line. We will adjust this in a later exercise.

11. Repeat steps 8 through 10, but this time, pick a point to the left of the alignment. Your results should appear as shown in Figure 15.7. Leave this drawing open to use in the next exercise.

FIGURE 15.7
The completed alignment with offsets in place

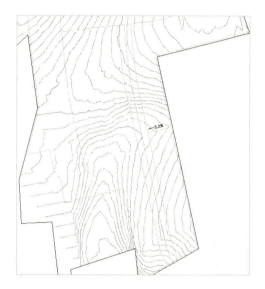

If you click the alignment now, you will select the feature line, but you will not see any grips available. This is because the feature line is dynamically linked to the design profile along the alignment and can't be modified. If either the alignment or the profile changes, the dynamic feature line will automatically update. Simply repeat the preceding procedure to create new offsets if needed. These three feature lines can be included in a new surface definition as breaklines, as discussed in Chapter 4.

Because dynamically linked feature line objects are slightly different, you'll look at them in our next exercise before using the name and style buttons to update all the feature lines:

1. This exercise uses the dynamic feature line created in the previous exercise. Pan or zoom to view the feature line along the Cabernet Court alignment.

2. Select the feature line.

3. Choose Feature Line Properties from the Modify panel to open the Feature Line Properties dialog, as shown in Figure 15.8. The information displayed on the Information tab is unique to the dynamic feature line, and you still have some level of control over the linking options.

FIGURE 15.8
The Feature Line Properties dialog for an alignment-based feature line

4. Deselect the Dynamic Link option and click OK. Notice the grips appear. The dynamic relationship between the feature line and the alignment has been severed.

5. Select Feature Line Properties from the Modify panel and notice the Dynamic Link options have disappeared.

6. Select the two feature lines opposite the alignment.

7. Choose Apply Feature Line Styles from the Modify panel. The Apply Feature Line Style dialog will appear.

8. Select the Corridor Ditch style from the Style drop-down list and click OK to dismiss the dialog.

Many people don't see much advantage in using styles with feature line objects, but there is one major benefit: linetypes. Zoom in on the feature lines on either side of the alignment and you'll see the Grading Ditch style has a dashed linetype. 3D polylines do not display linetypes, but feature lines do. If you need to show the linetypes in your grading, feature line styles are your friend.

In the Feature Line context menu activated by selecting a feature line, several more commands can be found on the Modify panel, as shown in Figure 15.9, that are worth examining before we get into editing objects.

FIGURE 15.9
The Modify panel on the context Feature Line tab

The Modify panel commands provide access to various properties of the feature line, the feature line style, and the feature line geometry as follows:

- The Feature Line Properties drop-down menu contains two commands. The first command, Feature Line Properties, is used to access various physical properties such as minimum and maximum grade. Only the name and style of the feature line can be modified on the Information tab, as shown in Figure 15.10. The second command, Edit Feature Line Style, is used to access various display characteristics such as color and linetype.

FIGURE 15.10
The Information tab of the Feature Line Properties dialog

- The Edit Geometry toggle opens the Edit Geometry panel on the Feature Line tab (see Figure 15.11). This panel will remain open until the Edit Geometry button is toggled off (it's highlighted when toggled on).

- The Edit Elevations toggle opens the Edit Elevations panel on the Feature Line tab (see Figure 15.12). This panel will remain open until the Edit Elevations button is toggled off (it's highlighted when toggled on).

FIGURE 15.11
The Edit Geometry panel on the Feature Line tab

FIGURE 15.12
The Edit Elevations panel on the Feature Line tab

- The Add To Surface As Breakline tool allows you to select a feature line or feature lines to add to a surface as breaklines.

- The Apply Feature Line Names tool allows you to change a series of feature lines en masse based on a new naming template. This tool can be helpful when you want to rename a group or just assign names to feature line objects. This tool cannot be used on a feature line that is tied to an alignment and profile.

- The Apply Feature Line Styles tool allows you to change feature line objects and their respective styles en masse. Many users don't apply styles to their feature line objects because the feature lines are found in grading drawings and not meant to be seen in construction documents. But if you need to make a global change, you can.

Once the Feature Line tab has been activated, the Quick Profile tool is available on the Launch Pad panel. The Quick Profile tool generates a temporary profile of the feature line based on user parameters found in the Create Quick Profiles dialog (Figure 15.13).

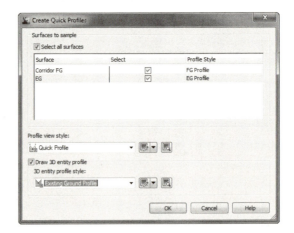

Figure 15.13
The Create Quick Profiles dialog

A few notes on this operation:

- Civil 3D creates a phantom alignment that will not display in Prospector as the basis for a quick profile. A unique alignment number is assigned to this alignment, so your number might be different from that shown.

- A closed feature line will generate a parcel when the quick profile is executed.

- Panorama will display a message to tell you that a profile view has been generated. You can close Panorama or move the Panorama palette out of the way if necessary.

Now that you've created a couple of feature lines, you'll edit and manipulate them some more.

Editing Feature Line Horizontal Information

Creating feature lines is straightforward. The Edit Geometry and Edit Elevations tools make them considerably more powerful than a standard 3D polyline. The Edit Geometry and Edit Elevations tools can be accessed by changing to the Modify tab and choosing Feature Line, or simply by selecting an existing feature line. Both tools are found on the Modify panel of the Feature Line tab. The Edit Geometry functions are examined in this section, and the Edit Elevations functions are described in the next section.

Grading revisions often require adding PIs, breaking apart feature lines, trimming, and performing other planar operations without destroying the vertical information. To access the commands for editing feature line horizontal information, first change to the Modify tab and choose Feature Line from the Design panel. Then choose Edit Geometry from the Modify panel to open the Edit Geometry panel. The first two tools are designed to manipulate the PI points that make up a feature line:

- The Insert PI tool allows you to insert a new PI, controlling both the horizontal and vertical design.

- The Delete PI tool removes a PI. The feature line will mend the adjoining segments if possible, attempting to maintain similar geometry.

The next few tools act like their AutoCAD counterparts, but understand that elevations are involved and add PIs accordingly:

- The Break tool operates much like the AutoCAD Break command, allowing two objects or segments to be created from one. Additionally, if a feature line is part of a surface definition, both new feature lines are added to the surface definition to maintain integrity. Elevations at the new PIs are assigned on the basis of an interpolated elevation.

- The Trim tool acts like the AutoCAD Trim command, adding a new end PI on the basis of an interpolated elevation.

- The Join tool creates one feature line from two, making editing and control easier.

- The Reverse tool changes the direction of a feature line.

- The Edit Curve tool allows you to modify the radius that has been applied to a feature line object.

- The Fillet tool inserts a curve at PIs along a feature line and will connect feature lines sharing a common PI that are not actually connected.

The last few tools refine feature lines, making them easier to manipulate and use in surface building:

- The Fit Curve tool analyzes a number of elevation points and attempts to define a working arc through them all. This tool is often used when the corridor utilities are used to generate feature lines. These derived feature lines can have a large number of unnecessary PIs in curved areas.

- The Smooth tool turns a series of disjointed feature line segments and creates a best-fit curve. This tool is great for creating streamlines or other natural terrain features that are known to curve, but there's often not enough data to fully draw them that way.

- The Weed tool allows the user to remove elevation points and PIs on the basis of various criteria. This is great for cleaning up corridor-generated feature lines as well.

- As discussed in detail earlier, the Stepped Offset tool allows offsetting in a horizontal and vertical manner, making it easy to create stepped features such as stairs or curbs.

By using these controls, it's easier to manipulate the design elements of a typical site while still using feature lines for surface design. In this exercise, you manipulate a number of feature lines that were created by corridor operations:

1. Open the `HorizontalFeatureLineEdits.dwg` file. This drawing is a continuation of the `SteppedOffset.dwg` file but has been populated with some more feature lines.

2. Pick the lower horizontal feature line (which is the Syrah Way road), as shown in Figure 15.14.

FIGURE 15.14
Picking the lower feature line on Syrah Way

3. Choose Edit Geometry from the Modify panel to enable the Edit Geometry panel.

4. Click the Break tool. Pick the lower feature line again.

5. Enter **F↵** to pick the first point of the break. When prompted to `Specify second break point or [First point]:`, enter **F↵**.

6. Using an Intersection osnap, pick the intersection of the horizontal feature line and the vertical feature line, as shown in Figure 15.15.

FIGURE 15.15
Using the Intersection object snap to select a point

7. Using a Nearest osnap, pick a point on the horizontal feature line, leaving a gap, as shown in Figure 15.16.

8. Pick the two vertical feature lines to activate grips on both lines. Notice the large number of grips. Press Esc.

9. Pick the left horizontal feature line that was previously broken and click the Weed tool on the Edit Geometry panel to display the Weed Vertices dialog.

10. Pick the feature line again when you see the prompt Select a feature line, 3d polyline or [Multiple/Partial]:. The Weed Vertices dialog will be displayed.

11. Watch the glyphs on the feature line and notice the number of vertices that will be removed from the feature line. Additionally, the glyphs on the feature line itself will change from green to red to reflect nodes that will be removed under the current setting, as shown in Figure 15.17. Click OK to complete the command and dismiss the Weed Vertices dialog.

FIGURE 15.16
The feature line after executing the Break command

FIGURE 15.17
Feature lines to be weeded

12. Using a standard AutoCAD Extend command, extend the horizontal feature lines that contain a gap.

13. Pick the left horizontal feature line and click the Trim tool from the Edit Geometry panel. A standard AutoCAD trim will not work in this case. Pick the horizontal feature line as the cutting edge and press ↵. Choose the vertical feature line when you see the prompt Select objects to trim:. Press ↵ to complete the command and review the results (see Figure 15.18). Press Esc.

FIGURE 15.18
The left horizontal feature line trimmed

14. Select the left horizontal feature line and click the Join tool on the Edit Geometry panel.

15. Select the left vertical feature line and press ↵. Notice the grips and that the two feature lines have been joined.

16. Select the Fillet tool. Enter **R**↵ at the command line to adjust the radius value. Enter **15**↵ (**4.57** m) to update the radius value.

17. Move your cursor toward the intersection of feature lines B and D until the glyph appears. Pick near the glyph to fillet the two feature lines.

18. Click the Edit Curve tool from the Edit Geometry panel and pick the curve when you see the prompt `Select feature line curve to edit or [Delete]:`. The Edit Feature Line Curve dialog opens, as shown in Figure 15.19.

FIGURE 15.19
The Edit Feature Line Curve dialog

19. Enter a Radius value of **34** (**10.36** m), as shown in Figure 15.20, and click OK to close the dialog. Press ↵ to exit the command.

When modifying the radius of a feature line curve, it's important to remember that you must have enough tangent feature line on either side of the curve segment to make a curve fit, or the program will not make the change. In that case, tweak the feature line on either side of the arc until there is a mathematical solution. Sometimes it is necessary to use the Weed Vertices tool to remove vertices and create enough room to fillet feature lines. In some cases, it may be necessary to plan ahead when creating feature lines to ensure that vertices will not be placed too closely together.

FIGURE 15.20
Filleted
feature lines

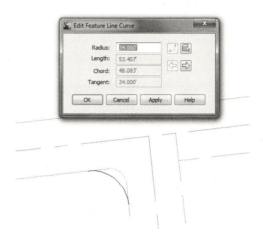

Editing Feature Line Elevation Information

To access the commands for editing feature line elevation information, first change to the Modify tab and choose Feature Line from the Design panel. Then choose Edit Elevations from the Modify panel to open the Edit Elevations panel. Moving across the panel, you find the following tools for modifying or assigning elevations to feature lines:

- The Elevation Editor tool activates a palette in Panorama to display station, elevation, length, and grade information about the feature line selected. Feature lines, survey figures, and parcel lines can be edited using this tool.

- The Insert Elevation Point tool inserts an elevation point at the point selected. Note that this point can control only elevation information; it does not act as a horizontal control point. Feature lines, survey figures, parcel lines, and 3D polylines can be edited using this tool.

- The Delete Elevation Point tool deletes the selected elevation point; the points on either side then become connected linearly on the basis of their current elevations. Feature lines, survey figures, parcel lines, and 3D polylines can be edited using this tool.

- The Quick Elevation Edit tool allows you to use onscreen cues to set elevations and slopes quickly between PIs on any feature lines or parcels in your drawing.

- The Edit Elevations tool steps through the selected feature line, much like working through a polyline edit at the command line. Feature lines, survey figures, parcel lines, and 3D polylines can be edited using this tool.

- The Set Grade/Slope Between Points tool sets a continuous slope along the feature line between selected points. Feature lines, survey figures, parcel lines, and 3D polylines can be edited using this tool.

- The Insert High/Low Elevation Point tool places a new elevation point on the basis of two picked points and the forward and backward slopes. This calculated point is simply placed at the intersection of two vertical slopes. Feature lines, survey figures, parcel lines, and 3D polylines can be edited using this tool.

- The Raise/Lower By Reference tool allows you to adjust a feature line elevation based on a given slope from another location. Feature lines, survey figures, parcel lines, and 3D polylines can be edited using this tool.

- The Set Elevation By Reference tool sets the elevation of a selected point along the feature line by picking a reference point, and then establishing a relationship to the selected feature line point. This relationship isn't dynamic! This button also acts as a flyout for the next three tools. Feature lines, survey figures, parcel lines, and 3D polylines can be edited using this tool.

- The Adjacent Elevations By Reference tool allows you to adjust the elevation on a feature line by coming at a given slope or delta from another point or feature line point. Feature lines, survey figures, parcel lines, and 3D polylines can be edited using this tool.

- The Grade Extension By Reference tool allows you to apply the same grades to different feature lines across a gap. For example, you might use this tool along the back of curbs at locations such as driveways or intersections. Feature lines, survey figures, parcel lines, and 3D polylines can be edited using this tool.

- The Elevations From Surface tool sets the elevation at each PI and elevation point on the basis of the selected surface. It will optionally add elevation points at any point where the feature line crosses a surface TIN line. Feature lines, survey figures, parcel lines, and 3D polylines can be edited using this tool.

- The Raise/Lower tool simply moves the entire feature line in the z direction by an amount entered at the command line. Feature lines, survey figures, parcel lines, and 3D polylines can be edited using this tool.

Quite a few tools are available to modify and manipulate feature lines. You won't use all of the tools, but at least you'll have some concept of what is available. The next exercises give you a look at a few of them. In this first exercise, you'll make manual edits to set the grade of a feature line segment:

1. Open the `EditingFeatureLineElevations.dwg` file.

2. Select one of the vertical feature lines. The Grading Elevation Editor in Panorama will open, as shown in Figure 15.21.

FIGURE 15.21
The Grading Elevation Editor

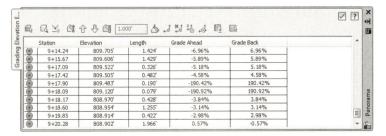

3. Click in the Grade Ahead column for the first PI. It's hard to see in the images, but as a row is selected in the Grading Elevation Editor, the PI or the elevation point that was selected will be highlighted on the screen with a small glyph.

4. You can use the Editor to make changes to the Station, Elevation, Length, Grade Ahead, and Grade Back. The exception is that you cannot edit stationing for primary geometry points, as indicated by the triangle glyph in the Grading Elevation Editor.

5. Click the green check mark in the upper-right corner to dismiss the Panorama. We will look at the Grading Elevation Editor more in-depth next.

Using the Grading Elevation Editor is the most basic way to manipulate elevation information.

The Grading Elevation Editor

We have discussed this very important editor and the plethora of tools available inside it. Many of the tools may seem redundant from the Feature Line Edit Elevations tools, but they are placed here for ease of use. Refer back to Figure 15.21 for a graphical display of the Grading Elevation Editor panorama.

- Clicking the Select tool allows you to select the object for editing.

- The Zoom To tool will do exactly as it says. If you have a station highlighted in the panorama and select the Zoom To tool, your plan view will be zoomed to that elevation point.

- The Quick Profile tool will create a quick profile based on the feature line selected. The Create Quick Profiles dialog will open and allow you to select what surface(s) you wish to display as well as what profile view style and 3D entity profile you want.

- Clicking the Raise/Lower tool will activate the Set Increment text box. This allows you to raise or lower selected elevation points, or if no elevation points are selected, it will raise or lower the entire feature line elevation points by the amount displayed in the text box.

- The Raise Incrementally/Lower Incrementally tools will raise or lower the elevation point or points by the amount listed in the Increment text box.

- The Set Increment tool works in tandem with the text box. You can type in a value that will be used by other tools for raising or lowering elevation point or points.

- The Flatten Grade Or Elevations tool will make all selected elevation points the value of the first selected point, or if no points are selected, the value of the first entry in the cells. When this tool is selected, the Flatten dialog will open asking if you want to flatten by constant elevation or grade.

- The Insert Elevation Point will let you select a spot on the feature line, and will create an intermediate elevation point. The Insert PVI dialog box will open, allowing you to fine-tune the station and enter an elevation.

- The Delete Elevation Point will delete a point or points that are highlighted in the editor. Note that this tool will only allow you to delete intermediate points.

◆ Clicking the Elevations For Surface tool will open the Select Surface dialog if there are multiple surfaces to choose from. If you have an elevation point or points selected, it will only affect those points, or if nothing is selected it will use all elevation points and drape them onto the selected surface.

 ◆ The Reverse direction tool will do exactly as it says; it will reverse the direction of the feature line.

 ◆ The Show Grade Breaks Only tool is a toggle (click, it's on and click, it's off) that will display only grade breaks on a feature line.

 ◆ The Unselect All Rows tool does exactly as it says; it will deselect any rows that have been highlighted for editing.

More Feature Line Editing Tools

Some of the relative elevation tools are a bit harder to understand, so you'll look at them in our next exercise and see how they function in some basic scenarios:

1. Open the `EditingRelativeFeatureLineElevations.dwg` file. This drawing contains a sample layout with some curb and gutter work.

2. Zoom to the ramp shown on the left-hand side of the intersection.

3. Select the feature line describing the ramp. Select Edit Elevations from the Modify panel to display the Edit Elevations panel if it isn't displayed already. Select Elevation Editor. The Panorama appears. Notice that the entire feature line is at elevation 0.000' (0.000 m). Click the green check mark in the upper right to close the Panorama. Press Esc to cancel the grips.

4. Select the feature line representing the flowline of the curb and gutter area.

5. Select the Adjacent Elevations By Reference tool. Civil 3D will prompt you to select the object to edit. You will edit the elevations along the feature line describing the ramp area.

6. Click the feature line describing the ramp area and Civil 3D will display a number of glyphs and lines to represent what points along the flowline it is using to establish elevations from, and it will prompt you for the elevation difference (you could also use a grade or slope).

7. Enter **0.5** (**0.15** m) at the command line and press ↵ to update the ramp elevations. Press ↵ again to exit the command, and press Esc to cancel grips.

8. Select the feature line describing the ramp. Select Elevation Editor, and the Panorama appears. Notice that the PIs now each have an elevation, as shown in Figure 15.22.

9. Click the green check mark in the upper right of the Panorama to dismiss it. Leave this drawing open.

Figure 15.22
Completed editing of the curb ramp feature line

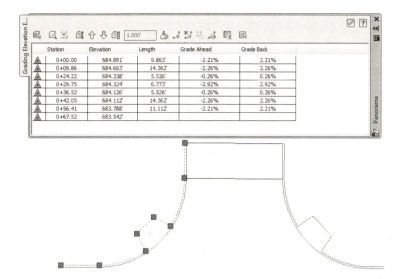

Next, you'll need to extend the grade along the line representing the flowline of the curb and gutter area on the left of the screen to determine the elevation to the south of your intersection on the right of the screen. You'll use the Grade Extension By Reference tool in the following exercise to accomplish this:

1. This exercise is a continuation of the previous exercise. Click the feature line representing the flowline of the curb and gutter area on the left side of the drawing.

2. Select the Grade Extension By Reference tool and select the flowline again when you see the prompt `Select reference segment:` (see Figure 15.23). This tool will evaluate the feature line as if it were three separate components (two lines and an arc). Because you are extending the grade of the flowline to the east and across the intersection, it is important to select the tangent segment, as shown in Figure 15.23.

Figure 15.23
Selecting the flowline of the curb and gutter section

3. At the `Select object to edit:` prompt, select the line representing the flowline of the curb and gutter area to the right of your screen, as shown in Figure 15.24.

4. At the `Specify point:` prompt, pick the PI at the left side of the tangent, as shown in Figure 15.25.

FIGURE 15.24
Selecting the flowline at the left side of the tangent segment

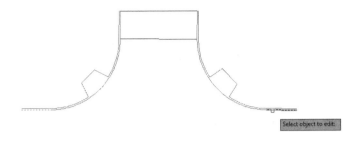

FIGURE 15.25
Selecting the PI at the left side of the tangent segment representing the flowline of the curb and gutter section

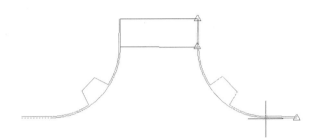

5. At the Specify grade or [Slope/Elevation/Difference] <2.21>: prompt, press ↵ to accept the default value of 2.21 (this is the grade of the flowline of the curb and gutter section to the left). Press ↵ to end the command, and press Esc to cancel grips.

6. Select the feature line representing the flowline of the curb and gutter section to the right of your screen to enable grips.

7. Move your cursor over the top of the PI, as shown in Figure 15.26, but do not click. Your cursor will automatically snap to the grip, and the grip will change color. Notice that the elevation of the PI is displayed on the status bar, as shown in Figure 15.26. This is a quick way to check elevations of vertices when modeling terrain. If your coordinates are not displayed, type the AutoCAD command **COORDS**, and set the value to **1** to display them. Leave this drawing open for the next exercise.

FIGURE 15.26
The x-, y-, and z-coordinates of the PI displayed on the status bar

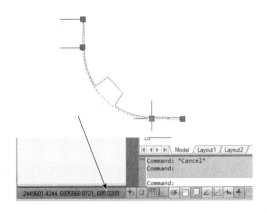

With the elevation of a single point determined, the grade of the feature line representing the curb and gutter section can be modified to ensure positive water flow. In the following exercise, you'll use the Set Grade/Slope Between Points tool to modify the grade of the feature line:

1. This exercise is a continuation of the previous exercise. Select the feature line representing the flowline of the curb and gutter section to the right of your screen. Select the Set Grade/Slope Between Points tool from the Edit Elevations panel.

2. When you see the prompt Specify the start point:, pick the PI as shown previously in Figure 15.25. The elevation of this point has been established and will be used as the basis for grading the entire feature line.

3. At the Specify elevation <685.021>: prompt, press ↵ to accept the default value. This is the current elevation of the PI as established earlier.

4. When you see the prompt Specify the end point:, pick the PI as shown in Figure 15.27.

FIGURE 15.27
Specifying the PI to establish the elevation at the flowline of the curb and gutter section

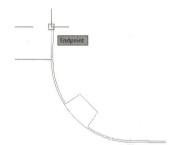

5. At the Specify grade or [Slope/Elevation/Difference]: prompt, type **2** and press ↵ to set the grade between the points at 2 percent. Do not exit the command.

6. When you see the prompt Select object:, select the feature line representing the flowline of the curb and gutter section to the right again.

7. When you see the prompt Specify the start point:, pick the PI (shown earlier in Figure 15.26) again.

8. At the Specify elevation <685.021> prompt, press ↵ to accept the default value. This is the current elevation of the PI as established earlier.

9. When you see the prompt Specify the end point:, pick the PI at the far bottom right along the feature line currently being edited.

10. At the Specify grade or [Slope/Elevation/Difference]: prompt, type **-2** and press ↵ to set the grade between the points at negative 2 percent. Press ↵ to exit the command but do not cancel grips.

11. Select the Elevation Editor tool from the Edit Elevations panel to open Panorama. Notice the values in both the Grade Ahead and Grade Back columns, as shown in Figure 15.28.

Figure 15.28
The grade of the feature line set to 2 percent

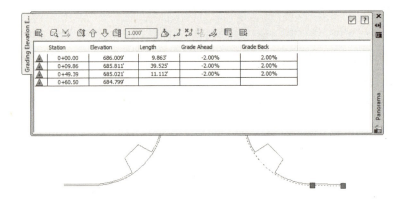

The possibilities are endless. In using feature lines to model proposed features, you are limited only by your creative approach. You've seen many of the tools in action, so you can now put a few more of them together and grade your pond.

Real World Scenario

Draining the Pond

You need to use a combination of feature line tools and options to pull your pond together, and get the most flexibility should you need to update the bottom area or manipulate the pond's general shape. Here is the method that was used when this pond was designed:

1. Open the PondDrainageDesign.dwg file. The engineer gave us a bit more information about the pond design, as shown here:

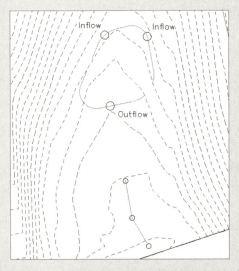

2. Click the pond basin outline and click the Insert Elevation Point tool.

3. Use the Center osnap to insert the elevation points in the center of each circle. Enter **779.5** (**237.59** m) as the elevation of each inflow, and accept the default elevation at the outflow. (You will change the elevation at the outflow in a moment.) Press ↵ to exit the command, and press Esc to cancel grips.

4. Draw a polyline, similar to the one shown here, from the northwestern inflow point, through the outflow, and snap to the endpoint of the drainage channel feature line. Be sure to place a PI at the center of the circle designating the outflow point. This polyline will be the layout for the pilot channel.

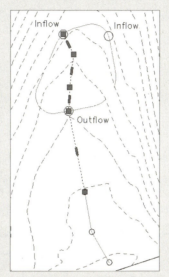

5. Change to the Home tab and choose Feature Line ➤ Create Feature Lines From Objects from the Create Design panel. Pick the polyline just drawn and press ↵. The Create Feature Lines dialog appears.

6. Click OK to complete the conversion, making sure that the Assign Elevations check box is left unchecked. Because this feature line was created in the same site as the bottom of the pond, the elevation of the PI at the southernmost inflow will reset to match the elevation of the endpoint of the pilot channel.

7. Select the feature line representing the pilot channel and pick the Fillet tool from the Edit Geometry panel.

8. Enter **R**↵, and enter **25**↵ (**7.62** m) for the radius.

9. Enter **A**↵ to fillet all PIs.

10. Press ↵ to exit the Fillet command.

11. Select the feature line representing to pilot channel. Click the Set Grade/Slope Between Points tool on the Edit Elevations panel.

12. Pick the PI at the inflow.

13. Enter **779.5**↵ (**237.59** m) to set this elevation.

14. Pick the other end of the feature line, as shown here. All the PIs will highlight, and Civil 3D will display the total length, elevation difference, and average slope at the command line.

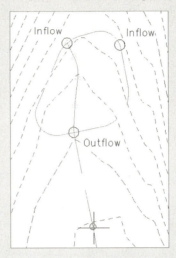

15. Press ↵ again to accept the grade as shown. This completes a linear slope from one end of the feature line to the other, ensuring drainage through the pond and outfall structure.
16. Press ↵ to exit the command. Press Esc to cancel grips.
17. Select the bottom of the pond and choose the Elevation Editor tool to open the Grading Elevation Editor in Panorama.
18. Click in the Station cells in the Grading Elevation Editor to highlight and ascertain the elevation at the outfall, as shown here. In this case, the elevation is 775.471′ (236.36 m). Some minor variation might occur depending on your pick points and the length of your pilot channel. Notice also that the icon for this point in the Panorama display is a white triangle, indicating that this point is a point derived from a feature line intersection. This is called a *phantom PI*.

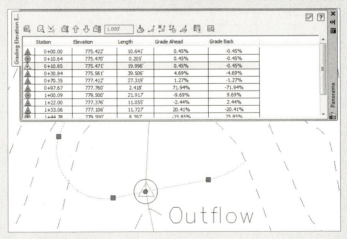

19. Click the Insert Elevation Point tool in the Panorama.
20. Using an intersection snap, set a new elevation point at the intersection of the pilot channel and the flowline. The Insert PVI dialog opens. Enter an elevation of **775.471′** (**236.36** m) or the elevation you ascertained earlier in the Elevation text box. This has to be added after the pilot channel has been created because the fillet process would tweak the location. Click OK to exit the Insert PVI dialog.
21. Close Panorama. A round grip is displayed at this new elevation point.
22. Select the pond bottom and click the Quick Elevation Edit tool on the Edit Elevations panel. Move your cursor over the PI at the inflow that we created the feature line from. Pick the PI. Enter **782** (**238.35** m) and press ↵ twice to exit the command.
23. Click the Set Grade/Slope Between Points tool on the Edit Elevations panel.
24. Move your cursor over the PI selected in step 22 once again and pick it. Press ↵ to accept the default elevation set previously as 782 (238.35 m).
25. Pick the PI at the outflow. Enter the elevation of **775.471** (**236.36** m).
26. Press ↵ to accept the grade. This sets the elevations between the one infall and the outfall PIs to all fall at the same grade.
27. Repeat steps 22–26 for the PI on the other infall (located on the northeast side of the pond) and the outfall. They will set the elevations between the other infall and the outfall to a constant slope.

The entire outline of the pond bottom is graded except the area between the two inflows, as shown here:

Low point between infalls

Because you want to avoid a low spot, you'll now force a high point:

1. Click the Insert High/Low Elevation Point tool.
2. Pick the PI near the northwestern inflow.
3. Pick the PI near the northeastern inflow as the endpoint. Enter **1.0**↵ as the grade ahead.

4. Enter **1.0↵** as the grade behind. A new elevation point will be created, as shown here:

5. Trim the feature line from inside the bottom of the pond using the Feature Line Trim tool.

By using all the tools in the Feature Lines toolbar, you can quickly grade elements of your design and pull them together. If you have difficulty getting all the elevations in this exercise to set as they should, slow down, and make sure you are moving your mouse in the right direction when setting the grades by slope. It's easy to get the calculation performed around the other direction—that is, clockwise versus counterclockwise. This procedure seemed to involve a lot of steps, but it takes less than a minute in practice.

There are roughly 25 ways to modify feature lines using both the Edit Geometry and Edit Elevations panels. Take a few minutes and experiment with them to understand the options and tools available for these essential grading elements. By manipulating the various pieces of the feature line collection, it's easier than ever to create dynamic modeling tools that match the designer's intent.

Labeling Feature Lines

Though it's not common, feature lines can be stylized to reflect particular uses, and labels can be applied to help a reviewer understand the nature of the object being shown. In the next couple of exercises, you'll label a few critical points on your pond design to help you better understand the drainage patterns.

Feature Line Labels

Feature lines do not have their own unique label styles but instead share with general lines and arcs. You can learn more about styles in Chapter 19, "Styles." The templates that ship with Civil 3D contain styles for labeling segment slopes, so you'll label the critical slopes in the following exercise:

1. Open the `Labeling Feature Lines.dwg` file if you closed it.
2. Change to the Annotate panel and choose Add Labels from the Labels & Tables panel.

3. In the Add Labels dialog, choose Line And Curve from the Feature drop-down, and then change Line Label Style to Grade Only and Curve Label Style to Radius Only, as shown in Figure 15.29.

FIGURE 15.29
Adding feature line grade labels

4. Click Add.
5. Pick a few points along the Pilot Channel feature line tangents to create labels, as shown in Figure 15.29.

Although it would be convenient to label the feature line elevations as well, there's no simple method for doing so. In practice, you would want to label the surface that contains the feature line as a component.

Grading Objects

Once a linear feature line is created, there are two main uses. One is to incorporate the feature line itself directly into a surface object as a breakline; the other is to create a grading object (referred to hereafter as simply a grading or gradings) using the feature line as a baseline. A grading consists of some baseline with elevation information, and a criteria set for projecting outward from that baseline based on distance, slope, or other criteria. These criteria sets can be defined and stored in grading criteria sets for ease of management. Finally, gradings can be stylized to reflect plan production practices or convey information such as cut or fill.

In this section, you'll use a number of methods to create gradings, edit those gradings, stylize the gradings, and finally convert the grading group into a surface.

Creating Gradings

Let's look at grading the pond as designed. In this section, you'll look at grading groups and then create the individual gradings within the group. Grading groups act as a collection mechanism for individual gradings and let Civil 3D understand the daisy chain of individual gradings that are related and act in sync with each other.

One thing to be careful of when working with gradings is that they are part of a site. Any feature line within that same site will react with the feature lines created by the grading. For that reason, the exercise drawing has a second site called Pond Grading to be used for just the pond grading.

1. Open the `GradingThePond.dwg` file.
2. Pick the pond bottom.
3. Right-click and choose Move To Site. The Move To Site dialog appears.
4. Choose Pond Grading from the Destination Site drop-down list. Click OK. This will avoid interaction between the pond banks and the pilot channel you laid out earlier.

5. With the feature line still selected, choose Grading Creation Tools from the Launch Pad panel. The Grading Creation Tools toolbar, shown in Figure 15.30, appears. This toolbar is similar to the one used in pipe networks. The left section is focused on settings, the middle on creation, and the right on editing.

FIGURE 15.30
The Grading Creation Tools toolbar

6. Click the Set The Grading Group tool to the far left of the toolbar to display the Site dialog. Choose the Pond site and click OK. The Create Grading Group dialog is displayed.
7. Enter **Pond** in the Name text box, as shown in Figure 15.31, and click OK. You'll revisit the surface creation options in a bit.

FIGURE 15.31
Assign the name Pond in the Create Grading Group dialog.

8. Click the Select Target Surface tool located next to the Select Grading Group tool. Select the EG surface.
9. Click the Select A Criteria Set tool to display the Select A Criteria Set dialog.
10. Select Pond Grading from the drop-down list and click OK to close the dialog.

11. Click the Create Grading tool on the Grading Creation Tools toolbar, or click the down arrow next to the Create Grading tool and select Create Grading, as shown in Figure 15.32.

FIGURE 15.32
Creating a grading using the 3:1 To Elevation criteria

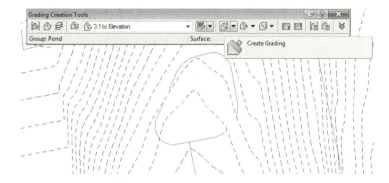

12. Pick the pond outline. If you get a dialog asking you to weed the feature line, select the Weed Feature Line option and click OK at the next dialog.

13. Pick a point on the outside of the pond to indicate the direction of the grading projections.

14. Press ↵ to apply the grading to the entire length of the pond outline.

15. Enter **784**↵ (**238.96** m) at the command line as the target elevation. The first grading is complete. The lines onscreen are part of the Grading style.

16. In the Select A Grading Criteria drop-down list on the Grading Creation Tools toolbar, select Flat Ledge.

17. Pick the upper boundary of the grading made in step 16, as shown in Figure 15.33.

FIGURE 15.33
Creating a daisy chain of gradings

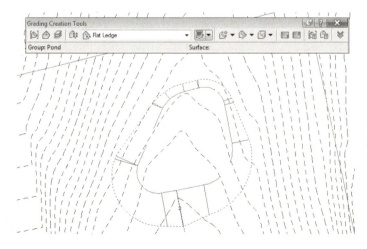

18. Press ↵ to apply to the whole length.
19. Enter **10**↵ (3.05 m) for the target distance to build the safety ledge.
20. In the Select A Grading Criteria drop-down list on the Grading Creation Tools toolbar, select 3:1 To Surface.
21. Pick the outer edge of the safety ledge just created.
22. Press ↵ to apply to the whole length. Press ↵ to exit the command. Your drawing should look similar to Figure 15.34.

FIGURE 15.34
Complete pond interior grading

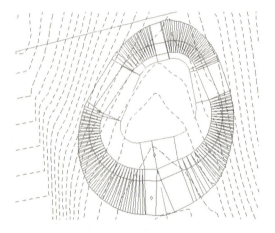

Each piece of this pond is tied to the next, creating a dynamic model of your pond design on the basis of the designer's intent. What if that intent changes? The next section describes editing the various gradings.

Editing Gradings

Once you've created a grading, you often need to make changes. A change can be as simple as changing the slope or changing the geometric layout. In this exercise, you'll make a simple change, but the concept applies to all the gradings you've created in your pond. Because the grading criteria were originally locked to make life easier, you'll now unlock them before modifying the daylight portion of the pond:

1. Open the EditingGrading.dwg file.
2. Pick one of the projection lines or the small diamond on the outside of the pond.
3. Choose the Grading Editor tool from the Modify panel. The Grading Editor in Panorama appears.
4. Enter **4:1** for a Fill Slope Projection, as shown in Figure 15.35.
5. Close Panorama. Your display should look like Figure 15.36. (Compare this to Figure 15.34 if you'd like to see the difference.)

FIGURE 15.35
Editing the Fill Slope value

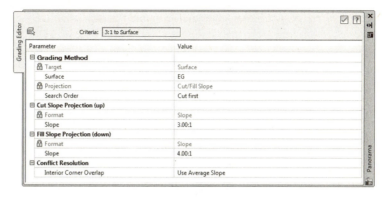

FIGURE 15.36
Completed grading edit

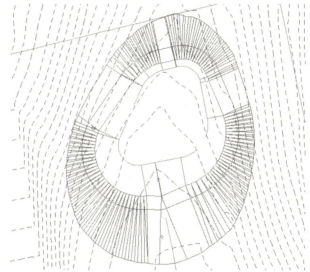

Editing any aspect of the grading will reflect instantly, and if other gradings within the group are dependent on the results of the modified grading, they will recalculate as well.

Creating Surfaces from Grading Groups

Grading groups work well for creating the model, but you have to use a TIN surface to go much further with them. In this section, you'll look at the conversion process, and then use the built-in tools to understand the impact of your grading group on site volumes.

1. Open the CreatingGradingSurfaces.dwg file.

2. Pick one of the diamonds in the grading group.

3. Select the Grading Group Properties tool from the Modify panel to display the Grading Group Properties dialog.

4. Switch to the Information tab if needed. Check the box for Automatic Surface Creation.

5. In the Create Surface dialog that appears, enter **Pond Surface** in the Name field.

6. Click in the Style field, and then click the ellipsis button. The Select Surface Style dialog appears. Select 1′ And 5′ (Prop) from the drop-down list in the selection box, as shown in Figure 15.37. Click OK twice to return to the Grading Group Properties dialog.

FIGURE 15.37
Creating a grading group surface

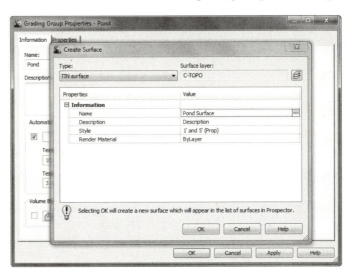

7. In the Grading Group Properties dialog, check the Volume Base Surface option and select the EG surface to perform a volume calculation. Click OK to dismiss the dialog.

You're going through this process now because you didn't turn on the Automatic Surface Creation option when you created the grading group. If you're performing straightforward gradings, that option can be a bit faster and simpler. There are two options available when creating a surface from a grading group. They both control the creation of projection lines in a curved area:

◆ The Tessellation Spacing value controls how frequently along an arced feature line TIN points are created and projection lines are calculated. A TIN surface cannot contain any true curves the way a feature line can because it is built from triangles. The default values typically work for site mass grading, but might not be low enough to work with things such as parking lot islands where the 10′ (3.05 m) value would result in too little detail.

◆ The Tessellation Angle value is the degree measured between outside corners in a feature line. Corners with no curve segment have to have a number of projections swung in a radial pattern to calculate the TIN lines in the surface. The tessellation angle is the angular distance between these radial projections. The typical values work most of the time, but in large grading surfaces, a larger value might be acceptable, lowering the amount of data to calculate without significantly altering the final surface created.

There is one small problem with this surface. If you examine the bottom of the pond, you'll notice there are no contours running through this area. If you move your mouse to the middle, you also won't have any Tooltip elevation as there is no surface data in the bottom of the pond. To fix that (and make the volumes accurate), you need a grading infill.

8. From the Modify tab and Modify panel, select Create Grading Infill.

9. Civil 3D will prompt you at the command line to select an infill area. Hover your cursor over the middle of the pond and the pond feature line created earlier will be highlighted, indicating a valid area for infill.

10. Click once to create the infill, and press ↵ to apply. Civil 3D will calculate, and Panorama may appear. Dismiss it. You should now have some contours running through the pond base area, as shown in Figure 15.38.

FIGURE 15.38
The pond after applying an infill grading

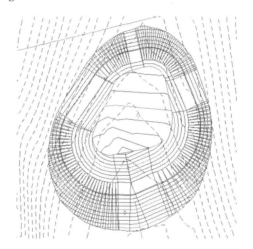

11. Zoom in if needed and pick one of the grading diamonds again to select one of the gradings. Make sure you grab one of the gradings and not the surface contours that are being drawn on top of them.

12. Select Grading Group Properties from the Modify panel to open the Grading Group Properties dialog.

13. Switch to the Properties tab to display the Volume information for the pond, as shown in Figure 15.39. This tab also allows you to review the criteria and styles being used in the grading group.

FIGURE 15.39
Reviewing the grading group volumes

This new surface is listed in Prospector and is based on the gradings created. A change to the gradings would affect the grading group, which would, in turn, affect the surface and these volumes. In the following exercise, you'll pull it all together:

1. Open the `CreatingCompositeSurfaces.dwg` file.

2. Right-click Surfaces in Prospector and select Create Surface.

3. In the Create Surface dialog, enter **Composite** in the Name text box. Click in the Style field, and then click the ellipsis to open the Select Surface Style dialog. Select the Contours 2′ And 10′ (Prop) option from the drop-down list box, and click OK.

4. Click OK to dismiss the Create Surface dialog and create the surface in Prospector.

5. In Prospector, expand the Surfaces ➢ Composite ➢ Definition branches.

6. Right-click Edits and select Paste Surface. The Select Surface To Paste dialog shown in Figure 15.40 appears.

FIGURE 15.40
Pasting surfaces together

7. Select EG from the list and click OK. Dismiss Panorama if it appears.

8. Right-click Edits again and select Paste Surface one more time.

9. Select Pond Surface and click OK.

10. Change the Corridor FG, EG, and Pond Surface display settings to No Display. The drawing should look like Figure 15.41.

Figure 15.41
Completed composite surface

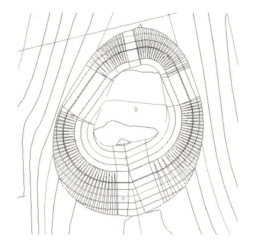

By creating a composite surface consisting of pasted-together surfaces, the TIN triangulation cleans up any gaps in the data, making contours that are continuous from the original grade, through the pond, and out the other side. With the grading group still being dynamic and editable, this composite surface reflects a dynamic grading solution that will update with any changes.

The Bottom Line

Convert existing linework into feature lines. Many site features are drawn initially as simple linework for the 2D plan. By converting this linework to feature line information, you avoid a large amount of rework. Additionally, the conversion process offers the ability to drape feature lines along a surface, making further grading use easier.

Master It Open the `MasteringGrading.dwg` file from the data location. Convert the magenta polyline describing a proposed temporary drain into a feature line and drape it across the EG surface to set elevations.

Model a simple linear grading with a feature line. Feature lines define linear slope connections. This can be the flow of a drainage channel, the outline of a building pad, or the back of a street curb. These linear relationships can help define grading in a model, or simply allow for better understanding of design intent.

Master It Add 100′ (30.48 m) radius fillets on the feature line you just created. Set the grade from the start of the feature line to the circled point to 5 percent and the remainder to a constant slope to be determined in the drawing. Draw a temporary profile view to verify the channel is below grade for most of its length.

Model planar site features with grading groups. Once a feature line defines a linear feature, gradings collected in grading groups model the lateral projections from that line to other points in space. These projections combine to model a site much like a TIN surface, resulting in a dynamic design tool that works in the Civil 3D environment.

Master It Use the two grading criteria just used to define the pilot channel, with grading on both sides of the sketched centerline. Calculate the difference in volume between them using 6:1 side slopes and 4:1 side slopes.

Chapter 16

Plan Production

So you've toiled for days, weeks, or maybe months creating your design in Civil 3D, and now it's time to share it with the world—or at least your corner of it. Even in this digital age, paper plan sets still play an important role. You generate these plans in Civil 3D using the Plan Production feature. This chapter takes you through the steps necessary to create a set of sheets, from initial setup, to framing and generating sheets, to data management and plotting.

In this chapter, you will learn to:

- Create view frames
- Edit view frames
- Generate sheets and review Sheet Set Manager
- Create section views

Preparing for Plan Sets

Before you start generating all sorts of wonderful plan sets, you must address a few concepts and prerequisites. Civil 3D takes advantage of several features and components to build a plan set. Some of these components have existed in AutoCAD and Civil 3D for years (for example, layout tabs, drawing templates, alignments, and profiles). Others are new properties of existing features (such as Plan and Profile viewport types). Still others are entirely new objects, including view frames, match lines, and view frame groups. Let's look at what you need to have in place before you can create your plotted masterpieces.

Prerequisite Components

The Plan Production feature draws on several components to create a plan set. Here is a list of these components and a brief explanation of each. Later, this chapter will explore these elements in greater detail:

Drawing Template Plan Production creates new layouts for each sheet in a plan set. To do this, the feature uses drawing templates with predefined viewports. These viewports have their Viewport Type property set to either Plan or Profile.

For the exercises in this chapter, the default location will refer to the `C:\Mastering Civil 3D 2012\CH 16\Final Sheets` location.

Object and Display Styles Like every other feature in Civil 3D, Plan Production uses objects. Specifically, these objects are view frames, view frame groups, and match lines. Before creating plan sheets, you'll want to make sure you have styles set up for each of these objects.

Alignments and Profiles   In Civil 3D, the Plan Production feature is designed primarily for use in creating plan and profile views. Toward that end, your drawing must contain (or reference) at least one alignment. If you're creating sheets with both plan and profile views, a profile must also be present.

Sample Lines and Sections   Creating section sheets requires a sample line group and a sheet template with section viewports associated.

With these elements in place, you're ready to dive in and create some sheets. The general steps in creating a set of plans are as follows:

1. Meet the prerequisites listed previously.
2. Create view frames.
3. Create plan/plan-profile sheets.
4. Create section view groups.
5. Create section sheets.
6. Plot or publish (hardcopy or DWF).

The next section describes this process in detail and the tools used in plan production. The Sheet Set Manager, which is found in basic AutoCAD, is an integral part of this process.

Using View Frames and Match Lines

When you create sheets using the Plan Production tool, Civil 3D first automatically helps you divide your alignment into sections that will fit on your plotted sheet and display at the desired scale. To do this, Civil 3D creates a series of rectangular frames placed end to end (or slightly overlapping) along the length of alignment, like those in Figure 16.1. These rectangles are referred to as *view frames* and are automatically sized and positioned to meet your plan sheet requirements. This collection of view frames is referred to as a *view frame group*. Where the view frames abut one another, Civil 3D creates *match lines* that establish continuity from frame to frame by referring to the previous or next sheet in the completed plan set. View frames and match lines are created in model space, using the prerequisite elements described in the previous section.

The Create View Frames Wizard

The first step in the process of creating plan sets is to generate view frames. Civil 3D provides an intuitive wizard that walks you through each step of the view frame creation process. Let's look at the wizard and the various page options. After you've seen each page, you'll have a chance to put what you've learned into practice.

You launch the Create View Frames wizard (Figure 16.2) by selecting the Create View Frames button on the Plan Production panel found on the Output tab of the Ribbon. The wizard consists of several pages. A list of these pages is shown along the left side, and an arrow indicates which page you're currently viewing. You move among the pages using the Next and Back navigation buttons along the bottom of each page. Alternatively, you can jump directly to any page by clicking its name in the list on the left. The following sections walk through the pages of the wizard and explain their features.

FIGURE 16.1
View frames and match lines

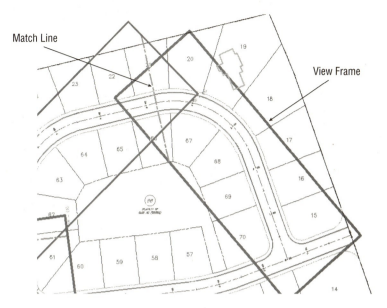

FIGURE 16.2
The Create View Frames wizard

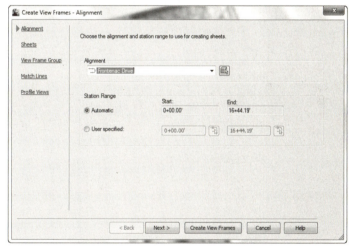

ALIGNMENT PAGE

You use the first page to select the alignment and station range along which the view frames will be created.

Alignment In the first section of this page, you select the alignment along which you want to create view frames. You can either select it from the drop-down menu or click the Select From The Drawing button to select the alignment from the drawing.

Station Range In the Station Range section, you define the station range over which the frames will be created. Selecting Automatic creates frames from the alignment Start to the alignment End. Selecting User Specified lets you define a custom range, by either keying in

Start and End station values in the appropriate box or by clicking the button to the right of the station value fields and graphically selecting the station from the drawing.

An example of selecting specific stations is if you have a subdivision that will be constructed in phases. You have designed an entire roadway but only need to create specific sheets for a specific phase.

SHEETS PAGE

You use the second page of the wizard (Figure 16.3) to establish the sheet type and the orientation of the view frames along the alignment. A plan production *sheet* is a layout tab in a drawing file. To create the sheets, Civil 3D references a predefined drawing template (with the file extension .dwt). As mentioned earlier, the template must contain layout tabs, and in each tab the viewport's extended data properties must be set to either Plan or Profile. Later in this chapter, you'll learn about editing and modifying templates for use in plan production.

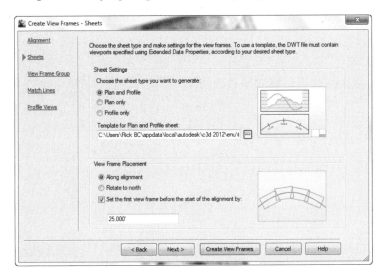

FIGURE 16.3
Create View Frames – Sheets

Sheet Settings In Civil 3D, the Plan Production feature provides options for creating three types of sheets:

Plan And Profile This option generates a sheet with two viewports; one viewport shows a plan view and the other shows a profile view of the section of the selected alignment segment.

Plan Only As the name implies, this option creates a sheet with a single viewport showing only the plan view of the selected alignment segment.

Profile Only Similar to Plan Only, this option creates a sheet with a single viewport, showing only the profile view of the selected alignment segment.

> **Graphics**
>
> Did you notice the nifty graphic to the right of the sheet-type options in Figure 16.3? This image changes depending on the type of sheet you've selected. It provides a schematic representation of the sheet layout to further assist you in selecting the appropriate sheet type. You'll see this type of graphic image throughout the Create View Frame wizard and in other wizards used for plan production in Civil 3D.

After choosing the sheet type, you must define the template file and the layout tab within the selected template that Civil 3D will use to generate your sheets. Several predefined templates ship with Civil 3D and are part of the default installation.

Clicking the ellipsis button displays the Select Layout As Sheet Template dialog. This dialog provides the option to select the DWT file and the layout tab within the template. Clicking the ellipsis button in that dialog lets you browse to the desired template location. Typically the default template location is:

```
C:\Users\<username>\AppData\Local\Autodesk\C3D2012\
enu\Template\Plan Production\
```

If you are on a network environment, your templates might be found in a common folder on the network.

After you select the template you want to use, a list of the layouts contained in the DWT file appears in the Select Layout As Sheet Template dialog (Figure 16.4). Here you can choose the appropriate layout.

FIGURE 16.4
Use the Select Layout As Sheet Template dialog to choose which layout you would like to apply to your newly created sheets.

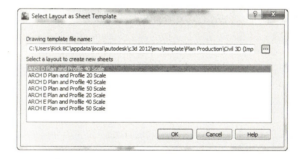

View Frame Placement Your view frames can be placed in one of two ways: either along the axis or rotated north. Use the bottom section of the Sheets page of the wizard to establish the placement.

Along Alignment This option aligns the long axes of the view frames parallel to the alignment. Refer to the graphic to the right for a visual representation of this option.

Rotate To North As the name implies, this option aligns the view frames so they're rotated to the north directions (straight up), regardless of the changing rotation of the alignment centerline. *North* is defined by the orientation of the drawing. Again, refer to the graphic.

If you want the north arrow to rotate according to the view twist of the viewport, the block that is being used for the north arrow must be included in the template and must be located over the plan viewport.

Set The First View Frame Before The Start Of The Alignment By Regardless of the view frame alignment you choose, you have the option to place the first view frame some distance before the start of the alignment. This option is useful if you want to show a portion of the site, such as an existing offsite road, in the plan view. When this option is selected, the text box becomes active, letting you enter the desired distance.

VIEW FRAME GROUP PAGE

You use the third page of the Create View Frames wizard (Figure 16.5) to define creation parameters for your view frames and the view frame group to which they'll belong. The page is divided into two sections: one for the view frame group and the other for the view frames themselves.

FIGURE 16.5
Create View Frames – View Frame Group

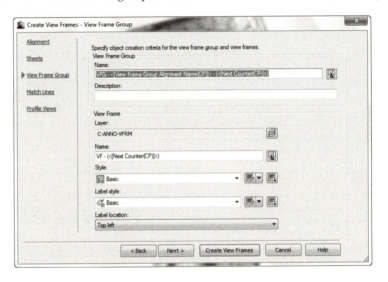

View Frame Group Use these options to set the name and an optional description for the view frame group. The name can consist of manually entered text, text automatically generated based on the Name Template settings (click the Edit View Frame Name button to open the Name Template dialog to adjust the name template), or a combination of both. In this example, the feature settings are such that the name will include manually defined prefix text (VFG-) followed by automatically generated text, which inserts the alignment name and a sequential counter number. For this example, this will result in a view frame group name of VFG – Frontenac CL - 1.

The Name Template dialog isn't unique to the Plan Production feature of Civil 3D. However, the property fields available vary depending on the features to be named. If you need to reset the incremental number counter, use the options in the lower area of the Name Template dialog (Figure 16.6).

Figure 16.6
The Name Template dialog. Use the options in the Incremental Number Format section to adjust automatic numbering.

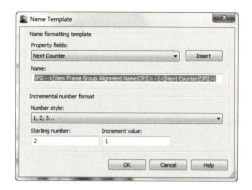

View Frame These options are used to set various parameters for the view frames, including the layer for the frames, view frame names, view frame object and label styles, and the label location. Each view frame can have a unique name, but the other parameters are the same for all view frames. Setting the view frame layer to no-plot will ensure your drawing does not end up plotting with unwanted rectangles.

Layer This option defines the layer on which the view frames are created. This layer is defined in the drawing settings, but you can override it by clicking the Layer button and selecting a different layer.

Name The Name setting is nearly identical in function to that for the view frame group name discussed earlier. In this example, the default name results in VF-1, VF-2, and so on.

Style Like nearly all objects in Civil 3D, view frames have styles associated with them. The view frame style is simple, with only one component: the view frame border. You use the drop-down menu to select a predefined style.

Label Style Also like most other Civil 3D objects, view frames have label styles associated with them. And like other label styles, the view frame labels are created using the Label Style Composer and can contain a variety of components. The label style used in this example includes the frame name and station range placed at the top of the frame.

Label Location The last option on this page lets you set the label location. The default feature setting places the label at the Top Left of the view frame. Other options include Top Center, Top Right, Bottom Left, Bottom Right, and so on.

All view frame labels are placed at the top of the frame. However, the term *top* is relative to the frame's orientation. For alignments that run left to right across the page, the top of the frame points toward the top of the screen. For alignments that run right to left, the top of the frame points toward the bottom of the screen. You can make the view frame label display along the frame edge closest to the top of the screen by using a large Y-offset value when defining your view frame label style.

Match Lines Page

You use the next page of the Create View Frames wizard (Figure 16.7) to establish settings for match lines. Match lines are used to maintain continuity from one sheet to the next. They're typically placed at or near the edge of a sheet, with instructions to "see sheet XX" for continuation. You have the option whether to automatically insert match lines. Match lines are used only for plan views, so if you're creating Plan And Profile or Profile Only sheets, the option is automatically selected and can't be deselected.

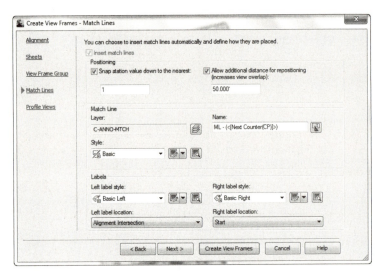

FIGURE 16.7
Create View Frames – Match Lines

Positioning The Positioning options are used to define the initial location of the match lines and provide the ability to later move or reposition the match lines.

Snap Station Value Down To The Nearest By selecting this option, you override the drawing station settings and define a rounding value specific to match-line placement. In this example, a value of 1 is entered, resulting in the match lines being placed at the nearest whole station. This feature always rounds down (snap station down as opposed to snap station up).

Allow Additional Distance For Repositioning Selecting this option activates the text box, allowing you to enter a distance by which the views on adjacent sheets will overlap and the maximum distance that you can move a match line from its original position.

Match Line The options for the match line are similar to those for view frames described on the previous page of the wizard. You can define the layer, the name format, and the match-line style.

Labels These options are also similar to those for view frames. Different label styles are used to annotate match lines located at the left and right side of a frame. This lets you define match-line label styles that reference either the previous or next station adjacent to the current frame. You can also set the location of each label independently using the Left and Right Label Location drop-down menus. You have options for placing the labels at the start, end, or middle, or at the point where the match line intersects the alignment.

Profile Views Page

The final page of the Create View Frames wizard is the Profile Views page (Figure 16.8). This page is optional and will be disabled and skipped if you chose to create Plan Only sheets on the second page of the wizard. The Plan Production feature needs to know what profile view and band set styles you intend to use for the profile views. This allows the correct positioning to be applied. Use the drop-down menus to select both the profile view style and the band set style.

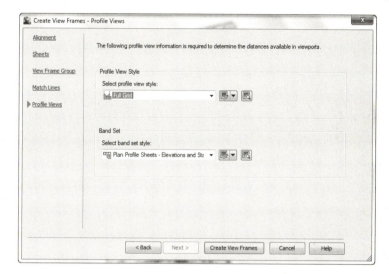

Figure 16.8
Create View Frames – Profile Views

Civil 3D has difficulty determining the proper extents of profile views. If you find that your profile view isn't positioned correctly in the viewport (for example, the annotation along the sides or bottom is clipped), you may need to create *buffer* areas in the profile view band set style by modifying the text box width. The _Autodesk Civil 3D NCS.dwt (both Imperial and Metric) file contains styles with these buffers created.

This last page of the wizard has no Next button. To complete the wizard, click the Create View Frames button.

Creating View Frames

Now that you understand the wizard pages and available options, you'll try them out in this exercise:

1. Open ViewFrameWizard.dwg, which you can download from this book's web page at www.sybex.com/go/masteringcivil3d2012. This drawing contains several alignments and profiles as well as styles for view frames, view frame groups, and match lines.

2. To launch the Create View Frames wizard, click the Create View Frames button on the Plan Production panel on the Output tab of the Ribbon.

3. On the Alignment page, select Frontenac Drive from the Alignment drop-down menu. For Station Range, verify that Automatic is selected, and click Next to advance to the next page.

4. On the Sheets page, select the Plan And Profile option.

5. Click the ellipsis button to display the Select Layout As Sheet Template dialog. Then, click the next ellipsis and browse to the `Plan Production` subfolder in the default template file location. Typically, this is:

   `C:\Users\<username>\AppData\Local\Autodesk\C3D2012\`
   `enu\Template\Plan Production\`

6. Select the template named `Civil 3D (Imperial) Plan and Profile.dwt`, and click Open.

7. A list of the layouts in the DWT file appears in the Select Layout As Sheet Template dialog. Select the layout named ARCH D Plan and Profile 20 Scale, and click OK.

8. In the View Frame Placement section, select the Along Alignment option. Then, select Set The First View Frame Before The Start Of The Alignment By. Note that the default value for this particular drawing is 30′. Click Next to advance to the next page.

9. On the View Frame Group page, confirm that all settings are as follows (these are the same settings shown previously in Figure 16.5), and then click Next to advance to the next page:

Setting	Value
View Frame Group Name	VFG - <[View Frame Group Alignment Name(CP)]> - (<[Next Counter(CP)]>)
View Frame Name	VF - (<[Next Counter(CP)]>)
Style	Basic
Label Style	Basic
Label Location	Top Left

10. On the Match Lines page, review the default settings and click Next to advance to the next page.

11. On the last page of the wizard, confirm that the settings are as follows (these are the same settings shown previously in Figure 16.8), and then click Create View Frames:

Setting	Value
Select Profile View Style	Full Grid
Select Band Set Style	Plan Profile Sheets - Elevations And Stations

The view frames and match lines are created and are displayed as a collection in Prospector, as shown in Figure 16.9.

The numbering for your view frames, view frame groups, and match lines may not identically match that shown in the images. This is due to the incremental counting Civil 3D performs in the background. Each time you create one of these objects, the counter increments. You can reset the counter by modifying the name template.

Editing View Frames and Match Lines

After you've created view frames and match lines, you may need to edit them. Edits to some view-frame and match-line properties can be made via the Prospector tab in the Toolspace palette. For both view frames and match lines, you can only change the object's name and/or style via the Information tab in the Properties dialog. All other information displayed on the other tabs is read-only.

You make changes to geometry and location graphically using special grip edits (Figure 16.10). Like many other Civil 3D objects with special editing grips (such as profiles and Pipe Network objects), view frames and match lines have editing grips you use to modify the objects' location, rotation, and geometry. Let's look at each separately.

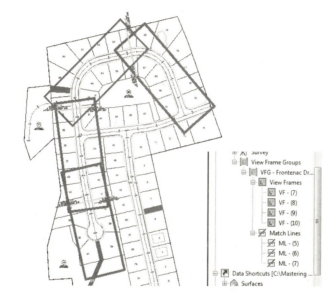

FIGURE 16.9
Finished view frames in the drawing and in Prospector

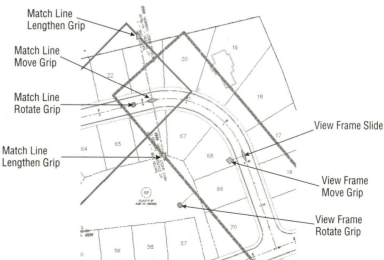

FIGURE 16.10
View frame and match line grips

View frames can be graphically edited in three ways. You can move them, slide them along the alignment, and rotate them as follows:

To Move a View Frame The first grip is the standard square grip that is used for most typical edits, including moving the object.

To Slide a View Frame Select the view frame to be edited, and then select the diamond-shaped grip at the center of the frame. This grip lets you move the view frame in either direction along the alignment while maintaining the orientation (Along Alignment or Rotated North) you originally established for the view frame when it was created.

To Rotate a View Frame Select the frame, and then select the circular handle grip. This grip works like the one on pipe-network structures. Using this grip, you can rotate the frame about its center.

> **DON'T FORGET YOUR AUTOCAD FUNCTIONS!**
>
> While you're getting wrapped up in learning all about Civil 3D and its great design tools, it can be easy to forget you're sitting on an incredibly powerful CAD application.
>
> AutoCAD features add functionality beyond what you can do with Civil 3D commands alone. First, make sure the DYN (Dynamic Input) option is turned on. This gives you additional functionality when you're moving a view frame. With DYN active, you can key in an exact station value to precisely locate the frame where you want it. Similar to view frame edits, with DYN active, you can key in an exact rotation angle. Note that this rotation angle is relative to your drawing settings (for example, 0 degrees is to the left, 90 degrees is straight up, and so on). Also, selecting multiple objects and then selecting their grips while holding Shift makes each grip "hot" (usually a red color). This allows you to grip-edit a bunch of objects at once, like sliding a group of view frames along the alignment.
>
> Whether you're learning Civil 3D yourself or training a group, it's a good idea to spend some time looking at the new AutoCAD features with every new release. You never know when you'll discover a nugget that cuts hours off your workday!

You can edit a match line's location and length using special grips. As with view frames, you can slide them along the alignment and rotate them. They can also be lengthened or shortened. Unlike view frames, they can't be moved to an arbitrary location.

To Slide a Match Line Select the match line to be edited, and then select the diamond-shaped grip at the center of the match line. This grip lets you move the match line in either direction along the alignment while maintaining the orientation (Along Alignment or Rotated North) that you originally established for the view frame. Note that match line can only be moved in either direction a distance equal to or less than that entered on the Match Line page of the wizard at the time the view frames were created. For example, if you entered a value of 50′ for the Allow Additional Distance For Repositioning option, your view frames are overlapped 50′ to each side of the match line, and you can slide the match line only 50′ in either direction from its original location.

To Rotate a Match Line Select the match line, and then select the circular handle grip. This grip works like the one on a view frame.

To Change a Match Line's Length When you select a match line, a triangular grip is displayed at each end. You can use these grips to increase or decrease the length of each half of the match line. For example, moving the grip on the top end of the match line changes the length of only the top half of the match line; the other half of the match line remains unchanged. See the sidebar "Don't Forget Your AutoCAD Functions!" for tips on using AutoCAD features.

The following exercise lets you put what you've learned into practice. Make sure Dynamic Input is active:

1. Continue working in the drawing from the previous exercise or open EditViewFramesAndMatchLines.dwg from this book's web page. This drawing contains view frames and match lines. You'll change the location and rotation of a view frame and change the location and length of a match line.

2. Select the lower middle view frame, which consists of stations 9+56 to 14+34, and select its diamond-shaped sliding grip. Slide this grip up so the overlap with the lower view frame isn't so large. Either graphically slide it to station 11+00 or enter **1100** in the Dynamic Input text box. Notice that the view frame label is updated with the revised stations.

3. Press Esc to clear your selection. Select the circular rotation grip on the lower view frame that consists of stations 14+34 to 16+44. Rotate the view frame slightly to better encompass the road. In the Dynamic Input text box, enter **8**. Then, press Esc to clear the grips from the view frame.

4. Now you'll adjust the match line's location and length. Select the Match Line, which is presently at Station 14+34, and then select its diamond sliding grip. Either graphically slide it to station 13+25 or enter **1325** in the Dynamic Input text box. Notice that the match line label is updated with the revised station.

5. Next, you'll lengthen the right side of the match line so it extends the width of the view frame. Select the triangular lengthen grip at the lower end of the Match Line, and either graphically lengthen it to 125′ or enter **125** in the Dynamic Input text box. Press Esc to exit the command.

Creating Plan and Profile Sheets

Civil 3D's Plan Production feature uses the concept of *sheets* to generate the pages that make up a set of plans. Simply put, *sheets* are layout tabs with viewports showing a given portion of your design model, based on the view frames previously created. The viewports have special properties set that define them as either Plan or Profile viewports. These viewports must be predefined in a template (DWT) file. You manage the sheets using the standard AutoCAD Sheet Set Manager feature.

The Create Sheets Wizard

After you've created view frames and match lines, you can proceed to the next step of creating sheets. Like view frames, sheets are created using a wizard. Let's look at the wizard and the various page options. After you've walked through each page, you'll have a chance to put what you've learned into practice.

You launch the Create Sheets wizard by switching to the Output tab and clicking the Create Sheets button on the Plan Production panel. A list of the wizard's pages is shown along the left side, and an arrow indicates which page you're currently viewing. You move among the pages using the Next and Back navigation buttons along the bottom of each page. Alternatively, you can jump directly to any page by clicking its name in the list on the left. Let's examine the wizard's pages and the features of each.

VIEW FRAME GROUP AND LAYOUTS PAGE

You use the first page of this wizard (Figure 16.11) to select the view frame group for which the sheets will be created. It's also used to define how the layouts for these sheets will be generated.

FIGURE 16.11
Create Sheets –
View Frame Group
And Layouts

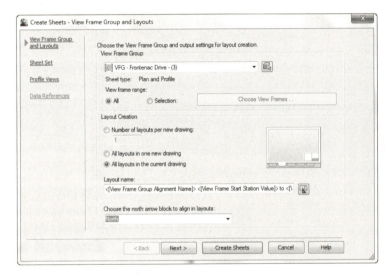

View Frame Group In the first section of this page, you select the view frame group either from the drop-down menu or by clicking the Select From The Drawing button to select the view frame group from the drawing. After you've selected the group, you use the View Frame Range option to create sheets for all frames in the group or only for specific frames of your choosing.

All Select this option when you want sheets to be created for all view frames in the view frame group.

Selection Selecting this option activates the Choose View Frames button. Click this button to select specific view frames from a list. You can select a range of view frames by using the standard Windows selection technique of clicking the first view frame in the range and then holding Shift while you select the last view frame in the range. You can also select individual view frames in nonsequential order by holding Ctrl while you make your view-frame selections. Figure 16.12 shows two of the three view frames selected in the Select View Frames dialog.

Figure 16.12
Select view frames by using standard Windows techniques.

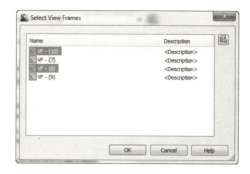

Layout Creation   In this section, you define where and how the new layouts for each sheet are created as well as the name format for these sheets, and you specify information about the alignment of the north arrow block.

There are three options for creating layout sheets: all the layout tabs are created in the current drawing (the drawing you're in while executing the Create Sheets wizard); all the new layouts are created in a new drawing file; or the layouts are created in multiple new drawing files, limiting the maximum number of layout sheets created in each file.

Number Of Layouts Per New Drawing   This option creates layouts in new drawing files and limits the maximum number of layouts per drawing file to the value you enter in the text box. For best performance, Autodesk recommends that a drawing file contain no more than 10 layouts. On the last page of this wizard, you're given the option to select the objects for which data references will be made. These data references are then created in the new drawings.

All Layouts In One New Drawing   As the name implies, this option creates all layouts for each view frame in a single new drawing. Use this option if you have fewer than 10 view frames, to ensure best performance. If you have more than 10 view frames, use the previous option. On the last page of this wizard you're given the option to select the objects for which data references will be made. These data references are then created in the new drawings.

All Layouts In The Current Drawing   When you choose this option, all layouts are created in the current drawing. You need to be aware of two scenarios when working with this option. (As explained later, you can share a view frame group via Data Shortcuts and reference it into other drawings as a data reference.)

- When creating sheets, it's possible that your drawing references the view frame group from another drawing (rather than having the original view frame group in your current drawing). In this case, you're given the option to select the additional objects for which data references will be made (such as alignments, profiles, pipes, and so on). These data references are then created in the current drawing. You select these objects on the last page of the wizard.

- If you're working in a drawing in which the view frames were created (therefore, you're in the drawing in which the view frame group exists), the last page of this wizard is disabled. This is because in order for you to create view frames (and view frame groups), the alignment (and possibly the profile) must either exist in the current drawing or be referenced as a data reference (recall the prerequisites for creating view frames, mentioned earlier).

Layout Name Use this text box to enter a name for each new layout. As with other named objects in Civil 3D, you can use the Name template to create a name format that includes information about the object being named.

Align The North Arrow Block In Layouts If the template file you've selected contains a north arrow block, it can be aligned so that it points north on each layout sheet. The block must exist in the template and be placed over the plan viewport. If there are multiple blocks, select the one you want to use from the drop-down menu.

> **WHERE AM I?**
>
> We strongly recommend that you set up the Name template for the layouts so that it includes the alignment name and the station range. This conforms to the way many organizations create sheets, helps automate the creation of a sheet index, and generally makes it easier to navigate a DWG file with several layout tabs.

CREATE SHEETS PAGE

You use the second page of the wizard (Figure 16.13) to determine whether a new or existing sheet set (with the file extension .dst) is used and the location of the DST file. The sheet name and storage location are also defined here. Additionally, on this page you decide whether to add the sheet-set file (with the file extension .dst) and the sheet files (with the file extension .dwg) to the project vault.

Sheet Set In this section of the page, you select whether to create a new DST file or add the sheets created by this wizard to an existing DST file.

New Sheet Set By selecting this option, you create a new sheet set. You must enter a name for the DST file and a storage location. By default, the sheet set is created in the same folder as the current drawing. You can change this by clicking the ellipsis and selecting a new location.

Add To Existing Sheet Set Selecting this option lets you select an existing sheet-set file to which the new sheets created by this wizard will be added. Click the ellipsis to browse to the existing DST file location.

Sheets You use the last section of the page to set the name and storage location for any new DWG files created by this wizard. On the previous page of the wizard, you had the choice of creating new files or creating the sheet layout in the current drawing. If you chose the latter option, the Sheets section on this page of the wizard is inactive. If you chose the former, here you enter the sheet file (DWG) name and the storage location.

CREATING PLAN AND PROFILE SHEETS | 663

FIGURE 16.13
Create Sheets –
Sheet Set

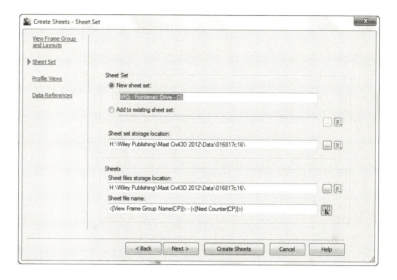

WHAT IS A SHEET?

This page can be a little confusing due to the way the word *sheets* is used. In some places, *sheets* refers to layout tabs in a given drawing (DWG) file. On this page, though, the word *sheets* is used in the context of Sheet Set Manager and refers to the DWG file itself.

PROFILE VIEWS PAGE

The next page of the wizard (Figure 16.14) lists the profile view style and the band set selected in the Create View Frames wizard. You can't change these selections. You can, however, make adjustments to other profile settings.

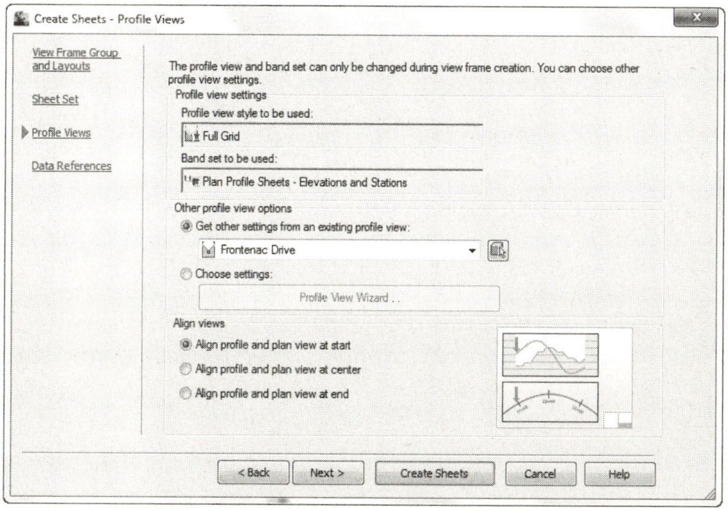

FIGURE 16.14
Create Sheets –
Profile Views

The Other Profile View Options section lets you modify certain profile view options either by running the Profile View wizard or by using an existing profile view in your drawing as an example. Regardless of what option you choose, the "other options" you can change are limited to the following:

- Split profile-view options from the Profile View Height page of the Profile View wizard
- All options on the Profile Display page
- Most of the Data Band Page settings
- Profile Hatch Options
- All settings on the Multiple Plot Options page

See Chapter 7, "Profiles and Profile Views," for details of each of these settings.

Data References Page

The final page of the Create Sheets wizard (Figure 16.15) is used to create data references in the drawing files that contain your layout sheets.

Figure 16.15
Select Create Sheets – Data References

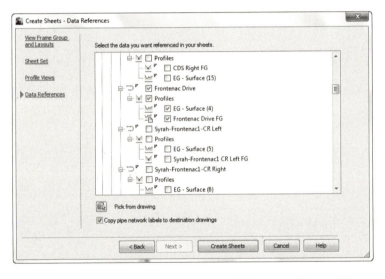

Based on the view frame group used to create the sheets and the type of sheets (plan, profile, plan and profile), certain objects are selected by default. You have the option to select additional objects for which references will be made. You can either pick them from the list or click the Pick From The Drawing button and select the objects from the drawing.

It's common to create references to pipe networks that are to be shown in plan and or profile views. If you choose to create references for pipe-network objects, you can also copy the labels for those network objects into the sheet's drawing file. This is convenient in that you won't need to relabel your networks.

Managing Sheets

After you've completed all pages of the wizard, you create the sheets by clicking the Create Sheets button. Doing so completes the wizard and starts the creation process. If you're creating sheets with profile views, you're prompted to select a profile view origin. Civil 3D then displays several dialogs, indicating the process status for the various tasks, such as creating the new sheet drawings and creating the DST file.

If the Sheet Set Manager isn't currently open, it opens with the newly created DST file loaded. The sheets are listed, and the details of the drawing files for each sheet appear in the Details area (Figure 16.16).

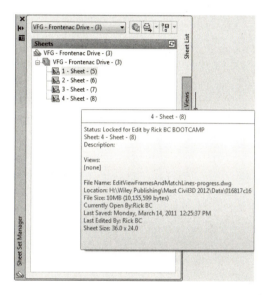

Figure 16.16
New sheets in the Sheet Set Manager

If you double-click to open the new drawing file that contains the newly created sheets, you'll see layout-sheet tabs created for each of the view frames. The sheets are named using the Name template as defined in the Create Sheets wizard. Figure 16.17 shows the names that result from the following template:

```
<[View Frame Group Alignment Name]> <[View Frame Start Station Value]>
to <[View Frame End Station Value]>
```

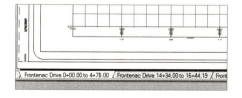

Figure 16.17
The template produces the Frontenac Drive tab names shown here.

To create the final sheets in this new drawing, Civil 3D externally references (XRefs) the drawing containing the view frames; creates data references (DRefs) for the alignments, profiles, and any additional objects you selected in the Create Sheets wizard; and, if profile sheet types were selected in the wizard, creates profile views in the final sheet drawing.

The following exercise pulls all these concepts together:

1. Open SheetsWizard.dwg, which you can download from this book's web page. This drawing contains the view frame group, alignment, and profile for Frontenac Drive. Note that the drawing doesn't have profile views.

2. To launch the Create Sheets wizard, click the Create Sheets button on the Plan Production panel on the Output tab of the Ribbon.

3. On the View Frame Group And Layouts page, confirm that View Frame Range is set to All and that Number Of Layouts Per New Drawing is set to 10. Click Next.

4. On the Sheet Set page, select the New Sheet Set option. For both Sheet Set File Storage Location and Sheet Files Storage Location, use the ellipsis to browse to C:\Mastering Civil 3D 2012\CH 16\Final Sheets, and click OK. Click Next.

5. On the Profile Views page, for Other Profile View Options, select Choose Settings and then click the Profile View Wizard button. The Create Multiple Profile Views dialog opens.

6. On the left side of the Create Multiple Profile Views dialog, click Profile Display Options to jump to that page.

    For the EG – Surface profile, scroll to the right and modify the Labels setting, changing it from Complete Label Set to _No Labels, as shown in Figure 16.18. After you select the label set, click OK; then, click Next to advance to the Data Bands page.

7. On the Data Bands page, change Profile2 to Frontenac Drive FG, as shown in Figure 16.19. Click Finish to return to the Create Sheets wizard. Click Next to advance to the Data References page.

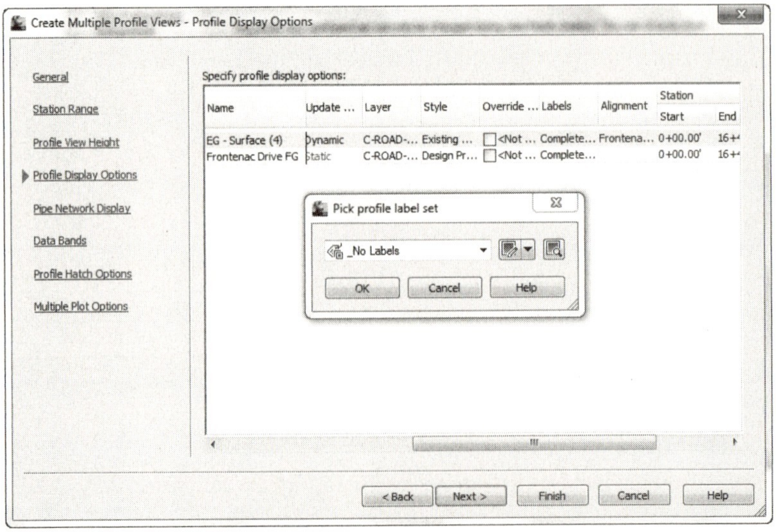

FIGURE 16.18
Change the labels for the EG – Surface profile

FIGURE 16.19
Set Profile2 to Frontenac Drive FG

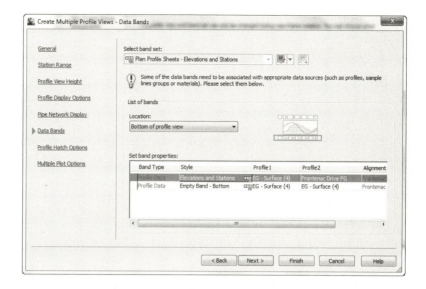

8. On the Data References page, confirm that Frontenac Drive and both of its profiles are selected. Click Create Sheets to complete the wizard.

9. Before creating the sheets, Civil 3D must save your current drawing. Click OK when prompted. The drawing is saved, and you're prompted for an insertion point for the profile view. The location you pick represents the lower-left corner of the profile view grid. Select an open area in the drawing, above the left side of the site plan. Civil 3D displays a progress dialog, and then Panorama is displayed with information about the results of the sheet-creation process. Close the Panorama window.

> **INVISIBLE PROFILE VIEWS**
>
> Note that the profile views are created in the current drawing only if you selected the option to create all layouts in the current drawing. Because you didn't do that in this exercise, the profile views aren't created in the current drawing. Rather, they're created in the sheet drawing files in model space in a location relative to the point you selected in this step.

10. After the sheet-creation process is complete, the Sheet Set Manager window opens (Figure 16.20). Click sheet 1, named Frontenac Drive 0 + 00.00 to 4 + 78.00. Notice that the name conforms to the Name template and includes the alignment name and the station range for the sheet. Review the details listed for the sheet. In particular, note the filename and storage location.

11. Double-click this sheet to open the new sheets drawing and display the layout tab for Frontenac Drive 0 + 00.00 to 4 + 78.00. Review the multiple tabs created in this drawing file. The template used also takes advantage of AutoCAD fields, some of which don't have values assigned.

FIGURE 16.20
The Sheet Set Manager once the sheet-creation process is complete

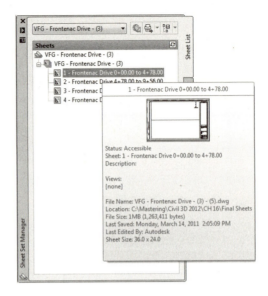

Creating Section Sheets

Similar to the process of creating plan and profile sheets, creating section sheets is a two-step process. First, you have to create a section view group that determines the layout, labeling, and styles of the profile views. Second, you have to generate the actual sheets and add them to the AutoCAD sheet set.

Creating Section View Groups

The process of creating the section view group is where you will determine how your section views will be laid out on the page, what labels will be used, and what styles will be used to represent the various components of the model. If you have any questions about section styles or section view styles, refer to Chapter 19, "Styles."

In this first exercise, you'll walk through setting up a basic section view group for the main road of our sample set:

1. Open MultipleSectionViews.dwg from the provided dataset. In this drawing, sample lines have been added along the Frontenac Drive alignment. These lines are sampling the existing and proposed surfaces.

2. From the Home tab's Profile And Section Views panel, select Section Views ➢ Create Multiple Views to display the Create Multiple Section Views wizard, shown in Figure 16.21.

3. Click Next to access the Section Placement Options page. Review the Section Placement Options at the top of the screen. The Production template and button allow you to navigate to your own sheet template that suits your needs. Change Template For Cross Section Sheet to ARCH D Section 50 scale. Change Group Plot Style to Plot By Page and click Next.

CREATING SECTION SHEETS | 669

4. The Offset range determines the spacing of the section views. Using the User Specified Offset option, change the Left Offset range to –20′. Change the Right Offset range to 20′. Click Next.

5. The options on the Elevation Range page help if you have extra tall sections, allowing you to set some limits manually. Click Next.

6. Change the Change Labels columns to reflect the options shown in Figure 16.22. Click Next.

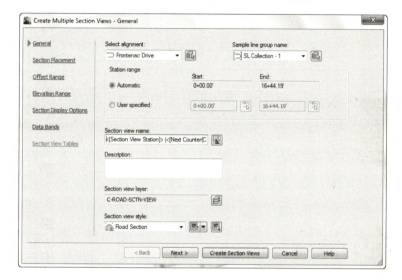

FIGURE 16.21
The General tab of the Create Multiple Section Views wizard

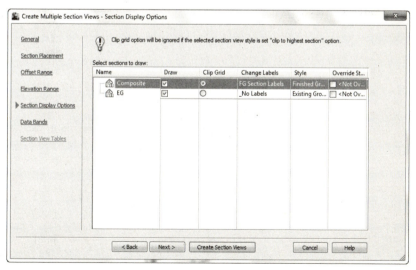

FIGURE 16.22
Changing label styles for the Multiple Section View wizard

670 | **CHAPTER 16** PLAN PRODUCTION

7. Select the Offsets Only band set from the drop-down list near the top of the screen. Click Create Section Views to exit the wizard and place your section views in the drawing.

8. Click a point to the east of the plan view to draw the section views and sheet outlines. Your drawing should look something like Figure 16.23.

FIGURE 16.23
The finished multiple section views operation

Now that you have a section view group, you can begin the process of actually creating section sheets for plotting.

Creating Section Sheets

Many long transportation projects such as highways, light-rail, or canals require the production of many section sheets. While Civil 3D could produce the views prior to 2012, the sheet creation process has improved greatly with this release. In this exercise, you'll convert a Section View Group into a collection of sheets and place them in a new sheet set.

1. Open the CreatingSectionSheets.dwg file. This file is the result of the previous exercise and contains the section view group for the Frontenac Drive alignment.

2. From the Output tab's Plan Production Panel, select Create Section sheets to display the Create Section Sheets dialog shown in Figure 16.24.

FIGURE 16.24
The Create Section Sheets dialog

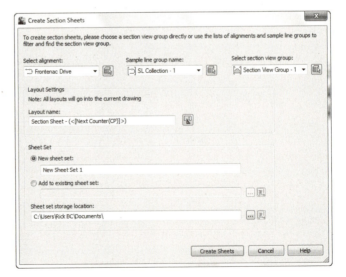

3. Review the settings and options shown. Note that in a drawing with more section view groups you would be able to select each one and create sheets quickly and easily.

4. Click Create Sheets to dismiss the dialog and generate sheets. A Warning dialog will appear warning you that the drawing will be saved. Click OK and Civil 3D will generate new layouts and sheets in a sheet set. The Sheet Set Manager will appear and the Create Section Sheets dialog will reappear.

5. Click Cancel to dismiss the dialog and switch to the Section Sheet – (3) layout tab. Your layout should look something like Figure 16.25.

FIGURE 16.25
A completed section sheet

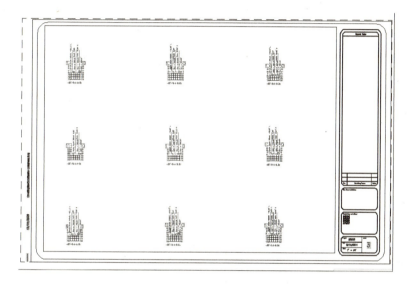

While there are still some tweaks to be made to any sheet, large portions of the mundane details are handled by the wizards and tools. There are some elements that you can modify to customize these details for your organization, and you'll look at those in the next section.

Supporting Components

The beginning of this chapter mentioned that there are several prerequisites to using the Plan Production tools in Civil 3D. The list includes drawing templates (DWT) set up to work with the Plan Production feature and styles for the objects generated by this feature. In this section of the chapter, you'll learn how to prepare these items for use in creating your finished sheets.

Civil 3D ships with several predefined template files for various types of sheets that Plan Production can create. By default, these templates are installed in a subfolder called Plan Production, which is located in the standard Template folder. You can see the Template folder location by opening the Files tab of the Options dialog, as shown in Figure 16.26.

FIGURE 16.26
Template files location

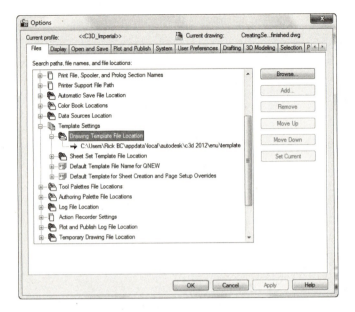

Figure 16.27 shows the default contents of the `Plan Production` subfolder. Notice the templates for Plan, Profile, and Plan And Profile sheet types. There are metric and imperial versions of each.

FIGURE 16.27
Plan Production DWT files

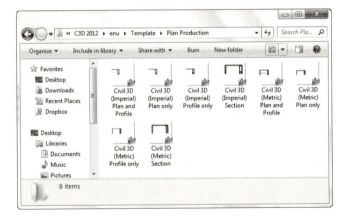

Each template contains layout tabs with pages set to various sheet sizes and plan scales. For example, the `Civil 3D (Imperial) Plan and Profile.dwt` template has layouts created at various ANSI and ARCH sheets sizes and scales, as shown in Figure 16.28.

The viewports in these templates must be rectangular in shape and must have Viewport Type set to Plan, Profile, or Section, depending on the intended use. You set Viewport Type on the Design tab of the Properties dialog, as shown in Figure 16.29.

FIGURE 16.28
Various predefined layouts in standard DWT

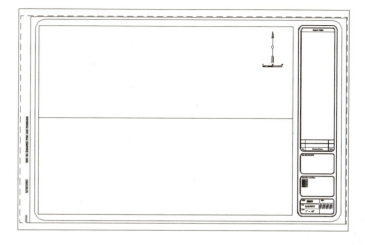

FIGURE 16.29
Viewport Properties – Viewport Type

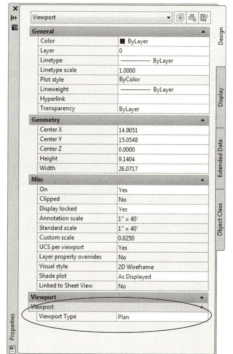

> **IRREGULAR VIEWPORT SHAPES**
>
> Just because the viewports must start out rectangular doesn't mean they have to stay that way. Experiment with creating viewports from rectangular polylines that have vertices at the midpoint of each side of the viewport (not just at the corners). After you've created your sheets using the Plan Production tool, you can stretch your viewport into irregular shapes.

The Bottom Line

Create view frames. When you create view frames, you must select the template file that contains the layout tabs that will be used as the basis for your sheets. This template must contain predefined viewports. You can define these viewports with extra vertices so you can change their shape after the sheets have been created.

> **Master It** Open the `MasteringPlanProduction.dwg`. Run the Create View Frames wizard to create view frames for the Bike Path alignment in the current drawing. (Accept the defaults for all other values.)

Edit view frames. The Edit View Frames command allows the user some freedom on how the frames will appear.

> **Master It** Open the `MasteringPlan Production1.dwg`, and move the VF- (9) view frame to Sta. 40+50 to lessen the overlap. Then adjust the match line so that it is now at Sta. 38+15.

Generate sheets and review Sheet Set Manager. You can create sheets in new drawing files or in the current drawing. Use the option to create sheets in the current drawing when a) you've referenced in the view frame group, or b) you have a small project. The resulting sheets are based on the template you chose when you created the view frames. If the template contains customized viewports, you can modify the shape of the viewport to better fit your sheet needs.

> **Master It** Continue working in the `MasteringPlanProduction1.dwg` file or you can open `MasteringPlanProduction2.dwg`. Run the Create Sheets wizard to create plan and profile sheets for Bike Path using the template `Mastering (Imperial) Plan and Profile.dwt`. (Accept the defaults for all other values.)

Create section views. More and more municipalities are requiring section views. Whether this is a mile-long road or a meandering stream, Civil 3D can handle it nicely via Plan Production.

> **Master It** Open `MasteringPlanProduction3.dwg`. Create Plan Production Sheets with the defaults using the methods you learned earlier.

Chapter 17

Interoperability

No man is an island, and it's the rare designer who works alone. Even in a one-person design team, breaking the project into finite elements, such as grading, paving, and utilities, makes sense from a plan production and management standpoint. To do so effectively, your design software has to understand and have some method for bringing in Land Desktop data as well as other third-party software which can result in better drawing management. Then you also need to take the Civil 3D design and give it to non-Autodesk clients. In this chapter, we'll look a variety of methods for importing and exporting data.

In this chapter, you will learn to:

- Create data shortcuts
- Import and export to earlier releases of AutoCAD
- Import and export to third-party CAD programs
- Use basic map queries to filter data

Data Shortcuts

Data shortcuts allow the cross-referencing of design data between drawings. To this end, the data is what is made available, and it's important to note that the appearance can be entirely different between the host and any number of data shortcuts. We'll use the term *data shortcut* or, more simply, *shortcut* when we discuss selecting, modifying, or updating these links between files.

There are two primary situations in which you need data or information across drawings, and they are addressed with external references (XRefs) or shortcuts. These two options are similar but not the same. Let's compare:

XRef Functionality An XRef is used when the goal is to get a picture of the information in question. XRefs can be changed by using the layer control, XRef clips, and other drawing-element controls. Though they can be used for labeling across files in Civil 3D 2012, the fact that you have to bring in the entire file to label one component is a disadvantage.

Shortcut Functionality A shortcut brings over the design information but generally ignores the display. Shortcuts only work with Civil 3D objects, so they will have their own styles applied. This may seem like duplicitous work because you have already assigned styles and labels in one drawing and have to do it again, but this functionality offers an advantage in that you can have completely different views of the same data in different drawings.

As noted, only Civil 3D objects can be used with shortcuts, and even then some objects are not available through shortcuts. The following objects are available for use through data shortcuts:

- Alignments
- Surfaces
- Profile data
- Pipe networks
- View frame groups

You might expect that corridor, parcel, assembly, point, or point group information would be available through the shortcut mechanism, but they are not. Parcels and corridors can only be accessed via XRef, but once they've been Xreffed, you can use the normal labeling techniques and styles. Now that you've looked at what objects you can tackle with shortcuts, you'll learn how to create them in the next section.

> **A Note about the Exercises in This Chapter**
>
> Many of the exercises in this chapter assume you've stepped through the full chapter. It's difficult to simulate the large number of variables that come into play in a live environment. To that end, many of these exercises build on the previous one. For the easiest workflow, don't close any files until the end of the chapter.
>
> Additionally, you'll have to make some saves to data files throughout this chapter. Remember that you can always get the original file from the chapter folder that you downloaded from www.sybex .com/go/masteringcivil3d2012.

Publishing Data Shortcut Files

Shortcut files are XML files that contain pointers back to the drawing containing the object in question. These shortcuts are managed through Prospector and are stored in a working folder. Creating shortcuts is a matter of setting a working folder, creating a shortcut folder within that folder, and then creating the shortcut files. We'll look at all three steps in this section.

As a precursor to making your first project, you should establish a typical folder structure. Civil 3D includes a mechanism for copying a typical project folder structure into each new project. Once you create a blank copy of the folder structure you'd like to have in place for your projects, you can use it as a starting point when creating a new project within Civil 3D.

1. Open Windows Explorer and navigate to `C:\Civil 3D Project Templates`.
2. Create a new folder titled **Mastering**.
3. Inside `Mastering`, create folders called **Survey**, **Engineering**, **Architecture**, and **Word**, as shown in Figure 17.1.

FIGURE 17.1
Creating a project template

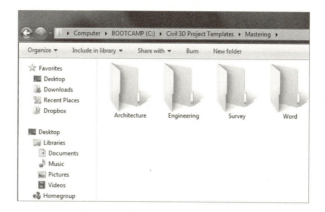

This structure will appear inside Civil 3D and in the working folder when a project is created. We've included a Word folder as an example of other, non–Civil 3D–related folders you might have in your project setup for users outside the CAD team, such as accountants or the project manager. If you have any files in these folders (such as a project checklist spreadsheet, a blank contract, or images), they will be checked into new projects as they are created.

THE WORKING AND DATA SHORTCUTS FOLDERS

You can think of the working folder as a project directory. The working folder will contain a number of projects, each with a shortcuts folder where the shortcut files reside. Each time you create a new shortcut folder within Prospector, you'll have the opportunity to create a full project structure. In this exercise, you'll set the working folder and create a new project:

1. Create a new blank drawing using any template.

2. Within Prospector, make sure the View drop-down list is set to Master View.

3. Right-click the Data Shortcuts branch and select Set Working Folder to display the Browse For Folder dialog shown in Figure 17.2.

FIGURE 17.2
Creating a new working folder

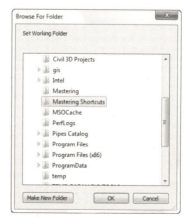

4. Click Local Disk (C:) to highlight it, and then click the Make New Folder button.

5. Type **Mastering Shortcuts** as the folder name and click OK to dismiss the dialog.

6. Right-click the Data Shortcuts branch in Prospector, and select New Data Shortcuts Project Folder to display the New Data Shortcut Folder dialog shown in Figure 17.3.

FIGURE 17.3
Creating a new shortcut folder (aka a project)

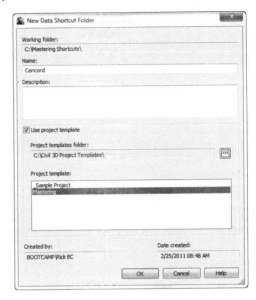

7. Type **Concord** for the folder name, and toggle on the Use Project Template option, as shown in Figure 17.3.

8. Select the Mastering folder from the list, and click OK to dismiss the dialog. Congratulations, you've made a new Civil 3D project! Notice that the Data Shortcuts branch in Prospector now reflects the path of the Concord project.

If you open Windows Explorer and navigate to C:\Mastering Shortcuts\Concord, you'll see the folder from the Mastering project template plus a special folder named _Shortcuts, as shown in Figure 17.4. One common issue that arises is that you may already have a project folder inside the working folder. This typically happens during some bidding or marketing work, or during contract preparation. If you already have a project folder established, it will not show up in Civil 3D unless there is a _Shortcuts folder inside it. To this end, you can manually create this folder. There's nothing special about it—it only has to exist.

FIGURE 17.4
Your new project shown in Windows Explorer

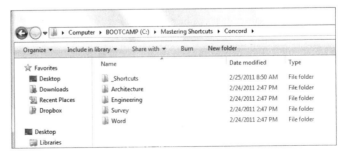

The *working folder* and *data shortcuts folder* are Civil 3D terms for a projects folder and individual projects. If you're familiar with Land Desktop, the working folder is similar to the project path, with various projects. When you publish or consume data from a project, that drawing will then be associated with the project being used, and from then on, your working folder will change to that project automatically.

> **ANOTHER WAY TO LOOK AT IT**
>
> One other option when setting up projects in Civil 3D with shortcuts is to set the working folder path to your particular project folder. Then you could have a CAD folder and within that would be the _Shortcuts folder, which would be the same for every project. This results in a folder structure like `H:\Projects\Project Name\CAD\_Shortcuts`, but some users may find this more useful and easier to manage, particularly if your company standards dictate that the top level of a project folder shouldn't include something like a _Shortcuts folder. Both methods are workable solutions—just decide on one and stick to it! We'll use the more conventional approach shown in the previous exercise throughout this text.

CREATING DATA SHORTCUTS

With a shortcut folder in place, it's time to use it. It's a best practice to keep the drawing files in the same location as your shortcut files, just to make things easier to manage. In this exercise, you'll publish data shortcuts for the alignments and layout profiles in your project:

1. Open the `CreatingShortcuts.dwg` file, which you can download from www.sybex.com/go/masteringcivil3d2012. This drawing contains samples of each shortcut-ready object, as shown in Figure 17.5.

FIGURE 17.5
The Creating Shortcuts drawing file

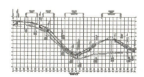

2. In Prospector, right-click the Data Shortcuts branch, and select Create Data Shortcuts to display the Create Data Shortcuts dialog shown in Figure 17.6.

FIGURE 17.6
The Create Data Shortcuts dialog

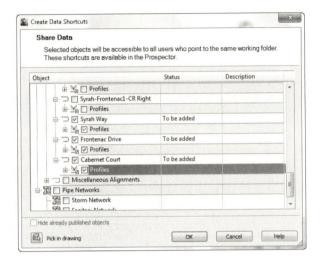

3. Check the EG Surface under the Surface option. Drill down through the Alignments options and check the three major alignments (Syrah Way, Cabernet Court, and Frontenac Drive). Notice that all the subitems will be selected. Leave the pipe networks and view frame groups unchecked. Note that the profiles associated with each alignment are also being selected for publishing.

4. Click OK to dismiss the dialog and create the data shortcuts.

5. Expand the branches under the Data Shortcuts heading as shown in Figure 17.7, and you should see all of the relevant data listed. The listing here indicates that the object is ready to be referenced in another drawing file.

FIGURE 17.7
Data shortcuts listed in Prospector

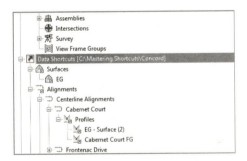

In Civil 3D 2008, you had to manually manage the XML data reference files. In Civil 3D 2012, Civil 3D manages these files for you. They are stored in the magic _Shortcuts folder as individual XML files. You can review these XML files using XML Notepad, but it's worth noting that the first comment in the XML file is PLEASE DO NOT EDIT THIS FILE!. In the past, some users found they needed to edit XML files to fix broken or lost references. This is no longer necessary with the addition of the Data Shortcuts Editor, which we'll look at later in the section "Updating and Managing References."

Using Data Shortcuts

Now that you've created the shortcut XML files to act as pointers back to the original drawing, you'll use them in other ways and locations. Once a reference is in place, it's a simple matter to update the reference and see any changes in the original file reflected in the reference object. In this section, you'll learn how create and explore these references, as well as update and edit them.

> **My Screen Doesn't Look Like That!**
>
> Many of the screen captures and paths shown in this chapter reflect the author's working folders during the creation of the data. Your screen will be different depending on how you have installed the data, network permissions, and so on. Just focus on the content and don't let the different paths fool you—you're doing the right steps.

Creating Shortcut References

Shortcut references are made using the Data Shortcuts branch within Prospector. In this exercise, you'll create references to the objects you previously published to the Concord project:

1. Open the `CreatingReferences.dwg` file. This is an empty file built on the Extended template.

2. In Prospector, expand Data Shortcuts ➤ Centerline Alignments ➤ Alignments.

3. Right-click the Cabernet Court alignment and select Create Reference to display the Create Alignment Reference dialog shown in Figure 17.8. Note that the profile data will be created automatically when you create the reference to the Cabernet Court alignment.

Figure 17.8
The Create Alignment Reference dialog

4. Set Alignment Style to Proposed and Alignment Label Set to Major Minor And Geometry Points, as shown in Figure 17.8.

5. Click OK to close the dialog.

6. Perform a zoom extents to find this new alignment.

7. Repeat steps 3 through 5 for the Syrah Way and Frontenac Drive alignments in the shortcuts list. When complete, your screen should look similar to Figure 17.9.

FIGURE 17.9
Completed alignment references

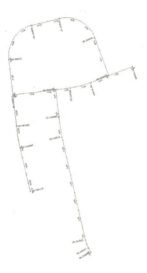

What's in a Name?

The name in the XML file doesn't seem to have much effect on the creation of references; however, it does seem to have an effect on their maintenance. Some users have reported issues with broken references when they changed the name of an object during the Create Reference part of the process. No one has a good feel for why it happens, but we don't recommend that you test it. Leave the name alone during the create reference step!

CERT OBJECTIVE

Each of these alignments is simply a pointer back to the original file. They can be stylized, stationed, or labeled, but the definition of the alignment cannot be changed. This is more clearly illustrated in a surface, so let's add a surface reference now:

1. Expand Data Shortcuts ➤ Surfaces.

2. Right-click the EG surface and select Create Reference to display the Create Surface Reference dialog.

3. Change the Surface Style setting to Contours 1 & 5 Background, and then click OK. Your screen should look like Figure 17.10.

4. Expand the Surfaces branch of Prospector and the EG surface, as shown in Figure 17.11.

FIGURE 17.10
Data referenced surface and alignments

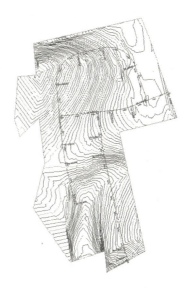

FIGURE 17.11
The EG surface in Prospector and the EG Surface Properties dialog

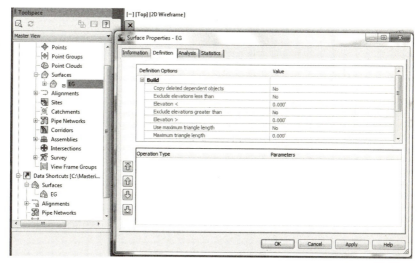

5. Right-click the EG surface and select Surface Properties.

   Here are a couple of important things to note:

   ◆ The small arrow next to the EG surface name indicates that the surface is a shortcut.

   ◆ There's no Definition branch under the EG surface. Additionally, in the Surface Properties dialog, the entire Definition tab is grayed out, making it impossible to change by using a shortcut.

6. Click OK to close the dialog.

Real World Scenario

DO I NEED TO KNOW THIS?

Yes, actually you do. Even if you are using Vault as your project-management scheme, there are times and situations where using data shortcuts is the only real solution. Let's look at a couple of cases where in spite of using Vault, we've used data shortcuts to pull the proverbial rabbit out of the hat.

When working with Vault, a drawing is typically attached to a particular project. Though this generally isn't a problem, it does limit your resources somewhat. When you have a multiphase project that has been divided in Vault, it is impossible to use the Vault mechanism to grab information from a drawing that is not part of the set project.

To get around this limitation, use a shortcut. Open the file containing the desired object and export a shortcut. Then, in the target file, you can import the data shortcut and reference the data accordingly. As the source file changes, the shortcut will update, keeping the two phases in synchronization.

Many firms just starting out with Civil 3D will try to put too much in one drawing. Since the program can handle it, why shouldn't they? This is fine until they run into a deadline at the end of the project. With no real sharing methodology in place, it becomes almost impossible to split up the work. Also, if you add everything to one drawing and the drawing is suddenly corrupted, then you have lost a lot of work.

Even if you don't want to get into sharing data at that point in the job, you can use the data shortcut mechanism to break out individual pieces to other files. Once a shortcut has been created in the target file, right-click the object name in Prospector and select Promote, as shown here:

The act of promoting a shortcut breaks the link with the original data and thus isn't a recommended practice, but sometimes the deadlines win. When it's time to get the project out the door, you do what works!

To make a reference into a live object in the current drawing, right-click its name and select Promote. This breaks the link to the source information and creates an editable object in the current drawing.

Now that you've created a file with a group of references, you can see how changes in the source drawing are reflected in this file.

Updating and Managing References

As it is, if the reference were just a copy of the original data, you'd have done nothing more than cut and paste the object from one drawing to another. The benefit of using shortcuts is the same as with XRefs: when a change is made in the source, it's reflected in the reference drawing. In this section, you'll make a few changes and look at the updating process, and then you'll learn how to add to the data shortcut listings in Prospector.

Updating the Source and Reference

When it's necessary to make a change, it can sometimes be confusing to remember which file you were in when you originally created a now-referenced object. Thankfully, you can use the tools in the Data Shortcut menu to jump back to that file, make the changes, and refresh the reference:

1. In Prospector, expand Data Shortcuts ➤ Alignments ➤ Centerline Alignments. Save your drawing.

2. Right-click Cabernet Court and select Open Source Drawing. You can also do this by selecting the object in the drawing window and right-clicking to access the context menu. The Open Source Drawing command appears on the context menu when a reference object is selected.

3. Make a grip-edit to Cabernet Court's northern end, dragging it up and further north, as shown in Figure 17.12.

Figure 17.12
Grip-editing Cabernet Court.

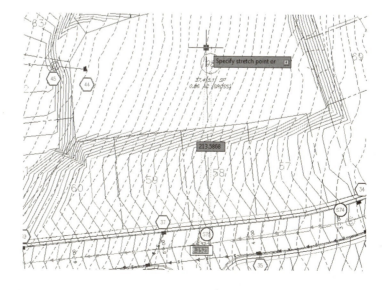

Once a change is made in the source drawing, Civil 3D will synchronize references the next time they are opened. In the current exercise, the reference drawing is already open. The following steps show you how the alert mechanism works in this situation.

4. Press Ctrl+Tab on your keyboard to change to the Creating References.dwg file and make it active, or right-click Creating References.dwg in Prospector and select Switch To. An alert bubble like the one shown in Figure 17.13 appears; it may take a few minutes. You can also choose Creating References ➢ Alignments ➢ Centerline Alignments in Prospector and you will see a series of warning chevrons.

FIGURE 17.13
Data Reference Change alert bubble

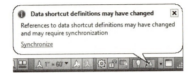

5. Click Synchronize in the alert bubble to bring your drawing current with the design file and dismiss the Panorama window if you'd like. Your drawing will update, and the Cabernet Court alignment will reflect the change in the source. If you do not get the bubble, you can also select individual references within Prospector and right-click them to access the Synchronize command. Or you can right-click on the data shortcut in Prospector and select Synchronize.

This change is simple enough and works well once file relationships are established. But suppose there is a change in the file structure of the source information and you need to make a change. You'll learn how next.

Managing Changes in the Source Data

Designs change often—there's no question about that—and using shortcuts to keep all the members of the design team on the same page is a great idea. But in the scenario you're working with in this chapter, what happens if new, additional alignment data is added to the source file? You'll explore that in this exercise:

1. Return to CreatingShortcuts.dwg.
2. Thaw the layer called Vino Lane. This is a polyline for a proposed road. It will appear as a thick red polyline.
3. Create a new alignment named **Vino Lane**. Your screen should look like Figure 17.14.
4. Save the file.
5. In Prospector, right-click Data Shortcuts and select Create Data Shortcuts to display the Create Data Shortcuts dialog.
6. Check the Hide Already Published Objects option in the lower left of the dialog to make finding the new object easier.
7. Check the Vino Lane alignment and click OK to dismiss the dialog.

8. Switch back to the CreatingShortcuts.dwg file, and add the Vino Lane alignment to your data shortcuts as in earlier examples.

9. Save and close the CreatingShortcuts.dwg file.

FIGURE 17.14
Vino Lane alignment

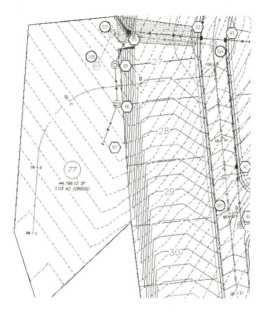

By using shortcuts to handle and distribute design information, you can keep adding information to the design as it progresses. It's important to remember that simply saving a file does not create new shortcut files for all of the Civil 3D objects contained within; they have to be created from the Data Shortcuts branch.

Fixing Broken Shortcuts

One of the dangers of linking things together is that eventually you'll have to deal with broken links. As files get renamed, or moved, the data shortcut files that point back get lost. Thankfully, Civil 3D includes a tool for handling broken links and editing links: the Data Shortcuts Editor. We'll explore that tool in this exercise. You are going to simulate one of the typical causes of broken shortcuts: that the file was modified.

1. There is a folder on the book's web page called MovedReference that should be copied to your local data folder (C:\Mastering\) for this exercise. Explode the Vino Lane alignment so that it is now a polyline.

2. Open the file RepairingReferences.dwg. This file contains a number of references pointing to a file that no longer exists (because it was moved), and Panorama should appear to tell you that five problems were found. Close the Panorama window.

3. In Prospector, select Repairing References ➢ Surfaces and you will see that EG has a warning chevron next to it.

4. Right-click EG and select Repair Broken References from the context menu to display the Choose The File Containing The Referenced Object dialog.

5. Navigate to the CreatingShortcuts.dwg file in the MovedReference folder. Click Open to dismiss the dialog. An Additional Broken References dialog will appear, as shown in Figure 17.15, offering you the option to repair all the references or cancel.

FIGURE 17.15
The Additional Broken References dialog

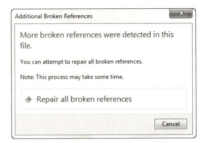

6. Click the Repair All Broken References button to close this dialog. Civil 3D will crawl through the file linked in step 4 and attempt to match the Civil 3D objects with broken references to objects in the selected drawing.

7. Perform a zoom extents, and your drawing should look like Figure 17.16, with a surface and four alignments.

FIGURE 17.16
Repaired references within an older file

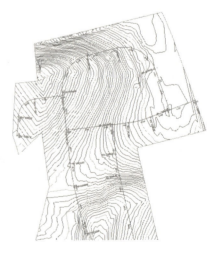

The ability to repair broken links helps make file management a bit easier, but there will be times when you need to completely change the path of a shortcut to point to a new file. To do so, you must use the Data Shortcuts Editor.

The Data Shortcuts Editor

The Data Shortcuts Editor (DSE) is used to update or change the file to which a shortcut points. You may want to do this when an alternative design file is approved, or when you move from preliminary to final design. In the following exercise, you'll change the Vino Lane alignment shortcut to an alternative design:

1. Make sure `RepairingReferences.dwg` is still open.

2. In Windows, choose Start ➢ All Programs ➢ Autodesk ➢ Autodesk Civil 3D 2012 ➢ Data Shortcuts Editor to load the DSE.

3. Select File ➢ Open Data Shortcuts Folder to display the Browse For Folder dialog.

4. Navigate to `C:\Mastering Shortcuts` and click OK to dismiss the dialog. Your DSE should look like Figure 17.17. (Your paths might be different from the author's.)

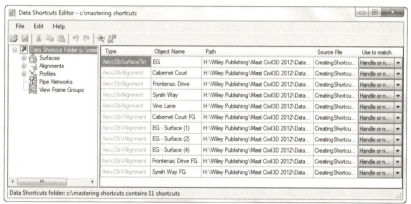

FIGURE 17.17
Editing the Concord data shortcuts

5. On the left, select the Alignments branch to display only the alignments.

6. Click on the Object Name column that presently says Vino Lane, and type **New Vino Lane**. This is the name of the alignment that we are replacing.

7. Click the Source File column on the New Vino Lane row, type `NewVinoLaneAlignment.dwg` as the Source filename, and press ↵. Unfortunately, you cannot browse. If you need to change the target file's path, you will need to change that manually as well.

8. Verify that the last column is set to Handle or Name.

9. Click the Save button in the DSE and switch back to Civil 3D. You should still be in the `RepairingReferences` drawing file.

10. Expand Data Shortcuts ➢ Alignments ➢ Centerline Alignments. Right-click Vino Lane and select Repair Broken Shortcut.

11. Scroll up within Prospector, and expand the Alignments branch.

12. Right-click Vino Lane and select Synchronize. The Object Change dialog may open; if so, select Update The Reference Name. You can also dismiss Panorama.

    Your screen should look like Figure 17.18, showing a completely different alignment for the Vino Lane alignment.

FIGURE 17.18
Completed repathing of the Vino Lane shortcut

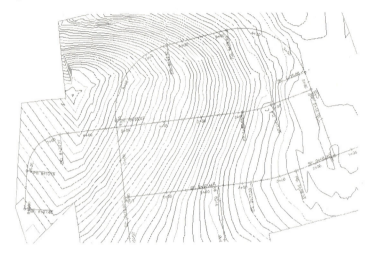

The ability to modify and repoint the shortcut files to new and improved information during a project without losing style or label settings is invaluable. When you use this function, though, be sure to validate and then synchronize.

> **WHAT ABOUT VAULT?**
>
> This section focused on the creation and use of data shortcuts. There is another method that is not discussed in this book that uses Vault. Many sources are available for finding out about Vault, such as James Wedding's Autodesk University presentation "How I Learned to Quit Worrying and Love the Vault," which can be found at http://au.autodesk.com/?nd=class&session_id=806. Note that to access any AU presentation, you must register. If you are a subscription customer, your registration is your username.

Playing Nicely in the Sandbox

Whether we want to admit it or not, there are many other CAD programs out there and, at some point, we may need to coordinate our efforts. It can be a request for a drawing from a third party who is not using Civil 3D, or even from a different department in your organization that is using other software.

In this section, we are going to discuss and demonstrate several of such cases—because we all have to play nicely in the proverbial sandbox.

Earlier Versions of Civil 3D or Land Desktop

The majority of requests that we receive have to do with third-party clients who are using an earlier release of Civil 3D or even Land Desktop. As you may (or may not) know, the Civil 3D releases are not backward compatible—that is, if you are working in Civil 3D 2012 and want to give your design to someone who is using Civil 3D 2009, you can't just give them your DWG file.

Even though Autodesk has a three-year cycle on drawing format, Civil 3D differs because of the new Civil 3D–specific items that are included each release.

There are a couple of ways to approach importing files and exporting your Civil 3D data, which we'll look at next.

XML It

> #### What Is LandXML?
>
> LandXML is not specific to Civil 3D. It is a consortium of fellow Civil users wanting to standardize a way of sharing files in a nonproprietary format.
>
> If you recall on the Import LandXML dialog, there was a drop-down list to select the version of LandXML. This is a result of that consortium. The latest as of this writing is the 1.2 schema. Most CAD programs have methods of importing and exporting LandXML files and this is one way to tackle that barrier.
>
> You can read up on LandXML at www.landxml.org/.

If you want to take your drawings and make them backward compatible, then LandXML is one option. In this example, a client needs our existing and proposed surfaces on the roads:

1. Open XMLPlan.dwg, which can be found at this book's web page.

2. From the Export panel on the Output tab, select Export To LandXML. The Export To LandXML dialog shown in Figure 17.19 opens.

Figure 17.19
The Export To LandXML dialog

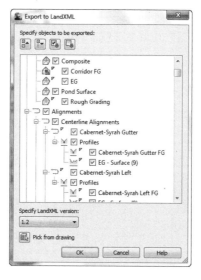

692 | **CHAPTER 17** INTEROPERABILITY

3. In the Export To LandXML dialog, click the Uncheck All icon. This setting does exactly as it says.

4. Click the Collapse The Tree icon, which will also do exactly as it says. These two tools are definite time-savers especially when you have the high number of items available in this dialog.

5. Expand the Surfaces section, select the Corridor FG and EG surfaces, and click OK.

6. In the Export XML dialog, navigate to the location where you wish to place the exported file. You can also rename the file. Remember the location as we will be looking for it in the next exercise. Click Save.

You have exported the files. Now let's make sure what you exported is what you want:

1. Start a new drawing.

2. In the Import panel on the Insert tab, select LandXML.

3. In the Import LandXML dialog (Figure 17.20), navigate to the location of the previously saved XML file. Click Open.

FIGURE 17.20
The Import LandXML dialog

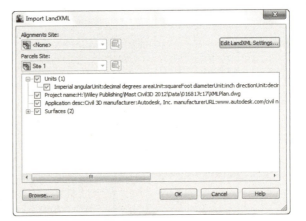

4. The Import LandXML dialog changes to reflect what it found in the file. You can select the parts that you want to be imported as well as edit any LandXML settings. Click OK to accept the defaults.

5. The file is imported into your drawing, as shown in Figure 17.21.

You will notice that it doesn't look quite like the file you used to export. This is because of the translation process that the XML has to do. In this case, the user will get sort of what they need. It should be good enough since the areas they are concerned about are correct (in this case, the road grades for existing and proposed).

FIGURE 17.21
The imported surfaces

> **LandXML Import and Alignments**
>
> In Chapter 6, "Alignments," you learned about the alignment constraints. New to Civil 3D 2012, when you import a LandXML file, you have the options to maintain the Free, Floating, or Fixed alignment tangency constraints.

eTransmit It

eTransmit has been around for some time. While it is not specific to Civil 3D, it is nonetheless another option for sending drawing files.

In this example, a third-party client wants our entire drawing file, but they are using AutoCAD Release 2000:

1. Open the `eTransmit.dwg` file.

2. From the AutoCAD icon in the upper left, click Publish ➢ eTransmit. You could also type **ETRANSMIT** at the command prompt. The Create Transmittal dialog opens (Figure 17.22).

3. Out of the box, Civil 3D includes just a standard transmittal setup. We want to add a new setup that will allow backward compatibility for AutoCAD entities. Click the Transmittal Setups button.

4. In the Transmittal Setups dialog, click the New button. In the New Transmittal Setup dialog, type a meaningful name, such as **R2000 Exploded**. This would indicate that it will take the files and turn them into AutoCAD Release 2000 format and also explode any AEC entries. Click the Continue button.

5. In the Modify Transmittal Setup dialog, the most important thing to select here is the drop-down in the File format section. Make sure it is set to AutoCAD 2000 Drawing Format With Exploded AEC Objects, as shown in Figure 17.23.

6. Click OK, and then close the Transmittal Setups dialog. Click OK in the Create Transmittal dialog.

7. Navigate to where you want the eTransmit file to reside. By default, eTransmit zips or packages the items into one file for ease of sending.

FIGURE 17.22
The Create Transmittal dialog

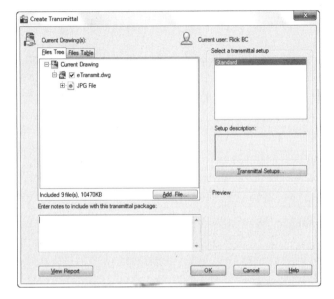

FIGURE 17.23
The modified settings for the Modify Transmittal dialog

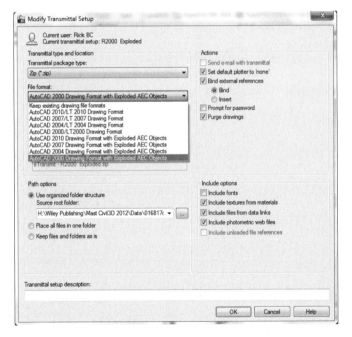

Note that using this method will turn any Civil 3D–specific objects into blocks. It is recommended that you explode the blocks into their native AutoCAD objects. This may be useful for the third party who is not using Civil 3D, but it's not useful for you to receive drawings and keep all the 3D objects.

It is beyond the scope of this book to go over every detail of the eTransmit dialog. You can find more information by clicking your friendly Help icon.

Land Desktop It

The predecessor to Civil 3D, Land Desktop can be imported into Civil 3D, and with a little finagling, the imported Land Desktop objects will show up as actual Civil 3D objects.

1. Open a new drawing using the default template.

2. From the Import panel on the Insert tab, select the Land Desktop tool. The Import Data From Autodesk Land Desktop Project dialog opens.

3. By default, everything is selected. Unlike the Import XML dialog shown in an earlier exercise, there are no controls for deselecting all; you must deselect and compress the selections yourself. For this example, you are just interested in the Existing and Proposed surfaces. Make the changes as shown in Figure 17.24 and click OK.

Figure 17.24
The Import Data From Autodesk Land Desktop Project dialog

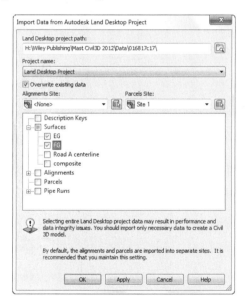

4. If everything went well, you should see no errors in the Import Data From Autodesk Land Desktop dialog. Click OK and perform a zoom extents to see your drawing, as shown in Figure 17.25.

Notice that even though you started a new drawing with no coordinate set, the Land Desktop project did have a coordinate set and translates when you import into Civil 3D.

FIGURE 17.25
The surfaces imported from Land Desktop

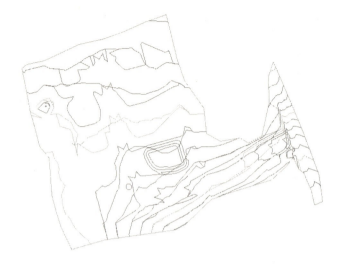

> **DON'T DO IT! YOU HAVE BEEN WARNED!**
>
> You may have the desire to select every object in the XML and Land Desktop import dialogs. Don't do it! Civil 3D gets very finicky when doing these imports and you most likely will crash.
>
> We like to think of these imports as a recipe. Mix slowly. So we recommend that you import one or two objects at a time, make the changes you want (styling, etc.), save, and then import some more. Or you can also keep the drawings separate, such as the EG in one drawing, alignments in another, and so forth.

EXPORTING TO AUTOCAD RELEASES

The ability to export to AutoCAD releases allows you to save your Civil 3D file as a plain AutoCAD release 2010, 2007, 2004, 2000, or release 14 format. This is also a one-way operation; it will explode any Civil 3D–specific objects. Here's how you do it:

1. From the AutoCAD menu, choose Export ➢ AutoCAD DWG and select one of the aforementioned release formats.
2. From the Export Drawing Name dialog, select a filename and a location and then click Save.

   By default, the filename will be the same one as the one you had originally opened. In addition, it will be prefixed by ACAD-.

Playing With Other Formats

In this day and age, to be competitive you must learn how to incorporate other CAD programs into Civil 3D. This section talks about these options.

Autodesk 3D Max Files

Many architectural firms use the popular Autodesk 3D Max format. Civil 3D has the capability of letting you import them.

In this example, a client who wishes to have a particular house on their newly purchased lot goes to the developer. In turn, the developer goes to the civil firm to show that house on their lot so they can visually see what it will look like before any dirt is moved. Here's how it works:

1. Open the `Import3DMax.dwg` file. This drawing shows our site along with existing and proposed rough grading.

2. From the Import panel on the Insert tab, select the Import tool. The Import File dialog opens.

3. Select 3D Studio (*.3ds) from the Files Of Type drop-down list.

4. Navigate to the `Pfaffle Residence.3ds` file (this file is courtesy of Emmanuel Garcia) and click Open. Remember, this and all other data files can be downloaded from www.sybex.com/go/masteringcivil3d2012.

5. In the 3D Studio File Import Options dialog, click the Add All button. The dialog is shown in Figure 17.26. Click OK to dismiss the dialog.

Figure 17.26
The 3D Studio File Import Options dialog

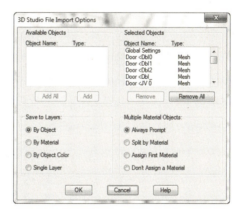

6. Click OK on the messages that come up and perform a zoom extents. The house model has been inserted around coordinates 0,0.

7. Using the AutoCAD Block command, block the house to make it easy to insert onto the lot. Choose the option to delete the block and then erase any leftover parts. Perform another zoom extents.

8. Zoom into your drawing so that Lot 4 is displayed as shown in Figure 17.27. The orange lines represent the maximum allowable building setback envelope by the municipality (also known as BSL).

FIGURE 17.27
Lot 4 shown with BSL

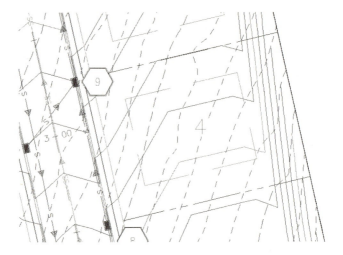

9. Insert the house block at the lower-left corner of the BSL. Since the house model came from an architect, you need to scale the entire block by $1/12$. Next, you need to rotate it into place, as shown in Figure 17.28.

FIGURE 17.28
The house rotated into place

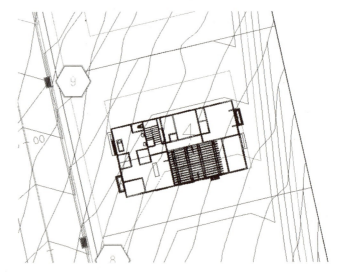

10. The house is not at the correct Z elevation. The desired building pad elevation is 800′. Click on the house block and type **MOVE**↵ ↵ **0,0,800**↵. The finished house along with some more grading is shown in Figure 17.29.

FIGURE 17.29
The finished house
(a) plan view
(b) 3D view

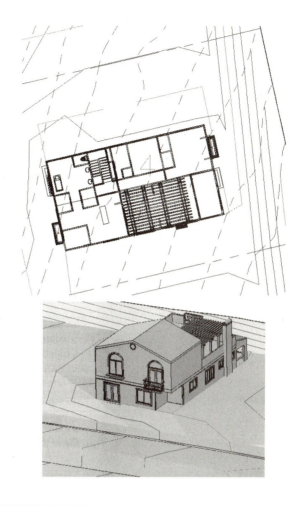

EXPORT TO CIVIL VIEW FOR 3DS MAX DESIGN

New to Civil 3d 2012 is the ability to dynamically export Civil 3D to Autodesk 3ds Max Design 2012. This tool can be found on the Export panel of the Output tab.

IMPORTING MICROSTATION FILES

Each release of Civil 3D provides a better method for importing and exporting MicroStation DGN files. Civil 3D 2012 is no exception. In this exercise, we have received a MicroStation DGN file that contains some roadwork:

1. Open a new drawing.
2. From the Insert tab and Import tab, select the Import tool. The Import File dialog opens.

3. At the bottom of the Import File dialog, make sure that the Files Of Type drop-down list is set to MicroStation (*.dgn).

4. Open the STPROP.dgn file (provided thanks to Matthew Anderson). Click Open. The Import DGN Settings dialog opens (Figure 17.30).

FIGURE 17.30
The Import DGN Settings dialog

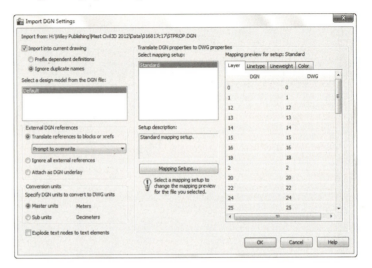

5. The Import DGN Settings dialog contains many options for handling the MicroStation file:

- At the top of the dialog, it shows you the complete path and filename that you are importing.

- When selected, the Import Into Current Drawing option allows you to insert the DGN file into your current drawing; if unchecked, it opens a new drawing and inserts it. When the box is checked, you have two choices:

  - The Prefix Dependent Definitions option sets a layer prefix, as shown in Figure 17.31 (left).

  - The Ignore Duplicate Names option sets the layers as MicroStation had them, as shown in Figure 17.31 (right).

- The External DGN References section contains three radio button options:

  - The Translate References To Blocks or XRefs will convert MicroStation references to blocks or XRefs; when it finds duplicates you can select a behavior via the drop-down list:

    - If you choose Prompt To Overwrite, when Civil 3D finds an occurrence it will prompt you to verify that you want to overwrite a block.

    - Overwrite Without Prompting is the brute-force method. It will ignore any duplicates and overwrite them regardless.

    - Do Not Overwrite will skip over any duplicates, leaving them intact.

- Ignore All External References will do exactly as indicated.
- Attach DGN As Underlay will also do exactly as indicated.

FIGURE 17.31
The Prefix Dependent Definitions (a) and the Ignore Duplicate Names (b) result in Layer Manager

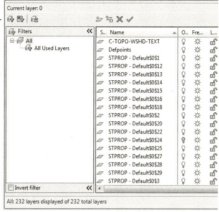

(a)   (b)

- If selected, the Explode Text Nodes To Text Elements check box will take a MicroStation text node and keep the text look and feel the way it was drawn in MicroStation. This is especially important if the text was drawn in MicroStation using a path. If the option is unchecked, the text will be converted to multiline text, destroying any MicroStation formatting that may have been in place.

- The Translate DGN Properties To DWG Properties section is the meat of the process. It is similar to the Layer Translator tool. You can select a mapping setup and to the right, it shows the current mapping. By default, there is no mapping.

    - Clicking the Mapping Setups button will open the DGN Mapping Setups dialog (Figure 17.32). You can create a new mapping setup, or rename, modify, or delete an existing one.

    - Clicking the New button opens the New Mapping Setup dialog (Figure 17.32). Here you name your new mapping setup.

- Clicking Continue opens the Modify DGN Mapping Setup dialog (Figure 17.33). You can manually enter mappings to layers, linetypes, lineweights, and colors. To import properties:

    - Clicking Add Properties From DGN File opens a dialog that lets you find a DGN file where it will import the layer properties.

    - Clicking Add Properties From DWG File opens a dialog that lets you find a DWG file where it will import the layer properties.

FIGURE 17.32
The process for DGN mapping setups

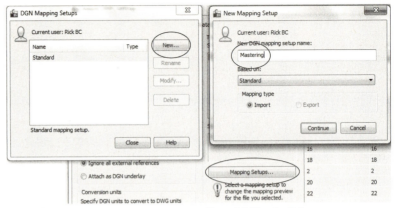

First, click Mapping Setups

FIGURE 17.33
The Modify DGN Mapping Setup dialog

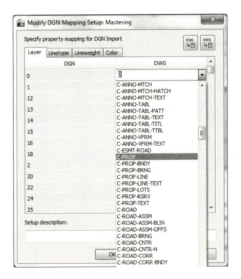

6. Click OK and then perform a Zoom All command. Your drawing should look similar to Figure 17.34. Notice that the drawing was placed on coordinates based off the original DGN file.

FIGURE 17.34
The completed DGN import

> **DGN as Underlay**
>
> New to Civil 3D 2012 is the capability to use DGN files as underlays. On the Reference panel of the Insert tab, select Attach. In the Select Reference dialog's Files Of Type drop-down list, select MicroStation DGN (*.dgn).

Exporting MicroStation Files

In this scenario, you will be giving our subdivision plan to a third-party client who uses MicroStation. It is a pretty straightforward process, although as you will see, there is a caveat:

1. Open the `Import3DMax.dwg` if it is not already open.
2. From the AutoCAD menu drop-down, select Export ≻ DGN File. The Export DGN File dialog opens.
3. Navigate to where you want the file to be placed. You can leave the default name (which is the same as the drawing), or enter your own. You can also select whether you need to save in V8 DGN or V7 DGN format.
4. Click Save.
5. The Export DGN Settings dialog opens. This is similar to the Import DGN Settings dialog we discussed earlier. Make the changes you desire and click OK.
6. You'll see a warning message informing you that there are objects in the drawing that cannot be exported to DGN files. Your choices are to accept that fact and send the drawing without the objects or to cancel the process.

Well, that is not helpful since those objects that cannot be exported are Civil 3D–specific files—which sort of defeats the purpose. So, what do you do? You have to perform a two-part operation. As we mentioned earlier, one of the methods for exporting to plain AutoCAD and then running the DGN export will work.

So, with a little effort, you can coordinate your drawings with others and indeed, Civil 3D CAN play nicely in the sandbox with others.

An Introduction to Map 3D

Since Civil 3D is built upon Map 3D, we would be remiss if we didn't mention its existence and some of its features. An entire book could be written on this subject, so only the basic parts and how they pertain to Civil 3D will be discussed here.

Map 3D is the geographical information system (GIS) part of Civil 3D. It lies there ready and willing to jump in and help you out. Still, many people don't know about its existence. Hold on to your hats for a whirlwind tour of Map 3D.

Where Is It?

Out of the box, Civil 3D starts up with a workspace called Civil 3D. This workspace contains all the tools necessary to work on Civil 3D. Down on the status bar, you will find the Workspace switcher. The flyout workspace choices are:

- Civil 3D
- 2D Drafting & Annotation
- 3D Modeling
- Planning And Analysis Workspace

The last option (Planning And Analysis Workspace) is the Map 3D Workspace. Clicking that option will transform your screen into the new tools on the tabs and panels shown in Figure 17.35.

Figure 17.35
The Planning and Analysis Workspace

Setup

To efficiently use Map 3D, you need to do some housework, such as a creating a drive alias and attaching drawing files. So let's get to it.

Drive Alias

The ability to attach drawings and other objects is invaluable to Civil 3D so that you can load only what you want to. To assist in this effort, setting up drive aliases will make life a bit easier for you. You can liken a drive alias to data shortcuts (which we covered earlier this chapter). Here's the procedure:

1. Start a new drawing using any template you wish.

2. Using the Workspace switcher (located at the bottom of your drawing screen), select Planning And Analysis Workspace. This switches the workspace to the Map 3D with its associated tabs and panels.

3. From the Data panel on the Home tab, select the Attach tool. The Select Drawings To Attach dialog opens.

4. Click the Create/Edit Alias icon. The Drive Alias Administration dialog opens. Here is where you will set an alias and the path to it.

5. In the Drive Alias Details section, type **Mastering** in the Drive Alias text box. Click the Browse button and navigate to the `C:\Mastering\` folder and click OK.

6. Back in the Drive Alias Administration dialog, click the Add button, as shown in Figure 17.36. Click Close. If an error message pops up, close it. The alias has been saved, as shown in Figure 17.37. You can repeat this process for as many aliases you want. Click OK to close this dialog.

FIGURE 17.36
The Drive Alias Administration dialog

FIGURE 17.37
The Select Drawings To Attach dialog

7. You can now copy your GIS files to this new location. Locate the GIS folder, which you can download from this book's web page. Copy the GIS folder and all of its contents into the `C:\Mastering\` folder. (You could have also done this step before creating the drive alias).

ATTACH DRAWINGS

Now that you have a drive alias (or multiple aliases) set up, you can attach drawing files to your empty drawing. These drawings are not just regular AutoCAD drawings; they contain features that you will pull from later. So, let's attach a drawing:

1. Reopen the Select Drawings To Attach dialog by clicking the Attach tool on the Data panel of the Home tab.

2. In the Look In section, use the drop-down list box to select your aliased drive. Note that the author's alias is pointing to a larger drive where the GIS data is stored. This brings up a good point: You could alias a location on the server for these objects since many of them can be quite large in size.

3. Double-click the `Counties` folder, select the US Counties object, click Add, and then click OK. Yes, you select US Counties even though you are dealing with only a portion of a county. This way, you can see how powerful the query options are in Map 3D that will be covered later.

4. The counties drawing has been attached, but nothing is showing on the screen. This is because, although the drawing is loaded, it is not visible. Another way to think about this is you do not want to see every county in the United States on your drawing. You want to look inside that drawing and pull out the data that you really need to see, which we will cover next.

Queries

This section is one of the most powerful reasons to learn about 3D Map: the ability to pull data from different sources (such as SHP files) and come up with a drawing to just show those.

1. In the Task pane, select the Data tool.

2. From the flyout menu, select Connect To Data. The Data Connect palette opens.

3. Click the SHP icon, navigate to the `Counties` folder, and select the `countyp020.shp` file.

4. Type **Counties** in the Connection Name text box. Your palette should look similar to Figure 17.38. Click Connect.

5. In the Data Connect palette, click the Add To Map flyout and select Add To Map With Query. The Create Query wizard opens.

6. You want to find York County in the state of Pennsylvania, so you will start large and narrow down. Click the Property tool and select State.

7. Click the Equals tool.

8. Click the Get Values tool, and in the filter box, type the letter **P**. This sets the filter to all states that begin with the letter P. Click the green button and then select PA from the list. You could have simply selected the green arrow and then scrolled to find Pennsylvania, but we wanted to show you the power of the filter.

9. Click the Insert Value tool. Your query thus far looks like Figure 17.39.

FIGURE 17.38
The completed Data Connect palette

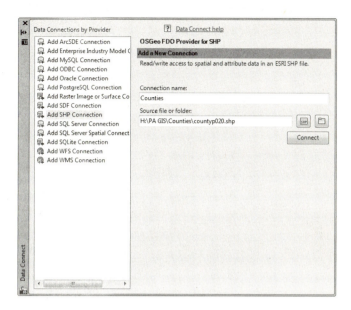

FIGURE 17.39
The query thus far

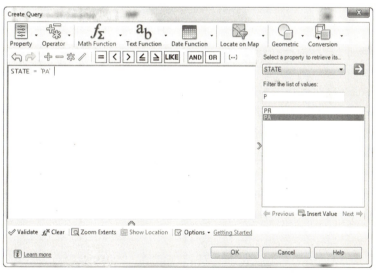

10. We need to add to the query by finding York County within the state of Pennsylvania. Click the Operator tool and select AND. This is a logical operator that means all conditions must be met in order to find a match.

11. Repeat steps 5–7 but look for the Property to query COUNTY and the value to query York County, as shown in Figure 17.40. Click OK.

12. Perform a zoom extents. Only the polygon for York County, Pennsylvania is displayed, although there are over 3,000 to choose from. In addition, there are five counties in the United States that are York County.

FIGURE 17.40
The finished query

13. Save your drawing as GISMAP.dwg.

Hopefully, you can see the power of Map 3D and how you can create custom reports like the following:

- What properties are affected by the proposed 50′ right-of-way that is going in on the highway?
- How many red maple (*Acer Rubrum*) trees are on your 36-acre lot?
- You need a coverage report of fire hydrants based off a cityscape, keeping in mind that hoses cannot go through buildings.

So you can see that the possibilities are pretty much endless as long as you have valid data, which can be procured from your municipal GIS department or even on the Internet.

SCRATCHING THE SURFACE

No, we have not gone in-depth at all with Map 3D—we haven't even scratched the surface! There could be books that would equal the size of this book just on the subject of Map 3D.

If our little exercise piqued your curiosity about Map 3D, then our object was fulfilled. Luckily, resources are available to help you learn more about the subject.

One such resource comes from James Murphy, who many consider to be a Map 3D guru. His blog, Map 3D and Murph's Law (map3d.wordpress.com), is a great source for Map 3D.

Don't forget the help and tutorials that are included with Civil 3D. As we mentioned earlier, Civil 3D is built on Map 3D and the help/tutorials are included with your Civil 3D.

The Bottom Line

Create data shortcuts.   The ability to load design information into a project environment is an important part of creating an efficient team. The main design elements of the project are available to the data shortcut mechanism, but some still are not.

> **Master It**   List the top-level branches in Prospector that cannot be shared through data shortcuts.

Import and export to earlier releases of AutoCAD.   Being able to work with outside clients or even other departments within your firm who do not have Civil 3D is an important part of the concept of playing nicely together.

> **Master It**   What are some methods for sharing your Civil 3D drawings with AutoCAD Release 2004 users?

Import and export to third-party CAD programs.   Exchanging drawings from other CAD formats is another important method that lets you work together on a common project, no matter what CAD program is being used.

> **Master It**   What tools can be used to receive CAD files from a firm using MicroStation?

Use basic map queries to filter data.   Map 3D is the program that Civil 3D is built on. In Map 3D there are numerous things that can be accomplished; using queries is one of them.

> **Master It**   Open `GISMAP.dwg`. This is the subdivision you have been working on for most of the book. Insert the PaMunicipalities shape file found in the `GIS\Municip_2011` folder. You only want to have the Felton municipality visible on the plan.

Chapter 18

Quantity Takeoff

The goal of every project is eventual construction. Before the first bulldozer can be fired up and the first pile of dirt moved, the owner, city, or developer has to know how much all of this paving, pipe, and dirt are going to cost. Although contractors and construction managers are typically responsible for creating their own estimates for contracts, the engineers often perform an estimate of cost to help judge and award the eventual contract. To that end, many firms have entire departments that spend their days counting manholes, running planimeters around paving areas, and measuring street lengths to figure out how much striping will be required.

Civil 3D includes a Quantity Takeoff (QTO) feature to help relieve that tedious burden. You can use the model you've built as part of your design to measure and quantify the pieces needed to turn your project from paper to reality. You can export this data to a number of formats and even to other applications for further analysis.

In this chapter, you will learn to:

- Open and review a list of pay items along with their categorization
- Assign pay items to AutoCAD objects, pipe networks, and corridors
- Use QTO tools to review what items have been tagged for analysis
- Generate QTO output to a variety of formats for review or analysis

Pay Item Files

Before you can begin running any sort of analysis or quantity, you have to know what items matter. Various municipalities, states, and review agencies have their own lists of items and methods of breaking down the quantities involved in a typical development project.

There are three main files associated with quantity takeoff in Civil 3D; the *pay item list*, the *pay item categorization file*, and the *formula file*. Each file serves a different purpose and is stored externally to the project. Once a file has been associated to a DWG, the path to the external file is stored with the DWG.

Pay Item List When preparing quantities, different types of measurements are used based on the items being counted. Some are simple individual objects such as light posts, fire hydrants, or manholes. Only slightly more complicated are linear objects such as road striping, or area items such as grass cover. These measurements are also part of the pay item list.

Pay Item Categorization File In any project, there can be thousands of items to tabulate. To make this process easier, most organizations have built up pay categories, and Civil 3D makes use of this system in a categorization file. Civil 3D includes an option to create favorite items that are used most frequently.

Formula File The formula file is used when the unit of measure for the pay item needs to be converted or calculated based on another value. For example, asphalt for a parking lot is usually paid for by the ton, but it is much easier to get the area out of CAD. You can set up a formula that converts square feet into tons, taking into account an assumed depth and density. Later in this chapter, you will create a formula that relates length to light poles along the road.

There is a key-in available if you need to disconnect your pay item and formula files from the drawing. At the command line you can type `DETACHQTOFILES`. The Undo command will not reconnect them if you type this in accidentally.

In the next section, you will learn how to connect the needed pay item files and create favorites.

Pay Item Favorites

Your Civil 3D file will "remember" which files it needs to correctly assess quantities. The *favorites list* is also stored as part of the DWG. The favorites list is a separate category that can be populated for quick access to commonly used pay items.

In this exercise, you'll look at how to open a pay item and categorization files and add a few items to the favorites list for later use:

1. Start a new drawing based on the template of your choice.

2. From the Analyze tab's QTO panel, select QTO Manager (see Figure 18.1) to display the QTO Manager palette.

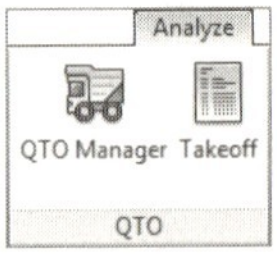

FIGURE 18.1
The QTO tools on the Analyze tab

3. Click the Open button at the top left of the palette to display the Open Pay Item File dialog.

4. From the Pay Item File Format drop-down list, select CSV (Comma Delimited).

5. Click the Open button next to the Pay Item File text box to display the Open Pay Item File dialog.

6. Navigate to the `Getting Started` folder and select the `Getting Started.csv` file. In Windows Vista and Windows 7, this file is found in `C:\ProgramData\Autodesk\C3D2012\enu\Data\Pay Item Data\Getting Started\`.

7. Click Open to select this pay item file.

8. Click the Open button next to Pay Item Categorization File to browse for the file.

9. Navigate to the `Getting Started` folder and select the `Getting Started Categories.xml` file. Click Open to select the file. Your display should look similar to Figure 18.2.

10. Click OK to close the Open Pay Item File dialog, and Panorama will now be populated with a collection of divisions. These divisions came from the `Getting Started Categories.xml` file.

11. Expand the Division 200 ➢ Group 201 ➢ Section 20101 branches, and select the item 20101-0000 CLEARING AND GRUBBING, as shown in Figure 18.3.

FIGURE 18.2
Open Pay Item File dialog

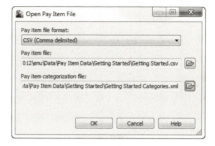

FIGURE 18.3
Selecting a pay item to add as a favorite

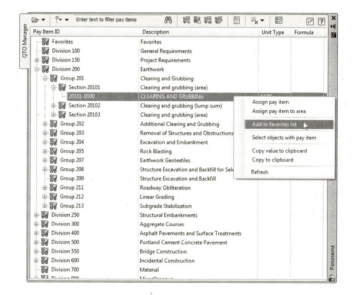

12. Right-click and select Add To Favorites List.

13. Expand a few other branches to familiarize yourself with this parts list. Add the following to your favorites in preparation for the next exercise:

    ◆ 60404-3000 Catch Basin, Type 3

    ◆ 61106-0000 Fire Hydrant

    ◆ 61203-0000 Manhole, Sanitary Sewer

    ◆ 63401-0600 Pavement Markings, Type C, Broken

14. Save the drawing as `QTO Practice.dwg` and keep it open for the next exercise.

    When complete, your QTO Manager should look similar to Figure 18.4.

Figure 18.4
A favorites list within the QTO Manager

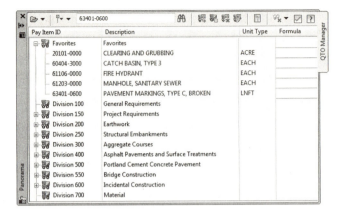

Civil 3D ships with a number of pay item list categorization files, but only one actual pay item list. This avoids any issue with out-of-date data. One commonly used categorization type found in the C:\ProgramData\Autodesk\C3D2012\enu\Data\Pay Item Data\CSI\ folder is MasterFormat. In the United States folder, you can also find categorization files for AASHTO, Federal Highway Administration, and several states' DOT formats. The pay item files that install with the software are intended to be examples. Contact your reviewing agency for access to their pay item list and categorization files if they're not already part of the Civil 3D product.

Once you have pay items to choose from, it's time to assign them to your model for analysis.

Searching for Pay Items

In the upper left of the QTO Manager palette is an extremely handy search tool. If you don't know what category an item is listed under, you can type it in the field to the left of the binoculars icon.

When you first click the binoculars icon to execute the search, it will appear as if nothing happened. That's because the categories are getting in the way of the listing. To see an uncategorized look at the list your search produced, select Turn Off Categorization from the drop-down to the left of the search field, as shown in Figure 18.5.

Figure 18.5
Choose Turn Off Categorization to see the results of your search.

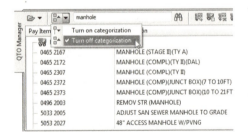

In the following exercise, you will use the search functionality to add items to your favorites list:

1. Continue working in the file that you created in the previous exercise.

2. If it is not already open, go to the Analyze tab and click the QTO Manager button to open it.

3. To the left of the search field, click the drop-down to select Turn Off Categorization.

4. In the search field of the QTO Manager type **grandi** and then click the binoculars icon. Your search should result in three types of trees. Right-click on the first item, FAGUS GRANDIFLORA, and select Add To Favorites List.

5. Using the same search technique, add the following item to the favorites list:

    63620-0500 Pole, Type Washington Globe No. 16

6. Save the drawing.

Keeping Tabs on the Model

Once you have a list of items that should be accounted for within your project, you have to assign them to items in your drawing file. You can do this in any of the following ways:

- Assign pay items to simple items like blocks and lines.
- Assign pay items to corridor components.
- Assign pay items to pipe network pipes and structures.

In the next few sections, you'll explore each of these methods, along with some formula tools that can be used to convert things such as linear items to individual quantity counts.

AutoCAD Objects as Pay Items

The most basic use of the QTO tools is to assign pay items to things like blocks and linework within your drawing file. The QTO tools can be used to quantify tree plantings, signposts, or area items such as clearing and grubbing. In the following exercise, you'll assign pay items to blocks as well as to some closed polylines. Be sure the `Getting Started.csv` and `Getting Started Categories.xml` files are loaded as described in the first exercise.

1. Open the `Acad Objects in QTO.dwg` file.

2. From the Analyze tab's QTO panel, select QTO Manager to display the QTO Manager palette if it is not already open.

3. Expand the Favorites branch, right-click on CLEARING AND GRUBBING, and select Assign Pay Item To Area, as shown in Figure 18.6.

FIGURE 18.6
Assigning an area-based pay item

4. Enter **O** at the command line and press ↵ to activate the Object option for assignment.

5. Click on the outer edge of the site, as shown in Figure 18.7. Notice the entire polyline highlights to indicate what object is being picked. The command line should also echo `Pay item 20101-0000 assigned to object` when you pick the polyline. Note that this will create a hatch object reflecting the area being assigned to the pay item.

FIGURE 18.7
Selecting a closed polyline for an area-based quantity

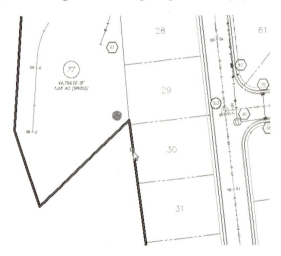

6. Press ↵ again to complete the selection.

7. Move the QTO Manager palette to the side and zoom in on an area of the road where you can see two or more of the blocks representing trees.

8. Select one of the tree blocks. To select all of the tree blocks in the drawing, right-click and pick Select Similar.

9. Back on the QTO Manager palette, under the Favorites branch, select 62606-0150 FAGUS GRANDIFLORA, AMERICAN BEECH.

10. Near the top of the palette, click the Assign Pay Item button. Notice the great tooltips on these buttons.

11. Press ↵ to complete the assigning.

12. Repeat steps 9–11 to assign pay items to the fire hydrant blocks in the road right of way areas.

The two assignment methods shown here are essentially interchangeable. Pay items can be assigned to any number of AutoCAD objects, meaning you don't have to redraw the planners' or landscape architects' work in Civil 3D to use the QTO tools.

Keep in mind that the pay item tag is saved with the block. This is a good thing if you assign a pay item to a block and copy the block because the copies will be automatically tagged. If you assign a pay item to an object and use the WBLOCK command to copy that object out of your current drawing and into another, the pay item assignment goes along as well. You'll find out how to unassign pay items after we discuss all the ways to assign them.

Pricing Your Corridor

The corridor functionality of Civil 3D is invaluable. You can use it to model everything from streams to parking lots to roads. With the QTO tools, you can also use the corridor object to quantify much of the project construction costs.

> **ASSEMBLIES AND QTO ARE RELATED**
>
> Be mindful of what parts of an assembly are available when creating quantity takeoff assignments in the code set style.
>
> If you plan to price an item based on the centerline of your road, make sure you have set Crown Point On Inside to Yes in the lane subassembly properties.
>
>
>
> Changes to QTO information will not cause your corridor to flag itself as out of date. Before you run a takeoff report, it is a good idea to rebuild your corridor to ensure the most up-to-date information is accounted for.

In this first example, you'll use the pay item list along with a formula to convert the linear curb measurement to an incremental count of light poles required for the project:

1. Open the `Corridor Objects in QTO.dwg` file.

2. From the Analyze tab's QTO panel, select QTO Manager to display the palette.

3. Expand the Favorites branch to display your 63620-0500 light standard pay item.

4. Scroll to the right to the Formula column and click in the empty cell on the Luminaire row. When you do, Civil 3D will display the alert box shown in Figure 18.8, which warns you that formulas must be stored to an external file.

5. Click OK to dismiss the warning, and Civil 3D will present the Select A Quantity Takeoff Formula File dialog. Navigate to your desktop, change the filename to **Mastering**, and click the Save button to dismiss this dialog.

   Civil 3D will display the Pay Item Formula: 63620-0500 dialog, as shown in Figure 18.9. (The expression will be empty when you first open it, but you'll take care of that in the next steps.)

FIGURE 18.8
Click in the Formula cell to display this warning dialog.

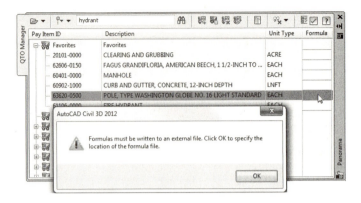

FIGURE 18.9
The completed pay item expression

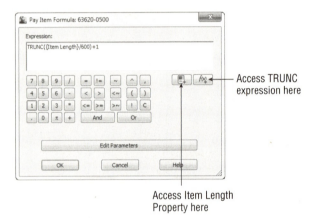

Access TRUNC expression here

Access Item Length Property here

Assume that you need a street light every 300 feet, but only on one side of the street. To do this, you add up all the lengths of curb, and then divide by 2 because you want only one half of the street to have lights. You then divide by 300 because you are running lights in an interval. Finally, you truncate and add 1 to make the number conservative.

6. Enter the formula using the buttons in the dialog or type it in to arrive at Figure 18.9; then click OK to dismiss the dialog. Note that the pay item list now shows a small calculator icon on that row to indicate a formula is in use.

   Now that you've modified the way the light poles will be quantified from your model, you can assign the pay items for light poles and road striping to your corridor object. This is done by modifying the code set, as you'll see in the next steps.

7. In the Toolspace Settings tab, select General ➤ Multipurpose Styles ➤ Code Set Styles ➤ All Codes. Right-click and select Edit to display the Code Set Style – All Codes dialog.

8. Change to the Codes tab, scroll down to the Point section, and find the row for Crown, as shown in Figure 18.10.

9. Click the truck icon in the Pay Item column, as shown in Figure 18.10, to open the Pay Item List dialog.

10. Expand the Favorites branch and select PAVEMENT MARKINGS, TYPE C, SOLID. Click OK to return to the Code Set Style – All Codes dialog. Your dialog should now reflect the pay item number of 63401-0500, as shown in Figure 18.11. Note that the tooltip also reflects this pay item.

FIGURE 18.10
Select the Crown row in the Point section in the Code Set Style – All Codes dialog.

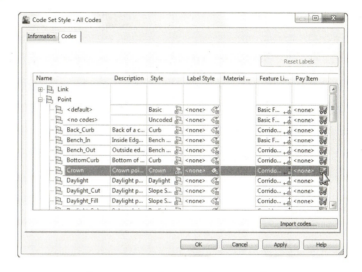

FIGURE 18.11
Crown point codes assigned a pavement marking pay item

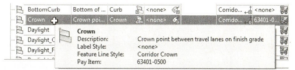

11. Find Back_Curb (which should be listed in the same area several rows above Crown), and repeat steps 9 and 10 to assign 63620-0500 to the Back_Curb code. Remember, this is your light standard pay item with the formula from the previous steps. Your dialog should look like Figure 18.12.

12. Switch to Prospector, and expand the Corridor branch. Right-click Main Corridor and select Properties to open the Corridor Properties dialog.

FIGURE 18.12
Completed code set editing for pay items

13. Change to the Codes tab and scroll down to the Point section. Notice that Back_Curb and Crown codes reflect pay items in the far-right column.

14. Click OK to close the dialog.

> **BEST PRACTICE: STORING A FORMULA FILE WITH THE PROJECT FILES**
>
> Every drawing in which you utilize QTO tools is intended to have its own, unique formula file. We recommend that when you create a new formula file, you store it in the same folder as the rest of the project.
>
> If you ever need to send a drawing to another firm and you want them to have your quantity takeoff formulas, use eTransmit to export everything they need to work with the drawing. The eTransmit command will attach the pay item file, the categorization file, and the formula file as part of the transmittal.

Corridors can be used to measure a large number of things. You've always been able to manage pure quantities of material, but now you can add to that the ability to measure linear and incremental items as well. Although we didn't explore every option, you can also use shape and link codes to assign pay items to your corridor models.

Now that you've looked at AutoCAD objects and corridors, it's time to examine the pipe network objects in Civil 3D as they relate to pay items.

Pipes and Structures as Pay Items

One of the easiest items to quantify in Civil 3D is the pipe network. There are numerous reports that will generate pipe and structure quantities. This part of the model has always been fairly easy to account for; however, with the ability to include it in the overall QTO reports, it's important to understand how parts get pay items assigned. There are two methods: via the parts lists and via the part properties.

ASSIGNING PAY ITEMS IN THE PARTS LIST

Ideally, you'll build your model using standard Civil 3D parts lists that you've set up as part of your template. These parts lists contain information about pipe sizes, structure thicknesses, and so on. They can also contain pay item assignments. This means that the pay item property will be assigned as each part is created in the model, skipping the assignment step later.

In this exercise, you'll see how easy it is to modify parts lists to include pay items:

1. Open the `Pipe Networks in QTO.dwg` file.

2. Change to the Settings tab, and expand Pipe Network ➢ Parts Lists ➢ Storm Sewer. Right-click and select Edit to bring up the Network Parts List – Storm Sewer dialog.

3. Change to the Pipes tab and expand the Storm Sewer ➢ Concrete Pipe part family. Notice that the far-right column is the Pay Item assignment column, similar to Code Sets in the previous section.

4. Click the truck icon in the 12-inch RCP row to display the Pay Item List.

5. Enter **12-Inch Pipe Culvert** in the text box, as shown in Figure 18.13, and press ↵ to filter the dialog. Remember that you can turn off categories with the button at the left of the text box.

FIGURE 18.13
Filtering and selecting the 12-inch pipe culvert as a pay item

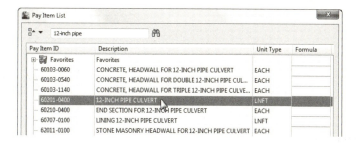

6. Select the 12-INCH PIPE CULVERT item as shown, and click OK to assign this pay item to the 12-inch RCP pipe part.

7. Repeat steps 4 through 6 to assign pay items to 15-, 18-, and 30-inch RCP pipes. Your dialog should look like Figure 18.14.

FIGURE 18.14
Completed pipe parts pay item assignment

8. Change to the Structure tab of the Parts List dialog.

9. Expand the Storm Sewer ≻ Concrete Rectangular Headwall branch.

10. Click the truck icon on the Headwall Up To 21 Inches row to display the Pay Item List dialog.

11. Search for Headwall and select CONCRETE, HEADWALL FOR 21-INCH PIPE CULVERT. Click OK to close the dialog.

12. Click the truck icon on the row of the Concentric Cylindrical Structure NF folder to display the Pay Item List dialog. By picking at the level of the part family, you will be assigning the same pay item to *all* sizes of that part family.

13. Expand the Favorites branch and select MANHOLE, TYPE 3. Click OK to close the dialog. Your parts list should now look like Figure 18.15.

FIGURE 18.15
Completed structure parts pay item assignment

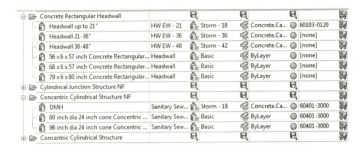

> ### HEADS UP ON PIPE AND STRUCTURE QTO ASSIGNMENTS
>
> Pay items change when pipes and structures change if the new part has a pay item assigned in the parts list. However, if you swap to a part that does not have a pay item assigned, Civil 3D drops the QTO tag.
>
> The best way to keep a proper count of your pipes and structures is to make sure every part in your network parts list has an appropriate pay item.
>
> If you graphically change a pipe property (such as its diameter), this will not cause a change in the pay item assignment. You should always use Swap Part to change sizes.
>
>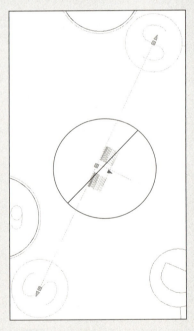
>
> This is not a defect, but it's something you definitely want to keep an eye on. So, if you do need to change a pay item assignment to a part that's already in the network, how do you do it? You'll find that out in the next section.

Pay Items as Part Properties

If you have existing Civil 3D pipe networks that were built before your parts list had pay items assigned, or if you change out a part during your design, you need a way to review and modify the pay items associated with your network. Unfortunately, doing so isn't as simple as just telling Civil 3D to reprocess some data, but it's not too complicated either. You simply remove the pay item association in place and then add new ones.

In this exercise, you'll add pay item assignments to a number of parts already in place in the drawing:

1. Open or continue working with the `Pipe Networks in QTO.dwg` file.

2. From the Analyze tab's QTO panel, select QTO Manager to display the QTO Manager palette.

3. Slide the QTO Manager to one side, and then select one of the manholes in the Storm Sewer Network (they have a D symbol on them).

4. Right-click and choose Select Similar, as shown in Figure 18.16. Nine manholes should highlight.

FIGURE 18.16
Use Select Similar to find all manhole structures.

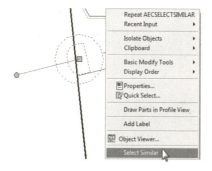

5. In the QTO Manager, expand Favorites and select MANHOLE, TYPE 3; then click the Assign Pay Item button.

6. At the command line, press ↵ to complete the assignment. You can pause over one of the manholes, and the tooltip will reflect a pay item now, in addition to the typical information found on a manhole.

Assigning pay items to existing structures and pipes is similar to adding data to standard AutoCAD objects. As mentioned before, the pay item assignments sometimes get confused in the process of changing parts and pipe properties, and they should be manually updated. To do so, you'll need to remove pay item data and then add it back in as demonstrated in this exercise:

1. Open or continue working with the `Pipe Networks in QTO.dwg` file. Pan to the northwest area of the site where the sanitary sewer network runs offsite.

2. Pause your cursor over the pipe. The tooltip will appear indicating the pipe information, including the pay item, as shown in Figure 18.17a.

3. Select the pipe, right-click, and select Swap Part, as shown in Figure 18.17b. This will display the Swap Part Size dialog.

FIGURE 18.17
Swapping out the part size (right)

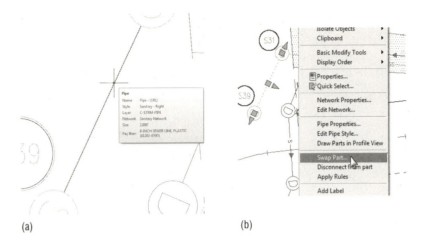

(a)   (b)

4. Select 12 Inch PVC from the list of sizes, and then click OK to dismiss the dialog.

5. Pause near the newly sized pipe and notice that the tooltip no longer shows the pay item, as shown in Figure 18.18.

FIGURE 18.18
The tooltip after the part has been swapped to a part with no assignment in the parts list

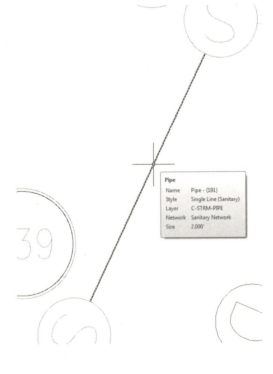

KEEPING TABS ON THE MODEL | **725**

6. Open the QTO Manager palette if it's not already open.

7. Enter **12-inch pipe** in the text box to filter the pay item list.

8. Select the 12-INCH PIPE CULVERT item in the list, right-click, and choose Assign Pay Item from the context menu.

9. Select the pipe you just modified to assign the pay item. The command line should echo `Pay item 60201-0400 assigned to object`.

10. Press ↵ to end the assignment command.

Next, you will remove an extraneous pay item assignment on the sanitary outfall:

11. Click the Edit Pay Items On Specified Object button in the QTO Manager.

12. Pick the downstream sanitary manhole connected to the pipe you edited in the previous step.

13. Select the FIRE HYDRANT row, and then click the red X in the upper right to remove this pay item from the structure, as shown in Figure 18.19.

FIGURE 18.19
Editing pay item assignments: deleting an erroneous fire hydrant

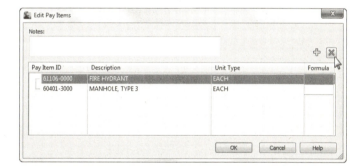

14. Click OK to dismiss the dialog.

15. Close the QTO Manager palette.

You might wonder why Civil 3D allows you to have multiple pay items on a single object. Linear feet of striping and tree counts can both be derived from street lengths; bedding and pipe material can both be calculated from pipe objects. You can also add related tasks to an item. For instance, a tree is usually a pay item by itself, but the labor to install the tree may be treated as a separate pay item.

You've now built up a list of pay items, tagged your drawing a number of different ways, updated and modified pay item data, and looked at formulas in pay items. In the next section, you'll make a final check of your assignments before running reports.

Highlighting Pay Items

Before you run any reports, it's a good idea to make a cursory pass through your drawing and look at what items have had pay items assigned and what items have not. This review will allow you to hopefully catch missing items (such as hydrants added after the pay item assignment was done), as well as see any items that perhaps were blocked in with unnecessary pay items already assigned. In this exercise, you'll look at tools for highlighting objects with and without pay item assignments:

1. Open the `Highlighting Pay Items.dwg` file.

2. From the Analyze tab's QTO panel, select QTO Manager.

3. In the QTO Manager, select Highlight Objects With Pay Items, as shown in Figure 18.20.

FIGURE 18.20
Turning on highlighting for items with pay items assigned

4. Pan to the middle of the drawing and note that some of the lot lines are still bright magenta instead of muted. This means they have a pay item assigned.

5. In the QTO Manager, select Highlight Objects Without Pay Items from the drop-down menu shown in Figure 18.20. Note that most of the alignments and lot lines are now highlighted.

6. In the QTO Manager, select Clear Highlight to return the drawing view to normal.

While objects are highlighted, you can add and remove pay items as well as any other normal AutoCAD work you might perform. This makes it easier to correct any mistakes made during the assignment phase of the process. Finally, be sure to clear highlighting before exiting the drawing, or your peers might wind up awfully confused when they open the file!

Inventorying Your Pay Items

At the end of the process, you need to generate some sort of report that shows the pay items in the model, the quantities of each item, and the units of measurement. This data can be used as part of the plan set in some cases, but it's often requested in other formats to make further analysis possible. In this exercise, you'll look at the Quantity Takeoff tool that works in conjunction with the QTO Manager to create reports:

1. Open the `QTO Reporting.dwg` file.

2. From the Analyze tab's QTO panel, select the Quantity Takeoff tool to display the Compute Quantity Takeoff dialog, shown in Figure 18.21.

    Note that you can limit the area of inquiry by drawing or selection; by sheets, if done from paper space; or by alignment station ranges. Most of the time, you'll want to run the full drawing. You can also report on only selected pay items if, for instance, you just want a table of pipe and structures.

3. Click Compute to open the Quantity Takeoff Report dialog, as shown in Figure 18.22.

FIGURE 18.21
The Quantity Takeoff tool with default settings

FIGURE 18.22
QTO reports in the default XSL format

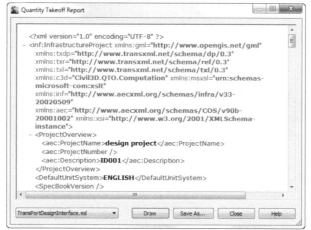

4. From the drop-down menu on the lower-left of the dialog, select Summary (TXT).xsl to change the format to something more understandable. At this point, you can export this data out as a text file, but for the purposes of this exercise, you'll simply insert it into the drawing.

5. Click the Draw button at bottom of the dialog, and Civil 3D will prompt you to pick a point in the drawing.

6. Click near some clear space, and you'll be returned to the Quantity Takeoff Report dialog. Click Close to dismiss this dialog, and then click Close again to dismiss the Compute Quantity Takeoff dialog.

7. Zoom in where you picked in step 5, and you should see something like Figure 18.23.

FIGURE 18.23
Summary Takeoff data inserted into the drawing

```
                              Summary Takeoff Report
                              ---------------------
Pay Item ID    Description                                                      Quantity    Unit
-----------    -----------                                                      --------    ----
15706-1100     SOIL EROSION CONTROL, INLET PROTECTION TYPE A                    37          EACH
20301-0080     REMOVAL OF BENCH                                                 1           EACH
60201-0400     12-INCH PIPE CULVERT                                             4246.817    LNFT
60401-3000     MANHOLE, TYPE 3                                                  10          EACH
60403-0100     INLET, TYPE 1                                                    37          EACH
61106-0000     FIRE HYDRANT                                                     10          EACH
61202-0700     8-INCH SEWER LINE, PLASTIC                                       3254.674    LNFT
61203-0000     MANHOLE, SANITARY SEWER                                          20          EACH
62606-0150     FAGUS GRANDIFLORIA, AMERICAN BEECH, 1 1/2-INCH TO 2-INCH CALIPER, BALLED AND BURLAPPED   8
63303-0900     SIGN, ALUMINUM PANEL, TYPE 3 SHEETING                            4           EACH
```

That's it! The hard work in preparing QTO data is in assigning the pay items. The reports can be saved to HTML, TXT, or XLS format for use in almost any analysis program.

Real World Scenario

CREATING YOUR OWN CATEGORIZATION FILE

One of the challenges in automating any quantity takeoff analysis is getting the pay item list and categories to match up with your local requirements. Getting a pay item list is pretty straightforward—many reviewing agencies provide their own list for public use to keep all bidding on an equal footing.

Creating the category file can be slightly more difficult, and it will probably require experimentation to get just right. In this example, you'll walk through creating a couple of categories to be used with a provided pay item list.

Note that modifying the QTO Manager palette affects all open drawings. You might want to finish the other exercises and come back to this when you need to make your own file in the real world.

1. Create a new drawing from the _AutoCAD Civil 3D (Imperial) template file.

2. From the Analyze tab's QTO panel, click the QTO Manager button to display the QTO Manager palette.

3. Click the Open Pay Item File button at the top left to display the Open Pay Item File dialog.

4. Verify that the format is CSV (Comma Delimited).

5. Browse to open the Mastering.csv file. Remember, all files can be downloaded from www.sybex.com/go/masteringcivil3d2012.

6. Click the Open button next to the Pay Item Categorization file text box to display the Open Pay Item Categorization File dialog.

7. Browse to open the Mastering Categories.xml file, and then click OK to dismiss the dialog. Your QTO Manager dialog should look like this:

8. Launch XML Notepad. (You can download this from Microsoft as noted in Chapter 17, "Interoperability.")
9. Browse and open the `Mastering Categories.xml` file from within XML Notepad.
10. Right-click the category branch, as shown here, and select Duplicate.

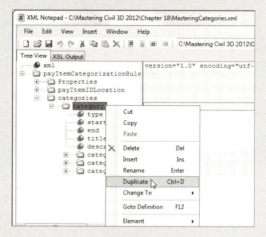

11. Modify the values as shown to match the following image.
12. Delete the extra category branches under the new branch by right-clicking on the two extra Category listings.

13. Save and close the modified `Mastering Categories.xml` file.
14. Switch back to Civil 3D's QTO Manager.

Next, you will need to reload the categorization file to view the changes.

15. Click the Open drop-down menu in the top left of the QTO Manager, and select Open ➤ Categorization File.

16. Browse to the `Mastering Categories.xml` file again and open it. Once Civil 3D processes your XML file, your QTO Manager should look similar to this:

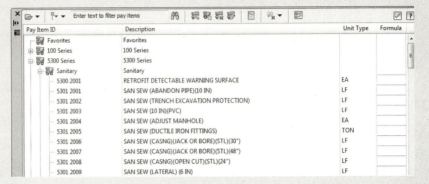

Creating a fully developed category list is a bit time-consuming, but once it's done, the list can be shared with your entire office so everyone has the same data to use.

The Bottom Line

Open and review a list of pay items along with their categorization. The pay item list is the cornerstone of QTO. You should download and review your pay item list and compare it against the current reviewing agency list regularly to avoid any missed items.

   Master It Add the 12-, 18-, and 21-Inch Pipe Culvert pay items to your favorites in the QTO Manager.

Assign pay items to AutoCAD objects, pipe networks, and corridors. The majority of the work in preparing QTO is in assigning pay items accurately. By using the linework, blocks, and Civil 3D objects in your drawing as part of the process, you reduce the effort involved with generating accurate quantities.

   Master It Open the `Mastering QTO.dwg` file and assign the CLEARING AND GRUBBING pay item to the project area.

Use QTO tools to review what items have been tagged for analysis. By using the built-in highlighting tools to verify pay item assignments, you can avoid costly errors when running your QTO reports.

   Master It Verify that the area in the previous exercise has been assigned a pay item.

Generate QTO output to a variety of formats for review or analysis. The Quantity Takeoff reports give you a quick understanding of what items have been tagged in the drawing, and they can generate text in the drawing or external reports for uses in other applications.

   Master It Calculate and display the amount of Type C Broken markings in the `Mastering Reporting.dwg` file.

Styles

Styles control the display properties of Civil 3D objects and labels. The creation of proper styles and settings can make or break your experience with Civil 3D.

Understanding and applying styles correctly can mean the difference between getting a job out in several hours and fighting with your CAD drawing for days.

This chapter is organized by style type. First, read the chapter to be introduced to the style concepts in a logical manner. Later, when you use this chapter as a reference to build styles, you will likely jump around to the examples that meet your needs.

Civil 3D styles and settings are stored in a DWT template file. At this point in the process, it would be beneficial to understand the role of the template file for base AutoCAD.

In this chapter, you will learn to:

- Override individual labels with other styles
- Create a new label set for alignments
- Apply a standard label set to profiles

Civil 3D Templates

Styles and settings should come from your template. Ideally, that template will have all the styles you need for the type of project you are working on. If you find you are constantly changing style settings, examine your workflow. Make sure to start new projects with a proper template. Use the new Import command located on the Manage tab to pull your preferred styles into drawings that you receive from outside sources.

There are hundreds of styles and settings in Civil 3D. The good news is that once you learn to create one or two styles, the rest should get easier. Autodesk has gone to great lengths to make the interface consistent between styles.

When you start a project with the correct template (DWT file), the repetitive task of defining the basic framework for your drawing is already completed. As you know, a DWT file contains the following:

- Units (architectural vs. decimal)
- Layers and respective linetypes
- Text styles
- Dimension and multileader styles
- Layouts and plot setups
- Block definitions

Civil 3D takes the base AutoCAD template and kicks it up several notches. In addition to the items just listed, a Civil 3D template contains the following:

- More specific unit information (international feet, survey feet, or meters)
- Civil object layers
- Ambient settings
- Label styles and formulas (expressions)
- Object styles
- Command settings
- Object naming templates
- Report settings
- Description key sets

Throughout this book, when you started a file from scratch, you used one of the templates that come with Civil 3D when you install it. There is good reason for this; the templates that come with Civil 3D may not be exactly what you want initially, but they are a great starting point when you are customizing your projects.

You must always start new projects with a proper Civil 3D template. If you receive a drawing from a non–Civil 3D user, and need to continue it in Civil 3D, you must import the styles and settings. Without suitable styles and settings, all of your object and label styles will show up with the name Standard, as shown in Figure 19.1. You do not want objects and labels to use the Standard style, as it is the Civil 3D equivalent of drawing on layer 0. Items that use the Standard style appear on layer 0 and will contain the most basic display settings.

FIGURE 19.1
A non–Civil 3D DWG will list all styles as Standard—which is a bad thing.

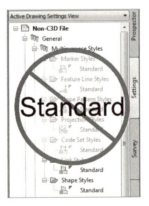

Importing Styles

Say someone sends you a drawing that was not done in Civil 3D. Perhaps it was exported from MicroStation, or initially created in Land Desktop, Carlson, or Eagle Point. You now have the task of creating Civil 3D objects, but making this task even more difficult is that there are no Civil 3D styles present. Perhaps the drawing was created in Civil 3D, but your organization's styles look completely different.

CIVIL 3D TEMPLATES | **733**

Import

Luckily, importing styles from an existing drawing or template has gotten much easier in Civil 3D 2012. You can find the new Import Styles button on the Styles panel of the Manage tab. When you click the button, you will be prompted to browse for the template (DWT) or drawing (DWG) that contains the styles you are looking for.

Once you select the file whose styles you will import, the dialog shown in Figure 19.2 appears.

Figure 19.2
Import Civil 3D Styles dialog

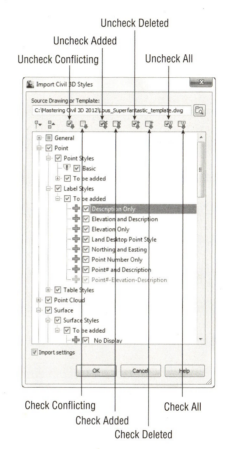

Notice the Import Settings option at the bottom of the dialog. This option is turned on by default. As you look through the list of styles to be imported, you will notice that some items are grayed out and can't be modified.

The grayed-out styles represent items in command settings. Styles referred to in command settings must be imported if the Import settings option is turned on. You will read about command settings in the upcoming section.

You will also notice items with a warning symbol such as Basic in Figure 19.2. The warning symbol indicates there is a style in the current drawing with the same name as a style in the batch to be imported. Use the Uncheck Conflicting button if you do not want styles in the destination drawing to be overwritten. If you leave these items checked on, the incoming styles "win." If you are not sure if there is a difference between the styles, pause your cursor over the style name and a tooltip will tell you what (if any) difference exists.

Use the Uncheck Added button if you only want styles with the same name to come in. Wherever possible, Civil 3D will release items in the To Be Added categories. In cases where a style is used by a setting, you will not be able to uncheck it unless you do not import settings.

A style in the current drawing will be deleted if the source drawing does not contain a style with the same name. Use the Uncheck deleted to prevent the style from being deleted.

Drawing Settings

The first stop when setting up a Civil 3D template is the Drawing Settings area. Access the Edit Drawing Settings command by selecting the Settings tab in Toolspace and right-clicking on the name of the drawing, as shown in Figure 19.3.

FIGURE 19.3
Accessing the drawing settings

Each tab in the Drawing Settings dialog controls a different aspect of the drawing. Most of the time, you'll pick up information for the Object Layers, Abbreviations, and Ambient Settings tabs from a companywide template. The drawing scale and coordinate information change for every job, so you'll visit the Units And Zone and Transformation tabs frequently.

Settings for Units And Zone, Object Layers, and Abbreviations are specified in this area and nowhere else. The Ambient Settings are set drawing-wide at this level but can be overridden "downstream" for object-specific control.

UNITS AND ZONE TAB

The Units And Zone tab (Figure 19.4) lets you specify metric or imperial units for your drawing. You can also specify the conversion factor between systems. If you choose a coordinate system with International or US Foot specified, this setting will become unavailable.

This tab also includes the options Scale Objects Inserted From Other Drawings and Set AutoCAD Variables To Match. The Set AutoCAD Variables To Match option sets the base AutoCAD angular units, linear units, block insertion units, hatch pattern, and linetype units to match the values placed in this dialog. As shown in Figure 19.4, you do want these checked on.

The scale that you see on the right side of the Units And Zone tab is the same as your annotation scale. You can change it here, but it is much easier to select your annotation scale from the bottom of the drawing window.

If you choose to work in assumed coordinates, you can leave the Zone set to No Datum, No Projection. To set the coordinate system for your locale, first set the category from the long list of possibilities. Civil 3D is a worldwide product; therefore, most recognized surveying coordinate systems (including obsolete ones) can be found in the options. Once your coordinate system has been established, you can change it on the Transformation tab if desired.

FIGURE 19.4
Units And Zone tab

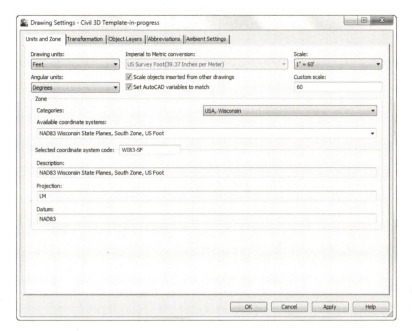

Try the following quick exercise to practice:

1. Open the drawing `Civil 3D Template-in-progress.dwg` file, which you can download from www.sybex.com/go/masteringcivil3d2012.

2. From the Settings tab of Toolspace, right-click on the name of the drawing and select Edit Drawing Settings. Switch to the Units And Zone tab.

3. Select USA, Wisconsin from the Categories drop-down menu.

4. Select NAD83 Wisconsin State Planes, South Zone, US Foot from the Available Coordinate Systems drop-down menu.

5. Save the drawing for the next exercise.

Transformation Tab

Earth is neither flat nor a perfect sphere. In reality, Earth is more of a squashed sphere, wider at the equator than at the poles. The mathematical approximation of the Earth's physical shape is called an *ellipsoid*. The ellipsoid is constantly being fine-tuned to increase accuracy for precise geographic modeling. To make matters even more complicated, the gravitational pull of Earth is not distributed evenly, creating a slight skew between what a surveyor measures and how it relates to the ellipsoid. The gravitational representation of Earth is referred to as the *geoid*.

Luckily, most engineers and surveyors do not need to worry about all this. Most survey-grade GPS equipment takes care of the transformation to local grid coordinates for you. In the United States, state plane coordinate systems already have regional projections taken into account. In the rare case that surveyors need to transform local observations from geoid to ellipsoid and ellipsoid to grid manually, the Transformation tab makes the computation quickly.

With a base coordinate system selected, you can do any further refinement you'd like using the Transformation tab. The coordinate systems on the Units And Zone tab can be refined to meet local ordinances, tie in with historical data, complete a grid to ground transformation, or account for minor changes in coordinate system methodology. These changes can include the following:

Apply Sea Level Scale Factor   This value is known in some circles as *elevation factor* or *orthometric height scale*. Sea Level Factor takes into account the mean elevation of the site and the spheroid radius that is currently being applied as a function of the selected zone ellipsoid.

Grid Scale Factor   At any given point on a projected map, there is a distortion between the "flat" measurement and the measurement on the ellipsoid. Grid Scale Factor is based on a 1:1 value, a user-defined uniform scale factor, a reference point scaling, or a prismoidal transformation in which every point in the grid is adjusted by a unique amount.

Reference Point   To apply Grid Scale Factor and Sea Level Factor correctly, you need to tell Civil 3D where you are on Earth. Reference Point can be used to set a singular point in the drawing field via pick or via point number, local northing and easting, or grid northing and easting values.

Rotation Point   Rotation Point can be used to set the reference point for rotation via the same methods as the Reference Point.

Specify Grid Rotation Angle   Some people may know this as the Convergence angle. This is the angle between Grid North and True North. Enter an amount or set a line to north by picking an angle or deflection in the drawing. You can use this same method to set the azimuth if desired.

Most engineering firms work on either a defined coordinate system or an arbitrary system, so none of these changes are necessary. Given that, this tab will be your only method of achieving the necessary transformation for certain surveying and geographic information system (GIS)–based and Land Surveying–based tasks. It should be noted that this is *not* the place to transform assumed coordinates to a predefined coordinate system. See Chapter 2, "Survey," to learn how to translate a survey.

OBJECT LAYERS TAB

Civil 3D and AutoCAD layers have a love–hate relationship with each other. Civil 3D is built on top of AutoCAD; therefore, all the objects do reside on layers. However, Civil 3D is not traditional CAD. Your surfaces, corridors, points, profiles, and everything else generated by Civil 3D are dynamic *objects* rather than simple lines, arcs, or circles.

When you created an alignment in Chapter 6, "Alignments," for example, you did not have to think about the current layer. You also were not concerned with which layer the polyline was on when you converted it to an alignment. This is because Civil 3D styles "push" objects and labels to the correct layer as part of their intelligence.

Layers are found in several areas of the Civil 3D template. The first location you will examine is the layers found in the Drawing Settings area. The layers listed here represent overall layers. For those of you who are familiar with AutoCAD blocks, it is useful to think of these layers in the same way as a block's insertion layer.

CIVIL 3D TEMPLATES | **737**

In the Object Layers tab, every Civil 3D object must have a layer set, as shown in Figure 19.5. Do not leave any object layers set to 0. An optional modifier can be added to the beginning (*prefix*) or end (*suffix*) of the layer name to further separate items of the same type.

FIGURE 19.5
Every object is on a layer; the corridor layer contains a modifier.

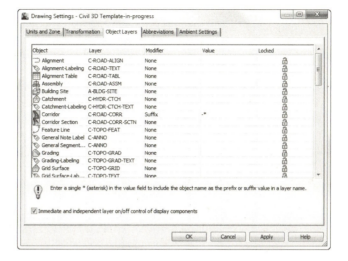

A common practice is to add wildcard suffixes to corridor, surface, pipe, and structure layers to make it easier to manipulate them separately. For example, if the layer for a corridor is specified to be C-ROAD-CORR and a suffix of -* (dash asterisk, as shown in Figure 19.5) is added as the modifier value, a new layer will automatically be created when a new corridor is created. The resulting layer will take on the name of the corridor. If the corridor is called 13th Street, the new layer name will be C-ROAD-CORR-13th Street. This new layer is created once, and is not dynamic to the object name. In other words, if you decide to rename "13th Street" to "Stephen Colbert Street," the layer remains C-ROAD-CORR-13th Street.

If the main layer name you are after does not exist in the drawing, you can create it as you work through the Object Layers dialog. Click the New button, as shown in Figure 19.6, and set up the layer as needed.

FIGURE 19.6
Click New to add a new layer.

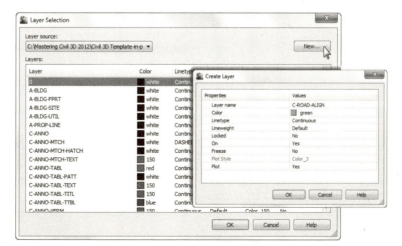

In the following exercise, you set object layers in a template:

1. Open the drawing `Civil 3D Template-in-progress.dwg` file, which you can download from this book's web page.
2. From the Settings tab of Toolspace, right-click on the name of the drawing and select Edit Drawing Settings. Switch to the Object Layers tab.
3. Click in the Layer field next to Alignment. Create a new layer called **C-ROAD-ALIGN**. Set the new layer as the layer for the Alignment object.
4. Set the layer for Building Site to A-BLDG-SITE.
5. Set the layer for Catchment-Labeling to C-HYDR-CTCH-TEXT.
6. For the corridor layer, keep the main layer as C-ROAD-CORR. Set the Modifier to Suffix. Set the modifier value to **-***. The asterisk acts as a wildcard that will add the corridor name as part of a unique layer for each corridor.
7. Scroll down to the pipe and structure listing. In this area you will create several new layers and add suffix information:

    - For Pipe, create a layer called **C-NTWK-PIPE** with a suffix of **-***.
    - For Pipe-Labeling, create a new layer called **C-NTWK-PIPE-TEXT**.
    - For Pipe And Structure Table, set the layer to **C-NTWK-PIPE-TEXT**.
    - For Pipe Network Section, create a new layer called **C-NTWK-XSEC**.
    - For Pipe Network Profile, create a new layer called **C-NTWK-PROF**.
    - Scroll down a bit further and create a new layer for Structure called **C-NTWK-STRC**. Add a suffix of **-***.
    - For Structure-Labeling, create a new layer called **C-NTWK-STRC-TEXT**.
    - Add a suffix to the Tin Surface object layer of **-***.

    Your layers and suffixes should now resemble Figure 19.7.

FIGURE 19.7
Examples of the completed layer names in the Object Layers tab

Object	Layer	Modifier	Value
Alignment	C-ROAD-ALIGN	None	
Alignment-Labeling	C-ROAD-TEXT	None	
Alignment Table	C-ROAD-TABL	None	
Assembly	C-ROAD-ASSM	None	
Building Site	A-BLDG-SITE	None	
Catchment	C-HYDR-CTCH	None	
Catchment-Labeling	C-HYDR-CTCH-TEXT	None	
Corridor	C-ROAD-CORR	Suffix	-*
Pipe	C-NTWK-PIPE	Suffix	-*
Pipe-Labeling	C-NTWK-PIPE-TEXT	None	
Pipe and Structure Table	C-NTWK-PIPE-TEXT	None	
Pipe Network Section	C-NTWK-XSEC	None	
Pipe or Structure Profile	C-NTWK-PROF	None	
Structure	C-NTWK-STRC	None	-*
Structure-Labeling	C-NTWK-STRC-TEXT	None	
Tin Surface	C-TOPO	Suffix	-*

CIVIL 3D TEMPLATES

8. Place a check mark next to Immediate And Independent Layer On/Off Control Of Display Components. This setting will allow you to use the On/Off toggle in Layer Manager to work with Civil 3D objects.

9. Click Apply and then OK. Save the drawing for use in the next exercise.

ABBREVIATIONS TAB

When you add labels to certain objects, Civil 3D automatically uses the abbreviations from this area to indicate geometry features. For example, left is L and right is R.

It is unusual to need to make many changes in this area, since Civil 3D uses industry-standard abbreviations wherever they are found. Civil 3D is also very customizable if you'd rather use VPI instead of PVI for Point of Vertical Intersection.

In most cases, changing an abbreviation is as simple as clicking in the Value field and typing a new one. Notice that the Alignment Geometry Point Entity Data section has a larger set of values and some special coding attached (as shown toward the bottom of Figure 19.8). These are more representative of other label styles. You will visit the Text Component Editor a little later in this chapter in the section "Label Styles."

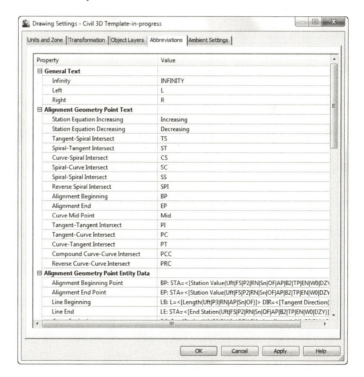

FIGURE 19.8
Customizable down to the letter, the Abbreviations tab

Ambient Settings Tab

Examine the settings in the Ambient Settings tab to see what can be set here. The main options you'll want to adjust are in the General category and the display precision in the subsequent categories.

The precision that you see in this dialog does not change the label precision. The precision you see here is the number of decimal places reported to you in various dialogs.

Being familiar with the way this tab works will help you further down the line, because almost every other settings dialog in the program works like the one shown in Figure 19.9.

FIGURE 19.9
Ambient settings at the main drawing level

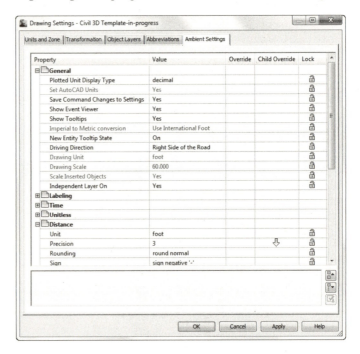

Plotted Unit Display Type Civil 3D knows you want to plot at the end of the day. In this case, it's asking how you would like your plotted units measured. For example, would you like that bit of text to be 0.25″ tall or ¼″ high? Most engineers are comfortable with the Leroy method of text heights (L80, L100, L140, and so on), so the decimal option is the default.

Set AutoCAD Units This displays whether or not Civil 3D should attempt to match AutoCAD drawing units, as specified on the Units And Zone tab.

Save Command Changes To Settings Set this to Yes. This setting is incredibly powerful but a secret to almost everyone. By setting it to Yes, you ensure that your changes to commands will be remembered from use to use. This means if you make changes to a command during use, the next time you call that Civil 3D command, you won't have to make the same changes. It's frustrating to do work over because you forgot to change one of the five things that needed changing, so this setting is invaluable.

Show Event Viewer Event Viewer is Civil 3D's main feedback mechanism, especially when things go wrong. It can get annoying, however, and it takes up valuable screen real estate (especially if you're stuck with one monitor!), so many people turn it off. We recommend leaving it on and pushing it to the side until needed.

Show Tooltips One of the cool features that people remark on when they first use Civil 3D is the small pop-up that displays relevant design information when the cursor is paused on the screen. This includes things such as station-offset information, surface elevation, section information, and so on. Once a drawing contains numerous bits of information, this display can be overwhelming; therefore, Civil 3D offers the option to turn off these tooltips universally with this setting. A better approach is to control the tooltips at the object type by editing the individual feature settings. You can also control the tooltips by pulling up the properties for any individual object and looking at the Information tab.

Imperial To Metric Conversion This setting displays the conversion method specified on the Units And Zone tab. The two options currently available are US Survey Foot and International Foot.

New Entity Tooltip State You can also control tooltips on an individual object level. For instance, you might want tooltip feedback on your proposed surface but not on the existing surface. This setting controls whether the tooltip is turned on at the object level for new Civil 3D objects.

Driving Direction Specifies the side of the road that forward-moving vehicles use for travel. This setting is important in terms of curb returns and intersection design.

Drawing Unit, Drawing Scale, and Scale Inserted Objects These settings were specified on the Units And Zone tab but are displayed here for reference and so that you can lock them if desired.

Independent Layer On This is the same control that was set on the Object Layers tab. Yes is the recommended setting.

The Ambient Settings for Direction offer the following choices:

- Unit: Degree, Radian, and Grad
- Precision: 0 through 8 decimal places
- Rounding: Round Normal, Round Up, and Truncate
- Format: Decimal, two types of DDMMSS, and Decimal DMS
- Direction: Short Name (spaced or unspaced) and Long Name (spaced or unspaced)
- Capitalization
- Sign
- Measurement Type: Bearings, North Azimuth, and South Azimuth
- Bearing Quadrant

When you're using the Bearing Distance transparent command, for example, these settings control how you input your quadrant, your bearing, and the number of decimal places in your distance.

Explore the other categories, such as Angle, Lat Long, and Coordinate, and customize the settings to how you work.

At the bottom of the Ambient Settings tab is a Transparent Commands category. These settings control how (or if) you're prompted for the following information:

Prompt For 3D Points   Controls whether you're asked to provide a z elevation after x and y have been located.

Prompt For Y Before X   For transparent commands that require x and y values, this setting controls whether you're prompted for the y-coordinate before the x-coordinate. Most users prefer this value set to False so they're prompted for an x-coordinate and then a y-coordinate.

Prompt For Easting Then Northing   For transparent commands that require Northing and Easting values, this setting controls whether you're prompted for the Easting first and the Northing second. Most users prefer this value set to False, so they're prompted for Northing first and then Easting.

Prompt for Longitude Then Latitude   For transparent commands that require longitude and latitude values, this setting controls whether you're prompted for Longitude first and Latitude second. Most users prefer this set to False, so they're prompted for Latitude and then Longitude.

DRAWING PRECISION VS. LABEL PRECISION

The precision and units that you are seeing in the Ambient settings and commands settings only affect how the values are reported to you in dialogs. These settings are not related to the number of decimal places or units displayed in labels.

You will work with labels later in this chapter in the section "Label Styles."

The settings that are applied here can also be changed at the object levels. For example, you may typically want elevation to be shown to two decimal places, but when looking at surface elevations, you might want just one. The Override and Child Override columns give you feedback about these types of changes. See Figure 19.10.

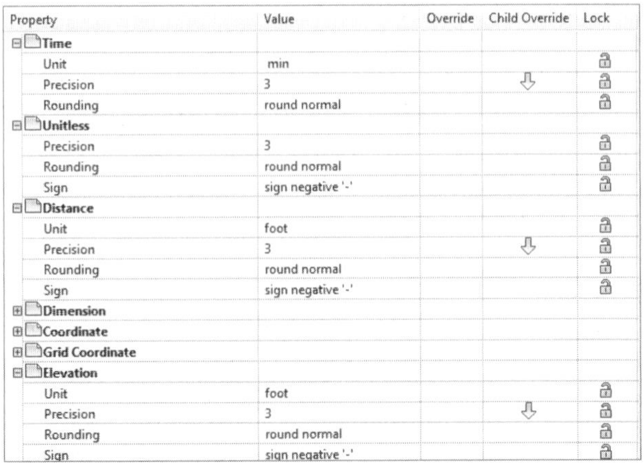

FIGURE 19.10
The Child Override indicator in the Time, Distance, and Elevation values

Property	Value	Override	Child Override	Lock
⊟ Time				
Unit	min			🔒
Precision	3		⬇	🔒
Rounding	round normal			🔒
⊟ Unitless				
Precision	3			🔒
Rounding	round normal			🔒
Sign	sign negative '-'			🔒
⊟ Distance				
Unit	foot			🔒
Precision	3		⬇	🔒
Rounding	round normal			🔒
Sign	sign negative '-'			🔒
⊞ Dimension				
⊞ Coordinate				
⊞ Grid Coordinate				
⊟ Elevation				
Unit	foot			🔒
Precision	3		⬇	🔒
Rounding	round normal			🔒
Sign	sign negative '-'			🔒

The Override column shows whether the current setting is overriding something higher up. Because you're at the Drawing Settings level, these are clear. However, the Child Override column displays a down arrow, indicating that one of the objects in the drawing has overridden this setting. After a little investigation of the objects, you'll find the override is in the Edit Feature Settings of the Profile View, as shown in Figure 19.11.

FIGURE 19.11
The Profile Elevation Settings and the Override indicator

Notice that in this dialog, the box is checked in the Override column. This indicates that you're overriding the settings mentioned earlier, and it's a good alert that things have changed from the general Drawing Settings to this Object Level setting.

But what if you don't want to allow those changes? Each Settings dialog includes one more column: Lock. At any level, you can lock a setting, graying it out for lower levels. This can be handy for keeping users from changing settings at the lower level that perhaps should be changed at a drawing level, such as sign or rounding methods.

Object Settings

If you click the Expand button next to the drawing name, you see the full array of objects that Civil 3D uses to build its design model. Each of these has special features unique to the object being described, but there are some common features as well. Additionally, the General collection contains settings and styles that are applied to various objects across the entire product.

The General collection serves as the catchall for styles that apply to multiple objects and for settings that apply to *no* objects. For instance, the Civil 3D General Note object doesn't really belong with the Surface or Pipe collections. It can be used to relate information about those objects, but because it can also relate to something like "Don't Dig Here!" it falls into the General category. The General collection has three components (or branches):

Multipurpose Styles   These styles are used in many objects to control the display of component objects. The Marker Styles and Link Styles collections are typically used in cross-section views, whereas the Feature Line Styles collection is used in grading and other commands. Figure 19.12 shows the full collection of multipurpose styles and some of the marker styles that ship with the product.

Toolspace is shown horizontally in Figure 19.13 and Figure 19.14 for illustration purposes. If you like the way this looks, you can also set your Toolspace like this by clicking the orientation toggle at the top. This will bring the preview area to the right of Toolspace.

FIGURE 19.12
General multipurpose styles and some marker styles

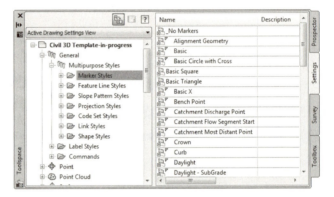

Label Styles   The Label Styles collection allows Civil 3D users to place general text notes or label single entities while still taking advantage of Civil 3D's flexibility and scaling properties. With the various label styles shown in Figure 19.13, you can get some idea of their use.

FIGURE 19.13
Line label styles

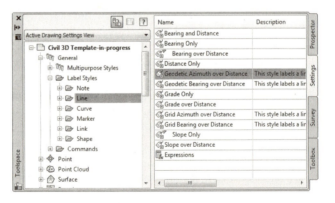

Label styles are a critical part of producing plans with Civil 3D. In the section "Label Styles" later in this chapter, you'll learn how to build a new basic label and explore some of the common components that appear in every label style throughout the product.

Commands   Almost every branch in the Settings tree contains a **Commands** folder. Expanding this folder, as shown in Figure 19.14, shows the typical long, unspaced command names that refer to the parent object.

FIGURE 19.14
Corridor command settings in Toolspace

Get Started with Object Styles

Before you get your hands on specific styles, it is nice to understand some general things all styles have in common.

There are several ways to enter the dialog. The easiest, most direct way into any style is from the Settings tab. Right-click on any style you see listed and select Edit, as shown in Figure 19.15.

FIGURE 19.15
Every style can be edited by right-clicking on the style name from the Settings tab of Toolspace.

You can also enter the styles' dialogs from the properties of any object by clicking the Edit button. Figure 19.16 shows the active styles on a point group, with the Edit buttons off to the right. Editing at this level affects all objects that use the style the same as it would if you had entered the style from the Settings tab. The downside to editing a style in this manner is that you will not immediately be able to click Apply to see your change. You'll need to exit the style dialog and click Apply at the object level before seeing your label or object update.

FIGURE 19.16
An object's properties reveal the active styles, as in this point group's properties.

Frequently Seen Tabs

All styles, regardless of type, have a few things in common:

Information Tab   The Information tab, shown in Figure 19.17, controls the name of the style.

The description is always optional, but we recommend that you add some information to the style. The description can be seen in tooltip form as you search through the Settings tab.

On the right side of the dialog you will see the name of the style, the date when it was created, and the last person who modified the style. These names are initially pulled from the Windows login information and only the Created By field can be edited.

Summary Tab   The Summary tab is exactly what it sounds like, that is, a summary of all other settings that exist in the style. The information from other tabs in list form as well as their override status is shown in Figure 19.18.

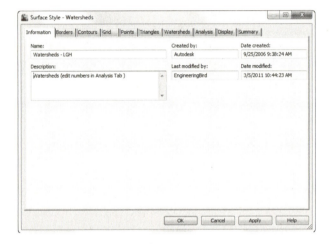

FIGURE 19.17
The Information tab exists for all object and label styles, including surface styles.

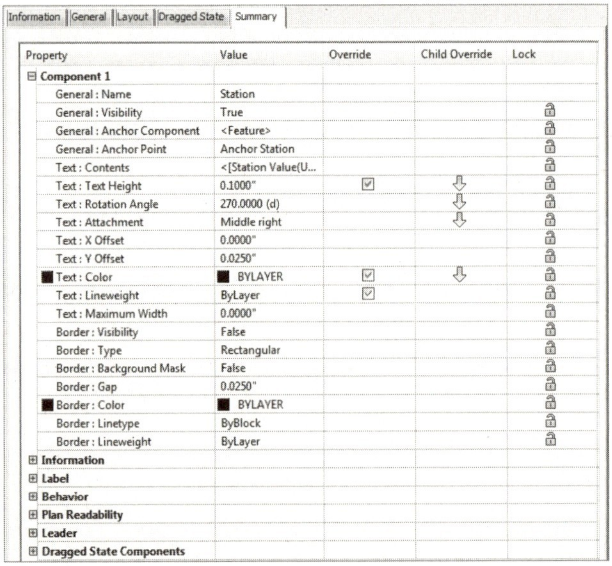

FIGURE 19.18
Summary tab of a label style

> **MAKING SENSE OF CHILD STYLES AND OVERRIDES**
>
> Civil 3D styles are customizable at several levels. In the Drawing Settings area, you saw overall settings that initially affect the entire drawing. The Drawing Settings are the highest level of the settings hierarchy (the main parent settings). You will find the same settings at the object level that can diverge from the overall drawing settings. Farther down the hierarchy are the command settings. The term "child" in Civil 3D refers to any style that can also be controlled at a higher level.
>
> For example, for most objects a precision of two decimal places (0.00) is adequate for your station work. However, for corridor creation it is advantageous to see more decimal places of precision. At the drawing level, keep the station precision to 0.00. At the Corridor level, right-click on the object to Edit Feature Settings.
>
>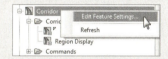
>
> Inside the Corridor Feature Settings, you will have a feeling of déjà vu from when you edited the Drawing Settings. All the same settings (and a few more object-specific ones) are here. The difference is that changing the setting here affects corridors only.
>
>
>
> The check mark in the Override column indicates that this setting differs from settings higher up in the hierarchy of settings.
>
> An arrow in the Child Override column indicates that further down the chain of command a style or setting differs. To force these subordinate styles to match the style of the parent, click the arrow so that a red X appears. The red X indicates that the change you make in the dialog you are looking at will get pushed to subordinate styles or settings.

Basic Object Styles

Object styles control the display of Civil 3D objects such as points, surfaces, alignments, pipes, sections, and so on.

Consider a Civil 3D surface. Sometimes you want to see the surface as contours, other times you want to see the surface as triangles, and sometimes you don't want to see it at all. Changing what you see is not a matter of freezing and thawing layers, but it is a matter of changing the active style. Think of the active style on an object as a wrapper. A surface model will have many potential wrappers depending on what you want to see. The underlying data does not change, but the representation of the data will change.

Object styles have quite a few things in common with each other. Almost every object style has a Display tab.

DISPLAY TAB

Inside the Display tab, you will see a list of the various components that can be displayed for the object you are working with. In Figure 19.19 you can see that a general-purpose marker contains only one component (the marker itself) and a surface model has components for contours, triangles, points, and more (as seen toward the bottom of Figure 19.19).

FIGURE 19.19
A simple marker style (top) and a more complex surface style (bottom)

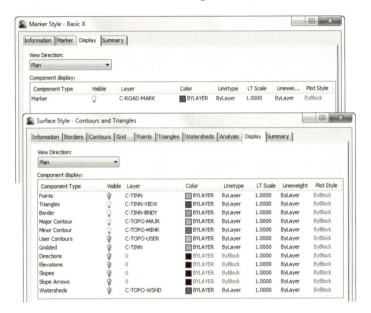

While other tabs in an object style control the specifics of *how* certain components look, the Display tab controls *if* the component displays at all. The visibility lightbulb indicates whether the component will be displayed when the style is applied to the object. In the bottom of Figure 19.19, you can see that triangles, border, major contours, and minor contours will all display for the Surface object style.

Each component will have a layer designation. You may be wondering, "Didn't we set a surface layer back in the object settings?" Yes, you did. Those were overall object layers. The layers we see here are component layers. Using the analogy of a base AutoCAD block, the Object layer can be thought of in the same way as a block insertion layer. The object component layers can be thought of in the same way as layers inside the block definition.

This is Civil 3D, of course, so the objects exist in three dimensions. View Direction controls the display of an object depending on how you are looking at it. Certain items, such as a profile view, are intended to only be seen in plan, so they do not have multiple view directions listed. A marker, on the other hand, can be shown in plan, model, profile, or section, as seen in Figure 19.20. Surface models make an appearance in plan, model, and section, as shown in Figure 19.20.

Marker Tab

Marker styles and point styles both contain a Marker tab (Figure 19.21). The Marker tab controls what symbol or block is used, its rotation, and how it should be sized when it is placed in the drawing.

There are three symbol types that can be used. Use Custom Marker and Use AutoCAD BLOCK Symbol For Marker are the best options. Use AutoCAD POINT For Marker is a poor choice because it uses the symbol specified by the DDPTYPE setting and is difficult to control. When you choose the Use Custom Marker option, you can select a combination of symbols to mix and match.

When you choose the Use AutoCAD BLOCK Symbol For Marker option, you will be able to access a listing of blocks in your drawing. If the block you are after does not yet exist in the drawing, you can right-click in the block listing and choose Browse, as shown in Figure 19.22.

FIGURE 19.20
View Directions for the marker style

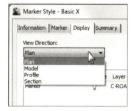

FIGURE 19.21
Marker tab for the Benchmark point style

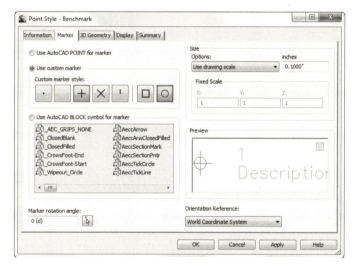

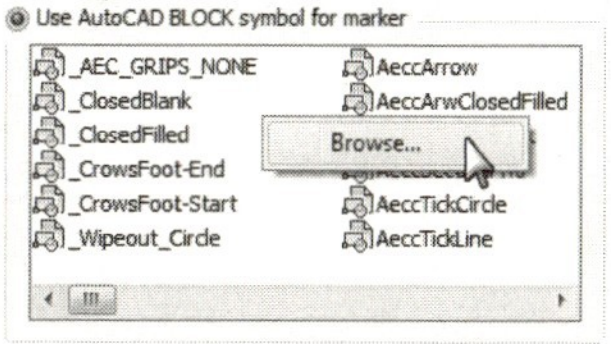

FIGURE 19.22
Right-click to browse for a block if it is not already defined in your drawing.

The Size options control how the symbol is scaled when inserted in the drawing (Figure 19.23). The most common options used for this are the Use Drawing Scale and Use Size In Absolute Units.

- Use Drawing Scale allows you to specify the plotted size of the symbol. The model space size of the symbol will be the size specified in the style multiplied by your annotation scale.

- Use Fixed Scale will scale the symbol based on the X, Y, and Z scale set in the style. This option will also use the Fixed scale factor in the description key set when used with a survey point.

- Use Size In Absolute Units is the option you will use in point styles that vary in size based on a survey description. For example, if this is a TREE symbol and you want the size of the symbol to reflect the trunk diameter in feet, set this to 1.

- Use Size Relative To Screen allows you to specify the size of the symbol as a percentage of your screen. This setting is annoying, because the marker will constantly change size as you zoom in or out. We do not recommend that you enable it.

FIGURE 19.23
Size Options for marker display

Orientation Reference (as seen in the bottom right corner of Figure 19.21) controls whether the symbol stays rotated with the view, world coordinate system, or, in the case of survey points, the object. In most cases, the Orientation Reference setting should be set to World Coordinate system.

CREATE A MARKER STYLE

Now get your hands on some object styles. You will start with simple styles and work your way up in complexity as this chapter progresses.

Markers are used in many places throughout Civil 3D. They are called from other styles to show vertices on civil objects, as label location marks, or as a marker to indicate the start of a flow line.

1. Open the `Civil 3D Template-in-progress.dwg` drawing file, which you can download from this book's web page.

2. From the Settings tab of Toolspace, choose General ➢ Multipurpose Styles ➢ Marker Styles. Right-click on Marker Styles and select New.

3. On the Information tab, name the style **VPI Marker**. Give the marker the description **Use to indicate vertical PI in profiles**. Click Apply. Your login name is listed in the Created By and Last Modified By fields.

GET STARTED WITH OBJECT STYLES | 751

4. Switch to the Marker tab. Set the marker type to Use AutoCAD BLOCK Symbol For Marker. In the block listing, highlight STA by clicking on it. Set the size to 0.2″. Leave all other Marker settings at their defaults.

5. Switch to the Display tab. For Plan, Model, Profile, and Section view directions, set the layer to C-ROAD-PROF-STAN-GEOM and set the color to BYLAYER.

6. Click OK to complete the marker style.

7. Save the drawing for the next exercise.

CREATE A SURVEY POINT STYLE

Survey Point styles contain many of the same options as Marker styles. As you work through the following example, you will perform many of the same steps as the previous exercise.

1. Continue working in the drawing `Civil 3D Template-in-progress.dwg` file, which you can download from this book's web page.

2. From the Settings tab of Toolspace, choose Point ➢ Point Styles. Right-click on Point Styles and select New.

3. On the Information tab, name the style **TRUNK**. Add a description of **Simple circle representing trunk diameter in inches**.

4. On the Marker tab, click the radio button Use Custom Marker. Click the blank marker option from the group of symbols on the left. Click to add the circle option on the right. Set the Size option to Use Size In Absolute Units. Set the size to **0.0833**. This value will scale down the symbol so the trunk diameter represents inches.

5. Switch to the Display tab. For Plan, Model, Profile, and Section view directions, set both the Marker and Label layers to V-NODE-TREE. Leave the other settings at their defaults.

   Hint! To save time in this step use the Shift key to multiselect the Marker and Label components, as shown in Figure 19.24. When you select the layer, it will apply to both components.

6. Click OK to complete the point style.

7. Save the drawing for the next exercise.

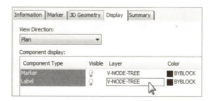

FIGURE 19.24
Use the Shift key on your keyboard as you click the components to multiselect.

Linear Object Styles

In this section, you will see some linear styles such as alignments, profiles, and parcels. Hopefully, you are already seeing that concepts from one type of style often apply to other types of object styles. For alignment and profile styles, this is especially true.

Both alignment and profiles styles have a Design tab, as shown in Figure 19.25. In the case of the alignment styles, the Enable Radius Snap option restricts the grip-edit behavior of alignment curves. If you enable this option and set a value of 0.5', the resulting radius value of curves will be rounded to the nearest 0.5'. In the case of profiles, the curve tessellation distance is a little more abstract. Curve tessellation refers to the smoothing factor applied to the profile when viewing it in 3D. Most users leave these settings at their default values.

Alignment and profile have very similar Markers tabs. This tab is where you can place markers (like the one you created in the first exercise of the chapter) at specific geometry points. Figure 19.26 shows the markers assigned to various locations along an alignment.

FIGURE 19.25
Design tabs exist in both alignment and profile object styles.

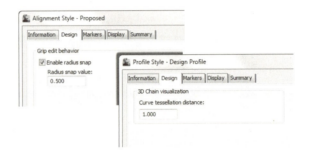

FIGURE 19.26
Markers for an alignment style

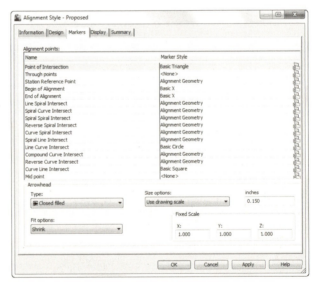

At the bottom of Figure 19.26 you see arrowhead information. Both alignments and profiles have the option of showing a direction arrow on each segment. Most people choose to omit this by turning the Arrow component off in the Display tab or setting the component to a layer that is set to No Plot. You should consider the latter, as knowing the direction of an alignment comes in handy when designing roundabouts, as you saw in Chapter 10, "Advanced Corridors, Intersections, and Roundabouts."

Figure 19.27 shows some commonly highlighted geometry points and components in an alignment. The markers in Figure 19.27 correspond to the markers displayed in Figure 19.26.

Figure 19.28 shows the Markers tab for the profile style. It looks pretty similar to the alignment Markers tab, don't you think?

Figure 19.29 shows the corresponding graphic for the markers used in Figure 19.28.

FIGURE 19.27
Example alignment with geometry highlighted by style

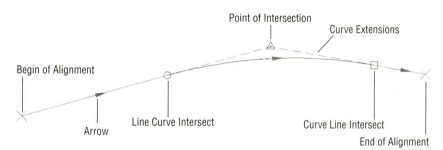

FIGURE 19.28
Profile Markers tab

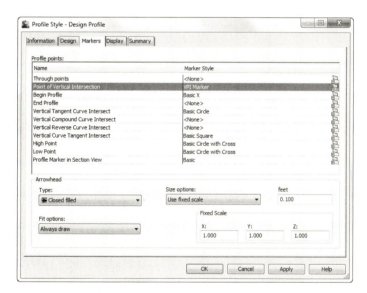

FIGURE 19.29
Profile geometry highlighted by style

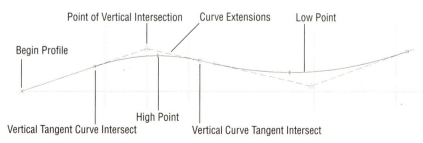

CREATE AN ALIGNMENT STYLE

Alignment styles can be helpful in identifying key design components, as well as showing the stationing direction. Use multiple alignment styles to visually differentiate centerline alignments from supplemental alignments such as offset alignments and curb return alignments.

In the following exercise, you will create a style that restricts radius grip edits to five-foot increments and displays basic alignment components.

1. Continue working in the `Civil 3D Template-in-progress.dwg` drawing file; if you have not completed the previous section exercises, open `Civil 3D template-in-progress2.dwg`. You can download either file from this book's web page.

2. From the Settings tab of Toolspace, choose Alignment ➢ Alignment Styles. Right-click on Alignment Styles and select New.

3. On the Information tab, name the style **Centerline**.

4. Switch to the Design tab and place a check mark next to Enable Radius Snap. Set the Radius Snap value to 5. Remember: To see the Help document about any dialog, press F1 on your keyboard to be taken directly to the help section for that topic.

5. Switch to the Markers tab. Set the Point Of Intersection marker by double-clicking the current value of <none>. Select PI Point from the marker listing and click OK. Set the Through Points marker to <None>. Set the Station Reference Point marker to <None>.

6. Switch to the Display tab. For the Plan View Direction, turn off the display components for Arrow, Line Extensions, and Curve Extensions. Use the Shift key to select Line, Curve, and Spiral together. Click the layer to set all three items to the layer C-ROAD-CNTR. Leave the Model and Section view directions at their defaults.

7. Click OK to complete the style. Save the drawing.

PARCEL STYLES

The parcel styles have several unique features that make them different from other styles.

In the Design tab of a parcel (shown in Figure 19.30a), you see parcel-specific options. A fill distance can be specified to place a hatch pattern along the perimeter of the parcel. This setting is useful to differentiate "special" parcels such as parks, limits of disturbance, or environmentally sensitive areas.

The fill distance indicates the width of the hatch pattern. On the Display tab, be sure to put the hatch component on its own layer so that it can be turned off independently from the parcel itself, as hatches tend to slow down the graphic. Figure 19.30b shows the parcel graphic resulting from the design settings shown.

FEATURE LINE STYLES

Feature lines are found in quite a few different places. They are created automatically as part of corridors, created as a result of grading, or can be created independently by the user. Because their scope crosses functionality, you will find feature line styles in the General ➢ Multipurpose Styles collection in Settings.

By definition, a feature line is a 3D object; therefore, its style can be controlled in plan, profile, and section. As shown in Figure 19.31, a feature line can use markers at geometry points.

FIGURE 19.30
Parcel style options and the resulting parcel graphic

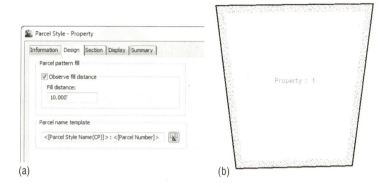

(a) (b)

FIGURE 19.31
Feature line profile marker options (left) and section option (right)

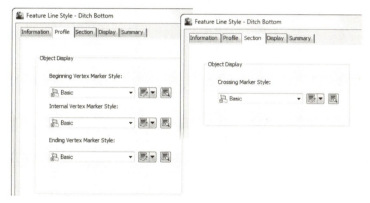

Surface Styles

Surface styles are the most widely used styles in any Civil 3D project. Depending on which style is active, certain editing options may be restricted. For example, you need to see triangle vertices before Civil 3D will let you use the Delete Point command.

Each tab in the Surface Style dialog corresponds to components listed in the Display tab. Remember, the Display tab only determines whether or not the component shows at all, but it does not control the specifics of how the component should be shown.

Once a surface is created, you can display information in a large number of ways. The most common so far has been contours and triangles, but these are the basics. By using varying styles, you can show a large amount of data with one single surface. Not only can you do simple things such as adjust the contour interval, but Civil 3D can apply a number of analysis tools to any surface:

Contours Allows the user to specify a more specific color scheme or linetype as opposed to the typical minor-major scheme. Commonly used in cut-fill maps to color negative colors one way, positive contours another, and the balance or zero contours yet another color.

Elevations Creates bands of color to differentiate various elevations. This can be a simple weighted distribution to help in creation of marketing materials, hard-coded elevations to differentiate floodplain and other elevation-driven site concerns, or ranges to help a designer understand the earthwork involved in creating a finished surface.

Direction Analysis Draws arrows showing the normal direction of the surface face. This is typically used for aspect analysis, helping site planners review the way a site slopes with regard to cardinal directions and the sun.

Slopes Analysis Colors the face of each triangle on the basis of the assigned slope values. While a distributed method is the normal setup, a common use is to check site slopes for compliance with Americans with Disabilities Act (ADA) requirements or other site slope limitations, including vertical faces (where slopes are abnormally high).

Slope Arrows Displays the same information as a slope analysis, but instead of coloring the entire face of the TIN, this option places an arrow pointing in the downhill direction and colors that arrow on the basis of the specified slope ranges.

User-Defined Contours Refers to contours that typically fall outside the normal intervals. These user-defined contours are useful to draw lines on a surface that are especially relevant but don't fall on one of the standard levels. A typical use is to show the normal pool elevation on a site containing a pond or lake.

In the following exercises, you will walk through the steps of creating or modifying surface styles.

Contour Style

The first style you create will be a contour display style. You will also manipulate the triangle display so that the elevations are more obvious when viewed in 3D.

1. Open the `Surface Styles.dwg` drawing file, which you can download from this book's web page.

2. From the Settings tab of Toolspace, choose Surface ➢ Surface Styles. Right-click Surface Styles and select New.

3. On the Information tab, name the style **Existing Contours**. Add a description indicating that this style will show 2′ minor contours and have a 5× exaggeration when viewed in 3D.

4. On the Contours tab, expand the Contour Intervals category. Set the Minor Interval to **2′**. Set the Major Interval to **4′**. The Contours tab will now look like Figure 19.32.

5. Switch to the Triangles tab. Change Triangle Display Mode to Exaggerate Elevation. Set Exaggerate Triangles By Scale Factor to **5.00** (see Figure 19.33).

6. On the Display tab, Shift+click to highlight all the components. Set all the component colors to BYLAYER. While all the components are still highlighted, set Linetype to ByLayer as well.

Step 5 rectifies a quirk with new surface styles. If you are a true CAD stickler, you will try to make as many items as possible set to BYLAYER. This approach greatly simplifies things if you need to change color or linetype. In some cases, setting a style component may not be practical. For surfaces, however, you can stick to your BYLAYER guns.

FIGURE 19.32
The Contours tab in the Surface Style dialog

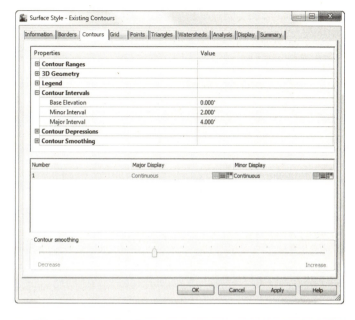

FIGURE 19.33
Exaggerate the elevations seen in the Object Viewer.

You will only be using the Minor Contour in this style, but later on you will copy this style and your efforts will be carried forward.

7. Click on the Minor Contour component so it is the only component highlighted. Turn this component's visibility on by clicking the lightbulb icon. Set the layer to C-TOPO-MINR. Turn off all other components for the Plan View Direction. Your Display tab should now resemble Figure 19.34.

8. Change your View Direction to Model. Verify that Triangles are the only component turned on. Set the Triangles layer to C-TINN. Set Color and Linetype both to BYLAYER.

To see the surface in the Object Viewer or in any other 3D view, you must have triangles set to display in Model, as shown in Figure 19.35.

9. Click OK to complete the style. Save the drawing.

10. Select the surface (shown as Border Only) and click Surface Properties from the context tab. Set Surface Style to Existing Contours. Click OK.

Figure 19.34
Minor Contour flying solo in plan

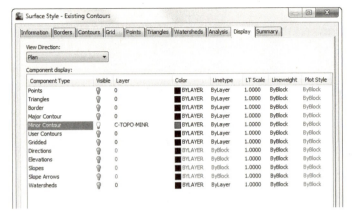

Figure 19.35
Triangle component set to display in Model

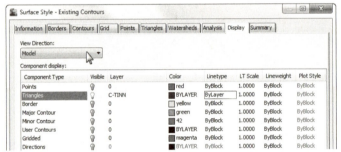

After the style is applied to the surface model, you should see simple contours in plan view (Figure 19.36a) and an exaggerated surface model in the Object Viewer (Figure 19.36b).

Figure 19.36
Your new style shown in plan and in Object Viewer

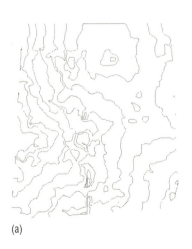
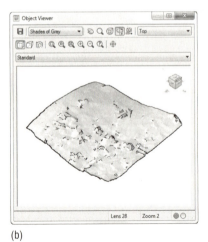

(a) (b)

Triangle and Points Surface Style

The next style you create will help facilitate surface editing. To work with the Swap Edge or Delete Line Surface edits, you must be able to see triangles. To work with the Delete Point, Modify Point, and Move Point commands, you must be able to see triangle vertices.

It is important to note that the points you see in the surface style do not refer to survey points. The points you are working with in this exercise are triangle vertices. The triangle vertices and survey points will initially be in the same locations for a surface built from points. However, as breaklines are added or edits are made to the triangle vertices, the surface model will differ from the original survey.

1. Continue working in the drawing file `Surface Styles.dwg`, which you can download from this book's web page.

2. From the Settings tab of Toolspace, choose Surface ≻ Surface Styles. Right-click the Existing Contours style you created in the previous exercise and select Copy, as shown in Figure 19.37.

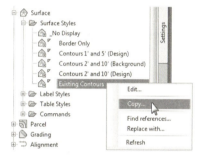

FIGURE 19.37
Copy is one of the options you will find when right-clicking directly on the style name.

3. On the Information tab, rename the surface style to **Surface Editing**. Add a description to reflect that this style will show points and triangles and will be used for editing purposes.

4. Switch to the Points tab. Expand Point Display and click the ellipsis for Data Point Symbol. Set the point type to the circle with a plus sign, as shown in Figure 19.38.

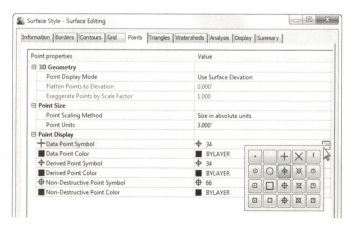

FIGURE 19.38
Change the point display so triangle vertices stand out.

5. Switch to the Triangles tab. Remove the exaggeration by setting the Triangle display mode back to Use Surface Elevation.

6. Switch to the Display tab. Use the light bulb icon to turn off visibility for Minor Contour, and turn it on for Points and Triangles. Create a new layer for the point display called **C-TINN-PNTS**. Make the new layer Red. Set the Points component layer to the new layer. Set the Triangles component to the C-TINN layer. Your Display tab will resemble Figure 19.39.

7. Click OK to complete the style. Use the same procedure from the previous exercise to apply the Surface Editing style to the surface. Your surface will now resemble Figure 19.40.

Component Type	Visible	Layer	Color	Linetype	LT Scale	Lineweight	Plot Style
Points		C-TINN-PNTS	BYLAYER	ByLayer	1.0000	ByBlock	ByBlock
Triangles		C-TINN	BYLAYER	ByLayer	1.0000	ByBlock	ByBlock
Border		0	BYLAYER	ByLayer	1.0000	ByBlock	ByBlock
Major Contour		0	BYLAYER	ByLayer	1.0000	ByBlock	ByBlock
Minor Contour		C-TOPO-MINR	BYLAYER	ByLayer	1.0000	ByBlock	ByBlock
User Contours		0	BYLAYER	ByLayer	1.0000	ByBlock	ByBlock
Gridded		0	BYLAYER	ByLayer	1.0000	ByBlock	ByBlock
Directions		0	BYLAYER	ByBlock	1.0000	ByBlock	ByBlock
Elevations		0	BYLAYER	ByBlock	1.0000	ByBlock	ByBlock
Slopes		0	BYLAYER	ByBlock	1.0000	ByBlock	ByBlock
Slope Arrows		0	BYLAYER	ByBlock	1.0000	ByBlock	ByBlock
Watersheds		0	BYLAYER	ByLayer	1.0000	ByBlock	ByBlock

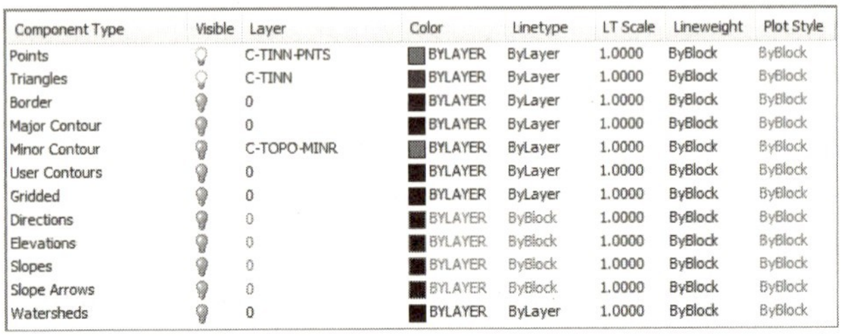

FIGURE 19.39 Change the Points and Triangles layers in the Display tab.

FIGURE 19.40 Triangles and points shown using the new surface style

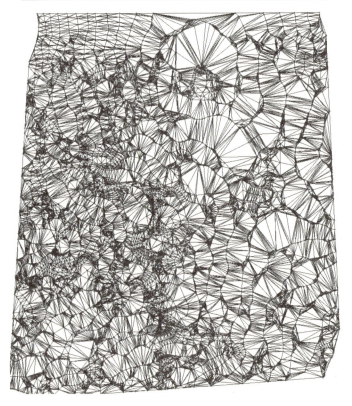

Watershed Analysis Style

The next style you will create in this chapter will be a watershed analysis style. Analysis styles are unique in several ways. To see the style applied to your design, you must run the analysis in the surface properties in addition to applying the style to the surface. The elevation and slope analysis styles do not let you set a layer for the components because the colors and behavior are set on the Analysis tab.

These distribution methods show up in nearly all the surface analysis methods. Here's what they mean:

Quantile This method is often referred to as an equal count distribution and will create ranges that are equal in sample size. These ranges will not be equal in linear size but in distribution across a surface. This method is best used when the values are relatively equally spaced throughout the total range, with no extremes to throw off the group sizing.

Equal Interval This method uses a stepped scale, created by taking the minimum and maximum values and then dividing the delta into the number of selected ranges. This method can create real anomalies when extremely large or small values skew the total range so that much of the data falls into one or two intervals, with almost no sampled data in the other ranges.

Standard Deviation Standard Deviation is the bell curve that most engineers are familiar with, suited for when the data follows the bell distribution pattern. It generally works well for slope analysis, where very flat and very steep slopes are common and would make another distribution setting unwieldy.

In the following exercise, you will create a surface style for watershed analysis. To apply the new style to the surface, you must also run the analysis.

1. Continue working in the drawing file `Surface Styles.dwg`, which you can download from this book's web page.

2. From the Settings tab of Toolspace, choose Surface ➤ Surface Styles. Right-click the Existing Contours style you created in the first exercise and select Copy.

3. On the Information tab, rename the surface style to **Watershed Analysis**. Add a description to reflect that this style will show watersheds and slope arrows.

4. Switch to the Watersheds tab. Expand the Boundary Segment Watershed area. Set the Use Hatching option to False. Expand the Multi-drain Watershed area. Set the Use Hatching option to False. Your Watersheds tab will now look like Figure 19.41.

5. Switch to the Analysis tab. Expand the Slope Arrows area. Change Scheme to Hydro. Change Arrow Length to 2′. The Analysis tab will now resemble Figure 19.42.

6. Switch to the Display tab. Turn off visibility for the Minor Contour component, and turn it on for Slope Arrows and Watersheds. Set the Watershed layer to C-TINN-VIEW. Click OK to complete the style.

7. Select the surface and open Surface Properties. Set Watershed Analysis as the surface style.

8. Switch to the Analysis tab. Set the Analysis type to Watersheds. Set the Merge Depressions value to 0.4. Place a check mark next to Merge Adjacent Boundary Watersheds. Click the down arrow to run the analysis. Your Surface Properties Analysis tab should now resemble Figure 19.43.

FIGURE 19.41
Changing the hatch options for watershed areas

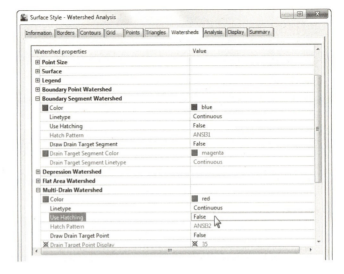

FIGURE 19.42
Set the color scheme and arrow length in the Analysis tab.

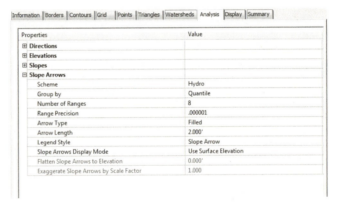

FIGURE 19.43
You must run the analysis in surface properties before the watershed style kicks in.

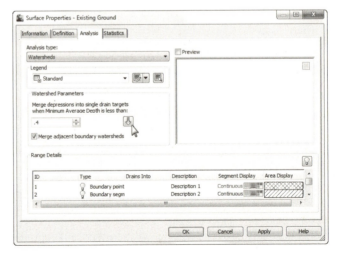

9. Click OK to close the surface properties. Your surface model should now resemble Figure 19.44.

FIGURE 19.44
The completed analysis style applied to the surface

ELEVATION BANDING

The last surface style you will create in this chapter is an elevation banding style:

1. Continue working in the drawing file `Surface Styles.dwg`, which you can download from this book's web page.

2. From the Settings tab of Toolspace, choose Surface ➢ Surface Styles. Right-click the Existing Contours style you created in the first exercise and select Copy.

3. On the Information tab, rename the surface style to **Elevations**. Add a description to reflect that this style will show elevation banding.

4. Switch to the Analysis tab, and expand the Elevations listing. Change Range Color Scheme to Rainbow, as shown in Figure 19.45.

FIGURE 19.45
Elevation analysis options

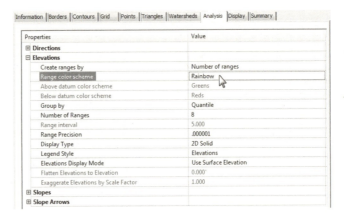

5. On the Display tab, turn off the display for all components except Elevations. Notice that you do not have a layer option for this item because the color and behavior is controlled by the analysis. Click OK to complete the style.

6. Use the same procedure from the previous exercises to apply the Elevations style to the surface. Your surface will now resemble Figure 19.46.

FIGURE 19.46
Pretty rainbow! The elevation style applied to the surface

Pipe and Structure Styles

In Chapter 13, you learned that the best way to manage pipes and structures was in a parts list. One of the functions of a parts list is to associate styles to pipes and structures. In this section, you will learn how to create the pipe and structure styles that are used by a parts list.

In your template, you will have multiple styles for the various parts lists. You will want to have separate styles for water systems, storm sewers, and sanitary sewers. Additionally, you may want to have separate styles for existing and proposed systems. The main difference between the styles for the different systems will be the layers you set in the Display tab.

Pipe Styles

It seems like no two municipalities want pipes displayed the same way on construction documents. Luckily, Civil 3D is very flexible in how pipes are shown. With a single pipe style, you can control how a pipe is displayed in Plan, Profile, and Section views. You can use multiple pipe styles to graphically differentiate larger pipes from smaller ones. This section explores all of the options.

The Plan Tab   The tab you see in Figure 19.47 controls what represents your pipe when you're working in plan view.

Figure 19.47
The Plan tab in the Pipe Style dialog

Options on the Plan tab include the following:

Pipe Wall Sizes   You have a choice of having the program apply the part size directly from the catalog part (that is, the literal pipe dimensions as defined in the catalog) or specifying your own constant or scaled dimensions.

Pipe Hatch Options   If you choose to show pipe hatching, this part of the tab gives you options to control that hatch. You can hatch the entire pipe to the inner or outer walls, or you can hatch the space between the inner and outer walls only, as shown in Figure 19.48.

FIGURE 19.48
Pipe hatch to inner walls (a), outer walls (b), and hatch walls only (c)

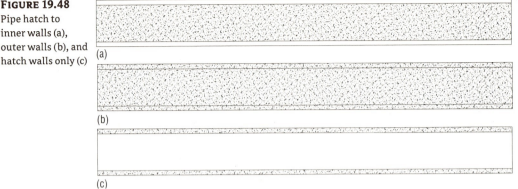

(a)

(b)

(c)

Pipe End Line Size   If you choose to show an end line, you can control its length with these options. An end line can be drawn connecting the outer walls or the inner walls, or you can specify your own constant or scaled dimensions. The pipes from Figure 19.48 are all shown with pipe end lines drawn to outer walls.

Pipe Centerline Options   If you choose to show a centerline, you can display it by the lineweight established in the Display tab, or you can specify your own part-driven, constant, or scaled dimensions. Use this option for your sanitary pipes in places where the width of the centerline widens or narrows on the basis of the pipe diameter.

The Profile Tab   The Profile tab (see Figure 19.49) is almost identical to the Plan tab, except the controls here determine what your pipe looks like in profile view. The only additional settings on this tab are the crossing-pipe hatch options. If you choose to display crossing pipe with a hatch, these settings control the location of that hatch.

The Section Tab   If you choose to show a hatch on your pipes in section, you control the hatch location on this tab (see Figure 19.50).

FIGURE 19.49
The Profile tab in the Pipe Style dialog

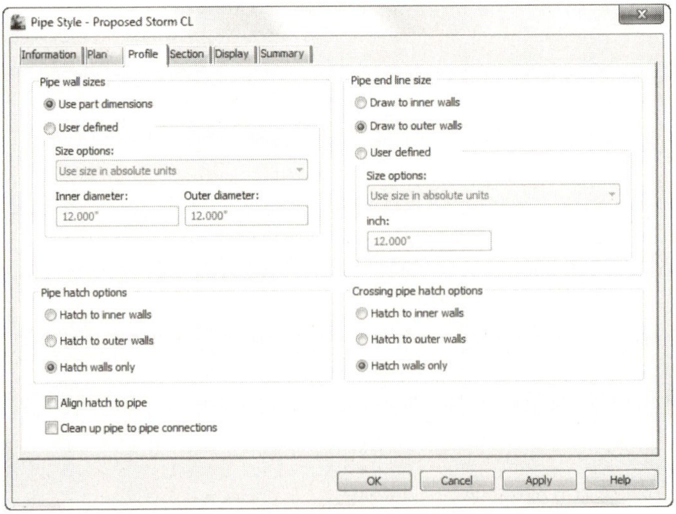

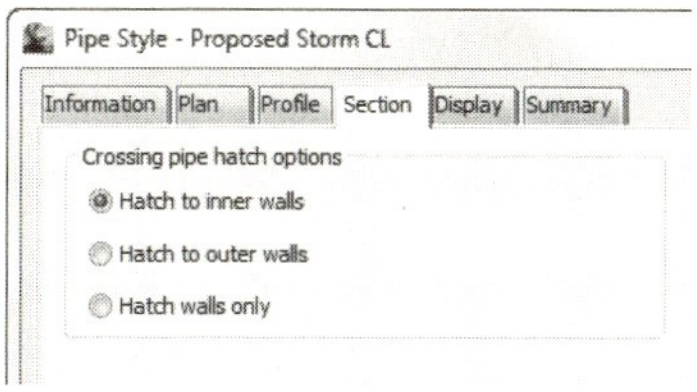

Figure 19.50
The Section tab in the Pipe Style dialog

In the examples that follow, you will create various types of pipe styles.

The first style is for a situation where the pipe must be shown in plan view with a single line, the thickness of which matches the pipe inner diameter. In profile, the pipe will show the inner diameter lines and in section, it will show as a hatch-filled ellipse.

1. Open the drawing `Pipe and Structure Styles.dwg`, which you can download from this book's web page.

2. From the Settings tab of Toolspace, choose Pipe ➢ Pipe Styles. Right-click Pipe Styles and select New.

3. On the Information tab, rename the style to **Proposed Storm CL**.

4. On the Plan tab, set the Pipe Centerline options to Specify Width. Set Specify Width to Draw To Inner Walls.

5. On the Profile tab, no changes are needed. On the Section tab, change the radio button to Hatch To Inner Walls.

6. On the Display tab, make these changes:

    ◆ In Plan View Direction, turn off all components except Pipe Centerline. Set the Pipe Centerline layer to C-STRM-PIPE.

    ◆ Change View Direction to Profile. Make Inside Pipe Walls the only visible component. Set the layer to C-STRM-PIPE.

    ◆ Change View Direction to Section. Set Crossing Pipe Inside Wall to the C-STRM-PIPE layer and make it visible. Set Crossing Pipe Hatch to the C-STRM-PIPE-PATT layer and make it visible. Turn off Crossing Pipe Outside Wall. At the bottom of the dialog, set Crossing Pipe Hatch to SOLID, as shown in Figure 19.51.

7. Click OK. The style is complete.

8. Select the rightmost pipe and open the Pipe Properties dialog. Change the style of the pipe to Proposed Storm CL. Click OK. Save the drawing to use in the next exercise.

Examine the pipe in plan, profile, and cross section. In plan, it should be a solid line. In section, it should be a filled pipe.

FIGURE 19.51
Hatch pattern display for the Section view direction

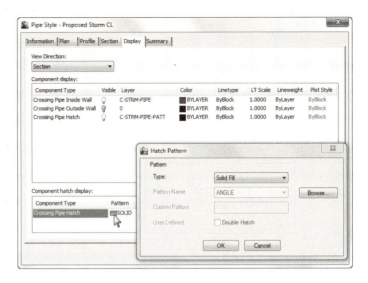

> ### HEY! WHY DOES MY PIPE LOOK LIKE AN OCTAGON?
>
> Civil 3D helps itself perform better on large drawings by knocking down the resolution of 3D curved objects, such as the pipe in the cross section view in the previous steps.
>
>
>
> The system variable you can use to make these pipes look nicer is Facet Deviation, or FACETDEV. The default FACETDEV value for any English unit drawing is 0.5 inches. In metric drawings, the default is 10 millimeters. The lower the FACETDEV value, the smoother the 3D curve.
>
> Note that another variable, FACETMAX, controls the maximum number of facets on any curved object. The Civil 3D default of 500 facets is usually more than enough to display a Civil 3D pipe smoothly.

In the next pipe style example, you will create a style that uses several options for hatching pipe walls:

1. Continue working in the drawing Pipe and Structure Styles.dwg.
2. From the Settings tab of Toolspace, select Pipe ➢ Pipe Styles. Right-click Pipe Styles and select New.
3. On the Information tab, rename the style to **Proposed Double-Wall Storm**.
4. On the Plan tab, verify that the Pipe Hatch options are set to Hatch Walls Only.

GET STARTED WITH OBJECT STYLES | 769

5. On the Profile tab, verify that the Pipe Hatch options are set to Hatch Walls Only.

6. On the Display tab, do the following:

    ◆ In Plan View Direction, the components that should be visible are Inside Pipe Walls, Outside Pipe Walls, Pipe End Line, and Pipe Hatch. Set all four layers to C-STRM-PIPE.

    ◆ In Profile View Direction, the components that should be visible are Inside Pipe Walls, Outside Pipe Walls, Pipe End Line, and Pipe Hatch. Set all four layers to C-STRM-PIPE.

    ◆ In Section View Direction, turn off Crossing Pipe Outside Wall. Set the layer for Crossing Pipe Inside Wall to C-STRM-PIPE.

7. Click OK to complete the style. Use the Pipe Properties dialog to apply the new style to the leftmost pipe.

8. Save the drawing to use in the next exercise.

The last pipe style example demonstrates how to create a crossing pipe override style:

1. Continue working in the drawing `Pipe and Structure Styles.dwg`.

2. From the Settings tab of Toolspace, select Pipe ➤ Pipe Styles. Right-click Pipe Styles and select New.

3. On the Information tab, rename the style to **Proposed Profile Crossing**.

4. Switch to the Display tab.

5. For Plan View Direction, turn off visibility for all components.

6. For Profile View Direction, turn off visibility for all components except Crossing Pipe Inside Wall. Set Crossing Pipe Inside Wall to the C-STRM-PROF layer.

7. Click OK. The style is complete.

8. Pan over to the profile view in the drawing. Click the profile view and select Profile View Properties from the context tab.

9. In the Profile View Pipe Networks tab, place a check mark in the Style Override column for Pipe - (15). Set the style for the style override to Proposed Profile Crossing, as shown in Figure 19.52.

10. Click OK.

FIGURE 19.52
Override the pipe style in Profile View Properties

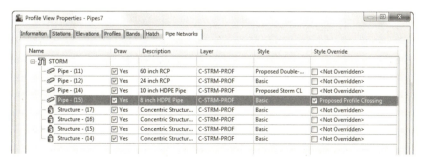

After all three pipe style exercises are complete, your profile view will resemble Figure 19.53.

FIGURE 19.53
The completed exercise in profile

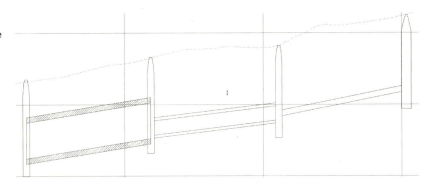

STRUCTURE STYLES

The following tour through the structure-style interface can be used for reference as you create company standard styles:

The Model Tab   This tab (Figure 19.54) controls what represents your structure when you're working in 3D. Typically, you want to leave the Use Catalog Defined 3D Part radio button selected so that when you look at your structure, it looks like your concentric manhole or whatever you've chosen in the parts list.

The Plan Tab   The Plan tab (Figure 19.55) enables you to compose your object style to match that specification.

FIGURE 19.54
The Model tab in the Structure Style dialog

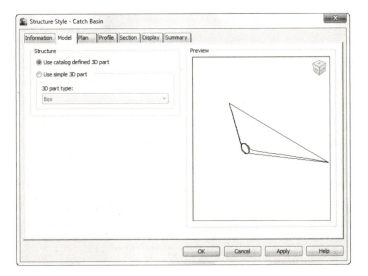

FIGURE 19.55
The Plan tab in the Structure Style dialog

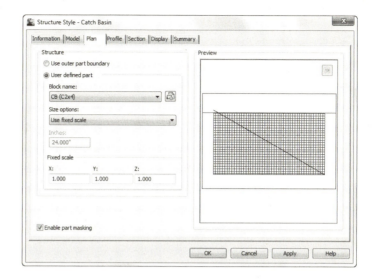

Options on the Plan tab include the following:

Use Outer Part Boundary This option uses the limits of your structure from the parts list and shows you an outline of the structure as it would appear in the plan.

User Defined Part This option uses any block you specify. In the case of your sanitary manhole, you chose a symbol to match the CAD standard.

Size Options The options in this drop-down are similar to what you see in other places in Civil 3D. Use Size In Absolute Units is a common way to represent a manhole at actual size. Use Drawing Scale will treat the object like an annotative block.

Enable Part Masking This option creates a wipeout or mask inside the limits of the structure. Any pipes that connect to the center of the pipe appear trimmed at the limits of the structure.

The Profile Tab Once you know what your structure must look like in the profile, you use the Profile tab (Figure 19.56) to create the style.

Options on the Profile tab include the following:

Display As Solid This option uses the limits of your structure from the parts list and shows you the mesh of the structure as it would appear in profile view.

Display As Boundary This option uses the limits of your structure from the parts list and shows you an outline of the structure as it would appear in profile view. You'll use this option for the sanitary manhole.

Display As Block This option uses any block you specify.

Size Options The options in this drop-down apply to most blocks placed by Civil 3D.

Enable Part Masking This option creates a wipeout or mask inside the limits of the structure. Any pipes that connect to the center of the pipe appear trimmed at the limits of the structure.

FIGURE 19.56
The Profile tab in the Structure Style dialog

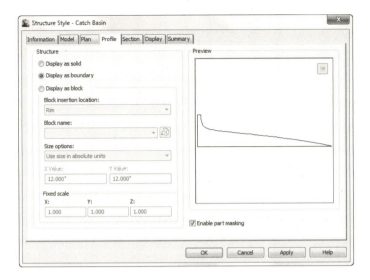

In the following exercise, you'll create a new structure style that uses a block in plan view to represent a catch basin. Because the block is drawn at actual size, you will use the size option Use Fixed Scale.

1. Continue working in the drawing `Pipe and Structure Styles.dwg`, which you can download from this book's web page. If you have not completed the pipe styles exercises, that's okay—completing the previous section is not required to continue.

2. From the Settings tab of Toolspace, select Structure ➢ Structure Styles. Right-click Structure Styles and select New.

3. On the Information tab, rename the style to **Catch Basin**.

4. On the Plan tab, in the Structure panel choose the User Defined Part radio button. Select the Block name CB(C2x4) from the list box. Set Size to Use Fixed Scale and leave the X, Y, and Z scale factors set to 1.

5. On the Display tab, set View Direction to Plan. Set the Structure layer to C-STRM-STRC.

6. Repeat step 5 for all remaining view directions (Model, Profile, and Section).

7. Click OK to complete the style.

8. Select the rightmost structure in the drawing. On the context tab, click Structure Properties. On the Information tab of the Structure Properties dialog, set the style to Catch Basin.

Blocks can be used anywhere a structure is shown. You have the option of inserting the blocks at Rim or Sump. In the case of a flared end section, there need to be blocks for both a left-facing view and right-facing view. In the next exercise you will edit an existing style to fix some "quirks."

1. Continue working in the drawing `Pipe and Structure Styles.dwg`. Completing the previous section is not required to continue.

2. Select the leftmost structure in plan view. Open the structure's properties and change the active style to Flared End Section. Click OK to close the Structure Properties dialog.

3. Select the structure and rotate it so the flared end opens to the left. (The structure will be smaller than the pipe.)

4. Zoom out and take a look at what is going on in profile view.

   The style is set incorrectly, causing the flared end section symbol to dive down to zero. A block is being used for the flared end section and looks pretty horrible. The block is even facing the wrong direction!

5. From the Settings tab of Toolspace, choose Structure ➤ Structure Styles. Right-click Flared End Section and select Edit.

6. On the Information tab, rename the style to **Flared End Section - L**.

7. Switch to the Profile tab. Choose Sump from the Block Insertion Location drop-down.

   Click the Browse button next to the Block Name option. Select the block from the Mastering dataset called FES-L.dwg. Set Size Options to Use Fixed Scale. Set the X and Y values to 4. Your flared end section Profile tab should now resemble Figure 19.57.

8. Click OK to complete the style.

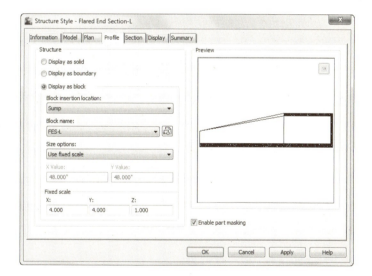

FIGURE 19.57
Flared end section done properly

Label Styles

The best design in the universe is not worth anything unless it is labeled properly. Civil 3D labels are smart objects that are dynamically linked to the object they are labeling. Civil 3D labeling is customizable to fit the needs of your design and local requirements.

When talking about labels in Civil 3D, keep in mind that you are not just talking about text. Labels can contain lines and blocks if desired. Label styles control the size, contents, and precision

CERT OBJECTIVE

of text. Label styles also control how leaders are applied when the label is dragged away from its initial position. You will even work with some labels that contain no text at all!

Like other settings in Civil 3D, there is a hierarchy that helps the user and the software decide which styles take precedence over other styles. There are also items that can be changed at a drawing-wide level and pushed out to object-level labels. Use this ability with caution, as most labels will use their own layers and behave differently depending on the object it is labeling.

In the first exercise, you will set all labels to use the same initial text style:

1. Open the drawing `Label Basics.dwg`, which you can download from this book's web page.

2. From the Settings tab of Toolspace, right-click on the name of the drawing and select Edit Label Style Defaults, as shown in Figure 19.58.

FIGURE 19.58
Accessing the global text settings

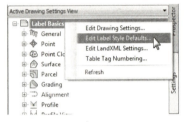

3. Expand the Label category and change Text Style to Arial. Click the arrow in the Child Override column to force all label styles in this drawing to use the same text, as shown in Figure 19.59.

FIGURE 19.59
Label placement options as shown at the drawing level

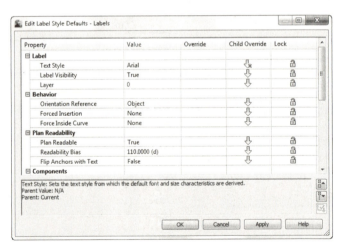

4. Do not make any more changes to this dialog, but examine the various options and settings. Click OK.

The options shown in Figure 19.59 include the following:

Text Style   This option refers to the base AutoCAD text style used in the label.

Layer   This option defines the layer on which the *components* of a label are inserted, not the layer on which the label *itself* is inserted. Think of labels as nested blocks. The label (the block) gets inserted on the layer on the basis of the object layers you saw earlier. The components of the label get inserted on the layer that is set here. This means a change to the specified layer can control or change the appearance of the components if you like. However, if you leave the component layers at 0, the overall object layer will control the look, making the object behave more like a traditional block.

Orientation Reference   This option controls how text rotation is controlled. Figure 19.60a shows the most common option where the label is aligned with the object. Figure 19.60b shows the text rotated to the view. Even though the view has been rotated, the text still appears parallel with the bottom of the screen. The last, and least used, option for text is the World Coordinate System option, shown in Figure 19.60c. The view is rotated, and the direction of the text rotates with the coordinate system.

FIGURE 19.60
Orientation reference options set to Object (a), View (b), and World Coordinate System (c)

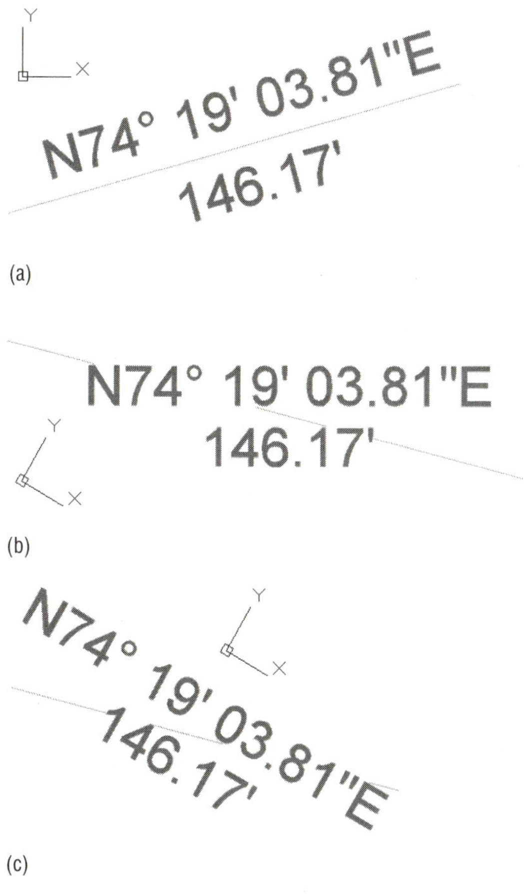

(a)

(b)

(c)

Forced Insertion This option makes more sense in other objects and will be explored further. The Forced Insertion feature allows you to dictate the insertion point of a label on the basis of the object being labeled. Figure 19.61 shows the effects of the various options on a bearing and distance label.

FIGURE 19.61
Forced Insertion options shown for parcel segments

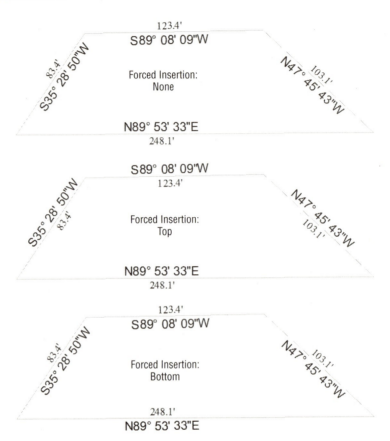

Plan Readable When this option is enabled, text maintains the up direction in spite of view rotation. This tends to be the "Ooooh, nice" feature that makes users smile. Rotating 100 labels is a tedious, thankless task, and this option handles it with one click.

Readability Bias This option specifies the angle at which readability kicks in. This angle is measured from the 0 degree of the x-axis that is common to AutoCAD angle measurements. When a piece of text goes past the readable bias angle, the text flips direction to maintain vertical orientation, as shown in Figure 19.62. The default Readability angle is 110. A common desired default is 90.1, which would flip any text just after passing a vertical direction.

Flip Anchors With Text Most users leave this obscure setting at its default, False, and never give it another thought. Figure 19.63a shows what happens to the text insertion points when this option is set to False and readability kicks in. Notice how the text flips to the bottom of the line it is labeling when this setting is set to True, as shown in Figure19.63b.

LABEL STYLES | 777

FIGURE 19.62
Plan-readable text shown on contours; note the direction difference between the top grouping and the bottom grouping.

FIGURE 19.63
Flip anchors with text? Best leave this alone!

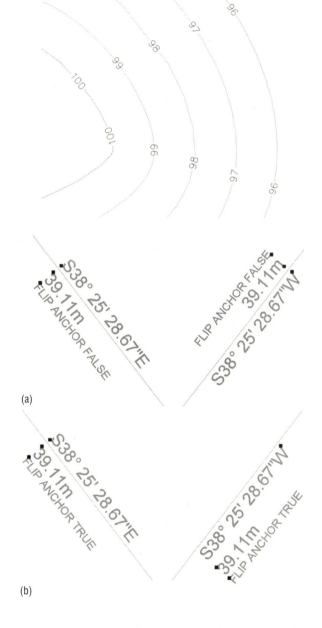

To get into a label style, find the appropriate label type in the Settings tab of Toolspace. Right-click the style you wish to edit (or copy) and select Edit, as shown in Figure 19.64.

FIGURE 19.64
Entering a label style for editing

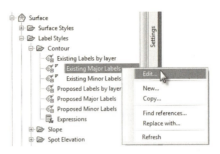

Inside the label styles themselves, you will see the same tabs regardless of the object you are labeling.

The General tab (Figure 19.65) has the label-specific version of the options we explored in the previous section.

The Layout tab (Figure 19.66) is the true heart of any label style. This is where you tell Civil 3D what to label and exactly how that label is to be attached to the feature you are labeling.

FIGURE 19.65
General tab for text placement

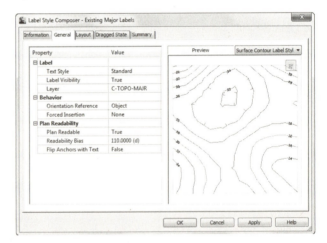

FIGURE 19.66
The Layout tab

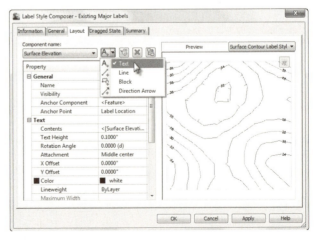

Each label can be made up of several components. A label component can be text, a block, or a line, and the top row of buttons controls the selection, creation, and deletion of these components:

Component Name   Choose the component you want to modify from the Component Name drop-down menu. The components are listed in the order in which they were created. Pay attention to which component is active when you make changes to properties on the layout tab. The changes you make to properties will only apply to the active component.

Create Component   The Create Component button lets you add new elements to enhance your labels. These components can be text, lines, blocks, ticks, direction arrows, or reference text. The options will vary depending on the object you are labeling. Not every option is available in every label.

The ability to label one object while referencing another (reference text) is one of the most powerful labeling features of Civil 3D. This is what allows you to label a spot elevation for both an existing and a proposed surface at the same time, using the same label. Alignments, COGO points, parcels, profiles, and surfaces can all be used as reference text.

Copy Component   The Copy Component button copies the component currently selected in the Component Name drop-down menu. This will be helpful when creating label styles that contain multiple pieces of similar information.

Delete Component   The Delete Component button deletes components. Elements that act as the basis for other components can't be deleted.

Component Draw Order   The Component Draw Order button lets you shuffle components up and down within the label. This feature is especially important when you're using masks or borders as part of the label.

You can work your way down the component properties and adjust them as needed for a label:

Name   This option defines the name used in the Component Name drop-down menu and when selecting other components. When you're building complicated labels, a little name description goes a long way.

Visibility   When set to True, this option means the component shows on screen. Why create a component that can't be seen? There are cool tricks you can do with styles, as you'll see in the section "Pipe Labels" later in this chapter.

Anchor Component and Anchor Point   Use these options to control text placement. Anchor Component lets you tell Civil 3D how text (or other label components) relate to the main object you are labeling. See Figure 19.67 and Figure 19.69 for a graphical explanation of how these options relate to each other. The circles represent anchor points and the squares represent attachment.

Text Height   This option determines the plotted height of the label. Text placed by Civil 3D is always annotative. Figure 19.68 shows two viewports showing some COGO points along a road. Even when the viewport scales differ, the text is the same size.

FIGURE 19.67
Closer look at the Layout tab options

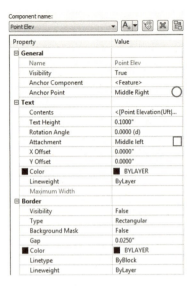

FIGURE 19.68
Civil 3D text is always annotative.

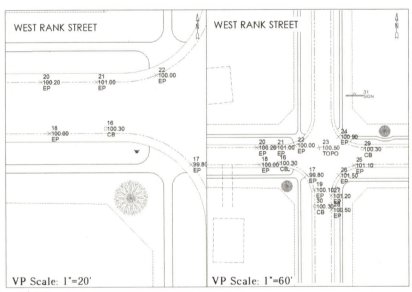

Rotation Angle, X Offset, and Y Offset These options give you the ability to refine the placement of the component by rotating or displacing the text in an *x* or *y* direction. Set your text as close as possible using the anchors and attachments and use the offsets as additional spacing.

Attachment This option determines which of the nine points on the label components bounding box are attached to the anchor point. See Figure 19.69 for an illustration.

Color and Lineweight These options allow you to force the color if desired. It's a good idea to leave these values set to ByLayer unless you have a good reason to change them.

FIGURE 19.69
Schematic showing the relationship between anchor points (circles) and attachments (squares)

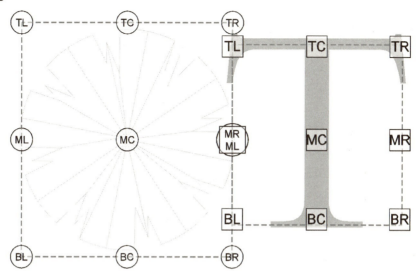

The final piece of the component puzzle is the Border option. These options are as follows:

Visibility   Use this option to turn the border on and off for the component. Remember that component borders shrink to the individual component; if you're using multiple components in a label, they all have their own borders.

Type   This option allows you to select a rectangle, a rounded rectangle (slot), or a circle border. Figure 19.70 shows examples of the three types of borders.

FIGURE 19.70
Border types shown on various surface label styles

Background Mask   This option lets you determine whether linework and text behind this component are masked. This option can be handy for construction notes in place of the usual wipeout tools. The surface labels in Figure 19.70 show the background mask in action.

Gap   This option determines the offset from the component bounding box to the outer points on the border. Setting this to half of the text size usually creates a visually pleasing border.

Linetype and Lineweight   These options give you control of the border lines.

After working through all the options for the default label placement, you need to set the options that come into play when a label is dragged. To do so, switch to the Dragged State tab, shown in Figure 19.71.

FIGURE 19.71
The Dragged State tab

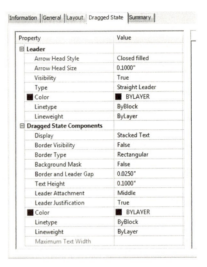

When a label is dragged in Civil 3D, it typically creates a leader, and text is rearranged. The settings that control these two actions appear on this tab. Here are some of the unique options:

Arrow Head Style and Size   These options control the tip of the leader. Note that Arrow Head Size also controls the tail size leading to the text object.

Type   This option specifies the leader type. Options are Straight Leader and Spline Leader. As of this writing, the AutoCAD multiple leader object can't be used.

Display   This option controls whether components rearrange their placement to a stacked set of components (Stacked Text) or maintain their arrangement as originally composed (As Composed).

Figure 19.72a shows an alignment label as it was originally placed. Figure 19.72b shows the same label in a dragged state with the Stacked Text option set. Figure 19.72c shows the label in a dragged state with the As Composed option set.

FIGURE 19.72
An alignment label as originally placed (a); Dragged State, Stacked Text (b); and Dragged State, As Composed (c)

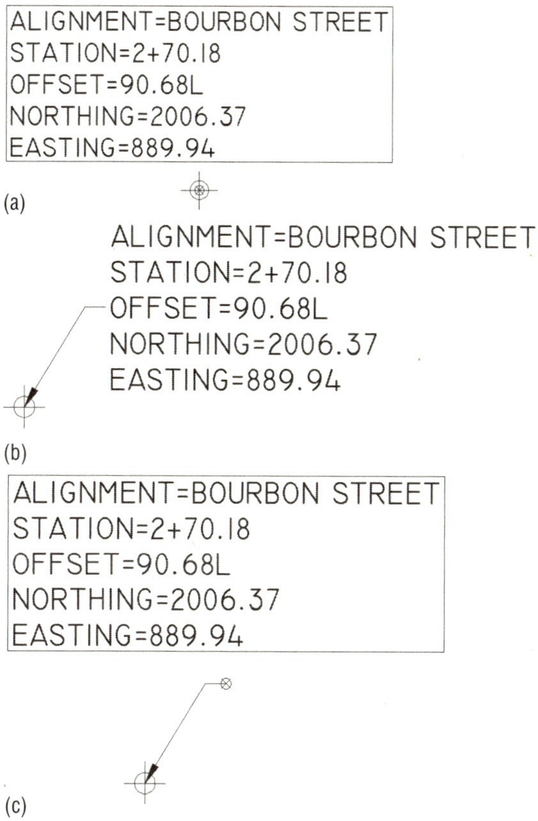

```
ALIGNMENT=BOURBON STREET
STATION=2+70.18
OFFSET=90.68L
NORTHING=2006.37
EASTING=889.94
```
(a)

ALIGNMENT=BOURBON STREET
STATION=2+70.18
OFFSET=90.68L
NORTHING=2006.37
EASTING=889.94

(b)

```
ALIGNMENT=BOURBON STREET
STATION=2+70.18
OFFSET=90.68L
NORTHING=2006.37
EASTING=889.94
```

(c)

GETTING TO KNOW THE TEXT COMPONENT EDITOR

Few dialogs in Civil 3D inspire expletives the way the Text Component Editor can. The interface is very logical and will do exactly what you tell it to do—not necessarily what you want it to do. Hopefully, with some explanation you will see the reasoning behind its behavior and master it!

To enter the Text Component Editor, from the Layout tab of any text style, click the ellipsis that pops up when you click in the Text Contents area.

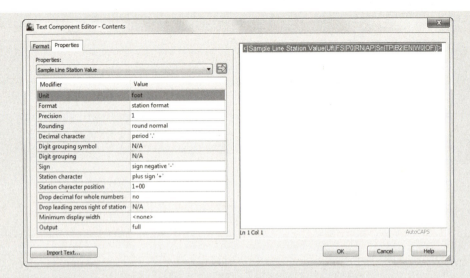

Within the Text Component Editor are two main areas. The left side is where you select what aspect of the Civil 3D object you want to pull into your label. The right side shows what is already in the label.

To modify an existing label, highlight the chunk of text on the right side of the dialog. Civil 3D's "smart" text will always highlight as a unit. Each of the cryptic codes in the label represents a setting for units, precision, or other ways the text can display. What appears on the left is a decoded list of what is currently part of your label. After making changes, don't forget to click the arrow to update the information on the right.

Before adding new text to a label, make sure you don't have any existing text highlighted on the right to avoid overwriting it.

The Properties list shows everything that can be labeled about the item you are working with. The length of the list varies depending on the object. Here you see the available properties for a contour (left) and a pipe network structure (right):

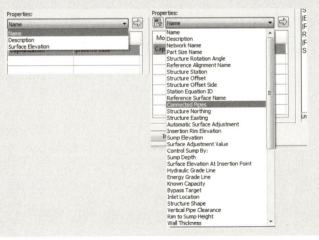

Once you pick what you'd like to label, you will set the units, precision, and any other special rounding or formatting you would like to see.

When everything is set, click the arrow next to the Properties list.

To further fine-tune the text, a Format tab sits inconspicuously behind the Properties list. Use this to add special symbols, or to override the color and text style.

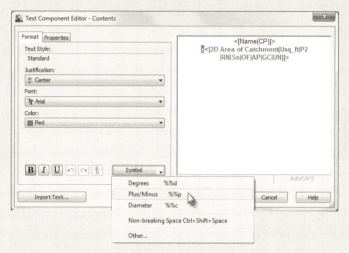

Inevitably, you will forget the arrow that adds or updates the text. You may even click it twice and end up with duplicates! Now that you've been given the heads-up you'll be able to laugh it off, knowing that it happens to even the most seasoned users.

General Note Labels

General Note labels are versatile, non-object-specific labels that can be placed anywhere in the drawing. There are several advantages to using these instead of base-AutoCAD MTEXT. Notes will leader off and scale the same as the rest of your Civil 3D text and, best of all, they can contain reference text.

In the following example, you will create an alternate parcel label that contains reference text:

1. Open the drawing file Example Labels.dwg, which you can download from this book's web page.

2. From the Settings tab of Toolspace, choose General ➢ Label Styles ➢ Note. Right-click on Note and select New.

3. On the Information tab, name the style **Easement Parcel Text**.

4. On the General tab, set Layer to C-PROP-TEXT. Set Orientation Reference to View.

5. On the Layout tab, click the field next to Contents under the Text category. Clicking this will cause an ellipsis button to appear. Click the ellipsis button to enter the Text Component Editor.

6. On the right side of the Text Component Editor, strike out the existing text and replace it with **Drainage Easement**, as shown in Figure 19.73. Click OK.

FIGURE 19.73
Entering the Text Component Editor for basic text

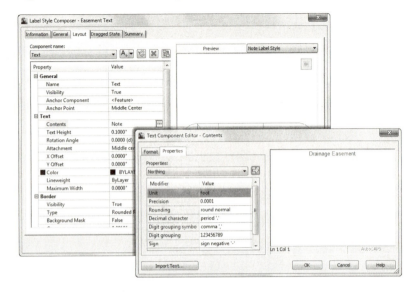

7. Back in the Layout tab, set Border Visibility to False.

8. Click the flyout next to Create Text Component and select Reference Text. You will be prompted to select the type of reference text, as shown in Figure 19.74. Select Parcel and click OK.

9. Rename the Reference Text.1 to **Parcel Area** and make the following changes:
    - Set Anchor Component to Text.
    - Set Anchor Point to Bottom Center.
    - Set Attachment to Top Center.
    - Click the Label Text field and enter the Text Component Editor.

10. Remove the existing label text. On the left side of the Text Component Editor, set the Properties drop-down to Parcel Area. Set Unit to Acre. Set Precision to **0.01**.

11. When you have set the properties, click the arrow icon to place the text in the right side of the editor. Add **ACRES** after the coding. The Text Component Editor will resemble Figure 19.75. Click OK.

FIGURE 19.74
Picking the reference type

12. You will still have question marks in the preview, but this is completely expected. Click OK to complete the command.

13. On the Annotate tab, click Add Labels. With Feature and Label Type set to Note, change Note Label Style to Easement Text. Click Add.

14. Click in the easement area in the southwest portion of the site. At the prompt `Select parcel for label style component Parcel Area:`, click the label on parcel 200. Press Esc on your keyboard to finish placing labels. Your completed and placed label should resemble Figure 19.76.

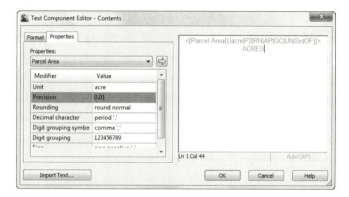

FIGURE 19.75
Adding "smart" text to the Text Component Editor

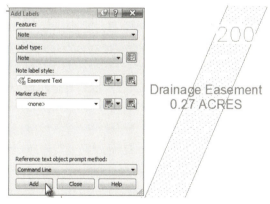

FIGURE 19.76
Your first label! Referencing parcel area.

Point Label Styles

If you have used other software packages for surveying work, odds are that you controlled point label text with layers, but Civil 3D is different. In Civil 3D, you must control what information is showing next to a point by swapping the label style applied to a group of points.

In the following exercise, you will create a new point label style. Your first point label style will show only Point Number and Description, so you will need to delete the default Elevation component.

1. Open the drawing `Point Labels.dwg`, which you can download from this book's web page.
2. From the Settings tab of Toolspace, choose Point ➢ Label Styles. Right-click Label Styles and select New.
3. On the Information tab, name the style **Point Number & Description**.
4. On the General tab, set the layer to V-NODE-TEXT. Leave all other General tab options at their defaults.
5. On the Layout tab, do the following:
   - Set the active component to Point Number. Change the anchor component to `<Feature>`. Set Anchor Point to Middle Right and set Attachment to Middle Left.
   - Change the Active component to Point Elevation. Click the red X to delete this component. You will receive a warning that reads "This label component is used an as anchor in this style or in a child style. Do you want to delete it?" Click Yes.
   - Change the active component to Point Description. Change the anchor component to Point Number. Change the anchor point to Middle Right. Change the attachment to Middle Left. Change the X Offset to **0.05"**.
   - Click OK to complete the label style.
6. In Prospector, locate the point group named TOPO. Right-click on TOPO and select Properties. Set the Point Label style to **Point Number & Description**. All the points in the group will change to resemble Figure 19.77.

FIGURE 19.77
Completed point label style

In the previous exercise, you created a simple new label style and made some modifications to the default components. In the following exercise, you will remove all the default components and add Northing and Easting values to the label using the Text Component Editor.

1. Continue working in the drawing `Point Labels.dwg`.
2. From the Settings tab of Toolspace, choose Point ➢ Label Styles. Right-click Label Styles and select New.
3. On the Information tab, name the style **Northing & Easting**.
4. On the General tab, set the layer to V-NODE-TEXT. Leave all other General tab options at their defaults.

5. On the Layout tab, do the following:

- Click the red X until all three default components are gone. When you are asked about deleting the anchor, click Yes.
- Click the button to create a new text component. Rename the component to **N-E**. Set the anchor point to Bottom Center. Set the attachment to Top Center.
- Click the Label Text field and enter the Text Component Editor by clicking the ellipsis. Highlight the default label text and delete it by pressing the Delete key on your keyboard.
- From the Properties list, select Northing. Set the Precision to **0.01** (two decimal places).
- Click the arrow to place the text to the right. After the label text, place an **N** as a static label after the Northing value.
- From the Properties list, select Easting. Set the Precision to **0.01** (two decimal places).
- Click the arrow to place the text to the right. After the label text, place an **E** as a static label after the Easting value.

The Text Component Editor will now resemble Figure 19.78.

FIGURE 19.78
The Northing & Easting label in progress

6. In Prospector, locate the point group named Group2. Right-click on Group2 and select Properties. Set the Point Label style to **Northing & Easting**. The labels will now resemble Figure 19.79.

FIGURE 19.79
Northing & Easting in the completed exercise

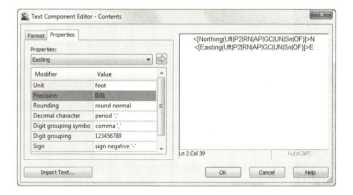

> **SANITY SAVING SETTINGS**
>
> In your Civil 3D template you will have your AutoCAD styles, linetypes, layers, Civil 3D styles, and a plethora of helpful goodies that make doing your job easier. There are a few drawing-specific CAD variables you may not have thought of that will improve your relationship with Civil 3D:
>
> **MSLTSCALE**   This variable stands for modelspace linetype scale. It is strongly, no, STRONGLY recommended that you have this set to 1 in your template. This setting makes line types react to your annotation scale. All your other Civil 3D objects are doing it, having your line types follow suit will help!
>
> **LAYEREVALCTL**   Set this to 0 to avoid the annoying pop-up that flags users when there are new layers in a drawing. Civil drafters are constantly using XRefs and inserting blocks, both of which cause the pop-up to occur.
>
> **GEOMARKERVISIBILITY**   This is the control for that weird thing in your drawing that appears any time a coordinate system is set up on your drawing. The geomarker looks like a block, but you can't select it and it keeps rescaling as you zoom. Set the GEOMARKERVISIBILITY variable to 0 in your template and the geomarker won't appear.
>
> **AUNITS**   This variable defines the angular units for the base CAD part of the world. Keep this set at 0 (decimal degrees) to help differentiate base AutoCAD angular entry from Civil 3D angular entry.

Line and Curve Labels

Many Civil 3D objects can have a bearing and distance label added to them. Anything from plain lines and polylines, to parcels and alignment tangent segments, can use nearly identical label types.

The examples in the following exercises will use parcels for labeling, but the tools can be applied to all other types of line labels.

SINGLE SEGMENT LABELS

In the following exercise, you will create a new line label style that uses default components. You will remove the direction arrow and change the display precision of the direction component.

1. Open the drawing file `Line and Curve Labels.dwg`, which you can download from this book's web page.
2. From the Settings tab of Toolspace, choose Parcel ➢ Label Styles ➢ Line. Right-click Line and select New.
3. On the Information tab, name the style **Parcel Segment**.
4. On the General tab, set the layer to C-PROP-LINE-TEXT. Leave all other General tab options at their defaults.
5. On the Layout tab, change the active component to Direction Arrow. Click the red X to delete this component.

6. Change the active component to Distance. Enter the Text Component Editor by clicking in the Text Contents area.

7. Highlight the Segment Length contents on the right by clicking on the text. All of the text should highlight as a unit.

8. On the left side of the Text Component Editor, change the Precision value to **0.01**.

9. Click the arrow to update the text. Click OK to dismiss the Text Component Editor. Click OK to complete the style.

10. Add labels to the parcels by selecting the Annotate tab. Click Add Labels and set Feature to Parcel. Set the label type to Single Segment. Set Line Label Style to Parcel Segment. Click Add and select several straight segments for labeling. The completed label will resemble Figure 19.80.

FIGURE 19.80
Your new bearing and distance label style in action on parcel segments

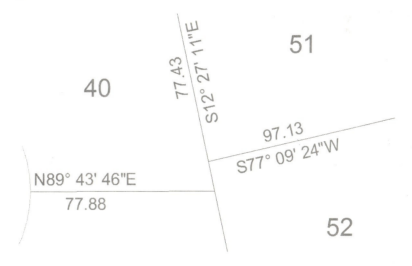

SPANNING SEGMENT LABELS

Looking at the labels you created in the previous exercise, you may notice that they stop at each parcel vertex. If the back property line is the same bearing, most plats show these as a single label with an overall length rather than many shorter lengths at the same bearing. To label these in Civil 3D, a separate label style is needed.

Spanning labels can be used in both line and curve parcel labels. In the following exercise, you will create line labels that span across multiple parcel segments:

1. Continue working in the drawing `Line and Curve Labels.dwg`.

2. From the Settings tab of Toolspace, choose Parcel ➢ Label Styles ➢ Line. Right-click Parcel Segments and select Copy.

3. On the Information tab, name the style **Spanning Segment**.

4. On the Layout tab, change the active Component Name to Table Tag. Change the Span Outside Segments setting to True.

5. Change to the Bearing component. Change the Span Outside Segments setting to True.

6. Change the active component to Distance. Change the Span Outside Segments setting to True. Click OK to complete the style.

7. Add labels to the parcels using the same technique you used in the previous exercise.

When you initially place a spanning label in the drawing, the overall length will not be reported unless the distance portion of the label is on the outside of a row of parcels. You may need to use the Flip Label command to see the spanning option take effect. Your completed label will resemble Figure 19.81.

FIGURE 19.81
A spanning label shown on the outside of parcel segments

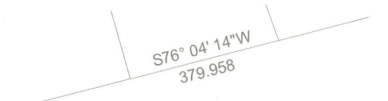

Curve Labels

In base AutoCAD, creating curve labels is a chore. If you want text to align to curved objects, it is no longer usable as traditional MTEXT. Luckily, Civil 3D gives you the ability to add curved text without compromising the usability.

In the following exercise, you will create a curve label style with a delta symbol and text that curves with the parcel segment:

1. Continue working in the drawing file `Line and Curve Labels.dwg`.

2. From the Settings tab of Toolspace, choose Parcel ➢ Label Styles ➢ Curve. Right-click on Curve and select New.

3. On the Information tab, rename the style to **Delta Length & Radius**.

4. On the General tab, set Layer to C-PROP-LINE-TEXT.

5. On the Layout tab, change the active component to Distance And Radius. Enter the Text Component Editor. Highlight the Segment Length property on the right and change the precision to **0.01**. Click the arrow to update the style.

6. Highlight the Segment Radius property and change the precision to **0.01**. Click the arrow to update the text.

7. Delete the comma that appears as static text in the Text Component Editor. Click OK.

8. Click the Create Text Component button and in the resulting window:
   - Rename the new text component to **Delta**.
   - Change Attachment to Bottom Center.
   - Change the Y offset to **0.025**.
   - Set Allow Curved Text to True.
9. Enter the Text Component Editor. Remove the default label text. Switch to the Format tab. Click the Symbol button and select Other.
10. You should now see the Character Map dialog (Figure 19.82). Browse through the dialog to find the Delta symbol. When you locate it, click Select and then click Copy.

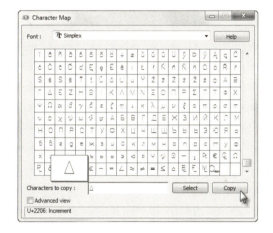

FIGURE 19.82
Browse for special symbols using the Windows Character Map.

11. Back in the Text Component Editor, on the right side of the dialog right-click and select Paste. Press Backspace on your keyboard if the cursor jumps to the next line of text. You should now see the delta symbol appear in the Text Component Editor.
12. Type in = as static text after the delta. Switch back to the Properties tab.
13. From the Properties drop-down, select Segment Delta Angle. Set the Format to DD° MM′ SS.SS″. Set the Precision value to 1 second. Click the arrow to place the text in the right side. Click OK to exit the Text Component Editor.
14. Click OK to complete the style.
15. Add labels to the parcels using the same technique you used in the previous exercise. Set the active curve label style to Delta Length And Radius.

    Your completed label when applied to the design should resemble Figure 19.83.

FIGURE 19.83
Completed curve labels with delta symbol and curved text

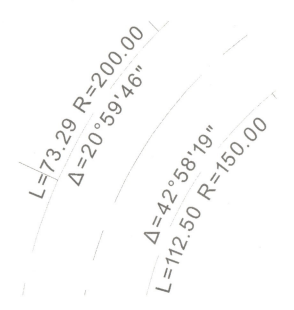

Pipe and Structure Labels

It seems like no two municipalities label their pipes and sewer structures exactly the same way. Luckily, Civil 3D offers lots of flexibility in how you label these items.

PIPE LABELS

Pipe labels have two separate label types: Plan Profile and Crossing Section. Both label types have many of the same options, but are used in different view directions.

In the following exercise, you will use a common "trick" in Civil 3D labels where a nonvisible component acts as an anchor to visible objects. In this case, you will use the flow direction arrow to force text to be placed at the ends and middle of the pipe regardless of the pipe length.

1. Open the drawing file `Pipe and Structure Labels.dwg`, which you can download from this book's web page.

2. From the Settings tab of Toolspace, choose Pipe ≻ Label Styles ≻ Plan Profile. Right-click Plan Profile and select New.

3. On the Information tab, name the style **Length Diameter Slope**.

4. On the General tab, set the layer to C-STRM-TEXT.

5. On the Layout tab, delete the existing Pipe Text component.

6. Click the Add New Component button and select Flow Direction Arrow; then adjust these settings:

    ◆ Set Visibility to False.

    ◆ Set Anchor Point to Top Outer Diameter.

    ◆ Set a Y offset of **0.1**.

7. Click the Add New Component button and select Text. Change these settings:
    - Rename the Text Component to **Length**.
    - Set Anchor Component to Flow Direction Arrow.1.
    - Set Anchor Point to Start.
    - Set Attachment to Bottom Left.
    - In the Text Component Editor, remove the default label text.
    - From the Properties list, set 2D Length - Center to Center Current.
    - Set Precision to 1 and Rounding to round up. This causes the pipe length value to round to the next highest whole foot.
    - Click the arrow to place the text in the editor.
    - Add a foot symbol after the text component.
    - Click OK.

8. With Length as the current component, click Copy. Then change these settings:
    - Rename the component to **Diameter**.
    - Change Anchor Point to Middle.
    - Change Attachment to Bottom Center.
    - Enter the Text Component Editor and remove the Length text.
    - From the Properties list, select Inner Pipe Diameter.
    - Set the Precision to **1** and click the arrow to add the text.
    - Add the inch symbol after the text component and click OK.

9. With Diameter as the current component, click Copy. Then change these settings:
    - Rename the component to **Slope**.
    - Set Anchor Point to End.
    - Set Attachment to Bottom Right.
    - Enter the Text Component Editor and set Pipe Slope as the current property.
    - Click the arrow to update the text and click OK.

10. Click OK to complete the style. Add the label to the pipe by selecting the Annotate tab and setting the Feature option to Pipe Network and Label Type to Single Part Plan. Set the Length Diameter Slope label style to Current and click Add. Your labeled pipe will resemble Figure 19.84.

FIGURE 19.84
Labeled pipe using the invisible arrow trick

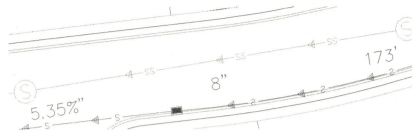

STRUCTURE LABELS

Structure labels are nifty because they have a component no other label style has: the Text For Each component. The number of pipes entering and exiting a structure will vary depending on where the structure is in a network, and the Text For Each option accommodates that unique feature.

> ### Real World Scenario
>
> #### ADDING EXISTING GROUND ELEVATION TO STRUCTURE LABELS
>
> In design situations, it's often desirable to track not only the structure rim elevation at finished grade, but also the elevation at existing ground. This gives the designer an additional tool for optimizing the earthwork balance.
>
> This exercise will lead you through creating a structure label that includes surface-reference text. It assumes you're familiar with Civil 3D label composition in general:
>
> 1. Continue working in the file Pipe and Structure Labels.dwg.
> 2. Locate the Structure Label Styles branch on the Settings tab of Toolspace.
> 3. Right-click Label Styles and select New. Set the options as follows:
>   - On the Information tab, name the label **Structure w Surface**.
>   - On the General tab, set the layer to C-STRM-TEXT.
>   - On the Layout tab is a default text component called Structure Text. Set the Y offset to **−0.25**.
>   - Click in the Contents box to bring up the Text Component Editor.
>   - Delete the <[Description(CP)]> text string, and then use the Properties drop-down menu to add the <Name> and RIM <Insertion Rim Elevation> (two decimal places).
> 4. Click OK to leave the Text Component Editor.
> 5. In the Label Style Composer, choose Reference Text from the Add Component drop-down menu.
> 6. In the Select Type dialog, choose Surface.

7. Rename the component from Reference Text.1 to **Existing Ground**.
8. Click in the Contents box to bring up the Text Component Editor.
9. Delete the "Label Text" text string, and then use the Properties drop-down menu to add the EG <Surface Elevation> (two decimal places).
10. Click OK to dismiss the Text Component Editor.
11. In the Label Style Composer, change Anchor Component for Existing Ground to Structure Text, change Anchor Point to Bottom Center, and change Attachment to Top Center.
12. Choose the Text For Each option from the Add Component drop-down menu.

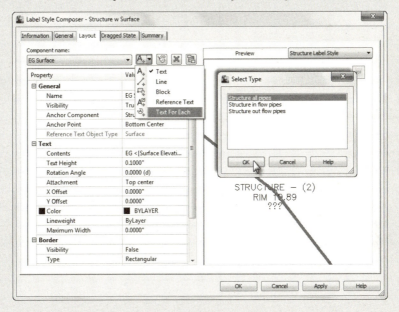

13. In the Select Type dialog, choose Structure All Pipes.
14. Click in the Contents box to bring up the Text Component Editor.
15. Delete the "Label Text" text string, and then use the Properties drop-down menu to add the INV <Connected Pipe Invert Elevation> (two decimal places).
16. Click OK to dismiss the Text Component Editor.
17. In the Label Style Composer, change Anchor Component for Text for Each.1 to Existing Ground and change Anchor Point to Bottom Center. Then, change Attachment to Top Center.
18. Click OK to dismiss the Label Style Composer.
19. Go into the drawing, and select one or more structure labels.

20. Select the Annotate tab and click Add Labels. Set the Feature option to Pipe Network and the label type to Single Part Plan. Set the structure label style to Structure w Surface and click Add.

21. Click the structure you want to label. You will immediately see a prompt at the command line that reads `Select surface for label style component EG Surface :`. Press ↵ to select EG from the surface listing. Your label is now complete.

Profile and Alignment Labels

Profile and alignment labels can take on many forms. On an alignment you may want to show labels every 100′ in addition to PC, PT, and PI information. On a profile, you will want tangent grades, curve information, and grade breaks.

For every element you wish to label, a separate style controls the look and behavior of that label.

Label Sets

A *label set* is a grouping of labels that apply to the same object. In lieu of having one big style that accounts for multiple aspects of an object, the labels are broken out into specific pieces to allow you more control.

Label sets come into play with alignments and design profiles. When you look at an alignment or profile and see labels, you are usually seeing multiple label styles in action.

Consider the alignment shown in Figure 19.85. How many labels are there on this alignment? The geometry points, the major ticks, the minor ticks, superelevation stations, and design speed are all separate labels.

How did those labels get here? When you first created the alignment or profile, one of the options was to specify a label set (as shown in Figure 19.86a). The labels from the set are applied as a batch to the object.

To edit which labels are applied to an alignment or profile, select Add/Edit Station Labels from the context tab (as shown in Figure 19.86b).

FIGURE 19.85
One alignment, five label styles in play

FIGURE 19.86
Specifying an alignment label set upon creation (a) and accessing the label list after creation (b)

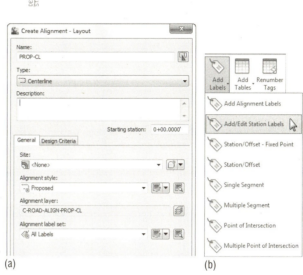

(a)   (b)

Label sets also dictate the placement of the annotation. An alignment label set controls the major and minor station labeling increment. A profile label set has a big job to do in helping you place labels in the correct location with respect to both the design and the profile view. In Figure 19.87, Dim Anchor Opt and Dim Anchor Val control whether the label is placed on the object or in relation to the graph.

FIGURE 19.87
Profile labels and placement

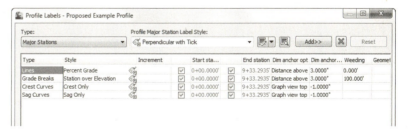

Alignment Labels

You'll create individual label styles over the next couple of exercises and then pull them together with a label set. At the end of this section, you'll apply your new label set to the alignments.

Major Station

Major station labels typically include a tick mark and a station callout. In this exercise, you'll build a style to show only the station increment and run it parallel to the alignment:

1. Open the `Alignment Labels.dwg` file.
2. Switch to the Settings tab, and expand the Alignment ➢ Label Styles ➢ Station ➢ Major Station branch.
3. Right-click the Parallel With Tick style and select Copy. The Label Style Composer dialog appears.
4. On the Information tab, type **Station Index Only** in the Name field.
5. Switch to the Layout tab.
6. Click in the Contents Value field, under the Text property, and then click the ellipsis button to open the Text Component Editor dialog.
7. Click in the preview area, and delete the text that's already there.
8. Click in the Output Value field, and click the down arrow to open the drop-down list.
9. Select the Left Of Station Character option, as shown in Figure 19.88 (you may have to scroll down). Click the insert arrow circled in the figure.

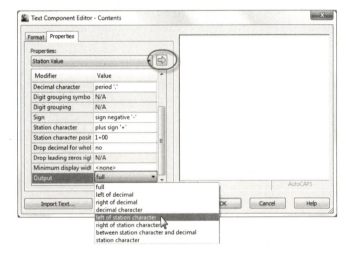

Figure 19.88
Modifying the Station Value Output value in the Text Component Editor dialog

10. Click OK to close the Text Component Editor dialog.
11. Click OK to close the Label Style Composer dialog.

The label style now shows in your label styles, but it's not applied to any alignments yet.

Geometry Points

Geometry points reflect the PC, PT, and other points along the alignment that define the geometric properties. The existing label style doesn't reflect a plan-readable format, so you'll copy it and make a minor change in this exercise:

1. Continue working in the `Alignment Labels.dwg` file.
2. Expand the Alignment ➢ Label Styles ➢ Station ➢ Geometry Point branch.
3. Right-click the Perpendicular With Tick And Line style, and select Copy to open the Label Style Composer dialog.
4. On the Information tab, change the name to **Perpendicular with Line**.
5. Switch to the General tab.
6. Change the Readability Bias setting to **90**. This value will force the labels to flip at a much earlier point.
7. Switch to the Layout tab and make these changes:
    - Set the Component Name field to the Tick option.
    - Click the Delete Component button (the red X button).
    - Click the Create New Line Component button.
    - Change the line angle to 90.
    - Change the component to Geometry Point & Station. Change the Anchor Component to Line.1. Change the Anchor Point to End. Change Rotation Angle to 0.
    - Click OK.

8. Click OK to close the Label Style Composer dialog. Save the drawing.

This new style flips the plan-readable labels sooner and includes a line with the label.

Alignment Label Set

Once you have several labels you wish to use on an alignment, it is time to save them as an alignment label set. Continue working in the `Alignment Labels.dwg` file.

1. Expand the Alignment ➢ Label Styles ➢ Label Sets branch.
2. Right-click Label Sets, and select New to open the Alignment Label Set dialog.
3. On the Information tab, change the name to **Paving**.
4. Switch to the Labels tab.
5. Set the Type field to the Major Stations option and the Major Station Label Style field to the Station Index Only style that you just created; then click the Add button.

6. Set the Type field to the Minor Stations option and the Minor Station Label Style field to the Tick option; then click the Add button.

7. Set the Type field to Geometry Points and the Geometry Point Label Style field to Perpendicular With Line. Then click the Add button to open the Geometry Points dialog, as shown Figure 19.89. Deselect the Alignment Beginning and Alignment End options as shown. Click OK to dismiss the dialog.

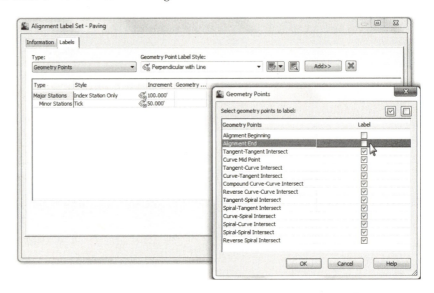

FIGURE 19.89
Deselecting the Alignment Beginning and Alignment End Geometry Point options

8. Three label types will now be shown in the Alignment Label Set dialog. Click OK to dismiss this dialog.

In the next exercise, you'll apply your label set to the example alignment and then see how an individual label can be changed from the set. Continue working in the `Alignment Labels.dwg` file.

1. Select the Example alignment on screen.

2. Right-click, and select Edit Alignment Labels to display the Alignment Labels dialog. This dialog shows which labels are currently applied to the alignment. Initially, it will be empty.

3. Click the Import Label Set button near the bottom of this dialog.

4. In the Select Style Set drop-down list, select the Paving Label Set and click OK. The alignment labels list populates with the option you selected.

5. Click OK to dismiss the dialog.

6. When you've finished, zoom in on any of the major station labels.

7. Hold down the Ctrl key, and select the label. Notice that a single label is selected, not the label set group.

8. Right-click and select Label Properties.

9. The Label Properties dialog appears, allowing you to pick another label style from the drop-down list.

10. Change the Label Style value to Parallel With Tick, and change the Flip Label value to True, as shown in Figure 19.90.

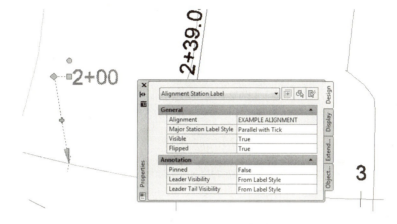

FIGURE 19.90
Modifying a single label's properties through base Auto-CAD properties

11. Press Esc to deselect the label item.

If you add labels to an alignment and like the look of the set, use the Save Label Set option. By using alignment label sets, you'll find it easy to standardize the appearance of labeling and stationing across alignments. Building label sets can take some time, but it's one of the easy, effective ways to enforce standards.

STATION OFFSET LABELING

Beyond labeling an alignment's basic stationing and geometry points, you may want to label points of interest in reference to the alignment. Station offset labeling is designed to do just that. In addition to labeling the alignment's properties, you can include references to other object types in your station-offset labels. The objects available for referencing are as follows:

- Other alignments
- COGO points
- Parcels
- Profiles
- Surfaces

In Chapter 10, you used special alignment labels that referenced other alignments to make adjusting your design easier. In this exercise, you will make a similar type of label. The label you create in the following exercise finds the intersection of two alignments.

1. Open the Advanced Alignment Labels.dwg drawing file.

2. On the Settings tab, expand Alignment ➢ Label Styles ➢ Station Offset.

3. Right-click Station And Offset, and select Copy to open the Label Style Composer dialog.
4. On the Information tab, change the name of your new style to **Alignment Intersection**.
5. Switch to the Layout tab. In the Component Name field, delete the Marker component.
6. In the Component Name field, select the Station Offset component.
7. Change the name to **Main Alignment**.
8. In the Contents Value field, click the ellipsis button to bring up the Text Component Editor.
9. Select the text in the preview area and delete it all.
10. Type **Sta.** in the preview area; be sure to leave a space after the period.
11. In the Properties drop-down field, select Station Value. Set the Precision to **0.01**.

12. Click the insert arrow in the Text Component Editor dialog; press the right arrow on your keyboard to move your cursor to the end of the line, and type one space.
13. In the Properties drop-down field, select Alignment Name.
14. Click the insert arrow to add this bit of code to the preview.
15. Click your mouse in the preview area, or press the right arrow or End key. Move to the end of the line, and type an equal sign (=). Your Text Component Editor should look like Figure 19.91.

FIGURE 19.91
The start of the alignment label style

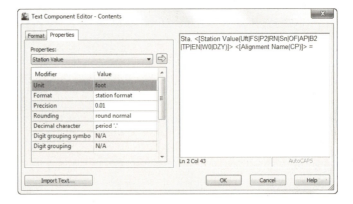

16. Click OK to return to the Label Style Composer dialog.
17. Under the Border Property, set the Visibility field to False.

18. Select Reference Text from the drop-down list next to the Add Component tool.
19. In the Select Type dialog that appears, select Alignment and click OK.
20. Change the name to **Intersecting Alignment**.

21. In the Anchor Component field, select Main Alignment.
22. In the Anchor Point field, select Bottom Left.
23. In the Attachment field, select Top Left. When you choose the anchor point and attachment point in this fashion, the bottom left of the Main Alignment text is linked to the top left of the Intersection Alignment text.
24. Click in the Contents field, and click the ellipsis button to open the Text Component Editor.
25. Delete the generic label text that currently appears in the preview area.
26. Type **Sta.** in the preview area; be sure to leave a space after the period.
27. In the Properties drop-down list, select Station Value.
28. Click the insert arrow in the Text Component Editor dialog, and then move your cursor using the right arrow on your keyboard or End key, and add a space.
29. In the Properties drop-down list, select Alignment Name.
30. Click the insert arrow to add this bit of code to the preview.
31. Click OK to exit the Text Component Editor, and click OK again to exit the Label Style Composer dialog.
32. Add the label to the drawing by selecting the Annotate tab and clicking Add Labels. Change the label settings to match those shown in Figure 19.92 and click Add. Watch the command line for placement instructions. You will be prompted to select the main alignment, intersection point, and offset. You will then be prompted to click the intersecting alignment.

FIGURE 19.92
Adding the new alignment label

33. Click the label to select it and reveal the grips. Select the square grip and drag it away from the current location to form a leader.

Your completed label should look like Figure 19.93.

FIGURE 19.93
The completed alignment label with reference text

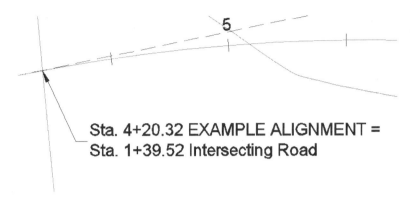

Profile Labels

It's important to remember that the profile and the profile view aren't the same thing. The labels discussed in this section are those that relate directly to the profile. This usually means station-based labels, individual tangent and curve labels, or grade breaks. You'll look at individual label styles for these components and then at the concept of the label set.

Profile Label Sets

As with alignments, you apply labels as a group of objects separate from the profile in the form of profile label sets. In this exercise, you'll learn how to add labels along a profile object:

1. Open the `Profile Labels.dwg` file.

2. Pick the blue layout profile (the profile with two vertical curves) to activate the profile object.

3. Select Edit Profile Labels from the Labels panel to display the Profile Labels dialog (see Figure 19.94).

FIGURE 19.94
An empty Profile Labels dialog

Selecting the type of label from the Type drop-down menu changes the Style drop-down menu to include styles that are available for that label type. Next to the Style drop-down menu are the usual Style Edit/Copy button and a preview button. Once you've selected a style from the Style drop-down menu, clicking the Add button places it on the profile. The middle portion of this dialog displays information about the labels that are being applied to the profile selected; you'll look at that in a moment.

4. Choose the Major Stations option from the Type drop-down menu. The name of the second drop-down menu changes to Profile Major Station Label Style to reflect this option. Set the style to Profile Station Labels in this menu.

5. Click Add to apply this label to the profile.

6. Choose Horizontal Geometry Points from the Type drop-down menu.

7. The name of the Style drop-down menu changes to Profile Horizontal Geometry Point. Select the Horizontal Geometry Station style, and click Add again to display the Geometry Points dialog shown in Figure 19.95. This dialog lets you apply different label styles to different geometry points if necessary.

FIGURE 19.95
The Geometry Points dialog appears when you apply labels to horizontal geometry points.

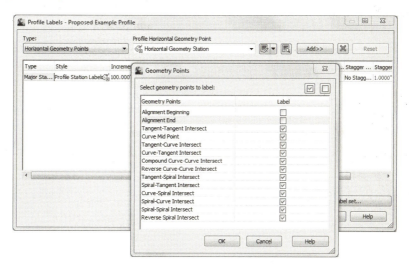

8. Deselect the Alignment Beginning and Alignment End rows, as shown in Figure 19.95, and click OK to close the dialog.

9. Click the Apply button. Drag the dialog out of the way to view the changes to the profile (see Figure 19.96).

10. In the middle of the Profile Labels dialog, change the Increment value in the Major Stations row to **50**, as shown in Figure 19.97. This modifies the labeling increment only, not the grid or other values.

11. Click OK to close the Profile Labels dialog.

FIGURE 19.96
Labels applied to major stations and alignment geometry points

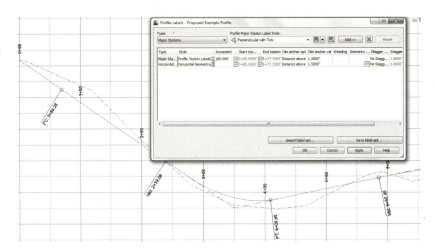

FIGURE 19.97
Modifying the major station labeling Increment

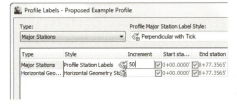

As you can see, applying labels one at a time could turn into a tedious task. After you learn about the types of labels available, you'll revisit this dialog and look at the two buttons at the bottom for dealing with label sets.

Line Labels

Line labels in profiles are typically used to convey the slope or length of a tangent segment. In this exercise, you'll add a length and slope to the layout profile:

1. Continue working in the `Profile Labels.dwg` file.
2. Switch to the Settings tab of Toolspace.
3. Expand the Profile ➢ Label Styles ➢ Line branches.
4. Right-click Percent Grade, and select the New option to open the Label Style Composer dialog and create a child style.
5. On the Information tab, change the name to **Length and Percent Grade**.
6. Switch to the Layout tab and make these changes:
    ◆ Change Attachment to Top Center.
    ◆ Set the Y-offset to **–0.025**.
    ◆ Set Background Mask to True.

7. Click in the cell for the Contents value, and click the ellipsis button to display the Text Component Editor.

8. Change the Properties drop-down menu to the Tangent Slope Length option and the Precision value to **0.01**, as shown in Figure 19.98.

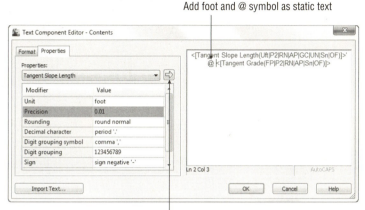

Figure 19.98
The Text Component Editor with the values for the Tangent Slope Length entered

Add foot and @ symbol as static text

Dont' forget the arrow!

9. Click the insert arrow, and then add a foot symbol, a space, an @ symbol, and another space in the editor's preview pane so that it looks like Figure 19.98.

10. Click OK to close the Text Component Editor, and click OK again to close the Label Style Composer dialog.

11. Pick the layout profile and select Edit Profile Labels from the Labels panel to display the Profile Labels dialog.

12. Change the Type field to the Lines option. The name of the Style drop-down menu changes to Profile Tangent Label Style. Select the Length And Percent Grade option.

13. Click the Add button, and then click OK to exit the dialog. The profile view should look like Figure 19.99.

> **Where Is That Distance Being Measured?**
>
> The *tangent slope length* is the distance along the horizontal geometry between vertical curves. This value doesn't include the tangent extensions. There are a number of ways to label this length; be sure to look in the Text Component Editor if you want a different measurement.

FIGURE 19.99
A new line label applied to the layout profile

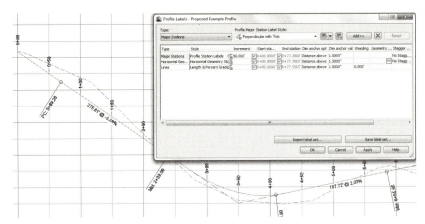

CURVE LABELS

Vertical curve labels are one of the most confusing aspects of profile labeling. Many people become overwhelmed rapidly because there's so much that can be labeled and there are so many ways to get all the right information in the right place. In this quick exercise, you'll look at some of the special label anchor points that are unique to curve labels and how they can be helpful:

1. Open the `Curve Profile Labels.dwg` file.

2. Pick the layout profile and select Edit Profile Labels from the Labels panel to display the Profile Labels dialog.

3. Choose the Crest Curves option from the Type drop-down menu. The name of the Style drop-down menu changes. Select the Crest Only option.

4. Click the Add button to apply the label.

5. Choose the Sag Curves option from the Type drop-down menu. The name of the Style drop-down menu changes to Profile Sag Curve Label Style. Select and add the Sag Only label style.

6. Click OK to close the dialog; your profile should look like Figure 19.100.

Most labels are applied directly on top of the object being referenced. Because typical curve labels contain a large amount of information, putting the label right on the object can yield undesired results. In the following exercise, you'll modify the label settings to review the options available for curve labels:

1. Continue working in the `Curve Profile Labels.dwg` file.

2. Pick the layout profile and select Edit Profile Labels from the Labels panel to display the Profile Labels dialog.

3. Scroll to the right, and change both Dim Anchor Opt values for the Crest and Sag Curves to Graph View Top.

4. Change the Dim Anchor Val for both curves to **–2.25"**, and click OK to close the dialog. Your drawing should look like Figure 19.101.

FIGURE 19.100
Curve labels applied with default Dim Anchor values

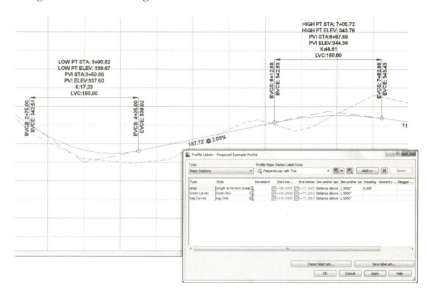

FIGURE 19.101
Curve labels anchored to the top of the graph

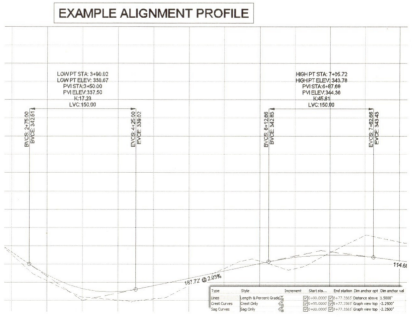

The labels can also be grip-modified to move higher or lower as needed. By using the top or bottom of the graph as the anchor point, you can apply consistent and easy labeling to the curve, regardless of the curve location or size.

THOSE CRAZY CURVE LABELS!

A profile curve label can be as intricate or simple as you desire. Civil 3D gives you many options for where along the curve feature you want your label component to appear. The following illustration shows where some of the commonly used curve locations are in a label.

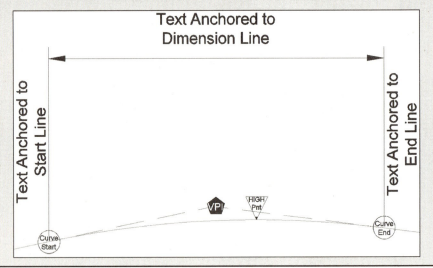

GRADE BREAKS

The last label style typically involved in a profile is a grade-break label at PVI points that don't fall inside a vertical curve, such as the beginning or end of the layout profile. Additional uses include things like water-level profiling, where vertical curves aren't part of the profile information or existing surface labeling. In this exercise, you'll add a grade-break label and look at another option for controlling how often labels are applied to profile data:

1. Open the `Grade Break Labels.dwg` file.
2. Pick the green surface profile (the irregular profile), and then select Edit Profile Labels from the Labels panel to display the Profile Labels dialog.
3. Choose Grade Breaks from the Type drop-down menu. The name of the Style drop-down menu changes. Select the Station Over Elevation style and click the Add button.
4. Click Apply, and drag the dialog out of the way to review the change.

   A sampled surface profile has grade breaks every time the alignment crosses a surface TIN line. Why wasn't your view coated with labels?

5. Scroll to the right, and change the Weeding value to **50'**. Click Apply. The profile labels should appear as shown in Figure 19.102.

FIGURE 19.102
Grade-break labels on a sampled surface

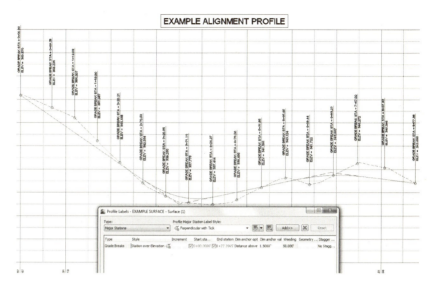

6. Click OK to dismiss the dialog.

Weeding lets you control how frequently grade-break labels are applied. This makes it possible to label dense profiles, such as a surface sampling, without being overwhelmed or cluttering the view beyond usefulness.

As you've seen, there are many ways to apply labeling to profiles, and applying these labels to each profile individually could be tedious. In the next section, you'll build a label set to make this process more efficient.

PROFILE LABEL SETS

Applying labels to both crest and sag curves, tangents, grade breaks, and geometry with the label style selection and various options can be monotonous. Thankfully, Civil 3D gives you the ability to use label sets, as in alignments, to make the process quick and easy. In this exercise, you'll apply a label set, make a few changes, and export a new label set that can be shared with team members or imported to the Civil 3D template. Follow these steps:

1. Open the `Profile Label Sets.dwg` file.

2. Pick the layout profile, and then select Edit Profile Labels from the Labels panel to display the Profile Labels dialog.

3. Click the Import Label Set button near the bottom of the dialog to display the Select Style Set dialog.

4. Select the Complete Label Set option from the drop-down menu, and click OK.

5. Click OK again to close the Profile Labels dialog and see the profile view. The label set you chose contains curve labels, grade-break labels, and line labels.

6. Pick the layout profile and select Edit Profile Labels from the Labels panel to display the Profile Labels dialog.

7. Click Import Label Set to display the Select Style Set dialog.

8. Select the _No Labels option from the drop-down menu, and click OK. All the labels from the listing will be removed.

In the next steps, you will add labels to the listing and save the listing as its own label set for future use.

9. Set the active type to Lines. Set Profile Tangent Label Style to Length & Percent Grade. Click Add.

10. Set the active type to Grade Breaks. Set Profile Grade Break Label Style to Station Over elevation. Click Add.

11. Set the type to Crest Curves. Set Profile Crest Curve Label Style to Crest Only. Click Add.

12. Set the type to Sag Curves. Set Profile Sag Curve Label Style to Sag Only. Click Add.

13. Set the Crest Curve and Sag Curve label types to use the Graph View Top as the Dim Anchor Opt. Set both Dim Anchor Val fields to **–1.5"**, as shown in Figure 19.103.

FIGURE 19.103
Four label types and dimension anchor settings in the label set to be saved

14. Click the Save Label Set button to open the Profile Label Set dialog and create a new profile label set.

15. On the Information tab, change the name to **Road Profile Labels**. Click OK to close the Profile Label Set dialog.

16. Click OK to close the Profile Labels dialog.

17. On the Settings tab of Toolspace, select Profile ➢ Label Styles ➢ Label Sets. Note that the Road Profile Labels set is now available for sharing or importing to other profile label dialogs.

Label sets are the best way to apply profile labeling uniformly. When you're working with a well-developed set of styles and label sets, it's quick and easy to go from sketched profile layout to plan-ready output.

Advanced Style Types

Now that you are familiar with the basics of object styles and label styles, you are ready to take your skills to the next level. The styles in the following section combine aspects of object styles and label styles.

You have a great deal of control over every detail, even ones that may seem trivial. Instead of being bogged down trying to understand every option, don't be afraid to try a "trial and error" approach. If you make a change you don't like, you can always edit the style until you get it right.

Table Styles

Civil 3D does a beautiful job of placing dynamically linked data tables that relate to your objects. The tables use the Text Component Editor to grab dynamic information from your objects. You also have control over fill colors, table headings, and how the data is sorted.

For the table style, the Data Properties tab contains all the column information. You can add columns by clicking the plus sign. You can remove columns by highlighting the column you want to remove and clicking the Delete button. Change column order by dragging them around and dropping them where you want them to go, as shown in Figure 19.104.

Figure 19.104
Modifying table columns

In the following exercise you will see the basic steps of modifying a table style. Now that you understand the ins and outs of the Label Style Composer, this procedure should be a breeze:

1. Open the `Tables.dwg` file. This file contains a parcel line table whose style you will modify.

2. Zoom into the parcel line table.

   Notice there are several things you will want to change. The line numbers are out of order, the directions have far too much precision, and the length does not display units. All that is about to change.

3. Click the table to select it. From the context tab, select Table Properties ➢ Edit Table Style.

4. Switch to the Data Properties tab.

   The Data Properties tab (Figure 19.105) is the main control area for all table styles. This is where you set behavior, text styles, and sizes for fields.

FIGURE 19.105
Data Properties tab for table styles

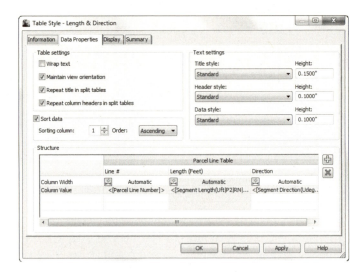

5. Place a check mark next to Sort Data. Set the sorting column to 1. Column 1 corresponds to the Column Value containing the Parcel Line Number. Set the order option to Ascending. The Ascending option will ensure that the parcel numbers are listed in the table from the lowest to highest value.

6. Double-click the column heading field for Length. Doing so opens a stripped-down version of the Text Component Editor. Headings and table titles are static text only; therefore only the text formatting tools are shown. Add **(Feet)** after the column heading to add a proper units heading to the column, as shown in Figure 19.106. Click OK.

FIGURE 19.106
Adding static text to a table column heading

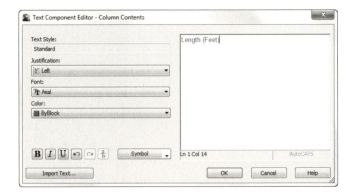

7. Double-click the Direction Column value field. Doing so opens the Text Component Editor, similar to what you've used in earlier exercises. Click on the text in the preview area so that it becomes highlighted. Change Precision to **1 Second**. Click the arrow to update the text. Click OK.

8. Click OK to complete the table style modifications.

Profile View Styles

When you are looking at a profile view that contains data, you are seeing many styles in play. The profiles themselves (existing and proposed) have a profile object style applied to them. The labeling has several styles applied to it, as you learned earlier in this chapter. Additionally, you will see profile view styles and band styles in action.

This section focuses on the profile view. A profile view controls many aspects of the display. The *profile view style* has a large role in determining:

- Vertical exaggeration
- Grid spacing
- Elevation annotation
- Title annotation

Figure 19.107 shows the basic anatomy of the profile view you have been working with throughout this book.

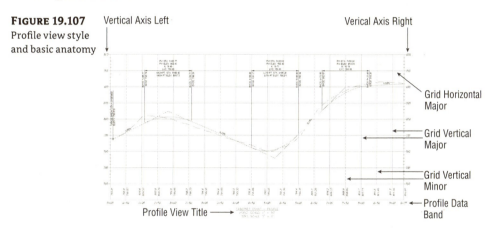

FIGURE 19.107
Profile view style and basic anatomy

In the example that follows, you will be making major modifications to a profile view style. The profile view that you will be practicing with does not contain any bands. Later on in this section, you will learn the ins and outs of band creation and modification.

1. Open the `Profile View Styles.dwg` file. This file contains a profile view with a very ugly style applied to it. You will be performing a complete makeover on this style.

2. Select the profile view by clicking anywhere on a grid line or axis. From the context tab, select Profile View Properties ➢ Profile View Style, as shown in Figure 19.108.

   Position the dialogs on your screen so you can make changes to the style and observe the changes in the profile view behind it.

3. On the Graph tab, change the Vertical exaggeration to **10**, as shown in Figure 19.109. Click Apply. If you can see your profile view in the background, you may be alarmed at the change. Don't worry—you are not even close to done!

FIGURE 19.108
Accessing the profile view style

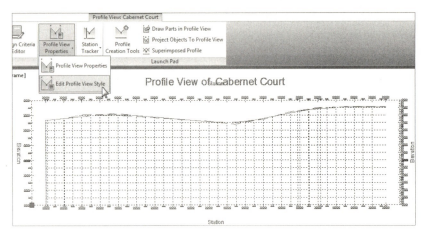

FIGURE 19.109
Change vertical exaggeration in the Graph tab of the Profile View Style dialog.

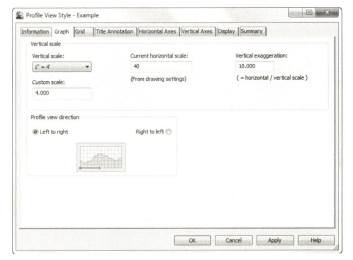

4. Switch to the Grid tab. Uncheck both Clip Vertical Grid and Clip Horizontal Grid options. Click Apply to observe the change. The vertical and horizontal grid lines will now extend to the entire length of the axes.

5. Change Grid Padding to **0.5** for Above Maximum Elevation and **0.5** for Below Datum. This setting will create additional space above and below the design data at least 0.5 times the vertical major tick interval (you will set the major tick interval later on). Click Apply to examine your changes.

6. Set all values for Axis offset to **0**. This will ensure that the axes and the grid lines coincide around the edges of the view. Click Apply to examine your changes. The settings on the grid tab should now match what is shown in Figure 19.110.

7. Switch to the Title Annotation tab. You will be working on the left side of this dialog only. Change the Text height to **0.4″**.

8. Click the Edit Text icon. In the Text Component Editor, remove all the text in the preview area. We will be starting over with a blank Text Component Editor.

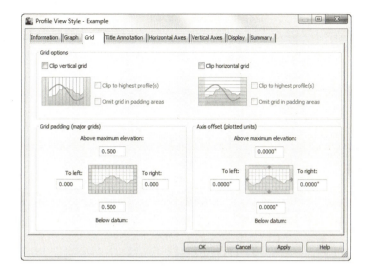

FIGURE 19.110
The Grid tab: so many options!

From the Properties menu, select Parent Alignment and make these changes:

- Set Capitalization to Upper Case.
- Click the arrow to add the text to the preview.
- After the dynamic text, add a space and type **PROFILE**.
- Click OK to dismiss the Text Component Editor.

9. Change the Y offset for the title position to **0.5**. Click Apply to examine your changes. If you do not see the changes, click OK and return to the dialog per Step 2. The Title Annotation tab should now match the settings shown in Figure 19.111.

   You will not bother to adjust any settings for the Axis title text on the right side of Figure 19.111 because the display will be turned off for all four of these possible elements.

10. Switch to the Horizontal Axes tab. Make sure the Axis To Control radio button is set to Bottom.

11. Under Major Tick Details, set Interval to **100′**.

12. For Tick Label Text, click the Edit Text icon. Highlight the dynamic station text that already exists in the preview by clicking it. Change the precision to **1**. Click the arrow to update the text. Click OK to exit the Text Component Editor.

13. Still under Major Tick Details, change Rotation to **90**. Change the X offset to **−0.05″** and the Y offset to **−0.25″**.

14. Under Minor Tick Details, set Interval to **50′**. No other changes are needed in the Minor Tick Details area. Click Apply to examine your changes. The Horizontal Axes tab should match the settings shown in Figure 19.112.

FIGURE 19.111
Working with the graph view title size and placement

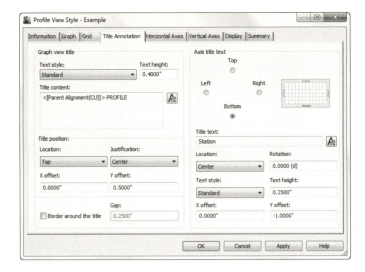

FIGURE 19.112
The bottom axis controls grid spacing.

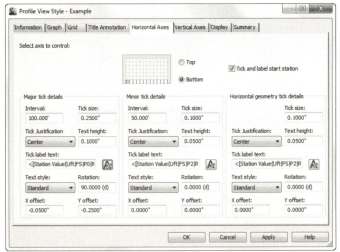

15. Switch to the Vertical Axes tab. With the Select Axis To Control radio button set to Left, click the Edit Text icon for the major tick. In the Text Component Editor:

    ◆ Highlight the dynamic Profile View Point Elevation text by clicking on it.

    ◆ Change Precision to **1**.

    ◆ Click the arrow to update the text.

    ◆ Click OK to exit the Text Component Editor.

16. Click the right radio button on the Vertical Axes tab to make it active. Click the Edit Text icon for the major tick. In the Text Component Editor:

    ◆ Highlight the dynamic Profile View Point Elevation text by clicking on it.

    ◆ Change Precision to **1**.

    ◆ Click the arrow to update the text.

    ◆ Click OK to exit the Text Component Editor.

17. Click Apply to examine your changes. Figure 19.113 shows the Vertical Axes tab as yours should look at this point in the exercise (the tab has been stretched out to show the Tick Label Text field contents).

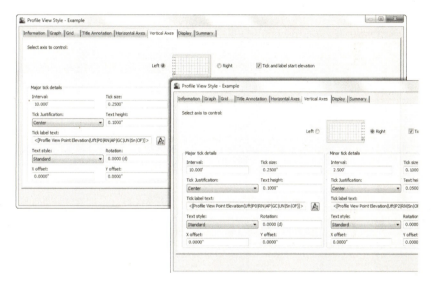

FIGURE 19.113 Don't forget to change the settings for both left and right axes in this tab.

18. Switch to the Display tab. For this illustration, you will not need to set the layers. Turn off the following components:

    ◆ Left Axis Title

    ◆ Left Axis Annotation Minor

    ◆ Left Axis Ticks Minor

    ◆ Right Axis Title

    ◆ Right Axis Annotation Minor

- Right Axis Ticks Minor
- Top Axis Title
- Top Axis Annotation Major
- Top Axis Annotation Minor
- Top Axis Ticks Major
- Top Axis Ticks Minor
- Bottom Axis Title
- Bottom Axis Annotation Minor
- Grid At Horizontal Geometry Point
- Top Axis Annotation Horizontal Geometry
- Top Axis Ticks Horizontal Geometry
- Bottom Axis Annotation Horizontal Geometry
- Bottom Axis Ticks Horizontal Geometry
- Grid At Sample Line Stations

19. Click OK to complete the profile view style. Your profile view should resemble Figure 19.114.

FIGURE 19.114
Profile view after the style is completed

WHAT DRIVES PROFILE VIEW GRID SPACING?

On the Horizontal Axes tab and Vertical Axes tab of the Profile View Style dialog you may notice that you can control opposing axes separately. Each tab has a toggle for Select Axis To Control; Top or Bottom and Left or Right.

It is the Bottom and Left options in the respective tabs that control grid spacing. For both axes, you will find options for Major Tick Details. The Interval values for the major ticks are the key to getting the grid spacing to look the way you want. Changes to either horizontal or vertical major tick intervals will affect the height and length of the Profile view, as well as grid spacing. Changing the Interval on Minor Tick Details will affect the grid spacing, but will not affect the aspect ratio of the profile view.

A good rule of thumb is to set your bottom horizontal major tick Interval value equal to the view's Vertical Exaggeration multiplied by the left vertical axis Interval value. For example, if you set the Vertical Exaggeration for the view to 10 on the Graph tab and your left vertical axis major tick Interval is 10′, the horizontal major tick Interval would be 100′.

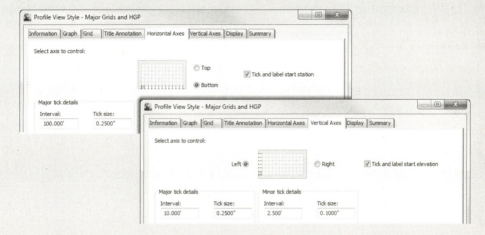

Even if you don't turn on the ticks or gridlines on these axes, the spacing increment will be reflected in the profile view.

PROFILE VIEW BANDS

Data bands are horizontal elements that display additional information about the profile or alignment that is referenced in a profile view. The most common band type is the profile data band. The other band types are not frequently used but can be helpful to designers by showing schematic representations of design as it relates to the profile stationing.

Bands can be applied to both the top and bottom of a profile view, and there are six band types:

Profile Data Bands   Display information about the selected profile. This information can include simple elements such as elevation, or more complicated information such as the cut-fill between two profiles at the given station. Figure 19.115 shows a typical profile data band.

FIGURE 19.115
Profile data band showing existing and proposed elevation in addition to major stations

Vertical Geometry Bands Create an iconic view of the elements making up a profile. Typically used in reference to a design profile, vertical data bands make it easy for a designer to see where vertical curves are located along the alignment. Figure 19.116 shows a typical vertical geometry data band.

FIGURE 19.116
Vertical geometry schematic shown in band

Horizontal Geometry Bands Create a simplified view of the horizontal alignment elements, giving the designer or reviewer information about line, curve, and spiral segments and their relative location to the profile data being displayed. Figure 19.117 shows a typical horizontal schematic data band.

FIGURE 19.117
Horizontal geometry shown as schematic in a data band

Superelevation Bands Display the various options for Superelevation values at the critical points along the alignment. Figure 19.118 shows an example of a superelevation data band.

FIGURE 19.118
Superelevation data in band form

Sectional Data Bands Displays information about the sample line locations, the distance between them, and other sectional-related information. In Figure 19.119 you can see a section data band showing sample line data.

FIGURE 19.119
Section data shown in a band

Pipe Data Bands Shows specific information about each pipe or structure being shown in the profile view. Figure 19.120 shows pipe data band with slopes and invert elevations.

FIGURE 19.120
Pipe data invert elevations and slope schematic in a band

Bands can be saved in *band sets*. Like alignment label sets and profile label sets, a band set determines which bands are applied to a profile view and will dictate the placement. The most common use for a band set is to create a single, viewable band and an additional nonvisible band for spacing purposes in plan and profile sheet generation.

In the following exercise, you will create a band that contains existing and proposed profile elevations:

1. Open the `Profile Bands.dwg` file. This file contains the profile view from the previous exercise and an empty band on the bottom of the view. You will be adding information to the profile data band.

2. From the Settings tab of Toolspace, select Profile View ➢ Band Styles ➢ Profile Data. Right-click the style called Example and select Edit.

3. On the Band Details tab, change Band Height to **1.0"**.

4. On the right side of the Band Details tab, highlight Major Station and click the Compose Label button, as shown in Figure 19.121.

FIGURE 19.121
Band Details tab

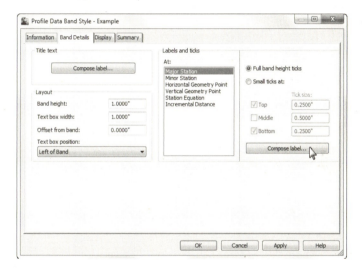

5. On the Layout tab, click Add New Text Component:

- Change the name to Station.
- Change Anchor Point to Band bottom.
- Change Attachment to Top Center.
- Change the Y offset to **−0.02"**.
- Enter the Text Component Editor by clicking on the field that currently reads Label Text. Click the ellipsis.
- In the Text Component Editor, remove Label Text and select Station Value from the Properties list.

- Set Precision to **1**.
- Click the arrow to add the dynamic text to the Preview window.
- Click OK to exit the Text Component Editor.

6. Click Add New Text Component again and make these changes:
   - Change the name to **Existing El**.
   - Change Anchor Point to Band Middle.
   - Set the rotation angle to **90**.
   - Change Attachment to Bottom Center.
   - Change the X offset to **–0.02"**.
   - Enter the Text Component Editor by clicking on the field that currently reads Label Text. Click the ellipsis.
   - In the Text Component Editor, remove Label Text and select Profile 1 Elevation from the Properties list.
   - Verify the Precision is **0.01**.
   - Click the arrow to add the dynamic text to the preview window.
   - Click OK to exit the Text Component Editor.

7. Click the Copy Text Component button and make these changes:
   - Change the name to **Proposed El**.
   - Change Attachment to Top Center.
   - Set the X offset to **0.02"**. Enter the Text Component Editor by clicking on the field that currently contains Profile 1 dynamic text. Click the ellipsis.
   - In the Text Component Editor, remove the Profile 1 text and select Profile 2 Elevation from the Properties list.
   - Verify the Precision is **0.01**.
   - Click the arrow to add the dynamic text to the preview window.
   - Click OK to exit the Text Component Editor.

8. Click OK to finish working with the Major Stations and return to the Band Details tab.
9. Switch to the Display tab. Locate the Minor Tick component and turn it off.
10. Click OK. Select the profile and select Profile View Properties from the context tab.
11. On the Bands tab, scroll over until you see the column for Profile 2. Set the value for Profile 2 to **Cabernet Court FG**. Click OK. The completed band should resemble Figure 19.122.

FIGURE 19.122
Text along the bottom of your profile view in the form of a band

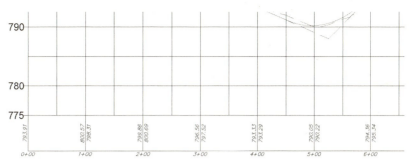

Section View Styles

Section view styles share many of the same concepts as creating Profile view styles. In fact, the Section View Style dialog has all the same tabs and looks nearly identical to the Profile View style.

In this section, you will walk through the creation of a section view style suitable for creating a section sheet:

1. Open the `Section Styles.dwg` file. This file contains section views created with the default settings for section views.

2. From the Settings tab of Toolspace, select Section View ➢ Section View Style ➢ Road Section. Right-click on Road Section and select Edit.

3. On the Grid tab, set the Grid padding settings to **0** for the Above Maximum Elevation and Below Datum options.

4. On the Display tab, turn off all components except:

    ◆ Graph Title

    ◆ Left Axis Annotation Major

    ◆ Right Axis Annotation Major

    ◆ Bottom Axis Annotation Major

5. Click OK. Your section views should resemble Figure 19.123.

6. Select one of the views by clicking on the station label. From the context tab, select Update Group Layout. The section views will rearrange to fit up to six per page.

FIGURE 19.123
Yes, this is correct! A very stripped-down section view.

Why the bare-bones section view? The section view grid will come from the group plot style, so the only information we really need is in this simple style.

Group Plot Styles

The driver behind how section sheets get laid out on a page is the *group plot style*. Upon section creation, the group plot style takes the Section template file and fits sections inside the paper-space viewport:

1. Continue working in the `Section Styles.dwg` file. (If you did not successfully complete the previous exercise, you can start with `Section Styles-2.dwg`.) This file contains section views created with the default settings.

2. From the Settings tab of Toolspace, select Section View ➢ Group Plot Styles. Right-click on Basic and select Edit.

3. On the Array tab, change the column spacing to **4"**. Change the row spacing to **2"**, as shown in Figure 19.124.

   These spacing changes will allow more cross sections per page and additional room for scooting the views up and right to fit on the page better in the upcoming steps.

4. Switch to the Plot Area tab for observation purposes. Leave all the settings as shown in Figure 19.125. This is where the grid spacing we will be using on our sheets comes from.

5. Switch to the Display tab. Under Component Display, turn on Major Horizontal Grid and the Major Vertical Grid. The components Minor Horizontal Grid and Minor Vertical Grid should be turned off. Set Color for the major grids to **9**. Set Print Area to Cyan and Sheet Border to Green. Your Display tab will look like Figure 19.126.

Figure 19.124
The Array tab controls section view spacing

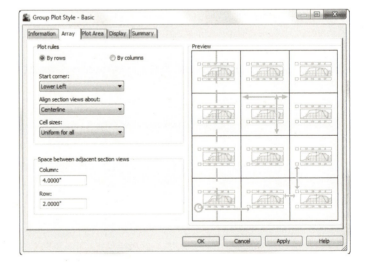

ADVANCED STYLE TYPES | 829

FIGURE 19.125
The Plot Area tab is where grid spacing on sheets comes from.

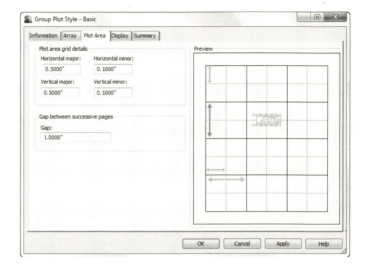

FIGURE 19.126
Group plot style display components

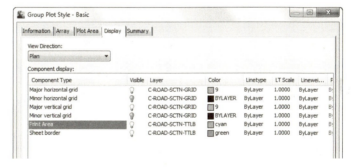

6. Click OK to complete the group plot style edits.

   Your section view sheets should be shaping up to the point where you could almost generate sheets. However, there is some text oozing out of the cyan line that represents the viewport border. In the next steps you will use a nonvisible section band to push the views onto the page.

7. Select any section view by clicking on the station label or the elevation labels. Select View Group Properties from the context tab. Click the ellipsis in the Change Band Set column, as shown in Figure 19.127.

FIGURE 19.127
Changing the band set in use for all section views

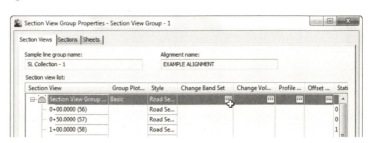

8. Verify the band type is set to Section Data. Set the band style as _NO DISPLAY. Click Add. Set the Gap distance to **0**, as shown in Figure 19.128. Click OK to dismiss the Section View Group Bands dialog, and click OK to exit the View Group Properties dialog.

FIGURE 19.128
Add the data band and set the gap to 0.

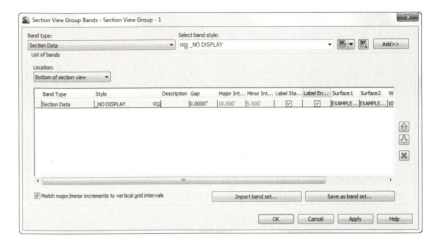

9. Click Update Group Layout. The views will reorganize and form only one column per page. Don't worry—next you will edit the _NO DISPLAY band style to fix this.

10. From the Settings tab of Toolspace, choose Section View ➢ Band Styles ➢ Section Data. Right-click on _NO DISPLAY and select Edit.

11. On the Band Details tab, change the band height to **0.2"**. Change the text box width to **0.2"**. The tab will resemble Figure 19.129.

FIGURE 19.129
Setting the band box size as a spacer

Even though the band is not visible, Civil 3D still accounts for this spacing when placing the views on the sheet. In this step, you are using this to your advantage.

12. Click OK to finish modifying the Data Band style.

13. Click the Update Group Layout button one last time. The cross section sheets should look like Figure 19.130.

FIGURE 19.130
The completed exercise

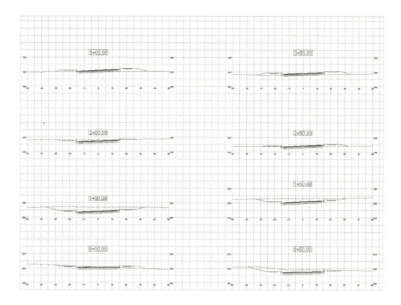

Code Set Styles

Code set styles determine how your assembly design will appear. Code set styles are used in many places. A code set style is in play when you first create your assembly. One is used in corridor creation and in the section editor. The most apparent use of a code set style is in section views like those that you created in the previous exercise.

Code set styles are a collection of many other styles. In a code set style you will find:

- Links and link label styles
- Points, point label styles, and feature line styles
- Shapes and shape label styles
- Quantity takeoff pay items
- Render materials for visualization tasks

It is helpful in naming your code set style to have the name of the set reflect its use. Multiple code set styles are needed because of different applications of its use. When you are designing an assembly, you may want to see more labels than when you are getting ready to plot the assembly in a cross-section sheet. Labeling that is useful in a cross-section sheet may obstruct your view of the design when working with it in the corridor cross-section editor.

The hardest part of working with code set styles is figuring out the name of the link or point you wish to label. Luckily, it is unusual for users to label shapes, so you won't need to worry about those. The names of each point or link can be found in the subassembly properties. Most of the links and points are logically named, but there's no harm in a little trial and error if you are not sure.

Shapes

Shapes are the areas that define materials. Because it is not common for people to label these materials in a section view, you will not be experiencing these in an exercise.

One heads-up, however: Resist the temptation to use a hatch pattern on shapes where multiple cross section views will be created. Solid fills and no patterns are your best bet to avoid performance issues and the annoying "Hatch pattern is too dense" warning.

Links and Link Labels

You learned in Chapter 8, "Assemblies and Subassemblies," that a link is the linear part of a subassembly. The object style for the link itself is very simple; just a single linear component. The label for a link is usually expressed as a percent grade or as a slope ratio.

In the following exercise, you will modify a code set style to apply link labels to an assembly.

1. Open the `Code Set Styles1.dwg` file. This file contains a corridor and cross-section views. Zoom into one of the cross section views so you can observe the changes as you apply them to the code set style.

2. From the Settings tab of Toolspace, choose General ➢ Code Set Styles. Right-click on All Codes and select Edit.

3. In the Link list, locate Pave. Click the Label Tag icon in the Label Style column, as shown in Figure 19.131.

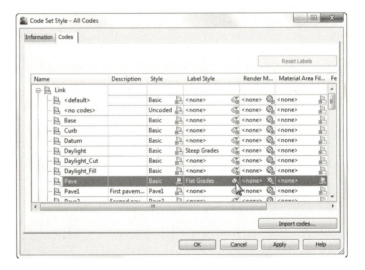

Figure 19.131
Adding labels to the link codes in the code set style

4. Select the Flat Grades label style and click OK. Click Apply to examine the change on the cross sections. You should see that the lanes now have slope information labeled.

5. In the Link list, locate Daylight. Click the Label Tag icon in the Label Style column and select Steep Grades as the label style. Click OK. Click OK to dismiss the All Codes label style and see what is happening with the cross section. The cross sections should resemble Figure 19.132.

FIGURE 19.132
Cross section with link labels applied to pave and daylight links

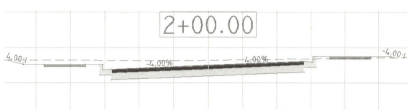

Points and Point Labels

A common frustration with new users of Civil 3D are the marker styles and their labels. For display in cross section views, you may not want points to display at all. In the following exercise, you will create a new code set style, modify point codes, and add more labels to the sections:

1. Continue working in the `Code Set Styles1.dwg` file or open `Code Set Styles2.dwg`.

2. From the Settings tab of Toolspace, choose General ➢ Code Set Styles. Right-click on All Codes and select Copy.

3. On the Information tab, rename the style to **All Codes-Plotting**.

4. On the Codes tab, locate the Points category. Locate the Back_Curb point and click the Tag icon in the Label Style column. Set the style to Offset Elevation. Click OK.

5. Repeat step 4 to set the label style for Sidewalk_Out to Offset Elevation.

6. Click the first point name, <default>. While holding down the Shift key on your keyboard, scroll down to the last point listing, Top_Curb. With all of the points selected, click the Tag icon and change the style to _No Markers. The Code Set Style dialog should resemble Figure 19.133.

7. Click OK. You do not see any changes to your cross sections yet because the style is not active.

FIGURE 19.133
Points set to _No Markers and labels set to Offset-Elevation

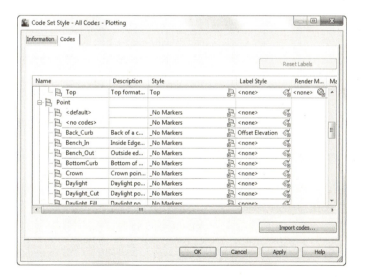

8. Select a section view and select View Group Properties from the context tab.

9. Change the style of the corridor to the All Codes-Plotting style you created, as shown in Figure 19.134 and click OK.

FIGURE 19.134
Setting the code set style current on the section views

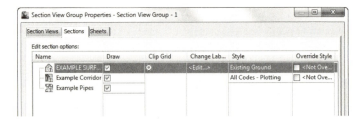

Your section view should now resemble Figure 19.135.

FIGURE 19.135
New code set style applied to the section view

The Bottom Line

Override individual labels with other styles. In spite of the desire to have uniform labeling styles and appearances between alignments within a single drawing, project, or firm, there are always exceptions. Using AutoCAD's Ctrl+click method for element selection, you can access commands that let you modify your labels and even change their styles.

Master It Open a new drawing based on the template of your choice. Create a copy of the Perpendicular With Tick Major Station style called Major With Marker. Change Tick Block Name to Marker Pnt. Replace some (but not all) of your major station labels with this new style.

Create a new label set for alignments. Label sets let you determine the appearance of an alignment's labels and quickly standardize that appearance across all objects of the same nature. By creating sets that reflect their intended use, you can make it easy for a designer to quickly label alignments according to specifications with little understanding of the requirement.

Master It Within the `Mastering Alignment Styles.dwg` file, create a new label set containing only major station labels, and apply it to all the alignments in that drawing.

Apply a standard label set to profiles. Standardization of appearance is one of the major benefits of using Civil 3D styles in labeling. By applying label sets, you can quickly create plot-ready profile views that have the required information for review.

Master It In the `Mastering Profile Styles.dwg` file, apply the Road Profiles label set to all layout profiles.

Appendix A

The Bottom Line

Each of The Bottom Line sections in the chapters suggests exercises to deepen skills and understanding. Sometimes there is only one possible solution, but often you are encouraged to use your skills and creativity to create something that builds on what you know and lets you explore one of many possible solutions.

Chapter 1: The Basics of AutoCAD Civil 3D

Find any Civil 3D object with just a few clicks. By using Prospector to view object data collections, you can minimize the panning and zooming that are part of working in a CAD program. When common subdivisions can have hundreds of parcels or a complex corridor can have dozens of alignments, jumping to the desired one nearly instantly shaves time off everyday tasks.

Master It Open BasicSite.dwg from www.sybex.com/go/masteringcivil3d2012, and find parcel number 18 without using any AutoCAD commands or scrolling around on the drawing screen.

Solution

1. In Prospector, expand Sites ➤ Proposed ➤ Parcels.
2. Right-click on Property:18 and select Zoom To.

Modify the drawing scale and default object layers. Civil 3D understands that the end goal of most drawings is to create hard-copy construction documents. By setting a drawing scale and then setting many sizes in terms of plotted inches or millimeters, Civil 3D removes much of the mental gymnastics that other programs require when you're sizing text and symbols. By setting object layers at a drawing scale, Civil 3D makes uniformity of drawing files easier than ever to accomplish.

Master It Change BasicSite.dwg from the 100-scale drawing to a 40-scale drawing.

Solution

1. Click on the Layout1 tab.
2. Select 1'=40 from the Scale list in the lower right of the application window.
3. Type **REA**↵ to regenerate and show the labels at the new scale.

Modify the display of Civil 3D tooltips. The interactive display of object tooltips makes it easy to keep your focus on the drawing instead of an inquiry or report tools. When too many objects fill up a drawing, it can be information overload, so Civil 3D gives you granular control over the heads-up display tooltips.

> **Master It** Within the same BasicSite drawing, turn off the tooltips for the Road A alignment.
>
> **Solution** Click the alignment in the drawing window or in Prospector, and bring up the Alignment properties. If the Properties palette method is used, select No from Show Tooltips in the Design tab. If the Alignment Properties dialog is used, select the Information tab and deselect the Show Tooltips check box.

Navigate the Ribbon's contextual tabs. As with AutoCAD, the Ribbon is the primary interface for accessing Civil 3D commands and features. When you select an AutoCAD Civil 3D object, the Ribbon displays commands and features related to that object. If several object types are selected, the Multiple contextual tab is displayed.

> **Master It** Using the Ribbon interface, access the Alignment Style Editor for the Proposed Alignment style. (Hint: it's used by the Road A alignment.)
>
> **Solution**
>
> 1. Select the Road A alignment to display the contextual Alignment Ribbon.
> 2. In the Alignment Properties menu, select the Information tab. Click the drop-down list next to the Proposed object style and select Edit Current Selection.

Create a curve tangent to the end of a line. It's rare that a property stands alone. Often, you must create adjacent properties, easements, or alignments from their legal descriptions.

> **Master It** Create a curve tangent to the end of the first line drawn in the first exercise that meets the following specifications:
>
> Radius: 200.00'
>
> Arc Length: 66.580'
>
> **Solution**
>
> 1. Select Lines/Curves ➢ Create Curves ➢ Curve From End Of Object.
> 2. Select the first line drawn (N 57° 06' 56.75' E; 135.441').
> 3. On the command line, press ↵ to confirm the radius.
> 4. On the command line, type **200.00**, and then press ↵.
> 5. Type **L** to specify the length, and then press ↵.
> 6. Type **66.580**, and then press ↵.

Label lines and curves. Although converting linework to parcels or alignments offers you the most robust labeling and analysis options, basic line- and curve-labeling tools are available when conversion isn't appropriate.

> **Master It** Add line and curve labels to each entity created in the exercises. Choose a label that specifies the bearing and distance for your lines and length, radius, and delta of your curve.

Solution

1. Change to the Annotate tab.

2. From the Labels & Tables panel, select Add Labels ➢ Line And Curve ➢ Add Multiple Segment Line/Curve Labels.

3. Choose each line and curve. The default label should be acceptable. If not, select a label, right-click and choose Label Properties. In the resulting AutoCAD properties box, you can select an alternative label.

Chapter 2: Survey

Properly collect field data and import it into Civil 3D. You learned best practices for collecting data, how the data is translated into a usable format for the survey database, and how to import that data into a survey database. You learned what commands draw linework in a raw data file, and how to include those commands into your data collection techniques so that the linework is created correctly when the field book is imported into the program.

Master It In this exercise, you'll create a new drawing and a new survey database and import the `MASTER_IT_C3.txt` file into the drawing. The format of this file is PNEZD (comma delimited).

Solution

1. Create a new drawing using a template of your choice.

2. On the Survey tab, create a new survey database.

3. Create a new network in the newly created survey database.

4. Import the `MASTER_IT_C3.txt` file and edit the options to insert both the figures and the points.

Set up description key and figure databases. Proper setup is key to working successfully with the Civil 3D survey functionality.

Master It In this exercise, you'll create a figure prefix database and a description key set.

Create a new description key set and the following description keys using the default styles. Make sure all description keys are going to V-Node:

- CL*
- EOP*
- TREE*
- BM*

Create a figure prefix database called MasterIt containing the following codes:

- CL
- EOP

Solution

1. Open a new drawing based on a Civil 3D template or continue working in the drawing from the previous exercise.

2. On the Settings tab, locate the description key area. Right-click description key sets and select New.

3. Add the description keys, including the asterisk as shown in the list.

4. Set the layer for each item in the table to V-NODE. Close the description key table and save the drawing.

5. In the Survey tab, right-click Figure Prefix Databases and select New. Create a new figure prefix database called `MasterIt`.

6. Add the CL and EOP codes to the list. Leave all options as default.

7. Import the file `Codetest.txt`. Your file should now contain linework that reflects your efforts.

Create and edit field book files. You learned how to create field book files using various data collection techniques and how to import the data into a survey database.

Master It In this exercise, you'll create a new drawing and survey database. Translate the database based on:

- Base Point **1**
- Rotation Angle of **10.3053°**

Solution

1. Create a new drawing and survey database and import the `traverse.fbk` file into a network.

2. Create construction lines to determine your reference angle.

3. Using the Translate Survey Network command, rotate the network.

4. Export a new FBK file.

5. Create another new drawing and survey database and import the new FBK file. Your drawing should look like `mastering2.dwg`, which you can download from this book's webpage, www.sybex.com/go/masteringcivil3d2012.

Manipulate your survey data. You learned how to use the traverse analysis and adjustments to create data with a higher precision.

Master It In this exercise, you'll use the survey database and network from the previous exercises in this chapter. You'll analyze and adjust the traverse using the following criteria:

- Use the Compass Rule for Horizontal Adjustment.
- Use the Length Weighted Distribution Method for Vertical Adjustment.
- Use a Horizontal Closure Limit of 1:25,000.
- Use a Vertical Closure Limit of 1:25,000.

Solution

1. Open a new drawing, create a new survey database and network, and import the `traverse.fbk` field book.
2. Create a new traverse from the four points.
3. Perform a traverse analysis on the newly created traverse, and apply the changes to the survey database.
4. You can find the results from the traverse analysis in the following files:
   - `batrav.trv.txt`
   - `fvtrav.trv.txt`
   - `antrav.trv.txt`
   - `trav.lso.txt`

Chapter 3: Points

Import points from a text file using description key matching. Most engineering offices receive text files containing point data at some point during a project. Description keys provide a way to automatically assign the appropriate styles, layers, and labels to newly imported points.

Master It Create a new drawing from `_AutoCAD Civil 3D (Imperial) NCS.dwt`. Revise the Civil 3D description key set to use the parameters listed here:

CODE	POINT STYLE	POINT LABEL STYLE	FORMAT	LAYER
GS	Basic	Elevation Only	Ground Shot	V-NODE
GUY	Guy Pole	Elevation and Description	Guy Pole	V-NODE
HYD	Hydrant (existing)	Elevation and Description	Existing Hydrant	V-NODE-WATR
TOP	Basic	Point#-Elevation-Description	Top of Curb	V-NODE
TREE	Tree	Elevation and Description	Existing Tree	V-NODE-TREE

Import the `Concord_PNEZD_SpaceDelim.txt` file from the data location, and confirm that the description keys made the appropriate matches by looking at a handful of points of each type. Do the trees look like trees? Do the hydrants look like hydrants?

Save the resulting file for use in the next exercises.

Solution

1. Select File ➢ New, and create a drawing from `_AutoCAD Civil 3D (Imperial) NCS Extended.dwt`.
2. Switch to the Settings tab of Toolspace, and locate the description key set called Civil 3D. Right-click this set and choose Edit Keys.

3. Delete any two keys in this set by right-clicking each one and choosing Delete.

4. Revise the remaining key to match the GS specifications listed under the "Master It" instructions for this exercise.

5. Right-click the GS key, and choose New. Create the four additional keys listed in the instructions. Exit the dialog.

6. On the Create Ground Data panel, select Points ➢ Point Creation Tools and then click the Import Points button on the toolbar. Navigate out to the `Concord_PNEZD_SpaceDelim.txt` file and click Open. Select PNEZD Space Delimited from the listing. Click OK.

7. Zoom in to see the points. Note that each description key parameter (style, label, format, and layer) has been respected. Your hydrants should appear as hydrants on the correct layer, your trees should appear as trees on the correct layer, and so on.

Create a point group.   Building a surface using a point group is a common task. Among other criteria, you may want to filter out any points with zero or negative elevations from your Topo point group.

Master It   Create a new point group called Topo that includes all points *except* those with elevations of zero or less. Use the DWG created in the previous exercise or start with `Master_It.dwg`.

Solution

1. In Prospector, right-click Point Groups and choose New.

2. On the Information tab, enter **Topo** as the name of the new point group.

3. Switch to the Exclude tab.

4. Click the Elevation check box to turn it on and enter **=<0** in the field.

5. Click OK to close the dialog.

Export points to LandXML and ASCII format.   It's often necessary to export a LandXML or an ASCII file of points for stakeout or data-sharing purposes. Unless you want to export every point from your drawing, it's best to create a point group that isolates the desired point collection.

Master It   Create a new point group that includes all the points with a raw description of TOP. Export this point group via LandXML and to a PNEZD comma-delimited text file.

Use the DWG created in the previous exercise or start with `Master_It.dwg`.

Solution

1. In Prospector, right-click Point Groups and choose New.

2. On the Information tab, enter **Top of Curb** as the name of the new point group.

3. Switch to the Include tab.

4. Select the With Raw Descriptions Matching check box and type **TOP** in the field.

5. Click OK, and confirm in Prospector that all the points have the description TOP.

6. Right-click the Top Of Curb point group, and choose Export To LandXML.

7. Click OK in the Export To LandXML dialog.

8. Choose a location to save your LandXML file, and then click Save.

9. Navigate out to the LandXML file to confirm it was created.

10. Right-click the Top Of Curb point group, and choose Export Points.

11. Choose the PNEZD comma-delimited format and a destination file, and confirm that the Limit To Points In Point Group check box is selected. Click OK.

12. Navigate out to the ASCII file to confirm it was created.

Create a point table. Point tables provide an opportunity to list and study point properties. In addition to basic point tables that list number, elevation, description, and similar options, you can customize point table formats to include user-defined property fields.

Master It Create a point table for the Topo point group using the PNEZD format table style. Use the DWG created in the previous exercise or start with `Master_It.dwg`.

Solution

1. Change to the Annotate tab, and select Add Tables ≻ Points.

2. Choose the PNEZD format for the table style.

3. Click Point Groups, and choose the Topo point group.

4. Click OK.

5. The command line prompts you to choose a location for the upper-left corner of the point table. Choose a location on your screen somewhere to the right of the project.

6. Zoom in, and confirm your point table.

Chapter 4: Surfaces

Create a preliminary surface using freely available data. Almost every land development project involves a surface at some point. During the planning stages, freely available data can give you a good feel for the lay of the land, allowing design exploration before money is spent on fieldwork or aerial topography. Imprecise at best, this free data should never be used as a replacement for final design topography, but it's a great starting point.

Master It Create a new drawing from the Civil 3D Extended template and bring in a Google Earth surface for your home or office location. Be sure to set a proper coordinate system to get this surface in the right place.

Solution

1. On the main menu, choose File ➢ New.
2. Select the _AutoCAD Civil 3D (Imperial) NCS.dwt or _AutoCAD Civil 3D (Metric) NCS.dwt file, and click Open.
3. Change to the Settings tab and right-click the drawing name to open the Drawing Settings dialog. Select an appropriate coordinate system.
4. In Google Earth, locate your home or office using the search engine.
5. In Civil 3D, change to the Insert panel.
6. On the Import panel, select Google Earth ➢ Google Earth Surface.

Modify and update a TIN surface. TIN surface creation is mathematically precise, but sometimes the assumptions behind the equations leave something to be desired. By using the editing tools built into Civil 3D, you can create a more realistic surface model.

Master It Modify your Google Earth surface to show only an area immediately around your home or office. Create an irregular-shaped boundary and apply it to the Google Earth surface.

Solution

1. Draw a polyline that includes the desired area.
2. Expand the Google Earth Surface branch in Prospector.
3. Expand the Definition branch.
4. Right-click Boundaries and select the Add option.
5. Select the newly created polyline and click Add to complete the boundary addition.

Prepare a slope analysis. Surface analysis tools allow users to view more than contours and triangles in Civil 3D. Engineers working with nontechnical team members can create strong meaningful analysis displays to convey important site information using the built-in analysis methods in Civil 3D.

Master It Create an Elevation Banding analysis of your home or office surface and insert a legend to help clarify the image.

Solution

1. Right-click the surface and bring up the Surface Properties dialog.
2. Change the Surface Style field to Elevation Banding (2D).
3. Change to the Elevation tab and run an Elevation analysis.
4. Click OK to close the Surface Properties dialog.

5. Select the surface to display the Tin Surface tab on the Ribbon.

6. On the Labels & Tables panel, click Add Legend Table. Enter **E** and then **D** at the command line and pick a placement point on the screen to create a legend.

Label surface contours and spot elevations. Showing a stack of contours is useless without context. Using the automated labeling tools in Civil 3D, you can create dynamic labels that update and reflect changes to your surface as your design evolves.

Master It Label the contours on your Google Earth surface at 1′ and 5′ (Design).

Solution

1. Change the Surface Style to Contours 1′ and 5′ (Design).

2. Select the surface to display the Tin Surface tab.

3. On the Labels & Tables panel, select Add Labels ➢ Contour – Multiple.

4. Pick a point on one side of the site, and draw a contour label line across the entire site. Repeat this step as needed to label the site appropriately.

Import a point cloud into a drawing and create a surface model. As point cloud data becomes more common and replaces other large-scale data-collection methods, the ability to use this data in Civil 3D is key. Intensity helps postprocessing software determine the ground cover type. While Civil 3D can't do postprocessing, you can see the intensity as part of the point cloud style.

Master It Import an LAS format point cloud `Denver.las` into the Civil 3D template of your choice. As you create the point cloud file, set the style to Scaled Color Intensity - Blue. Use a portion of the file to create a Civil 3D surface model.

Solution

1. Start a new file by using the default Civil 3D template of your choice. Save the file before proceeding as **Denver USA.dwg.**

2. Select the Point Clouds collection on the Prospector tab of Toolspace and right-click.

3. Select Create Point Cloud from the menu. The Create Point Cloud wizard is displayed. Name the point cloud **Denver**. Set Style to Scaled Color Intensity - Blue. Click Next.

4. Click the plus sign to browse for the LAS file. Select `Denver.las`. Click Finish. This file contains 4.7 million data points, so be patient while the file imports (this may be a good time to grab some coffee).

5. Select the point cloud to view the context Ribbon. Click Add Points To Surface.

6. Name the surface whatever you would like. Set the style to Scaled Color Intensity - Blue. Click Next.

7. Set the Region Option to Window. Click Define Region in Drawing. Create a window around the western half of the point cloud.

Chapter 5: Parcels

Create a boundary parcel from objects. The first step to any parceling project is to create an outer boundary for the site.

> **Master It** Open the `MasteringParcels.dwg` file, which you can download from www.sybex.com/go/masteringcivil3d2012. Convert the polyline in the drawing to a parcel.

> **Solution**
>
> 1. From the Home tab's Create Design panel, select Parcel ➢ Create Parcel From Objects.
>
> 2. At the `Select lines, arcs, or polylines to convert into parcels or [Xref]`: prompt, pick the polyline that represents the site boundary. Press ↵.
>
> 3. The Create Parcels – From Objects dialog appears. Select Subdivision Lots; Property; and Name, Square Foot and Acres from the drop-down menus in the Site, Parcel Style, and Area Label Style selection boxes, respectively. Keep the default values for the remaining options. Click OK to dismiss the dialog.
>
> 4. The boundary polyline forms parcel segments that react with the alignment. Area labels are placed at the newly created parcel centroids.

Create a right-of-way parcel using the right-of-way tool. For many projects, the ROW parcel serves as frontage for subdivision parcels. For straightforward sites, the automatic Create ROW tool provides a quick way to create this parcel. Cul-de-sacs serve as a terminal point for a cluster of parcels.

> **Master It** Continue working in the `Mastering Parcels.dwg` file. Create a ROW parcel that is offset by 25′ on either side of the road centerline with 25′ fillets at the parcel boundary. Then add the circles representing the cul-de-sac as a parcel.

> **Solution**
>
> 1. From the Home tab's Create Design panel, select Parcel ➢ Create Right Of Way.
>
> 2. At the `Select parcels`: prompt, pick your newly created parcels on screen. Press ↵ to stop picking parcels. The Create Right Of Way dialog appears.
>
> 3. Expand the Create Parcel Right Of Way parameter, and enter **25′** in the Offset From Alignment text box.
>
> 4. Expand the Cleanup At Parcel Boundaries parameter. Enter **25′** in the Fillet Radius At Parcel Boundary Intersections text box. Select Fillet from the drop-down menu in the Cleanup Method selection box.
>
> 5. Click OK to dismiss the dialog and create the ROW parcels.
>
> 6. Trim the two circles at the ROW line to create arcs.
>
> 7. From the Home tab's Create Design panel, select Parcel ➢ Create Parcel From Objects.
>
> 8. Pick the two arcs and accept the default settings. Two new parcels are created.
>
> 9. Pick a label at one of the newly created parcels, and from the Parcel tab and Modify panel, select Parcel Layout Tools.

10. Select the Delete Sub-Entity tool and pick the two ROW lines and ROW arc, leaving the outer arc alone. The cul-de-sac is created and is part of the ROW parcel.

11. Repeat steps 9 and 10 for the other cul-de-sac arc.

Create subdivision lots automatically by layout. The biggest challenge when creating a subdivision plan is optimizing the number of lots. The precise sizing parcel tools provide a means to automate this process.

Master It Continue working in the `Mastering Parcels.dwg` file. Create a series of lots with a minimum of 8,000 square feet and 75′ frontage.

Solution

1. From the Home Tab's Create Design panel, select Parcel ≻ Parcel Creation Tools.

2. Expand the Parcel Layout Tools toolbar.

3. Change the value of the following parameters by clicking in the Value column and typing in the new values:

    Default Area: **8000 Sq. Ft.**

    Minimum Frontage: **75′**

4. Change the following parameters by clicking in the Value column and selecting the appropriate option from the drop-down menu:

    Automatic Mode: On

    Remainder Distribution: Redistribute Remainder

5. Click the Slide Line – Create tool. The Create Parcels – Layout dialog appears.

6. Select Subdivision Lots, Single Family, and Name Square Foot & Acres from the drop-down menus in the Site, Parcel Style, and Area Label Style selection boxes, respectively. Keep the default values for the rest of the options. Click OK to dismiss the dialog.

7. At the `Select Parcel to be subdivided:` prompt, pick the Property: 1 label for your property parcel.

8. At the `Select start point on frontage:` prompt, use your Endpoint osnap to pick the point of curvature along the ROW parcel segment.

9. The parcel jig appears. Move your cursor slowly along the ROW parcel segment, and notice that the parcel jig follows the segment. At the `Select end point on frontage:` prompt, use your Endpoint osnap to pick the point of curvature along the ROW parcel segment.

10. At the `Specify angle at frontage:` prompt, type **90**. Press ↵.

11. At the `Accept Result:` prompt, press ↵ to accept the lot layout.

12. Repeat steps 5 through 11 for the other property parcels, if desired. Note that the parcels are not going to line up properly. For extra credit, fix them to your liking.

Add multiple parcel-segment labels. Every subdivision plat must be appropriately labeled. You can quickly label parcels with their bearings, distances, direction, and more using the segment labeling tools.

Master It Continue working in the `MasteringParcels.dwg` file. Place Bearing Over Distance labels on every parcel line segment and Delta Over Length And Radius labels on every parcel curve segment using the Multiple Segment Labeling tool.

Solution

1. From the Annotate tab, select Add Labels ➢ Parcels ➢ Add Parcel Labels.
2. In the Add Labels dialog, select Multiple Segment, Bearing Over Distance, and Delta Over Length And Radius from the drop-down menus in the Label Type, Line Label Style, and Curve Label Style selection boxes, respectively.
3. Click Add.
4. At the `Select parcel to be labeled by clicking on area label or [CLockwise/COunterclockwise]<CLockwise>:` prompt, pick the area label for each of your single-family parcels.
5. Press ↵ to exit the command.

Chapter 6: Alignments

Create an alignment from a polyline. Creating alignments based on polylines is a traditional method of building engineering models. With Civil 3D's built-in tools for conversion, correction, and alignment reversal, it's easy to use the linework prepared by others to start your design model. These alignments lack the intelligence of crafted alignments, however, and you should use them sparingly.

Master It Open the `MasteringAlignments-Objects.dwg` file, and create alignments from the linework found there.

Solution From the Home tab and Create Design panel, select Alignment ➢ Create Alignment From Objects. Select the lines and arc.

Create a reverse curve that never loses tangency. Using the alignment layout tools, you can build intelligence into the objects you design. One of the most common errors introduced to engineering designs is curves and lines that aren't tangent, requiring expensive revisions and resubmittals. The free, floating, and fixed components can make smart alignments in a large number of combinations available to solve almost any design problem.

Master It Open the `MasteringAlignments-Reverse.dwg` file, and create an alignment from the linework on the right. Create a reverse curve with both radii equal to 200 and with a pass-through point at the intersection of the two arcs.

Solution

1. Trace both lines with Fixed Segments.
2. Use the Floating Curve (From Entity With Passthrough Point) tool to draw an arc from the endpoint of the line with a pass-through point at the intersection of the two sketched arcs.
3. Use the Floating Curve (From Entity With Passthrough Point) tool to fillet the floating curve created in step 2 and the last fixed segment using a 200′ radius.

Replace a component of an alignment with another component type. One of the goals in using a dynamic modeling solution is to find better solutions, not just a solution. In the layout of alignments, this can mean changing components out along the design path, or changing the way they're defined. Civil 3D's ability to modify alignments' geometric construction without destroying the object or forcing a new definition lets you experiment without destroying the data already based on an alignment.

Master It Convert the reverse curve indicated in the `MasteringAlignments-Rcurve.dwg` file to a floating arc that is constrained by the following segment. Then change the radius of the curves to 150′.

Solution

1. Select the indicated alignment, and right-click to edit alignment geometry.
2. Select the Alignment Grid View tool.
3. Starting with the first segment, click in the Tangency Constraint field and change it to Constrained By Next (Floating). Repeat for the other segments except the last one, which cannot be modified because it is dependent on the previous constraint.
4. Change radius of the two curves to 150′.

Create alignment tables. Sometimes there is just too much information that is displayed on a drawing, and to make it clearer, tables are used to show bearings and distances for lines, curves, and segments. With their dynamic nature, these tables are kept up-to-date with any changes.

Master It From the `Mastering Alignments-Table.dwg`, make a line table, curve table, and segment table. Use whichever style you want to accomplish this.

Solution For lines:

1. Click on the alignment.
2. From the Alignment contextual tab and Labels & Tables panel, select Add Tables ➢ Add Line
3. Using the Pick On Screen icon at the bottom of the dialog, select the line segments of the alignment. If a warning comes up regarding child styles, select the "Convert all selected label styles to tag" mode. Click OK.
4. Place the table anywhere on your drawing.
5. The bearings and distances are now replaced by tag labels.

For curves:

1. Click on the alignment.
2. From the Alignment contextual tab and Labels & Tables panel, select Add Tables ➢ Add Curve.
3. Using the Pick On Screen icon at the bottom of the dialog, select the curve segments of the alignment. If a warning comes up regarding child styles, select the "Convert all selected label styles to tag" mode. Click OK.
4. Place the table anywhere on your drawing.

5. The bearings and distances are now replaced by tag labels.

For segments:

1. Click on the alignment.
2. From the Alignment contextual tab and Labels & Tables panel, select Add Tables ➢ Add Segments.
3. In the By Alignment section, select the alignment you want to label. Click OK.
4. Place the table anywhere on your drawing.
5. The bearings and distances are now replaced by tag labels.

Chapter 7: Profiles and Profile Views

Sample a surface profile with offset samples. Using surface data to create dynamic sampled profiles is an important advantage of working with a three-dimensional model. Quick viewing of various surface slices and grip-editing alignments makes for an effective preliminary planning tool. Combined with offset data to meet review agency requirements, profiles are robust design tools in Civil 3D.

Master It Open the `MasteringProfile.dwg` file and sample the ground surface along Alignment A, along with offset values at 15' left and 25' right of the alignment.

Solution

1. On the Home tab's Create Design panel, select Profiles ➢ Create Surface Profile.
2. Click the Add button to add the EG surface.
3. Check the Sample Offsets check box and enter **15, –25** in the box below the sample offsets and then click the Add button.
4. Click the Draw In Profile button and in the Create Profile View dialog, click the Create Profile View button.
5. Place the profile anywhere on the drawing.

Lay out a design profile on the basis of a table of data. Many programs and designers work by creating pairs of station and elevation data. The tools built into Civil 3D let you input this data precisely and quickly.

Master It In the `Mastering Profiles.dwg` file, create a layout profile on Alignment B with the following information:

STATION	PVI ELEVATION	CURVE LENGTH
0+00	812.76	
1+45	818.59	250'
5+22	794.48	

Solution

1. Create a profile for Alignment B.
2. On the Home tab's Create Design panel, select Profiles ➢ Profile Creation Tools.
3. Select the Profile B Grid and press ↵ to accept the value in the Create Profile - Draw New dialog.
4. In the Profile Layout Tools palette, set the L-value of the Curve settings to **250'**.
5. Use the Draw Tangents With Curves tool and the Transparent Commands toolbar to enter station-elevation data.

Alternatively, you can import a text file.

Add and modify individual components in a design profile. The ability to delete, modify, and edit the individual components of a design profile while maintaining the relationships is an important concept in the 3D modeling world. Tweaking the design allows you to pursue a better solution, not just a working solution.

Master It In the `MasteringProfile.dwg` file, on profile B insert a PVI at Sta 3+52, Elevation 812. Modify the curve so that it is 100' and then, add a 175' parabolic vertical curve at the newly created point.

Solution

1. Pick the Design profile and from the Modify Profile panel, select the Geometry Editor tool.
2. In the Profile Layout Tools palette, select the Insert PVI tool.
3. Using the Profile Station elevation transparent command, select the profile grid, enter **352** for the station and **812** for the elevation. Press Esc three times.
4. Back in the Profile Layout Tools palette, select the Profile Grid View tool.
5. In the Profile Entities panorama, change the Profile Curve Length field to **100**↵.
6. In the Profile Layout Tools palette, select the Draw Fixed Parabola By 3 Points drop-down and choose More Free Vertical Curves ➢ Free Vertical Parabola (PVI Based).
7. Pick the newly created PVI and enter **175** for Curve Length. Press ↵ twice.

Apply a standard label set. Standardization of appearance is one of the major benefits of using Civil 3D styles in labeling. By applying label sets, you can quickly create plot-ready profile views that have the required information for review.

Master It In the `MasteringProfile.dwg` file, apply the Road Profile Labels set to all layout profiles.

Solution

1. Pick the layout profile, right-click, and select the Edit Labels option.
2. Click Import Label Set, and select the Road Profile Labels set.

Chapter 8: Assemblies and Subassemblies

Create a typical road assembly with lanes, curbs, gutters, and sidewalks. Most corridors are built to model roads. The most common assembly used in these road corridors is some variation of a typical road section consisting of lanes, curb, gutter, and sidewalk.

Master It Create a new drawing from the DWT of your choice. Build a symmetrical assembly using LaneInsideSuper, UrbanCurbGutterValley1, and LinkWidthAndSlope for terrace and buffer strips adjacent to the UrbanSidewalk. Use widths and slopes of your choosing.

Solution

1. Create a new drawing from the DWT of your choice.
2. From the Home tab's Create Design Panel, select Assembly ➢ Create Assembly.
3. Name your assembly and set styles as appropriate.
4. Pick a location in your drawing for the assembly.
5. Locate the Basic tab on the tool palette.
6. Click the LaneInsideSuper button on the tool palette. Use the AutoCAD Properties palette, and follow the command-line prompts to set the LaneInsideSuper on the left and right sides of your assembly.
7. Repeat the process with UrbanCurbGutterValley1, LinkWidthAndSlope, and UrbanSidewalk. Complete this portion of the exercise by placing a final LinkWidthAndSlope on the outside of the UrbanSidewalk. (Refer to the "Subassemblies" section of this chapter for additional information.)
8. Save the drawing for use in the next Master It exercise.

Edit an assembly. Once an assembly has been created, it can be easily edited to reflect a design change. Often, at the beginning of a project, you won't know the final lane width. You can build your assembly and corridor model with one lane width and then later change the width and rebuild the model immediately.

Master It Working in the same drawing, edit the width of each LaneInsideSuper to 14' (4.3 m), and change the cross slope of each LaneInsideSuper to –3.08%.

Solution

1. Select both lane subassemblies. Be sure these are the only two elements selected.
2. Click Properties.
3. In the Advanced Parameters, change the width to 14' (4.3 m). Note that width will be listed twice. The topmost width reports the default value. You will change the second occurrence.
4. Change the slope to –3.08%.
5. Save the drawing for use in the next Master It exercise.

Add daylighting to a typical road assembly. Often, the most difficult part of a designer's job is figuring out how to grade the area between the last engineered structure point in the cross section (such as the back of a sidewalk) and existing ground. An extensive catalog of daylighting subassemblies can assist you with this task.

Master It Working in the same drawing, add the DaylightMinWidth subassembly to both sides of your typical road assembly. Establish a minimum width between the outermost subassembly and the daylight offset of 10′ (3 m).

Solution

1. Locate the Daylight tab on the tool palette.
2. Click the DaylightMinWidth button on the tool palette. Use the AutoCAD Properties palette, and follow the command-line prompts to set the DaylightMinWidth on the right side of your assembly.
3. Press Esc on your keyboard to complete the command. Select the right daylight subassembly.
4. From the context Ribbon, select Mirror Subassemblies.
5. Click the outermost left point on the LinkWidthAndSlope link. You should now have daylighting subassemblies visible on both sides of your assembly.

Chapter 9: Basic Corridors

Build a single-baseline corridor from an alignment, profile, and assembly. Corridors are created from the combination of alignments, profiles, and assemblies. Although corridors can be used to model many things, most corridors are used for road design.

Master It Open the `Mastering Corridors.dwg` file. Build a corridor on the basis of the Project Road alignment, the Project Road Finished Ground profile, and the Project Typical Road Assembly.

Solution

1. From the Home tab's Create Design panel, select Corridors ➢ Create Simple Corridor.
2. Pick the Project Road alignment, the Project Road Finished Ground profile, and the Project Typical Road Assembly. Target the Existing Ground surface.

Use targets to add lane widening. Targets are an essential design tool used to manipulate the geometry of the road.

Master It Open `Mastering Corridor Widening`. Set the Lane - L to target the USH 10 Left 16 alignment and set Lane - R to target the USH 10 Right alignment.

Solution

1. Open the Corridor Properties dialog and switch to the Parameters tab.
2. At the regions level, scroll over and click the targets ellipsis.

3. In the Target Mapping dialog, click the width alignment for Lane - R. Select USH 10 Right 16.000 and click Add. Click OK.

4. Click the width alignment for Lane - L. Select USH 10 Left 16.000 and click Add. Click OK.

5. Click OK to dismiss the Target Mapping dialog box. Click OK to dismiss the Corridor properties and allow the corridor to build.

Create a corridor surface. The corridor model can be used to build a surface. This corridor surface can then be analyzed and annotated to produce finished road plans.

Master It   Continue working in the `Mastering Corridors.dwg` file or open `Mastering Corridors - Step 1 Complete`. Create a corridor surface from Top links.

Solution

1. Open the Corridor Properties dialog and switch to the Surfaces tab.
2. Click Create Surface and then the plus (+) sign to add Top links.
3. Click OK to dismiss the dialog. The corridor and surface will build.

Add an automatic boundary to a corridor surface. Surfaces can be improved with the addition of a boundary. Single-baseline corridors can take advantage of automatic boundary creation.

Master It   Continue working in the `Mastering Corridors.dwg` file or open `Mastering Corridors - Step 2 Complete`. Use the Automatic Boundary Creation tool to add a boundary using the Daylight code.

Solution

1. Open the Corridor Properties dialog and switch to the Boundaries tab.
2. Right-click on the surface entry and click Add Automatically ➢ Daylight.
3. Click OK to dismiss the dialog. The corridor and surface will build.

Chapter 10: Advanced Corridors, Intersections, and Roundabouts

Create corridors with noncenterline baselines. Although for simple corridors you may think of a baseline as a road centerline, other elements of a road design can be used as a baseline. In the case of a cul-de-sac, the EOP, the top of curb, or any other appropriate feature can be converted to an alignment and profile and used as a baseline.

Master It   Open the `Mastering Advanced Corridors.dwg` file, which you can download from www.sybex.com/go/masteringcivil3d2012. Add the cul-de-sac alignment and profile to the corridor as a baseline. Create a region under this baseline that applies the Typical Intersection assembly.

Solution

1. Select the corridor, right-click, and choose Corridor Properties. Switch to the Parameters tab.
2. Click Add Baseline. Choose Cul de Sac EOP in the Pick A Horizontal Alignment dialog.
3. In the Profile column, click inside the <click here ...> box. Choose Cul de Sac EOP FG in the Select A Profile dialog.
4. Right-click the new baseline. Choose Add Region. Select Intersection Typical in the Select An Assembly dialog box.
5. Click OK to leave the Corridor Properties dialog and build the corridor.

Add alignment and profile targets to a region for a cul-de-sac. Adding a baseline isn't always enough. Some corridor models require the use of targets. In the case of a cul-de-sac, the lane elevations are often driven by the cul-de-sac centerline alignment and profile.

Master It Continue working in the `Mastering Advanced Corridors.dwg` file. Add the Second Road alignment and Second Road FG profile as targets to the cul-de-sac region. Adjust Assembly Application Frequency to 5', and make sure the corridor samples are profile PVIs.

Solution

1. Select the corridor, right-click, and choose Corridor Properties. Switch to the Parameters tab.
2. Click the Target Mapping button in the appropriate region.
3. In the Target Mapping dialog, assign Second Road as the transition alignment and Second Road FG profile as the transition profile. Click OK to leave the Target Mapping dialog.
4. Click the Frequency button in the appropriate region. Change the Along Curves value to 5' and the At Profile High/Low Point value to Yes. Click OK to exit the Frequency To Apply To Assemblies dialog.
5. Click OK to leave the Corridor Properties dialog and build the corridor.

Use the Interactive Boundary tool to add a boundary to the corridor surface. Every good surface needs a boundary to prevent bad triangulation. Bad triangulation creates inaccurate and unsightly contours. Civil 3D provides several tools for creating corridor surface boundaries, including an Interactive Boundary tool.

Master It Continue working in the `Mastering Advanced Corridors.dwg` file. Create an interactive corridor surface boundary for the entire corridor model.

Solution

1. Select the corridor, right-click, and choose Corridor Properties. Switch to the Boundaries tab.
2. Select the corridor surface, right-click, and choose Add Interactively.

3. Follow the command-line prompts to add a feature line–based boundary all the way around the entire corridor.

4. Type **C** to close the boundary, and then press ↵ to end the command.

5. Click OK to leave the Corridor Properties dialog and build the corridor.

An example of the finished exercise can be found in `Mastering Advanced Corridors Finished.dwg` at the book's web page.

Chapter 11: Superelevation

Apply superelevation to an alignment. Civil 3D has very convenient and flexible tools that will apply safe, correct superelevation to an alignment curve.

Master It Open `Master Super.dwg` file, which you can download from www.sybex.com/go/masteringcivil3d2012. Set the design speed of the road to 20 miles per hour and apply superelevation to the entire length of the alignment. Use AASHTO 2004 Design Criteria with an eMax of 6%.

Solution

1. Select the alignment and choose Alignment Properties from the context tab.

2. On the Design Criteria tab, place a check mark next to Use Criteria Based Design. Set the design criteria file and superelevation eMax from the right side of the dialog box. Set the design speed to 20mph on the left side. Click OK.

3. From the context tab, click Superelevation ➢ Calculate/Edit Superelevation. Click Calculate Superelevation Now.

4. Step through the superelevation wizard, taking all the defaults for pivot, shoulder, and attainment. Click Finish. You should now have superelevation applied to the design with no overlap.

Create a superelevation view. Superelevation views are a great place to get a handle on what is going on in your roadway design. You can visually check the geometry as well as make changes to the design.

Master It Continue working in the drawing from the previous exercise. Create a superelevation view for the alignment.

Solution

1. Select the alignment and choose Superelevation ➢ Create Superelevation View.

2. In the Create Superelevation View dialog box, toggle off Left Outside Shoulder and Right Outside Shoulder.

3. Set Left Outside Lane Color to Blue. Set Right Outside Shoulder Color to Red. Click OK.

4. Place the view in the drawing.

Chapter 12: Cross Sections and Mass Haul

Create sample lines. Before any section views can be displayed, sections must be created from sample lines.

> **Master It** Open `sections1.dwg` and create sample lines along the alignment every 50'.

> **Solution** Using the Cabernet Court alignment, sample all data except the datum surface. Create sample lines by station range and set your sample line distance to 50'.

Create section views. Just as profiles can only be shown in profile views, sections require section views to display. Section views can be plotted individually or all at once. You can even set them up to be broken up into sheets.

> **Master It** In the previous drawing, you created sample lines. In that same drawing, create section views for all the sample lines.

> **Solution** Using the Create Multiple Sections command, create section views for all sample lines. Use the Plot All option, and don't add any labels to the views.

Define materials. Materials are required to be defined before any quantities can be displayed. You learned that materials can be defined from surfaces or from corridor shapes. Corridors must exist for shape selection and surfaces must already be created for comparison in materials lists.

> **Master It** Using `sections4.dwg`, create a materials list that compares EG with Cabernet Court Top Road Surface.

> **Solution** To create this materials list, you will need to use Cut and Fill criteria. The EG surface will be Surface2, and the Datum will be Cabernet Court Top Road Surface.

Generate volume reports. Volume reports give you numbers that can be used for cost estimating on any given project. Typically, construction companies calculate their own quantities, but developers often want to know approximate volumes for budgeting purposes.

> **Master It** Continue using `sections4.dwg`. Use the materials list created earlier to generate a volume report. Create an XML report and a table that can be displayed on the drawing.

> **Solution** Use the Cut And Fill table style to display the volume calculations on the drawing. It should be set to dynamically update in the event that the quantities change (if the profile is adjusted or the alignment is moved).

Chapter 13: Pipe Networks and Part Builder

Create a pipe network by layout. After you've created a parts list for your pipe network, the first step toward finalizing the design is to create a Pipe Network By Layout.

> **Master It** Open the `MasteringPipes.dwg` file. Choose Pipe Network Creation Tools from the Pipe Network drop-down in the Create Design panel on the Home tab to create a sanitary sewer pipe network. Use the Finished Ground surface, and name only structure and pipe label styles. Don't choose an alignment at this time. Create 8" (20.32 cm) PVC pipes and concentric manholes. There are blocks in the drawing to assist you in placing manholes. Begin at the START HERE marker, and place a manhole at each marker location. You can erase the markers when you've finished.

Solution

1. From the Home tab's Create Design panel, select Pipe Network ➢ Pipe Network Creation Tools.

2. In the Create Pipe Network dialog, set the following parameters:

    Network Name: **Mastering**

    Network Parts List: Sanitary Sewer

    Surface Name: Corridor FG

    Alignment Name: <none>

    Structure Label Style: Plan (Sanitary)

    Pipe Label Style: Name Only

3. Click OK. The Pipe Layout Tools toolbar appears.

4. Set the structure to SMH and the pipe to 8 Inch PVC. Click Draw Pipes And Structures, and use your Insertion osnap to place a structure at each marker location.

5. Press ↵ to exit the command.

6. Move the structure labels. Select a marker, right-click, and choose Select Similar. Click Delete.

Create an alignment from network parts and draw parts in profile view. Once your pipe network has been created in plan view, you'll typically add the parts to a profile view on the basis of either the road centerline or the pipe centerline.

Master It Continue working in the `MasteringPipes.dwg` file. Create an alignment from your pipes so that station zero is located at the START HERE structure. Create a profile view from this alignment, and draw the pipes on the profile view.

Solution

1. Select the structure labeled START HERE to display the Pipe Networks contextual tab and select Alignment From Network on the Launch Pad panel.

2. Select the last structure in the pipe run. Press ↵ to accept the selection.

3. In the Create Alignment dialog, name the Alignment **Mastering** and make sure the Create Profile And Profile View check box is selected. Accept the other defaults, and click OK.

4. In the Create Profile dialog, sample both the EG and Corridor FG surfaces for the profile. Click Draw In Profile View.

5. In the Create Profile View dialog, click Create Profile View and choose a location in the drawing for the profile view. A profile view showing your pipes appears.

Label a pipe network in plan and profile. Designing your pipe network is only half of the process. Engineering plans must be properly annotated.

> **Master It** Continue working in the `MasteringPipes.dwg` file. Add the Length Description And Slope style to profile pipes and the Data With Connected Pipes (Sanitary) style to profile structures. Add the alignment created in the previous exercise to all pipes and structures.
>
> **Solution**
>
> 1. Change to the Annotate tab and select Add Labels.
> 2. In the Add Labels dialog, change Feature to Pipe Network, and then change Label Type to Entire Network Profile. For pipe labels, choose Length Description And Slope; for structure labels, choose Data With Connected Pipes (Sanitary). Click Add, and choose any pipe or structure in your profile.
> 3. Drag or adjust any profile labels as desired.
> 4. Expand Pipe Networks ➢ Networks ➢ Mastering and select Pipes.
> 5. Select all pipes in Prospector, right-click on Reference Alignment, and select Edit.
> 6. Choose the Mastering alignment.
> 7. Repeat steps 5 and 6 but choose Structures.

Create a dynamic pipe table. It's common for municipalities and contractors to request a pipe or structure table for cost estimates or to make it easier to understand a busy plan.

> **Master It** Continue working in the `MasteringPipes.dwg` file. Create a pipe table for all pipes in your network.
>
> **Solution**
>
> 1. Change to the Annotation tab, and select Add Labels ➢ Add Tables ➢ Add Pipe.
> 2. In the Pipe Table Creation dialog, make sure your pipe network is selected. Accept the other defaults, and click OK.
> 3. Place the table in your drawing.

Chapter 14: Storm and Sanitary Analysis

Create a catchment object. Catchment objects are the newest object type that Civil 3D can use to determine the area and time of concentration of a site. You can create a catchment from surface, but in most cases you will use the option to create from polyline.

> **Master It** Create a catchment for predeveloped conditions on the site.
>
> **Solution**
>
> 1. Open the file `Mastering Catchment Creation.dwg`.
> 2. On the Analyze tab, choose Catchments ➢ Create Catchment Group.

3. Name the new catchment group **Predeveloped** and click OK.

4. On the Analyze tab, select Catchments ➢ Create Catchment From Object.

5. Select the red closed polyline that represents the catchment. When prompted for the flow path, select the polyline that runs through the site. Be sure to select it on the north part of the site (the uphill side).

6. In the Create Catchment dialog, name the catchment **Basin A**. Leave all other styles and settings at their defaults and click OK. You now have a completed catchment.

Export pipe data from Civil 3D into SSA. Civil 3D by itself cannot do pipe flow or runoff calculations. For this reason, it is important to export Civil 3D pipe networks into the Storm And Sanitary Analysis portion of the product.

Master It Verify your pipe export settings and then export to SSA.

Solution

1. Open the file `Export to SSA.dwg`.

2. Select the Settings tab, then choose Pipe Networks, and click Commands.

3. Double-click the EditInSSA option.

4. Expand the Storm Sewers Migration Defaults area.

5. Click the field for part matching defaults. Click the ellipsis to open the Part Matchup Settings dialog.

6. Examine the Import tab and verify that the settings make sense. Switch to the Export tab and examine your options for bringing files back into Civil 3D. No changes are needed. Click OK.

7. Click OK to exit the command settings.

8. Switch to the Analyze tab in Civil 3D. Click Edit In Storm And Sanitary Analysis.

9. Click OK to export the Highway Drainage storm network.

10. When Storm Sewers launches, click OK to create a new project.

11. Click No to saving the log file. Switch to the Plan View tab.

12. Double-click one of the structures. Make sure it contains elevation data.

13. Save the SSA file and exit.

Chapter 15: Grading

Convert existing linework into feature lines. Many site features are drawn initially as simple linework for the 2D plan. By converting this linework to feature line information, you avoid a large amount of rework. Additionally, the conversion process offers the ability to drape feature lines along a surface, making further grading use easier.

Master It Open the `MasteringGrading.dwg` file from the data location. Convert the magenta polyline describing a proposed temporary drain into a feature line and drape it across the EG surface to set elevations.

Solution

1. From the Home tab's Create Design panel, select Feature Lines ➢ Create Feature Lines From Objects.
2. Pick the polyline.
3. Toggle the Assign Elevations check box on.
4. Select the EG surface in the Assign Elevations dialog.
5. Click OK twice to close the dialogs and return to your model.

Model a simple linear grading with a feature line. Feature lines define linear slope connections. This can be the flow of a drainage channel, the outline of a building pad, or the back of a street curb. These linear relationships can help define grading in a model, or simply allow for better understanding of design intent.

Master It Add 100′ (30.48 m) radius fillets on the feature line you just created. Set the grade from the start of the feature line to the circled point to 5 percent and the remainder to a constant slope to be determined in the drawing. Draw a temporary profile view to verify the channel is below grade for most of its length.

Solution

1. Display the Feature Line toolbar's Edit Geometry panel, and select the Fillet tool.
2. Pick the feature line.
3. Enter **R↵** to change the radius.
4. Enter **100↵** (30.48 m) for the radius.
5. Enter **A↵** to fillet all the points.
6. Toggle on the Edit Elevations panel and select Insert Elevation Point.
7. Use the Center osnap to pick the center of the circle. Press ↵ to accept the elevation.
8. From the Edit Elevations panel, select the Set Grade Between Points tool.
9. Pick the feature line. Pick the PI near the start of the feature line.
10. Press ↵ to accept the elevation. Pick the elevation point at the circle's center.
11. Enter **-5↵** to set the grade.
12. Pick the feature line again.
13. Pick the elevation point created near the circle. Press ↵ to accept the elevation.
14. Pick the PI at the downstream end of the channel. Press ↵ to accept the elevation.
15. Press ↵ to exit the command.

16. Pick the feature line. Right-click and select Quick Profile.
17. Select only the EG surface and Layout from the drop-down list in the 3D Entity Profile Style selection box.
18. Click OK and pick a point on the screen to draw a profile view.

Model planar site features with grading groups. Once a feature line defines a linear feature, gradings collected in grading groups model the lateral projections from that line to other points in space. These projections combine to model a site much like a TIN surface, resulting in a dynamic design tool that works in the Civil 3D environment.

Master It Use the two grading criteria just used to define the pilot channel, with grading on both sides of the sketched centerline. Calculate the difference in volume between using 6:1 side slopes and 4:1 side slopes.

Solution

1. From the Home tab's Create Design panel, select Grading ➢ Create Grading to activate the Grading Creation Tools toolbar. Set Surface to EG.
2. Click the Set The Grading Group tool.
3. Check the Automatic Surface Creation option.
4. Check the Volume Base Surface option, select the EG surface, and click OK.
5. Click OK to accept the surface creation options.
6. Change Grading Criteria to Grade To Distance.
7. Click the Create Grading tool and pick the feature line.
8. Pick the left or right side. Press ↵ to model the full length.
9. Type **5′** (**1.52** m) for the distance.
10. Press ↵ to accept 2 percent grade.
11. Pick the main feature line again and grade the other side.
12. Change Grade to Surface Criteria and grade both left and right sides, accepting the default values.
13. Right-click to complete the gradings.
14. Pick one of the diamonds in the grading. Right-click and select Grading Group Properties.
15. Switch to the Properties tab and note the volume (approximately 424 Cu. Yd. (12 Cu. M.) Net Fill). Click OK.
16. Right-click the diamond representing the Grade To Surface grading.
17. Click the Grading Editor tool.
18. Change Slopes to 4:1. Click OK.

19. Repeat for the other side of the channel.

20. Pick the diamond again and select Grading Group Properties. The new net volume is approximately 293 Cu. Yd. (8.29 Cu. M.) The difference is approximately 131 Cu. Yd. (3.71 Cu. M.).

Chapter 16: Plan Production

Create view frames. When you create view frames, you must select the template file that contains the layout tabs that will be used as the basis for your sheets. This template must contain predefined viewports. You can define these viewports with extra vertices so you can change their shape after the sheets have been created.

Master It Open the `MasteringPlanProduction.dwg`. Run the Create View Frames wizard to create view frames for the Bike Path alignment in the current drawing. (Accept the defaults for all other values.)

Solution

1. Change to the Output tab, and then select Plan Production Tools ≻ Create View Frames.

2. On the Alignment page, select Bike Path from the Alignment drop-down menu. Click Next.

3. On the Sheets page, select the Plan And Profile option. Click the ellipsis button to display the Select Layout As Sheet Template dialog. In this dialog, click the ellipsis button, browse to `C:\Mastering Civil 3D 2012\CH 16\Data`, select the template named `Mastering (Imperial) Plan and Profile.dwt`, and click Open.

4. Select the layout named ARCH D Plan And Profile 20 Scale, and click OK.

5. Click Create View Frames.

Edit view frames. The Edit View Frames command allows the user some freedom on how the frames will appear.

Master It Open the `MasteringPlan Production1.dwg`, and move the VF- (9) view frame to Sta. 40+50 to lessen the overlap. Then adjust the match line so that it is now at Sta. 38+15.

Solution

1. Click on the VF- (9) view frame.

2. Make sure you have Dynamic Input on. Click on the triangular grip. Type **4050↵**. The view frame is now centered on Sta. 40+50. Press Esc to clear the selection.

3. Click on the Match Line 8 to show its grips.

4. Click on the triangular grip. Type **3815↵**. The match line is now centered better between the two view frames.

Generate sheets and review Sheet Set Manager. You can create sheets in new drawing files or in the current drawing. Use the option to create sheets in the current drawing when a) you've referenced in the view frame group, or b) you have a small project. The resulting sheets are based on the template you chose when you created the view frames. If the template contains customized viewports, you can modify the shape of the viewport to better fit your sheet needs.

Master It Continue working in the `MasteringPlanProduction1.dwg` file or you can open `MasteringPlanProduction2.dwg`. Run the Create Sheets wizard to create plan and profile sheets for Bike Path using the template `Mastering (Imperial) Plan and Profile.dwt`. (Accept the defaults for all other values.)

Solution

1. Change to the Output tab, and then select Plan Production Tools ➢ Create Sheets.
2. On the View Frame Group And Layouts page, under the Layout Creation section, select All Layouts In The Current Drawing, then click Create Sheets.
3. Click OK to save the drawing.
4. Click a location as the profile origin.
5. Dismiss the events Panorama.
6. In Sheet Set Manager, double-click the sheet named 1- Sheet - (1).

Create section views. More and more municipalities are requiring section views. Whether this is a mile-long road or a meandering stream, Civil 3D can handle it nicely via Plan Production.

Master It Open `MasteringPlanProduction3.dwg`. Create Plan Production Sheets with the defaults using the methods you learned earlier.

Solution

1. From the Home tab, and Section Views panel, select Section Views ➢ Create Multiple Views.
2. Click Create Section Views.
3. Select a spot above the centerline profile. The multiple section views are created.
4. From the Output tab and Plan Production panel, select Create Section Sheets.
5. Click Create Sheets. You will have to save your drawing before the process ends.

Chapter 17: Interoperability

Create data shortcuts. The ability to load design information into a project environment is an important part of creating an efficient team. The main design elements of the project are available to the data shortcut mechanism, but some still are not.

Master It List the top-level branches in Prospector that cannot be shared through data shortcuts.

Solution   Points, point groups, sites, corridors, assemblies, subassemblies, and survey pieces are all unavailable as shortcuts.

Import and export to earlier releases of AutoCAD.   Being able to work with outside clients or even other departments within your firm who do not have Civil3D is an important part of the concept of playing nicely together.

Master It   What are some methods for sharing your Civil 3D drawings with AutoCAD Release 2004 users?

Solution   In no particular order, exporting to LandXML, eTransmitting with the option to save to Release 2004 with exploded AEC objects, or exporting to AutoCAD Release 2004 are some of the methods.

Import and export to third-party CAD programs.   Exchanging drawings from other CAD formats is another important method that lets you work together on a common project, no matter what CAD program is being used.

Master It   What tools can be used to receive CAD files from a firm using MicroStation?

Solution   The Import tool on the Insert tab and Import panel will allow you to import MicroStation (DGN) files. You could also request an XML file, which imports just fine into Civil 3D via the LandXML tool found on the Import panel of the Insert tab.

Use basic map queries to filter data.   Map 3D is the program that Civil 3D is built on. In Map there are numerous things that can be accomplished; using queries is one of them.

Master It   Open `GISMAP.dwg`. This is the subdivision you have been working on for most of the book. Insert the PaMunicipalities shape file found in the `GIS\Municip_2011` folder. You only want to have the Felton municipality visible on the plan.

Solution

1. Make sure you are in the Planning and Analysis Workspace by clicking the Workspace switcher at the bottom of the drawing area.

2. Make sure the Task pane is also open by typing **MAPWSPACE↵ ON↵**.

3. In the Task pane, click the Data tool and select Connect To Data. The Data Connect palette opens.

4. Click the SHP icon, navigate to the `C:\Mastering\Municip_2011` folder, select the `PaMunicipalities2011_01` shape file, and click Open.

5. Back in the Data Connect palette, click Connect.

6. Select Add To Map With Query from the drop-down list.

7. Click the Property icon and select MUNICIPAL1.

8. Select the Equal icon.

9. Click the Get Values icon and, type **FELTON** in the Filter Values text box, and press the green arrow.

10. Click Insert Value and then click OK.

11. Use the Zoom Extents and Draw Order commands on the shape file to move it to the background.

Chapter 18: Quantity Takeoff

Open and review a list of pay items along with their categorization. The pay item list is the cornerstone of QTO. You should download and review your pay item list and compare it against the current reviewing agency list regularly to avoid any missed items.

Master It Add the 12-, 18-, and 21-Inch Pipe Culvert pay items to your favorites in the QTO Manager.

Solution

1. Open the QTO Manager palette (it does not matter which drawing is open).
2. Enter 12-Inch Pipe in the text box to filter. Turn the categorization option off to more easily see the filter results.
3. Right-click on the 12-INCH PIPE CULVERT line and select Add To Favorites.
4. Repeat for the other sizes.

Assign pay items to AutoCAD objects, pipe networks, and corridors. The majority of the work in preparing QTO is in assigning pay items accurately. By using the linework, blocks, and Civil 3D objects in your drawing as part of the process, you reduce the effort involved with generating accurate quantities.

Master It Open the `Mastering QTO.dwg` file and assign the CLEARING AND GRUBBING pay item to the project area.

Solution

1. Open the QTO Manager palette.
2. Expand the Favorites branch and select the CLEARING AND GRUBBING item.
3. Right-click and select Assign Pay Item To Area.
4. Switch to the Object option by typing O at the command line.
5. Select the polyline representing the limits of disturbance of the site.

Use QTO tools to review what items have been tagged for analysis. By using the built-in highlighting tools to verify pay item assignments, you can avoid costly errors when running your QTO reports.

Master It Verify that the area in the previous exercise has been assigned a pay item.

Solution

1. Turn on Highlight Objects With Pay Items in the QTO Manager palette.
2. Pan and review that the polyline in question is highlighted and not shown in black, indicating no pay item assignment.

Generate QTO output to a variety of formats for review or analysis. The Quantity Takeoff reports give you a quick understanding of what items have been tagged in the drawing, and they can generate text in the drawing or external reports for uses in other applications.

Master It Calculate and display the amount of Type C Broken markings in the `Mastering Reporting.dwg` file.

Solution

1. From the Analyze tab's QTO panel, select the Takeoff Command, and click OK to run the report with default settings.

2. In the lower-left corner of the screen, change the report style to `Summary(TXT).xsl`.

3. The calculated amount for Type C Broken pavement markings should be 3594.486′.

Chapter 19: Styles

Override individual labels with other styles. In spite of the desire to have uniform labeling styles and appearances between alignments within a single drawing, project, or firm, there are always exceptions. Using AutoCAD's Ctrl+click method for element selection, you can access commands that let you modify your labels and even change their styles.

Master It Open a new drawing based on the template of your choice. Create a copy of the Perpendicular With Tick Major Station style called Major With Marker. Change Tick Block Name to Marker Pnt. Replace some (but not all) of your major station labels with this new style.

Solution

1. On the Settings tab, expand the Alignment ➢ Label Styles ➢ Station ➢ Major Station branch.

2. Right-click Perpendicular with Tick, and select Copy.

3. Change the name to **Major with Marker**.

4. Change to the Layout tab.

5. Change to the Tick component.

6. Change AeccTickLine Block Selection to Market Pnt.

7. Click OK to close the dialog.

8. Open the AutoCAD Properties dialog.

9. Ctrl+click a major station label.

10. Change the style to Major With Marker.

Create a new label set for alignments. Label sets let you determine the appearance of an alignment's labels and quickly standardize that appearance across all objects of the same nature. By creating sets that reflect their intended use, you can make it easy for a designer to quickly label alignments according to specifications with little understanding of the requirement.

> **Master It** Within the `Mastering Alignment Styles.dwg` file, create a new label set containing only major station labels, and apply it to all the alignments in that drawing.

> **Solution**
>
> 1. On the Settings tab, expand the Alignments ➢ Label Styles ➢ Label Sets branch.
> 2. Right-click Major And Minor Only, and select Copy.
> 3. Change the name to **Major Only**.
> 4. Delete the Minor label on the Labels tab.
> 5. Right-click each alignment, and select Edit Alignment Labels.
> 6. Import the Major Only label set.
> 7. Repeat for each label.
>
> Solutions may vary!

Apply a standard label set to profiles. Standardization of appearance is one of the major benefits of using Civil 3D styles in labeling. By applying label sets, you can quickly create plot-ready profile views that have the required information for review.

> **Master It** In the `Mastering Profile Styles.dwg` file, apply the Road Profiles label set to all layout profiles.

> **Solution**
>
> 1. Pick the layout profile, right-click, and select the Edit Labels option.
> 2. Click Import Label Set, and select the Road Profile Labels set.
> 3. Repeat this procedure for all layout profiles.

Appendix B

AutoCAD Civil 3D Certification

Autodesk certifications are industry-recognized credentials that can help you succeed in your design career—providing benefits to both you and your employer. Getting certified is a reliable validation of skills and knowledge, and it can lead to accelerated professional development, improved productivity, and enhanced credibility.

This Autodesk Official Training Guide can be an effective component of your exam preparation. Autodesk highly recommends (and we agree!) that you schedule regular time to prepare, review the most current exam preparation roadmap available at www.autodesk.com/certification, use Autodesk Official Training Guides, take a class at an Authorized Training Center (find ATCs near you here: www.autodesk.com/atc), take an assessment test, and use a variety of resources to prepare for your certification—including plenty of hands-on experience.

To help you focus your studies on the skills you'll need for these exams, the following tables show the objective and in which chapter you can find information on that topic—and when you go to that chapter, you'll find certification icons like the one in the margin here.

Table B.1 is for the Autodesk Certified Associate Exam and lists the topic, exam objectives, and chapter where the information is found. Table B.2 is for the Autodesk Certified Professional Exam. The topics and exam objectives listed in the table are from the Autodesk Certification Exam Guide.

These Autodesk exam objectives were accurate at press time; please refer to www.autodesk.com/certification for the most current exam roadmap and objectives.

Good luck preparing for your certification!

TABLE B.1: Certified Associate Exam Topics and Objectives

Topic	Objective	Mastering AutoCAD Civil 3D 2012 Chapter
User Interface	Navigate the user interface	1
	Use the functions on the Prospector tab	1
	Use functions on the Settings tab	19
Styles	Create and use object styles	19
	Create and use label styles	19
Lines and Curves	Use the Line and Curve commands	1
	Use the Transparent command	1

TABLE B.1: Certified Associate Exam Topics and Objectives *(CONTINUED)*

TOPIC	OBJECTIVE	MASTERING AUTOCAD CIVIL 3D 2012 CHAPTER
Points	Create points using the Point Creation command	3
	Create points by importing point data	3
	Use point groups to control the display of points	3
Surfaces	Create and edit surfaces	4
	Use styles and settings to display surface information	19
	Use styles to analyze surface display results	4
Parcels	Create parcels using parcel layout tools	5
	Select parcel styles to change the display of parcels	19
	Select styles to annotate parcels	5
	Create alignments	6
Profiles and Profile Views	Create a surface profile	7
	Create a layout profile	7
	Create a profile view	7
Corridors	Design and create a corridor	9
	Derive information and data from a corridor	10
	Design and create an intersection	10
Sections and Section Views	Create and analyze sections and section views	9
Pipe Networks	Design and create a pipe network	13
Grading	Design and create a grading model	15
	Create a grading model feature line	15
Managing and Sharing Data	Use data shortcuts to share and manage data	17
Plan Production	Generate a sheet set using plan production	16
Survey	Use description keys to control the display of points created from survey data	3
	Use figure prefixes to control the display of linework generated from survey data	2

TABLE B.2: Certified Professional Exam Topics and Objectives

Topic	Objective	Mastering AutoCAD Civil 3D 2012 Chapter
Styles	Create and use object styles	19
	Create and use label styles	19
Points	Use point groups to control the display of points	3
Surfaces	Use styles and settings to display surface information	4
	Create a surface by assembling fundamental data	4
	Use styles to analyze surface display results	4
Parcels	Design a parcel layout	5
Alignments	Design a geometric layout	6
Profiles and Profile Views	Design a profile	7
	Create a profile view style	19
Corridors	Design and create a corridor	9
	Derive information and data from a corridor	10
	Design and create an intersection	10
Sections and Section Views	Create and analyze sections and section views	9
Pipe Networks	Design and create a pipe network	13
Managing and Sharing Data	Create a data sharing setup	17
Plan Production	Create a sheet set	16
Survey	Create a topographic/boundary drawing from field data	4

Index

Note to the reader: Throughout this index **boldfaced** page numbers indicate primary discussions of a topic. *Italicized* page numbers indicate illustrations.

Symbols and Numbers

* (asterisk) as wildcard, 94
    for layers, 737
2D boundaries, creating and evaluating, 63
2D Wireframe, 140
3D points, prompt for, 10
3D polyline, extracting from corridor feature line, 385
3D Studio File Import Options dialog, *697*, 697

A

abbreviations, **739**, *739*
    settings, **8**
Action Recorder panel, 39
Active Drawing View, 2
ActiveX controls, for viewing part catalog, 508
Add Automatically boundary tool, for corridor, 374
Add Boundaries dialog, 126
Add Contour Data dialog, 111, *111*
Add Distances tool, 68
Add Fixed Curve (Three Points) tool, 214
Add From Polygon tool, for corridor, 374
Add Interactively boundary tool, **430–433**
    for corridor, 374
Add Items tool, 571
Add Labels dialog, *35*, 35, 197, *198*, 239–240, *240*, 267, *267*, 496, *805*
    for intersection, 424, *425*
Add Point File dialog, *113*, 113–114
Add Points To Surface wizard, *158–159*
Add Rule dialog, *519*, 519, 524
Add To Surface As Breakline tool, 620
Additional Broken References dialog, 688, *688*
Adjacent Elevations By Reference tool, 626, 628
AECC_POINTs, converting to Civil 3D points, *77*, 77

Alignment Design Check Set dialog, 220, *220*
Alignment From Corridor utility, 385
Alignment From Network tool, 560
Alignment Intersection label style, *239*
Alignment Labels dialog, 237, *237*, 802
Alignment Layout Parameters dialog, 222, 225–226, *226*
Alignment Layout Tools toolbar, *212*
Alignment Properties dialog
    Constraint Editing tab, 228, *229*
    Design Criteria tab, *235*, 235, *236*
    Information tab, 232
    Masking tab, 210, *211*
    Point Of Intersection tab, 228, *228*
    for renaming objects, 232
    Station Control tab, 234, *234*
alignment segment table, **243**
alignment tables, **240–243**
alignment target, 362, *363*
alignments
    added to site, *188*
    best fit from lines and curves, **215–217**, *217*
    changing components, **229–230**
    component-level editing, **225–226**
    constraints, **226–228**, *227*
    corridor feature lines extracted to produce, 359
    creating, **207–222**
        by alignment, **211–215**
        from line, arc, or polyline, **208**, **210–211**
    creating design constraints and design check sets, **219–222**
    creating shortcut references, **681–685**
    data shortcuts for, 679
    editing geometry, **222–230**
        grip editing, *223*, **223–224**, *224*
        tabular design, **224–225**
    grip-editing, *251*

information in profile, 247
intersections of, label for, 803
label sets for, *799*, **801–803**
    applying, 802–803
labels for, **236–240**
    reference text in, *806*
    styles, **799–814**
as objects, **230–243**
for pipe networks, creating for parts, **548–549**
placement on site, 162, *163*
Plan Production and, 648
reversing, 233
for roundabouts, **445–450**
    with slip lane, *450*
and sites, **205**
stationing, **233–235**, *234*
styles, 209, *752*, 752
    creating, **754**
    station offset labeling, **803–805**
superelevation and, **469–470**
types, **206–207**
Ambient Settings, **8–13**, *9*
    in civil 3D templates, *740*, **740–743**
Analyze tab on ribbon
    Design panel, Edit in Storm And Sanitary
        Analysis, 599
    Ground Data panel
        Catchments, Create Catchment From
            Object, 579
        Catchments, Create Catchment Group,
            577, *578*
    QTO panel
        QTO Manager, *712*, 712, 715
        Quantity Takeoff, 726
    Volumes And Materials panel, 145, 377
        Compute Materials, 489, 491
        Mass Haul, 498
        Total Volume Table, 490
        Volume Report, 493
anchor points
    for curve labels, 274
    relationship with attachments, *781*

Angle-Distance Transparent command, 68
    icon, *37*
angle, for creating line, 22, *22*
Angle Information tool, 68
Annotate tab on Ribbon
    Add Labels, Alignment, Alignment
        Labels, 238
    Add Tables, Alignment, 241
    Labels & Tables panel, 35, *35*
        Add Labels, 196, 197, 241, 267, 304, 496, 636
        Renumber Tags, 201
annotation scale, 734
annotations, *780*
    in profile views, **304–305**
AOR (axis of rotation) subassembly, **465**, *465*
Applications panel, 39
Apply Feature Line Names tool, 620
Apply Feature Line Styles tool, 620
Apply Rules tool, 559
Apply To X-Y field, for description key, 93
Apply To Z field, for description key, 93
arch pipe, adding to part catalog, **518**
arcs
    Best Fit for, *32*, 32
    creating alignment from, **210–211**
area labels, splitting on 2 layers, **196**
arrow
    invisible for labeled pipe, 794, *796*
    for label, 782, *783*
    for offset alignment grip, 211
    in styles, 752–753
ASCII text file, for survey database source, 51
assemblies, 309. *See also* subassemblies
    building, **312–329**
    center-point marker for, 328, *328*
    of corridors, **349**, *349*
    creating
        for intersection, *414*, **414**
        for nonroad uses, **327–329**
    with daylight subassemblies, *337*
    editing, **325–327**
    for intersection, 402, *403*

offset, *340*, **340**
organizing within drawings, **344**
pre-cooked, **312**, *313*
and Quantity Takeoff, 717
road, creating, *313*, **313–318**
for roundabouts, *453*, 453–454
storing completed on tool palette, **343–344**
without superelevation, *459*
tools for organizing, **342–344**
assembly offset, for corridors, **433–437**
Assembly Properties dialog
    Construction tab, *326*, 326–327, *327*, 328
    Information tab, 326
assembly sets, **403–404**
    changing, *316*
Assign Elevations dialog, 615
asterisk (*) as wildcard, 94
    for layers, 737
Astronomic Direction Calculator, *60*, *61*
Asymmetric curve type, 253
Attach Multiple Entities command, **34**, *34*
attached parcel segments, *174*, **174–175**
    sliding, *182*
attachment point, default, for subassemblies, 321
attachments, relationship with anchor points, *781*
AUNITS variable, 790
AutoCAD
    attribute extraction, for converting outside program point blocks, **77–80**
    exporting, **696**
    functions for plan sets, 658
    geometry to create parcels, 167
    visual styles, 140
AutoCAD angular units, vs. Surveyor Units, 67–68
AutoCAD BLOCK Symbol For Marker, 749
AutoCAD drawing units, 740
AutoCAD Map product, 38
AutoCAD Object Properties Manager (OPM) palette, *231*, 231
AutoCAD objects, as pay items, **715–716**
AutoCAD point entities, converting, **81**
AutoCAD POINT For Marker, 749

AutoCAD Properties palette, 313
    for assemblies, *316*, 316
    Design tab, 316, 328
    Dimension Anchor Option, *564*
AutoCAD units, 11
autocomplete feature, for commands, 39
Autodesk 3D max files, importing, **697–698**, *698*, *699*
Autodesk Map 3D, 582
Automatic Begin On Figure Prefix Match option, 47
Automatic Layout, for parcels, **176**
Automatic Mode parameter, 176
Automatic Surface Adjustment, 553
Automatic Surface Creation option, for grading group, 642
axis of rotation (AOR) subassembly, *465*, **465**
Azimuth Distance transparent command, icon, *37*

B

background mask, for label components, 782
backward lanes, troubleshooting, 421
Balanced, in Mass Haul diagram, 498
Band Set dialog, Information tab, *306*, 306
band sets, 825
    for profile view labels, **305–307**
    profile views after importing, *307*
banding. *See also* data bands
    elevations, *139*, **139–140**
banking, 459. *See also* superelevation
baseline profile, for corridors, 355, *356*, 357
baselines
    adding to corridor for intersecting road, 412–413
    for corridors, **349**
    for grading, 637
    for intersection, *402*, **415–419**, *420*
    multiple, for cul-de-sac, **392–393**, *393*
    multiregion, for corridors, **390–391**
`Basic Site.dwg` file, 5
BasicCurbandGutter subassembly, 314

BasicLane subassembly, 362, 435, *435*
BasicShoulder subassembly, 324, *324*
BasicSideSlopeCutDitch subassembly, 338, *339*
BasicSidewalk subassembly, *318*
Bearing Distance transparent command, 36, 741
    icon, *37*
Begin Full Super (BFS), *461*
Begin Shoulder Rollover (BSR), 460, *460*
Behavior, for alignment table, 241
benchmark point style, Marker tab, *749*
berms, designing to contain swales, 339
best fit, alignments from lines and curves, **215–217**, *217*
Best Fit Entities, **31–33**
    arc, 32, *32*
    lines, *32*, **32**
best practices
    for parcel creation, **185–193**
    point groups to control point display, **90**
    for site topology interaction, **161–164**
BFS (Begin Full Super), *461*
bike path, *433*, 433, *434*
    with assembly offset, 434–437
Block command (AutoCAD), 697
blocks, structure style using, 772
Blunder Detection/Analysis, 57, *58*
border
    adding for surface, **123–127**
    adjusting with grips, *125*
    for label components, 781
Borrow, in Mass Haul diagram, 498
boundaries, 105
    for catchment, 576
    for corridor surface, **372**, *373*, **374–377**
        troubleshooting, 376
    data clip, 126
    of structures, in profile view, 771
boundary parcel, creating, **167**, *168*
Bounded Volumes tool, 149
bowties, troubleshooting, *437*, **437–440**, *438*
branching, for feature lines, 358
Break tool, for feature lines, 621

breaklines, 44, 105, **122–123**
    and building surface from links, 371
    crossing, 118, 123, 438
    in intersection design, 429
    nondestructive, 122, *126*, 126
    proximity, 118
bridges, assembly for, **327–329**
Browse For Folder dialog, 677, *677*
BSR (Begin Shoulder Rollover), *460*, 460
buffers, 190
    in profile view band set, 655
bypass link, 592

C

CAD programs, **690–703**
    earlier versions of Civil 3D or Land Desktop, **691–696**
    eTransmit, **693–695**
    importing files, to SSA, 590
    Land Desktop, **695**
CAD Standards panel, 39
Calculate Superelevation screen
    Attainment, 470, *471*
    Lane, 469, *470*
    Roadway Type, *469*, 469
    Shoulder Control, 469, *470*
camber, 459. *See also* superelevation
cant, 459. *See also* superelevation
Carlson
    data collectors, 52
    survey blocks, 78
catchment areas, GIS data and, **582–584**
catchment boundary, 576
catchment flow path, 576
catchment groups, 575
Catchment object, changes to, 7
catchments, **575–581**
    creating, 575, **575–576**
    delineating, 577–581, *580*
    options, **577–581**
categorization files, creating, **728–730**
ceiling value for surface, 118

center-point marker, for assembly, *328*, 328
Centerline alignment type, 209
Change Flow Direction tool, 559
channel assembly, **327–329**, *329*, *334*, *335*
Channel subassembly, 378
Character Map dialog, *793*, 793
child styles, and overrides, 747
Circular curve type, 253
circular grip, 223, 255
Civil 3D
    earlier versions, **691–696**
    modeling components, 7
    tutorials for creating custom structures, 512
    underlying engine, **38–39**
Civil 3D Line and Arc tool (CGLIST), 67
Civil 3D objects, order for placement, refining, 213
Civil 3D subassemblies tool palette, *310*
cliff effect, troubleshooting, *421*, 421–422
Clip Horizontal Grid option, 818
Clip Vertical Grid option, 818
closed polygon, 189, *190*
    for parcels, 186
Code field, for description key, 92
Code Set Style - All Codes dialog, *719*
    Codes tab, *832*
coding diagram, *322*, **322**
COGO (coordinate geometry) points, 153
    vs. survey points, **72**
color
    of label component, 780
    scheme for contours, 136
columns in tables, adding or deleting, 815
commands
    autocomplete feature for, 39
    saving changes to settings, 11, 740
    settings for, 14, *14*, 745
Component Draw Order, for label, 779
component layers, 748
component-level editing
    for alignments, **225–226**
    for profiles, **264–265**

Composite Volume palette, 377
compound curves, creating, **29**, *29*
Compute Materials dialog, *489*
Compute Quantity Takeoff dialog, 726, *727*
Connect and Disconnect Part Tools tool, 559
connection marker, for pipe networks, *536*
constraints, for alignments, **226–228**, *227*
Content Builder, *516*
Continuous Distance tool, 68
Contour Analysis tool, 148–149
contours
    basics, **137–140**
    color scheme or linetype for, 136
    labeling, **150–151**, *151*
    smoothing, vs. surface smoothing, 138
    in surface styles, 755, **756–758**, *757*
control points, in survey database, 52
Convergence angle, 736
Convert AutoCAD Line And Spline tool, 260
Conveyance Link dialog (SSA), 591, 594
conveyance links, overland, 599–600
coordinate geometry (COGO) points, 153
    vs. survey points, **72**
Coordinate Geometry Editor, *63*, **63–64**
    traverse report from, *65*
coordinate line commands, **18–20**, *19*
coordinate system
    of DEM file, 106
    setting, 734
    setting up drawing for specific area, 6
    transformation, 75
    troubleshooting, 369
    warning of no setting, 790
COORDS command (AutoCAD), 630
COPY command, for profile views, 295
Copy Profile Data dialog, *266*, *266*
Copy Profile tool, 260
Corridor Extents As Outer Boundary method, 374
Corridor Modeling Catalog, 309, *310*, **310–311**, 323, 331
Corridor Properties dialog, 375
    Boundaries tab, 374, *374*
    Feature Lines tab, *358*, 358–359

Hide Boundary setting, 433
Parameters tab, *354*, 354, 355, 357, *357*, 411, 417, *417*
Surfaces tab, *376*, 376, *478*
corridor stations, 355
corridor surfaces, 423, *424*, 429–430
adding boundary, **372**, *373*, **374–377**
creating, *369*, **369–377**
from feature lines, **371**, *372*, **372**
from link data, 370–371, *372*, **372**
refining, **429–430**, *430*
troubleshooting, **376–377**
Corridor tool palette, daylight subassemblies, *335*
corridors, 353, **477–478**, *478*. *See also* intersections; roundabouts
in 3D view, *347*
anatomy of, *351*
assembly offset, **433–437**
baseline of, **349**
baseline profile for, 355, *356*, 357
basics, **347–348**
checking and fine-tuning, *423*, **423–429**
completed model, *443*
creating, **348–358**
creating sample lines along, **481–482**
creativity in using, **383–384**
for cul-de-sac, **392–400**
EOP design profiles, **393–394**
with multiple baselines, **392–393**, *393*
putting pieces together, **394–397**, *395*, *396*, *397*
daylighting for, troubleshooting display, 357–358
editing sections, *367*, **367**
feature lines, 348, **350–354**, *358*, **358–361**
with guardrails, 390–391, *391*
intersection modeled with, *348*
lane widening, targets using alignments and profiles, 362–364, *363*
lot-grading feature line integration with, **441–443**, *442*
materials from shapes, 489
multiregion baselines, **390–391**
non-road, **378–379**, *379*

Parameter Editor for, *368*
for pipe trenches, 329, *380*, 380–381, *381*
pricing, **717–720**
properties modification, **366**
rebuilding, **354–355**, *355*
and surface update, 375
regions, **349**
for roadside swale, **384–390**
Section Editor for, *437*
stream modeled with, *348*
troubleshooting problems, **355–358**
volume calculation, **377–378**
cost estimates. *See* Quantity Takeoff (QTO) feature
Cover And Slope rule, 521, *522*, **533–534**
Cover Only rule, 521, *522*
Create Alignment From Objects dialog, 208, *208*, **209**
for swale, 385, *386*
Create Alignment - Layout dialog, *212*, 212, 213, 216, 217–218, 221, *221*
Create Alignment Reference dialog, *681*, 681
Create Alignments From Objects dialog, *360*
Create Assembly dialog, *312*, 312, 315
for channel, 327–328
Create Best Fit Arc command, **32**, *32*
Create Best Fit Entities menu, *31*
Create Best Fit Line command, *32*, **32**
Create Best Fit Parabola command, **33**, *33*
Create Catchment From Object dialog, *579*, *579*
Create Catchment From Surface dialog, *578*, *578*
Create Cropped Surface dialog, *128*, 128–129, *129*
Create Curve Between Lines command, **26–27**, *27*
Create Curve From End of Object command, **29**, *29*
Create Curve on Two Lines command, **27**, *27*
Create Curve Through Point command, **27**, *28*
Create Curves commands, *26*
Create Data Shortcuts dialog, *680*
Create Feature Line From Alignment dialog, *616*
Create Feature Line From Stepped Offset tool, 612, 616–617, *617*
Create Feature Line tool, 612
Create Feature Lines dialog, *613*, 613, 615, 633
Create Feature Lines From Alignment tool, 612

Create Feature Lines From Corridor tool, 612
Create Feature Lines From Objects tool, 612
Create Full Parts List tool, 559
Create Grading Group dialog, *638*, 638
Create Interference Check dialog, 565, *565*
Create Interference Check tool, 560
Create Intersection wizard
    Corridor Regions page, 403–404, *404*, 409, *409*
    General page, *405*, 405
    Geometry Details page, *406*
Create Line By Angle command, *22*, **22**
Create Line By Azimuth command, *22*, 22
Create Line By Bearing command, **21**, *21*
Create Line By Deflection command, **22**, *22*
Create Line By Grid Northing/Easting command, **20**
Create Line By Latitude/Longitude command, **20**
Create Line By Northing/Easting command, **20**
Create Line By Point Name command, **20**
Create Line By Point Object command, **20**
Create Line By Point # Range command, **18–19**, *19*, 20
Create Line By Side Shot command, **23–24**, *24*
Create Line By Station/Offset command, **22–23**, *23*
Create Line command, 18
Create Line Extension command, *24*, **24**
Create Line From End of Object command, *25*, **25**
Create Line Perpendicular from Point command, **25**, *25*
Create Line Tangent from Point command, **25**, *25*
Create Mass Haul Diagram dialog, 498
    Balancing Options, *500*, **501**
    General page, 498, *499*, **500**
    Mass Haul Display Options, 498, *499*, **501**
Create Multiple Curves command, **28**, *28*
Create Multiple Profile Views wizard, 286–287, *287*, *666*, 666
    Data Bands page, *667*
Create Multiple Section Views wizard
    General page, *669*
    Section Display Options page, *669*
Create Offset Alignments dialog, *210*
Create Parcel From Objects tool, 167, 189k, *190*
Create Parcels - From Objects dialog, 167

Create Parcels - Layout dialog, 167
Create Parts List tool, 559
Create Pipe Network dialog, *528*, 528–529
Create Pipe Network From Object command, 536
Create Point Cloud wizard, *156*
    summary page, *157*
Create Points dialog, **72**, *73*, **84**, 84
Create Points toolbar, **81–85**, *82*
    Alignment Point-Creation Options, 83, *83*
    Interpolation Point-Creation Options, *83*, 83–84
    Intersection Point-Creation Options, *82*, 82
    Miscellaneous category, 82
    Slope Point-Creation Options, *84*, 84
    Surface Point-Creation Options, 83, *83*
Create Profile dialog, 252, *252*
    for swale centerline, *387*
Create Profile - Draw New dialog, *360*
Create Profile From Surface dialog, 246–247, *247*, *248*, *281*, *284*, *285*, 548–549, *549*
Create Profile View wizard, *282*, 282, 288, 549, 562
    General page, *549*
Create Quick Profiles dialog, *620*
Create Reverse or Compound Curves command, **29**, *29*
Create Right Of Way dialog, *171*
Create Roundabout dialog
    Approach Refs, *446*
    Circulatory Road, *446*
    Draw Slip Lane, 449, *449*
    Islands, *447*
    Markings And Signs, 447, *448*
Create ROW tool, *170*, **170–171**, *172*, 190
Create Sample Line Group dialog, 480, *480*, 482, *482*
Create Sample Lines - By Station Range dialog, *481*, 481
Create Section Sheets dialog, *670*, 670–671
Create Section View Wizard
    Data Bands page, 486, *486*
    Elevation Range page, *485*
    General page, 484, *484*
    Offset Range page, 484, *485*
    Section Display Options page, 485, *486*
    Section View Tables page, 486, *487*

Create Sheets Wizard, **659–664**, 666
    Data References, *664*, **664**
    Profile Views page, *663*, **663–664**
    Sheet Set, **662**, *663*
    View Frame Group And Layouts, *660*, **660–662**
Create Simple Corridor dialog, *352*
Create Surface dialog, 146, *147*, 642
Create Total Volume Table dialog, 490, *490*
Create Transmittal dialog, 693, *694*
Create View Frames wizard, **648–655**
    Alignment page, *649*, **649–650**
    Match Lines page, **653–654**, *654*
    Profile Views page, *655*, **655**
    Sheets page, *650*, **650–652**
    View Frame Group page, *652*, **652–653**
Create Widening command, 211
Creating Grading tool, 639, *639*
criteria-based design, 252
Criteria dialog, 565, *566*
critical points, 153
cross-referencing design data. *See* data shortcuts
Cross-Section Area, *603*
cross sections, 477
cross-slope, 459. *See also* superelevation
crossing breaklines, 118, 123, 438
crossing pipe override style, 769, *769*
Crossing Section label, 794
CUI (custom user interface), 39
cul-de-sac
    corridors for, **392–400**
        EOP design profiles, **393–394**
        with multiple baselines, **392–393**, *393*
        putting pieces together, **394–397**, *395*, *396*, *397*
    creating, **172–173**, *173*
    troubleshooting, *398*, **398–400**, *399*, *400*
culvert crossing, 384
culverts, 586
curb ramp feature line, *629*
Curb Return alignment type, 209
Curb Return Fillets assembly, *418*
curb subassemblies, **324–325**
    and superelevation, 464

Curve Calculator, *34*, **34–35**
curve labels, *274*, **274–275**
    location for, 275, *276*
    styles, **792–793**, *794*
Curve Settings tool, 260
curve table, *203*
curve tags, for table creation preparation, **200–201**
curves, **26–36**
    adding labels, **35–36**
    best fit alignments from, **215–217**, *217*
    and frontage offset, 179
    label styles, **790–793**
    labels, styles, **810–812**, *811*
    re-creating deed using, **30–31**, *31*
    reverse curves
        creating, **217–219**, *218*, *219*
        removing, 229–230
    standard, **26–29**
custom marker, 749
custom user interface (CUI), 39
Customization panel, 39
cut-fill analysis, 146

D

data bands
    profile view styles for, **823–826**
    on profile views, **298–302**, *300*
data clip boundaries, 126
data collector market, manufacturers, 52
Data Extraction tool, *78*, 78–80, *79*, *80*
data shortcuts, **675–690**
    creating, **679–681**
    displaying, 2
    fixing broken, **687–688**
    folders for, **677–679**
    in Prospector, 4
    publishing files, **676–681**
    setting up projects in, 679
    using, **681–690**
    value of knowing, 684
Data Shortcuts Editor (DSE), 687, *689*, **689–690**
Data Shortcuts panel, 39

database setup for Survey, **41–64**
    defaults, **41–42**
    equipment database, **43–44**
    figure prefix database, **44–46**
    linework code set database, **46**
    survey database, **47–50**
datum links, schematic of those connected to form surface, *371*
datum surfaces, overhang correction for, 370–371
Davenport, Cyndy, 516
daylight subassemblies, *335*, **335–336**
    alternative, **338–339**
    when to ignore parameters, 337
DaylightBasin subassembly, *339*, 339
daylighting
    for corridor, troubleshooting display, 357–358
    with generic links, **334**
DaylightInsideROW subassembly, 336
DaylightMaxOffset subassembly, 314, *315*, 318, *318*
DaylightToROW subassembly, *338*, 338
Decimal Degrees, for AutoCAD angular units, 68
`Deed Create Start.dwg` file, 30
deed, creation using Line and Curve tools, **30–31**, *31*
default attachment point, for subassemblies, 321
default layer, for points, **72**, *73*
deflection angle, **22**
Deflection Distance transparent command, icon, *37*
DeKalb Rational hydrology methods, 589
Delaunay triangulation, 103
Delete Elevation Point tool, 625
    for Grading Elevation Editor, 627
Delete Entity tool, 261, 265
Delete Line Surface, triangles for, 759
Delete PI tool, for feature lines, 621
Delete Pipe Network Object tool, 534
Delete PVI tool, 260
Delete Sub-Entity tool, 182, *183*, *184*
deleting
    connection between points, 129
    from cul-de-sac, *173*

data files from Coordinate Geometry Editor, 63, *64*
    parcel segments, **182–185**
    point from surface, 130
    surface boundary, 127, *127*
    unassigned subassemblies, 344
    water drop path, 574
delimited text file, importing points from, 72
Delta symbol, 36
delta symbol, 793
    in styles, 792
DEM (Digital Elevation Model) files, 105
    surfaces from, **106–108**
depression contours, 138
depth label, 305, *305*
DescKey Editor, in Panorama window, 94
Description Key Set, 91, *92*
    activating, **95–96**
    creating, *93*, **93–96**
    editing, *94*
Description Key Sets Search Order dialog, *96*
description keys, **91–96**
    and layers, 96–97
    vs. point groups, **94**
design check sets, creating alignment with, **219–222**
design checks, vs. design criteria, 221
design constraints, creating alignment with, **219–222**
Design Criteria Editor, 462, *462*
    adding example data, *464*
design criteria files, **461–463**
    adding design speed, *463*, 463
    user-defined, 461
design criteria, vs. design checks, 221
design profiles, reversing alignment and, 233
design speeds, assigning, **235**
DETACHQTOFILES command, 712
DGN files, 699
    as underlays, 703
DGN Mapping Setup dialog, *702*
diameters for pipe, 541
diamond grip, for sample line, 479, *479*

Digital Elevation Model (DEM) files, 105
    surfaces from, **106–108**
Dim Anchor Opt value, 275
direction
    for alignment, 233
    ambient settings for, 9–10, 741
    in survey database, 53
direction analysis, 136
    in surface styles, 756
direction arrow, in styles, 752–753
direction-based line commands, **21–25**
discharge point, for catchment, 576
display styles, for Plan Production, 647
documentation
    process, 222
    requirements, 240
    for sanitary sewer planning, 505–507, *506*
drainage
    basins, 575. *See also* catchments
    for roundabout, **444**, *444*
draining pond, **632–636**
draping image, 109
Draw Fixed Vertical Curve By Three Points
    tool, 265
Draw Order Icons tool, 558
Draw Pipes And Structures tool, 535
Draw Tangent-Tangent With No Curves tool, 167
Draw Tangent tool, 260
Draw Tangents With Curves tool, 260
drawing objects, 105. *See also* objects
drawing precision, vs. label, 9
Drawing Scale, settings, 11
drawing settings, **5–6**
    in civil 3D templates, *734*, **734–743**
        Abbreviations tab, *739*, **739**
        Ambient Settings, *740*, **740–743**
        Object Layers, **736–739**, *737*
        Object Settings, **743–745**
        Transformation settings, **735–736**
        Units And Zone settings, **734–735**, *735*
    number of layouts per new drawing, 661
Drawing Settings dialog, *5*
    Units And Zone settings, *106*

drawing templates, 647, 671–672
    displaying, 2
    in Prospector, *4*, **4–5**
Drawing Unit, settings, 11
drawings. *See also* exercise files
    cross-referencing design data between. *See*
        data shortcuts
    organizing assemblies within, **344**
    precision, vs. label precision, 742
    sending, eTransmit for, **693–695**
DRAWORDER command, 387
Drive Alias Administration dialog, *705*
drive alias, for Map 3D, **704–705**
Driving Direction setting, 11
    and side of road, 741
DSE (Data Shortcuts Editor), 687, *689*, **689–690**
dump site, for Mass Haul diagram, 502, *503*
duplicate segments, troubleshooting, *187*, 187
DWG files. *See* exercise files
DWT template file. *See* template files
dynamic input (DYN)
    pipe networks and, **542–543**, 547
    for view frames, 658
dynamic model, corridor as, 354
dynamically linked feature line objects, 617–618

E

earthwork, 144
    calculating, 489
    depth labels for, 305
    elevation analysis settings, *147*
    volume analysis for, **145–146**
    volume calculation for, 377–378
easements, 190
Easements site, 164
EATTEXT command, 78–80
edge of pavement (EOP) profile, creating,
    393–394
Edit Command Settings dialog, 95, *95*
Edit Curve tool, for feature lines, 621
Edit Drawing Settings command, 734
Edit Elevations panel, *619*, 619, 625
    for feature lines, 620–621

Edit Feature Line Curve dialog, *624*, 624
Edit Geometry panel, *619*, 619
    Edit Curve tool, 624
    for feature lines, 620–621, 622
Edit In Storm And Sanitary Analysis tool, 560
Edit Interference Style tool, 560
Edit Label Style Defaults dialog, *774*, 774–775
Edit Linework Code Set dialog, 47
Edit Material List dialog, *492*
Edit Parcel Properties dialog, 194, *195*
Edit Parts List tool, 559
Edit Pipe Network tool, 559
Edit Pipe Style tool, 559
Edit Sample Lines dialog, 482, *483*
Edit Structure Style tool, 559
Edit Table Style tool, 571
editing
    Drawing Settings, 5
    survey points, 51
Elevation Editor tool, 625, 631
elevation factor, 736
elevation points, 615
elevations, 245–281
    adjusting in profile views, *296*, **296–297**
    analysis, *148*, 148
    banding, *139*, **139–140**
        color, 136
        in surface styles, **763–764**, *764*
    Child Override indicator, 12, *12*
    editing for grading feature lines, **625–628**
    excluding those less than floor, 118
    inheriting by point, 82
    locking, 269
    matching profile, **267–269**
    range for section view, 485
    setting for inlets, 594–595, *595*
    sump control by, 553
    in surface styles, 756
elevations for points
    changing, **90–91**
    prompt for, 73, *73*
Elevations For Surface tool, for Grading Elevation Editor, 628
Elevations From Surface tool, 91, 626
ellipsoid, 735
End Normal Crown (ENC) station, 459, *460*
Endpoint-Location grip, 541
entity, layout profiles by, **257–258**
Environmental Protection Agency Storm Water Management Model (EPA-SWMM) method, 589
EOP (edge of pavement) profile, creating, **393–394**
Equal Interval distribution method, for surface analysis, 761
equipment database, **43–44**
Equipment Database Manager, 44, *44*
Equipment Properties dialog, 43–44
Erase tool (AutoCAD), 182, *183*, 184
    for pipe network parts, 541
error tolerance for survey database, default settings, 43
ESRI Shape files (SHP), 582
eTransmit, **693–695**, 720
ETRANSMIT command, 693
Event Viewer, displaying, 11, 741
Excel. *See* spreadsheets
exercise files
    Acad Objects in QTO.dwg, 715
    Advanced Alignment Labels.dwg, 803
    Alignment Labels.dwg, 801
    AlignmentFromNetworkParts.dwg, 548
    AlignmentLabels.dwg, 237
    AlignmentProperties.dwg, 230, 233
    AlignmentReverse.dwg, 217
    AlignmentReverseEdit.dwg, 229
    AlignmentsBestFit.dwg, 216
    AlignmentsbyLayout.dwg, 212
    AlignmentSegments.dwg, 239, 241
    AlignmentsFromPolylines.dwg, 208
    AOR Assembly.dwg, 466
    ApplyingProfileLabels.dwg, 270
    AutoCAD Civil 3D (Imperial) NCS.dwt, 518
    Basic Site.dwg file, 5
    C3DtoSSA.dwg, 599

Catchment-1.dwg, 577
ChangeAreaLabel.dwg, 194
CheckingAlignments.dwg, 221
Civil 3D Template-in-progress.dwg, 735, 738, 750, 751
Code Set Styles1.dwg, 832, 833
Completed Roundabout Corridor example.dwg, 457
ComponentEditingProfiles.dwg, 264
Concord Commons Corridor.dwg, 431
Concord Commons.txt file, 113
Convert LDT Points.dwg, 81
Corridor Objects in QTO.dwg, 717
Corridor Pipe Trench.dwg, 380
Corridor Stream.dwg, 378–379
Corridor Surface Volume.dwg, 377
Corridor Surface.dwg, 375
Corridor Swale.dwg, 385, *386*
CreateBoundaryparcel.dwg, 167
CreateFreeForm.dwg, 180
CreateROWParcel.dwg, 171
CreateSite.dwg, 165
CreateSubdivisionLots.dwg, 177
CreatingChecks.dwg, 219
CreatingCompositeSurfaces.dwg, 644
CreatingFeatureLines.dwg, 613, 615
CreatingGradingSurfaces.dwg, 641
CreatingPipeTables.dwg, 567
CreatingReferences.dwg, 681
CreatingSectionSheets.dwg, 670
CreatingShortcuts.dwg, *679*, 679, 686
Cul-de-sac_Design.dwg, 394
Cul-de-sac_Profile.dwg, 393
CulDeSacBlock.dwg, 172
Curve Profile Labels.dwg, 810
Deed Create Start.dwg, 30
DeleteSegments.dwg, 184
DescriptionKeys and Layers.dwg, 96
DynamicProfiles.dwg, 249
EditingAlignments.dwg, 224, 225
EditingFeatureLineElevations.dwg, 626
EditingGrading.dwg, 640
EditingPipesPlan.dwg, 546
EditingRelativeFeatureLineElevations.dwg, 628
EditViewFramesAndMatchLines.dwg, 659
eTransmit.dwg, 693
Extract Feature Line.dwg, 359
Feature Line Target.dwg, 441
figure prefix library.dwg, 45
GE_Surface.dwg, 109
Getting Started Categories.xml, 712
Grade Break Labels.dwg, 812
Grade Break Profile Labels.dwg, 276
GradingThePond.dwg, 638
GripEditingProfiles.dwg, 261
Highway 10 Criteria.dwg, 463
Highway 10 Super.dwg, 469
Highway 10 with Bike Path_Assembly.dwg, 434
Highway 10 with Bikepath_Corridor.dwg, 436
HorizontalFeatureLineEdits.dwg, 622
Import3DMax.dwg, 697, 703
InterferenceStart.dwg, 564
Intersection.dwg, 404
Labeling Feature Lines.dwg, 636
LayoutProfiles-PVI.dwg, 252
LayoutProfiles-Transparent.dwg, 255
Line and Curve Labels.dwg, 790, 791
Manual Intersection_Adjustment.dwg, 427
Manual Intersection_Assembly.dwg, 414
Manual Intersection.dwg, 412
Manual Intersection_Labels.dwg, 424
Manual Intersection_Refining.dwg, 429
Manual Intersection_Surface.dwg, 423
Manual Intersection_Targets.dwg, 416
Mapcheck-start.dwg, 61
MassHaul.dwg, 498
Mastering Point Creation.dwg, 84
Mastering Point Display.dwg, 90
Mastering Point Groups.dwg, 88
Mastering Points.dwg, 76
Mastering.csv file, 728
Multi-Region Corridor.dwg, 390
MultipleSectionViews.dwg, 668

Mystery Plat.dwg, 78
ObjectProjection.dwg, 292
OtherProfileEdits.dwg, 266
ParameterEditingProfiles.dwg, 262
Parcel Export.dwg, 582
Pipe and Structure Labels.dwg, 794
Pipe and Structure Styles.dwg, 767, 769, 772, 796
Pipe Network in QTO.dwg, 720, 723
Pipes-Exercise1.dwg, 535
Pipes-Exercise2.dwg, 537
Pipes-Exercise3.dwg, 561
PipeTableStaticDynamic.dwg, 570
Point Cloud.dwg, 157
Point Labels.dwg, 788
Point Table.dwg, 97
PondDrainageDesign.dwg, 632
Profile Bands.dwg, 825
Profile Label Sets.dwg, 813
Profile Labels.dwg, 806, 808
Profile View Styles.dwg, 817
ProfilefromFile.dwg, 260
ProfileSampling.dwg, 246
ProfileViewBands.dwg, 299
ProfileViewBandSets.dwg, 305
ProfileViewLabels.dwg, 304
ProfileViewProperties.dwg, 294, 296
ProfileViewProperties1.dwg, 298
ProfileViews.dwg, 281, 283
ProfileViewsModify.dwg, 303
ProfileViewsSplit.dwg, 284
ProfileViewsStaggered.dwg, 284, 286
QTO Reporting.dwg, 726
RepairingReferences.dwg, 689
RoadsMatchProfiles.dwg, 267
Roundabout Layout.dwg, 445
Section Styles.dwg, 827, 828
sections1.dwg, 482, 483, 489
sections2.dwg, 483
sections4.dwg, 489
sections5.dwg, 493
sections7.dwg, 496
SegmentLabels.dwg, 197, 198
SheetsWizard.dwg, 666
Simple Corridor.dwg, 351
Soil Borings.dwg, 98
SSA-Rain.spf, 604
SSA-Reports.spf, 606, 607
StackedProfiles.dwg, 288
SteppedOffset.dwg, 616
Storing Subassemblies and Assemblies.dwg, 342
Subassembly Practice.dwg, 332
Subassembly Practice.dwg, 336
SuperimposeProfiles.dwg, 290
Surface Styles.dwg, 756, 759, 761, 763
SurfaceAnalysis.dwg, 139
SurfaceBreaklines.dwg, 122
SurfaceContours.dwg, 137
SurfaceCrop.dwg, 128
SurfaceEdits.dwg, 131
SurfaceFromPolylines.dwg, 111
SurfaceLabeling.dwg, 150
SurfacePoints.dwg, 124
SurfaceProperties.dwg, 118
SurfaceSimplifying.dwg, 133
SurfaceSlopeLabeling.dwg, 152
SurfaceVolumeGridLabels.dwg, 154
SurfaceVolumes.dwg, 146, 149
Tables.dwg, 815
Target Practice - Complete.dwg, 366
Target Practice.dwg, 362, 366
ViewFrameWizard.dwg, 655
WaterDrop.dwg, 573
WetlandsParcel.dwg, 167
XMLPlan.dwg, 691
exercises
    deed creation using Line and Curve tools, **30–31**, *31*
    deed labels, **35–36**, *36*
    on figure prefix database, 45–46
    on importing survey data, 49–50
exiting drawing, Section Editor and, 369
expansion factor for soil, 491
exploding objects, 60
Export DGN File dialog, 703

Export Drawing Name dialog, 696
Export To LandXML dialog, *691*, 691–692
Export To Storm Sewers dialog, 599, *599*
exporting
    AutoCAD, **696**
    database settings, 42
    MicroStation files, **703**
    pipes to Storm and Sanitary Analysis (SSA), **584–585**
    SSA data into CAD, 607–609
    superelevation tables, *472*, 472
Express tool, 560
extended properties, default settings, 43
extents of profile views, troubleshooting, 655
external references (XRefs), 675
Extract Objects From Surface utility, 123–124

F

face of surface, 103
FACETDEV system variable, 768
FACETMAX system variable, 768
FBK (Field Book Format) files, 52
Feature Line Properties dialog, Information tab, 617–618, *618*, *619*
feature lines. *See also* grading feature lines
    in corridors, 348, **350–354**, *351*, *358*, **358–361**
    creating surface from, **371**
    creating surface from links and, **372**, *372*
    curb ramp, *629*
    editing
        horizontal information, **620–624**
        tools, **628–632**
    labels for, **636–637**
    moving, 442
    storm drainage pipe network creation from, **536–538**, *537*
    styles, **754–755**, *755*
    on subassemblies, 320
    as width and elevation target, **440–443**
Feature Lines From Corridor utility, 385
FG (finished ground) profile, 415
figure groups, in survey database, 48
figure prefix database, **44–46**

Figure Prefix Database Manager, 44–45, *45*
`figure prefix library.dwg` file, 45
figures, in survey database, 48
File menu (SSA)
    Export, Hydraflow Storm Sewers File, 607, 608, *609*
    Import, Layer Manager, 590, *591*
files
    creating profile from, **259–260**, *260*
    format for importing points, 74–75, *75*
    sharing, LandXML for, 691
Fill Slope value, editing, *641*
Fillet tool, for feature lines, 621
filter, for point groups, 87
finished ground (FG) profile, 415
Fire Hydrant object, 292, *293*, 293
Fit Curve tool, for feature lines, 621
Fixed Property selection field, for Curve Calculator, 34, *35*
Fixed Rotation field, for description key, 93
Fixed Scale Factor field, for description key, 93
fixed segments, *206*, 206
Fixed Vertical Curve command, 258
flared end section, style for, 772–773, *773*
flat areas, minimizing, 130
Flatten Grade Or Elevations tool, for Grading Elevation Editor, 627
Flip Anchors with Text option, for label styles, *776*, 777
Flip Label function, Ctrl+click to access, 238
Floating Curve tool, *214*, *215*
floating segments, 206, *206*
floor for surface, excluding elevations less than, 118
floor Thickness, 548
flow direction
    of pipes, *533*, 533
        changing, **539**
        preview, *538*
flow diversion node, in SSA project, 587
flow path segments, 576
flow paths, locating, 573
flow segment, *581*
    creating, 581

folders
    for data shortcut files, 676, **677–679**
    for part catalog, 507–508
    for plan template, 651
    for survey database, 47–48, *48*
forced insertion, for label styles, 776, *776*
Format field, for description key, 92–93
formula file, 711, **712**
    storing with project files, 720
Free Form Create tool, *180*, **180–181**, *181*
Free Haul, in Mass Haul diagram, 498
free segments, 206, *207*
frequency
    of assembly application to corridor design, 350
    in corridor assembly, *351*
frequency ellipsis, 397
friends, and marked points, *340*, **340–341**
frontage offset, and curves, 179

G

gap, for label components, 782
gapped profile views, creating, **286–287**, *288*
General collection, for object settings, 13–14, 743–745
General Multipurpose Styles collection, shape styles from, 302
General Note labels, **785–787**
General Segment Total Angle, 36
generic links, daylighting with, **334**
Generic Subassembly Catalog, 331
Generic Subassembly tool palette, *332*, **332**
Geodetic Calculator, 60–61, *61*
geoid, 735
geomarker, 790
GEOMARKERVISIBILITY variable, 790
geometry bands, *824*, 824
Geometry Editor, 225
geometry points, **237–238**, *753*
    styles for alignment labels, **801**
Geometry Points dialog, *271*, 271, *807*, *807*
Geometry Points To Label In Band dialog, 299–300, *300*
Getting Started - Catalog Screen dialog, 514, *514*
    tools, 514, *515*
GIS (geographical information system) data. *See also* Map 3D
    catchment areas and, **582–584**
    folder, 705
    Surfaces from, **114–117**, *117*
Google Earth
    surfaces from, **108–109**, *109*
    Timespan tool, 558
grade breaks, *276*, **276–277**, *277*
    label styles, **812–813**, *813*
Grade Extension By Reference tool, 626, 629–630, *629–630*
grading
    creating, **637–640**
    editing, **640–641**, *641*
    and feature line editing, 621
Grading Creation Tools toolbar, 638, *638*
Grading Elevation Editor, 626, **626–628**
grading feature lines, **611–637**
    corridor feature lines extracted to produce, 359
    creating, **612–620**
    editing elevation information, **625–628**
    extracting, 385
    tools, **611–612**
Grading Group Properties dialog, 641–643
    Properties tab, 643–644, *644*
grading groups, 637
    creating surfaces from, **641–645**, *643*
grading infill, 643, *643*
grading objects, **637–645**
Grading site, 164
grass, coefficients for long, 581
grid labels, 153–154, *154*
Grid Northing Easting transparent command, icon, 37
Grid Padding, 818
Grid Rotation Angle, 7, 736

Grid Scale Factor, 6, 736
grid spacing, for profile views, 823
grips
    adjusting border with, *125*
    alignment, *251*
        editing, *223*, **223–224**, *224*
    for feature lines, 615
    for fine-tuning superelevation stationing, *474*
    for label movement, 812
    for match lines, *657*
    for offset alignments, 211
    for pipe ends, 541, *541*
        in profile view, *554*
    Point-Rotation, 86, *86*
    for profile editing, *261*, **261–262**
    for PVI-based layout profiles, 255
    for sample line, *479*, 479
    structure, 540, *540*
    for structure label in profile, 563
    type on layout profile, *254*
    for vertical movement pipe edits, *552*, **552–554**, *554*
    for view frames, *657*
ground elevation, adding to structure labels, **796–798**
Group Plot Style dialog
    Array tab, *828*, 828
    Display tab, 828, *829*
    Plot Area tab, 828, *829*
group plot styles, **828–830**
groups, names, changing, *326*, **326–327**, *327*
guardrail
    corridors with, 390–391, *391*
    options in Daylight subassemblies, 337
gutter slope, in cross-section, *600*

H

hatches
    for parcels, 754
    for pipe walls, 768–769
    for pipes, 765
    for profile views, *302*, **302**
    for section view direction, *768*
    for watershed areas, *762*

HEC-1 for hydrology calculations, 589
help
    for subassemblies, *311*, **311**, 313, **319–322**, *322*
        coding diagram, **322**, *322*
    tutorial for custom structures, 512
Help icon, 3
hidden assemblies, 344
hiding boundaries, 125
hierarchy
    for styles, *784*
Home tab on Ribbon
    Create Design panel
        Alignment, Alignment Creation Tools, 212, 213, 216, 221
        Alignment, Create Alignment From Objects, 208
        Alignment, Create Offset Alignments, 210
        Assembly, Create Assembly, 312, 315, 327
        Corridor, 350–351
        Corridor, Create Corridor, 394
        Create Best Fit Profile tool, 259
        Create Feature Line, 611, *612*
        Feature Line, Create Feature Lines From Alignment, 616
        Feature Line, Create Feature Lines From Objects, 615
        Intersection tool, 401
        Intersections Add Turn Slip Lane, 449
        Intersections, Create Intersection, 405
        Intersections Roundabout, 445, *445*
        Parcel, 172
        Part Builder, 514
        Pipe Network, Pipe Network Creation Tools, 528
        Pipe Network Pipe Network Creation Tools, 535
        Profile, Create Superimposed Profile, 290
        Profile, Create Surface Profile, 281
        Profiles, Create Profile, 260
        Profiles, Profile Creation Tools, 252
    Create Ground Data panel, 81, 84
        Import Survey Data, 48, 49
        Points, 72
        Surfaces, 113

Data panel, Attach tool, 705
Draw panel, 18
Modify panel
    Feature Line tab, *618*, 618–620
    Match Properties, 306
Palettes panel, Tool Palettes, 309
Profile & Section Views panel
    Profile View, Project Objects To Profile View, 292
    Sample Lines, 480, 482
    Section Views, Create Section View, 484
Profile And Section View panel, Section Views, Create Multiple Views, 668
horizontal geometry bands, 299, *824*, 824
horizontal information, for feature lines, **620–624**
Hydraflow, vs. Storm and Sanitary Analysis, 588
hydraulic calculations, 573
hydraulically most distant point, for catchment, 576
Hydrographs tool, 560
hydrology, 573
    methods, **588–598**
        rainfall assumptions, 604

I

icons, in Prospector, 2–3
IDF (intensity-duration-frequency) curve, *596*, 596
    for rainfall computation, 589
Immediate And Independent Layer On/Off Control Of Display Components setting, 8
Imperial units
    conversion to metric, 11
    coordinate settings for DEM import, *106*
    specifying, 6
Import Civil 3D Styles dialog, *733*, 733–734
Import Data From Autodesk Land Desktop Project dialog, 695, *695*
Import DGN Settings dialog, 700, *700*
Import Gis Data Wizard
    Connect To Data screen, 114, *115*
    Data Mapping, 116, *117*
    Geospatial Query, 116, *116*
    Object Options, 114, *115*
    Schema And Coordinates, 115, *116*
Import LandXML dialog, *692*
Import Points dialog, *74*, 75
Import Survey Data dialog
    Import Options page, *50*
    Specify Data Source page, *49*
importing
    Autodesk 3D max files, **697–698**, *698*, *699*
    CAD file, to SSA, 590
    GIS data, *117*
    MicroStation files, **699–702**, *702*
    point cloud, **155–156**
    points
        from Land Desktop database, 77
        from text file, 72, 74, **74–76**
    styles, **732–734**
    survey data, 48
        exercise, 49–50
    text file, point group creation from, 88
Independent layer, 11
inlets
    connecting subbasin to, *592*
    importing into SSA, 599
    setting elevations, 594–595, *595*
    in SSA project, 587
    subbasins assigned to, *592*
Inquiry commands, **66–68**
Insert dialog, for CulDeSacBlock, *172*
Insert Elevation Point tool, 625, 632
    for Grading Elevation Editor, 627
Insert High/Low Elevation Point tool, 626
Insert PI tool, for feature lines, 621
Insert PVI Tabular tool, 260
Insert PVI tool, 260, 265, 427
Insert tab on Ribbon, Import panel, 692
    Google Earth, 108
    Points From File, 76
    Storm Sewers, *609*, 609
inserting, Delta symbol, 36
intensity-duration-frequency (IDF) curve, 596, *596*
    for rainfall computation, 589

interface for Civil 3D 2012, **1–17**. *See also* Panorama window; Toolspace
interference check, between storm pipe and sanitary pipe networks, **564–567**, *565*
Interference Check Properties, 560
interference cloud, 566
interference marker
    in 3D view, *567*
    in plan view, *566*
Interference Properties tool, 560
Internet Explorer
    for viewing US Imperial Pipes catalog, 510
    for volume XML report display, *488*
Internet, free surface information from, **105–109**
intersecting stations, on profiles, 268–269
Intersection Curb Return Parameters dialog, 406–407, *407*, *408*
Intersection Lane Slope Parameters dialog, *407*, *408*
Intersection Offset Parameters dialog, *406*
Intersection wizard, **402–410**
intersections, **401–430**, *420*
    of alignments, label for, 803
    assemblies for, 402, *403*
    assembly creation for, **414**, *414*
    baselines for, *402*
    baselines, regions, and targets for, **415–419**, *420*
    labels for, 424–426, *425*
        centerline design label, *426*
    manually modeling, *412*, **412–413**, *413*
    sketch of required baselines, 401
    troubleshooting, **421–422**
inventorying, pay items, **726–728**, *728*
inward branching, for feature lines, 358
Irregular Cross Sections dialog, *600*, *601*
ISD files, 155
Isolate Objects tool, 558
Item Preview Toggle icon, 3

J

jigs, 254, *256*, 431–433
    parcel, 178, *178*

Join tool, for feature lines, 621
junction node, in SSA project, 587

K

K value, 253
Kirpich method of time of concentration, 597
kriging, 130

L

label precision, vs. drawing, 9
label sets, **236–238**
    for alignment, **801–803**
        applying, 802–803
    for alignments, **798–799**, *799*
    for profiles, **279–281**, *280*
        styles, **806–807**, **813–814**
Label Style Composer, 797
    General tab, *778*
    Layout, *797*
    Layout tab, *778*, 778–779, *780*
label styles, 744
    for view frames, 653
Label Styles collection, 13, *14*
labels
    adding to lines and curves, **35–36**
    for alignments, **236–240**
    components for, 779
    for curves, styles, **810–812**, *811*
    for feature lines, **636–637**
    General Note, **785–787**
    grid of point labels, 153–154, *154*
    for intersection, 424–426, *425*, *426*
    intersection profile/surface comparison, *427*
    layout profiles with, *254*
    leader for, *782*, *783*
    for lines, styles, **808–809**, *810*
    for links, styles, **832**
    modifying single label properties, *803*
    Northing & Easting, 788–789, *789*
    for parcel areas, **193–195**, *194*
    for parcel segments, **196–203**, *197*
        multiple segments, **197–199**, *198*
        spanning segments, *199*, **199–200**

for pipe networks, 558, **560–564**, *561*
for points, styles, **787–789**, **833–834**, *834*
precision, vs. drawing precision, 742
for profile views, styles, **304–307**
for profiles, **270–281**
    applying, **270–272**, *272*
    curve labels, *274*, **274–275**
    grade breaks, *276*, **276–277**, *277*
    line labels, *273*, **273**
    station labels, **272–273**, *273*
    styles, 806
reference text in, 779
in section views, 495
for segments, **239–240**
for slopes, **152–153**
station offset, **238–239**
style for alignments, **799–814**
styles, **773–796**
    grade breaks, **812–813**, *813*
    of lines and curves, **790–793**
    for pipes and structures, **794–795**
for surface points, **151–154**
for surfaces, **150–154**
    contours, **150–151**, *151*
tables as alternative, 567
Land Desktop
    converting points from, **77**, **81**
    earlier versions, **691–696**
    Import dialog, minimizing object selection, 696
    importing into Civil 3D, **695**
    migrating parcels from, 188
    point objects, converting, **81**
    surfaces imported from, *696*
LandXML, 691
LandXML file, 76, 105
    importing, and alignment tangency constraints, 693
    for survey database source, 51
    for volume report, 492
lane slope parameters, 407, *408*
lane subassemblies, **323–324**
LaneBrokenBack subassembly, *324*, 324
LaneInsideSuper subassembly, 323, *323*

LaneOutsideSuper subassembly, 313, *314*, *317*, *323*, 362, 414
LaneParabolic subassembly, *323*, 323–324
LaneSuperelevationAOR subassembly, 465, *466*
LAS file format, **155–159**
Latitude Longitude transparent command, icon, *38*
latitude, prompt for, 10
Launch Pad panel, *384*, 384–385
    Grading Creation Tools, 638
Layer field, for description key, 93
layers, 72
    Civil 3D vs. AutoCAD, 736
    description keys and, 92, 96–97
    for figures, 45
    for label components, 775
    vs. point groups, for display control, **90**
    for points, default, **72**, 73
    setting in template, *738*, 738
    settings, **7–8**, *8*
    splitting area labels on 2, **196**
    for view frames, 653
LAYISO command, 96
layout profiles, **251–261**
    adding length and slope, 808–809
    best fit, **259**
    creating, 251
        from file, **259–260**, *260*
    data shortcuts for, 679
    edited curve length, *264*
    by entity, **257–258**
    grip types on, *254*
    with labels, *254*
    PVI-based, grips for, 255
`LayoutProfiles-Entity.dwg`, 257
layouts for sheets, 661
LC (Level Crown), *460*, 460
leader, for label, 782, *783*
least squares analysis, default settings, 43
legend table, for slopes, 141–142, *142*
Leica, data collectors, 52
Length Check rule, 522, *523*
Leroy method of text heights, 740
Level Crown (LC), *460*, 460

LiDAR (Light Detection and Ranging), for point cloud collection, 154
Line And Arc Information tool, 66–67, *67*
line commands, 18–25, *19*
    coordinate, **18–20**, *19*
    direction-based, **21–25**
    re-creating deed using, **30–31**, *31*
line table, creating, **242**
linear object styles, **752–754**
lines
    adding labels, **35–36**
    best fit alignments from, **215–217**, *217*
    Best Fit for, *32*, **32**
    creating alignment from, **210–211**
    label styles, **790–793**
    labels, *273*, **273**
        styles, **808–809**, *810*
    sample, *479*, **479–483**
linetypes
    for contours, 136
    for feature line objects, 618
    for label component border, 782
lineweight
    of label component, 780
    of label component border, 782
linework code set database, **46**
link data, for creating corridor surfaces, 370–371
Link Styles collection, 13
links
    in corridor assembly, 348, *351*
    creating surface from feature lines and, *372*, **372**
    generic, **331–334**, *332*
        daylighting with, **334**
    labels, styles, **832**
    to marked points, **341**
    overlapping, 437
    in SSA project, 587
    styles, **832**
    in subassembly, **319**, *319*
LinkSlopesBetweenPoints subassembly, 435
LinkSlopetoSurface subassembly, **334**, *335*
LinkSlopeToSurface subassembly, 436
LinkToMarkedPoint2 subassembly, 341
LinkWidthandSlope subassembly, 332
LinkWidthAndSlope subassembly, 441, 442
LIST command (AutoCAD), 67
List Slope tool, 66
ListAvailablePointNumbers command, 76
location, assembly to flag, 340
locking elevations, 269
longitude, prompt for, 10
lot-grading feature line, integrating with corridor model, **441–443**, *442*
lot lines, 45
Low Shoulder Match (LSM), *460*, 460
LSM (Low Shoulder Match), *460*, 460

M

macros, adding to Toolbox, 15–16
Major Grids style, for profile views, *303*
major station labels, 236
    styles, **800**
Major Tick Details
    for grid spacing, 823
    for profile view style, 819
Manage tab on Ribbon, **39**
Manning's roughness, 581
Map 3D, **703–708**
    attaching drawings, **706**
    drive alias for, **704–705**
    queries, **706–708**, *707*, *708*
    setup, **704–706**
Mapcheck Report, 61–62, *62*
marked points
    and friends, *340*, **340–341**
    linking to, **341**
    subassemblies with, 341
marker points
    in corridor assembly, 348, *351*
    and feature line, 350
    in subassembly, 319, *319*, **320**, *320*
marker styles
    creating, **750–751**
    View Direction for, *749*
Marker Styles collection, 13

markers, for alignment style, *752*
MarkPoint assembly, *435*, 435
mask, for parts, 771
Mass Haul, **496–502**
Mass Haul diagram, *497*, **497–498**
    creating, **498–499**
    editing, **501–502**
    with free haul diagram, *502*
Mass Haul Line Properties dialog, 501, *502*
master coordinate zone, for database, 43
Master View mode, 2
match lines, *649*
    editing, **657–659**
    grips for, *657*
    length changes, 659
    for view frames, 648, 653–654, *654*
matching transparent commands, **38**
materials list
    adding soil factors, **491–493**
    creating, **489**
    volume calculation, **488–493**
maximum curve radius, for vertical curves, 259
Maximum Pipe Size Check rule, 519, *520*
measurement types, default settings, 43
median islands for roundabouts, 454, *455*
Merge Network tool, 559
merging pipe networks, 538
Metric Pipes catalog, 507
Metric Structures catalog, 507
metric units
    imperial conversion to, 11
    specifying, 6
MicroStation files
    exporting, **703**
    importing, **699–702**, *702*
MicroSurvey, data collectors, 52
migrating parcels, from Land Desktop, 188
Minimum Radius Tables from AASHTO, 462
Minor Tick Details, for profile view style, 819
Mirror Subassemblies tool, 318, 333
Miscellaneous alignment type, 209
mistakes. *See also* troubleshooting
    with subassemblies, correcting, 317

modeless toolbar, 252
modelspace linetype scale, 790
Modified Rational hydrology methods, 589
Modify commands
    for pipes, 541
    for structures, 540
Modify DGN Mapping Setup dialog, *702*
Modify panel
    Edit Elevations panel, 628
    Grading Group Properties tool, 641–643
Modify Point option, 130
Modify Profile panel, Geometry Editor, 266
Modify tab on ribbon
    Design panel, Assembly, 310
    Modify panel, Create Grading Infill, 643
Modify Transmittal dialog, 693, *694*
Move a View Frame grip, 658
Move Point option, 130
Move PVI tool, 260
Move To Assembly option, for subassemblies, 317
Move To Site dialog, 638
moving
    feature lines, 442
    parcel segments, *187*
MSLTSCALE variable, 790
MTEXT, 785
multiple curves, creating, 28, **28**
multipurpose styles, *13*, 13, *744*, 744
multiregion baselines, for corridors, **390–391**
Murphy, James, 708

N

names
    of alignments, 209, 415
    of assemblies, changing, 326
    of feature lines, 614
    of groups, changing, *326*, **326–327**, *327*
    of objects, changing, **230–233**
    of pipe networks, 528
    of points
        vs. description, 74
        prompt for, *73*, 73

of profiles, 415
for sheets, 665
of subassemblies, 315, *349*
changing, *326*, **326–327**, *327*
of view frame group, 652
in XML files, 682
National Resources Conservation Service (NRCS), 576
Natural Neighbor Interpolation (NNI), 130
for surface smoothing, 132, *133*
natural vertices, 192, *192*
spanning labels and, 199
Net Cut values, in Mass Haul diagram, 497, *497*, *501*, 501
Net Fill values, in Mass Haul diagram, 497, *497*
network groups, in survey database, 48
Network Layout Tools toolbar, *529*, **546–547**
Network Properties tool, 559
New Data Shortcut Project Folder dialog, *678*, *678*
New Local Survey Database dialog, 59
New Point Cloud Database - Processing In Background dialog, 156
New Transmittal Setup dialog, 693
NNI (Natural Neighbor Interpolation), 130, 132, *133*
_NO DISPLAY band style, 830
nodes, in SSA project, 587
non-control points, in survey database, 53
non-road corridors, **378–379**, *379*
nondestructive breakline, 122, 126, *126*
<none> site, 165
Northing & Easting label, 788–789, *789*
Northing Easting transparent command, icon, 37
NRCS (National Resources Conservation Service), 576
Null Assembly, 404
null structures, in pipe network, 527

O

Object Settings, **13–14**
object styles, **745–773**
child styles and overrides, 747
Display tab, *748*, **748**
Information tab, *746*, 746
linear, **752–754**
Marker tab, *749*, **749–750**
for Plan Production, 647
Summary tab, 746, *746*
Object Viewer, 558
for corridor, 390
exaggerated surface model, *758*
objects. *See also* drawing objects
alignments as, **230–243**
creating catchment from, 577
exploding, 60
feature line dynamically related to, 615
grading, **637–645**
multiple pay items on single, 725
parent layer of, 8
projecting, **291–293**, *293*, *296*
references to other, 238
renaming, **230–233**
settings, **743–745**
Offset alignment type, 209
from centerline alignment, 210
offset assemblies, *340*, **340**
advantages, 434
and axis of rotation superelevation assembly, 465
offset range, for section views, 484
Offset tool (AutoCAD), 190
one-point slope labels, *152*, 152
Open Drawings branch of Prospector, *3*, **3–4**
Open Pay Item File dialog, 712, *713*, 728
Open Source Drawing command, 685
open space, 190
areas dedicated for, Free Form Create tool for, *180*, 180–181, *181*
Options dialog, Files tab, 671, *672*
orientation reference, in label styles, *775*, 775
Orientation Reference setting, 750
Origin Point, in Mass Haul diagram, 498
Ortho mode, 269, 295
orthometric height scale, 736
outer boundaries, 125
outfall node in SSA project, 587
elevation and settings, *595*

output parameters, for subassemblies, **321–322**
Output tab on ribbon
    Export panel, Export Civil Objects To SDF, 582
    Plan Production, Create Sheets, 660
    Plan Production panel
        Create Section Sheets, 670
        Create Sheets, 666
        Create View Frames, 648, 655
outward branching, for feature lines, 358
Over Haul, in Mass Haul diagram, 498
overhang correction, for datum surfaces, 370–371
overland conveyance links, in SSA, 599–600
overlapping segments, problems from, 192, *193*
Override column for settings, 12, *12*
overrides, and child styles, 747

P

page setup
    creating, 483
    group plot style and, 828
Panorama Display Toggle icon, 3
Panorama window, **17**
    for alignment, 225
    for Best Fit line, *32*, 32
    for Best Fit parabola, *33*, 33
    Composite Volume palette, 377
    Composite Volumes tab, *145*, 145
    creating and saving custom views, 225
    DescKey Editor in, 94
    to edit alignment segments, 222
    editing curve length, *264*
    Events tab, 249
    modifying curve properties, 262
    Pipe Network Vistas via, *534*, 534
    for point editing, 86, *86*, 91, *91*
    Profile Entities tab, 263
    Tangency Constraint field, 227
    tangency constraints in, 227, *227*
parabola, Best Fit for, *33*, 33
Parameter Editor, for corridor, *368*
parameters
    for subassemblies, 309
    target and output, for subassemblies, **321–322**

parametric parts, in pipe network catalogs, 513
Parcel Area Label Style dialog, 169
parcel areas, labels for, **193–195**, *194*
parcel jig, *178*, 178
Parcel Layout Tools toolbar, *175*, *176*
parcel segments
    constructing, with appropriate vertices, **192–193**
    creating table for, **202–203**
    forming parcels from, *185*, **185–186**
    labels for, **196–203**, *197*, **239–240**
        multiple segments, **197–199**, *198*
        spanning segments, *199*, **199–200**
    moving, 187
    spanning labels for, **790–791**, *791*
    troubleshooting duplicate, *187*, 187
parcel styles, **754**, *755*
parcels, 161
    alignment crossing to divide, *162*
    Automatic Layout for, **176**
    creating and managing
        best practices, **185–193**
        boundary parcel, **167**, *168*
        cul-de-sac, **172–173**, *173*
        right-of-way parcel, *170*, **170–171**
        wetlands parcel, **168–169**, *169*
    editing, by deleting segments, **182–185**
    Free Form Create tool, *180*, **180–181**, *181*
    migrating from Land Desktop, 188
    for moving data to SHP format, 582–583
    reacting to site objects, **186–189**
    separation of soil boundary segment from subdivision lot, 163–164, *164*
    sites, creating and managing, **161–166**
    sizing for new, **175**
    subdivision lot parcels with precise sizing tools, **174–180**
    typical property boundary, *162*
parent layer, of objects, 8
Part Builder, 508, **512–513**
    adding part size using, **516–517**
    closing, 515
    Content Builder, *516*
    exploring part families, *515*, **515–516**

orientation, **513–518**
tools, 515
part catalog for pipe network, **507–512**
    adding arch pipe to, **518**
    backup, 513
    parametric parts in, 513
    part rules, **519–525**
        pipe rules, **521–523**
        structure rules, **519–521**
    Pipes domain, 507, **510–512**, *511*
    sharing custom part, **518**
    Structures domain, 507, **508–510**
    supporting files, **512**
part masking, 771
Part Matchup Settings dialog, Export tab, *585*, 585–586
partcatalogregen command, 518
parts list
    assigning pay items in, **720–721**
    pipe and structure styles for, 765
Parts List tool (Network Layout Tools toolbar), 532
pasting surfaces, 130, *644*, 644–645
pay item categorization file, 711
Pay Item List dialog, 717
pay items
    assigning to items in drawing file, 715
    AutoCAD objects as, **715–716**
    highlighting, **725–726**, *726*
    inventorying, **726–728**, *728*
    multiple on single object, 725
    as part properties, **723–725**
    pipes and structures as, **720–725**
    searching for, **714–715**
PC (point of curve), 27
peer-road intersection, 404–410, *410*, *411*
PennDOT drainage design manual, 581
performance, point clouds and, 157
perpendicular line, *25*, **25**
phantom PI, 634
Pick Label Style dialog, 280
pick point, jigs to help document, 254
Pipe Connection marker, *532*
pipe data bands, 299

Pipe-Diameter grip, 541
Pipe Drop Across Structure rule, *520*, 520–521
pipe endpoint edits, dynamic input (DYN) and, 543
Pipe-Length grip, *541*
pipe length grip, 541
    dynamic input (DYN) and, 542
Pipe Network Catalog Settings dialog, 513, *513*
Pipe Network Properties dialog, 529
    Information tab, 529, *529*
    Layout Settings tab, 530, *530*
    Profile tab, *530*, 530
    Section tab, 530, *531*
    Statistics tab, 530, *531*
Pipe Network Vistas tool, 534
pipe networks. *See also* sanitary sewer network
    adding to sample line group, 495
    alignments, creating for parts, **548–549**
    assigning pay items to existing structures and pipes, 723–725
    changing, and pipe tables, **570**
    composite finished grade surface for, 531
    dynamic input (DYN) and, **542–543**
    editing, **539–547**
        with context menu, **545**, *545*
        with Network Layout Tools toolbar, **546–547**
        in plan view, **540–541**
        tabular edits, **544–545**
        toggling numbers, 547
    exploring, *526*, **526–527**
    flow direction, changing, **539**
    invisible arrow for labeled pipe, 794, *796*
    label styles, **794–795**
    labels, **560–564**, *561*
    merging, 538
    object types, **527**
    part catalog, **507–512**. *See also* part catalog for pipe network
    parts list, **526**
    pay items changes and, 722
    pipes and structures as pay items, **720–725**
    placing parts in, **532–534**
    planning, **505–507**

profile views, *550*
    drawing parts in, **550–558**
    removing part, **554–555**
Pipe Networks tab
    Analyze panel, **560**
    General Tools panel, **558**
    Labels & Tables panel, **558**
    Launch Pad panel, **560**
    Modify panel, **559**
    Network Tools panel, **559–560**
Pipe Properties tool, 559
pipe rule sets, creating, **524–525**
Pipe Style dialog
    Plan tab, *765*
    Profile tab, 766, *784*
    Section tab, 766, *767*
pipe style, overriding, 557
Pipe Table Creation dialog, 568, *568*
pipe table creation dialog, 570
pipe tables
    creating, **567–571**, *568, 569*
    network changes and, **570**
Pipe To Pipe Match rule, 522–523, *523*
pipe trenches
    assembly for, 327–329, **329–331**
    corridor for, *380*, 380–381, *381*
pipes, 527
    data bands, 824, *824*
    distortion on profile views, *551*, 551–552
    exporting to Storm and Sanitary Analysis (SSA), **584–585**
    flow direction of, 533, *533*
    lengthening, 547
    shapes in default catalogs, 510
    showing those that cross profile view, *555*, **555–558**
    stormwater network profile, plotting hydraulic grade line against, 259
    styles, **765–769**
Pipes domain, 507, **510–512**, *511*
pivot point
    center crown of road as, 464
    for superelevation, **466–468**, *468*

Place Lined Material parameter, 337
Plan Production, 647. *See also* sheets
    prerequisite components, **647–648**
    supporting components, **671–672**
    view frames and match lines, **648–659**
Plan Profile labels, 794
Plan Readable option, for label styles, 776
plan sets
    AutoCAD functions for, 658
    preparation, **647**
    steps in creating, 648
plan view
    interference marker in, *566*
    pipe style for, 767
    projecting objects from, 290
Plan viewports, 659
Planning And Analysis Workspace, *704*
    changing, 582
plans, sheet with viewport for, 650
Plotted Unit Display Type, 11
plotted unit display type, 740
plotting scale of drawing, 6
point cloud surfaces, **154–159**
    creating, **157–159**
point description, **20**
Point File Format - Create Group dialog, *74*
Point File Format dialog, *75*
point files, 105
point groups, *87*, **87–90**, 105
    for controlling point display, **90**
    creating, 88
    vs. description keys, **94**
    for style control, 92
    for precedence control, 90
Point Groups dialog, 124
Point Inverse option, in Inquiry command, 66, *67*
Point Name transparent command, icon, *38*
Point Number transparent command, icon, *38*
Point Object transparent command, icon, *38*
point of tangency (PT), 27
Point-Rotation grip, 86, *86*
Point Style field, for description key, 92
Point Style Label field, for description key, 92

point tables, **97**
    creation options, 97, *98*
Point to Point tool, 144
POINTCLOUDDENSITY value, 157
points
    adding to surface, 130
    anatomy of, *71*, **71–72**
    assigning numbers, 76
    basic editing, **85–91**
        Panorama and Prospector for, **86**, *86*
    changing elevations, **90–91**
    COGO vs. survey, **72**
    converting from Land Desktop, SoftDesk, and other sources, **77**, **81**
    creating, **72–85**
    creating curve through, **27**, *28*
    critical, 153
    default layer for, **72**, *73*
    importing, from text file, *74*, **74–76**
    labels, styles, **787–789**
    names, **20**
        vs. description, 74
    removing connection between, 129
    settings, 72
    styles, **833–834**, *834*
    surface approximation from, **113–114**
    user-defined properties, **98**
Points From Corridor utility, 385
points of vertical intersection (PVIs), 251
    locking, 264
polygon, closed, 189, *190*
    for parcels, 186
Polyline From Corridor utility, 385
polylines
    converting to catchment area, 577
    converting to feature lines, 614–615, *615*
    creating alignment from, **210–211**
    and sample lines, 481
    surface approximation from, **111–112**, *112*
    trimmed/extended, 190
pond, draining, **632–636**
pop-up, avoiding, 790
pre-cooked assemblies, **312**, *313*

precision
    drawing vs. label, 9, 742
    Survey User settings for, 43
predefined viewports, 647
Preview Area Display Toggle icon, 3
pricing corridor, **717–720**
printing. *See* plan production
Process Linework dialog, 50
profile attachment points, for structure labels, 563, **563–564**
profile curve label, location, 812
Profile Data Band Style dialog, Band Details tab, 825, 825
profile data bands, 298
Profile From Corridor utility, 385, 387
Profile Grid Box tool, 261
Profile Grid View tool, 263
Profile Labels dialog, 270–272, *271*, *272*, *806*, 810
    Import Label Set, 279, 280
    for station labels, 272, *273*
Profile Layout Parameter dialog, 262–263, *263*
Profile Layout Parameters tool, 261
Profile Layout Tools toolbar, 252, 265–267
    Delete Entity tool, 261, 265
    Draw Fixed Vertical Curve By Three Points tool, 265
    Insert PVI tool, 260, 265
profile plot, from SSA, 607, *608*
Profile Plot Options dialog, *608*
Profile Properties dialog, Profile Data tab, 355, *356*
Profile Station Elevation command, 256
Profile Style dialog, Markers tab, *753*
Profile View Properties dialog
    Bands tab, 299, *299*, 300–301, *301*
    Elevations tab, *296*, 296–297, *297*
    Hatch tab, *302*, 302
    Information tab, *294*, 303
    Pipe Networks tab, *555*, *557*, 557
    for removing pipe network parts, 554–555, *555*
    Stations tab, 294–295, *295*

Profile View Style dialog
  Display tab, 821–822
  Graph tab, *818*
  Grid tab, *819*
  Horizontal Axes, 819, *820*
  Title Annotation, *820*
  Vertical Axes tab, 820, *821*
Profile viewports, 659
profile views, 249, **281–290**, *282*
  after importing band set and matching properties, *307*
  bands, *827*
  with completed style, *822*
  creating
    gapped, **286–287**, *288*
    manually, **283**
    for pipe networks, *550*
    during sampling, **281–283**
    stacked, **288–290**, *289*, *290*
    staggered, **284**, *285*, **286**, *288*
  data bands, **298–302**, *300*
    troubleshooting, *300*, 300–301, *301*
  drawing parts in, *550*, **550–558**
  editing, **294–307**
    properties, **294–303**
    station limits adjustments, **294–295**
  elevation adjustments, *296*, **296–297**
  extents, troubleshooting, 655
  grid spacing for, 823
  grips for pipe end in, *554*
  hatches, *302*, **302**
  invisible, 667
  label styles, **304–307**
    band sets, **305–307**
  manually created gap between, *295*
  organizing, 415
  pipe distortion on, *551*, 551–552
  position in viewport, 655
  vs. profiles, 270, 281, 303
  projecting objects from plan view into, 290
  removing part, **554–555**
  showing pipes that cross, *555*, **555–558**, *556*
  splitting, **283–290**, 296–297, *297*

  creating manually limited, **284**
styles, **303**, **817–826**
  accessing, *818*
  anatomy of, *817*
  bands, **823–826**
vertical exaggeration for, 817, *818*
vertical movement editing using grips, *552*, **552–554**, *554*
profiles, 246. *See also* layout profiles
  alignment information in, 247
  associating with main alignment, 387
  corridor feature lines extracted to produce, 359
  creating, surface sampling for, **246–251**
  data bands, 823, *824*
  direction for reading, 251
  display options, **297–298**
  editing, **261–267**
    component-level editing, **264–265**
    grip profile editing, *261*, **261–262**
  intersecting stations on, 268–269
  label sets for, **279–281**, *280*
    styles, **806–807**, **813–814**
  labels, **270–281**
    applying, **270–272**, *272*
    curve labels, *274*, **274–275**
    grade breaks, *276*, **276–277**, *277*
    line labels, **273**, *273*
    station labels, **272–273**, *273*
    styles, 806
  layout, **251–261**
  matching elevations, **267–269**
  for offset assembly, 434
  Plan Production and, 648
  vs. profile views, 270, 281, 303
  for roundabouts, **451–453**, *452*
  sheet with viewport for, 650
  staggering, **277–279**, *277–279*
  structure style for, 771
  styles, *752*, 752
  superimposing, **290–291**, *291*
  utilities, **290–293**
    object projection, **291–293**, *293*, *296*

project data, access to, 2
project files, storing formula file with, 720
Project Objects To Profile View dialog, *292*, 292
Project Options dialog (SSA), 586, *587*, *597*, *601*
projecting objects, **291–293**, *293*, *296*
Projection View Properties dialog, 293
projects
    displaying, 2
    setting up with shortcuts, 679
Projects branch, in Prospector, 4
prompts
    controlling, 10
    for point elevations, names and descriptions, *73*, **73**
properties
    of points, user-defined, **98**
    of profile views, **294–303**
    of surfaces, **118–121**
Properties dialog, Design tab, *672*, *673*
Properties list, and label options, 784
Properties tool, for pipe networks, 558
Prospector tab in Toolspace
    alignment and profile in, *361*
    for alignment style, *232*, 232
    Area Label Style column, *195*, 195
    assemblies in, 344
    changing alignment properties, 230
    for changing group of parcel area labels, 195
    context-sensitive menus, *3*, 3–4
    creating site, 165
    Data Shortcuts, 4
    Drawing Templates branch, *4*, **4–5**
    Open Drawings branch, *3*, **3–4**
    pipe networks branch in, *544*, 544–545
    for point groups, 88
    Projects branch, 4
    and Ribbon, 114
    shortcuts managed through, 676
    view frames in, *657*
    to view Points collection, *87*
proximity breaklines, 122
    converting to standard, 118
PT (point of tangency), 27

public utility lots, areas dedicated for, Free Form Create tool for, *180*, 180–181, *181*
publishing data shortcut files, **676–681**
PVI-based layout profiles, grips for, 255
PVI-based layouts, 251, **252–256**
PVI or Entity-Based tool, 261
PVI stations, adding, *428*
PVIs (points of vertical intersection), 251
    definition, 259
    locking, 264

Q

QTO. *See* Quantity Takeoff (QTO) feature
QTO Manager, favorites list in, *714*
Quantile distribution methods, for surface analysis, 761
Quantity Takeoff (QTO) feature, 711
    and assemblies, 717
    pay item files, **711–715**
        pay item favorites, **712–714**
    pricing corridor, **717–720**
    searching for pay items, **714–715**
Quantity Takeoff Report dialog, 726, *727*
queries, in Map 3D, **706–708**, *707*, *708*
Quick Elevation Edit tool, 625, 635
Quick Profile tool, 620
    for Grading Elevation Editor, 627
Quick Select (QSelect) tool, 558

R

Rain Gauge dialog, *604*, 604
rainfall
    information, 581, 604–605
    location, 602
    runoff, 588–589
Rainfall Designer dialog, 604, *605*
Raise Incrementally/Lower Incrementally tools, for Grading Elevation Editor, 627
Raise/Lower By Reference tool, 626
Raise/Lower PVI dialog, *266*, 266
Raise/Lower PVI tool, 260
Raise/Lower tool, 626
    for Grading Elevation Editor, 627

Rational method for computing rainfall
    runoff, 588–589
RC (Reverse Crown), 460, *460*
Reactivity Mode, for alignment table, 241
Readability Bias option, for label styles, 776, *777*
Realign Stacks tool, 571
rear lot lines, from precise sizing tools, 191
Rebuild command, for corridor, 354–355, *355*
red triangle grip, 255
Reference Point, *7*, 736
reference text
    alignment label with, *806*
    in General Note labels, 785
    in labels, 779
Refill Factor value, 491
regions
    in corridors, **349**
    for intersection, **415–419**, *420*
    for roundabouts, *454*
    setting stations for, 417
Regression Data window, 216–217, *217*
Remainder Distribution parameter, 176
Remove Items tool, 571
Rename Parts tool, 559
rendering materials, 129
Renumber/Rename Parcels dialog, 194
Replace Items tool, 571
Report Quantities dialog, 493
reports
    of pay items, 726
    from Storm and Sanitary Analysis, **606–609**
    in Toolbox, 14
Reports toolbar (SSA), *606*
Resolve Crossing Breaklines tool, 123
retaining walls, assembly for, **327–329**
Reverse Crown (RC), 460, *460*
reverse curves
    creating, *29*, **29**, **217–219**, *218*, *219*
    removing, 229–230
Reverse direction tool, for Grading Elevation
    Editor, 628
Reverse Label function, Ctrl+click to access, 238
Reverse tool, for feature lines, 621
reversing alignments, 233

Ribbon, 2, **17**, *18*. See also tab names
    for COGO points vs. survey points, 72
    Manage tab, **39**
    and Prospector tab in Toolspace, 114
right of way, for model, 210
right-of-way parcel, creating, *170*, **170–171**
Rim Insertion Point grip, 553
road assembly, creating, *313*, **313–318**
Roads and Lots site, 164
roadside swales
    assembly for, **434**, *434*
    berms to contain, 339
    Create Profile dialog for centerline, *387*
    creating alignment and profile for, **384–390**
    grading feature lines for, 613–614
roadway, specification for superelevation, *469*
Rotate a View Frame grip, 658
rotating match lines, 658
rotation angle
    of label component, 780
    tracking, 542
Rotation Direction field, for description key, 93
Rotation Parameter field, for description key, 93
Rotation Point, *7*, 736
rotational grip, for pipe network structure, 540
roundabouts, **443–457**
    alignments, **445–450**
        layout, *448*
        layout with slip lane, *450*
    assemblies for, *453*, 453–454
    center design, **450–451**, *451*
    corridors for, 444
    drainage, **444**, *444*
    finishing touches, **454–457**, *455*
    profiles, **451–453**, *452*
    regions and targets, *454*
    turn lane for, 447

S

Sample Line Creation Methods tool, 480–481
sample line creation tool, 384
sample line group
    adding pipe network to, 495
    creating, 478, 480

editing swath width of, **483**
in views, **483–484**
Sample Line Group Properties dialog, *494*, 495
Sample Line Tools toolbar, *480*, 480, 483
sample lines, creating along corridor, **481–482**
samples, lines, *479*, **479–483**
sampling
    after section view creation, **493–495**
    creating best surface for, 477
    creating profile views during, **281–283**
sanitary sewer network
    adding to sample line group, *494*
    creating, **527–538**
        with layout tools, 528
        with network layout creation tools, **529–536**
        parameters, **528–529**
    design criteria files, 505
    interference check between storm pipe and, **564–567**, *565*
saving command changes to settings, 11
Scale Inserted Objects, settings, 11
Scale Parameter field, for description key, 93
scale, setting for, 734
Schematic of Top links, connecting to form surface, *371*
screen, splitting for plan and profile editing, 249, *250*
SCS (Soil Conservation Service), 576
SDF files, creating, 582
SDTS (Spatial Data Transfer Standard), 106
sdts2dem program, 106
Sea Level Scale Factor, 6, 736
searching for pay items, **714–715**
Section Data Band Style dialog, Band Details tab, *830*
Section Editor, 367, *367*, *368*
    for corridor, *437*
    Station Selection panel, 367
section methodology, for earthwork, 144
section sheets
    creating, **668–671**, *671*
    styles for, 827

Section Sources dialog, 494
Section View Group Bands dialog, 830, *830*
Section View Group Properties dialog, *829*
    Sections tab, *834*
section view groups, creating, **668–670**
section views, 484
    annotations, **495–496**, *496*
    arranged to plot by page, *483*
    hatches, *768*
    object projection, 487
    sampling after creation, **493–495**
    styles, **827–831**
sectional data bands, 299, *824*, 824
segments. *See* parcel segments
Select A Quantity Takeoff Formula File dialog, 717
Select A Sample Line Group dialog, 489, 491
Select Alignment tool (Network Layout Tools toolbar), 532
Select Drawings To Attach dialog, 705, *705*, 706
Select Label Style dialog, 195, 232
Select Layout As Sheet Template dialog, 651, *651*
Select PVI tool, 261
Select Similar tool, 558
Select Style Set dialog, for label set, 279
Select Surface Style dialog, 642
Select Surface To Paste dialog, *644*, 644–645
Select Surface tool (Network Layout Tools toolbar), 531
Select tool, for Grading Elevation Editor, 627
Selection Rule, 241
sending drawing files, eTransmit for, **693–695**
Set Elevation By Reference tool, 626
Set Grade/Slope Between Points tool, 625, 631
Set Increment tool, for Grading Elevation Editor, 627
Set Network Catalog tool, 559
Set Pipe End Location rule, 523, *524*
Set Slope Or Elevation Target dialog, *419*, *441*
Set Sump Dept rule, 521, *521*
Set The Grading Group tool, 638
Set Width Or Offset Target dialog, *364*, 364, *418*, *441*, 441

SetAutoCAD Variables To Match option, 6
Settings tab. *See* Toolspace, Settings tab
setups, in survey database, 53
sewer specifications, 505
shape styles, 302
shapes
    in corridor assembly, *351*
    in part catalog, 508, *509*
    styles, **832**
    in subassembly, *319*, 319, *321*, **322**
sharing files, LandXML for, 691
sharing workload, 10
Sheet Set Manager, 665, *665*, 667, *668*
sheet style, page setup in, 483
sheets, 659
    creating, **659–664**
    managing, **665–667**
    for Plan Production, 650
    what it is, 663
shortcut references
    creating, **681–685**
    updating and managing, **685–690**
shortcuts. *See* data shortcuts
shots, 71. *See also* points
shoulder subassemblies, **324–325**
    and superelevation, 464
ShoulderExtendAll subassembly, *325*, 325
ShoulderExtendSubbase subassembly, 325, *325*
Show Grade Breaks Only tool, for Grading
    Elevation Editor, 628
SHP (ESRI Shape) files, 582
shrinkage factor for soil, 491
Side Shot transparent command, icon, *38*
single-section views, creating, **484–487**, *487*
single segment label styles, **790–791**, *791*
site, for figures, 45
site geometry problems, avoiding, 164
site objects, parcels reacting to, **186–189**
Site Properties dialog, 613
siteless alignment, 165
sites
    alignments and, 162, *163*, **205**
    best practices for topology interaction,
        161–164

creating and managing, **161–166**
    feature lines and, 612
Slide a View Frame grip, 658
Slide Line - Create tool, **177–179**, *179*
sliding
    match lines, 658
    PVI grips, 255
slope arrows, 137, *143*, 143
slopes, **141–143**, *142*
    of corridor, control of, 415
    labels, **152–153**
    legend table, 141–142, *142*
slopes analysis, 136
    in surface styles, 756
"smart" text, adding to Text Component
    Editor, *787*
smooth surface, 130
Smooth Surface dialog, 132
Smooth tool, for feature lines, 621
smoothing surfaces, *132*, **132–133**
    vs. contour, 138
snaps, and profiles, 305
SoftDesk, converting points from, **77**, **81**
soil boring data, user-defined properties for,
    98–100
soil boundary parcel segment, separation from
    subdivision lot parcel, *163*, 163–164, *164*
Soil Conservation Service (SCS), 576
    TR-55 method, 589
soil factors, adding to materials list, **491–493**
solid, displaying structure as, 771
source data, managing changes in, **686–687**
source drawing, updating, **685–686**
spanning labels
    for pipes, 563
    for segments, **790–791**, *791*
    styles for, 199
    troubleshooting, 200
Spatial Data Transfer Standard (SDTS), 106
split-created vertices, 192, *192*
    spanning labels and, 199
Split Network tool, 559
Split Region tool, *390*, 390
splitting profile views, **283–290**, 296–297, *297*
    creating manually limited, **284**

spot grade, 151
spreadsheets
   from SSA reports, 606
   volume report for, 493
square feature line grip, 615
square grip, 223
   for sample line, *479*, 479
SRTM (Shuttle Radar Topography Mission)
   data, 108
SSA. *See* Storm and Sanitary Analysis (SSA)
stacked profile views, creating, **288–290**, *289*, *290*
staggered profile views, creating, **284**, *285*, **286**, *288*
staggering, profiles, *277–279*, **277–279**
Standard Deviation, for surface analysis, 761
standard surfaces, 104
standards
   for design criteria files, 461
   label sets for, 238, 803
   matching available parts to, 509
   matching available pipes to, 511–512
Starting Station, for alignment, 209
Static Mode tool, 571
static text, adding to table column heading, *816*
station equation, 234
station labels, **272–273**, *273*
station offset labeling, **238–239**
Station Offset transparent command, icon, *38*
station range, for plan set, 649–650
stationing, alignments, **233–235**, *234*
stationing reference point, 234
stations
   corridor, 355
   intersecting, on profiles, 268–269
   limits adjustments for profile views, **294–295**
   offset labeling, **803–805**
   setting for regions, 417
SteppedOffset tool, for feature lines, 621
storage nodes, in SSA project, 587
storing, formula file with project files, 720
Storm and Sanitary Analysis (SSA)
   catchments, **575–581**
   example of use, 590–598
   exporting pipes to, **584–585**

guided tour, **586–587**
hints for working in, 590
vs. Hydraflow, 588
hydrology methods, **588–598**
importing CAD file, 590
inlet vs. manhole, 585–586
launching, 599
Plan view, 599
profile plot from, 607, *608*
running reports from, **606–609**
Water Drop command, **573–575**, *574*
storm crossing, invert shown in profile, *556*
storm drainage pipe network
   creating, from feature lines, **536–538**, *537*
   interference check between sanitary pipe an, **564–567**, *565*
   plotting hydraulic grade line against profile, 259
Storm Sewers tool, 560
stormwater-management facilities, areas dedicated for, Free Form Create tool for, *180*, 180–181, *181*
Stormwater Management site, 164
street-name labels, creating, 239–240
Structure Connection marker, *532*
Structure drop-down list (Network Layout Tools toolbar), 532, *532*
structure grips, 540, *540*
structure labels
   adding ground elevation, **796–798**
   profile attachment points for, *563*, **563–564**
   styles, **794–795**, *796*
Structure Properties dialog, Part Properties tab, *546*
Structure Properties tool, 559
structure rule sets, creating, **524–525**
structure rules, **519–521**
Structure Style dialog
   Model tab, 770, *770*
   Plan tab, 770–771, *771*
   Profile tab, 771, *772*
structure styles, **770–773**
   with blocks, 772
Structure Table Creation dialog, *568*, 570

structures, 527
Structures domain, 507, **508–510**
styles. *See also* object styles
    for alignment, 209, *752*, 752
        creating, **754**
    for alignment labels, **799–814**
        geometry points, **801**
        major station labels, **800**
    for alignments, station offset labeling, **803–805**
    civil 3D templates and, **731–745**
        drawing settings, **734–743**
        importing styles, **732–734**
    for curve labels, **792–793**, *794*, **810–812**, *811*
    description keys control of, 92
    direction arrow in, 752–753
    DWT template file for storing, 731
    editing, *745*
    for feature lines, 618, **754–755**, *755*
    for figures, 45
    group plot, **828–830**
    hierarchy for, *784*
    for labels, 236, 653, **773–796**
        grade breaks, **812–813**, *813*
        of lines and curves, **790–793**
        for pipes and structures, **794–795**
    for line labels, **808–809**, *810*
    links, **832**
    parcel, **754**, *755*
    for pipes and structures, **765–773**
    point group control of, 88–89
    for point labels, **787–789**
    for points, 71
    points and point labels, **833–834**, *834*
    for profile labels, 806
    for profile view labels, **304–307**
        band sets, **305–307**
    for profile views, 303, **817–826**
        anatomy of, *817*
        bands, **823–826**
    for section views, **827–831**
    shapes, **832**
    for spanning labels, 199
    for spanning segment labels, **790–791**, *791*
    structure, **770–773**
    for surfaces, **136–144**, **755–764**
        elevation banding, **763–764**, *764*
        watershed analysis, **761**, *762*, *763*
    for tables, 241, *815*, **815–816**
    for view frames, 653
    warning symbol for, 733
Styles panel, 39
Sub-Entity Editor tool, 225
    for changing constraints, 227
subassemblies, **309–311**
    adding to tool palette, *311*, **311**
    commonly used, **323–325**
        lane, **323–324**
        shoulder and curb, **324–325**
    customized, on tool palette, **342–343**, *343*
    daylighting with generic links, **334**
    default attachment point for, 321
    help for, *311*, **311**
    with marked points, 341
    mistakes, correcting, 317
    name changes, *326*, **326–327**, *327*
    specialized, **331–339**
        generic links, **331–334**, *332*
    target parameters and output parameters, **321–322**
    targets and, 361–362
    unassigned, 344
Subassemblies Properties dialog, 325
subbasins
    assigned to inlets, *592*
    connecting to inlet, *592*
    preparation of data, *598*
    in SSA project, 587
subdivision lot parcels. *See also* parcels
    creating with precise sizing tools, **174–180**
Subdivision Tangent design check, results, *220*
sump depth, 548, 553
super assemblies, **464–465**, *465*
superelevation, 384
    applying to design, **469–472**
        alignment, **469–470**
    assembly without, *459*
    attainment equations, 462

critical stations and regions calculated by
Civil 3D, *462*
data bands, *824*, 824
pivot point for, **466–468**, *468*
potential uses, 474
preparation, **459–465**
    alignments, **464**
        design criteria files, **461–463**
preserving manual edits, 472
super assemblies, **464–465**, *465*
table with glyphs, *471*
transition station overlap, **471–472**, *472*
views, *473*, **473**, 474
superelevation bands, 299
Superelevation View command, 307
Superimpose Profile Options dialog, 290, *291*
superimposing profiles, **290–291**, *291*
surface cropping, **127–129**
Surface Definition tab, reordering build operation on, *121*, 121
surface points, labels, **151–154**
surface profiles, for roundabout design, *452*
Surface Properties dialog, 126
    Analysis tab, 140, *141*, 148
    Information tab, 120
    Statistics tab, 135
Surface Style dialog, 755
    Contours tab, *137*, 137–138, *757*
    Display tab, *758*
    Points tab, *759*
Surface tab on ribbon, Analyze panel, Water Drop command, 573
surfaces, 247. *See also* corridor surface
    adding border, **123–127**
    approximations, **109–114**
        from points or text files, **113–114**
        from polyline information, **111–112**, *112*
    basics, **103–104**
    comparing, **144–149**
    contours, basics, **137–140**
    creating, **104–117**
    creating best for sampling, 477
    creating catchment from, 577
    creating from grading groups, **641–645**, *643*
    deleting boundary, 127, *127*
    detail level, 135–136, *136*
    from GIS data, **114–117**, *117*
    from Google Earth, **108–109**, *109*
    grid labels, 153–154, *154*
    imported from Land Desktop, *696*
    information on Internet, **105–109**
    labels for, **150–154**
        contours, **150–151**, *151*
    manual edits, **129–135**
        point and triangle editing, *131*, **131**–132
        simplifying, **133–135**, *134*
        smoothing, *132*, **132–133**
    materials from, 489
    pasting, 130
    for pipe networks, 529
    point cloud, **154–159**
        creating, **157–159**
    raise/lower, 130
    refining and editing, **117–135**
        additions, **121–135**
        properties, **118–121**
    sampling for creating profile, **246–251**
    slopes, **141–143**, *142*
    styles, **755–764**
        contours in, **756–758**, *757*
        elevation banding, **763–764**, *764*
        triangle and points, **759–760**, *760*
        watershed analysis, **761**, *762*, *763*
    styles and analysis, **136–144**
    volume, **146–149**
Surfaces Properties dialog, 121, *121*
Survey Command window, default settings, 43
survey database, **47–50**, *53*
    details, **52–60**
    settings, *42*
        defaults, **41–42**
    source of data, **51–52**
    traverses section, **54–57**
survey network
    manipulating data, **58–60**
    in survey database, 48
    translating, 59–60

Survey palette, 14
    database setup, **41–64**
        equipment database, **43–44**
        figure prefix database, **44–46**
        linework code set database, **46**
        survey database, **47–50**
survey point groups, in survey database, 49
survey point styles, creating, **751**
survey points
    vs. COGO points, **72**
    editing, 51
    in survey database, 48–49
Survey tab on Ribbon
    Inquiry panel, *66*, **66–68**
    Modify panel
        Geometry Editor, 225
        Parcel Layout Tools, 183
        Renumber/Rename, 193
        Surface tab, Surface Properties, 139
        Translate Database, 58, *58*
Survey User Settings dialog, 41, *42*, 43
surveying coordinate system, 6
surveyor measurement, 735
Surveyor Units, vs. AutoCAD angular
    units, 67–68
swale. *See* roadside swales
Swap Edge option, 130
    triangles for, 759
Swap Part command, *545*
Swap Part tool, 559
Swing Line - Create tool, **179–180**
symbol, size options, 750, *750*

T

Table Creation dialog, *202*, 202–203, *241*, 241, **570**
    for line table, 242, *242*
Table panel, **570–571**
Table Properties tool, 571
Table Style dialog, Data Properties tab,
    815–816, *816*
Table Tag Numbering dialog, 201, *202*

tables
    alignment, **240–243**
    alignment segment, **243**
    creating
        curve tags to prepare for, **200–201**
        for parcel segments, **202–203**
    curve table, *203*
    line, creating, **242**
    for pipe networks, 558
    section view settings, 486, *487*
    styles for, *815*, **815–816**
    superelevation, exporting, 472, *472*
    types, 240
    volume, **490**, *491*
tabular design, for alignment, **224–225**
tag-only style, for segments to be in table, 200
Tangent By Best Fit dialog, *216*, 216
tangent slope length, 809
Tangent - Tangent (With Curves) tool, *212*
Target Mapping dialog, 351, 357, 397
    for corridor, *352*, 352
    for cul-de-sac, 398
    for intersection, 422
    for regions, *391*
target parameters, for subassemblies, **321–322**
targets, **361–367**
    alignments and profiles, **362–364**
    for intersection, **415–419**, *420*
    multiple, **415–416**, *416*
    for roundabouts, *454*
    on subassemblies, 320
TDS, data collectors, 52
Technical Release (TR)-55 document, 576–577, 581
template files
    contents, 790
    creating for projects, *677*
    for drawings, 671–672
    for generating sheets, 651
    non-Civil 3D, 732, *732*
    for projects, creating, *677*
    setting object layers in, 738, *738*
    for storing styles, 731
    for viewports, 659, 672

Tessellation Angle value, 642–643
Tessellation Spacing value, 642
test database, creating, 41
text, adding static to table column heading, *816*
Text Component Editor, 36, **783–785**, *784*, *786*, *809*
    adding "smart" text, *787*
    symbols, 793
    for tables, 815
text file
    breakline information from, 122
    importing, point group creation from, 88
    importing points from, 72, *74*, **74–76**
    surface approximation from, **113–114**
text height, for label, 779
thickness, absence for surface, 104
time of concentration
    for catchment, 576
    Kirpich method of, 597
Time Series dialog, for rain, *605*
TIN files, 105
Tin Surface tab on Ribbon
    Labels & Tables panel, Add Labels, 150, 152
    Surface Tools, Drape Image, 109
TIN (Triangulated Irregular Network), 33, 103, *104*
    reducing data amount processed, 130
Toggle Upslope/Downslope tool, *533*, 533, 539
tool palettes
    adding subassemblies to, **311**, *311*
    customized subassemblies on, **342–343**, *343*
    opening, 309
    storing completed assembly on, **343–344**
Tool Properties dialog, for customized subassemblies, *343*
toolbar
    expanding, 72
    modeless, 252
Toolbox, *15*
    editing, **15–16**
Toolspace, **1–14**, *2*
    horizontal display, 744
    Prospector tab, **2–5**. *See also* Prospector tab in Toolspace
        Open Drawings branch, *3*, **3–4**

Settings tab, **5–14**, 72, *745*
    Abbreviations, **8**
    Alignment, Design Checks, 219
    Ambient Settings, **8–13**, *9*
    Drawing Settings, **5–6**
    General, Label Styles, Note, 785
    Object Layers, **7–8**, *8*
    Object Settings, **13–14**
    Parcel, Label Styles, Curve, 792
    Parcel, Label Styles, Line, 791
    Pipe, Label Styles, Plan Profile, 794
    Point, Label Styles, 788
    Section View, Group Plot Styles, 828
    Section View, Section View Style, Road Section, 827
    Structure Rule set, 524
    Structure, Structure Style, 772
    Transformation, **6–7**, *7*
    Units And Zone, **6**
Survey, 14. *See also* Survey palette
Toolbox, **14**, *15*
    editing, **15–16**
Tooltips, 11
    displaying, 741
    for quadrant, bearing and distance, *21*
Topo point group, creating, *89*
Transformation settings, **6–7**, *7*
transition station overlap, **471–472**, *472*
Transmittal Setups dialog, 693
transparent commands, **36–38**
    matching, **38**
    settings for, 742
    toolbar, *37*
Transparent Commands category, for Ambient settings, 10
Transparent Commands toolbar, *255*
    to create layout profile, *257*
    Profile Station Elevation command, 255–256, 262
transportation design
    design speed in, 235
    PVI layout in, 252
traverse analysis, 63
    default settings, 43
    results, *57*

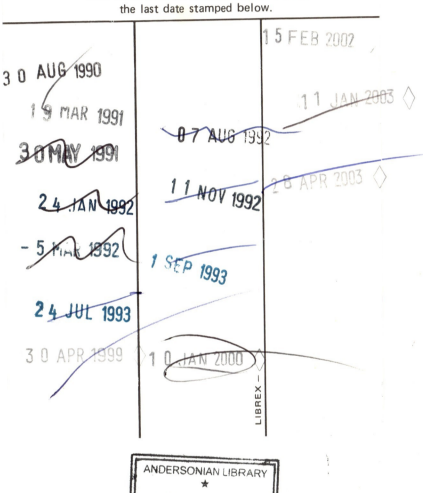

This publication is based on presentations from the Wheat Industry Utilization Conference held in San Diego, CA, on October 7—8, 1988. To make the information available in a timely and economical fashion, this book has been reproduced directly from typewritten copy submitted in final form to the American Association of Cereal Chemists by the editor of the volume. No editing or proofing has been done by the Association.

Library of Congress Catalog Card Number: 89-84430
International Standard Book Number: 0-913250-68-6

©1989 by the American Association of Cereal Chemists, Inc.

All rights reserved.
No portion of this book may be reproduced in any form, including photocopy, microfilm, information storage and retrieval system, computer database or software, or by any means, including electronic or mechanical, without written permission from the publisher.

Copyright is not claimed in any portion of this work written by United States Government employees as a part of their official duties.

Reference in this volume to a company or product name by personnel of the U.S. Department of Agriculture or anyone else is intended for explicit description only and does not imply approval or recommendation of the product to the exclusion of others that may be suitable.

Printed in the United States of America

American Association of Cereal Chemists
3340 Pilot Knob Road
St. Paul, Minnesota 55121, USA